AF889911

Pyrolysis and Gasification of Biomass and Waste

Proceedings of an Expert Meeting

Strasbourg, France, 30 September – 1 October 2002

Edited by A V Bridgwater

Sponsored by
EC Altener, CIRAD & CEA

Supported by
PyNe & GasNet

© Aston University, Bio-Energy Research Group and CPL Press, UK
May 2003

All rights reserved. No part of this publication may be reproduced, stored in a retrieval system or transmitted in any form or by any means, electronic, mechanical, photocopying, recording or otherwise, without the prior written permission of the copyright holders.

Published in the United Kingdom by:

CPL Press, Liberty House, The Enterprise Centre
New Greenham Park, Newbury, Berks RG19 3UP, UK
Tel: +44 1635 817408 Fax: +44 1635 817409
Email: press@cplbookshop.com Website: www.cplpress.com

for:

Aston University, Bio-Energy Research Group
Aston Triangle, Birmingham B4 7ET, UK
Email: a.v.bridgwater@aston.ac.uk Fax: +44 121 359 6814

Sponsored by:

European Commission DG Energy and Transport Altener Programme, CIRAD and CEA

Supported by:

PyNe and GasNet
Website: www.thermonet.co.uk

This book includes material that may contain the personal views of the authors and does not necessarily express the views of the editor, members and/or participants in PyNe or GasNet, Aston University Bio-Energy Research Group, the European Commision, CIRAD or CEA.

No responsibility is assumed by the authors, editor or any of the above-mentioned organisations for any injury and/or damage to persons or property as a matter of products liability, negligence or otherwise, or from any use or operation of any methods, products, instructions or ideas contained in the material herein.

CPL Press Online Bookshop

Numerous publications covering pyrolysis, gasification and many other aspects of renewable energy, as well as the life and chemical sciences, are available at: **www.cplbookshop.com**

Reprinted in the UK by Lightning Source Ltd, August 2008

ISBN 978-1-872691-77-0

Foreword

This Expert Meeting was organised by the PyNe and GasNet networks to assist the European Commission, particularly DG TREN, to formulate a strategy for the research, development and demonstration of advanced thermal processes for realisation of renewable energy from biomass.

The presentations included overviews of the European Commission and US policies and programmes; status reports on the major thermal conversion technologies of pyrolysis and gasification as applied to biomass and waste, together with an overview of commercial requirements for successful translation of research and development into commercial practice; and a series of 11 case studies from major industrial players that provided authoritative examples of how industry is developing and implementing these advanced conversion technologies and in particular the range of problems that were encountered and how they were addressed. The overviews and case studies gave a sound basis for the six workshops that followed, each led by a chairman and supported by a panel. Each workshop presented their conclusions and recommendations at the wrap-up session at the conclusion of the meeting, and provided a report with conclusions and recommendations all of which are included at the end of this book.

These proceedings provide a comprehensive set of papers on pyrolysis and gasification of biomass and waste from fundamental research through to commercial plants all of which were presented as posters, case studies or overviews during the meeting. The R&D community in particular used the occasion to demonstrate their active involvement in contributing efficient and environmentally sound science and technology to underpin the industrial realisations that will follow as the European policies for renewable energy and particularly bio-energy become implemented.

The Organisers were very grateful for sponsorship of the meeting by the European Commission DG TREN, CIRAD and CEA. They would particularly like to thank Mrs. Julie Ellen of Aston University for her considerable help in organising the meeting.

Tony Bridgwater
Harrie Knoef

January 2003

Contents

The European Commission perspective in Biomass and Waste Thermochemical Conversion

K. Maniatis*, G. Guiu,° J. Riesgo*

**Directorate General for Energy and Transport*
°Directorate General for Research and Technological Development
European Commission
Rue de la Loi, 200, 1049 Brussels

ABSTRACT: This paper presents an overview on the European Community policy and actions on renewable energy sources with particular emphasis on bioenergy and the thermochemical conversion routes. The overview covers the present energy situation in the European Union and outlines the contribution of renewable energy sources. It shows that bioenergy has the highest potential and it is the only one that can directly replace fossil fuel based energy. The paper then concentrates on the conversion technologies of gasification and pyrolysis and provides the current status of the research and demonstration projects supported under the Fifth Framework Programme of the European Commission. The status of the conversion technologies is then reviewed in terms of market deployment and commercialisation aspects.

CONTRIBUTION OF RENEWABLE ENERGY SOURCES

According to the recent Commission's Green Paper on the security of supply, current energy demand in the EU is covered 80% by fossil fuel, of which over 40 % are oil. Renewable energy sources today have a share of around 6% of total energy consumption, although there is a target to raise this share to 12% by 2010. The level of energy-related pollution caused by today's energy mix (where our main concern now is CO_2) is too high. The European Energy mix in 1999 is presented in Fig. 1 where it can be seen that the contribution of renewables is at present 6%.

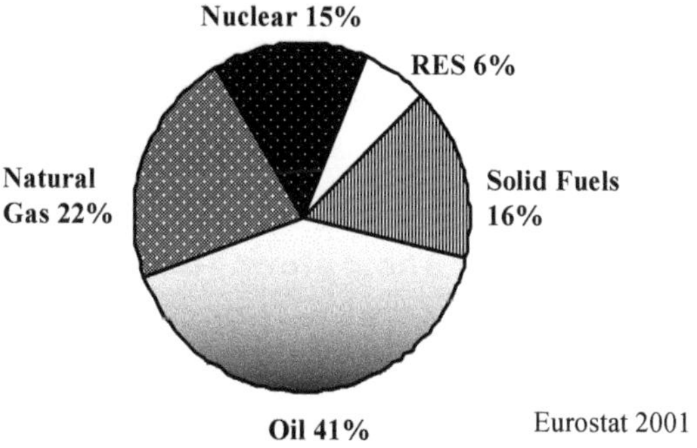

Figure 1: The European energy mix in 1998

While the rate of energy intensity has actually fallen the last decade in the EU, due to improved energy efficiency and structural change in the energy sector and industry as well, energy consumption per capita has continued to increase, along with the rate of global oil consumption. The EU energy consumption and CO_2 emissions change between 1990-1999 is graphically represented in Fig. 2.

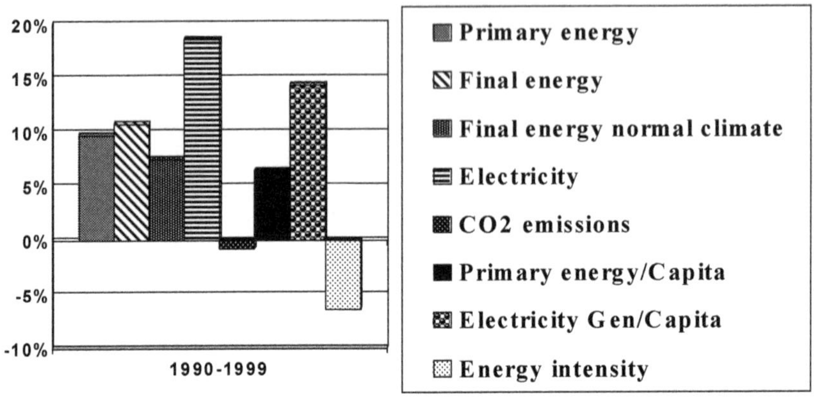

Figure 2: EU Energy consumption and CO_2 emissions % change between 1990-1999

An additional problem for the European Union, mainly due to continued use of oil, is its degree of import dependency in energy. It has increased from around 40% in 1985 to around 50% today. It is expected to continue to rise, possibly reaching 70% in the next 20 years if appropriate measures are not taken. Moreover, 50% of EU oil is now imported from a small number of OPEC member countries. Energy Consumption and production forecasts for Europe (of 30 assuming that all candidate countries become members of the European Union) can be seen in Fig. 3.

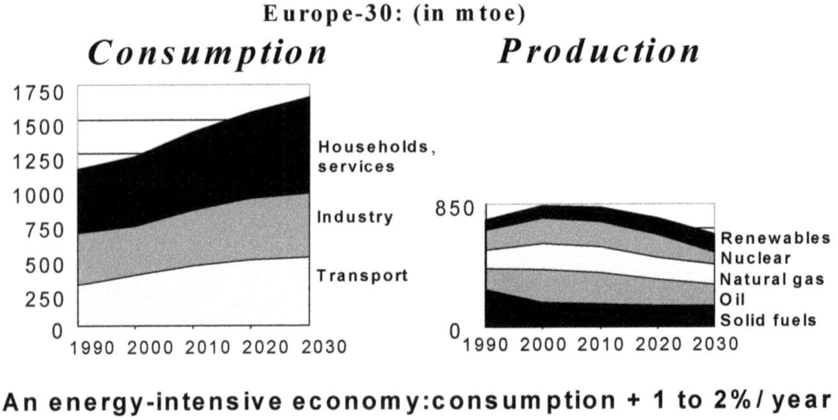

Figure 3: Energy Consumption and production forecasts for Europe (of 30)

It is obvious that a robust energy policy has to address this unsatisfactory situation and envisage reversing this trend for reasons of security of energy supply. Any such policy has to carefully examine the role of renewable energy sources that are indigenous energy sources and directly replace imports.

Figure 4 shows the contribution of the various renewable energy sources to the Union's energy mix. Even though in this Figure large-scale hydropower is also included, it is manifested that bioenergy (energy from biomass and waste) has the highest contribution amongst all renewables. Bioenergy is globally recognised as the renewable resource that will make the most significant contribution for sustainable energy in the near to medium term (1).

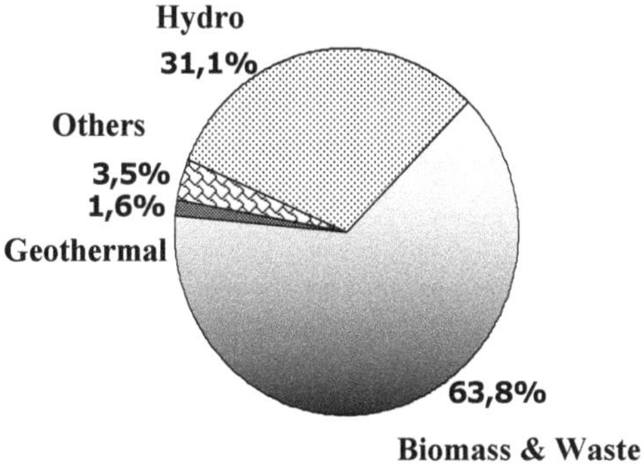

Figure 4: Bioenergy's contribution to RES energy

POLICY INTRUMENTS

Achieving sustainability requires that the Community, Member States and local and regional governments continue to play an active role in energy policy and demonstrate a strong political will to create the market conditions which favour sustainability. Efficient end-use technology and renewables technology should be given level playing fields by internalising external costs or compensating for the fact that this hasn't yet been done. To some extent this is being done, a fact made evident by recent Commission initiatives in energy efficiency and renewables that are briefly described below. On the supply side, the *Internal Market Directives* have led to price reductions of electric power for some customers of up to 30%. However, it should also be noted that falling electricity prices may render new and less developed technologies, including many renewables, and Combined Heat and Power (CHP) less attractive, due partially to external costs not being fully internalised in the price of energy. The Commission has therefore taken, and will be taking, additional measures.

DIRECTIVE ON ELECTRICITY FROM RENEWABLE ENERGY SOURCE

This *Directive on Electricity from Renewable Energy Source (RES-e)* (2001/77/EC) (2) sets indicative targets for Member State shares of renewables in their electricity mix balances. It also establishes a guarantee of origin certificate, rules concerning access for renewables to the grid and fair and transparent connection and transmission costs. The Directive also has a favorable impact on energy efficiency. Meeting Member State targets will increase the share of renewables-produced electricity from 14% to 22% for the EU as a whole.

DIRECTIVE TO PROMOTE THE PENETRATION OF BIOFUELS FOR TRANSPORT

The Commission has proposed a *Directive to promote the penetration of biofuels for transport applications* (3). This initiative includes an obligation for Member States to introduce legislation and to take the necessary measures to ensure that a minimum share of transport fuel sold nationally is biofuels. The target is a 2% share in 2005 to be gradually increased to 5.75% by 2010. However, the Council position is that these targets shouldn't be mandatory but indicative. It is expected that an agreement similar to that of the RES-e directive will be finally reached between the Council and the European Parliament by which the Member States will have to justify any deviation from these targets.

The proposed Directive is part of a more comprehensive package on alternative fuels that includes a Communication on Alternative fuels for Transport and the proposed amendment of

Directive 92/81/EEC, in favour of tax differentiation for biofuels. The directive covers all biofuels, both liquid and gaseous as well as those that are available in the market now such as bioethanol, biodiesel & biogas and those that could become available in the near to medium term such as bio-hydrogen, bio-methanol, bio-dimethylether, bio-Fischer-Tropsch etc.

OTHER RELATED DIRECTIVES

A *Directive on CHP* is being considered to complement and strengthen existing measures to promote CHP, in line with the Community target of doubling the share of CHP in EU electricity generation from 9% in 1994 to 18% by 2010. The proposal would cover both small-scale and large CHP, with special provisions for renewables and "quality CHP". Definitions, certification, grid access and connection cost issues, procedures and permits will also be covered. National CHP targets may be considered.

A *Directive on Energy Demand Management* is also being considered to encourage utilities to market energy-efficient and environmentally friendly heating, cooling, lighting and motive power instead of simply selling kilowatt hours.

THE EUROPEAN COMMISSION'S FRAMEWORK PROGRAMMES

The European Commission is now working extensively to increase the deployment of renewables energy technologies and applications. Barriers to a well functioning market are being removed whenever possible. This is being done with the help of Community technology development programmes, such as the *RTD Framework Programme*, and with pro-active support programmes such as *ALTENER.* Regulations, directives and other actions to promote energy efficiency and renewables are being set forth to develop an Energy Intelligent Europe with a system of energy production, distribution and consumption that aims at being safe, secure, cost-effective, and sustainable.

The emphasis on technology support has changed significantly over the various programmes and years following the pattern of technology development and the identification of the most promising technologies in terms of the accelerated penetration of bioenergy into the EU energy markets. In the Fifth Framework Programme (4) that ended on 31.12.2002 care was also taken to identify the most promising technologies for the future generation of technologies in combination to societal, economic and industrial demands. In the Sixth Framework Programme (FP6) that followed FP5, the emphasis has been in combining the European capacity in infrastructure and scientific and technical knowledge into a more comprehensive structure with closer cooperation between activities at national and European level (5) aiming to maintain and strengthen the international leading role of the European Research Area.

The sector of Energy from Biomass and Waste has been recognised as a very important RES sector and has received significant support during all European Commission programmes (6). Within the numerous technologies and market applications thermochemical conversion has received dedicated support with priority so far given to gasification, since this conversion technology is considered to be closer to the market after combustion technologies that are already competing in the EU energy markets (7). Strong support has been given to pyrolysis technologies too, however, so far there has been little interest from the market for demonstration projects. It is nevertheless expected that in the future more pyrolysis demonstration projects will emerge.

The research and demonstration projects supported by the European Commission in the Framework of FP5 for the period 1999 to 2001 are summarised in the Appendix to this paper. The Appendix is not complete as by the time this paper was written the contract negotiations for the proposals short-listed for support after the evaluation process of the last call of FP5 (the deadline for submission of applications was 18 December 2001) had not been completed yet and most of the contracts had not been signed. The Appendix is also not exhaustive as there can be other research contracts such as Cooperative Research (CRAFT) which are precursors to actual research projects.

In summary during the first three calls of FP5 24 projects on thermochemical biomass and waste conversion focusing on applied research and/or demonstration were supported as shown in Table 1.

Table 1: Summary of Thermochemical Bioenergy Projects Supported in FP5
(first three programme calls only)

Gasification		Pyrolysis		Dissemination
Research	Demonstration	Research	Demonstration	
11	7	2	3	1

STATUS OF THERMOCHEMICAL TECHNOLOGIES

Although numerous projects have been supported in the framework of the European Commission programmes as well as national programmes commercialisation of both technologies is progressing at a low pace not only in Europe but also globally. The status of the gasification and pyrolysis technologies has been reviewed recently (8). Although it is recognised that significant progress has been achieved in trying to address practically all barriers to technological development, full commercial implementation has not been attained yet. However it must be pointed that this is also partly due to non-technical barriers such as the liberalisation of the electricity markets in the EU which has oblige the utilities to look for the cheapest fuels in order to maintain their competitive prices at the expense of renewables.

Status of Gasification

A detailed review of the status of the gasification projects in the members of IEA Bioenergy Task 33 'Thermal Gasification of Biomass' is annually updated by NOVEM (9). This was further updated by the GasNet (10) to include activities in other European Countries that do not participate in IEA Bioenergy.

A careful analysis of these reviews shows that there are still two main problems that hinder the reliable and trouble free operation of gasification systems, a) a thorough systems approach to a gasification facility that will address all individual sub-systems as well as their interaction as a whole and b) the gas cleaning with main problem the efficient and economically viable tar elimination.

Gasification is characterised by a continuous and constant operation of all subsystems from the fuel feeding to the generation of power and/or heat, and unless all subsystems operate efficiently the gasification plant is rendered inoperable. The collective experience of problems in this area is numerous though not properly published in the literature. However, most gasification pilot or demonstration plants have faced problems with peripheral equipment to the gasifier rather than the gasifier itself. Indeed areas such as constant feeding, fouling of heat exchange surfaces, tar elimination, wastewater treatment and disposal of effluents and emissions of NOx continue to present barriers to trouble free operation.

In spite of all efforts that have been dedicated to tar elimination over the last 30 years this remains a problem area for most technologies under development with the exception of co-firing and pressurised operation where the fuel gas is maintained at temperatures above 300 °C avoiding tar condensation till its complete combustion. Advances in catalytic tar conversion though significant, still are far from commercial status.

Unless both areas will be critically addressed and the remaining problems of the various processes under development and/or demonstration will be resolved gasification will remain a technology in development.

By the end of 2002 there have been renewed efforts to restart the only two existing IGCC projects, the ARBRE plant in the UK (see GD4 in Appendix) and the Värnamo plant in Sweden with the support of the European Commission via FP5 (for both projects proposals were received and short-listed for support in the last call of FP5 and thus are not included in the Appendix). If all progresses well, both plants are expected to be re-commissioned in the Autumn of 2003. The Brazilian IGCC project has also been support (see GD2 in Appendix) and recent developments are promising that this project could also finally enter the implementation stage. The success of the Lahti co-firing plant (supported by THERMIE) has been disseminated in Ruen, Belgium, and the problems facing the Amer plant in the Netherlands (supported by THERMIE) have been overcome by bypassing a section of the gas conditioning before firing the gas into the coal boiler. Progress has also been achieved with the operation of the Güssing plant in Austria (supported by the Regional Funds of the EC) which is a facility producing medium calorific value gas based on steam gasification with indirect heating. Success was also achieved at the Varkaus bubbling fluidized bed gasifier for the gasification of tetra-pack waste and separation of aluminium (this project was not supported by EC funds) (11).

There is renewed attention to gasification due to the possibility to produce synthesis gas and subsequently liquid biofuels or bio-hydrogen. The eventual markets for such bio-products are enormous on a global scale. The main challenge to be solved with a long-term perspective is to generate a clean synthesis gas that can be used in existing catalytic process for its conversion to bio-methanol, bio-ethanol, bio-Fischer-Tropsch, bio-dimethylether and or other bio-chemicals.

Status of Pyrolysis

Bridgwater has reviewed the status of pyrolysis technologies and projects in these proceedings (12) and therefore no further detailed analysis will be carried out by the authors. However, it is worth noting that from the three pyrolysis demonstration projects in the Appendix, two are based on slow pyrolysis processes and one on flash pyrolysis. Since the projects are ongoing no further information may be given at this stage.

REFERENCES

1. Maniatis K. & Tustin J. (2002) The IEA Bioenergy Research Network: Promoting Bioenergy for Sustainable Development, in Parz W., Spitzer J., Maniatis K., Kwant K., Helm P. & Grassi A., Editors, *Proceedings Twelfth European Biomass Conference*, ETA Florence & WIP Minich.
2. Directive 2001/77/EC, On the Promotion of Electricity produced from Renewable Energy Sources in the Internal Electricity Market, http://europa.eu.int/scadplus/leg/en/lvb/l27035.htm
3. Proposal for Directive COM(2002)508, On the Promotion of Biofuels for Transport, http://europa.eu.int/comm/energy/en/fa_3_en.html
4. Fifth framework programme of the European Community for research, technological development and demonstration (RTD) activities for the period 1998 to 2002, OJ L 26, 1.2.1999, For Workprogramme see European Commission Decision C (2000) 670. http://www.cordis.lu/fp5/
5. Specific programmes implementing the Framework Programme 2002-2006 of the European Community for research, technological development and demonstration activities, COM(2001)279, http://www.cordis.lu/fp6/
6. Maniatis K. & Kotronaros, A. (2002) The Bioenergy Policy of the European Commission, in Parz W., Spitzer J., Maniatis K., Kwant K., Helm P. & Grassi A., Editors, *Proceedings Twelfth European Biomass Conference*, ETA Florence & WIP Munich.

7. Maniatis K. (2001), Progress in biomass gasification: an overview, in Bridgwater A.V. ed. (2001), *Progress in Thermochemical Biomass Conversion*, Blackwell, Oxford, UK. (2001), pp. 1–32.
8. Bridgwater A.V. & Maniatis K. (2003) The Production of Biofuels by the Thermochemical Processing of Biomass, Chapter 10, in M.D. Archer & J. Barber ed. *Photo-conversion of Solar Energy Molecules to Global Photosynthesis*, in print, Imperial College Press.
9. *Status of Gasification in countries participating in the IEA*, (2001), ed. Kwant K. & Knoef H., Novem
10. *Status of Gasification in countries participating in the IEA and GasNet activity*, (2002), ed. Kwant K. & Knoef H., Novem.
11. Kurkela E. (2003), Status of biomass & waste gasification in Finland, in these proceedings.
12. Bridgwater A.V. (2003) in these proceedings.

Appendix: Thermochemical Bioenergy Projects Supported in the Framework of FP5 (first three programme calls only).

N°	Title	Contract & Acronym	Summary	Contractors	Status
GR1	Biomass gasification for CHP with dry gas cleaning and regenerative heat recovery	ERK5-CT-1999-003 DRY GAS CLEANNING	The objective of the project is to develop, integrate and prove a dry gas cleaning system on an existing biomass gasification CHP plant (250 kWth). The innovative idea of this project is a newly developed dry gas cleaning and heat recovery system. The dry gas cleaning concept produces no waste water and, intends to solve and eliminate the tar problems associated with operation of biomass gasifiers by recirculating the tars back into the gasifier. The ultimate objective is to create a new generation of energy efficient and environmentally friendly gasification systems.	1. CIRAD. 2. DK-TEKNIK Energy & Environment 3. Thomas Koch Energi A/S	Ongoing
GR2	Straw gasification for CO-combustion in large CHP plants	ERK5-CT-1999-004 STRAWGAS	The project will develop and test gasification of straw in a circulating fluid bed gasifier (CFBG). The gas is intended as supplementary fuel in a modern coal-fired power plant, which also supplies district heat (CHP). The project includes development and test of a straw feeding system, of gas cleaning equipment and of procedures to ensure reuse of the residues. The project is preparation for a demonstration gasifier which will have a thermal input of 100 MWth and use app. 150.000 tons of straw per year. The energy efficiency from straw to electricity will be 40%. The project can lead to demonstration of an efficient and clean method to use straw for power production in a large scale. The method is clean, since the straw-gas is burnt in a boiler with modern flue-gas cleaning equipment.	1. EK Energi Power Company Lrd. 2. Foster Wheeler Energia OY	Ongoing
GR3	Development of selective catalytic oxidation "Sco" technology and other high temperature NH3 removal processes for gasification power plant	ERK5-CT-1999-020 AMMONIA REMOVAL	The objective of this research is to develop two main technologies for ammonia removal in gasification plants. First, technology based on the use of nickel monolith catalyst for ammonia removal will be studied in order to estimate the feasibility of this process in real gasification conditions. Secondly, new Selective Catalytic Oxidation (SCO) process will be studied in fundamental as well as in more applied level in order to optimise the process and to make a decision about further testing of the SCO technology in pilot scale.	1. Technical Research Centre of Finland 2. Universidad Complutense deMadrid 3. Sydkraft 4. Abo Akademi 5. Foster Wheeler Energie 6. EK Energi Power Company 7. Univesrity of Leeds	Ongoing

N°	Title	Contract & Acronym	Summary	Contractors	Status
GR4	Development of a standard method (protocol) for the measurement of organic contaminants "tars" in biomass producer gases	ERK6-CT-1999-20002 TAR PROTOCOL	Measurement methods and definitions for organic contaminants in biomass gasification based CHP (combined heat and power) systems are numerous and non-uniform. As result, comparison of data and definition of clear maximum allowable concentrations for these contaminants is problematic. This forms an obstacle for market introduction of the CHP systems as the contaminants can cause damage or an unacceptable level of maintenance. The objective of the project is to remove this obstacle by developing a standard measurement method (Protocol) which is accepted and used by parties working on biomass gasification and has entered the standardisation procedure at CEN. This work continues an initiative started by IEA, EU and US-DoE to develop such a Protocol and will be based on the two draft Protocols that resulted from this initiative. The new Protocol will be optimised and tested thought R&D activities co-ordinated from the project. As acceptance and use by others is considered to be very important, dissemination and internalisation forms an essential part of this project. It will be accomplished by the use of Internet, Conference contributions, involvement of Rewiers and a Workshop.	1. ECN 2. CIRAD 3. Univ.Comp. de Madrid 4. Termiska Processer AB 5. Vienna Univ. of Techn. 6. Foster Wheeler Energia 7. Danish Techn. Inst. 8. National Renewable Energy Laboratory 9. Enerkem Technologies 10. Kungliga Tekniska Hoegskolan 11. National Technical University of Athens 12. Technical Research Centre of Finland 13. Universite Catholique de Louvain 14. Lurgi Mrtallurgie GmbH 15. CRE Group Ltd. 16. Verenum	Ongoing
GR5	Sewage sludge gasification for CHP applications	ENK5-CT-2000-050 CHP SEWAGE GASFN	In the EU, constraints on sewage sludge disposal routes are growing as former options, such as marine disposal, landfill, incineration and agricultural uses, are under increasing environmental, legislative and economic pressure. Gasification of sewage sludge for heat and power generation in CHP applications is an attractive concept as it has the potential to provide an environmentally acceptable, efficient and economically viable means of generating energy from a waste disposal problem. The overall objective of the project is to examine the key technical, environmental and economic issues for sewage sludge gasification in CHP applications and to establish the commercial viability of the process.	1. EMC Environment Engineering Ltd. 2. Imperial College of Science, Technology & Medicine 3. University of Ulster 4. Universidad de Zaragoza 5. Technical Research Centre of Finland 6. Sistemas de Transferencia de Calor, S.A.	Ongoing

N°	Title	Contract & Acronym	Summary	Contractors	Status
GR6	Tar decomposition by novel catalytic hot gas cleaning methods	ENK5-CT-2000-305 NOVACAT	The successful implementation of biomass and waste gasification for power production depends on the availability of low-cost, simple, reliable and high performance gas cleaning systems. Present solutions for gas cleaning do not fulfil these conditions. Hence, the project concentrates on improving gas cleaning systems. 1. Develop and improve nickel-activated filter substrates and screen the performance; 2. Determine the stability of the catalytic filter in extended time test with real biomass gasification gas and scale-up the preparation technique; 3. Determine the performance of nickel monolith catalyst, catalytic filter and fluid bed dolomite catalyst combined to fluidised bed gasifier; 4. Determine long-term stability of a nickel monolith catalyst in real gasification gas, characterise the catalyst and develop and test in pilot-scale an optimised complete gas cleaning train.	1. Technical Research Centre of Finland 2. Universidad Complitense de Madrid 3. Vrije Universitetit Brussel 4. Sereco Biotest S.S. 5. Maridiana SRL 6. Condens OY	Ongoing
GR7	Biomass Heatpipe Reformer	ENK5-CT-2000-311 BIOHPR	The project is aimed at the developing, constructing and testing of two 200 kW prototypes of the Heatpipe Reformer. The work will mainly focus on indirectly heated gasifiers - the realization of extremely high heat fluxes into a small sized gasification reactor. The application of heat pipes for this purpose will be tested and demonstrated. The excellent heat transfer characteristics of the heat pipes combined with high heat transfer rates between fluidized bed and heat pipe surface allow to design a very compact, highly integrated apparatus for steam gasification. Both prototypes will be tested with wood pellets, straw pellets and pellets produced from cotton stalk residues. Long-term experiments (72 hours) are planned to test the long-term stability of the heat pipes and to identify technical problems. The requirements of small scale gas turbines and fuel cells will be considered.	1. Technical University Munich 2. University of Stuttgart 3. National Technical University of Athens 4. Deutsche Montan Technologie 5. Luft und Feuerungstechnik 6. Saar Energie	Ongoing

N°	Title	Contract & Acronym	Summary	Contractors	Status
GR8	The influence of tar composition and concentration on fouling, emission and efficiency of micro and small scale gas turbines by combustion of biomass derived low calorific valued gas	ENK5-CT-2000-313 TARGET	Gasification of biomass and combustion in a gas engine or turbine is the most efficient way of power production. However, for gas turbine stars in the LCV gas are a big problem and can result in fouling, increased emissions and failures during operation. In order to avoid those problems, intensive gas cleaning is up to now recommended because limits and effects in gas turbine combustion are not known. This makes biomass gasification less cost competitive then combustion systems. The focus of this project is on the effect of tars on fouling and emissions of turbines and micro-turbines and especially on interactions between gasifier, gas cleaning and (micro) turbine. Only by an integrated approach it will be possible to provide customised & cost competitive biomass IGCC systems.	1. Delft University of Technology 2. Universitaet Stuttgart 3. Alstom Power Technology Ltd. (SW) 4. Kungliga Tekniska Hoegskolan 5. Host B.V. 6. Alstom Power Technology Ltd. (UK)	Ongoing
GR9	Biomass-gasification and fuel-cell coupling via high-temperature gas clean-up for decentralised electricity generation with improved efficiency	ENK5-CT-2000-314 CLEAN ENERGY FROM BIOMASS	The project is aimed at studying, developing and testing the integration of Biomass Gasification and Molten Carbonate Fuel Cell for distributed power generation. The research activities include the assembly and operation of a pilot plant (made of an internally-circulating, catalytic fluidised-bed gasifier, a novel hot gas clean-up system and a 125 kWe MCFC) and an ancillary workprogramme in key areas, which includes cold modelling of the gasifier, detailed simulation of the overall system and catalyst upgrading. The programme is addressed at fully assess the technical feasibility of economically viable units for wide scale industrial applications in the short/medium term.	1. Universita Degli Studi L'Aquila 2. Schumacher Umwelt und Trenntechnik 3. Universite Louis Pasteur	Ongoing
GR10	A new approach for the production of a hydrogen rich gas from biomass – an absorption enhanced reforming process	ENK5-CT-2001-545 AER-GAS	The Objectives of this project are: -Investigate and improve a new, efficient and less expensive process for clean conversion of solid biomass in a single step fluidized bed AER reactor. -Convert biomass into a gas containing > 80Vol. % Hydrogen and < 300 mg/m3 tar. -Develop and improve catalytic absorbent bed material with high mechanical and chemical stability. -Regenerate bed material by combustion of AER residuals.	1. Organisation for Solar Energy and Hydrogen Research. 2. Foundation for Research and Technology-Hellas 3. Proplan Ltd. 4. University of Cyprus 5. Vienna University of Technology	Ongoing

N°	Title	Contract & Acronym	Summary	Contractors	Status
			-Reduce tar content by optimization of operation conditions and bed composition. -Investigate range of organic feedstock that can be processed. -Combine AER gasification and absorbent regeneration in one plant consisting of a fast internally circulating fluidized bed (FICFB). -Design 1MW and 50 MW AER biomass conversion plant including gas purification. -Outline an AER biomass plant for market introduction with less than 800 Euro/kW investment and estimate electricity and H2 production efficiencies and costs.	5. University of Stuttgart 6..Paul Scherrer Insitut 7. Ive Weimer	
GR11	Biomass and Waste conversion in Supercritical Water for Production of Renewable Hydrogen	ENK6-CT-2001-555 SUPERHYDROGEN	The main objectives of this project is to develop the innovative SuperCRitical Water (SCW) gasification process for cost-effective (< 12E/GJH_2) conversion of wet biomass and bio-waste into clean, renewable, compressed hydrogen with an energy efficiency to pure hydrogen exceeding 60%. Integrated parts of the SCW process development are: -The development of a preparation method to convert wet feedstocks of different origin into a high-solids content (up to 30wt%), pumpable slurry -the development of a multi- functional, catalytic, membrane reactor to produce hydrogen with a purity >98 vol % (Grade A), and obtain at least 70% conversion of CO and CH_4 -The basic engineering of the overall process and a detailed cost estimate(more or less 30%)	1. Biomass Technology Group 2. University of Twente 3. University of Warwick 4. Dytech Corporation LTD 5. Netherlands Organisation for Applied Scientific , Research 6. UHDE Hochdrucktecnik 7. Sparqle International BV	Ongoing
GD1	Upgrade to CHP of a Biomass-Fired District Heating Plant in Denmark, Using a Combination of Gasification, a Novel Gas Treatment Process	NNE5/240/1999 BIOCURE	The main objectives are: - To demonstrate an updraft, fixed bed gasifier, designed specifically to gasify wood chips; - To demonstrate an innovative thermal/catalytic cracking system to clean the raw gas, to make it suitable to fuel a 1 MWe gas engine/generator set gas; - To demonstrate operational reliability and economic potential, high overall efficiency and low environmental impact.	1. Shell International Renewables 2. Lemvig Varmevaerk 3. Jenbacher 4. Wellman Process Engineering 5. A/S Dansk Shell	Abandoned

N°	Title	Contract & Acronym	Summary	Contractors	Status
GD2	EU-Brazilian Industrial Demonstration of Gasification to Electricity	NNE5/489/1999 EU-BR-IDGE	This proposal has as its goal to demonstrate biomass integrated gasification, gas turbine combined cycle (BIG-GT) technology in the largest installation this far by constructing a large, 32 MW net electrical, power plant in Brazil for demonstration and commercial operation based on, eucalyptus wood from plantations. In the plant tests will also be performed with agricultural residues (e.g. sugar cane bagasse) available locally in large quantities. The objectives of the R&D part of this combined projects are: 1. To establish the technical and other requirements necessary, e.g. fuel recovery technology and cost, pilot plant and prototype testing; 2. To proceed into a large scale demonstration of a power plant BIG-GT plant fuelled by an agro fuel, bagasse; 3. To access the huge energy and emission reduction potential of the sugar cane industry.	1. Termiska Processer AB 2. Centro de Tecnologia Copersucar 3. Thomas Koch Energi A/S 4. Consórcio SER -	Ongoing
GD3	Demonstration of a 500 KW Fixed Bed CHP Gasifier system Including a Novel Tar Cracker	NNE5/528/1999 DEMO-GAS	The main objective is to demonstrate, monitor and optimise fixed bed gasification systems. The demonstration plant should establish the warranty conditions for a commercial product in the capacity range of 500 to 2500 KWe. Specific objective are: (i) To scale-up the fixed bed gasifier and the catalytic tar cracker from 30KW to 500 KW electric output (16x); (ii) To combine the different components developed within different R&D projects into an integrated system; (iii) To produce CHP efficiently (electric efficiency 36%) with minimal environmental impact (no harmful ash, tar and condense water production); (iv) To use as fuel a difficult biomass residue; a woody biomass waste stream from the composting process which has a low density and moreover not uniform; (v) To establish warranty conditions for commercial fixed bed gasification systems.	1. N.V. Edon Group 2. BTG Biomass Technology Group B.V. 3. Kara Energy Systems B.V. 4. Sork Thermeq B.V. 5. Jenbacher AG	Cancelled

N°	Title	Contract & Acronym	Summary	Contractors	Status
GD4	Completion of the ARBRE Plant with the Typhoon Gas Turbine	NNE5/20065/1999 ARBRE-TYPHOON	The primary objective of the project is to demonstrate that BIGCC technology has reached industrial reliability and can therefore be scaled up and replicated internationally. This will be achieved by demonstrating conclusively that: - short rotation forestry (SFR) can be successfully established in the EU under market conditions to produce the renewable fuel resource for power plants; - fluidized bed gasification of biomass followed by catalytic cracking of tars has been scaled up to industrial size operations successfully; - the Typhoon gas turbine has been successfully modified to burn low calorific value fuel gases and now commercial guarantees can be provided. The project also aims to demonstrate operational reliability and economic viability with very high overall efficiency and very low environmental impact.	1. Kelda Group Plc. 2. Termiska Processer AB 3. Alstom Power UK Ltd.	Ongoing
GD5	13 MW CHP Plant Based on Biomass Gasifier with Gas Engines	NNE5/124/2000 BGGE	The partners will demonstrate and optimise a novel CHP plant based on biomass gasification on a full scale (13 MW thermal input). In this plant two biogas, fuelled engines produce electricity (4 MWe) from gasified biomass. While the district heat (8 MWth) is produced in various heat exchangers. The BGGE CHP plant will be built at an existing CHP plant in Lemvig, Denmark.	1. FLS Miljoe a/s 2. Energy Power Resources 3. Jenbacher 4. Carbona 5. Lemvig Varmevaerk	Ongoing Request to change location of project
GD6	Testing the Gasification of Secondary Fuels for High-efficiency Energy Conversion	NNE5/285/2000 MULTIFUEL GASIFICATION	The objective of the proposal is to test the applicability of very diverging secondary fuels in a circulating fluid bed gasification system including gas cooling and gas cleaning followed by combustion of the product gas in a boiler. Examples of secondary fuels to be considered are: poultry litter, sewage sludge, paper sludge and other industrial sludges, plastic and paper derived fuel, oil sludges from tanker cleaning, waste plastics, different biomass products. Final goal is CO_2 emissions reduction and/or re-use of low quality waste materials.	1. N.V. EPZ 2. Netherland Energy Research Foundation 3. Lurgi Envirotherm GmbH	Abandoned

N°	Title	Contract & Acronym	Summary	Contractors	Status
GD7	A 1MWe Biomass Fluidised Bed Gasifier Power Plant with Catalytic Conversion of Tars	NNE5/312/2000 BIO- GASCAT - POWER	The overall objective of this project is to support the realisation of Bioenergy in the EU by demonstrating an innovative technology, which can be applied for the medium scale applications (0.5 - 5 MWe) for which a very large market is known to exist. The project will also have a significant effect on the socio-economic aspects of sustainable development and will be assisting the EU towards meeting its Kyoto obligations. The project will demonstrate the efficient gasification of biomass residues to a fuel gas, the innovative and efficient catalytic conversion of tars, the continuous & stable combustion of the clean fuel gas in a gas engine of nominal rating of 1 MWe, the operational reliability of the complete plant and significantly lower environmental impact compared to fossil fuels. Also, the project will solve a major environmental & disposal problem concerning biomass residues, which is faced by numerous local authorities in the regions of the EU.	1. Envitec SA 2. Vrije Universiteit Brussel 3. Universidad Complutense de Madrid 4. Municipality of Nea Makri 5. Jenbacher AG	Ongoing
PR1	Pyrolysis oil for heat generation: verification of a second-generation pyrolysis process	ERK5-CT-1999-011 PYROHEAT	Among the biomass-derived biofuels, pyrolysis oil (PO) appears to be the most cost effective. Pilot plants installed in Europe suffer of many defects typical of young technologies. Moreover, the oils have unfavourable fuel properties. The proposals intends to advance the pyrolysis technology developing new concepts for the pyrolysis process and making technological improvements in the pilot plant that will used to assess the concepts. The use of the oil in small and large-scale thermal boilers is foreseen to verify the fitness of PO to substitute heavy and light fossil fuels and to assess the environmental impact.	1. Enel Produzione SPA 2. Technical Research Centre of Finland 3. ASUR IA 4. CCT S.R.L. - 5. Fortum Oil and Gas OY 6. UGM SPA	Ongoing
PR2	Catalyst Development For Catalytic Biomass Flash Pyrolysis Producing Promising Liquid Bio-Fuels	ENK6-CT-2001-510 BIOCAT	Biomass flash pyrolysis is the most promising process for the production of renewable liquid fuels. Large scale applications are not foreseen shortly, because of the high upgrading costs of the liquids that are necessary (e.g. polymerisation, high viscosity) for their effective handling and processing as a renewable fuel. Catalytic biomass flash pyrolysis (CBFP) is a	1. Grace GmbH & CO KG 2.Dechema Gesellschaft Fuer Chemische Technik & Biotechnologie E.V. 3. Foundation for Technical & Industrial Research at the	Ongoing

N°	Title	Contract & Acronym	Summary	Contractors	Status
			promising and economical process combining pyrolysis and catalytic upgrading in the same reactor. Commercial catalysts were found inappropriate for CBFP leading to high water and coke yields. Development of tailored catalysts for CBFP within this project will substantially improve the liquids fuel quality (less water; higher heating values; increased content of hydrocarbons and valuable chemicals, ~30 wt% each).	Norwegian Institute of Technology 4. Kungliga Tekniska Hoegskolan 5. Adhesives Research Institute Ltd. 6. Biomass Technology Group BV	
PD1	Pyrolytic Advanced Process for Energy and Recycling	NNE5/144/2000 PAPER	The purpose of PAPER is to demonstrate a pyrolysis plant running on a industrial scale to take advantage of the energy generated by waste derived from the production of recycled paper and which is applicable to any waste of similar characteristics. The waste is generated during the industrial processes of breaking down and purifying the paper pulp obtained from used paper. 80% of the cellulose is recovered and the waste generated consists fundamentally of polyethylene, aluminium (from liquid board packaging) and a small amount of fibre. The industrial pyrolysis plant will be designed to treat 36.000 ton per year of waste in order to produce steam that will be used directly in the manufacture of recycled paper or will be sent to condensation steam turbine generating electricity.	1. Alier S.A. 2. UM Engineering SA 3. Tetra Pak France 4. Institut Catala C'Energia	Ongoing
PD2	Demonstration of a Flash Pyrolysis Plant	NNE5/233/2000 DEMO-PYROLYSIS	The main objective is to construct and operate a 50 t/d wood flash pyrolysis plant. The technical objectives are to establish: 1. Flash pyrolysis process characteristics, such as the plant availability, validation of environmental impact, exact manpower requirements and skills, accurate capital costs (+/-15%) and technical specifications. 2. The warranty conditions for the pyrolysis oils, with respect to oil quality, its consistency and the production costs; 3. Production of 1,000 tons of pyrolysis oil to combust in a district-heating boiler.	1. BTG Biomass Technology Group B.V. 2. Ansaldo Ricerche s.r.l 3. ASM Brescia S.P.A. 4. Kara Energy Systems B.V. 5. AS Enere. A.	Ongoing

N°	Title	Contract & Acronym	Summary	Contractors	Status
PD3	Tyre Pyrolysis Process for Cement Manufacturing and Tyre-to-energy Plants	NNE5/271/2001 TYREPYRO	The aim of this project is to demonstrate a patented tyre pyrolysis process with energy and by-product recovery, first in a cement plant and subsequently in a full scale 17 MWe tyre-to-energy power plants. Scrap tyre is undesirable for landfill, but has very high calorific value, and is due to the high natural rubber content a renewable energy source. In tyre pyrolysis process scrap, tyres are partially gasified producing fuel gas and residual saleable char. In the tyre-to-energy plant the fuel gas is completely combusted in a secondary chamber and is passed through the steam boiler used in the steam cycle electric power plant. The ultimate objective of this project is to demonstrate tyre-to-energy plant's long-term technical and economical viability.	1. FLS Miljoe a/s 2. Energy Power Resources Ltd. 3. Carbona Incorporated 4. BPI Projects Ltd.	Ongoing
DIS	Network Cluster on Thermal Biomass Conversion Implementation	NNE5/168/2000 THERMONET	This project has established a cluster of two Networks on thermal processing of biomass for fuels and electricity. One network will address gasification and the other pyrolysis. Each network will have its own work programme, but both will have a common focus of addressing commercialisation issues and providing support for more rapid and more effective implementation of all the technologies in the market place. There will be joint tasks that involve both Networks and provide common approaches including pyrolysis of waste for gas production, market assessments, economics, legislative impacts, health and safety, bio-fuel standards and incentives for bio-energy. The Networks will function through a regular programme of meetings and workshops that will address a range of technical and economic issue that affect those industries that are developing and potentially using these conversion processes. Outputs include a regular newsletter, websites and technical reports that address all the issues under consideration. The Cluster Co-ordinator will provide administrative support to manage the Network while an Executive Group of the two Network Leaders will oversee the integration and interactions	1. Aston University 2. Biomass Technology Group B.V.	Ongoing

N°	Title	Contract & Acronym	Summary	Contractors	Status
			between the two networks. The output will be the establishment of a comprehensive trans-European forum on thermal biomass conversion that will cover all the technical and economic issues.		

Key to Table:

G = Gasification
P = Pyrolysis
R = Research Project
D = Demonstration Project
DIS = Dissemination/Network Project

The Future for Advanced Thermal Conversion Technologies – The Commercial Perspective

IP Burdon
PB Power, Amber Court, William Armstrong Drive
Newcastle upon Tyne, NE4 7YQ, UK

ABSTRACT: This overview attempts to explain the way in which new and advanced technologies find a place in the world's electricity supply industries. It discusses the commercial realities facing a designer or developer of such technologies and highlights the gap between the "testbed" prototype and the fully-proven "turnkey" scheme; a gap which has to be bridged to get the technology into the market. It suggests that Governments which enter into commitments under the Kyoto Protocol have responsibilities to support and encourage the new technologies which will make those commitments attainable in a free market economy.

INVOLVEMENT IN ADVANCED THERMAL CONVERSION ENERGY TECHNOLOGIES

PB Power is involved with many of the so-called renewable energy technologies including wind, hydro and energy-from-waste schemes. The latter schemes often involve the utilisation of a biomass feedstock, particularly in mass-burn or fluidised bed combustors. More advanced technologies, such as gasification and pyrolysis schemes, have been evaluated by us, either on behalf of a project developer who may wish to have an independent view of their potential, or on behalf of a Bank who may be contemplating the provision of debt finance to a specific project.

RESEARCH, DEVELOPMENT AND DEMONSTRATION

PB Power, as an organisation, generally acts, firstly, as a technical advisor on electricity supply and, secondly, as a project manager on large capital schemes eg power stations and transmission and distribution networks. About one-half of our corporate turnover is attributable to the former activity, the balance being on activities associated with project management. We undertake, therefore, very little **practical** research and development ie we neither design nor build experimental or prototype devices for research or manufacturing development purposes. We do, however, undertake "desk-top" research, ie investigation into the performance, characteristics and costs of the newer, less-conventional, energy technologies as part of our technical advisory activities mentioned above.

Demonstration of performance is a **critically** important activity in the development of new technologies. It almost goes without saying that new technologies are brought to market because they:

- are more efficient at converting primary resources (such as fuel) into useful and versatile energy such as electricity;
- have less of an adverse impact on the environment in which they have to operate;
- are less expensive in terms of either or both of their capital costs and their ongoing operation and maintenance costs.

One or more of these benefits have to be demonstrated to be applicable to the new technology if it is to find a buyer in the market place.

THE CURRENT MARKET PLACE

In many countries in the developed world, and in some countries in the developing world, there has been a relatively rapid shift in the last 10-15 years away from the old order of the national electricity supply industry being a department of the State where the taxpayer effectively supported the difference between the costs that the industry incurred and the revenue that it earned from sales of electricity. The new order takes the form of either the sale of the entire assets of the industry by the issue of shares (equity) or the solicitation by the State (who maintains ownership of the majority of the industry) of offers of private-sector participation in certain parts of the industry (most likely investment in the electricity generation part of the industry). This latter arrangement is generally adopted where the State has neither the means nor the inclination to invest in new assets. Long-term arrangements for the purchase of electricity from the private generator are generally offered as an inducement to participate in such an arrangement. Commercial arrangements will be put in place to ensure that the private generator honours his commitments to provide electricity when required.

Where the entire industry is offered for sale (as was the case in the UK in 1990), all generation assets are owned by plcs and the generators sell their output to electricity retailers who are also plcs. Similar strict commercial arrangements exist between producer and retailer in this model as in the limited privatisation exampled above as being typical of the developing world. It is important that the commercial realities of such strict commercial arrangements are appreciated in the consideration of the place for advanced technologies in the real world.

TECHNOLOGY RISK

When the world's electricity supply industries were, for the majority of the developed and the developing world, state-owned utilities, they were funded from the Government exchequer and the risk of a new technology failing to perform in accordance with expectations were rarely quantified in financial terms – the exchequer would bear any losses.

In the developed world eg in the UK prior to 1990 and in France up to the present time, many new technologies were introduced as experiments on a limited scale with the exchequer bearing the costs of the associated research and development. Sometimes such expenditure was proven in the long-term as money well spent, often it was not. The whole of the nuclear power programme in the UK and in France was funded in this way.

The privatised industries, however, take an entirely different view; risk, whether it be technology risk or any other of the risks they face in their day-to-day operations have to be quantified and offset or managed in some way. High risks have very little or no place in the highly-efficient and cost-effective private electricity supply industries of the 21st Century. Neither do these industries have large budgets for research, development or demonstration. Their profit margins do not provide for the allocation of funds for such activities and neither would their shareholders see any rationale for such activities.

Accordingly, the new companies in electricity supply have little or no time for the unproven technologies which are radically different from those which they are intended to replace.

ESTABLISHING CONFIDENCE

It almost goes without saying that, in a changing world, the demand for new energy technologies which are more efficient, cleaner or less-expensive will increase from industries which are striving to maintain a competitive advantage in a free market situation. Opportunities exist, therefore, to bring forward new ideas and new technologies to fulfil this latent demand. Demonstration projects play an essential part in establishing confidence and allowing forward predictions of performance to be made and incorporated into business models. This begs the question as to who will bear the cost of the demonstration. The gap between the "test bed" and the full-scale "turn-key" commercial scheme is vast and the costs that can arise in bridging that gap are huge. The appendix to this paper gives more details of this aspect of prototype technologies.

DEMAND PULL

It is arguable that the genuine demand for electricity produced from renewable resources, is small, even though it is widely acknowledged that there is a potential threat to climate stability from global warming and that there is a connection between the ever-increasing man-made CO_2 emissions into the atmosphere and climate change. The population of the market place are, in the main, disinterested in these esoteric scientific concepts and are much more interested in the price of the product on offer, ie the kWh of electricity, than the environmental costs associated with its production. In any event, apart from the nuclear part of the electricity supply industry, these external costs are not borne by the producers of electricity anyway and so the true **total** costs are not passed on to the end-user. Thus the much-praised market mechanisms associated with unfettered trading of the biggest (by value) commodity bought and sold in the world today are failing.

Conscious of obligations arising under the Kyoto Protocol, many Governments are distorting this free market by imposing obligations on the retailers of the commodity ie the electricity supply plcs to secure a proportion of their sales from "green" or zero-carbon sources. In the UK this will amount to 10 per cent of total electricity sales by 2010. Thus there is an intervention in the supply chain between the electricity producer and the retailer to induce the introduction of renewable energy technologies. This arrangement has been in existence since April 2002 but there is little, if any, evidence that it is encouraging the development and application of the new low-carbon technologies or projects.

Rather than distort the market mechanism to try to force zero-carbon technologies into the production sector, a more efficient mechanism could be adopted whereby the external costs associated with traditional producers were attributed to them as a "carbon/environmental tax" to allow the conventional price mechanism to operate more efficiently. The current situation in UK electricity supply in which major non-carbon producers ie the nuclear power generators, are denied Climate Change Levy Exemption Certificates on their production, demonstrates the absurdity of the current "market".

CONCLUSION

The aggressively competitive organisations which characterise the electricity supply industries in most of the developed world will continue to strive for competitive advantage and seek to win market share by adopting more efficient and more environmentally-friendly technologies. The drivers emanating from the Kyoto accord will provide encouragement, if not compulsion, to the adoption of cleaner or "greener" technologies for the production of electricity.

In order to satisfy the sometimes conflicting demands for fuel conversion processes which, on the one hand, are more efficient and cleaner but, on the other hand, are more technologically advanced and perhaps more complex, process manufacturers will have to demonstrate that the risks involved are small or that they can be managed satisfactorily without any sacrifice of cost, efficiency or environmental impact.

Demonstration projects will need to be established, perhaps at full-scale, to prove to sceptical developers and investors that the technology is capable of delivering the benefits that are claimed for it.

Funding of such demonstration projects is unlikely to come from the private-sector companies whose object in life is to sell electricity and satisfy shareholders. The signatories to the Kyoto Protocol have a further responsibility to ensure that their commitments can be achieved in practice. This they can do by providing support for advanced technologies until they are securely established in the aggressive commercial marketplace that they were responsible for creating in the first place.

APPENDIX

Test Bed to Turnkey – The Introduction of New Thermal Renewable Energy Technologies

IP Burdon

ABSTRACT: The step between a laboratory-scale research project and a full-scale, commercially-viable, scheme which will attract financial support from investors is large. Considerable effort needs to be expended by the developer to demonstrate that the risks involved in moving from test-bed into a competitive, environmentally-sensitive, real world are mitigated to the maximum extent and borne by the project participants most able to control them. The paper considers a generic fuel-energy conversion process and describes the commercial climate in which such a project has to develop, grow and survive. The trends in the world's electricity supply industry, from state-controlled utilities to liberalised, commercial, regimes in which competition plays an important part are explained. Key ingredients for a successful project are identified and the way in which independently-owned power (ie electricity) projects are structured is described. The requirements of the financiers of such projects are highlighted. The author presents his views from the standpoint of an "independent engineer"; his organization often being employed in such a role by project financiers to appraise energy projects employing novel technologies. It is anticipated that the information given in the paper will be of use to individuals or organisations who are seeking financial support to take an advanced-technology process from a proven design concept to commercial operational service.

INTRODUCTION

Commercial viability is a fundamental requirement for any process or project which is intended to supply useful energy into the competitive market places which, nowadays, characterise the electricity supply industries in many parts of the world.

Most independent power projects which compete in these markets are financed "off-balance sheet" ie they rely on debt provided by banks and similar financial institutions to provide 70-85 per cent of the project's funding. The banks will require all risks associated with the project to be identified and mitigated or accepted by others; they will have a thorough appraisal of the project's technical and commercial aspects conducted by independent engineers and lawyers with all "due diligence".

This paper describes the project development process and attempts to highlight the commercial climate in which an energy project, based on a thermal conversion process, has to become established and survive. The detail of the thermal conversion process itself, ie the "technical" process, is not relevant to the conception or the understanding of the project development process, ie the "financial" process, although the former process will have a major influence on the project's risk profile which will come under intense scrutiny in the latter process.

A GENERIC PROCESS MODEL

Almost all thermochemical biomass conversion processes can be considered and thus analysed as a "black box" with a feedstock input and an energy output (either thermal or electrical). Useful co-products will arise in the operation of the process (eg sulphur, acetic acid, methanol, even food flavouring!) from which revenue may be earned. Relatively useless by-products (eg ash, frit and other emissions to air, land and water) which arise may incur a cost in their disposal if a market for their use cannot be found.

Certain defined design and operational parameters will be intimately associated with the black box and these are shown on the diagrammatic representation of the overall process shown in Fig. 1. The parameters can be considered of two types:

(1) "Hard" - ie those which are amenable to scientific analysis and, therefore, fairly precise calculation or prediction.

(2) "Soft" - which may be largely qualitative and, accordingly, subject to opinion and be affected by judgement or simple prejudice.

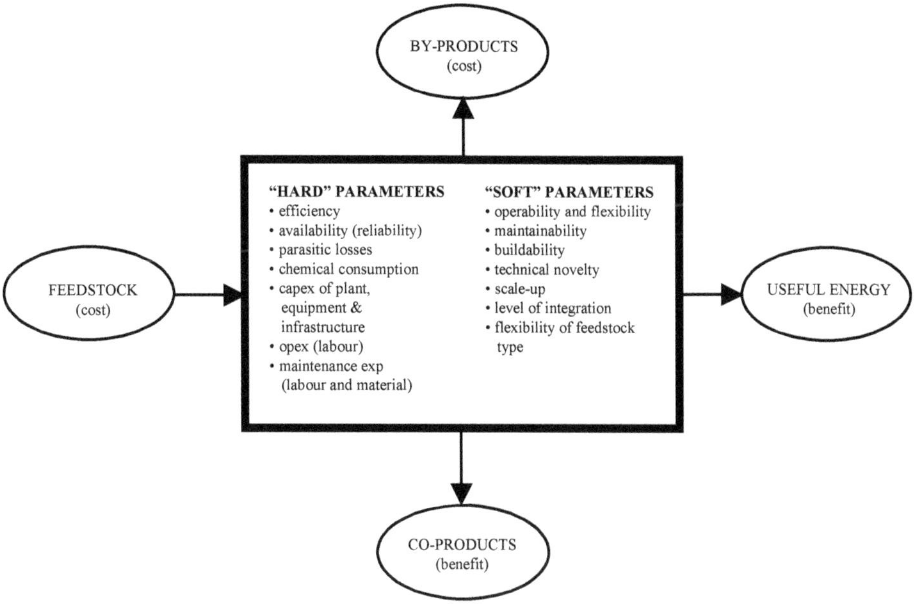

Figure 1: Black-box approach to project evaluation and risk assessment.

All of the hard parameters will need to be taken into account in the financial model of the scheme which will evaluate the cashflows, in terms of costs and revenues to calculate the financial return on the investment in the project.

Two vitally important hard parameters associated with any conversion process, which have a critical effect on project viability, are efficiency and availability. These are defined as follows:

$$\textbf{efficiency} = \frac{\textbf{useful electricity exported}}{\textbf{heat value of feedstock}}$$

$$\textbf{availability} = \frac{\textbf{period time - maintenance time - forced outage time}}{\textbf{period time}}$$

NB efficiency should be measured at defined ISO-type conditions and forced outage time includes time at reduced capacity and the attainment of high availability is critical in maximizing the energy produced from the process and, therefore, the revenue.

The financial model is characterized, in summary, as follows:

(1) Based on selected technology
(2) Includes all project development costs
(3) Includes costs of the site, insurances, working capital and contingency allowance
(4) Allows for inflation and taxes
(5) Gives IRR/NPV, cashflow and cover ratios (ie the ratio of net project revenue to debt service costs) on project investment.

The model invariably involves many iterations before a satisfactory combination of project parameters is assembled; it will set the scene for the business case which the investor or the financier will consider prior to committing funds to the project. Further discussion on the financial model is given in the section on risks and critical issues later.

THE REAL WORLD

Once upon a time, electricity supply was a monopoly activity, the industry's assets were owned by the state in most countries or, where they were not state-owned, such as in the US or Germany or Japan, they were heavily regulated.

In the 1980s, fundamental changes took place in the state-owned sector with the so-called "privatization" of the industry. First in Chile, then in the UK, the assets of the publicly-owned electricity authorities were vested in companies which issued equity shares to the general public. This change in ownership was coupled with a fundamental shift in the way electricity was produced and sold. New, cheaper, technologies such as gas-fired, combined cycle, gas turbine plant, which could be built more quickly than the traditional coal-fired steam turbine plant, and which is less labour intensive in its operation, was introduced extensively in the UK. The separate producers of electricity competed with each other on price (per kWh) to win market share. It is alleged, with some justification in the UK, that such competition has reduced the price to the end-user.

In such a highly competitive market place, expensive technology, and that which is unproven or of a small-scale, will find it very difficult to compete with tried and tested CCGT plant fired on plentiful and (currently) cheap natural gas.

In the developing world, the requirement is for large increases in reliable capacity to provide power for industrial development, new townships and agriculture. Governments in such countries, in most parts of the world, are encouraging private sector developers to invest in the national electricity system and sell power to the distribution companies for onward supply to consumers. Again, conventional technology, utilising lowest cost, often indigenous, fuel such as coal, is a powerful imperative in the thinking of private sector investors.

Thus, it is suggested, expensive complex technology will be hard pushed to find a secure place in a market place where the emphasis is on low capital cost and technical efficiency for large-scale power production.

One area which, however, does offer some chance of success for advanced conversion technology lies in the field of renewable energy. The ability of a process to accept a feedstock which is sustainable, such as biomass, or can be considered, to all intents and purposes, as a renewable such as waste arisings eg sewage sludge, municipal solid waste or food waste, from large centres of population, can be beneficial for technological development in the context of a national renewable energy policy.

Such a policy manifests itself, for example, in the UK at the present time where Government has set targets for 5 per cent of all electricity production to be from renewable energy sources by 2003, rising to 10 per cent in 2010.

Thermal conversion of biomass will play a part in meeting these commitments, along with wind energy, hydro, etc. Premiums paid for renewable energy from such sources can compensate for the inevitable high capital cost of the technology and the disadvantages arising from dis-economies of scale.

A further important issue to be considered in a "real world" environment is not only competition in the market place for the output of the conversion process (ie electricity or heat) but competition between rival technologies for the attention and sponsorship of project developers. Factors which are likely to be considered by developers in selecting a particular technology will include:

(1) Previous operating experience - where the "technical" process has been developed from laboratory prototypes, data will need to be gathered on the performance of the process in terms of its thermal efficiency in recovering useful energy from the feedstock and the production of by-products (and co-products if relevant), particularly discharges to air, water and land which will be of keen interest to the environmental regulatory agency. Inevitably, such operating experience will, generally, relate to a small scale plant and predictions on performance and operating experience will have to be made where scaling-up is envisaged (see below).

(2) Scale of process – those investing in a project where significant scaling-up is needed to attain critical "commercial" mass will require evidence that such a scale-up is a practical proposition and that the relevant factors have been taken into account, both from a mechanical engineering aspect and from the thermochemical conversion standpoint. Sometimes, a multiple parallel-flow process may be adopted to avoid having to stretch the design too far. The risks inherent in the scaling-up will be of special interest to those involved in the independent project appraisal.

(3) Current status of technology – the development status of the technology is interrelated with the feedstock type to a large extent. For example, whilst the gasification of a coal feedstock is a relatively well understood and well tried process, the gasification of many biomass-type wastes is not, certainly outside of a research laboratory. The question generally asked under this heading in the appraisal process is "how long has the technology been around and where can we see it working?"

(4) Cost/MW of capacity – given that the process will be competing for developers funds against a number of other technologies, there will have to be significant advantages, in terms of efficiency or environmental emissions, to be gained from relatively expensive processes if the attention of a developer is to be caught.

(5) Potential application – it should be remembered that a conversion process can, fundamentally have one of three potential uses:-
    - electricity generation
    - feedstock utilization or disposal
    - production of co-products.

Each of these aspects needs to be quantified in the project evaluation process.

FINANCING A PROJECT

The majority of independent energy projects which have been promoted in countries such as the UK, which have a non-parastatal electricity supply industry, have been funded on the so-called "project finance" basis. Under such an arrangement (which was developed originally in connection with North Sea oil projects in the 1960s) the majority (in the region of 80 per cent) of the finance required to implement the project comprises debt borrowed over a period of, typically, 12-15 years. Bond financing can extend this period to 25 years in certain cases. The debt is repaid from the net revenues of the project. The remainder of the finance required for the project comprises equity provided by the project sponsors.

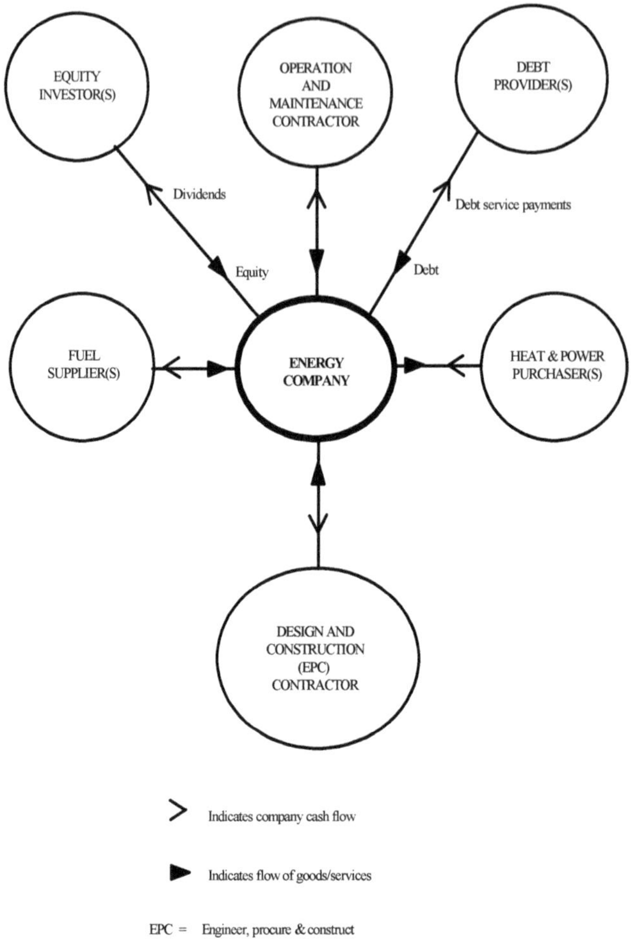

Figure 2: Contractual arrangement for independently financed energy conversion project.

SPECIAL PURPOSE COMPANY

In order to provide a vehicle to take the project forward on a project finance basis, a special purpose company is established by the project sponsors. This company (the "energy company") becomes the counter-party to the various agreements for goods and services which, essentially, comprise the project (see Figure 2). The primary agreements in a typical renewable energy project are:

(1) Fuel or feedstock supply
(2) heat and electricity off-take
(3) design and construction of the facility
(4) operation and maintenance.

There are other secondary agreements covering lease or sale of land, finance etc as well as statutory consents such as planning, integrated pollution control, consents to build a power station, licence to generate electricity etc.

The primary agreements are written in such a way that each complements one another eg the specification and quantity of feedstock supplied corresponds with that required by the plant to be built under the design and construction (D&C) contract to deliver the electrical energy output specified in the electricity off-take contract. At the same time, the obligations contained within the operation and maintenance (O&M) contract require specific levels of fuel conversion efficiency and plant availability to be attained by the operator in line with the terms and conditions of the electricity off-take contract.

The essential elements of each of these primary contracts are summarized below:

Fuel supply

Specific quantities of feedstock, of a defined quality, are to be made available under specified conditions at a particular price. A "gate fee" ie a payment from the feedstock supplier to the project company is generally paid in the case of a project in which the disposal of a waste (eg a by-product from a food processing industry) is a fundamental purpose of the project. Some form of "put or pay" arrangement is generally included in the contract in such circumstances. Where the feedstock has an intrinsic value eg it can be used as a fuel in other processes (and, thus, there is some competition between procurers), the independent assessors of the project will require to see evidence of its long term, reliable, delivery in sufficient quantity to satisfy the needs of the project. Some measure of price stability will be necessary, for example by linking fuel purchase price to recognized and appropriate cost indices. Diversity of supplier is generally desirable, whether there is a gate fee payable by the supplier or the feedstock has to be paid for by the project company, so as to mitigate the risk of a single supplier going out of business. The duration of the fuel supply agreement usually exceeds the term of the finance.

Electricity off-take

An electricity off-take contract in a renewable energy project in many countries which offer support to renewable energy is usually a pro-forma agreement with one or other of the generating companies (or perhaps electricity supply companies) for a fixed term sometimes at a premium price. In the UK, for example, the Non-Fossil Fuel Obligation under the 1989 Electricity Act obliged the electricity distribution companies to procure a certain amount of the electricity that they supplied to their customers from renewable sources. Similar obligations exist under the recently introduced Utilities Act.

Design and construction

The D&C contract in any project financed on a project finance basis is, invariably, based on a turnkey arrangement. This ensures that a single contractor assumes an undivided responsibility for completion date, final cost and performance of the facility. Liquidated damages are payable by the contractor for delay in completion or shortfall in performance. Such a contract is based on a carefully-written performance specification issued by the project company against which the contractor submits a fixed price tender in competition with other contractors. The contract is generally arranged so that the turnkey contractor accepts all the major construction risks.

Operation and maintenance

Given that the payments made to the O&M contractor are likely to be a small fraction of the revenues earned by the project as a whole from electricity sales, it is not usually possible to secure damages from the O&M contractor for failure to achieve set performance targets for output (ie efficiency and availability) which are commensurate with the value of the lost electricity output. The essence of an O&M contract, therefore, is the establishment of, and commitment by the O&M contractor to, an agreed quality assurance system covering the activities of the contractor such as organization, documentation, safety, training, reporting, planning etc. Regular audits of the contractor's activities and procedures give confidence to the owner that the contractor is likely to achieve the required performance targets. There may also be financial incentives included in the contract to encourage the O&M contractor to meet the desired level of performance and, conversely, penalties for non-achievement.

Consents

Certain consents will be required by the project owner or operator prior to putting the facility into use. These, typically, include in the UK for example Integrated Pollution and Prevention Control Authorization (IPPCA) for thermal conversion processes, water abstraction licences for hydro electricity schemes etc. Assuming the use of proven technology and competent D&C and O&M contractors, such consents may not be difficult to obtain, unless the project is sited in a particularly environmentally sensitive area. In the case of pyrolysis processes, it will be necessary to convince the environmental regulators that the organic constituents of a pyrolysis oil-fuel (some of which might be carcinogenic) are destroyed in the overall energy production process.

An environmental statement is required under planning legislation in certain countries for all electricity generation projects and consultation with appropriate organizations may be necessary as part of the procedure for obtaining planning permission.

RISKS AND CRITICAL ISSUES

The main categories of risk associated with a renewable energy project are listed in the Appendix. The majority of risks associated with a given project can be covered in the contractual arrangements described in the preceding section. In this way, risks are allocated to the party who is best able to deal with them. The construction contractor, for instance, will typically be called upon to provide a guarantee of the performance of the plant, with appropriate damages for non-performance which protects other participants from financial loss. Project financiers will wish to see a suitable spreading of risks and mitigation of them (perhaps by insurance) over all the project participants. It is highly unlikely that the banks responsible for the provision of debt to the project will agree to bear any of the risks themselves.

RISK ANALYSIS

The interests of lenders, insurers and contractors require a very careful analysis of risk in the context of the various contractual obligations that go to make up the structure of a project.

Matters which would typically be regarded by lenders as representing a potential risk include:

(1) Technology - lenders will expect an independent engineer to report on the technical specification for the plant and will require evaluation of the appropriateness and track record of the relevant technology in the proposed application. The developer may be enthusiastic about the potential benefits of a new technology whereas the lender would regard untried features as an unacceptable risk;

(2) Contractors - lenders' interests extend beyond just the identity of the principal contractors. They will consider their reputation and financial strength in assessing whether there is any risk of default in relation to any function which is material to the project's success;

(3) Security of cash flow - lenders will analyse all factors which may have a bearing on the ability of the project to maintain the project cash flow comfortably beyond the prospective term of the borrowing including, for instance, as mentioned in the previous section, the security of the fuel supply and contingency arrangement for alternative fuel;

(4) statutory authorizations - lenders will check on all factors affecting the consents and permits necessary for the commencement and continuation of the project.

The success of the developer in structuring and obtaining finance for projects is closely linked with their ability to identify areas of risk and to design technical solutions and allocate contractual responsibility so as to create the greatest comfort in the light of perceived risk. Much of this can be achieved by seeking initial expert advice in each relevant field so that there is a coherent strategy from an early stage. Some of the relevant issues which relate to risk analysis include:

(1) Financial model - a crucial test is to see how the economics of a project stand up to the perceived areas of risk. The development of any project will involve the production of a financial model which shows how the sensitivities of the various parameters at risk affect the economics. This is a process which will be followed by any prospective lender in relation to a funding proposal.

(2) Contractual framework - the contractual framework for any project will need to take account of the risk factors which are relevant to the specific features of the project. It involves careful planning on the identity of the contracting parties and assessment of their capacity to discharge their obligations, the scope of the positive obligations under each contract and the effect of limitations of liability.

(3) Financial structure - the funding structure will require a detailed evaluation of the capital requirement and the balance between equity or "risk" capital and loans. The proposed "gearing" will indicate the potential reward for the providers of equity and will also be an important indication in any proposal for project finance of the way in which the developers have considered the allocation of risk. Lenders like to see proposals which show that these principles have been understood and will gain confidence in the developers if the proposed financial structure of the project takes them into account.

(4) Insurance - insurance cover will be available for many of the areas of perceived risk and it is important that the scope of the cover is planned with reference to the specific detail of the project. This will, in any case, be a requirement of the lender. Specialist insurance brokers can assist in creating an effective insurance programme covering relevant interests with an indication of the sort of scope of cover that lenders will wish to see.

As a final note on this subject, the difference between risk and uncertainty is frequently misunderstood and the terminology, accordingly, mis-applied. The following note gives a clear enunciation of the difference:

> *Risk has its place in a calculus of probabilities. It applies to a specific course of action. The risk of an action is the likelihood that it will produce an* unwanted *result. Risk lends itself to quantitative expression, as when we say that the chances of failing to strike oil in a field are better than fifty-fifty or that the chances of finding a defective part in a batch are two out of a hundred. In the framework of benefit-cost analysis, the risk of an innovation is how much we stand to lose if we fail, multiplied by the probability of failure.*
>
> *Uncertainty is quite another matter. A situation is uncertain when it requires action but resists analysis of risks. A gambler takes a risk in an honest game of blackjack when, knowing the odds, he calls for another card. The same gambler, unsure of the odds and of the honesty of the game, is in a situation of uncertainty. He can act, but he cannot estimate the risks or rewards of his action. Even so, he operates in a risklike situation because, at any rate, he has two well-defined alternatives - to call for another card or not to call for it. He must act even though relevant alternatives are undefined. He must invent what to do. He has no way of calculating with any precision the risks of action. He has only rough guidelines of skill and experience to help him. (1)*

CONCLUSIONS

Translating thermal conversion research into a full-scale, practical, process plant in which all technical risks are understood and capable of quantification is bound to be a difficult task, but every new development has to start somewhere.

The commercial world of electricity supply, in which the thermal conversion process hopes to find an application, is changing rapidly in the developed world and the support and sustenance that prototype processes might have received in years gone by from central authorities is now vanishing.

In the developing world, the requirement is generally for tried and tested, reliable and easily maintained technology capable of providing a supply of electricity utilizing indigenous fuels. Advanced technologies, not long out of the research stage, are unlikely to satisfy their demands.

The drive to utilize more non-fossil fuels in electricity production, in order to meet climate change accords, is likely to provide the main stimulus to technological development of new thermal conversion processes which can utilize biomass or waste products as a feedstock in the developed world.

The financing of such projects, unless heavily underwritten as a stimulus to their commercial application from grant support of their capital cost from public funds, will require intensive analysis of the intrinsic technical risks of the process. The satisfaction of such a risk analysis will require the production of comprehensive data (and its intensive scrutiny) on operational performance and design, particularly in relation to scaling-up from the test-bed to the turnkey, commercial, installation.

ACKNOWLEDGEMENTS

Thanks are due to the Directors of PB Power Ltd (which now includes the power businesses of both Merz and McLellan and Kennedy & Donkin) for permission to publish this paper. It is emphasized that the opinions expressed are entirely those of the author and in no way reflect the corporate attitudes or views of PB Power.

BIBLIOGRAPHY

1. *A guide to finance for the UK renewable energy industry.* ETSU for the DTI, London, 1997.

REFERENCES

1. Schon DA (1967) *Technology and Change: The New Heraclitus*, Delacorte Press, New York

APPENDIX

CATEGORIES OF RISK IN A TYPICAL RENEWABLE ENERGY PROJECT

Major category	Specific aspect
Sponsor	• track record and reputation as a developer • degree of financial commitment • competing interests • insurability
Construction	• time, cost, quality • sponsor's ability to arrange construction • experience and standing of contractor • terms of main construction contract - fixed price/turnkey - penalties/bonuses - performance warranties and damages - insurances - incentives - force majeure - commissioning - subcontracting • competency of contractor's organization
Completion	• site conditions • infrastructure • competence of project manager • labour availability and suitability
Technology	• previous applications • competitive options • robustness • flexibility • adequacy of guarantees and warranties • achievability of performance targets • maintainability
Fuel supply	• availability • specification • price and exposure to cost increases • delivery and deliverability • terms and duration of fuel supply agreement
Operation and maintenance	• experience of operator • adequacy of infrastructure • maintenance of performance levels • access to replacement supplies and parts • cost constraints • exposure to cost increases • quality management system • appropriate maintenance schedules • terms of O&M agreement • emission monitoring

Major category	Specific aspect
Electricity and heat off-take	• quantity • product reliability • capacity charge/energy charge • market price levels and stability • competition • indexation • term of contract • predictability of heat and electricity loads • take or pay • penalties for non-supply • status of off-taker • potential for export off-site
Political	• energy policy • legislation • taxation • environmental controls • consents and authorizations

Fast Pyrolysis of Biomass: Technical Requirements for Commercialisation

A.V. Bridgwater
Bio-Energy Research Group, Aston University, Birmingham, B4 7ET, United Kingdom

ABSTRACT: Fast pyrolysis of biomass for liquid fuels is moving steadily towards commercialisation. Several companies now have substantial pilot plants or demonstration plants operating including BTG, ENEL, Ensyn, Dynamotive, Fortum and Wellman, which cover a range of reactor technologies. The interest is driven by the logistical advantage of a liquid fuel that can be stored and/or transported, and in several countries by enhanced electricity or liquid fuel prices such as the Non Fossil Fuel Obligation in the UK and untaxed liquid fuels in Germany. All EU countries now have some incentives to encourage renewable electricity. The UK in particular has offered contracts for 69 MWe to be generated by fast pyrolysis of biomass at a significantly enhanced price.

As the technologies are scaled up to commercial size, a number of technical issues will have to be resolved including provision of process heat; heat transfer; hydrodynamics (for fluid bed systems); maintenance of yields; separation of char; biomass preparation; scale-up; feeding; back-mixing in fluid bed systems; control of secondary reactions; and product quality definition, monitoring and control. These are described and discussed in relation to their impact on commercial developments and the direction of further RD&D.

INTRODUCTION

Several companies now have substantial pilot plants or demonstration plants operating including BTG, ENEL, Ensyn, Dynamotive, Fortum and Wellman, which cover a range of reactor technologies including fluid beds, transported beds, and rotating cone reactors (1). The technology is steadily progressing to a stage where large scale demonstration and commercial plants can soon be implemented. There is also extensive research and development both in the science and technology of fast pyrolysis as well as in the development of applications for heat, power and chemicals (2).

SOME KEY FACTS ABOUT FAST PYROLYSIS

- Fast pyrolysis is a relatively new technology,
- Fast pyrolysis has only been developed for the last 20 years compared to more than 200 years for gasification, and more than 2000 years for combustion,Fast pyrolysis gives high yields of liquid that can be stored and/or transported,
- The characteristics of the liquid can be compared to crude oil in the early years after its discovery,
- The liquid is a combustible fuel but is not miscible with conventional liquid fuels,
- The liquid can be used for production of chemicals both directly and indirectly,
- The liquid energy yield is about 70% compared to a typical cold gas efficiency of 75% from gasification,
- Applications include energy applications such as fuel for heat and power and an energy carrier, and chemicals applications including a source of hydrogen and transport fuels, a source of speciality and bulk chemicals,
- Pyrolysis should be viewed as complementary to gasification as a liquid is produced that can be stored and transported to the point of use and can be used intermittently as well as

continuously (see below), whereas producer gas from gasification cannot be economically stored or transported, so is better suited to base load applications.

CENTRALISATION AND DECENTRALISATION

The singular advantage of fast pyrolysis as a thermochemical renewable energy process is in delivering a liquid that can be readily stored and/or transported is illustrated in Figure 1 and Figure 2. Figure 1 depicts collection of biomass from several sources, delivering it a large central fast pyrolysis unit. The advantages include maximising the economies of scale in conversion of biomass to liquid, which significantly reduces the bio-oil cost. Figure 2 depicts de-centralised bio-oil production delivering liquid to a central point for further processing of distribution.

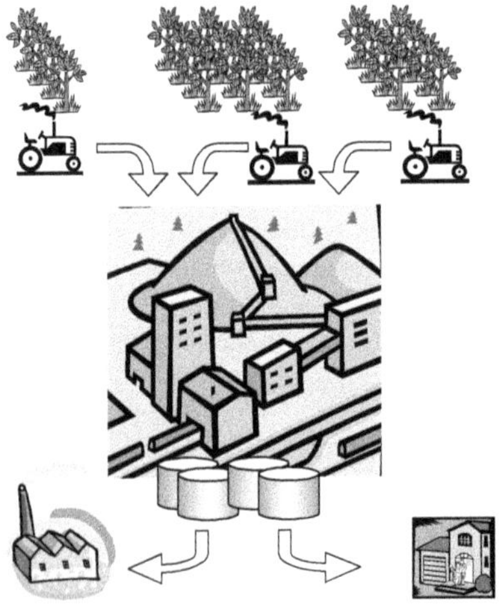

Figure 1: Delivery of biomass to a central pyrolysis of biomass and distribution of bio-oil

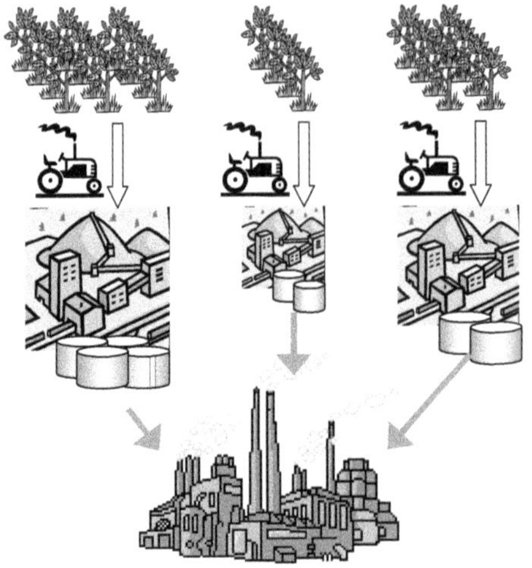

Figure 2: Distributed pyrolysis of biomass and delivery of bio-oil to a central processing point, e.g. for upgrading or a bio-refinery.

CHALLENGES AND OPPORTUNITIES

As the technology of fast pyrolysis and applications for the product bio-oil have developed, so has knowledge and awareness of the technical and economic challenges to be addressed. *Table 1* summarises the main challenges and opportunities for process development. This is not an exhaustive list, but serves to illustrate the range and type of challenges facing the industry and the contributions that the R&D community can play in aiding successful commercialisation.

Table 1: Challenges and opportunities for the successful development of fast pyrolysis

Feed production	**Challenges**	**Opportunities**
	Feed selection Feed characteristics	Improve feed specifications and establish relationships between feed and bio-oil characteristics
	Harvesting	Improve harvesting methods to give feed characteristics that are more focussed on conversion process requirements, such as particle size and shape

Feed production	Challenges	Opportunities
	Seasonality	Develop multi-feed processes to provide year round operation and minimise storage requirements
	Use of wastes	Key parameters to be addressed are: Quality; Consistency; Contamination; Cost.
	Reduce feed cost	Improve all steps in the biomass production chain to minimise delivered feed cost and overall product cost
Feed preparation	**Challenges**	**Opportunities**
Handling & storage	Minimise cost and uncertainty	Transfer technology from the pulp and chipboard and similar industries
Drying	Energy integration	Provide a high level of cost effective energy integration with heat recovery through design and optimisation
Comminution	Chipping – energy and financial cost	Improve biomass harvesting techniques (see above), Minimise energy and financial cost
	Grinding – energy and financial cost	Improve harvesting techniques, Minimise energy and financial cost
	Particle shape and size	Improve harvesting techniques (see above)
Fast pyrolysis	**Challenges**	**Opportunities**
Reactor and reaction	Heat transfer rates and heat transfer area	Fundamental research to improve design methods and modelling, Design and optimisation of reactor, heaters and associated operations, Selection of optimum fuel and method of heating
	Char separation	Improved reactor design through modelling and innovation
	Lignin cracking	Fundamental research into primary and secondary reactions, leading to improved reactor design and process design, Process improvement and optimisation
	Secondary reactions	Fundamental research into primary and secondary reactions, leading to improved reactor design and process design, Process improvement and optimisation
Char separation	Cyclones	Improved design and optimisation
	Vapour filtration	Applied research and development on hot vapour filters, Design and optimisation of filtration sub-system for minimise yield loss and maximum char removal
	Liquid filtration	Applied research and development to evaluate and develop liquid filters
Liquid collection	Quench	Improve design through modelling and applied research, Develop relationship between bio-oil quality and quench design and process operation
	ESP	Develop relationship between bio-oil quality and quench design and process operation, Improve design and operation of subsystem, including evaluation of alternatives
	Aerosols	Applied research to improve demisters as alternative to ESP
	Low vapour partial pressure of vapours.	Review technology and design to improve liquid and aerosol capture, Innovation in technology development

Fast pyrolysis	Challenges	Opportunities
Liquid quality	Stability	Fundamental research to understand stability and relate it to process parameters, technology and feedstock, Improve process and process control
	Viscosity	Fundamental research to relate viscosity to process parameters, technology and feedstock, Improve process and process control
	Solids	Improve process and process control, Develop filtration (see above) Develop tolerant applications for bio-oil with solids, Establish relationship between char and quality (stability, viscosity etc.),
	Norms and Standards	Need to be developed and validated,
	Environment, health and safety	Need to be thoroughly reviewed and appropriate standards defined and authorised
	Consistency	Process design, Process operation, Norms and standards
	Mixing	Establish compatibility of bio-oils from different feedstocks and technologies
	Upgrading	Improve catalysts, Improve process, Improve yields from upgrading processes, Reduce costs, Develop new upgrading processes and routes such as emulsions, and bio-oil reactions with suitable solvents such as alcohols
Scale up	**Challenges**	**Opportunities**
	Design and optimisation	Successfully apply the results from laboratory, pilot plant and demonstration facilities to commercial scale plants, Use of modelling
	Heat transfer	Design and test suitable systems for providing process heat in large scale fast pyrolysis processes, Integrate energy usage
	Feeding	Develop, test and apply reliable feeders
	Yield	Maintain liquid yield at higher capacities
Cost reduction	**Challenges**	**Opportunities**
	Scale	Design, Development, Optimisation, Learning
	Learning	Improved designs and operating procedures will emerge from experience with research, development and demonstration activities, Support is needed for long term operation of existing facilities to provide experience, More development, demonstration and commercialisation
	Modular construction	Offers possibilities for cost reduction and factory assembled skid mounted modules
	Innovation	Encouragement of innovation at all levels and stages of a fast pyrolysis system and applications for the products will aid successful development and implementation

APPLICATIONS

A wide range of applications have been investigated and successfully demonstrated as feasible as summarised in *Table 2* (3, 4, 5). Only production of food flavours is currently commercially practiced by Red Arrow in the USA (6). The focus of most research and development to date has been on energy applications due to the driving force of political commitments to reduce greenhouse gas emissions, with renewable energy targets of 10-12% by 2010 being adopted by many western countries.

While energy applications are currently the main driving force, the much higher value of bio-oil as a chemical resource means that there has been extensive work carried out on applications for bio-oil in the chemicals area, of which resins or adhesives for the wood panel industry has been of particular interest. The production of any chemical may lead to by-products or residues, such as the fuel oil remaining after production of food flavourings at Red Arrow, which is used for heat and power production (7). It is the potential offered for both chemicals and fuels production that has led to the concept of a bio-refinery (8) being proposed to optimise the process and products.

Table 2: Examples of applications for bio-oil

SPECIFIC ENERGY USAGE
Fuel for power,
Fuel for heat including co-firing,
Energy carrier,

ORTHODOX FUELS
Diesel
Fuel oil
Gasoline
Reformulated gasoline
Olefins
Novel fuels e.g. emulsions

COMMODITIES
Emission control reagent – bio-lime
Hydrogen
Fatty acids
Food flavourings
Glyoxal
Hydroxyacetaldehyde
Levoglucosan
Oxychemicals
Phenols
Polyphenols
Slow release fertiliser
Wood preservative

CHEMICALS
Acetic acid
Acetol
Adhesives and resins
Alkanes and alkenes
Anhydrosugars
Anisole
Aromatics
Aryl ethers
Calcium acetate
Carboxylic acids
Cresols

Table 3 summarises the main challenges and opportunities for successful development of applications for bio-oil. Again this is not an exhaustive list, but serves to illustrate the range and type of challenges facing the industry and the contributions that the R&D community can play in aiding successful commercialisation.

The main challenge to date has been matching the characteristics of bio-oil to the specification and requirements of the application. A key factor in the successful exploitation of bio-oil, at least in the short term, will be minimising the extent of downstream processing or upgrading that is needed. Use of whole oil or at least a major fraction will minimise both cost and by-product disposal requirements.

Co-firing is currently seen as a particularly attractive short commercial opportunity that will enable companies to take advantage of the incentives available for renewable electricity without the uncertainties of developing a power generation unit. The opportunities are summarised in *Figure 3.*

Table 3: Challenges and opportunities for the successful development of applications for bio-oil from fast pyrolysis of biomass

Applications	
Match product quality to application	Develop fast pyrolysis and bio-oil to be as consistent, as well defined and characterised, and as close to industrial needs as possible, Develop applications to accept products from an improved fast pyrolysis process
Power – engines and turbines	Fundamental research, Applied research, Demonstration
Heat	Fundamental research, Design burners and combustors, Modelling, Applied research, Demonstration
Standards	Establish and register bio-oil as an approved substance
Co-firing	Establish feasibility, Establish design, Modelling for design, prediction and optimisation, Process optimisation, Bio-oil quality and matching applications requirements are very important
Energy carrier	Develop a process to deliver a bio-oil char slurry with as high an energy efficiency as possible, Develop handling and transport systems, Develop applications for the product such as gasification
Bio-refinery	Explore and evaluate opportunities for integrated fuels and chemicals production
Transport fuels	Develop catalysts, Develop and evaluate process, Evaluate products, Carry out technical and economic analysis to ascertain potential
Upgrading	Explore and evaluate processes and products, Minimise requirements, Use whole bio-oil where-ever possible, Minimise by-products and waste streams, Innovation and explore new ideas
New applications	To be explored
Markets and costs	To be explored

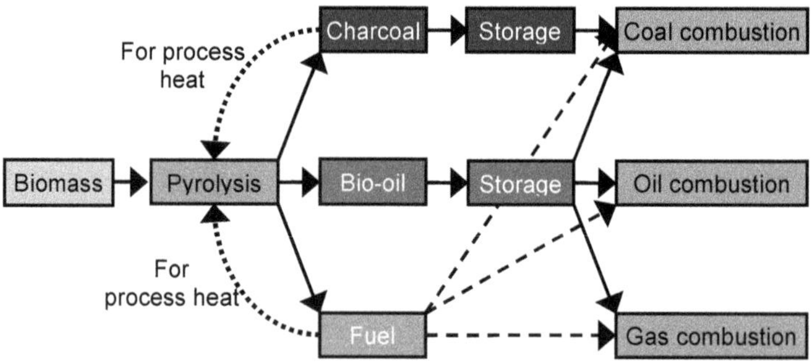

Figure 3: Opportunities for co-firing bio-oil and other pyrolysis products.

THE FAST PYROLYSIS AND BIO-OIL INDUSTRY

A complete fast pyrolysis process for electricity production is constituted from three main steps shown in *Figure 4.* A comparable figure for chemicals production can be described. In both cases the main steps can be separated by time and place. Each main stage is made up from series of process steps of which the main elements are shown, and each stage is very roughly of similar capital cost at small scales of operation. However, as scale increases, the economies of scale benefit pyrolysis and preparation more than power generation, which takes an increasingly large proportion of the total capital cost. It is important that complete systems are properly specified and characterised, and the interfaces between biomass production and pyrolysis and between pyrolysis and applications for power, heat and chemicals are properly considered.

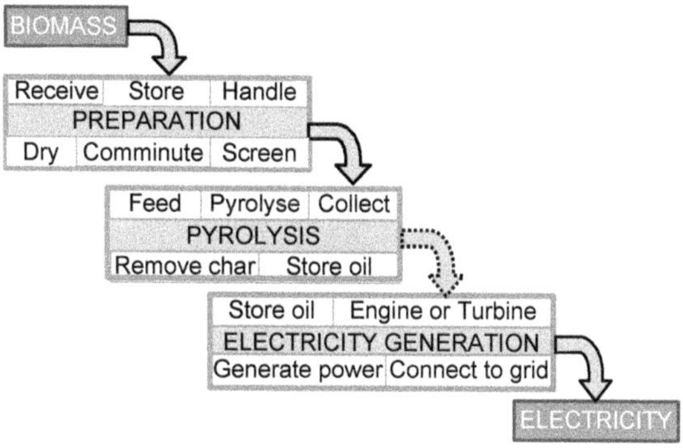

Figure 4: Complete bio-energy system based on fast pyrolysis to electricity

BARRIERS

Fast pyrolysis technologies have to overcome a number of technical and non-technical barriers before industry will implement their commercialisation. The technical barriers have been summarised above, while some of the more important non-technical barriers are summarised below.

Economics is a critical barrier as all bio-fuels have to compete with fossil fuels. For those countries who have made commitments to reduce fossil fuel derived carbon emissions, the current disincentive to implementation of bio-energy on simple cost grounds will have to be overcome - no company is going to invest in ventures that are guaranteed to lose money, regardless of the environmental benefits that may accrue. Industry will only invest in technology that has an acceptable return at an acceptable risk (9). Acceptable returns will come from incentives to capital expenditure of product purchase, or to dis-incentives to orthodox options by, for example taxation on fossil fuels or legislation, etc. This is one of the major roles that governments have to play.

There is widespread approval of the interest in, and move towards, renewable energy and bio-energy - as long as it is not near me! Increasing attention will need to be placed on "selling" the idea to the populace where the plant is to be built, and this problem could be exacerbated by the need to build smaller plants and thus many more plants than conventional power stations and refineries. Early plants will have the curiosity factor and may enjoy popularity as an attraction by itself or as part of an attraction of a "green" site. However the increasing acceptance of bio-fuels such as pellets in many parts of Europe should lead to greater acceptance of the value and virtues of these technologies.

For industry to implement renewable energy technologies in order that commitments made to mitigate greenhouse gases can be met, investment has to be attractive. Without some fiscal incentives (or disincentives against fossil fuels), companies will only invest in those projects that are sufficiently profitable and most of these will be in niche markets and special opportunities. As commented above, only governments can create the necessary instruments.

Economies of scale are a vital feature of the development of industry and technology in which the larger a process can be built, the cheaper it becomes. This is particularly important in the energy and process industries that will shoulder the responsibility for technology development and implementation. However in the bio-energy industry, biomass is a diffuse resource, which has to be harvested over large areas. A modest 10 MWe power station operating at a modest efficiency of 35% will require about 40000 t/year of wood on a dry basis, which will require about 4000 ha of land or 40 square km. This could reduce to 2000 ha or 20 square km if the promise of high yielding short rotation forestry is realised. Neither figure makes any allowance for non-productive land. There are therefore finite sizes that bio-energy processes can be built in considering the costs and logistics of transporting biomass to a processing plant. The maximum size that has been suggested in Europe ranges from 30 to 80 MWe in the short to medium term, and 100 to 150 MWe in North America. This places a practical upper limit on the benefits of scale.

Investors are generally risk averse and always prefer low risk investments, but if risks have to be accepted, then an appropriately higher return is expected. Technology developers can do much to minimise technical risk and this topic has been thoroughly described and discussed (9).

The established energy suppliers and providers have considerable investments in conventional energy systems and will always seek to maximise their returns and maintain their competitive edge. Most major energy companies have their own programmes of supporting renewable energy, but there have always been concerns over the extent to which they will seek to protect their interests.

CONCLUSIONS

Fast pyrolysis offers the advantage of a high efficiency one-step process for giving high yields of a liquid that can be directly or indirectly used for energy and/or chemicals. There is still much work needed on process and product development, but the attraction of a liquid fuel with commensurate advantage of easy and low cost storage and transport is becoming widely appreciated.

REFERENCES

1 Bridgwater AV and Peacocke GVC, "Fast pyrolysis processes for biomass", Sustainable and Renewable Energy Reviews, 4 (1) 1-73 (Elsevier, 1999) ISSN 1-364-0321
2 Bridgwater AV, Czernik S, Piskorz J, "An Overview of Fast Pyrolysis Technology", pp 977-997 in Progress in Thermochemical Biomass Conversion, Ed. Bridgwater AV, (Blackwell, Oxford, UK 2001) ISBN 0-632-055533-2
3 Bridgwater AV, "Catalysis in thermal biomass conversion", Applied Catalysis A, 116, (1-2), pp5-47, 1994
4 Bridgwater AV, "Production of high grade fuels and chemicals from catalytic pyrolysis of biomass", Catalysis Today, 29, 285-295 (1996) ISSN 0-920-5861
5 Czernik S and Bridgwater AV
6 Underwood, G., "Commercialisation of fast pyrolysis products", in Biomass Thermal Processing, Eds. Hogan, E., Robert, J., Grassi, G. and Bridgwater A. V., CPL Scientific Press, 1992, pp. 226-228.
7 Manitowac
8 Radlein D, PyNe Newsletter No. 5, March 1998 pp 12-13 (Aston University, UK)
9 Thornley P. and Wright E., "Evaluation Of Bio-Energy Projects", in Bridgwater AV (Ed), 'Fast pyrolysis of biomass – A handbook Volume 2' (CPL Press, Newbury, UK, 2002)

Polyethylene Pyrolysis in a Conical Spouted Bed Reactor

R. Aguado, M. Olazar, S. Alvarez, P. González, J. Bilbao
Departamento de Ingeniería Química, Universidad del País Vasco, Apartado 644, E-48080 Bilbao, Spain

ABSTRACT: Pyrolysis is considered as the more suitable method for upgrading waste plastics in order to recover monomers and to obtain feedstock and fuel. The importance of the technological development of large scale pyrolysis units is evident in view of the growing amount of post-consumer plastic wastes in developed countries. In this work, a conical spouted bed reactor is proposed for the kinetic study of polyethylene pyrolysis at high temperatures (500 – 600 °C), in order to obtain the parameters for designing these units. Due to the inherent properties of this fluid-solid contact method, the spouted bed technology may be a solution to the problems arising in fluidized beds with sticky solids, as it avoids particle agglomeration and bed defluidization. Moreover, the coating of inert particles is more uniform, giving way to greater productivity and improved heat transfer. The kinetic results and yields obtained in this work are evidence of the good performance of this original reactor.

INTRODUCTION

Pyrolysis (or thermal degradation) of polyolefins and other polymeric materials was originally carried out with the aim of ascertaining the structure of the material [1,2]. The results obtained in these studies showed the great interest of pyrolysis as a way of upgrading and recycling of post-consumer plastics, due to the fact that it allows for recovering monomers and obtaining feedstock and fuel [3], which are easily introduced in the market. In the developed countries, where plastic consumption is high and an increase of 4% per year is expected, the waste plastics is a serious problem that requires a fast solution. The environmental regulations prohibit their upgrading by combustion and other recycling ways have not been technologically developed at large scale. Taking into account that thermal degradation is applicable to all the plastics and several large scale units have been developed, pyrolysis may be the most interesting way for upgrading of this kind of waste.

In the upgrading of polyolefins by pyrolysis two alternative methods may be followed. In order to maximize the yield of gaseous products, flash pyrolysis at high temperatures is the more widely studied [4,5,6,7]. To maximize the yield of olefins ultrapyrolysis has been proposed because it allows for reducing the residence time of the gaseous product [5,8,9]. On the other hand, and in order to minimize energy requirements, the formation of wax (primary product in pyrolysis) at low temperatures is also attractive [10]. On the basis of its composition, wax is a raw material used by the petrochemical industry for obtaining fuels by thermal or catalytic cracking, which have been improved in recent years with the view of using alternative raw materials (of similar characteristics to wax). Furthermore, given the treatment capacity of these units, when the wax is fed included in the habitual feed it only represents a small proportion.

The more developed technology for plastic pyrolysis in the literature is the bubbling fluidized bed reactor [4,11-16], which has advantages (isothermicity and shorter residence time of the gaseous product) over rotary kilns [17,18] and does not need to operate at high pressures, as does the stirred tank reactor [19]. Anyway, the operation at large scale of this kind of reactor have defluidization problems due to particle agglomeration. The main difficulties are due to the low thermal conductivity of the polymeric materials and to the handling of fused plastic.

With the aim of improving the gas-solid contact and of avoiding defluidization, other technologies have also been proposed, such as the circulating fluidized bed reactor [5,8], the tubular reactor [20], the screen reactor [21], the conical rotary reactor [22,23], the circulated-spheres reactor [24] and a reactor with stirred heat medium particles [25].

In this paper a new reactor, which is a conical spouted bed reactor, is proposed for the upgrading of polyethylene by pyrolysis at high temperatures. The conical spouted bed reactor has several advantages over fluidized beds: simpler design because no distributor plate is needed, lower pressure drop, cyclic circulation of the particles, high heat and mass transfer between gas and solid (whose circulation is mainly counter-current), short gas residence time and great versatility as far as gas flow is concerned, which allows for a vigorous solid movement. In view of these properties, the conical spouted bed reactor performs well in the pyrolysis of vegetable biomass [26,27] and in catalytic polymerization [28,29].

In order to improve the design and simulation of the reactor and to establish the optimum process conditions, a knowledge of pyrolysis kinetics is of great importance. Most of the kinetic studies in the literature have been carried out using thermogravimetric analysers (TGA), and they correspond to temperatures lower than 450 °C, which means that their extrapolation for the design of reactors that operate at higher temperatures is speculative. Anyway other techniques, such as screw reactors, screen heaters [21,30] or laminar flow reactors [20], have been used in order to overcome heat and mass transfer limitations of conventional TGA equipment. Great differences in the kinetic results obtained in the literature are observed, which are attributable to several experimental factors, such as the material used (particle size and molecular weight of the polymer) or the operating conditions (heating ramp, gas used and its flow rate, heat transfer rate between sample and crucible)[31]. Another factor that justifies the difference in the results is the methodology followed for treating the experimental results in order to calculate the kinetic parameters [32].

EXPERIMENTAL

The materials studied are: low density polyethylene (LDPE) and high density polyethylene (HDPE). These materials are supplied by Dow Chemical (Tarragona, Spain) and their properties are set out in Table 1.

Table 1: Properties of the materials studied

material	mol wt, g mol^{-1}	polydispersity	density, kg m^{-3}
LDPE	92200	5.13	923
HDPE	46200	2.89	940

Figure 1 shows a diagram of the laboratory set-up provided with a conical spouted bed reactor. As the plastic material enters the reactor, it melts onto the sand particles and coats them. The vigorous cyclic movement of the particles allows for obtaining a uniform coating of the particles without agglomeration problems.

Fig. 2 shows an scheme of the reactor where its characteristic dimensions are defined: H_T= 0.340 m; H_c = 0.205 m, γ = 28 °; D_c = 0.123 m; D_i = 0.02 m; D_o = 0.01 m. These dimensions have been established from previous hydrodynamic studies carried out in the spouted bed regime in conical contactors [33-35] and bearing in mind the versatility of the equipment. When the inert gas flow rate is increased, the gas-solid contact regime evolves from the conventional spouted bed to a dilute spouted bed (which characteristizes conical contactors) and the gas residence time decreases from a few seconds to approximately 20 ms.

Temperature is measured by means of three thermocouples placed at different radial positions in the reactor and provided with free vertical movement. Bed isothermicity is noteworthy and it is attained due to vigorous particle circulation.

Reaction temperature is reached by means of two electric resistances covered with ceramic material. One resistance is located within the tube for heating the inert gas before entering the reactor and the second one surrounds the conical section of the reactor. Temperature in these two positions is measured by two thermocouples. The two heating zones are thermally insulated.

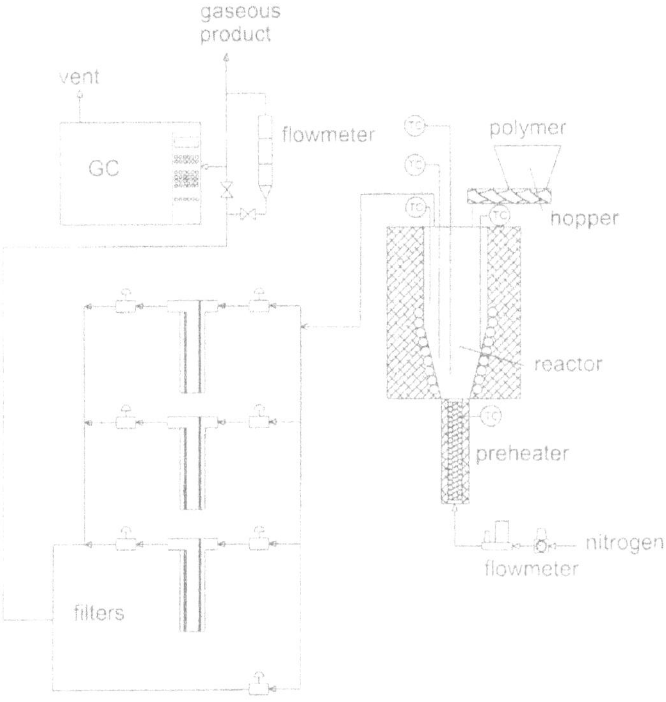

Figure 1: Diagram of the pyrolysis equipment.

The pyrolysis product stream is transported by the inert gas towards a set of three 25 μm filters made of sintered steel placed in parallel for collecting the wax. In this set of three parallel filters, Figure 1, only one filter is operating and, once the time fixed for each kinetic measurement has elapsed, the product stream is re-routed to another filter and the used filter is removed and weighed in order to ascertain the amount of wax retained. Following this procedure, samples are obtained every 10 s until total conversion of the feed. The non-condensable gases of the product stream (under the reaction conditions, all products except wax) are analysed by gas chromatography (Hewlett Packard 6890) using a BPX5 capillary column of 50 m.

The runs for the kinetic study have been carried out at 450, 500, 550 and 600 °C by feeding 1g of plastic material (with a particle size of approximately 1 mm) and using 30 g of sand (with a size between 0.63 and 1 mm). The reaction times are between 1 s and 1080 s (sufficient time for reaction to be complete at 500 °C).

In order to obtain sufficient amount of wax for its analysis, continuous operation has been carried out at 450 °C by feeding 0.1 g min^{-1} of plastic.

The average molecular weight has been measured by gel permeability chromatography (Hewlett-Packard HPLC 1100) at 40 °C by using a detector based on a refraction index and PL Caliber commercial software. Wax is dissolved into trichlorobenzene, which is the mobile phase with a flow rate of 1 ml min^{-1}. The column used is a Tosohaas TSK-Gel GMHHR-M of 30 cm length and of 7.8 mm internal diameter, which is filled with a mixture of polystyrene/divinylbenzene. Calibration has been carried out by means of a polyethylene kit supplied by American Polymer.

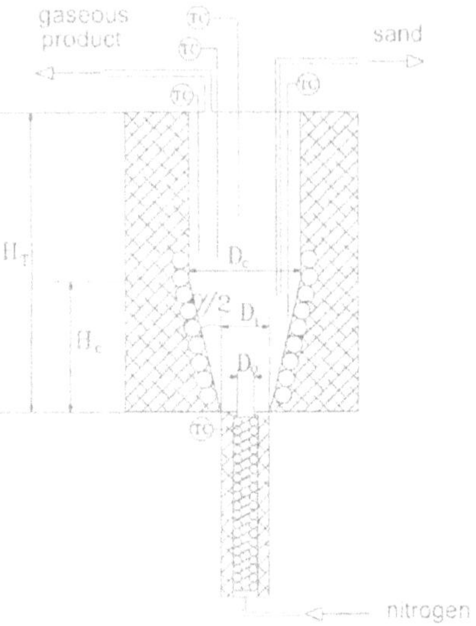

Figure 2: Scheme and design parameters of the conical spouted bed reactor.

The structure of the wax has been studied by FTIR spectrophotometry (Nicolet 740) by dissolving the sample in trichlorobenzene and depositing it on a KBr tablet.

RESULTS

HYDROCARBON AND WAX YIELD

Polyolefin pyrolysis proceeds via a random degradation mechanism, yielding a broad product spectrum (C_1 to C_{50}) [36]. The primary devolatilization reaction yields mainly wax-like materials with a typical carbon number in the range of 20-50. These products are subsequently cracked in secondary reactions to produce hydrocarbons in which olefinic monomers prevail. Due to the high reactivity of these gaseous products, they follow ternary reactions to form more stable compounds, mostly aromatics (mainly benzene and toluene). Other products are methane and hydrogen and, when residence time of the product stream is long, coke is formed by progressive condensation of aromatics [37].

The wax yields obtained in the discontinuous runs are set out in Table 2. The remaining products are gases and volatile compounds. The decrease in the yield of wax when temperature is increased is almost linear. No significant differences are observed between the results for the two types of polyethylene (being the maximum yield 80 wt% at 450 °C) and for polypropylene the yield of wax is higher, 92 wt%. The yield of wax obtained for polyethylene pyrolysis is lower than that obtained by Predel and Kaminsky [38] in a fluidized bed reactor at 510 °C, which is between 85 and 88 wt% when polyolefins are continuously fed into the reactor. Nevertheless, it is higher than the value of 71 wt% obtained by Chaala et al. by vacuum pyrolysis [39].

It is noteworthy that the wax yield may be increased by decreasing temperature. The temperature corresponding to the maximum wax yield, 450 °C, is the lowest used with the aim of ensuring a high level of bed isothermicity, which in turn ensures the formation of wax with uniform composition and reproducibility of kinetic results. Likewise, the wax yield may also be increased by modifying the contactor geometry. Thus, as cone angle is decreased, residence time distribution corresponding to the gaseous stream that carries the wax is narrower [40], which minimizes secondary reactions. Nevertheless, the angle used is suitable for a versatile gas-solid contact, which allows for changing process conditions (amount of sand, plastic feed, size distribution of sand and plastic, linear velocity of inert gas).

Table 2: Yields (wt%) of the products obtained by polyethylene pyrolysis.

material	compound	500 °C	550 °C	600 °C
	wax	69.0	57.0	50.8
	methane	5.5	4.1	3.6
	ethene	1.6	9.8	11.4
	C_3	1.1	7.0	6.9
LDPE	1-butene	2.6	6.6	8.0
	C_2-C_4	7.9	4.9	6.2
	C_5-C_8	11.3	10.2	12.6
	aromatics	1.0	0.4	0.5
	wax	68.1	61.0	48.9
	methane	5.9	4.3	4.2
	ethene	2.8	10.7	15.5
	C_3	2.0	3.9	8.4
HDPE	1-butene	2.9	3.4	5.4
	C_2-C_4	6.6	9.7	6.4
	C_5-C_8	10.7	6.1	10.8
	aromatics	1.0	0.9	0.4

The wax obtained in this study is mustard coloured with a melting point of 70 °C. The elemental composition (C, 86 wt%; H, 14 wt%) corresponds to long chain alkanes. The molecular weights determined by gel permeability chromatography are: wax from LDPE, 13950 g mol^{-1}; HDPE, 9214 g mol^{-1}.

In view of the FTIR analysis of the waxes obtained [41], the spectra are very similar to each other and also very similar to those obtained in the literature for commercial paraffinic wax and to those obtained by vacuum pyrolysis of electric cables based on polyethylene [39].

Taking into account that the implementation of residual plastic pyrolysis at large scale requires treatment of all kinds of plastic materials without selection, it is interesting to study the composition of the hydrocarbon stream formed in the pyrolysis of polyolefins at low temperatures. In this strategy of performing pyrolysis at low temperatures in order to obtain wax selectively, these hydrocarbons will leave the reactor together with the product stream stemming from the pyrolysis of the other plastic materials. The characteristics of the conical spouted bed reactor are especially suitable for the kinetic study of the formation of individual hydrocarbons.

In the chromatogram corresponding to the product stream obtained in LDPE pyrolysis, 31 compounds are present in significant quantities, from which 21 account for 98 wt%. These compounds, ordered according to increasing retention time, are the following: methane, ethylene, ethane, propylene, propane, isobutane, 1-butene, 1,3-butadiene, 2 methyl 1-butene, 2 methyl 2-butene, 1-pentene, n-pentane, cyclopentene, 1-hexene, n-hexane, 1-heptene, n-heptane, 1-octene, benzene, toluene and styrene.

In the product stream of HDPE pyrolysis 18 compounds account for more than 98 wt% and they are the same as those for LDPE, except for 2 methyl 2-butene, n-heptane and 1-octene. The products of PP pyrolysis that account for more 98 wt % are the same as those for LDPE, except for 2 methyl 2-butene.

Table 2 also shows the yields of individual components (methane and ethylene) and of the lumps corresponding to C_3 (propane and propylene), n-butenes, C_2-C_4 (ethane, butane, isobutane e isobutene), C_5-C_8 (paraffins, linear and branched light olefins and naphthenes) and aromatics (benzene, toluene and styrene). It is noteworthy that the yield of ethylene and propylene increases as temperature is increased, which is evidence that the optimum temperature for maximizing olefin production (alternative strategy to that of maximization of wax production) is higher than the range studied in this paper, as has been proven by several authors [5,6,7].

KINETIC RESULTS

Due to the simplicity of its design and its versatility in process conditions (such as gas linear velocity and solid particle size), the conical spouted bed reactor is an interesting alternative to other technologies described in the literature for plastic pyrolysis, such as the fluidized bed reactor [4,6,11,12, 14-16,42], the tubular reactor [9,37], the rotary kiln [17,18], the mixing reactor [19], the rotating cone reactor [22,23], the internally circulating fluidized bed reactor [5,8], the circulated-spheres reactor [24] and a reactor with stirred heat medium particles [25]. The different technologies aim to circumvent the difficulties of the process, amongst which the following are noteworthy:

(1) Difficulty in handling plastics due to their physical nature (irregular texture and size, sticky).
(2) Heat transfer requirements due to the endothermal nature of the reaction.
(3) The complexity of the reaction scheme, which implies the need for an accurate control of the residence time of primary pyrolysis products in order to optimize product distribution.

The results of conversion vs. time are shown in Figure 3 for HDPE as an example and they are obtained by monitoring the formation of wax and gases. The results for the other material are qualitatively similar.

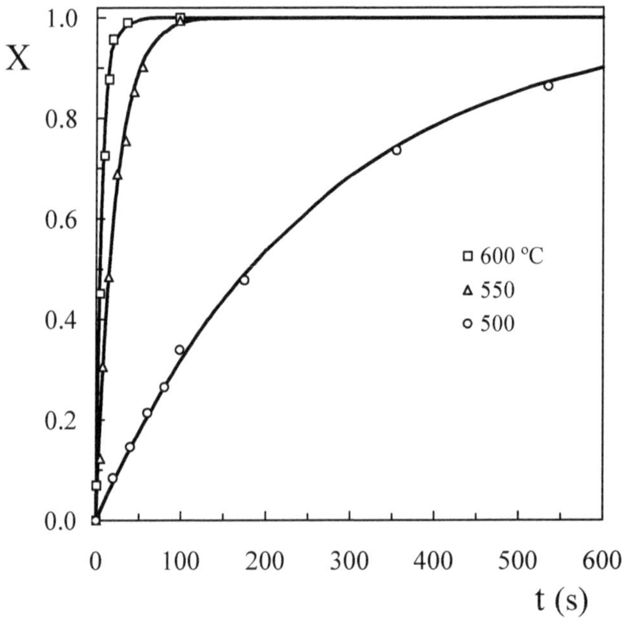

Figure 3: Evolution of HDPE conversion with pyrolysis time in the spouted bed reactor at different temperatures.

The results of Figure 3 suitably fit a first order kinetics [43]. The values of the kinetic constants and of frequency factors and activation energies, corresponding to the two materials, are set out in Table 3. These latter values are shown with a 95 % confidence interval.

Table 3: Kinetic parameters of the pyrolysis of the two materials.

material	T, °C	k , s^{-1}	E, kJ mol^{-1}	k_o, s^{-1}
	500	0.00404		
LDPE	550	0.04503	213 (±14)	$1.09\ (\pm 0.07)\cdot 10^{12}$
	600	0.17620		
HDPE	500	0.00385		
	550	0.04417	208 (±13)	$4.74\ (\pm 0.31)\cdot 10^{11}$
	600	0.15245		

The results of evolution with time of the yield of waxes, individual components and lumps of the gas stream for each one of the materials pyrolysed have been fitted to a first order kinetics [41] and the values of the kinetic constants for formation of individual products have been calculated. From these constants, the values of frequency factor and activation energy have been calculated, which are set out in Table 4 with a 95% confidence interval.

Table 4: Kinetic parameters of the formation of gaseous stream components and waxes in the pyrolysis of the two materials.

material	compound	E, kJ mol^{-1}	k_o, s^{-1}
LDPE	wax	214±19	$9.23(\pm 0.83)\cdot 10^{11}$
	methane	174±17	$6.06(\pm 0.61)\cdot 10^{9}$
	ethene	230±21	$1.55(\pm 0.14)\cdot 10^{13}$
	propylene	137±12	$3.73(\pm 0.34)\cdot 10^{7}$
	1-butene	129±13	$1.27(\pm 0.13)\cdot 10^{7}$
	C_2-C_4	200±16	$2.27(\pm 0.18)\cdot 10^{11}$
	C_5-C_8	200±18	$2.53(\pm 0.23)\cdot 10^{11}$
	aromatics	213±19	$1.28(\pm 0.12)\cdot 10^{12}$
HDPE	wax	225±18	$4.46(\pm 0.37)\cdot 10^{12}$
	methane	186±17	$4.17(\pm 0.38)\cdot 10^{10}$
	ethene	181±18	$1.78(\pm 0.18)\cdot 10^{10}$
	propylene	151±12	$2.05(\pm 0.16)\cdot 10^{8}$
	1-butene	122±12	$3.57(\pm 0.36)\cdot 10^{6}$
	C_2-C_4	198±20	$1.93(\pm 0.19)\cdot 10^{11}$
	C_5-C_8	163±13	$1.13(\pm 0.09)\cdot 10^{9}$
	aromatics	185±17	$2.02(\pm 0.19)\cdot 10^{10}$

PERFORMANCE OF THE CONICAL SPOUTED BED REACTOR

In the pyrolysis of plastics defluidization by solid agglomeration is a serious problem, which has difficult solution when a mixture of post-consumer plastics containing PVC is fed into a bubbling fluidized bed [44]. In the deterioration of the particle mixture and evolution to bed blockage, the mechanism of the initial step of inert particle (sand) coating with fused plastic is crucial. Subsequently, the tendency of the particles to agglomerate depends on their stickiness, momentum and contact surface.

Concerning the first step of coating, when the particles of polyolefin are introduced into the bed, their external surface becomes softer but their inside is cold due to the low thermal conductivity. The softened surface is the responsible for the formation of an agglomerate made up of a plastic particle surrounded by sand particles. When the whole plastic particle reaches the softening temperature, its material is distributed by coating the surrounding sand particles, so that if the thickness of the layer that coats the particles is lower than a critical value the sand particles do not fused. This situation corresponds to a low flowrate of plastic in the feed and beyond this value agglomerates grow irreversibly and instead of sand coating total blockage of the bed or defluidization is the result. This growing mechanism depends on the properties of the plastic material [45].

The good behaviour of the conical spouted bed reactor is quantified on the basis of the critical thickness of the fused plastic that may be handled. The critical thickness is the value above which the particles fuse when they collide and is a function of the stickiness (viscosity) and of the momentum of the colliding particles. In continuous operation it will depend on the kinetics of the pyrolysis. From the hypotheses of Arena and Mastellone [45] the following expression has been deduced for the critical thickness:

$$\delta_{crit} = \frac{r}{2}\left[\exp\left(\frac{m v_0}{\mu \pi r^2}\right) - 1\right] \tag{1}$$

Eq. (1) shows the dependency of the critical thickness with mass, m, radius, r, and velocity, v_0, of the inert particles coated with plastic and with the viscosity of the material that coats the sand, μ. For the experimental conditions (total mass of sand, 30 g; mass of plastic in the feed, 1 g; sand particle diameter, between 0.6 and 1.2 mm; sand density, 2600 kg m^{-3}; minimum velocity of the particles in the annular zone, 0.35 m s^{-1} [46]), the critical thickness predicted by eq. (1) is 250 μm [47]. In this calculation, the viscosity of the plastic has been taken at its fusion temperature, 16 poises, because it is in the initial step of fusion when the stickness of the plastic is maximum.

In the fluidized bed, the velocity of the particles is of few cm s^{-1} and under these conditions the critical thickness of the fused plastic is between 4 and 7 μm. A fact that should be taken into account is that, although gas velocity is increased, particle velocity hardly changes because most of the excess gas crosses the bed in the bubble phase. Nevertheless, in the conical spouted bed, particle velocity is increased by increasing gas velocity up to the dilute spouted bed, in which bed voidage is uniform (between 0.9 and 0.99) and the particle descending velocity is almost equal to the ascending velocity [48].

Eq. (1) has been used to calculate the maximum amount of plastic that can be fed (corresponding to the critical thickness) without bed agglomeration, for different amounts of sand in the reactor. Fig. 4 shows these results for different nitrogen flowrates. When the nitrogen flowrate is increased, the bed is in the regime of spouted bed (beds made up of high amounts of sand) or in the regime of dilute spouted bed (beds made up of low amounts of sand). For intermediate amounts of sand the bed is the transition between spouted bed and dilute spouted bed as velocity is increased from 1.05 u_{ms} to 2.50 u_{ms}.

In Fig. 4 it is observed that, for a given amount of sand in the reactor, as gas velocity is increased the amount of plastic that may be fed increases almost linearly, which is a consequence of particle velocity increase. Consequently, the ratio of plastic/sand in the feed may be as high as 0.03 for the experimental reactor that may treat 100 g of sand in a spouted bed regime, which corresponds to 6.2 cm of stagnant bed height.

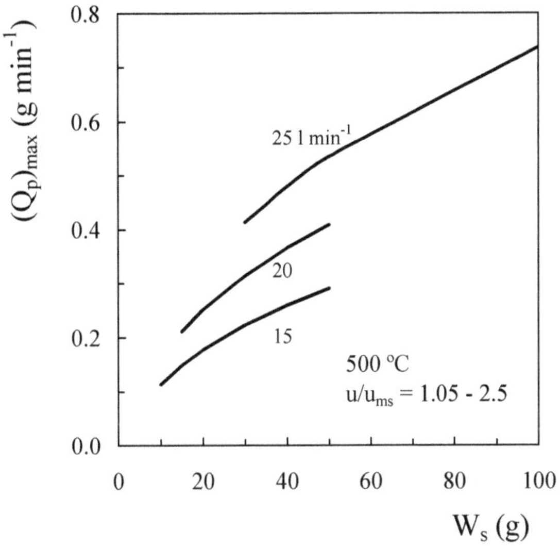

Figure 4: Influence of the inert gas flowrate on the maximum amount of HDPE fed at 500 °C.

CONCLUSIONS

The conical spouted bed reactor performs well in the kinetic study of polyolefin pyrolysis at high and low temperatures. This good performance is due to fast and uniform coating of sand particles with melted polyolefins. Pyrolysis occurs in the thin layers of melted plastic, which minimizes the diffusional restrictions and allows for reaching an isothermal regime and avoids particle agglomeration. The good conditions for bed isothermicity and the fact that no particle agglomeration was observed, which is due to the vigorous gas-solid contact, have anabled the calculation of the kinetic parameters for formation of both wax and hydrocarbons (individual components and lumps) in the 500-600 °C range.

In view of the results, the conical spouted bed reactor may be an alternative to the fluidized bed reactor for obtaining olefins by polyolefin pyrolysis and, moreover, it is also suitable for obtaining high yields of wax, which is a raw material that is handled easily in commercial cracking units, either by steam cracking (the fraction of boiling point lower than 500 °C) or by catalytic cracking in FCC units (the heavy fraction dissolved in the habitual gas-oil feed). Consequently, it may be interesting to carry out this transformation of polyolefins into waxes near the places where post-consumer plastic wastes are collected and, once the waxes are obtained, they are easily transported to commercial units for their processing at larger scale. This strategy allows for using existing industrial technology and upgrading plastic wastes, resulting in the same products as from present petrochemical processes.

It should be noted that the conditions used in this paper have been chosen in order to ensure the bed isothermicity required for the kinetic study and not for maximizing production, which is conditioned by the maximum plastic flow rate in the feed that may coat the sand and by the sand amount required for this purpose (which conditions the volume of the reactor). Nevertheless, given the capacity of conical spouted beds for scaling up in physical operations and chemical processes, the study of aspects related to scaling up in the future is interesting in order to ascertain the feasibility of residual plastic pyrolysis in conical spouted bed reactors at large scale.

ACKNOWLEDGEMENTS

This work was carried out with the financial support of the University of the Basque Country (Project UPV069.310-G34/98) and of the Ministry of Science and Technology of the Spanish Government (Project PPQ2001-0780).

NOTATION

D_c, D_i, D_o	diameter of the cylindrical section, of the bed bottom, and of the gas inlet, respectively, m
E	activation energy, kJ mol^{-1}
H_c, H_T	height of the conical section and total height of the reactor, m
k	kinetic constant, s^{-1}
ko	frequency factor, s^{-1}
m	particle mass, kg
r	sand particle radius, m
T	temperature, °C
t	time, s
u, u_{ms}	gas velocity and gas velocity at minimum spouting, m s^{-1}
W_s, W_p	sand mass and plastic mass at zero time, kg
W, W_g, W_w	plastics mass, (gases+ volatile) mass and wax mass at t time, respectively, kg
X	conversion, mass fraction

Greek Letters

δ_{crit}	critical thickness of the plastic layer, m
γ	cone angle, deg
μ	viscosity of the melted plastic, kg m^{-1} s^{-1}
v_0	particle velocity, m s^{-1}

REFERENCES

1. Sugimura, Y.; Tsuge, S., *Macromolecules* 1979, *12*, 512.
2. Jaradat, Q. M.; Saleh, F. Y.; Lewis, R. V.; Daugherty, K. E. *J. Anal. Appl. Pyrolysis* 1990, *17*, 169.
3. Sasse, F.; Emig, G. *Chem. Eng. Technol.* 1998, *21*, 777.
4. Kaminsky, W.; Rössler, H. *Chemtech* 1992, *February*, 108.
5. Sodero, S. F.; Berruti, F.; Behie, L. A. *Chem. Eng. Sci.* 1996, 51, 2805.
6. Kaminsky, W.; Schlesselmann, B.; Simon, C. *J. Anal. Appl. Pyrolysis* 1995, *32*, 19.
7. Batelle Memorial Institute; Paisley, M. A.; Litt, R. D. *PCT Int. Appl. WO 92 04.423 (Cl. C10853/00)*, March 19, 1992.
8. Lovett, S.; Berruti, F.; Behie, L. A. *Ind. Eng. Chem. Res.* 1997, *36*, 4436.
9. Cozzani, V.; Nicolella, C.; Rovatti, M.; Tognotti, L. *Ind. Eng. Chem. Res.* 1997, *36*, 342.
10. Kastern, H.; Kaminsky, W. *Hydroc. Process.* 1995, *May*, 109.
11. Stammbach, M. R.; Hagenbucher, R.; Kraaz, B.; Richarz, W. *Chimia* 1988, *42*, 252.
12. Scott, D. S.; Czernik, S. R.; Piskorz, J.; Radlein, A. G. *Energy & Fuels* 1990, *4*, 407.
13. Kaminsky, W. *Adv. Polym. Technol.* 1995, *14*, 337.
14. Conesa, J. A.; Font, R.; Marcilla, A.; García, A. N. *Energy & Fuels.* 1994, *8*, 1238.
15. Conesa, J. A.; Marcilla, A.; Font, R. *J. Anal. Appl. Pyrolysis* 1994, *30*, 101 .
16. Conesa, J. A.; Font, R.; Marcilla, A. *Energy & Fuels* 1997, *11*, 126.
17. Sueyoshi, H.; Kitaoka, Y. *Hydroc. Process.* 1972, *51*, 161.
18. Schulman, B. L.; White, P. A. *ACS Symp. Ser.* 1978, *76*, 274.

19. Matsumoto, K.; Kurisu, S.; Oyamoto, T. In *Proceedings of the Conversion of Refuse to Energy Conference. IEEE: New York,* 1975, pp. 538-543.
20. Westerhout, R. W. J.; Kuipers, J. A. M.; van Swaaij, W. P. M. *Chem. Eng. Sci.* 1996, *51*, 2221.
21. Westerhout, R. W. J.; Balk, R. H. P.; Meijer, R.; Kuipers, J. A. M.; van Swaaij, W. P. M. *Ind. Eng. Chem. Res.* 1997, *36*, 3360.
22. Westerhout, R. W. J.; Waanders, J.; Kuipers, J. A. M.; van Swaaij, W. P. M.. *Ind. Eng. Chem. Res.* 1998, *37*, 2293.
23. Westerhout, R. W. J.; Waanders, J.; Kuipers, J. A. M.; van Swaaij, W. P. M. *Ind. Eng. Chem. Res.* 1998, *37*, 2316.
24. Bockhorn, H.; Hentschel, J.; Hornung, A.; Hornung, U. *Chem. Eng. Sci.* 1999, *54*, 3043.
25. Masuda, T.; Kushino, T.; Matsuda, T.; Mukai, S. R.; Hashimoto, K.; Yoshida, S. *Chem. Eng. J.* 2001, *82*, 173.
26. Aguado, R.; Olazar, M.; San José, M. J.; Aguirre, G.; Bilbao, J., *Ind. Eng. Chem. Res.* 2000, *39*, 1925.
27. Olazar, M.; Aguado, R.; Barona A.; Bilbao, J., *AIChE J.* 2000, *46*, 1025.
28. Olazar, M.; San José, M. J.; Zabala, G.; Bilbao, J. *Chem. Eng. Sci.* 1994, *49*, 4579.
29. Olazar, M.; Arandes, J. M.; Zabala, G.; Aguayo, A. T.; Bilbao, J. *Ind. Eng. Chem. Res.* 1997, *36*, 1637.
30. Darivakis, G. S.; Howard, J. B.; Peters, W. A. *Combust. Sci. Tech.* 1990, *74*, 267.
31. Conesa, J. A.; Marcilla, A.; Caballero, J. A.; Font, R. *J. Anal. Appl. Pyrolysis* 2001, *58-59*, 617.
32. Carrasco, F., Pagés, P., In *Proceedings of the 7th Mediterranean Congress of Chemical Engineering*, Barcelona, Spain, 1997, p. 244.
33. Olazar, M.; San José, M. J.; Aguayo, A. T.; Arandes, J. M.; Bilbao, J. *Ind. Eng. Chem. Res.* 1992, *31*, 1784.
34. Olazar, M.; San José, M. J.; Aguayo, A. T.; Arandes, J. M.; Bilbao, J. *Chem. Eng. J.* 1993, *51*, 53.
35. San José, M. J.; Olazar, M.; Aguayo, A. T.; Arandes, J. M.; Bilbao, J. *Chem. Eng. J.* 1993, *51*, 45.
36. Seeger , M.; Gritter, R. J., *J. Polym. Sci.* 1977, *15*, 1393.
37. Westerhout, R. W. J.; Kuipers, J. A. M.; van Swaaij, W. P. M.. *Ind. Eng. Chem. Res.* 1998, *37*, 841.
38. Predel, M.; Kaminsky, W. *Polym. Degrad. Stab.* 2000, *70*, 373.
39. Chaala, A.; Darmstadt, H.; Roy, C. *J. Anal. Appl. Pyrolysis* 1997, *39*, 79.
40. Olazar, M.; San José, M. J.; Peñas, F. J.; Arandes J. M.; Bilbao, J. *Ind. Eng. Chem. Res.* 1994, *33*, 3267.
41. Aguado, R., Olazar, M., San José, M.J., Gaisán, B., Bilbao, J., *Energy & Fuels*, 2002, in press.
42. Kaminsky, W. Possibilities and Limits of Pyrolysis. *Makromol. Chem., Macromol. Symp.* 1992, *57*, 145.
43. Aguado, R., Olazar, M., Gaisán, B., Prieto, R., Bilbao, J., *Ind. Eng. Chem. Res.*, 2002, *41*, 4559.
44. Sinn, H., Kaminsky, W., Janning, J. *Angewante Chemie*, 1976, *88*, 737.
45. Arena, U., Mastellone, M.L., *Chem. Eng. Sci.*, 2000, *55*, 2849.
46. Olazar, M., San José, M .J., Alvarez, S., Morales, A., Bilbao, J. *Ind. Eng. Chem. Res.*, 1998, *37*, 4520.
47. Aguado, R., Olazar, M., San José, M.J., Alvarez, S., Bilbao, J., In *Proceedings of the 17th International Symposium on Chemical Reaction Engineering*, 2002, in press.
48. San José, M. J., Olazar, M., Alvarez, S., Bilbao, J. *Ind. Eng. Chem. Res.*, 1998, *37*, 2553.

Optimising Reaction Conditions Relative to Product Slates in Aqueous Pressurised Pyrolysis of Biomass and Waste Samples

T. Barth

Department of Chemistry, University of Bergen, Allégaten 41, N- 5007 Bergen, Norway

ABSTRACT: As part of the JOULE project "Hydrocarbons from Biomass" (JOR3-CT97-0176, 1998-2000), the effect of temperature, residence time, water-to-biomass ratio and catalyst addition has systematically been investigated for a number of biomass and waste organic materials. Experimental designs and multivariate interpretation are used to express the quantitative relationships between the experimental factors and the yields. The whole product slate has been analysed and quantified for each pyrolysis experiment, i.e. oil phase (with sub-fractions), gas, aqueous compounds and coke/residue for each experiment, and used to establish the experimental designs. Results for the design testing effects of temperature, duration and water/biomass ratio with a clean softwood pulp sample are presented. Additional designs compare the effects of adding sodium hydroxide and potassium hydroxide to the system. Without base, coke dominated the product slate. A high biomass loading and a high water-to biomass ratio increases the yields of organic liquids ("oil"), while the temperature and duration were not significant in the ranges tested. With addition of a base, the oil yields increases dramatically, and the coke can be completely eliminated, but a high water-to-biomass ratio is still important. The results for a design on degraded waste biomass are given for comparison. This is an organic rich sludge from a chemical sewage cleaning plant, which is much less sensitive to the amount of water and gives high yields of oil phase products under all conditions. Addition of a base did not change the product slate significantly. The methodology of experimental designs and multivariate interpretation is very suitable for finding optimal reaction conditions with regard to the product slate, and for adjusting the conditions to different raw materials. Aqueous pyrolysis is shown to be a good technique for producing high-quality products from biomass and waste.

INTRODUCTION

Pyrolysis and related thermal conversion methods have the potential to convert biomass and bio-sourced waste into a wide range of exploitable products, and thus improve sustainability of fuels and chemicals production and use. The products of greatest interest have been perceived as the fuels. However, biomass also contains a number of constituents that could be substitutes for high volume petroleum derived chemicals. The type and value of the products that can be obtained will depend both on the initial biomass composition and the reaction conditions in the conversion process.

Optimal exploitation of biomass therefor requires development of conversion processes for a variety of raw materials and products. Pyrolysis techniques are the most widely used conversion methods, and have the potential for producing a wide slate of products depending on the raw material type and pyrolysis conditions. However, specific procedures that give easily refined products need to be developed to make extensive exploitation of biomass derived product as replacement for petroleum derived products practical and economically interesting.

APPROACH

As part of the JOULE project "Hydrocarbons from Biomass" (JOR3-CT97-0176, 1998-2000), pyrolysis in the presence of water in closed reactors has been used to convert different biomass and waste types into chemically simpler, preferably fluid products. This type of pyrolysis was selected to simulate the natural conversion processes that produce petroleum from sedimentary organic matter, with the aim of producing high yields of petroleum compatible fluid product (1,2). Similar techniques have been applied to biomass and waste conversion with promising results before (3), but have recently not been given much attention.

The effect of temperature, residence time, water-to-biomass ratio and catalyst addition has been systematically investigated using experimental designs and multivariate interpretation to express the quantitative relationships between the experimental factors and the yields (4). The whole product slate has been quantified for each pyrolysis experiment, i.e. oil phase (with sub-fractions), gas, aqueous compounds and coke/residue. Molecular analysis by GC-MS (Gas chromatography-Mass spectrometry) has been performed on the major liquid and gas phase products. The experimental procedures are described in more detail in references 5 and 6. The sample work-up follows a standard fractionation procedure used in petroleum analysis, and the corresponding fraction names are used here: Asphaltenes for liquid products insoluble in excess hexane, oil for the hexane soluble fraction, hydrocarbons for the fraction of the deasphalted oil that is eluted from a silica column with 10 % DCM (dichloromethane) in hexane, and polars for the fraction eluted with 50 % methanol, 50 % DCM).

The experimental designs are used to quantitatively relate the product distribution to the experimental conditions using multivariate linear regression models. The models are used to predict "best" conditions for the product of interest and thus optimise the product slate for each biomass type. The results presented here have been selected from a database of more than 200 experiments performed during the "Hydrocarbons from biomass" project.

RESULTS AND DISCUSSION

PRODUCT DISTRIBUTION

The results from a selection of experiments using a thermo-mechanical softwood pulp (TMP) as raw material are shown in Figs. 1-3, while selected results from experiments using sewage sludge from a chemical sewage treatment plant as raw material is given in Fig. 4. The experimental conditions and product distributions for the experiments shown here are listed in detail in Appendix 1.

In Figure 1, the experiments TMP-3, TM-5 and TMP-7 are from pyrolysis with a limited amount of added water. The water was added to dry biomass to provide either half or the double of the saturation pressure at the pyrolysis temperature. No gas phase analysis was done for these experiments. The yield of fluid products is low, and the recovery is below 50 % by weight, mainly due to the low oxygen content of the products. The TMP has an elementary composition of 46.6 % C, 6.4 % H, N below detection level and O 47.17 % (by difference), and an ash content of 14.5 mg/g., so the complete recovery of hydrocarbons and coke with no oxygen would be about 50 % by weight. The hydrocarbon fraction yields are too low to be seen in the figure, so the whole oil fraction values are given.

Based on interpretation of the initial designs, the amount of water was increased to the maximum possible in the reactor (9 ml for a 21 ml reactor). The results are shown in the TMP-10 experiment, where coke yields are strongly reduced. However, much of the increase is in the asphaltene fraction, which is not a preferred product. The yields of water soluble short chain organic acids also increase, and although this was not initially intended as a target product, it has a good potential for exploitation.

Figure 2 shows examples of the range of variation obtained when 1 M sodium hydroxide is used instead of distilled water. Gas phase composition is included in the data. Under the high pH conditions of these experiments, additional CO_2 will be dissolved in the aqueous phase, but this has not been quantified. The recovery is up to 80 % by weight of the original dry biomass, partly due to the large contribution from the aqueous acids that contain an appreciable amount of oxygen. Additional non-ionic aqueous products have been observed (see below), but have not been systematically quantified. In many of the experiments, the polar, petroleum-compatible fraction is the major non-aqueous product. This is an easily exploitable product fraction (together with coke/charcoal), and further optimisation may increase the yield even more. The addition of NaOH thus is positive for the product slate from this woody biomass.

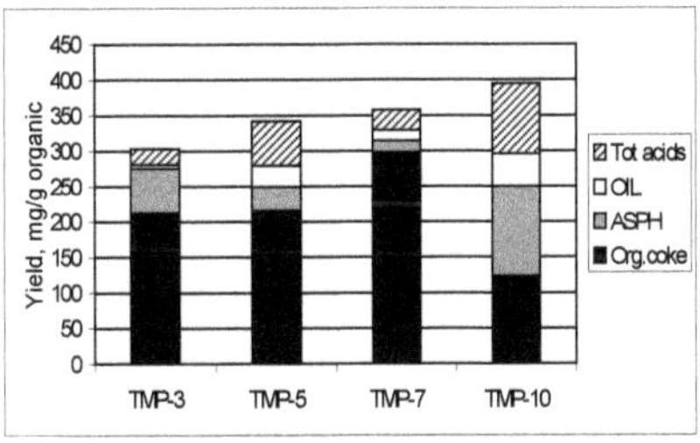

Figure 1: Major products from pyrolysis of woody biomass (TMP) with water (reaction conditions given in Appendix 1).

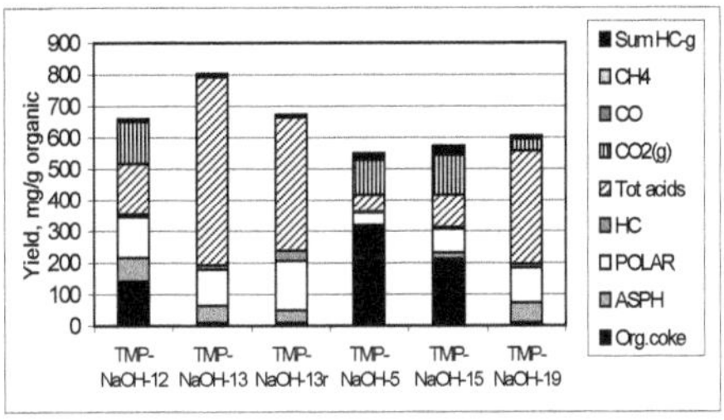

Figure 2: Product distribution from selected experiments with TMP and NaOH as catalyst.

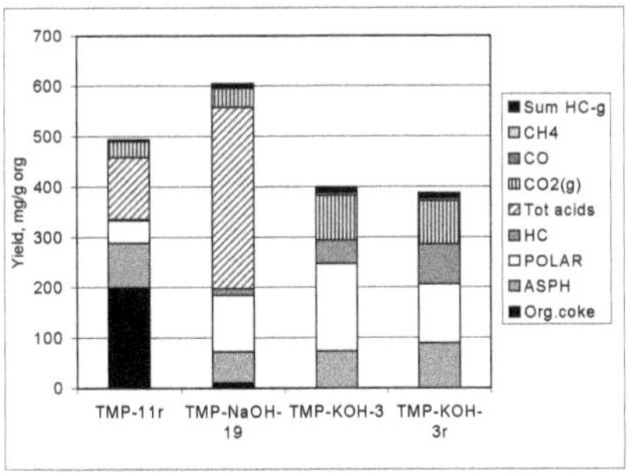

Figure 3: Comparison of product distributions from TMP with water, NaOH and KOH.

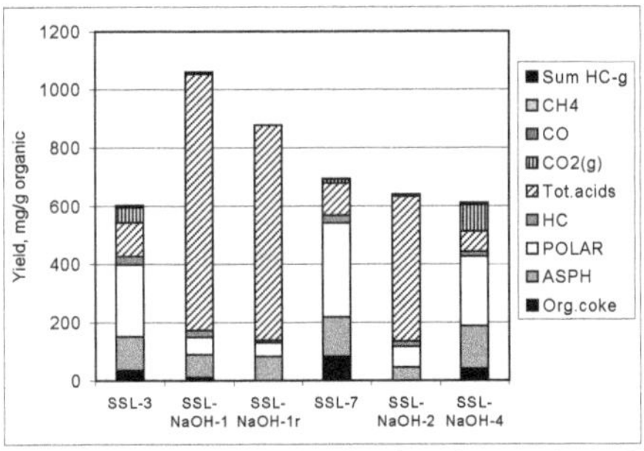

Figure 4: Product distribution from selected experiments with sewage sludge

Some experiments substituting potassium hydroxide (KOH) for the NaOH have been made, and are compared with experiments with water only and with NaOH at the same experimental conditions in Figure 3. Even though analysis of the aqueous phase is missing, the KOH obviously gives more fluids, more hydrocarbon fraction and less coke than for any other conditions. KOH is therefore the preferred catalyst for further optimisation.

In Figure 4, selected experiments using sewage sludge from a chemical sewage treatment plant are shown for comparison. This material contains high levels of iron oxides used as a flocculent for the organic colloids in the wastewater (46.3 % inorganic components). The sewage sludge gives higher fluid yields and lower asphaltene and coke yields than TMP at most conditions and is less sensitive to addition of both water and bases then the woody biomass. It is clearly a better raw material for fluid fuel production by aqueous pyrolysis than the woody biomass, but the high ash content may pose technical problems.

MODELS OF DEPENDENCE ON REACTION CONDITIONS

The regression models are based on a larger number of experiments than listed in Appendix 1, but the complete data table is too large to be included here, and the results presented in the table represent the range of variation observed. The experimental conditions used are given in Table 1 together with the type of designs. Fractional factorial and full factorial designs were used, with the experimental variables set to a "high" and "low" level (3). Some replicate experiments were included to check reproducibility, so the number of experiments is larger than the number required for a complete design. The data table showing the experimental variable values and the yields was analysed using multivariate models: Principal component analysis (PCA) for data exploration and validation of the systematic structure in the data, and PLS (Partial Least Squares/Projection to Latent Structures) to quantitatively model the dependence of the product yields on the experimental conditions. This is expressed as regression coefficients for each variable relative to each yield.

Table 1: Designs and experimental variables for the TMP and sewage sludge experiments.

	Biom.	**Design**	**Temp.**		**Duration**		**Biom. load.**		**Water load.**		**Cat. conc.**	
	type		deg. C		hours		g dry biom.		ml		M NaOH	
		factorial	Low	High	Low	High	Low	High	Low	High	Low	High
Design 1*	TMP	$2^{(4-1)}$	360	380	6	17(48)	1.5	3.0	50 %	200 %	none	
Design 2	TMP	$2^{(5-2)}$	360	380	6	17	1.5	3.0	2.2	4.4	0.5	1.0
Design 3	TMP	$2^{(2)}$	const. =360		const.=17		0.5	1.0	const. =9		0.0	1.0
Design 4	SSL	$2^{(3-1)}$	360	380	6	17	0.5	1.0	const. =9		none	
Design 5	SSL	$2^{(3-1)}$	const. =360		6	17	const.=1		4.4	9.0	0.0	1.0

*: 10 ml stainless steel tubing reactors. Other designs used 21 ml Parr 316 stainless steel reactors.

Table 2 gives the regression coefficients for each product or product group relative to the experimental conditions, and the correlation coefficient R for the resulting linear model of product yields as a function of the experimental variables. The fit of the model to the experimental values is given as the correlation coefficient R for the line fitting the modelled value to the measured value. For some products, and notably the hydrocarbon fraction yields, the model did not fit well, and cannot be used for predictions. This is marked in the table as "non-linear", and the model parameters are only an indication of the major trend.

For the woody biomass, the catalyst concentration is the most important factor for the oil and hydrocarbon yields. Addition of sodium hydroxide (NaOH) in design 3, and increased concentration of NaOH in design 2, increases the yields significantly. This is seen as a large, positive value for the regression coefficients in these designs in Table 2. The coke yield decreases correspondingly in Design 2, and there is no coke produced in Design 3. Asphaltene yields decrease in Design 3 relative to the experiments with water only. The yield of aqueous acids increases with the base addition in all the designs, including the sewage sludge (SSL) in Design 5. The basic conditions result in "trapping" of acids as carboxylate ions in the water phase, and this seems to reduce the further thermal conversion. A similar mechanism may exist for the carbon dioxide, but the lack of data on aqueous carbonic acid concentrations makes interpretation difficult.

Water and biomass loadings are the next most important experimental parameters. A high water content increases the yields of the organic liquids from TMP and reduces the coke, both with and without added base. The biomass loading gives more variable results, and possibly goes through an optimal value for oil phase product generation, given the presence of sufficient water.

Table 2: Regression coefficients for the models.

Product	Temp.	Duration	Bm loading	W. loading	Conc., cat.	Constant	Cor. coeff, R
Design 1: Softwood pulp (TMP), no catalyst (8 exp.)							
Coke	-0.2	1.4	16.4	-31.2		318.8	0.892
Oil	0.1	0.0	8.7	1.1		-26.7	0.794
Asphalt	-0.9	-1.0	-4.1	12.4		361.0	0.95
Hydrocarbons	0.0	0.0	1.5	0.3		-4.2	0.993
Polars	0.0	0.0	7.2	1.6		-12.4	0.739
Aq. acids	1.5	0.0	-5.0	15.6		-519.9	0.889
Design 2: TMP with aqueous NaOH (11 exp.)							
Coke	0.1	1.6	26.4	-39.6	-714.0	0.4	0.969
Oil	-0.6	-1.2	-14.5	21.9	37.8	249.3	0.986
Asphalt	-0.6	-1.5	-5.8	7.4	1.2	241.4	0.944
Hydrocarbons	0.0	0.1	-1.4	1.3	3.0	6.4	non-linear
Polars	-0.6	-0.8	-14.9	18.7	29.9	258.3	0.981
Aq. acids	-0.1	-0.9	-22.1	50.5	123.7	-44.8	non-linear
CO_2	0.7	-1.4	15.0	0.0	14.1	-175.2	0.639
Methane	0.1	0.1	0.6	0.0	2.2	-29.2	0.911
C_2-C_6 gases	0.2	0.5	0.9	0.3	6.6	-71.9	0.851
Design 3: TMP with aqueous NaOH at high water to biomass ratio (6 exp.)							
Oil			27.3		75.7	30.2	0.956
Asphalt			-21.1		-58.5	126.6	0.933
Hydrocarbons			2.8		7.7	0.7	0.809
Polars			21.2		58.9	33.8	0.975
CO_2			-13.1		-36.3	58.6	non-linear
Methane			0.1		0.3	1.8	non-linear
Design 4: Sewage sludge, no catalyst (4 exp.)							
Oil	-2.2	-4.3	-73.8			1218.6	0.951
Asphalt	-1.5	-3.0	-51.7			766.2	0.845
Hydrocarbons	-0.1	-0.3	-4.6			82.8	non-linear
Polars	-2.4	-4.6	-80.0			1257.1	0.906
Aq. acids	-0.4	-0.7	-12.2			210.6	non-linear
CO_2	0.9	1.8	30.3			-319.0	0.987
Methane	0.1	0.2	3.4			-39.6	0.962
Design 5: Sewage sludge with aqueous NaOH (4 exp.)							
Oil		-9.8		-19.7	-135.6	565.0	0.999
Asphalt		-4.3		-5.0	-38.2	204.9	0.997
Hydrocarbons		-0.5		-1.0	-7.1	44.0	non-linear
Polars		-9.5		-17.4	-121.6	508.3	0.99
Aq. acids		5.3		22.0	145.1	-142.1	0.971
CO_2		1.7		-10.6	-64.9	130.8	0.992
Methane		0.1		-0.1	-0.7	2.0	0.984

The models in general show that the temperature and duration of the experiments are not of major importance for the yields within the ranges tested. The exception is a minor trend of increasing hydrocarbon fraction yields with time, correlating with decreased asphaltene and increased coke yields. This can be interpreted as a disproportion of asphaltenes to hydrocarbons and coke over time.

The sewage sludge is very different from the TMP, and the fluid yields are generally highest at the lower temperature, with the lower level of water and no base in the range tested here. However, the number of data points is limited, so the models represent a smaller range of variation. The data definitively demonstrates that it is necessary to optimise the conversion conditions for each raw material type.

PRODUCT MOLECULAR COMPOSITION

Identification of individual molecular species in the different product fractions is a challenging task, and has been routinely performed only for selected fractions where the quantification procedure has made it necessary. This is the case for the gas phase products, which were quantified by gas chromatographic analysis with individual calibration curves for all major components, and for the short chain fatty acids in the water phase, which have been analysed by isotachophoresis (7), with quantification of C_1-C_8 alkanoic acids based on individual calibration curves. From the other fractions, only selected representative samples have been analysed by gas chromatography combined with mass spectrometry (GC-MS), using suitable derivatisation procedures for the polar fractions. The results are briefly summarised here, but will be presented in grater detail in later publications.

The non-polar hydrocarbon fraction is complex and not well suited for GC separation. Cyclic compounds seem to dominate. Methyltetraline and other tetraline homologues have been identified, but most of the components at present are of unknown structure. Some aromatic hydrocarbons, e.g. phenanthrene, have also been found. In addition, the n-alkane series from C_{10} to C_{33} have been found in low concentrations in many samples. This could support the hypothesis that the conversion could in part be similar to natural hydrocarbon generation, but the hydrocarbon yields from the pyrolysis are very low compared to the natural situation.

Polar fraction compositions are strongly dependent on the raw material and in many cases well defined, but again only selected sample have been analysed. For example, the TMP-NaOH-19 experiment gave a cleanly phenolic polar fraction, with the major components consisting of dimethyl-phenols, trimethylphenols, methoxyphenols and some bicyclic phenols with a saturated 5-ring condensed to the aromatic ring of the phenol. These are interpreted as products from decomposition of the lignin component of the woody biomass. The sewage sludge polar fractions, e.g. SSL-NaOH-1 below, contains minor amounts of the same phenols and a full range of alkanoic and alkenoic acids from C9 to C20, with maximum yields of hexadecanoic acid. These acids can be related to the lipid components of the sewage sludge, and could be used for diesel fuel after conversion to methyl esters.

The main products in the water phase are the alkanoic acids in the C_2-C_6 range. Acetic acid is the dominating species. Additional GC-MS analysis of selected samples with regard to non-ionic components showed significant amounts of low molecular weight ketones and alcohols in the C_4-C_6 range, with mainly cyclic structures.

CONCLUSIONS

Pyrolysis in an aqueous medium shows interesting potentials for use in conversion of biomass and waste organic matter into simple organic product groups that can be used as petroleum substitutes or petroleum product substitutes with a limited degree of refining. The presence of water during the reaction is very important for the product quality, resulting in high yields of the low molecular weight, petroleum compatible components.

Woody biomass needs a large excess of water to prevent coke from being the major product. Even so, coke and asphaltenes (heavy tars) dominate the products when run in an unmodified water phase. Aqueous acids, especially acetic acid, can be produced in high yields depending on the amount of water. The fluid products have low oxygen contents, and the oil is petroleum compatible.

Addition of sodium hydroxide reduces or prevents coke production. The oil yields increase significantly, and hydrocarbons are produced in moderate yields. The aqueous acids can be major products. The total yields by weight increase due to higher oxygen content in the aqueous products.

Using potassium hydroxide (KOH) instead of NaOH increases the oil yield further, and the hydrocarbon fraction yields are considerable. This shows that the cation in itself acts as a catalyst, in addition to the pH effect of the base.

Sewage sludge shows a very different pattern. The oil yields are high at all conditions, and the organic coke is low, but there are considerable amounts of inorganic residues. Adding NaOH or KOH does not increase the oil or hydrocarbon yields, but does increase the yield of aqueous acids.

These results demonstrate the considerable potential of the aqueous thermal processing of biomass and organic wastes as a source of non-fossil carbon based products. However, it will be necessary to choose one type of raw material and optimise the reaction conditions relative to the target product slate to establish an efficient conversion process in each application.

ACKNOWLEDGEMENT

Anne Eva Borgund and Reidun Myklebust are thanked for their enthusiastic work on the project. Carla Sofia Aureliano is thanked for the polar fraction analysis. JOULE is gratefully acknowledged for the funding. The project partners, Geoff Abbott (University of Newcastle) and Karsten Pedersen (Danish Technological Institute/Chemidan), are thanked for inspiring co-operation.

REFERENCES

1. Barth, T. (1999) Similarities and differences in hydrous pyrolysis of biomass and source rocks. *Organic Geochemistry* 30, 1495-1508.
2. Elliott, D.C., Sealock, L.J. Jr., Baker, E.G. (1993) Chemical processing in high-pressure aqueous environments. 1. Historical perspectives and continuing developments. Industrial and Engineering Chemistry Research 32, 1535-1541.
3. Michels, R., Landais, P., Torkelson, B.E., Philp, R.P. (1995) Effects of effluents and water-pressure on oil generation during confined pyrolysis and high-pressure hydrous pyrolysis. *Geochimica et Cosmochimica Acta* 59, 1589-1604.
4. Carlsson, R. (1992) Design and optimisation in organic synthesis. Elsevier, Amsterdam.
5. Barth T., Borgund A.E., Hopland A.L. (1989) Generation of organic compounds by hydrous pyrolysis of Kimmeridge oil shale - Bulk results and activation energy calculations. *Organic Geochemistry* 14, 69-76.
6. Borgund, A.E and Barth, T, (1999) Effects of base catalysis on the product distribution from pyrolysis of woody biomass in the presence of water. *Organic Geochemistry* 14, 1517-1526.
7. Barth, T. (1987) Quantitative determination of volatile carboxylic acids in formation water by isotachophoresis. *Anaytical.Chemistry* 59, 2232-2237.

Appendix 1: **Experimental conditions and products for the experiments shown in Figs. 1-4.**

Name	Temp C	Duration h	Biomass			Catalyst I		Water ml	Recov. water g	Tot. Rec. Water %	Org. coke mg/g.org	EXTR. mg/g.org	OIL mg/g.org	ASPH mg/g.org
			type	amount g	org. %	type	conc. M							
TMP-3	360	6	TMP	1.624	98.55			1.9	0.708	37.3	215.2	83.2	5.6	63.3
TMP-5	380	6	TMP	2.952	98.55			2.2	0.512	23.3	218.3	72.7	29.8	34.3
TMP-7	380	17	TMP	2.95	98.55			1.2	0.651	54.3	302.4	35.8	15.1	16.4
TMP-10	360	17	TMP	0.494	98.55			9	10.677	118.6	123.5	191.6	46.2	126.5
TMP-10r	360	17	TMP	0.506	98.55			9	9.576	106.4	128.6	185.4	53.9	121.7
TMP-11	360	17	TMP	1.05	98.55			9	9.435	104.8	166.8			
TMP-11r	360	17	TMP	1.014	98.55			9	11.82	131.3	198.4	163.8	46.8	89.0
TMP-NaOH-5	380	17	TMP	2.99	98.55	NaOH	1	2.2	2.206	100.3	313.3	66.9	47.7	6.6
TMP-NaOH-12	360	6	TMP	1.546	98.55	NaOH	1	4.4	4.27	97.0	141.5	233.4	139.6	74.8
TMP-NaOH-13	360	17	TMP	0.51	98.55	NaOH	1	9	8.947	99.4	9.3	234.7	118.0	54.3
TMP-NaOH-13r	360	17	TMP	0.5	98.55	NaOH	1	9	9.317	103.5	9.5	268.5	180.7	38.7
TMP-NaOH-15	380	17	TMP	3.04	98.55	NaOH	1	4.4	4.78	108.6	212.4	138.8	93.7	19.1
TMP-NaOH-19	360	17	TMP	1.021	98.55	NaOH	1	9	9.181	102.0	10.3	244.7	151.4	61.7
TMP-KOH-3	360	17	TMP	1.001	98.55	KOH	1	9	9.115	101.3	0.0	237.6	141.9	73.1
TMP-KOH-3r	360	17	TMP	1.001	98.55	KOH	1	9	8.411	93.5	0.0	234.6	129.4	88.4
SSL-3	360	17	KSLAM	0.503	57.2			9	6.905	76.7	37.2	444.7	292.3	114.4
SSL-7	360	17	KSLAM	0.5	57.2			9	7.5	83.3	83.6	501.0	385.1	135.1
SSL-NaOH-1	360	17	KSLAM	0.52	57.2	NaOH	1	9	8.524	94.7	12.4	476.0	101.5	77.0
SSL-NaOH-1r	360	17	KSLAM	0.544	57.2	NaOH	1	9	7.96	88.4	0.0	220.7	123.4	83.5
SSL-NaOH-2	360	17	KSLAM	1.028	57.2	NaOH	1	9	3.783	42.0	0.0	552.6	88.5	46.2
SSL-NaOH-4	360	6	KSLAM	3.063	57.2	NaOH	0.5	2.2	0.826	37.5	41.7	470.0	301.1	145.7

Appendix 1, continued.

Name	HC mg/g.org	POLAR mg/g.org	HAc M	HAc mg/g.org	Tot. acids M	Tot. acids mg/g.org	P(gas) bar	CO_s(gas) mg/g.org	H_s mg/g.org	CO mg/g.org	CH_4 mg/g.org	Sum HC-g mg/g.org	Mass sum mg/g.org
			(initial water base)									(C_2-C_6)	(extr. base)
TMP-3	0.9	4.0	0.17	12.1	0.19	22.3							320.8
TMP-5	3.3	26.5	0.57	25.9	0.84	62.6							353.7
TMP-7	3.0	11.6	0.52	12.8	0.70	28.4							366.6
TMP-10	4.6	43.6	0.05	51.0	0.05	98.4	2.9	91.38	2.00	0.05	3.36	6.60	514.8
TMP-10r	1.0	52.5		0.0			2.8	29.35	1.69	0.02	1.03	2.47	346.9
TMP-11			0.07	35.0	0.21	179.1	5.2	34.65	1.47	0.03	1.34	2.16	384.1
TMP-11r	2.2	45.7	0.07	38.9	0.14	123.4	4.6	31.09	1.79	0.04	1.22	2.05	519.9
TMP-NaOH-5	3.3	39.2	0.53	23.6	0.74	54.7	16.7	111.29		0.24	5.25	15.92	567.6
TMP-NaOH-12	9.5	129.6	0.32	55.8	0.57	161.9	10.0	132.48		0.21	2.29	6.30	678.1
TMP-NaOH-13	13.3	115.1	0.15	162.2	0.34	600.0	3.5	0.41		0.04	2.57	6.67	853.7
TMP-NaOH-13r	32.4	157.2	0.13	140.3	0.24	426.6	3.5	0.29	5.04		2.06	6.20	713.2
TMP-NaOH-15	5.5	76.3	0.48	42.7	0.72	103.5	18.2	127.73	0.00	0.20	6.63	21.14	610.4
TMP-NaOH-19	12.7	111.6	0.21	111.1	0.41	361.4	6.5	37.08	17.15	0.07	2.58	6.63	662.8
TMP-KOH-3	47.3	173.4					7.0	88.92	3.80	0.18	4.70	10.40	341.8
TMP-KOH-3r	80.4	116.7					7.0	86.80	3.78	0.18	4.73	10.04	336.4
SSL-3	29.6	247.1	0.0	38.1	0.04	116.9	1.4	50.91	0.45	0.07	1.61	5.22	569.4
SSL-7	26.4	323.5	0.0	32.1	0.04	110.1	1.6	14.02	0.48	0.00	0.37	1.55	580.0
SSL-NaOH-1	21.9	61.2	0.1	147.5	0.29	880.5	2.9	0.87	4.92	0.00	2.12	3.99	986.6
SSL-NaOH-1r	8.8	47.3	0.1	110.0	0.26	737.5	3.1						642.6
SSL-NaOH-2	17.6	72.1	0.1	79.2	0.33	497.4	2.7	0.29	2.19	0.00	1.96	5.09	844.5
SSL-NaOH-4	17.4	239.5	0.3	24.8	0.55	69.2	7.2	91.73	0.92	0.02	1.92	4.20	607.

Abbreviations used in Appendix 1

Temp: Temperature; Conc.: Concentration; Recov.water: Recovered water from decantation and water separation during extraction; Org. coke: Carbon in solid residue calculated from measured TOC values; EXTR.: Extractable non-volatile liquids; OIL: Hydrocarbon-soluble fraction of the extract; ASPH: Fraction of the extract precipitated after 40 v/v hexane dilution; HC: Hydrocarbon fraction from separation of oil on a silica column; POLAR: Polar fraction form separation of oil on a silica column; HAc: Acetic acid; Tot.acids: Sum of all ionic components in the water phase; P(gas): Gas overpressure in vessel at room temperature; $CO_2(g)$: Amount of carbon dioxide recovered in the gas phase; H_2: Amount of hydrogen gas recovered in the gas phase; CO: Amount of carbon monoxide gas recovered in the gas phase; CH4: Amount of methane gas recovered in the gas phase; Sum HC-gas C2-C6: Sum of the amounts of all hydrocarbon gases in the 2-6 carbon range recovered in the gas phase; Mass sum: Sum of all components recovered from the experiment, using the crude extract values to represent the liquids.

A New Technology for Fast Pyrolysis of Biomass, Development of the PyRos Reactor.

E.A. Bramer[a] and G. Brem[a,b]

[a]University of Twente, Laboratory of Thermal Engineering, P. O. Box 217, 7500 AE Enschede, The Netherlands

[b]TNO, P. O. Box 342, 7300 AH Apeldoorn, The Netherlands

ABSTRACT: A new reactor type for flash pyrolysis of biomass and other solid materials is presented. In this novel reactor, denoted the PyRos reactor, the flash pyrolysis process in a cyclonic reactor and high temperature gas cleaning by the rotational particle separator are integrated in one unit. In this way a high quality bio-oil can be produced.

The biomass and the inert heat carrier are introduced as particles into the cyclone. By centrifugal force the particles are transported to the periphery of the cyclone where the pyrolysis process takes place. The evolved vapors are transported rapidly to the centre of the cyclone and leave the cyclone by the rotating filter were the solid particles are removed. Immediately after the pyrolysis reactor the vapors are quenched with cooled bio-oil. The quenching of the vapors and the separation of droplets and aerosols are done in a second rotational particle separator. In this apparatus also the very small aerosols can be separated. These particles are difficult to remove by common demister techniques. This paper gives a description of the process, the pilot plant and some results on heat transfer and particle residence time are reported.

INTRODUCTION

In the Netherlands the target for the use of renewable energy sources is set at 10 % of the total energy demand in the year 2020. The contribution of biomass and waste is 42 % (120 PJ) of the total demand of renewable energy. At the moment approx. 32 PJ is produced by the combustion or incineration of biomass and waste. In the last decade the research on more sophisticated techniques like gasification and pyrolysis was intensified. For a successful implementation of biomass in our energy production system high efficient and economical competitive conversion techniques must be developed.

Fast pyrolysis is a thermo-chemical process that converts biomass into a liquid product: bio-oil. This bio-oil can be burned in a diesel engine, a gas turbine, a furnace or be used as a feedstock for chemical plants such as FT-synthesis. The main product within the fast pyrolysis process is a fuel with a high density and heating value. This product can be transported and stored easily. This gives the possibility to decouple the primary biomass conversion process and the prime mover, regarding the time, location, and scale of the process. The fast pyrolysis technology can have relatively low investment costs and high energy efficiencies compared to other processes, especially on a small scale.

In practice about 40-75% of the (dry) biomass is converted into oil vapour, which after condensation yields the liquid bio-oil. About 10-20% of the biomass is converted into char (solid porous carbon particles) and 10-30% is converted into a noncondensable gas. The water content of the fuel usually is limited to about 15 wt-%. Present fast pyrolysis reactors are based on fluidized bed, entrained flow, ablative vortex, rotating cone technology, ablative rotating blade and vacuum pyrolysis [1]. Most of the reactor types are pilot plant or demonstration stage.

A major problem of the present reactors for flash pyrolysis is the quality and the stability of the produced oil, both are strongly affected by the char/ash content of the bio-oil. Besides the known problems concerning solid particles in oil, the char fines will catalyze the repolymerization reactions inside the bio-oil resulting in a higher viscosity. Furthermore the char transports the alkali from the biomass into the bio-oil as the alkali dissolve in this oil (pH = 2-3). The char can be removed by after-treatment of the condensed products like filtering the oil. By liquid filtration of the char, the alkali content of the oil could only be reduced from 325 ppm down to 190 ppm this in contrast to the option of hot gas cleaning of the oil vapor, the alkali content is dropped then down to 2.2. ppm [2]. During hot gas filtration however problems will occur with increased pressure drop over the filter with time. The char or, after oxidative regeneration of the filter, the ash will block the filter passage [3]. The char layer on the surface of the filter will also act as a catalyst for secondary cracking of tar, the yield of tar can be reduced by 10 to 20 % by the char layer on the filter [4]. For both, oil and hot gas filtration additional equipment and capital investments are required.

University of Twente and TNO, a Dutch contract research organization, have invented and developed a novel reactor in which the fast pyrolysis and the char removal from the vapors take place simultaneously: PyRos reactor, patented by TNO NL99/00688.

The paper gives a description of the rotational particle separator as a filter for particle separation, followed by the application of the rotational particle separator as a reactor for the pyrolysis process. Experimental research on the heat transfer and solid residence time on small lab-scale is reported. The paper is concluded with a small description of the pilot plant currently in test phase.

THE ROTATIONAL PARTICLE SEPARATOR

In recent years a new technique for separating particles and aerosols from gases is developed by the Thermal Engineering group of the University of Twente. This is done in close cooperation with different companies. Present this patented technique (RPS, US-patent US5073177) is commercial available as a gas cleaning device for industry and as a portable air cleaner for domestic use.

The rotational particle separator (RPS) consists of a cyclonic pre-separator for coarse particles and a rotating filter element for the separation of the small particles of micron or submicron size. The core of the RPS is a rotating filter element that consists of a large number of small axial channels, which rotate as one body around a common central axis. Solid and/or liquid particles present in the gas flowing through the channels are centrifuged towards the outer collecting walls: see Fig. 1.

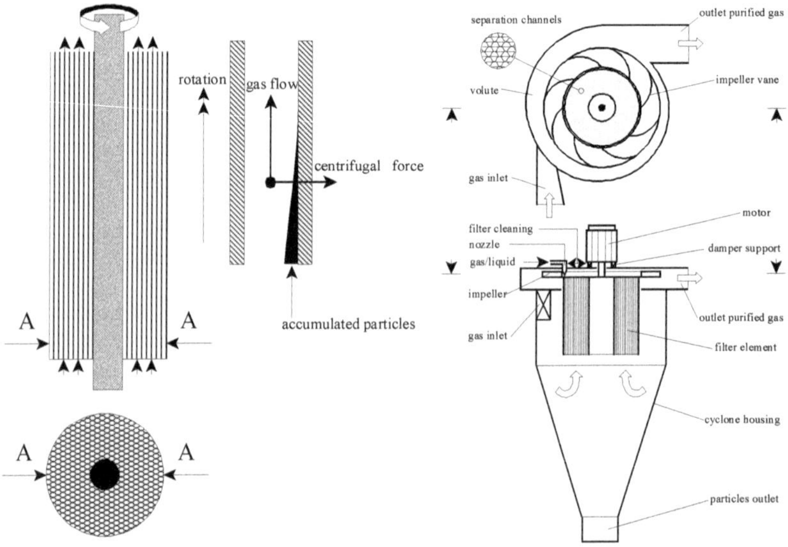

Figure 1: The Rotational particle separator.

As a result of the action of centrifugal forces, ' van der Waals' forces and/or forces due to surface tension, particles remain at the wall. The purified gas leaves the channels at the top while the particles material collected at the walls can be removed periodically by means of a cleaning system. The flushing fluid is pressed by a nozzle mounted at the top of the filter element, at high velocity through the channels in reverse direction. Particles are blown out as clusters or agglomerates and are separated by the cyclone and leave at the system bottom exit. The nozzle is moving in radial direction over the filter element during rotation to ensure that all the channels will be cleaned.

The radial height of the channels is small, up to a few millimeters, so that the radial distance to be traveled by a particle is small. At limited circumferential speed of the filter element (some tens of meters per second) and relatively large axial gas velocities (a few meters per second) particles of (sub) micron size can still reach the outer collecting wall for sufficient axial length of the channels (up to about one meter). The diameter of the filter element can amount to one meter allowing gas flows of 10.000 m^3/hr. Length and height of the channels are dimensioned such that the pressure loss is limited: up to some hundreds of Pascal.

To design the RPS theoretical separation efficiency will be used. If the flow in the separation channels is laminar the migration velocity can be calculated from the equilibrium between the centrifugal force and the Stokes force, resulting in:

$$u_p = \frac{\Omega^2 * R * d_p^2 * \rho_p}{18 * \mu_g}$$

Where, Ω is the angular velocity of the centrifuge, R is the distance between the channel and the rotation axis , d_p is the particle diameter, ρ_p particle density and μ_g the viscosity of the gas.

Whether a particle reaches the outer collecting wall of the channel depends on the residence time in the channel and the radial distance to be traveled by the particle.

The smallest particle which is just able to reach the outer collecting wall with 100% probability is the particle which in the available time moves over a radial distance corresponding to the height of the channel. The residence time of a particle in the channel is:

$$t_r = \frac{L}{v_p}$$

With L is the length of the filter element and v_p is the axial velocity of the particle. The maximum migration time of a particle, reaching the outer wall of the channel is:

$$t_m = \frac{d_c}{u_p}$$

With d_c is the channel diameter and u_p is the radial velocity of the particle.

The particle will reach the wall if the migration time is less than or equal to the residence time. So this will lead to the smallest particle collected for 100%.

$$d_{p,100\%} = \left(\frac{18.\mu.v_p.d_c}{(\rho_p - \rho_g).\Omega^2.R.L} \right)^{1/2}$$

In case of an optimum distribution of the axial gas velocity, i.e. $v_p \div R$, an expression for $d_{p.100\%}$ can be derived which is applicable to the entire filter element:

$$d_{p,100\%} = \left(\frac{27.\mu.\phi.d_c}{(\rho_p - \rho_g).\Omega^2.L.\pi.(1-\varepsilon)(R_o^3 - R_i^3)} \right)^{1/2}$$

With, ϕ is the gas flow through the filter element, ε = reduction of the effective cross-sectional area of the element due to wall thickness of the channel R_o is the outer radius of the filter element and R_i is the inner radius of the filter element.

From the previous considerations it follows those particles with a diameter larger than $d_{p.100\%}$ are collected with an efficiency of 100 %. Particles with a diameter less then $d_{p.100\%}$ can also be separated in the channels. The probability to reach the collection wall depends on the radial position when entering the channel.

In Figure 2 the collection efficiency is plotted versus the particle diameter, made dimensionless by $d_{p.100\%}$ and a theoretical efficiency curve. The measurements were executed with different types of dust and different types of the RPS, for more details see [5].

As we can see from this figure that very small particles can be separated at an efficiency with this technique and that the efficiency can be predicted rather well.

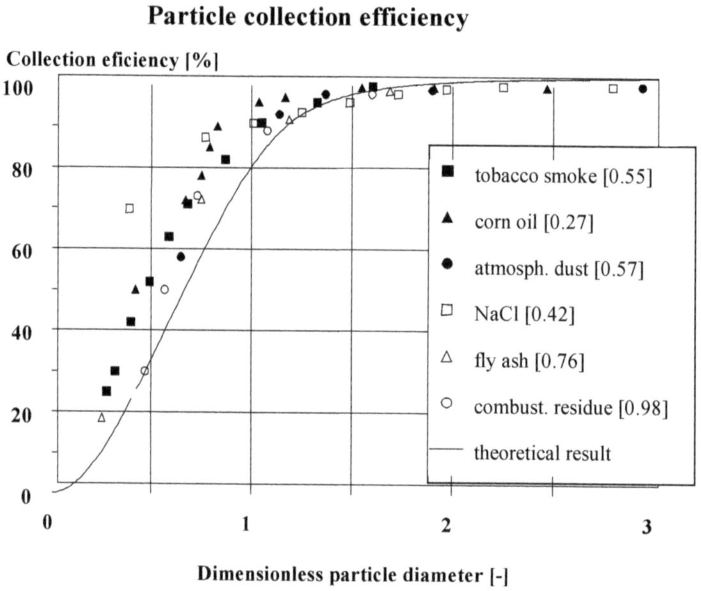

Figure 2: The collection efficiency as a function of the dimensionless particle diameter for different RPS types and dust types, the number between the brackets is the $d_{p.100\%}$ in microns for each dust type.

This high collection efficiency for small particles and the possibility to manufacture the apparatus from high-grade material make the RPS an excellent tool for high temperature gas cleaning. The Rotational Particle Separator can be a used for the particle separation in the off-gases of gasification or pyrolysis units to meet the restrictions of the equipment off-stream or to produce high quality oil.

In this project however we developed a new pyrolysis reactor where the pyrolysis process is integrated in the particle separator, the PyRos reactor.

PYROS REACTOR

REACTOR DESCRIPTION

The novel reactor, denoted by PyRos -reactor, integrates fast-pyrolysis and high temperature gas cleaning in one unit, to produce particle free bio-oil. The hot gas filtration is based on the principle of the rotational particle separator [5].

Cyclone or vortex reactors were used before for pyrolysis of biomass by Lede [6] and Diebold [7]. In both systems the main part of the heat is supplied by heating the walls of the reactor. The reactor of Lede, operating at approx. 1000 °C, had problems with blocking of the reactor at higher biomass flow rates. In the vortex reactor of Diebold very high inlet velocities (up-to 400 m/s) for particles and the transport gas (steam or nitrogen), this results wear of the reactor and short residence time for the particles. For complete conversion of the particles a high recycle rate back to the reactor is needed.

The PyRos reactor described here consists of a cyclonic reactor with a rotating particle separator placed in the so called vortex finder of this reactor. A scheme of the novel reactor is given figure 3. The heat supply to the reactor is through adding preheated inert material (sand) to the reactor. The inert over biomass ratio is typically around 20. Besides the function of heat source the material also wipes the surface of the cyclone so build-up of a layer of reaction product is prevented. By the high inert-biomass ratio the inert is the predominant factor in the transport of the biomass in the reactor, so the biomass particle residence time can be controlled by the inert flow.

The biomass is fed into the cyclone via a transport medium consisting of an inert gas with a solid that acts as heat carrier. By the centrifugal force in the reactor the particles are swept out of the gas stream and are sliding along the cyclone wall downwards to the particle exit. In the reactor the biomass and the heat carrier are in direct contact, so efficient heat transfer can take place. During the transport downwards in the reactor the biomass particles are dried, heated up and devolatilized. The average process temperature is 450-550 °C. The typical gas residence time in the reactor is 0.5 to 1 sec, so secondary cracking reactions of tars in the reactor can be reduced. In the gas outlet, in the middle of the cyclone reactor, the fine solid particles are removed from the oil vapor by the rotational particle separator. In this separator the micron particles are captured in a bundle of channels with small diameters, rotating around a common axis. The collected particles stick to the channel walls until they periodically removed on-line and leave the reactor with the main solid stream at the lower outlet of the cyclone.

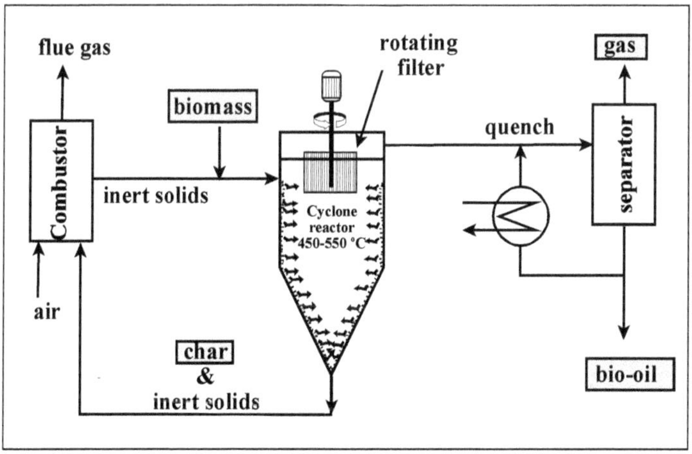

Figure 3: A simplified scheme of the PyRos-process.

Features of the PyRos reactor are:

- High heat transfer rates to the biomass particles.
- Short gas residence time, typically <1 second.
- Controllable particle residence times to have the possibility to convert also larger biomass particles.
- Particle free vapour production.
- Integration of reactor and particle filter.
- Compact/low cost reactor.

To develop the PyRos reactor a bench scale test unit with a throughput of 1 kg/hr biomass was constructed. In this cyclone reactor an experimental program is carried to study the influence of the reactor configuration and the most important parameters of the process, temperature, gas and solid residence time particle size of the biomass on the process performance. The first results were very promising a stable operation for many hours was possible and liquid yields of more than 50 wt.% on dry feed were achieved. The heating value of the oil was comparable to the supplied biomass and oil had a low viscosity.

The next step in the scale-up of the process is a pilot plant with a throughput of 30 kg/hr. This installation is erected at the laboratories of the Thermal Engineering group of the University of Twente. The installation is at the moment in the test phase. This is a fully integrated pilot plant with pyrolysis reactor, biomass feeding and transportation system, hot gas particle separation, char/gas combustion for inert particle heat-up, condenser for quick oil separation and a oil separator/demister. See below for more details about this pilot plant.

To scale-up the PyRos reactor we need more information about the processes taking place inside the cyclone reactor. On lab-scale heat transfer and particle residence time in a cyclone are studied separately.

PARTICLE RESIDENCE TIME IN THE CYCLONE REACTOR

One of the goals of the project is to develop a pyrolysis reactor which is able to convert millimeter sized particles with a high bio-oil yield. The economics of the process will be strong dependent on the cost for pretreatment of the biomass. The energy consumption and so grinding cost will increase exponential with a decreasing particle size when the feedstock particle size has to be reduced below 1 – 2 mm [7]. It can be calculated that the heat up time and so the conversion time of 2 mm particles is in the order of a few seconds. For complete conversion of the biomass particle the residence time in the cyclone reactor should exceed the reaction time to avoid more complicated recycle systems.

A cyclone is normally used as a gas-solid separator and not as a reactor for gas-solid reactions. In literature different models are available to describe the flow behavior and the particle separation in cyclones. Almost all of these models have in common that the description of the path of the particles stops when the particles reach the wall. In our reactor we use relative large particles, they are separated almost immediately after entering the cyclone and flow downwards along the cyclone wall. In that case the available models will fail. The forces on the particles sliding along the cyclone wall are:

- gravity (F_g)
- friction between particle and wall (F_f and F_w)
- friction between gas and particle (F_d)
- centrifugal forces (F_{cf})
- normal force on the wall (F_n)

It is clear that the forces are dependent on position in the cyclone, cylindrical or conical part and on the angle of the cone, but also on the movement of direction of the particle with respect to the direction of the gas flow, see Figure 4a and 4b.

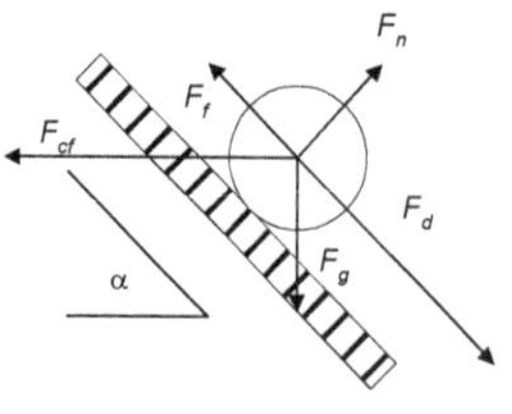

Figure 4a: Forces acting on a particle by the tangential movement.

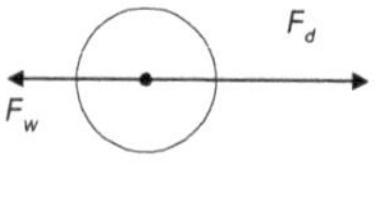

Figure 4b: Forces acting on a particle by the axial movement.

From observations in glass cyclones it is known that the particles in the cylindrical part act independent from each other. For individual particles it is impossible to leave the cyclone at the bottom exit, the centrifugal force acting on one particle is that high that the particles are forced upwards when they are in the conical part. Particles are concentrated in the transition between the cylindrical and the conical part. A particle can only travel downwards within strings of particles. Within this string the drag force of the gas on the particle is reduced and so the centrifugal force will decrease. This whole flow behavior of the gasses and the particles and the interaction between each other is very complex and modeling of this type of two phase flow is not yet fully understood.

In the past a few research groups have performed experimental research on the residence time of particles in cyclones, Kang[8], Kemp [9] and Lede [10]. Residence times ranging from tenths of seconds up to 40 seconds. Also different correlations have been derived which are predicting the solid residence time. Unfortunately, these correlations are only valid within a certain operating range and usually also for only one specific type of cyclone. There is also a great discrepancy in the trends on the influence of the different parameters on the residence time of particles.

An experimental research was started at the Thermal Engineering Group to investigate the influence of the different process parameters and cyclone geometry on the particle residence time in the cyclone. The experimental set-up is given in the figure 5. The cyclone is made of glass to be able to visual observe the flow behavior. The cyclone is operated in continuous mode, a constant feed of inert solid material (in this case sand) and gas (air) to the cyclone.

In literature different methods for the measurement of the residence time are suggested [11]. Most of these methods were rather time consuming or were not able to distinguish between the sand and the biomass particles. Here a new technique, using magnetic induction of coils by tracer particles is described. The idea is that two electrical coils are placed around the exit and the entrance of the cyclone. The coils are connected to an electrical bridge and an electrical amplifier. Particles containing some iron were used as tracer and were injected batch-wise at the cyclone entrance. These batches of tracer will influence the magnetic field in the coils when they passing the coil, thereby causing a potential difference. These variations are recorded with a data acquisition card at a high frequency. The advantage is that a measurement is not time consuming, the resolution is very high at high sampling rates, no interference of the flow behavior in the cyclone and it is possible to measure the biomass wood residence time separate. Furthermore it is possible to measure under all operating conditions, like high gas velocities and high solid loadings, as long as enough material is used to give a detectable signal. In Figure 5 a schema of the experimental setup is given.

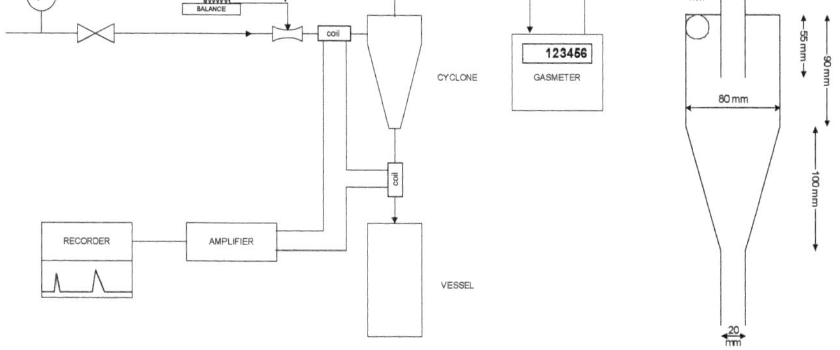

Figure 5: Experimental setup for the measurement of the particle residence time in a cyclone.

In Figure 6 the measured residence time as a function of the inlet air velocity is given for two different sand feed rates. As can be seen from this figure the particle residence time in this small cyclone is approximately 0.6 to 0.7 sec and can be increased by increasing the gas flow rate. A

Also the sand feed rate is of importance, if the sand feed rate is increased the mean residence time of the tracer particles decreases.

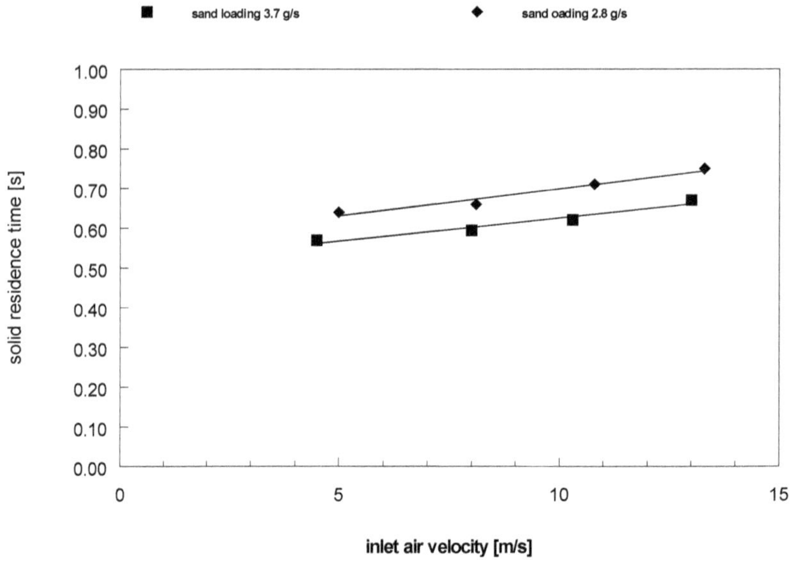

Figure 6: Influence the inlet air velocity on the tracer particle residence time for two different sand loadings, sand diameter is 0.2-0.63 mm and tracer particle diameter is 0.2-0.5 mm.

In Table 1 a summary of the experimental results of the particle residence time measurements is given. It can be seen that the particle residence time can be increased by increasing the inlet gas velocity and the sand particle diameter or decreasing the sand flow. These results have to be verified in a large cyclone.

Table 1: A summary of the experimental results of the particle residence time measurements, + = increase, - = decrease

Parameter increase of:	Observed change in residence time
Inlet gas velocity (by flow)	+
Sand flow rate (loading)	-
Sand diameter	+
Tracer particle diameter	-
Tracer particle density	+

HEAT TRANSFER IN A CYCLONE

One of the demands for flash pyrolysis is the high heat up rate of the biomass particles to insure complete conversion and an acceptable particle residence time in the reactor. In the cyclone reactor there is a direct contact between the sand particles (heat source), the hot vapors and the biomass particles. For these reasons together with good mixing of the solids we can expect heat transfer rates comparable to fluidized beds.

In literature there is only information available for heat transfer in cyclones in the case the heat is flowing from hot walls to particles and gas or vice wards, [11], [12].

In the PyRos reactor there are two different separate solid streams, hot sand and biomass particles to be heated up. The heat can be supplied by different mechanism to the biomass particles e.g. by direct contact with the wall, contact to the heated sand, convection by the gases around the particles.

To get more information about the heat transfer mechanism in a cyclone, experiments were conducted in the small glass cyclone also used for the particle residence time measurements. The cyclone was operating in a continuous mode at room temperature. Small traces of preheated particles at a known temperature were introduced into the cyclone. The outlet temperature of the particles was detected by using the fluoroptic measurement technique. This technique was developed for heat transfer measurements in the Rotating Cone reactor by Wagenaar [14] and Janse [15]. The heat transfer coefficient was calculated from the particle temperature and the residence time of the particles. For these experiments biomass particles could not be used because their low internal heat conductivity. In that case temperature gradients will occur inside the particles and the surface temperature is not a measure for the amount of heat hold by the particles. A special polymer, PEEK, was chosen to simulate the biomass. This polymer has a density lower than the sand particles and comparable to the biomass particles so more or less the same flow behavior can be expected. The thermal conductivity is however more than five times that of biomass. For small particles ($d_p < 0.5$ mm) the temperature will not vary over the diameter and a simple model of the lumped heat capacity can be used to calculate the heat transfer to the particles.

First experiments were carried out on the heat transfer in the glass cyclone. The influence of particle loading, particle size of the sand and cyclone inlet velocity was studied.

The heat transfer coefficients measured were in the order of 600 to 1400 W/m^2K. The heat transfer could be improved by increasing the sand load and the inlet air velocity. One of the first experimental results is given in Figure 7. At higher inlet air velocities the slip velocity of the particles will be higher improving the conductive gas-solid heat transfer. At higher inlet velocities also mixing of the particles will be improved resulting in a higher heat transfer rate. By increasing the particle load the sand to biomass contact will intensively enhancing the heat transfer.

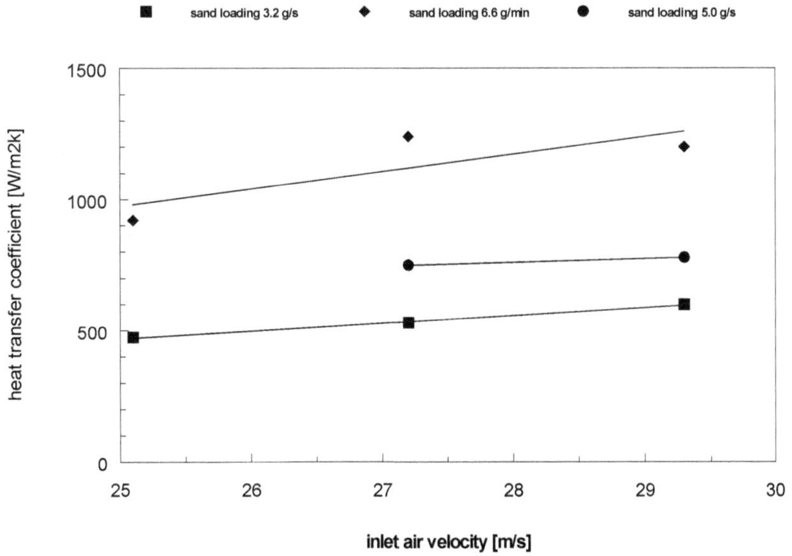

Figure 7: Measured heat transfer coefficient as a function of the inlet air velocity for different sand loadings. PEEK particles 0.3-0.4 mm sand particles 0.5-0.65 mm.

THE PYROS PILOT PLANT

A pilot plant with a biomass through-put of 30 kg/hr is assembled and erected at the University of Twente. A schematic view of the pilot plant is given in figure 8. The pyrolysis reactor in this pilot plant is a high temperate cyclone with a rotational particle separator mounted in the center of this cyclone. The whole reactor is rather compact, the cyclone has an internal diameter of 40 cm and a total height of 40 cm.

The biomass is supplied from a hopper by a screw feeder with a controllable rotational speed. The sand is preheated by combustion of the produced char and non condensable gases from the pyrolysis process. The combustion process takes place in an already existing fluidized bed oven. The sand flow rate back to the PyRos reactor is controlled by a screw feeder. The non condensable gases are used as a carrier/transportation agent for the sand and biomass into the reactor. The transport gases are set at pressure by means of a side channel blower. This type of compressor is simple of construction, high reliable and almost insensitive for small amounts of liquid droplets in the gas. After the PyRos reactor the product gases and vapors are transported by insulated lines to the quench. In the quench the gases are cooled down very quickly by spraying a large excess of cold bio-oil into the hot gases. The process of quenching is done in a second Rotational Particle Separator. The cooled bio-oil is introduced by means of nozzle at elevated pressure. The spray nozzles are mounted at different levels in the centre of the cyclone. The nozzles produce a mist of small droplets (approx. 100-200 micron). The small particles produce a large surface area for efficient heat transfer and condensation. The droplets and condensed vapors are separated by the cyclone of the RDS. The small aerosols will be separated by the rotating filter in the quencher outlet. The liquid separated by the rotating filter element can flow downwards out of the filter by gravity, for high viscous oils the element can be cleaned by a nozzle with compressed inert gas. The oil separated by the cyclone and the filter element will flow downwards in a thin film to the liquid outlet of the quench. In this manner a very compact and efficient quencher is obtained.

At the moment the different part of the pilot plant are tested en evaluated. In the coming months first experiments with wood pyrolysis are planed.

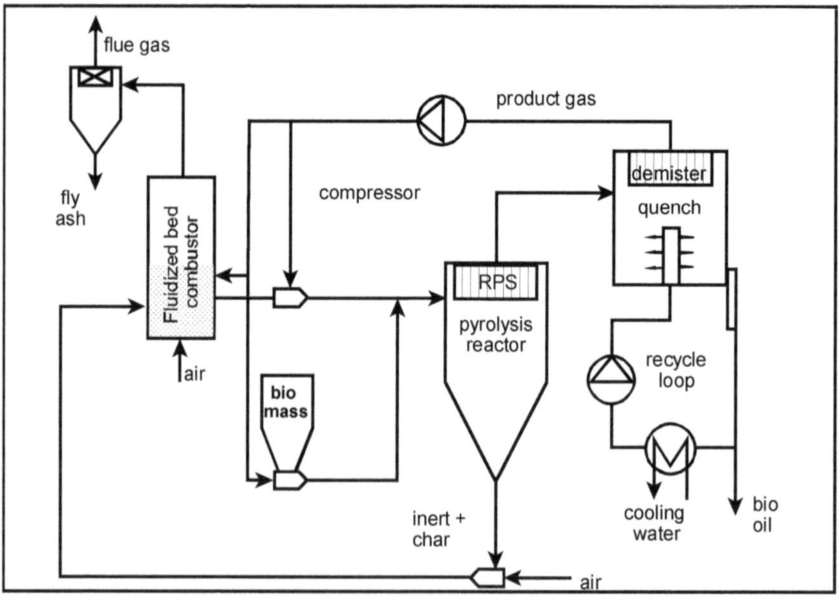

Figure 7: The PyRos pilot plant with 30 kg/hr biomass input.

CONCLUSIONS

- A very compact new type of reactor for flash pyrolysis of biomass is developed.
- The PyRos reactor has an internal particle separator, to produce a high quality bio-oil.
- An oil yield of more than 50 % is already achieved in a small laboratory scale cyclone.
- The particle residence time in the reactor can be controlled by the process conditions.
- High heat transfer rates can be achieved in a cyclone reactor.

REFERENCES

1. Bridgwater, A.V., Peacocke, G.V.C., 2000, Fast pyrolysis processes for biomass, Renewable and Sustainable Energy reviews 4, pp. 1-73.
2. Diebold, J.P.,Bridgwater, A.V., 1999, Overview of fast pyrolysis of biomass for the production of liquid fuels, fast pyrolysis of biomass, A handbook, ed. by A.V. Bridgwater, CLP-press, pp. 14-32.
3. Scahill, J., Diebold, J.P., Feek,C. , 1997, Removal of residual char fines from pyrolysis vapors by hot gas filtration. Developments in thermochemical biomass conversion. ed. by A.V. Bridgwater, Blackie Academic & Professional, pp. 253-266.
4. Bridgwater, A.V., Czernic, S., Pizkorz, J., 2001, An overview of fast pyrolysis. Progress of thermochemical biomass conversion, ed. by A.V. Bridgwater, Blacky Science, pp. 977-997.
5. Brouwers, J.J.H., Rotational particle separator: A new method for separating fine particles and mist from gases. Chemical engineering and technology, 19 (1996) pp.1-10.
6. Lede, J. Cyclone: a multifunctional reactor for the fast pyrolysis of biomass, Industrial and Engineering Chemistry Research, Volume 39, Issue 4, 2000, pp. 893-903.
7. Diebold, J.P., Scahill, J.W., 1997, Improvements in the Vortex Reactor Design, Developments in thermochemical biomass conversion. ed. by A.V. Bridgwater, Blackie Academic & Professional, pp. 242-252
8. Kang, S.K., Kwon, T.W., Kim, S.D., 1989, Hydrodynamic Characteristics of Cyclone Reactors', Powder Technology, 58, p211-220
9. Kemp, I.C., Frankum, D.P., Abrahamson, J., Saruchera, T.,1998 Solids Residence Time and Drying in Cyclones, International Drying Congress, Greece
10. Lédé, J., Li, H.Z., Soulignac, F, Villermaux,J., 1989, 'Le cyclone réacteur: Partie II: Mesure Directe de la Distribution des Temps de Séjour de la Phase Solide – Lois d'Extrapolation', The Chemical Engineering Journal, Vol 42, pp 103-117,
11. Lede, J.; Souglignac, F.; Villermaux, J., 1987, Measurement of Solid Particle Residense Time in a Cyclone Reactor: A Comparison of Four Methods, Chemical Eng. Process., 22, pp 215-222.
12. Lede, J.; Li, H.Z.; Souglignac, F.; Villermaux, J., 1992, Le Cyclone Reacteur. 4: Mesure de l'efficacite des transfer de chaleur entre les parois et les phases gazeuze et solide, The Chemical Engineering Journal, 48, pp. 83-99.
13. Szekely, J., Carr, R. 1966, Heat transfer in a cyclone, Chemical Engineering Science, 21, pp. 1119-1132.
14. Wagenaar, B., 1994,The rotating cone reactor, PhD-thesis, University of Twente.
15. Janse, A.M.C., 1998, A Heat Integrated Rotating Cone Reactor System, PhD-thesis University of Twente.

Opportunities for Bio-oil in European Heat and Power Markets

JG Brammer, AV Bridgwater
Bio-Energy Research Group, Aston University, Birmingham B47ET, UK
M Lauer
Joanneum Research, Elisabethstrasse 5, A-8010 Graz, Austria

ABSTRACT: This EC-funded project (ALTENER Contract 1030/C/00-015/2000) aims to develop and implement a methodology for identifying opportunities for the use of biomass fast pyrolysis liquid (bio-oil) in the heat and power markets of Europe, by means of a comparative economic and environmental assessment. The methodology is based on a "competitiveness factor" - the ratio of the total annual cost of meeting a heat and/or power demand at a particular location by conventional sources (i.e. grid electricity, and heat from a gas or oil boiler), to that of meeting the same demand from a bio-oil technology. A range of standard bio-oil applications have been identified, and location-specific data have been obtained by questionnaire from members of the ALTENER-funded Pyrolysis Network (PyNe) representing 15 European countries. Competitiveness factors are being calculated for each standard application in each country. Also, competitiveness factors are being calculated for comparable non-bio-oil biomass technologies, and also for any non-standard applications identified by the PyNe members as being of particular relevance to their country. Greenhouse gas emission mitigation is also being calculated for each application, so that environmental as well as economic competitiveness can be assessed. Supporting assessments of bio-oil markets and non-technical barriers are included. This paper presents the methodology employed, together with interim results including sensitivity studies.

INTRODUCTION

The production of bio-oil by fast pyrolysis of biomass is a technology on the boundary between development and demonstration for both heat and power applications [1]. Compared to combustion and gasification of biomass, fast pyrolysis has the advantage that the intermediate energy carrier (the bio-oil) is a liquid fuel which may be stored and transported economically, so that biomass conversion and heat and/or power generation are independent and can be carried out at different locations, times and scales. This allows exploitation of economies of scale and minimisation of environmental impact.

Bio-oil can be used as a fuel in boilers for heat, and diesel engines or gas turbines for electricity or CHP production. However, the attractiveness of bio-oil systems compared with alternatives will depend on many factors, and will vary with the application, the scale, and importantly the location and its associated economic and logistical factors, for example biomass availability and price, conventional energy prices, incentives, industrial traditions etc.

The objective of the present project, funded by the European Commission ALTENER programme (Contract 1030/C/00-015/2000), is to develop a methodology for the evaluation of opportunities for biomass fast pyrolysis liquid (bio-oil) technologies in the heat and power markets of Europe, and to evaluate a range of potential applications in 12 European countries. Bio-oil technologies are being compared with both conventional energy solutions and with other biomass technologies for the same heat and power duty. The project is utilising the resources of the PyNe network for biomass pyrolysis (sponsored by the European Commission and IEA Bioenergy) to collect the necessary data.

APPROACH

The evaluation of opportunities for bio-oil technologies is carried out by a quantitative assessment of competing alternatives for the provision of heat and/or power in a set of standard applications in each of the 12 countries, and also by a number of less specific analyses relating to overall markets and non-technical barriers. The standard applications comprise heat only, power (electricity) only and combined heat and power (CHP) cases, and are specified in terms of the heat or power rating in MW. Some non-standard applications may also be included.

For each application and location, the economic competitiveness of one or more bio-oil technologies is quantified using a newly developed methodology. Also for each application and location, the comparative environmental performance of the bio-oil technologies is also quantified using a standard measure.

The evaluation also includes reviews of the current status of non-technical barriers to the adoption of bio-oil technologies, and an assessment of potential market volumes in each of the 12 countries.

ECONOMIC COMPETITIVENESS

Competitiveness in terms of advantage in economics compared to other options for providing the same service is one of the most important issues for the implementation of new technologies. Thus economic competitiveness is the principal focus of the present project.

Competitiveness is a relative issue. The overall costs of a technology meeting a given duty should be at the same level or lower than the overall costs of a competing technology meeting the same duty. The relation of the overall costs of two competing technologies depend very much on the specific application and location, and it is necessary to to consider items such as output, fuel prices, labour costs, taxes, incentives, availability and capacity factor.

COMPETITIVENESS FACTOR

In order to provide a measure of competitiveness, a non dimensional index "CF" (competitiveness factor) is introduced. This is defined as the ratio of the total annual cost of meeting the required duty using "conventional" energy to that of meeting the same duty using the new or proposed technology.

$$CF = \frac{\text{total annual cost: preferred conventional alternative}}{\text{total annual cost: proposed technology}}$$

The value of CF indicates the competitive situation of the proposed technology in a specific application. If CF>1, the proposed technology is economically superior to the preferred conventional alternative. If the preferred conventional alternative is standardised across all applications, then it is possible to compare technologies that are very different in scale and/or service provided. In this project the preferred conventional alternative is specified for all applications as:

- ***Electricity*** Supplied from the local electricity grid according to the commercial terms offered by local electricity supply companies
- ***Heat*** Supplied from a dedicated on-site boiler burning either natural gas or fuel oil (depending on local availability and tariffs)

The total annual cost for dedicated generating plant is calculated by adding investment and operating costs, following the guideline VDI 2067 prepared by the Association of German Engineers [2]. Cost related to investment are calculated as an annuity (a constant annual payment in real terms over the lifetime of the project, equivalent to borrowing the total investment cost at a specified real interest rate). Cost related to operation include fuel, personnel, utilities, maintenance and other overheads. The total annual cost for purchase of utility electricity is a simple product of price and consumption.

DATA ACQUISITION

For each specified application, the calculation of the total annual cost of the proposed technology and of the preferred conventional alternative requires a combination of technology-related data and location-related data.

Technology-related data

These are data associated with plant and equipment providing a given heat and/or power output, and are assumed to be constant across Europe:

- investment cost
- operation and maintenance cost (excluding feed and labour)
- labour requirement (hours per year)
- feed requirement per annum (i.e. overall efficiency)

Data have been obtained from manufacturers, previous studies and published literature, where necessary updated to reflect technology changes and cost inflation.

Location-related data

These are data specific to the location of the application, and will vary across Europe:

- annual biomass availability and cost
- biomass type, moisture and heating value
- interest rate on capital loans
- cost of labour
- price of utility-supplied electricity and oil or gas

Data are being obtained by questionnaire to the PyNe representatives in the 12 countries. The questionnaire has been designed to ensure consistency of scope in the data requested, and a strong emphasis is placed on the provision of source references.
To aid comparability, the questionnaire requires that the principal biomass feedstock must be virgin wood of some kind. However, provision is made for the specification of additional biomass feedstocks if they are of particular local importance.

APPLICATIONS

The standard applications comprise heat only, power (electricity) only and combined heat and power (CHP) cases at a range of scales. For each application, one or more bio-oil technologies are evaluated, and also one or more non-bio-oil biomass technologies for comparison.

The complete set of standard applications and associated technologies are given in Table 1 (bio-oil technologies) and Table 2 (non-bio-oil biomass technologies). Rated heat output is given as MW_{th}, rated electrical output as MW_e. In heat or CHP cases, it is assumed that the heat is supplied as hot water, typically for space heating purposes. CHP applications include some rated by electrical ouptut and some rated by heat output to assist comparisons, as heat to power ratios can vary widely between technologies.

Table 1: Standard applications and associated technologies - bio-oil

Technology	Product	Rated Output
bio-oil boiler	heat	0.5, 1.0, 2.0 MW_{th}
bio-oil IC engine (dual fuel diesel)	electricity	1.0, 5.0 MW_e
bio-oil IC engine (dual fuel diesel)	CHP	1.0, 5.0 MW_e
bio-oil IC engine (dual fuel diesel)	CHP	2.0, 10.0 MW_{th}
bio-oil gas turbine (simple cycle)	electricity	5.0, 15.0 MW_e
bio-oil gas turbine (simple cycle)	CHP	5.0, 15.0 MW_e
bio-oil gas turbine (simple cycle)	CHP	10.0, 25.0 MW_{th}
bio-oil gas turbine (combined cycle)	electricity	15.0 MW_e
bio-oil gas turbine (combined cycle)	CHP	15.0 MW_e
bio-oil gas turbine (combined cycle)	CHP	25.0 MW_{th}
bio-oil boiler (Rankine cycle)	CHP	15.0 MW_e
bio-oil boiler (Rankine cycle)	CHP	25.0 MW_{th}

Table 2: Standard applications and associated technologies - non-bio-oil biomass

Technology	Product	Rated Output
biomass boiler	heat	0.5, 1.0, 2.0 MW_{th}
biomass gasifier IC engine (SI)	electricity	1.0, 5.0 MW_e
biomass gasifier IC engine (SI)	CHP	1.0, 5.0 MW_e
biomass gasifier IC engine (SI)	CHP	2.0, 10.0 MW_{th}
biomass gasifier gas turbine	electricity	5.0 MW_e
biomass gasifier gas turbine	CHP	5.0, 15.0 MW_e
biomass gasifier gas turbine	CHP	10.0, 25.0 MW_{th}
biomass IGCC	electricity	15.0 MW_e
biomass IGCC	CHP	15.0 MW_e
biomass IGCC	CHP	25.0 MW_{th}

In the questionnaire issued to PyNe members, the opportunity is given to specify additional application and technology combinations if considered locally important.

BIO-OIL PRODUCTION

As indicated earlier, one of the key advantages of bio-oil-based technologies over other biomass conversion technologies lies in the storability and transportability of bio-oil. This allows the bio-oil production operation to be quite independent of the utilisation operation. Although bio-oil does exhibit slow property changes over time, these are inconsequential for periods of 12 months or more provided the oil has met certain quality standards and correct handling and storage procedures are followed. Hence it is reasonable to assume that bio-oil would be a traded commodity in much the same way as fossil fuel oil.

Therefore in the present study it is assumed that the bio-oil production facility is located in a region of high biomass availability, and takes all the available biomass from an area of radius 100 km. The size of area has been limited to reflect the high transportation costs of raw biomass. Bio-oil is then produced and stored, with an associated cost of production. The bio-oil utilisation technologies specified for the standard applications are then assumed to purchase bio-oil at this cost. The complete system is depicted in Figure 1.

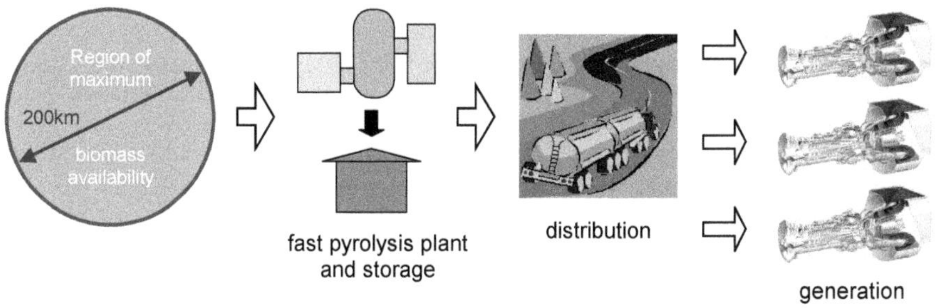

Figure 1: Bio-oil production concept.

RESULTS ANALYSIS

Results will be presented as charts of competitiveness factor, allowing comparison between countries, between technologies and between applications. Although values of CF may not exceed unity in many cases, cases where values of CF approach most closely to unity will indicate the most promising opportunities for focussed support measures or research and development.

Sensitivity analyses will also be conducted in which the effect of variation in key parameters on the value of CF will be presented. This is of particular value where high levels of uncertainty surround a current parameter value, or where a significant change in parameter value is anticipated over time.

ENVIRONMENTAL PERFORMANCE

Environmental performance is evaluated through the consideration of greenhouse-gas emissions mitigation potential. Information on specific greenhouse gas emissions rates will be taken either from the GEMIS databank [3] or subsequent work based on it, and updated and adapted to the specific needs of the different bio-oil and other technologies considered.

The difference in greenhouse-gas emissions between the competing conventional technology and the proposed technology is defined in terms of the initial biomass energy input, so that specific results for potential greenhouse-gas emissions mitigation (kg/GJ) are produced. By using results from the economic competitiveness assessment, specific costs for the mitigation of greenhouse gases can be calculated.

NON-TECHNICAL BARRIERS

As part of the project, reviews are being conducted of the current status relating to non-technical barriers to the adoption of bio-oil technologies. The most important of these are considered to be bio-oil specifications and standards, and bio-oil health and safety issues. The reviews involve both literature surveys, and direct contact with the key players involved.

For each of the non-technical barriers, consideration will be given to the importance of the barrier in impeding deployment, and to the specification of measures for overcoming the barrier and their likely success.

- ***Specifications and standards*** The requirements for effective operation of bio-oil fuelled equipment will be set against the ability of bio-oil producers to produce a consistent product.
- ***Health and safety*** Particular attention will be paid to toxicity issues, and to the development of safe handling procedures if deemed necessary.

MARKET ASSESSMENT

An assessment will be carried out of the heat, power and CHP markets across Europe, using data supplied by the PyNe members. This information will then be combined with the results of the other parts of the study to arrive at the following:

- the market volume realisable in each country and market sector by bio-oil technologies assuming non-technical barriers remain
- the market volume realisable in each country and market sector by bio-oil technologies assuming non-technical barriers are overcome

CURRENT STATUS

The project commenced on 1st June 2001 and is due to finish 31st May 2003. The current status (September 2002) is as follows:

1. ***Economic competitiveness*** The questionnaire has been designed and issued to PyNe members - most responses have been received and data is under review. Technology cost and performance data have been collected and are currently being assessed. First competitiveness results will be ready before end 2002.
2. ***Environmental performance*** A carbon balance for bio-oil production is complete. System carbon budgets are being calculated. First results will be ready before end 2002.
3. ***Non-technical barriers*** The reviews are being carried out by the leaders of the relevant PyNe topics. Sub-contracts have been issued. Reviews will be complete by early 2003.
4. ***Market assessment*** This is the final project task and will not commence until other activities are complete. The assessment will be delivered at the end of the project.

REFERENCES

1. Bridgwater A.V.(1999) An Introduction to Fast Pyrolysis of Biomass for fuels and Chemicals. In: *Fast Pyrolysis of Biomass; a Handbook.* pp. 1-13. CPL Scientific Publishing Services Ltd., Newbury.
2. VDI (1983) *Richtlinie VDI 2067.* VDI Gesellschft Technische Gebäudeausrüstung. Beuth Verlag, Berlin.
3. GEMIS (1994) *Gesamt-Emissions-Modell Integrierter Systeme, Version 2.1.* Öko-Institut, Darmstadt

Full-scale Application of Sewage Sludge Pyrolysis: Experience from the Subiaco Plant in Australia

TR Bridle
Environmental Solutions International Ltd, 21 Teddington Road, Burswood Western Australia 6100
Jean-Marie Rovel
Ondeo-Degrémont, 183, Avenue du 18 Juin 1940, 92508 Rueil Maimaison Cedex, France

ABSTRACT: Management of sewage sludge via "publicly acceptable" methods is becoming increasingly difficult primarily due to health and environmental concerns with respect to reuse of the product in agriculture. Consequently thermal processes are gaining popularity with significantly increased interest being shown in pyrolysis and gasification processes, due to their "non-incineration status".

One such process is the **ENERSLUDGE™** technology which has been developed and commercialised by Environmental Solutions International Ltd (ESI). The world's first commercial **ENERSLUDGE™** plant is located at the Subiaco Wastewater Treatment Plant (WWTP) that was recently handed over to the client, the Water Corporation of Western Australia. Extensive design knowledge and operational experience has now been accumulated from this commercial pyrolysis facility and future applications of the technology will benefit immensely from the lessons learned and experience gained from the Subiaco facility.

ESI has recently signed an exclusive license agreement with Ondeo-Degrémont for use of this technology in France and we are jointly bidding for work in France and elsewhere.

INTRODUCTION

The **ENERSLUDGE™** technology has been developed and commercialised over the past 15 years, with extensive pilot plant demonstration programs conducted in both Australia and Canada.[1, 2, 3] A contract to build the first commercial plant was awarded to the ESI-Clough Joint Venture, by the Water Corporation of Western Australia in late 1996. The contract called for the design, construction, commissioning and operation of a 25 dry tpd sludge pyrolysis plant, based on ESI's patented **ENERSLUDGE™** conversion process, at the Subiaco WWTP in Perth, Western Australia.

Construction of the facility commenced in early 1997 and commissioning of sludge dewatering commenced in January 1998. Since the plant comprises numerous unit operations, which are designed to operate in a cascade mode, commissioning of the entire plant was conducted in phases and continued to mid 2000.

The ESI-Clough Joint Venture was responsible for plant operations and maintenance until hand-over of the integrated facility to the client in June 2001. This included a nine month operational period of the fully integrated facility to generate the required operational performance data. The process knowledge and environmental data reported in this paper was generated during this nine month operational period.

DESCRIPTION OF THE SUBIACO PLANT

Unlike other biomass pyrolysis processes, the Subiaco facility comprises a number of unit operations that operate in cascade mode, as shown in Figure 1. The plant processes all of the raw primary (RPS) and thickened excess activated sludge (TEAS) generated at the Subiaco WWTP, which currently amounts to between 16 and 20 dry tpd.

Both RPS and TEAS are strained to remove most of the hair and long fibres in the sludge. This unit operation which removes about 2% of the dry sludge mass is required to improve the dried pellet bulk density and minimise blockages in the dryer system. The strained sludges are then blended, by storage / mixing for up to 6 hours, to produce a relatively consistent feedstock for downstream processes. Dewatering is conducted using solid bowl centrifuges, generating a sludge cake of between 26 and 28% TS. This is then dried in an indirectly heated flue gas drum dryer, producing pellets, with a nominal TS of 95%.

The dried sludge pellets are then fed to the conversion (pyrolysis) process, shown schematically in Figure 2. The pellets are first heated to 450°C in the absence of oxygen at a pressure of 2 to 5 kPag producing char and a raw pyrolysis gas. The char and gas are re-contacted in the second reactor, also operated at 450°C to facilitate the catalysed vapour phase reactions which refine the raw gas, producing mainly hydrocarbons. The catalysts necessary (alumino silicates and heavy metals) are inherently present in all sewage sludges.

The refined vapours are then condensed to 50°C in a direct contact spray condenser using cooled reaction water (RW) as the cooling medium. The char is cooled to 70°C and then used as the main fuel in the Hot Gas Generator (HGG). The condensed products are separated in an oil-water separator, with the oil stored prior to use as a fuel off-site, predominately in industrial boilers for steam raising. The RW is cooled using once-through secondary effluent and then used to condense the incoming vapours in the spray condenser.

The RW generated by the process is also combusted in the HGG. The vapours not condensed at 50°C, called non-condensed gas (NCG) is the third fuel combusted in the HGG. The reactors are heated externally, using hot flue gas, from LPG combustion.

The HGG combusts the three low grade fuels (char, RW, NCG) to produce all the energy needed to dry the sludge. The flue gas, at a nominal 850°C transfers heat to a closed-loop of drying air in an air-to-air heat exchanger. Cooled flue gas from the heat exchanger is then cleaned in a venturi scrubber and SO_2 scrubber, before discharge to the atmosphere, via the plant stack.

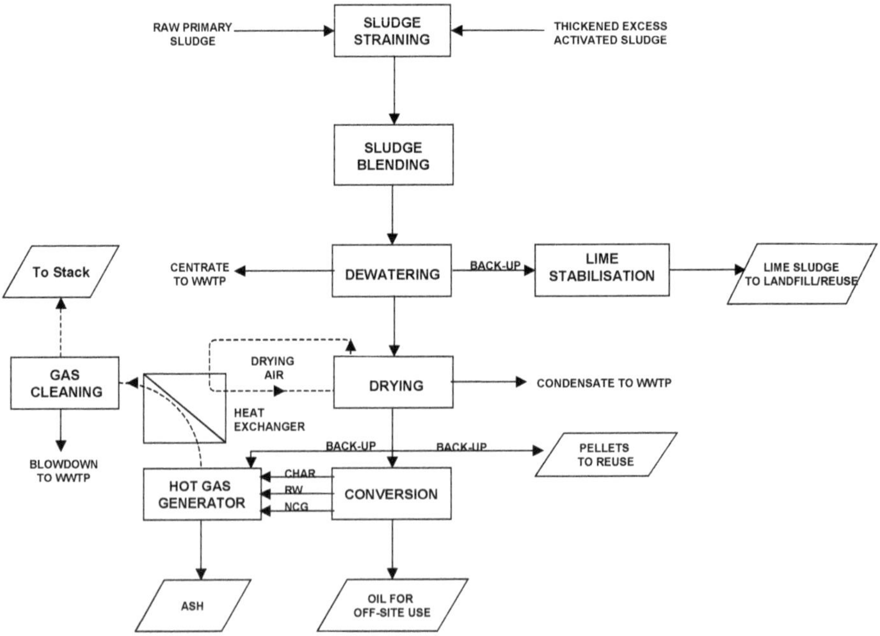

Figure 1: Integrated Plant Process Flow Diagram.

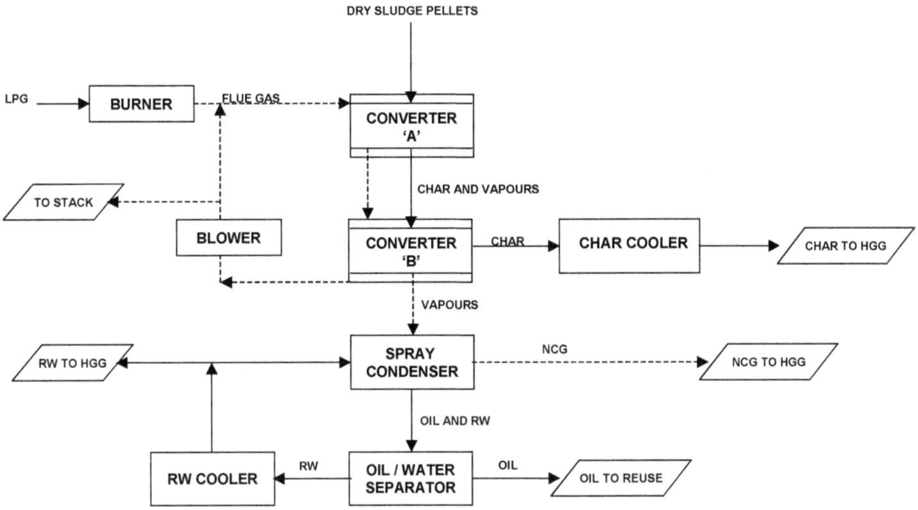

Figure 2: Conversion Plant Process Flow Diagram.

For periods when the conversion system is down, the plant is designed to operate as an autogeneous sludge dryer. In this mode about 40 to 50% of the dried sludge is combusted in the HGG to provide the thermal energy needed in the dryer. The remaining 50 to 60% of dried pellets are then available for beneficial reuse on agricultural land. The inerts in the sludge pellets or char are recovered as ash from the HGG. This ash can be used as an ingredient in concrete products / bricks. However, at Subiaco, the material is currently being landfilled.

PROCESS PERFORMANCE SUMMARY

Routine operations of the integrated plant commenced in October 2000 and the ESI-Clough Joint Venture operated the plant until June 2001, when hand-over to the Water Corporation occurred. During this operational period the required short-term and long-term Performance Testing of the integrated facility was conducted.

The overall process performance of the conversion system has been essentially as predicted. Average product yields and energy densities, as measured from December 2000 to June 2001 are shown in Table 1.

Table 1: Average Conversion Product Data.

Product	Yield (%)	GCV (MJ / kg)	% of Sludge Energy
Oil	29	30	45
Char	43	18	40
NCG	14	15	11
RW	14	6	4

The data has confirmed that all of the energy available in the sludge is recovered in the four conversion products.

The water content of the oil is slightly higher than predicted with the result that the oil gross calorific value (GCV) is slightly lower than predicted. However, this does not limit use of the oil as a fuel. In fact, to date over 300 tonnes of oil has been successfully used by industry, predominately for steam raising in boilers.

ENERGY RECOVERY

Extensive operating data has been obtained from the integrated plant during both the short-term (3 by 6 hours) and long-term (30 days) performance test periods. In addition, during the earlier performance testing of the dryer and HGG, data on the energy requirements for drying and the energy efficiency of the HGG were determined.[4, 5] The critical parameters verified during this early testing was that the gross energy requirements for sludge drying is equal to or less than 3.6 GJ/tonne of water evaporated and that the gross efficiency of the HGG (gross thermal energy in flue gas divided by input fuels GCV) was 83%. The overall thermal efficiency of both the HGG and the dryer heat exchanger was determined to be about 72%. That is, the nett energy requirement for sludge drying (i.e. the actual energy transferred to the dryer) is 2.6 GJ/tonne water evaporated.

Based on this data, and that measured from the converters (Table 1) it is evident, that provided the sludge is dewatered to between 25 and 26% TS, there is sufficient energy in the conversion byproducts (char, NCG and RW) to provide all the energy needed to dry the sludge. A thermal energy balance for the integrated plant, when processing 15 dry tpd of sludge, is shown in Figure 3. As has been mentioned previously the conversion reactors are heated using LPG. When taking this energy into account, it can be seen that the nett energy output of the plant is 116 GJ/d when processing 15 dry tpd, equivalent to 7.7 GJ/t sludge processed. This is understood to be the only thermal process which generates this level of export energy from sewage sludge. If the oil is combusted in an internal combustion engine with an efficiency of 35%, the process will generate 925 kWh/t sludge processed.

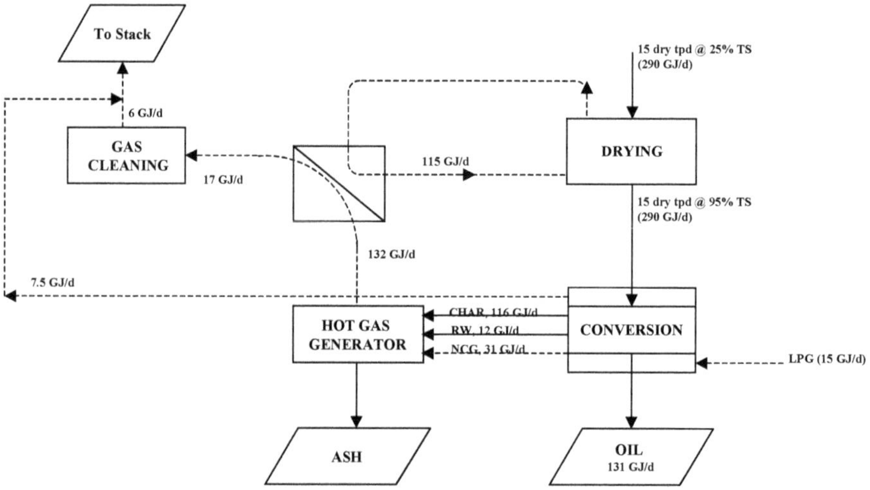

Figure 3: Energy Balance for Integrated Plant.

ENVIRONMENTAL ISSUES

The process provides numerous environmental benefits compared to other sludge management systems, the major ones being identified below:

- Minimum trucking of products off-site
- Complete control of pathogens and viruses
- Greenhouse gas credits
- Complete control of heavy metals, and
- Extremely low levels of gaseous contaminant emissions.

The last three environmental issues will be discussed in more detail, based on the operating results generated from the integrated plant for the period December 2000 to June 2001.

GREENHOUSE GAS CREDITS

The total greenhouse gas (CO_2) emissions from all sludge management systems are very similar. This is so, since ultimately, all the carbon is driven off as CO_2. However, what is important in the greenhouse gas debate, is the emission generated from the combustion of non-renewable fossil fuels. Consequently, the CO_2 emissions from this source, for various sludge management options, varies very significantly. Because the **ENERSLUDGE™** process uses very little fossil fuels for thermal energy and because it is a nett exporter of energy, it has very low emissions of what are called "new" greenhouse gas emissions. An extensive comparisons of greenhouse gas emissions for the most commonly used sludge management systems has been conducted in-house by ESI. Results from this study shows that the technology provides a greenhouse gas benefit of between 300 and 1300 kg CO_2/kg dry sludge processed, compared to other sludge management systems. These are significant CO_2 credits and if the technology were applied widely, these credits would help countries achieve their Kyoto Accord commitments.

CONTROL OF HEAVY METALS

Substantial monitoring of the integrated plant has confirmed the unparalleled control of heavy metals previously demonstrated during pilot plant studies.[6] Data generated during the three short-term performance tests, each of 6 hour duration, are shown in Table 2.

Table 2: Fate of Heavy Metals in the ENERSLUDGE™ Process.

Heavy Metal	Sludge Concentration (mg/kg)	Distribution in Conversion Products				% of Sludge Metal in Ash
		Oil	Char	NCG	RW	
Cadmium	1.5	<1	>99	<1	<1	>65
Copper	900	<1	>99	<1	<1	>99
Chromium	35	<1	>99	<1	<1	>99
Mercury	2.5	5	5	0.5	2	0
Nickel	15	<1	>99	<1	<1	>99
Lead	50	<1	>99	<1	<1	>95
Zinc	550	<1	>99	<1	<1	>99

Sludge drying has no impact on sludge heavy metals since temperatures are too low to volatilise any of the metals present. The first thermal process to have any effect on heavy metals is the conversion process. Operating at 450°C, under reducing conditions, the only heavy metal volatilised is mercury. As can be seen in Table 2, about 95% of the mercury is vaporised in the reactors, and then condensed in the spray condenser, with about 7.5% of the mercury remaining in the recovered oil, RW and NCG. Most of the mercury (over 85%) is recovered in the process sludge generated in the oil-water separation and oil polishing unit operations. About 1kg of this mercury rich oily sludge is generated per tonne of dried sludge processed. The **ENERSLUDGE™** process is the **only** thermal process with this unique ability to remove essentially all of the mercury from the fuels generated. Consequently mercury removal is **not** required in the flue gas cleaning operation, producing both environmental and economic benefits.

As shown in Table 2, essentially all of the other heavy metals remain in the char product, since at 450°C, they are not volatile. Upon combustion of the char, NCG and RW in the HGG, almost all the metals are quantitatively classified to the ash. A small quantity of the semi-volatile metals cadmium and lead are vaporised in the HGG, and report to the stack.

Extensive testing of the ash produced, during both pilot plant programs and that from the commercial Subiaco plant, has clearly shown that the metals are speciated as non-leachable silicates and oxides.[7] Leachability testing of Subiaco ashes, using the Toxicity Characteristic Leaching Procedure (TCLP) has also confirmed this finding. Typical TCLP leach results and the WA criteria for disposal in a Class III landfill [8], are shown in Table 3. As can be seen all criteria are readily met. With the exception of arsenic and selenium, all other metal values meet the stringent Class II landfill criteria.

Table 3: Ash Leachability Data (mg/L).

Metal	Subiaco Ash Result	WA Class III Landfill Criteria
Arsenic	0.13	0.7
Silver	0.01	10
Barium	0.03	70
Cadmium	<0.01	0.2
Copper	<5	200
Chromium	0.01	5
Iron	0.01	-
Mercury	<0.01	0.1
Nickel	0.10	2
Lead	0.07	1
Selenium	0.40	1
Zinc	4	500

PLANT GASEOUS EMISSIONS

As has been demonstrated above the unparalleled control of heavy metals and the previously demonstrated control of organochlorine components [(6)] means that very simple and low cost gas cleaning equipment is required to achieve world Best Environmental Practise emission standards. Of most significance is the destruction of organochlorine compounds present and removal of mercury in the conversion process. Thus the traditional costly mercury and dioxin removal processes are not required during gas clean-up. A simple venturi scrubber for particulate removal and packed column scrubber for SO_2 removal is all that is required for gas cleaning.

Emission testing of the plant was performed by external contractors / laboratories during the Performance Testing and results are shown in Table 4. As can be seen all the plant license parameters were readily met. Most pleasing was the extremely low dioxin emission, which is more than 300 times lower than the stringent European standards.

Table 4: Plant Emission Levels and Standards (mg/Nm^3).

Parameter	Plant Emission Value	Plant License Value	TA Luft (17 BImSchV) Standard
Particulates	12	75	30
SO_x	<36	-	200
CO	45	100	50
HF	Not detected	5	4
HCl	19	50	60
Cd	0.02	0.03	0.05
Cu	0.36	2.0	0.5
Cr	<0.007	0.5	0.5
Hg	0.008	0.05	0.05
Ni	0.11	0.5	0.5
Pb	0.08	0.5	0.5
Tl	0.0001	0.05	0.05
Zn	0.1	1.0	-
Dioxins (ng TEQ/Nm^3)	0.00032	-	0.1

LESSONS LEARNED FROM AUSTRALIA'S FIRST SLUDGE PYROLYSIS PLANT

Numerous lessons have been learned from operation of Australia's first sludge pyrolysis plant and the more important and notable are discussed below. A more detailed account of the lessons learned can be found elsewhere (9).

EFFECT OF SLUDGE QUALITY ON PLANT PERFORMANCE

The first important lesson learned was the impact of the variable sludge properties on the up-front and supposedly "proven" components of the plant, namely dewatering, drying and HGG operations. These unit operations are more sensitive to sludge quality parameters than the actual pyrolysis / conversion part of the plant. Sludge quality parameters such as Volatile Solids (VS) content, ratio of the two sludges (RPS and TEAS) and sludge phosphorus content had major impacts on plant operability and costs. Major upgrades of the sludge dryer and HGG were required to handle the greater than anticipated variability in these sludge properties (4, 5, 10). The interaction between the "liquids" and "solids" handling trains at the WWTP are now much better understood. To maximise performance of the integrated pyrolysis plant, more stringent control of the WWTP, with respect to these sludge quality parameters, is required. This understandably places additional and possibly new constraints on the operator of the WWTP.

OPERATOR EXPERIENCE

The second important lesson learned was the need for operators with industrial experience, who understand the interactions between the various unit operations, to be "comfortable" with thermal processes and be familiar with the relatively fast process response times required. All the operators at Subiaco now come from an industrial operations background, and the integrated plant is manned by one operator per shift. As operational experience was gained, plant uptime and efficiencies increased markedly.

PYROLYSIS PERFORMANCE

Pleasingly, the conversion (pyrolysis) reactors have performed extremely well. This part of the process has proven to be thermally robust and not nearly as sensitive to sludge quality issues as the other thermal unit operations in the plant. The reactors are not sensitive to changes in operating parameters. One major change was made to the process configuration, namely change of the vapour condensers from indirect to direct spray condensers. The original shell-and-tube condensers (vapour on tube-side) experienced plugging problems, which could not be satisfactorily resolved by changing operating conditions. Since this change, the condenser has operated satisfactorily, but still requires cleaning about once per month. The oil water separation equipment has proven adequate, but does require routine desludging.

Two issues require improvements yet to be made. The first is the sealing of the char discharge from Reactor B. An improved lock-hopper system, or a "plug-forming" screw conveyor, as is successfully used on the sludge feed to the reactor, are proposed. The second issue is the management of emergency vents from the reactor system, when the HGG "slams-shut". Currently, the NCG, which continues to be generated for about 30 minutes after feed to the reactors is stopped, is quenched and then routed to the existing plant odour control scrubber. This system does not provide adequate odour control and a number of alternate systems are being evaluated.

OCCUPATIONAL HEALTH AND SAFETY SYSTEMS

An important lesson learned is that it is now recognised that for a pyrolysis plant the Occupational Health and Safety (OH&S) systems, procedures and practices required are more akin to those applicable to the hydrocarbon processing industry. A much greater emphasis on OH&S procedures and practices is required relative to those normally used in conventional sludge management systems, especially in the areas of equipment isolation and purging procedures.

PLANT MAINTENANCE

Maintenance on the plant has proved to be more difficult and costly than expected, predominately because the "fast-track" nature of the contract fixed the building size before all the equipment had been selected. Consequently the chosen equipment had to be designed into the available space, with the result that access for maintenance of some components of equipment is limited. In addition some of the equipment finally selected and delivered by suppliers was, in hindsight, not as suitable as it may have been, which has added to maintenance costs. Also the variability of the sludge quality, particularly the intermittent high phosphorus levels, increased maintenance on the HGG.

PLANT ECONOMICS

A wealth of operational and cost data has now been gained from the plant operating in both the back-up "autogenous sludge dryer" and pyrolysis mode. The most cost-effective mode of operation largely depends on the costs / revenues from the products generated by the plant. For example, currently in WA there is an environmentally sound market for beneficial reuse of the dried pellets on farmland relatively close to the WWTP, which thus makes operation of the plant in autogenous drying mode the most cost effective mode and the plant is currently operating in this mode. Should the dried pellets require a long transport distance, or indeed there be a limited or no market for the pellets, as was the case when the plant was designed, then operation in conversion mode is the more attractive mode.

The economics of operation in conversion mode are clearly influenced by the end market and revenue gained for the oil. If the oil is used off-site for industrial purposes, with no revenue, then it is more cost effective to operate the plant as an autogenous dryer, while the present outlet for the pellets is available.

This economic data for pyrolysis plants has indicated the importance of "value adding" through using the oil on-site for power generation. This option results in the oil having a value exceeding 30¢/L, making operation of the plant in this mode very cost effective and the most competitive thermally based sludge treatment process. ESI is currently investigating this option for future plants.

CONCLUSIONS

Australia's first commercial- scale sewage sludge pyrolysis plant has clearly demonstrated the energy recovery and environmental benefits offered by the technology, as detailed below:

1. All of the thermal energy needed for sludge drying is provided by the process, eliminating the need for costly fossil fuels
2. In addition the plant has demonstrated a nett energy export of 7.7 GJ/t of dry sludge processed.
3. Greenhouse gas (CO_2) credits of between 300 and 1300 kg/kg dry sludge processed have been demonstrated compared to other sludge management systems
4. Heavy metals are completely controlled, with mercury being removed from the conversion products. Essentially all other metals are retained as non-leachable species in the ash product.
5. Plant gaseous emissions readily meet the Plant License Limit using simple and inexpensive gas cleaning equipment. Notably dioxin emissions are 300 times lower than the stringent European standard.

Many engineering design, process, operational and maintenance lessons have been learned from this world-first facility and improvements to the facility are ongoing. Hard "real-world" economic data is now available regarding the installation and operation of this technology in Australia. Future applications of the technology, both in Australia and indeed around the world, will benefit immensely from the lessons learned and the experience gained from the Subiaco facility.

REFERENCES

1. Bridle, T.R. and Campbell, H.W., (1986). "Oil from Sludge: A Technology Update", Proceedings 9th Annual AQTE Conference, Montreal.
2. Bridle, T.R., Hertle, C.K., and Luceks, T., (1989). "The Oil-from-Sludge (OFS) Technology: A Cost Effective Sludge Management Option". *Water*, Vol 16, No 1, pp 30 to 33.
3. Gough, D.E., Hammerton, I.F., Bridle, T.R. and C.K. Hertle, (1991). "Demonstration of the Oil-from-Sludge Technology at the Malabar Sewage Treatment Plant", Proceedings, 14th Federal AWWA Conference, Perth.
4. Bridle, T.R., Molinari, L., Skrypski-Mantele, S., Ye., P, and Mills, J., (1999). "Commissioning the Subiaco **ENERSLUDGE**^{TM} Plant Part 2: Sludge Drying, Energy Recovery and Gas Cleaning". Proceedings 18th AWWA Federal Convention.
5. Bridle, T.R and Skrypski-Mantele, S., (2000)., "Assessment of the **ENERSLUDGE**^{TM} Process for the Management of Biosolids in the European Community, Proceeding 5th European Biosolids and Organic Residuals Conference, Wakefield, UK.
6. Bridle, T.R., Hammerton, I.F. and Hertle, C.K., (1990). "Control of Heavy Metals and Organochlorines using the Oil-from-Sludge Process". *Wat.Sci. Tech*, Vol 22, No , pp249-258.
7. Bridle, T.R., Côte, P.L., Constable T.W. and Fraser, J.L., (1986)."Evaluation of Heavy Metal Leachability from Solid Wastes", Proceedings 13th Biannual IAWPRC Conference, Rio de Janeiro.
8. Western Australian Department of Environmental Protection (1996). "Landfill Waste Classification and Waste Definitions"
9. Bridle, T.R. and Unkovich, I (2002). Critical Factors for Sludge Pyrolysis in Australia *Water*.29(4), pp43-48.
10. Bridle, T.R., Casey, T.G. and Domurad, M.K., (2002). "Converting Sewage Sludge to Fuel Oil", *Water* 29(2) pp55-58.

Biomass Pyrolysis for Liquid Co-Firing in Existing Oil Power Plants: can Italy be a Potential Market?

David Chiaramonti, Francesco Martelli
Dept of Energetics "S.Stecco", Faculty of Mechanical Engineering, Florence, Italy
Gianluca Tondi
ETA S.r.L., Florence, Italy
Massimo Bonini, Piero Baglioni
CSGI – Faculty of Chemistry, Florence, Italy

ABSTRACT: Even if the amount of Heavy Fuel Oil (HFO) used in large scale power plants for electricity production shows a decreasing trend, HFO still covers a significant share for fossil fuel consumption for electricity generation: in the framework of the measures aimed at reducing GHG and pollutant emissions, pyrolysis oil (PO) co-fired with HFO in large scale conventional thermal power plants could therefore be considered. The proposed approach offers a number of advantages: it does not require the construction of new biomass power plants, but only the adaptation of the existing ones; the efficiency of these large scale plants (~38 %) is considerably greater than state-of-the-art bioenergy plants; even considering PO-HFO emulsions composed by small amounts (e.g. 2-5%) of PO in HFO, the market potential for PO is very high, given the present and future total consumption of HFO, and the contribution to meet both the Kyoto targets and the requirements national regulations high as well. In the present paper the potential application of PO as biofuel in large scale thermal power plant is considered, the market in Italy discussed and preliminary figures estimated for PO-HFO emulsions.

OIL CONSUMPTION FOR POWER GENERATION IN EU AND ITALY

Oil consumption in the EU [1], which reached a maximum of 820 Mt/y in 1973 and 830 Mt/y in 1979, and a minimum of 662 Mt/y in 1985, is smoothly but steadily increasing: it was 760 Mt/y in 2001. Germany (132 Mt/y), France (96 Mt/y), Italy (93 Mt/y), United Kingdom (76 Mt/y), Spain (73 Mt/y) and Netherlands (50 Mt/y) are the six EU countries with the greatest oil consumption. As far as regional consumption is concerned, oil covered approximately 40 % of EU consumption, similarly to North-America, Africa and Asia-Pacific: Middle East and South America had a greater oil consumption (approximately 50 %), while the former Soviet Union had the lowest figure (less than 20 %).

Despite a decreasing trend, a considerable share of this amount of oil is still used for energy production in large scale power plant: nevertheless, the mix of fossil fuels, nuclear and renewables for energy production varies significantly amongst the different EU Countries.

At world level, electricity generation from thermal plants covers the major part (Fig. 2) of the total amount of electricity generated [2]. The oil share passed from 24.6 % in 1973 to 8.5 % in 1999: however, in absolute terms, this meant a reduction of only 251 TWh/y, from 1,506 TWh/y in 1973 to 1,255 TWh/y in 1999 (Fig. 3).

It is interesting to observe that electricity generation from nuclear energy in the EU, despite the reduction in nuclear programmes in various EU Countries, increased from 575 TWh (1985) to 850 TWh (end of 90's), representing more than one third of the total EU gross electricity production [3].

The European Union self-sufficiency in terms of energy production and consumption decreased from approximately 60 % (mid of '80s) to 50 % (2000), even if the situation in the various EU countries is very different (Ireland, Italy and Portugal depend on import for more than 80 %, while Denmark, Netherlands and United Kingdom for less than 20 %). The EU15 energy consumption had a 10 % increase during the last ten years.

Similarly, EU self-sufficiency in terms of electricity generation has also been decreasing: from more than 80 % (mid of '80s) to less than 70 % (2000).

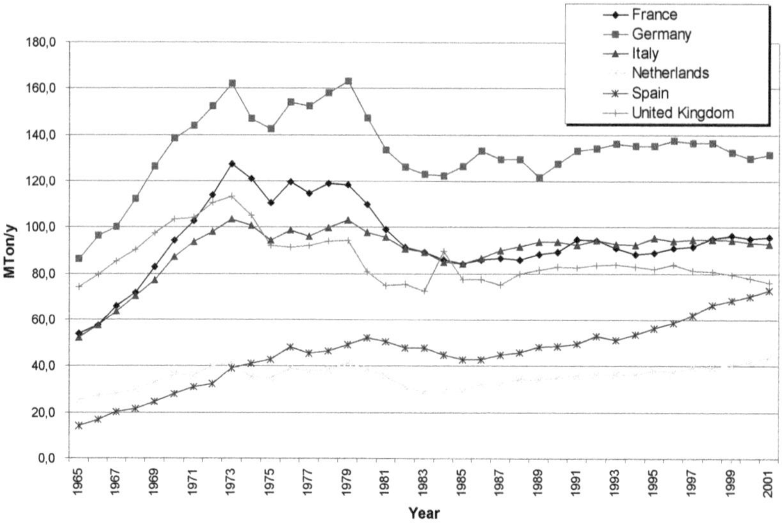

Figure 1: The six EU countries with the greatest oil consumption (Source: BP).

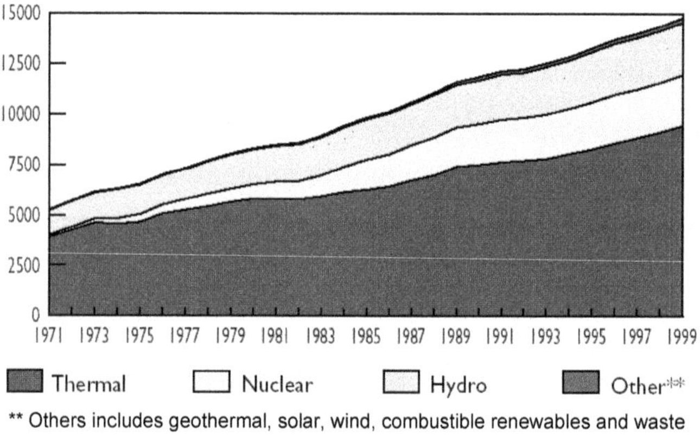

Figure 2: Electricity generation (excludes pump storage) by fuel in TWh/y (Source: IEA).

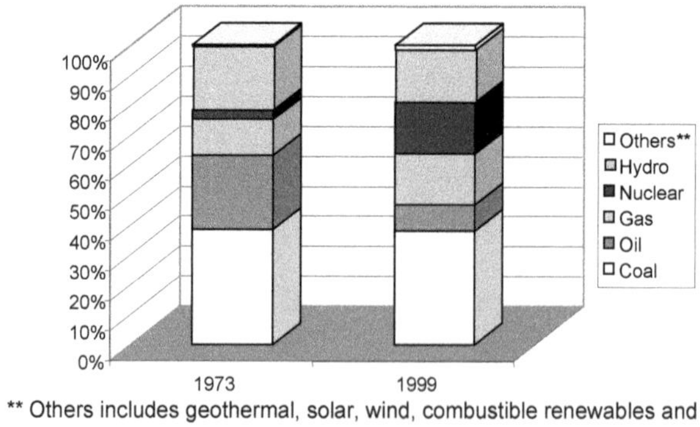

Figure 3: Fuel shares of electricity generation (Source: IEA).

The total gross electricity production [3] in EU-15 increased from 1,690.7 TWh/y in 1980 to 2,575.8 TWh/y in 2000. The most considerable increase in the same period were recorded by Portugal (+ 190 %), Ireland (+ 120 %) and Finland (+110 %).

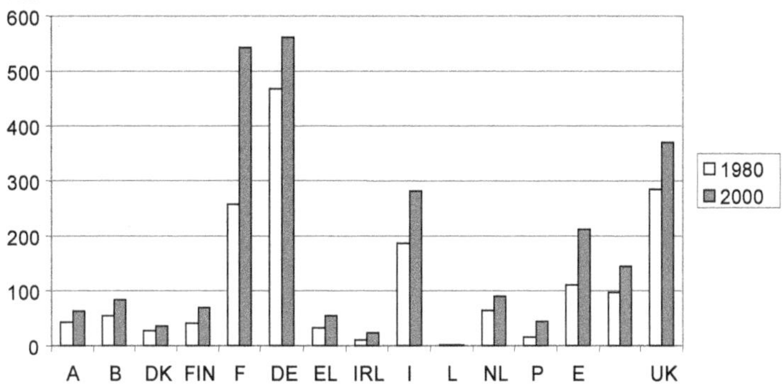

Figure 4: Total gross electricity production in EU countries (Source: Autorità per l'Energia Elettrica ed il Gas).

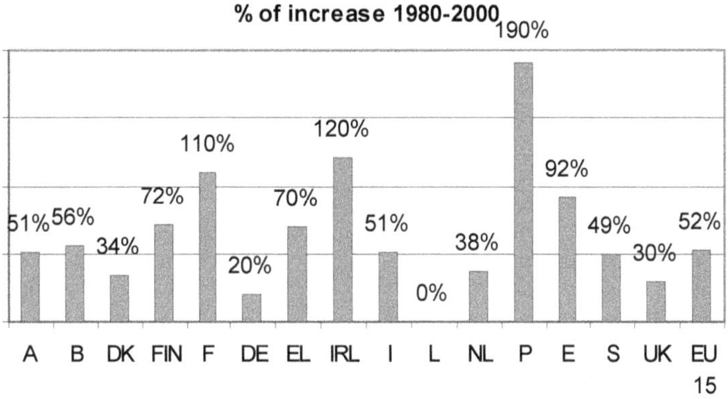

Figure 5: Total gross electricity production: % of increase 1980 – 2000 (Source: Autorità per l'Energia Elettrica ed il Gas).

As far as regards the incidence of electricity production from oil fuel in the various countries [2], in 1999 Italy ranked 3rd (Table 1), immediately after Japan and United States, and in the same position as Mexico. Italy was, during the same time, increasing very rapidly the share of electric energy production from natural gas.

The electricity generation from thermal plants using oil products in Italy was 85,878 GWh in 2000 (not including natural gas). In the same year, the consumption of fossil fuels for electricity generation was 19,532 thousand tons [4]. The incidence of electricity production on the internal gross energy consumption was equal to 34 % in 1999 (i.e. 64.8 MTEP on a total of 148.8 MTEP).

Forecasts on electricity demand in Italy give rising figures: from 298,5 TWh/y in 2000 and 305,4 TWh/y in 2001, the electricity demand is expected to increase to 345 TWh/y and 400 TWh/y in 2005 and 2010 respectively. According to recent estimations, oil should represent, at that time, 11 % of the national energy sources (it was 35 % in 1999) [5, 6].

Table 1: Electricity production from fossil fuels in 1999 (Source: IEA).

Coal	TWh/y	Oil	TWh/y	Gas	TWh/y
United States	2,204	Japan	176	United States	613
P.R.China	964	United States	121	Russia	359
India	397	Italy	91	Japan	234
Germany	286	Mexico	91	United Kingdom	141
Japan	224	Saudi Arabia	77	Italy	87
South Africa	187	P.R.China	49	Islamic Rep.of Iran	86
Russia	161	Russia	41	Germany	55
Australia	158	Chinese Taipei	40	Thailand	53
Poland	135	Iraq	29	Malaysia	51
Canada	109	Spain	24	Netherlands	49
Rest of the World	982	Rest of the World	510	Rest of the World	803
WORLD	**5,627**	**WORLD**	**1,249**	**WORLD**	**2,531**

As presented in Fig. 6 and 7, oil consumption for electricity generation is constantly decreasing in Italy, in favour of natural gas (which however showed an unexpected reduced increase, while coal use augmented thanks to its low price) [3, 5]. Nevertheless, oil consumption in Italy for electricity production remains significant in absolute terms, and future trends and forecasts [6] on electric energy demand and fuel consumption confirm this scenario, despite the intention of ENEL to convert oil power stations to less expensive fossil fuels in the shortest possible time. Therefore, it could make sense to consider heavy fuel oil for liquid co-firing biomass derived fuels, at least in the next ten years.

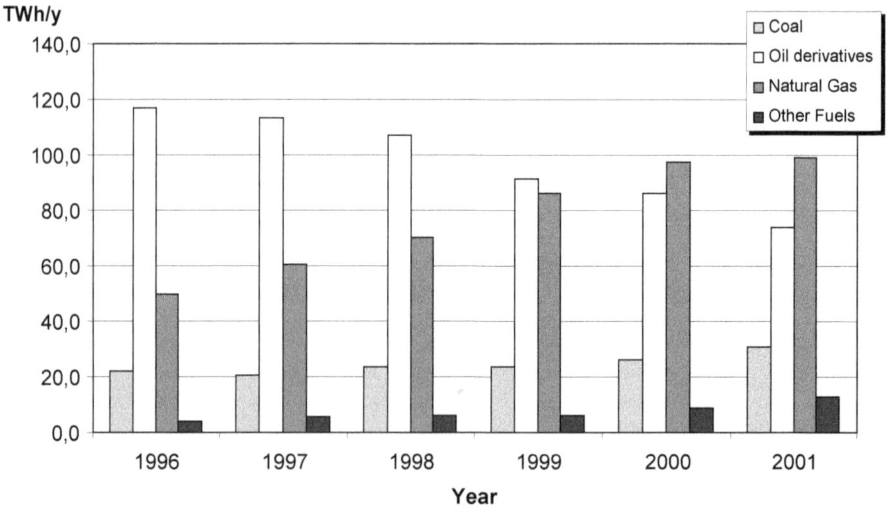

Figure 6: Energy production from thermal power plants in Italy by source (Source: Autorità per l'Energia Elettrica ed il Gas).

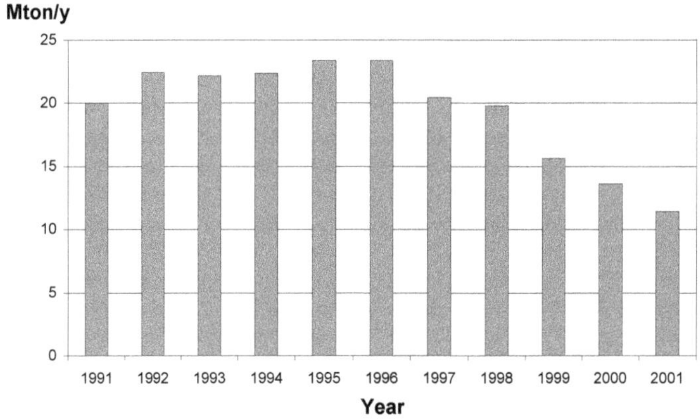

Figure 7: Oil product consumption in thermal plants in Italy (Source: Ministero dell'Economia e delle Finanze, Dipartimento del Tesoro).

EMISSION REDUCTION AND BIOELECTRICITY IN EU AND ITALY

EU is taking an active role on the environmental issue, defining programmes and plans for emission reduction. Several recent EU Protocols and Directives deal with pollutant emissions from large fossil fuel plants. Among these:

- the Goteborg Protocol identifies emission limits for SO_2, NO_x, NH_3 and VOC (Volatile Organic Compounds) [7]
- the Directive 2001/81/CE of the European Parliament and Council, which imposes to Member States to reduce further the emissions of the previously mentioned four classes of pollutants, defining emission limits slightly lower than those identified by the Protocol [8]
- the Directive 2001/80/CE of the European Parliament and Council about pollutant emissions from large combustion plants, which intervene on the instruments necessary to achieve the objectives. It is important to remark that this Directive modifies the emission regulations for large scale plants (including thermal power plants above 50 MW) on the basis of the fuel used and, for the first time, introduces severe emission limits to existing old plants (before 1987) [9]

As far as regards greenhouse gas (GHG) emissions, the Kyoto Protocol represents the cornerstone of the environmental policy of the EU. It will be ratified after the recent Johannesburgh Conference 2002, in which Russia and the P.R.China stated their intention to sign the Protocol. As soon as the Kyoto Protocol will be in force (i.e. 90 days after its ratification), the European Union will be committed to reduce its greenhouse gas emission by 8 % with respect to 1990 figures within the year 2010 [3]. But which is the situation at EU level nowadays?

During the period 1990-2000, EU emissions were reduced by 3.6 %, even if in 2000 a slight increase of 0.3 % was registered with respect to 1999. However, the situation in the various Member States is very different. The most significant contribution to GHG emission reduction was given by Germany and United Kingdom. On the contrary, Italy (that is supposed to reduce its emission level by 6.5 %) has increased them by 4.9 % [3].

The European Union has proposed or adopted several Directives related to the final goal of reducing GHG emissions, as the one on Emission Trading (COM/2001/0581 def. COD.2001/0245) or the Directive 2001/77/CE, on the promotion of the electricity production from renewable energy source. Moreover, the scenario is made even more complex if the Directives on the liberalisation of the electricity (Directive 1996/92/CE) and the gas (Directive 1998/30/CE) markets are considered (however, these will not be discussed in the present paper).

As a consequence of the liberalisation of these strategic markets, and with the aim of promoting renewable energies, in Italy the Ministry of Industry issued a Decree (nr 79/99) which established the "Green Certificate" mechanism: the Decree imposes a 2 % share of renewable energy to producers or importers of electric energy (above a minimum threshold). This renewable energy has to be produced in new power plants, i.e. entered in operation after the 1st of April 1999. This decree opens new opportunity for renewable energy projects in Italy, including co-firing.

Moreover, recently a further Decree [10, 11] increased the amount of the renewable energy eligible for the Green Certificate scheme from 10 % to 50 % of the green electricity generated in co-firing. This new framework therefore establishes more favourable economic conditions for projects on co-firing.

As far as regards CO_2 emissions per fuel at world level, these are presented in fig. 8: in 1999, oil was responsible for 41.7 % of CO_2 emissions. Projections and outlook from IEA forecast oil responsible for 36.9 % and 37.0 % in 2010 and 2020 respectively.

If we look into more details on the CO_2 emissions from oil for electricity production in Italy [7], in 1990 fuel oil combustion emitted 66,621 thousand tons of CO_2 on a total amount of 123,770 thousand tons from the whole electric sector (i.e. 53.4 %). In 1999, oil still accounted for 53,595 thousand tons of CO_2 on a total amount of 129,784 thousand tons of CO_2 emitted (i.e. 41.3 %). With the present trend, by 2010 fuel oil is expected to be responsible for 49,611 thousand tons of CO_2.

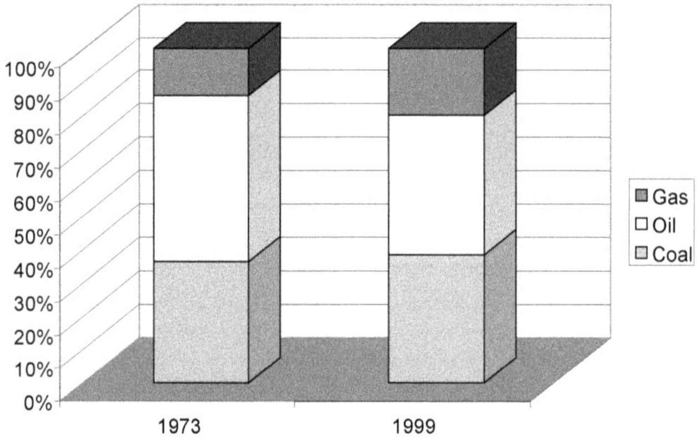

Figure 8: 1973 and 1999 fuel shares of CO_2 emissions (Source: IEA).

Analysing the specific emission factors of large scale power plants in Italy, as reported in Table 2, it can be seen that specific emissions from oil plants are lower only than those from coal plants.

Table 2: Carbon dioxide emission factors per fuel and type of plant (Source: ENEA).

Type of plant	**Power MW_{el}**	**Average electric efficiency**	**CO_2 production Kg CO_2/kWh**
Oil power plant	600	0.38	0.711
Coal power plant	600	0.36	0.946
Natural gas power plant	600	0.43	0.467
Combined cycle natural gas power plant	600	0.50	0.402
Cogeneration by counter-pressure combined cycle power	32	0.72	0.278
Cogeneration by counter-pressure combined cycle power	15	0.70	0.287

If Kyoto targets are to be met by Italy, in 2010 the total emission of CO_2 should not be greater than 116 million tons, i.e. an average emission factor of 0.387 $kgCO_2/kWh$ should be attained. Given the present value of 0.532 $kgCO_2/kWh$, it appears evident how far Italy is from achieving this goal [12]. Not considering contributions from renewable energies, even the conversion of the existing power plants to non-cogenerated combined cycle natural gas power plant would not allow to meet this goal: only co-generation would be compatible with Kyoto targets.

Summarising, Kyoto goals will be very difficult to achieve for Italy even considering power plant modernisation. A set of different and co-ordinated measures will have to be taken, such as the use of renewable energies as well as the implementation of other actions in the field of rational use of energy and energy saving. As oil consumption in Italy, in the medium terms, will continue to provide a significant contribution to national GHG emissions, co-firing heavy fuel oil and biomass derived fuels (as pyrolysis oils) could offer a short-term option to meet the national goals and comply with the existing and recently established environmental regulations.

PYROLYSIS FOR CO-FIRING

There are several reasons to consider pyrolysis oil and heavy fuel oil co-firing in the medium term, both from the fuel market as well as from the environmental point of view. Pyrolysis oil co-firing could be effectively used in large power stations, provided that its cost is competitive. Two different approaches could be adopted: in fact, pyrolysis oil could be used as pure fuel or emulsified with fossil fuel oil. This choice should be carefully evaluated and justified by specifically designed experimental work. Furthermore, PO could be produced at the same site as the power plant, or transported: similarly, emulsification could occur at the production site or at the power plant site. In the present paper the case of PO-HFO emulsion use in large scale power plants is preliminary evaluated.

Pyrolysis oil could be used directly as fuel in large power stations, provided that a dedicated burner and fuel feeding system is installed, corrosion resistant materials are adopted and the combustion characteristics are adjusted to the use of this biofuel. Combustion properties of biomass derived pyrolysis oil PO (or bio-crude-oil, BCO) are well known, and specifications for biomass derived oils in substitution to heavy fuel oils have been elaborated [13]. Handling and storage precautions have to be taken, in order to combine the characteristics of pyrolysis oils with the typical needs of large scale power plants, which requires huge amount of very stable fuels. A first choice could be the production of PO in a decoupled way to the power plant (even if produced in the same site as the power station). In this case a very stable pyrolysis oil is required, and long term storage arranged. Otherwise, a continuous in-line PO production system should be arranged. This second option, if apparently preferable because of the reduced storage and stability problems, will limit the overall plant flexibility, and generate a greater integration between the pyrolysis plant and the power plant. A final choice will depend on several aspects such as local conditions, technology available, feedstock delivery, etc.: however, as biomass pyrolysis is still in a demonstration phase and not yet fully proven at commercial scale (at least for energy production), decoupling between PO production and energy production seems preferable, as it will allows for more flexibility during plant operation (i.e. interruptions for maintenance and set-up of the pyrolysis unit).

The use of emulsions [14] would meet - at least from the technical point of view - the needs of this integrated configuration. PO-HFO emulsion would allow PO storage, and easier handling and transportation, maintaining good combustion properties of the fuel.

TECHNO-ECONOMIC EVALUATION: THE CASE OF EMULSIONS

The base-case plant is a 600 MWel heavy fuel oil plant, having 38 % total electric efficiency. It is assumed to operate for 7,884 hours per year (90 % availability), producing 4,730 GWh/y of electricity and consuming 1,101,287 t/y of heavy fuel oil. We also assumed that the plant maintains its production capacity, i.e. that the combustion chamber will accept higher amounts of lower calorific value fuels, so to produce the same amount of energy and maintain the same efficiency. In fact, typical biomass derived fuels lower heating value (LHV) is approximately 54 % w/w of the HFO LHV. The reference typical chemical composition of pyrolysis oil and heavy fuel oil used for the calculation is by A.V.Bridgwater *et* al [15], and reported in Table 3. Details on the emulsification process between pyrolysis oils from biomass and liquid fossil fuels are given in [14].

It has been assumed to supply an amount of PO-HFO emulsion sufficient to produce the same electric energy per year as in the HFO base-case. The emulsions under investigation were 20 %, 5 % and 2 % w/w of PO in HFO. As a first approximation, the calorific value of the emulsifier has been considered negligible. The properties of the emulsions are given in the Table 4.

Table 3: Pyrolysis oil and heavy fuel oil characteristics (Source: [10]).

	BCO	Heavy Fuel Oil
Flash point °C	66	100
Pour Point °C	-27	21
Viscosity (cSt @ 15 °C)	13	351
Elemental analysis (% w/w)		
H	6,40	11,80
C	48,50	86,10
O	42,50	-
N	-	-
S	-	2,10
Ash	0,13	0,03
Water weight (% w/w)	20,50	0,1
Density @ 15 °C (kg/litre)	1,220	0,963
HHV (kJ/kg)	22.746	41.768
LHV (kJ/kg)	17.500	40.700
LHV (kJ/litre)	21.350	39.194
Acidity (Ph)	3	-
PO/HFO v/v	54%	
PO/HFO w/w	43%	

Table 4: PO content, lower heating value and density of the three emulsions.

Emulsion 20 %	
PO content (% w/w)	20,0%
LHV (kJ/kg)	35.979
Density (kg/dm3)	1,012
Emulsion 5 %	
PO content (% w/w)	5,0%
LHV (kJ/kg)	39.520
Density (kg/dm3)	0,975
Emulsion 5 %	
PO content (% w/w)	2,0%
LHV (kJ/kg)	40.228
Density (kg/dm3)	0,968

The share of green power and energy produced from biomass depends on the type of emulsion considered. For the three emulsions reported in Table 4, it corresponds to 58 MWel, 13 MWel and 5 MWel respectively for the three selected cases (20 %, 5 % and 2 % w/w emulsions). It corresponds to the hypothetical size of a 100 % biomass plant producing the same amount of energy per year. It is worth to remark that, in the case of the 20 % w/w emulsion, the size of this plant would be extremely relevant, as well as the assumptions made on its availability (90 %).

The use of emulsion in the HFO base-case plant would require modifications to existing pumps and feeding systems, as well as the adaptation of storage tanks and supply systems. As a first estimation, these costs have been assumed equal to 20 M€, 18 M€ and 16 M€ respectively. As far as regards the emulsification unit, on the basis of previous small-scale (50 kg/h) prototypes [9], a preliminary estimation on the plant size, its energy consumption (function of its size) and the investment cost has been performed. It has to be remarked that these two assumptions on costs are the most critical ones, as the project dimension is significantly different from any previous experience: moreover, any long-term test with emulsions is available yet. The cost for the emulsification unit has been considered equal to 5 M€ for all the three cases. Amortisation costs have been estimated (20 years, 8 % interest rate).

In Table 5, the amount of emulsion and PO needed is given. The efficiency of the PO plant has been assumed constant for the three cases and equal to 67 %.

Table 5: Amount of emulsions needed and characteristics of the PO plant.

		20%	**5%**	**2%**
Tot amount of emulsion needed	**t/y**	1,245,807	1,134,180	1,114,212
Emuls.syst.	**t/h**	158	144	141
PO needed	**t/y**	249,161	56,709	22,284
PO Plant Production capacity	**t/h**	47	11	4
Total Efficiency of the PO plant		67 %	67 %	67 %
Dry Feedstock in	**t/y**	371,883	84,640	33,260

Finally, as the present scheme requires large amounts of PO, a cost of 118 €/kg_{PO} has been used, according to estimations given in [15] for future conditions. It has been assumed to buy the PO from a producer independent from the power plant owner. The emulsification process is instead assumed to be performed within the battery limits of the power plant. Operation and maintenance costs of the power plant fed with emulsions have been considered the same as in the case of pure heavy fuel oil, and equal to 0.004 €/kWh [12]. Under these assumptions, the economic analysis gives the following main results:

- The operating costs per year related to the green electricity produced are equal to 40 M€, 13.5 M€ and 8.6 M€ for the 20 %, 5 % and 2 % cases respectively.
- The PO cost accounts for 73 %, 50 % and 31 % respectively of the yearly costs of the green electricity produced by the system.
- The costs of the emulsification units are almost constant in both terms of investment as well as operating costs, since the total amount of emulsion to be treated does not change significantly in the three cases even if the amount of PO in HFO decreases significantly. Therefore, differently from the PO cost, the relative importance of this cost component tends to increase as the amount of PO in HFO decreases: from 11 % to 29 % to 45 % for the three emulsions 20 %, 5 % and 2 % respectively.

The green electricity production costs for the three different emulsions is given in Figure 9. The reason why it increases as the amount of PO in HFO decreases is clear if we look at the cost composition of the green kWh (Fig. 10).

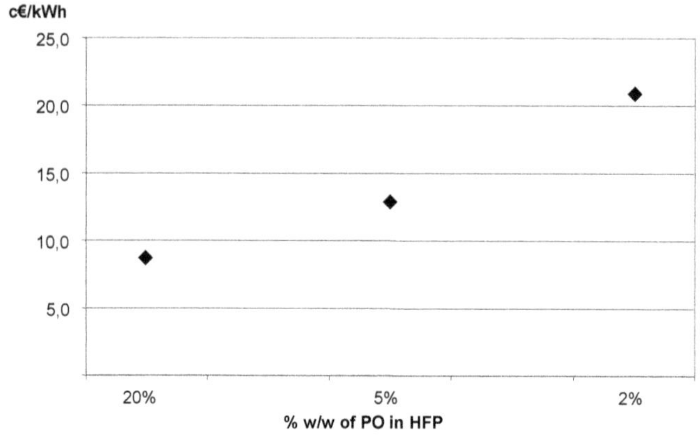

Figure 9: Green electricity production costs.

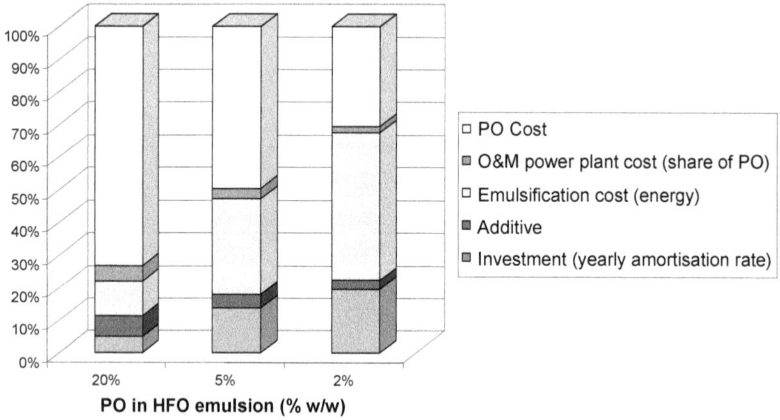

Figure 10: Components of green electricity production costs.

In fact, PO costs represent 73 % of electricity production cost in case of a 20 % w/w emulsions and only 31 % in case of a 2 % w/w emulsion, while the costs associated to technology adaptation and emulsification remains almost constant in absolute terms, i.e. increase from 5 % (20 % w/w) to 20 % (2 % w/w). The composition of electricity production cost is given Figure 11.

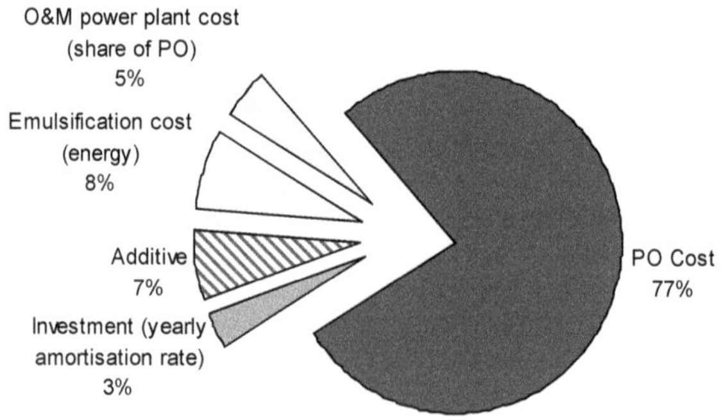

Figure 11: Components of the green electricity production costs
(30 % w/w PO in HFO emulsion, 2 % w/w of HFO consumption per year).

Based on the results of this preliminary analysis, a second possible approach has been developed. It has been assumed that:

- The total amount of PO fed into the 600 MWel HFO power station is equal to 2 % w/w of the yearly HFO consumption
- This amount of PO is emulsified with HFO at 30 % w/w, and fed as a 30 % w/w emulsion to the HFO power plant in a separate burner

This approach will maintain the emulsification costs at a low level, as these are mainly determined by the total amount of emulsion to be treated per unit of time (i.e. the amount of PO and HFO emulsified in the system per hour). At the same time, an overall contribution of 2 % w/w to the overall fuel consumption is maintained as in the previous case.

The economic analysis of this scheme gives, compared to the previous case in which PO is emulsified with the total flow of HFO at 2 % w/w, the following results:

- The size of the emulsification unit is greatly reduced: from 141 t/y to 9 t/h, while the amount of PO emulsified is the same (22,284 t/y).
- The yearly operating cost is significantly reduced as well, from 8.6 M€/y to 3.4 M€/y.
- The electricity production cost is reduced as well, from 20.9 c€/kWh to 8.2 c€/kWh.
- The cost for the PO represents 78 % of the yearly operating and financial costs of the plant.

Comparing these figures with predicted electricity production costs for 100 % biomass plants based on pyrolysis, it is very interesting to observe how these forecasts [15], using future system conditions and based on pyrolysis oil and diesel engines at production capacity level of 5 MWel, indicate a cost of approximately 9 c€/kWh (pyrolysis oil combined with Diesel engine has the lowest electricity production costs at scales up to 5 MWel).

With respect to the composition of the electricity production cost, as already mentioned PO cost is the main cost component (77.6 %).

As far as regards the sensitivity of this second scheme to variations in the PO cost or in the costs for technology adaptation and emulsification, as expected the system is more sensitive to the first rather than to the second (technology-related) item, as shown in Figure 12.

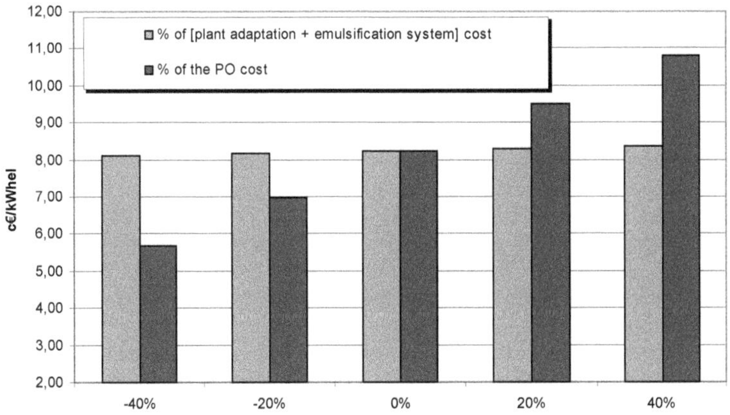

Figure 12: Sensitivity analysis on green electricity production costs.

Finally, the sensitivity of the scheme to green pricing has been studied. As already mentioned, only 50 % the green electricity produced in co-firing in Italy is eligible for green certificates (i.e. 8.418 c€/kWh) [10]. This share has been recently increased, as it was 10 % until March 2002. The analysis showed that the economic balance between costs and revenues is obtained at approximately 50 %, which is the value established by the current law. However, it seems reasonable to foresee a further increase of this share during the next years, to harmonise Italy with the other EU Countries. It is worth to remark that carbon-trading financial mechanisms (e.g. Joint Implementation, Clean Development Mechanisms) could also be activated for this type of projects in EU and abroad.

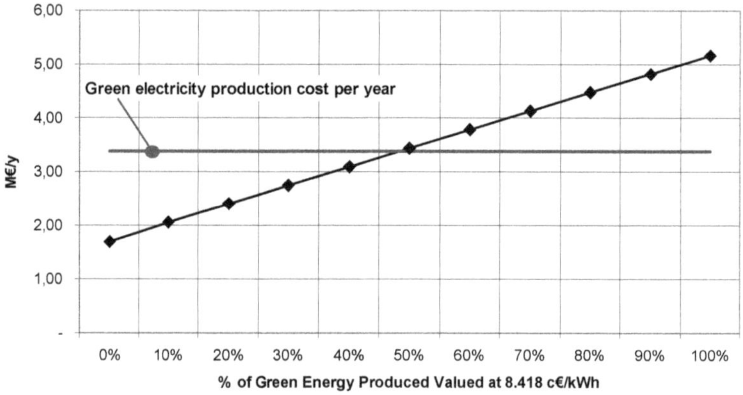

Figure 13: Sensitivity analysis on green energy pricing for co-firing in Italy.

Considering the environmental aspects, the reduction in CO_2 emissions is proportional to the amount of biofuel used (pure or emulsified) with respect to the amount of HFO. Similar considerations can be done for SOx emission. With respect to NOx emission reduction, it is reasonable to expect significant improvements even on this issue: however, the nitrogen content of the feedstock should be carefully taken into account in the analysis.

CONCLUSIONS

The use of biomass PO in large HFO power plants would offer a very important opportunity to address not-niche markets with biofuels. The analysis of the energy scenario in Italy showed that in the next years fuel oil could maintain a significant market share for power production in large scale plants, and that it could still provide a significant contribution to electricity production.

The use of biomass pyrolysis emulsions co-firing with pure HFO in large power plant is technically possible, economically interesting and environmentally sound.

From the technical point of view, the adaptation of existing combustion systems of large scale HFO plants can be performed with success. The use of PO in large scale plants will offer significant advantages in terms of energy conversion efficiency, as the total electrical efficiency of these large plants is considerably higher than state-of-the-art biomass plants. Moreover, these HFO power plants are operated by skilled personnel that could be trained on PO (pure or emulsified) use in a relatively short time.

From the economic point of view, the cost of the biofuel is still the main factor in the kWhel production cost: moreover, a full agreement on the estimation of the future PO production cost [15] has not yet been reached among experts in biomass PO. The costs for the adaptation of combustion systems in the power plant, as well as the costs associated to the construction of an in-line emulsification system of this size need more detailed verification, as the extrapolation from laboratory or prototype systems is very difficult, and results should be better assessed. This study should also aim at assessing the differences in using pure PO or emulsified PO in large scale burners: nevertheless, it has to be remarked that emulsions offer significant advantages in terms of fuel stability (and therefore storage) and handling [14].

From the environmental point of view, the use of biomass PO, pure or emulsified with HFO, will provide significant benefits not only in terms of CO_2 emission reduction (as requested in the EU and Italian legislation), but also in terms of SO_x and NO_x (that, from the regulatory point of view, could be even more interesting, as recent laws apply new emission limits also to existing large scale plants).

ACKNOWLEDGMENTS

Authors wish to acknowledge Ing.Neri, ENEL, for the support provided in the analysis of the regulatory framework in Italy.

REFERENCES

1. BP (2002) *Statistical Review of World Energy*, www.bp.com/centres/energy
2. International Energy Agency (2001) *Key World Energy Statistics from IEA* www.iea.org.
3. Autorità per l'Energia Elettrica ed il Gas (2002) *Relazione Annuale sullo stato dei servizi e sull'attività svolta", presentazione del Presidente dell'Autorità* www.autorita.energia.it
4. GRTN (2002) *Dati Storici* www.grtn.it
5. Ministero dell'Economia e delle Finanze, Dipartimento del Tesoro (2002) *Quaderno dei Prodotti Petroliferi: Consumi e Prezzi Internazionali, Maggio 2002* www.dt.tesoro.it
6. La Repubblica, Affari e Finanza (8 July 2002) *Rapporto Energia.*
7. United Nation Economic Commission for Europe UNECE (1999) *Goteborg Protocol: Convention on Long Range Transboundary Air Pollution*, www.unece.org
8. *Directive 2001/81/EC of the European Parliament and of the Council of 23 October 2001 on National Emission Ceilings for Certain Atmospheric Pollutants.* In: Official Journal of the European Communities L 309, 27.11.2001, pp. 22-30
9. *Directive 2001/80/EC of the European Parliament and of the Council of 23 October 2001 on the Limitation of Emissions of Certain Pollutants into the Air from Large Combustion Plants.* In: Official Journal of the European Communities L 309, 27.11.2001, pp. 1-21
10. Ministero delle Attività Produttive (2002) *Decreto 18 marzo 2002, "Modifiche e integrazioni al decreto del Ministro dell'industria, del commercio e dell'artigianato, di concerto con il Ministro dell'ambiente, 11 novembre 1999, concernente «direttive per l'attuazione delle norme in materia di energia elettrica da fonti rinnovabili di cui ai commi 1, 2 e 3 dell'art. 11 del decreto legislativo 16 marzo 1999, n. 79»".* In: *Gazzetta Ufficiale della Repubblica Italiana* n. 71 del 25-03-2002.
11. Ministero dell'Industria del Commercio e dell'Artigianato (1999) *Decreto 11 novembre 1999, "Direttive per l'attuazione delle norme in materia di energia elettrica da fonti rinnovabili di cui ai commi 1, 2 e 3 dell'articolo 11 del decreto legislativo 16 marzo 1999, n. 79".* In: *Gazzetta Ufficiale della Repubblica Italiana* n. 292 del 14-12-1999.
12. Vignati S. (2002) Il sistema di generazione elettrica italiano dopo i recenti accordi sul protocollo di Kyoto. *Energia, Ambiente e Innovazione*, **2/2002**, 41-50.
13. Diebold J.P., Milne T.A., Czernik S., Oasmaa A., Bridgwater A.V., Cuevas A., Gust S., Huffman D., and Piskorz J., (1999) *Proposed Specifications for Various Grades of Pyrolysis Oils.* In : *Fast Pyrolysis of Biomass: a Handbook* (Ed. by Bridgwater A., Czernik S., Diebold J., Meier D., Oasmaa A., Peacocke C., Piskorsz J. And Radlein D.), pp 102-114. CPL press.
14. Baglioni P., Chiaramonti D., Bonini M., Soldaini I. (2000) *Bio Crude Oil /Diesel Oil Emulsification: Main Achievements of the Emulsification Process and Preliminary Results of Tests on Diesel Engines.* In: IEA Bioenergy, *Progress in Thermochemical Biomass Conversion*, (Ed.by A.V.Bridgwater) Blackwell Science Ltd, Oxford.

15. Bridgwater A.V., Toft A.J., Brammer J.G. (2002) A Techno-Economic Comparison of Power Production by Biomass Fast Pyrolysis with Gasification and Combustion. *Renewable and Sustainable Energy Reviews*, Elsevier, 6, 181-248.

Compact Power puts Pyrolysis at the Heart of Sustainable Developments Optimising Value Recovery

Nic Cooper
Compact Power Limited, Hydro House, St. Andrews Road, Avonmouth, Bristol BS11 9HZ, UK

ABSTRACT: Compact Power has developed a total waste solution for local communities using pyrolysis, gasification and high temperature oxidation. Its advantages are based on the ability to provide an economic and effective process for a wide range of wastes, to recover up to 80% of the energy value of the waste fuel as usable heat and power, and to operate to the highest environmental standards.

The process can facilitate the recovery of materials, principally carbon and inert residues, in the form of valuable by-products. As such it constitutes a form of "thermal recycling".

The technology has been developed over a period of 8 years and is now being demonstrated in a commercial plant running in Avonmouth in Bristol, UK. The operations are carried out under an Integrated Pollution Prevention and Control Authorisation from the Environment Agency and emissions are continuously monitored and reported daily in accordance with the authorisation. Results show that operations are well within permitted limits and in some respects new standards are being demonstrated.

The process holds all material for nearly 1½ hours at high temperature in a closed system during which it is first pyrolysed in the absence of oxygen to release the hydrocarbon gases and leave carbon and inert residues. The carbon is either captured for conversion into high value carbon products, or reacted with steam and air to produce hydrogen and carbon monoxide. The gases are then taken forward for combustion through high temperature oxidation at 1250° C for more than 2 seconds. Energy recovery is achieved through a steam boiler and steam turbine. Emissions meet the highest standards and dioxins are undetectable.

The ability of the plant to process multiple waste streams simultaneously gives the potential for optimizing a total waste solution for a community. The combination of the technology with DryVac sludge drying plant provides an economic solution for highly aqueous wastes and represents a technical breakthrough for wastewater treatment. An integrated closed system returns clean water to the system and valuable materials are recovered from the solid content.

Compact Power seeks to maximise the opportunities for sustainable development around its projects by collaborating with complementary activities that can use the heat and power and other by-products from the Compact Power process. It is, for example, developing the capacity to produce industrial quality carbon black from processed tyres. Similarly the Compact Power technology for production of activated carbon from sewage sludge will offer the water industry the opportunity to recycle a high value product for its operations from its own waste stream.

In the field of conversion of biomass, including a wide range of agricultural wastes containing high concentration of fertilisers, the Compact Power technology is designed to provide some practical and economic solutions for the developed world which will also provide some much needed answers for developing countries.

INTRODUCTION

HISTORY

The business of Compact Power was established in 1992 by Nic Cooper and Professor John Sharpe, an expert on thermal power plant. The founders identified a niche in the market for waste-to-energy plant, appropriately engineered to provide economic answers for local integrated waste management facilities with minimal environmental impact. It is based on a modular concept to give the best options for matching the variable requirements of truly sustainable solutions.

The company has developed a proprietary technology for processing a wide range of wastes based on the processes of pyrolysis, gasification and high temperature oxidation. These processes permit the highest relevant environmental standards to be achieved and offer the potential for maximising the reuse and recycling of materials. The high level of energy recovery in the form of heat and power enables the technology to support complementary processes and promote job creation at local level.

A full scale prototype plant was built in 1994 based on a single pyrolyser tube with a throughput capacity of 3,000 tonnes a year; and between 1994 and 1997 a series of trials were conducted on a variety of wastes, including dewatered sewage sludge, MSW, clinical waste, waste from vegetable oil refining, tyre crumb, and coal slurry. These tested the core thermo-chemical processes and provided data on the performance of the plant and on emissions and solid residues. The results indicated that the emissions fell within or close to all relevant standards without any air pollution control equipment and that the plant could achieve the highest environmental standards with minimum conventional flue-gas remediation.

Subsequently a commercial plant was built at Avonmouth in Bristol with double the throughput capacity, and Integrated Pollution Prevention and Control Authorisation was obtained from the Environment Agency in October 2001. Since then the plant has been taking clinical and municipal waste and demonstrating the performance of the plant in normal operating conditions.

The company is also pursuing further technology developments with the object of creating applications for the re-processing of waste materials to create useful and, in some cases high value, industrial products. Currently the particular focus is on the production of activated carbon and carbon black.

THE TECHNOLOGY

Compact Power combines the thermo-chemical processes of pyrolysis, gasification and high temperature oxidation.

Compact Power converts up to 80% of the energy value of the waste into readily usable steam with a wide range of applications such as power generation and combined heat and power ("CHP"), including process heat for industrial purposes (such as the DryVac dewatering process), chilling and refrigeration requirements using absorption chillers, and water de-salination plant.

Solid wastes are converted into simple gases and used to generate heat and power. The plant is effectively a gas production plant integrated with a gas fired steam boiler and power plant. The plant is a closed system and the only emissions are the exhaust gases, which are well within the relevant emissions standards, and the ash which typically is high quality bottom ash which may be suitable for uses in construction.

The plant design permits accurate control and optimal conditions for each separate stage of the process. Provision can be made within the same facility for the handling and processing of a range of different wastes simultaneously within an integrated waste management strategy which optimises the recycling, reprocessing and energy recovery potential of each waste stream.

PYROLYSIS

The waste is initially fed into a system of tubes in which pyrolysis takes place over a period of about 30 minutes. The material is heated in the absence of oxygen. In the temperature range 400°C to 800°C the organic materials break down into "pyrogas" (low molecular weight hydrocarbon gases), leaving carbon residues and inert materials such as ash, metals and glass, all of which pass into the gasification chamber.

GASIFICATION

This is the process by which the residual carbon is reacted with steam at high temperature to produce hydrogen and carbon monoxide which are both combustible gases, and carbon dioxide. This is the conventional "water gas" reaction and takes about 1 hour.

High temperature oxidation

The pyrogas and the water gas pass into a thermal oxidizer which is maintained at a constant temperature of 1250°C. Sufficient air is introduced for complete oxidation of all gases and the design of the chamber ensures that the gases and any small particles of solid matter that may come through in the gas stream remain at these temperatures for at least two seconds. These conditions are designed to minimise the emission of pollutants and to meet the highest environmental standards.

The exhaust gases from the thermal oxidizer provide the heat for the process, and the plant is self-sustaining even with wastes which have relatively low calorific value. However auxiliary fuel is used to ensure that the temperature conditions in the thermal oxidizer are maintained at all times and to enable the plant to maintain the required heat and power output even if the energy value of the waste is inadequate or if the processing of the waste is interrupted for any reason.

The exhaust gases then pass through a boiler where the heat is transferred to the steam and the temperature of the gases is rapidly reduced to below 250°C. The steam is then available as a convenient energy source for heat and power applications.

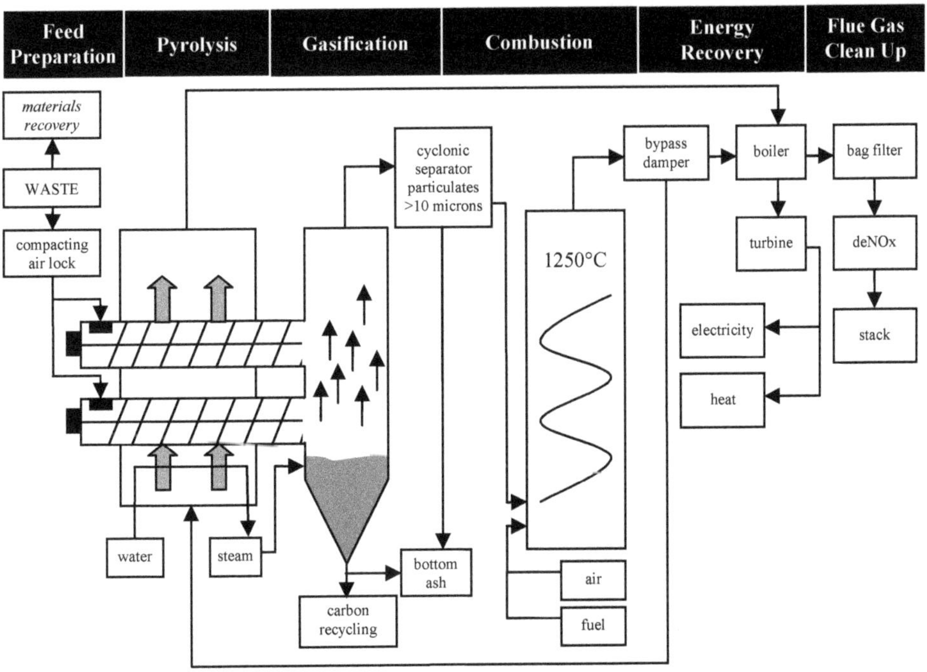

Figure 1: Process diagram of Compact Power technology.

EMISSIONS CONTROL AND ASH

The process is designed to minimise the risk of dioxins arising and the gas clean-up system has the capability to reduce any residual dioxins in the flue gas. Similarly such nitrogen oxides ("NO_X") as are produced are brought well within all relevant standards by the de-NO_X unit included in the flue gas clean-up system.

Any solid residues will be removed from the plant. Ash will be recycled to the extent that this is practicable or otherwise sent to landfill. The amount of ash will depend on the characteristics of the waste fuel.

PLANT DESIGN

Compact Power's designs are based on the principle of compact, modular units of plant which are used to produce integrated systems for waste to energy and CHP applications. The items of plant are designed to be transportable so that they can be modified or relocated to meet changing needs and varying waste streams.

The multi-tube pyrolysis system is based on 2 tube units each with a 1 tonne per hour throughput capacity. Each 2 tube unit feeds a single gasifier. There is a standard 2 tube (MT2) system or a standard 8 tube (MT8) configuration made up of 4 x 2 tube units. The gas line / thermal oxidizer (G) is sized to serve an MT2 or an MT8. Facilities are designed around combinations of these units, so that a unit designed to process 64,000 Tpa would comprise 2 MT8 units and 2 oxidisers [2xMT8 (2G)].

Typical configurations are as follows:

- The MT2 unit which can process up to 8,000 tonnes per annum and is appropriate for lower volume difficult wastes which command relatively higher gate fees such as clinical waste;
- The MT8 unit which in the case of MSW has a capacity of up to 32,000 tonnes per annum and can generate up to 2MW of electricity (net of plant requirements) and 10.4MW of reusable heat.
- The 2xMT8(2G) with a capacity of up to 64,000 Tpa of mixed waste and can generate up to 4.2MWe and 20.8MW of heat.
- The 2XMT8(3G) with a capacity of up to 64,000 Tpa of mixed waste including highly calorific waste such as tyres and can generate up to 7MWe and 31.2 MW of heat.

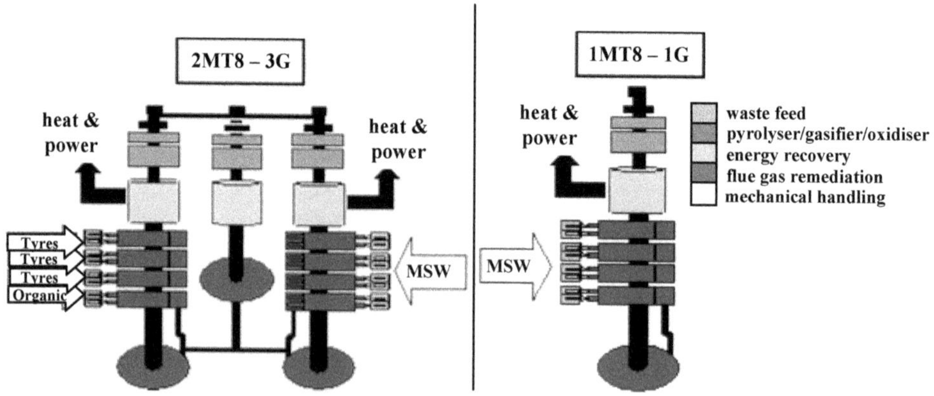

Figure 2: Modularity of Compact Power plant.

The plant is designed to provide economies in operation and maintenance with features that minimise the risk of unscheduled stoppage. A principal design criterion is the achievement of the environmental objective set by the U.K. Environment Agency to use Best Available Techniques ("BAT") to avoid or minimise harmful pollutants from any process. The combined effect of pyrolysis, gasification, high temperature oxidation and advanced NO_x treatment minimises pollution to such an extent that we believe that the process can offer the best environmental options for waste disposal.

Compared with a typical mass burn incinerator, the system has very low environmental impact in terms of air pollution and discharges of residues to land. There is no aqueous waste effluent and the plant has a low visual impact compared with a typical mass burn incinerator.

Table 1: Emissions data.

Mg/Nm^3 11% 0_2@ 273K & 101.3kPa	EU Directive Limits s: spot / d: daily ½ hour		Compact Power
Particulates	10 (d)	30	0.2
VOC's as carbon	10 (d)	20	trace
NO ($NO + NO_2$)	200 (d)	400	<37
HCl	10 (d)	60	2
HF	1 (d)	4	<0.1
SO_2	50 (d)	200	<25
CO	50 (d)	150	trace
Cd & Tl	0.05 (s)		0.006
Mercury	0.05 (s)		0.006
Pb Cr Cu Mn Ni As Sb Co V Sn	0.5 (s)		0.006
Dioxins TEQ ng/Nm^3	0.1 (s)		<0.003
N_2O	30 (d)		trace
NH_3	10 (d)	20	<1

The ability of the plant to process multiple waste streams simultaneously gives the potential for optimizing a total waste solution for a community. The combination of the technology with DryVac sludge drying plant provides an economic solution for highly aqueous wastes and represents a technical breakthrough for wastewater treatment. An integrated closed system returns clean water to the system and valuable materials can be recovered from the solid content.

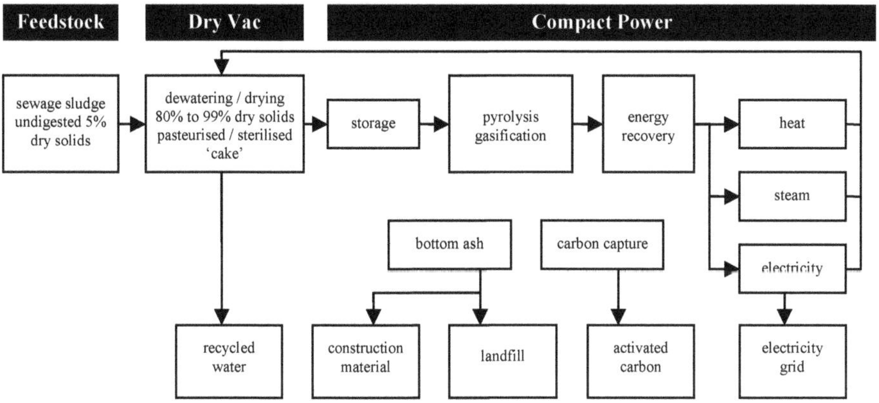

Figure 3: Process flow of combined Compact Power / DryVac application.

Compact Power seeks to maximise the opportunities for sustainable development around its projects by collaborating with complementary activities that can use the heat and power and the other by-products of the Compact Power process. It is, for example, developing the capacity to produce industrial quality carbon black from processed tyres. Similarly the Compact Power technology for production of activated carbon from sewage sludge will offer the water industry the opportunity to recycle a high value product for its operations from its own waste stream.

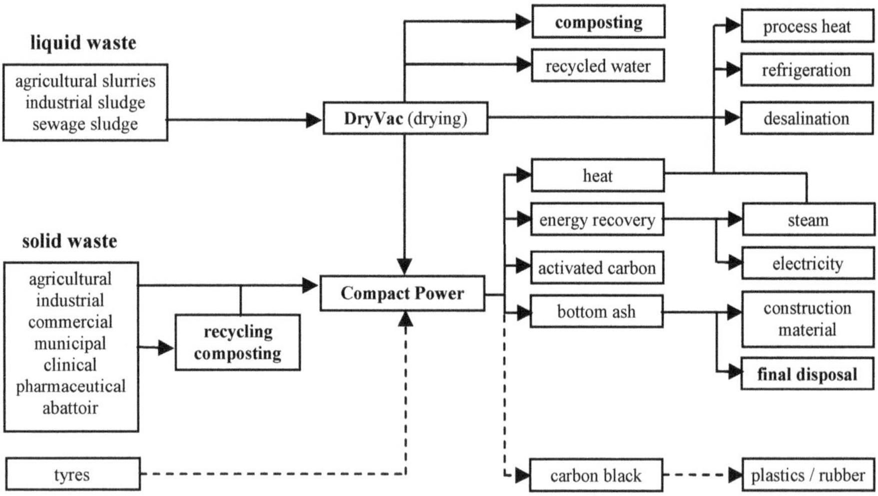

Figure 4: Process flow of typical integrated waste management facility.

In the field of conversion of biomass, including a wide range of agricultural wastes, the Compact Power technology is designed to provide some practical and economic solutions for the developed world which will also provide some much needed answers for developing countries.

Case studies will be presented of different applications of the technology in project situations.

CONCLUSION

Compact Power has a "hub and spoke" approach to project development in which it can offer a total waste solution for a community at the heart of an integrated infrastructure facility. This can provide an economic process for relevant municipal, commercial and industrial wastes, recycling materials where practically and economically feasible, recycling water from aqueous wastes and contributing renewable energy to local distributed power systems. Environmental objectives of reducing carbon emissions can be met by converting carbon residues.

The Compact Power process is conceived more as a thermal re-processing of waste with optimised energy recovery than as a means of disposal and the company's business is dedicated to realizing this potential for truly sustainable solutions in partnership with industries and the communities which they serve.

Heated Grid Flash Pyrolysis of *Miscanthus* with In-situ Infrared Spectrometry Species Analysis and Comparison with FG-DVC Biomass Model Simulations

W. de Jong , A. Slabbekoorn
Section Thermal Power Engineering, Faculty of Mechanical Engineering, Delft University of Technology, Mekelweg 2, NL-2628 CD, Delft, The Netherlands
J. Guo , A. Veefkind
Section Gas Dynamics, Faculty of Applied Physics, Eindhoven University of Technology, P.O. Box 513, NL-5600 MB Eindhoven, The Netherlands

ABSTRACT: The desire to develop robust and comprehensive models to predict the product evolution and yield of biomass pyrolysis requires advanced methods for experimental validation. In this research the biomass version of the Functional Group - Depolymerization, Vaporization and Cross-linking (FG-DVC) pyrolysis model developed by Advanced Fuel Research Inc. (AFR) has been validated by flash pyrolysis experiments with *Miscanthus giganteus* in a heated grid reactor. The pyrolysis gas yield has been analysed by means of infrared absorption spectroscopy with an infrared tuneable diode laser. The research was focused on the yield of CO, CO_2 and NH_3 at a heating rate of 280-320K/s and a final temperature of 1050-1400K. The experiments show exceptionally low CO_2 and CO yields, indicating considerable yields of primary products like primary tars and acetic acids. The evolution of these primary products into light gasses by secondary reactions most likely didn't occur in the grid reactor experiments. The model predicts well the qualitative trends of the CO and CO_2 formation. However, the yield prediction is not accurate at high heating rates. The straightforward extrapolation that the model uses to predict the pyrolysis product yield at high heating rates, based on kinetic input data that was determined by Thermo-Gravimetric experiments with Fourier Transform Infrared analysis (TG-FTIR) at low heating rates (10-100 K/min), is found to be precarious. The heating rate dependent character of secondary reactions and occurrence of these during TGA pyrolysis experiments is believed to be the reason for this.

INTRODUCTION

The increased interest in the production of liquid or gaseous fuels from biomass and the co-firing of biomass waste streams with coal arises mainly from the environmental benefits of the CO_2 neutral and renewable qualities of biomass energy sources. Inevitably, this has resulted in more and more research on the thermal conversion processes of biomass, in which pyrolysis plays an important first role. Besides the composition of the biomass feedstock, the pyrolysis conditions in terms of temperature, heating rate, atmosphere and residence time also have a major impact on the product yield of biomass pyrolysis. This has led to the desire for a reliable model to predict the pyrolysis product yield as a function of the biomass characteristics and the process conditions. Advanced experimental methods are required to validate and further develop these models.

FG-DVC MODEL

In the past several pyrolysis models for solid fuels, mostly coal, have been developed as reported in [2-4, 7]. A typical one-stage multi reaction model is the comprehensive Functional Group - Depolymerization, Vaporization and Cross-linking (FG-DVC) model described in [2,12,13], originally developed for coal pyrolysis and recently adapted for biomass feedstock. The model has been developed by Advanced Fuel Research Inc. (AFR, USA). The performance of this biomass version has been validated extensively with TG-FTIR experiments at heating rate of 10-100 K/min with a variety of biomass feedstock as described in [1,2]. The model considers certain functional groups or pools in the feedstock, which lead to the formation of specific product species. The evolution rate of a specific species during pyrolysis is modelled with the first order equation (1.1):

$$\frac{dY_i}{dt} = -k_i Y_i \qquad (1.1)$$

in which the left hand part is the evolution rate of a specific species (e.g. a carboxyl group into CO_2) and Y_i size the pool size of this species in the biomass. The rate constant k_i is determined by the Arrhenius expression shown in (1.2):

$$k_i = A_i \cdot e^{\left(\frac{-(E_i \pm \sigma_i)}{RT}\right)} \qquad (1.2)$$

where A_i is the pre-exponential factor, E_i is the activation energy and σ_i the width of distribution in activation energies. A Gaussian distribution in activation energies is assumed. The kinetic parameters of equation (1.2) are determined by accurate TG-FTIR experiments at heating rates of 10 to 100 K per minute. The technique usually employed to determine these kinetic parameters is the non-isothermal method based on the temperature at which the rate of volatile evolution is maximum (T_{max}). T_{max} is measured at the heating rates 10, 30 and 100 K/min and the activation energy E_i en pre-exponential factor A_i are extracted from equation (1.3).

$$Ln\left(\frac{M}{T_{\max}^2}\right) = Ln\left(\frac{A_i R}{E_i}\right) - \left(\frac{E_i}{RT_{\max}}\right) \qquad (1.3)$$

where M represents the varying heating rate.

The model uses straightforward extrapolation to simulate pyrolysis at different heating rates and temperature, using these constants determined at three heating rates, i.e. 10, 30 and 100 K/min. In this research the model is validated with pyrolysis experiments at high heating rates (280-320K/s) and temperature (1050-1400K) in order to verify the validity of the extrapolation that the model uses. These experimental conditions are typical conditions of fluidised bed gasifiers and other thermal equipment used for biomass conversion.

The results of the TG-FTIR analysis of *Miscanthus giganteus* and Wood pellets, used in this research are shown in Table 1. The kinetic parameters used in the FG-DVC model are shown in Table 2.

Table 1: Product yields (wt% daf [2]) of the TG-FTIR analysis of *Miscanthus giganteus* pellets and Wood pellets at different heating rates.

Sample	Wood	Wood	Wood	*Miscanthus*	*Miscanthus*	*Miscanthus*
Heating rate K/min (K/s)	10 (0.17)	30 (0.5)	100 (1.7)	10 (0.17)	30 (0.5)	100 (1.7)
Ash [1]	0.0	0.0	1.0	1.5	3.8	-
Moisture [1]	5.6	6.3	6.6	5.0	2.5	6.0
Volatile Matter (from TGA)[2]	86.2	86.2	86.2	80.2	79.4	81.9
Char [2]	13.8	13.8	13.8	19.8	20.6	18.1
Tar [2]	37.2	38.8	48.4	18.1	21.9	28.3
Methane (CH_4) [2]	1.27	1.30	1.15	1.18	1.14	1.16
Water (pyr.) (H_2O) [2]	10.6	16.0	13.6	17.7	18.5	20.6
Carbon Monoxide (CO) [2]	7.48	6.77	5.10	8.55	7.15	5.65

Sample	Wood	Wood	Wood	*Miscanthus*	*Miscanthus*	*Miscanthus*
Carbon Dioxide (CO_2) [2]	5.90	5.08	4.88	10.4	9.43	9.11
Ethylene (C_2H_4) [2]	0.15	0.23	0.09	0.52	0.27	0.13
Hydrogen Cyanide (HCN) [2]	0.09	0.09	0.05	0.22	0.14	0.11
Ammonia (NH_3) [2]	0.00	0.02	0.03	0.13	0.10	0.11
Isocyanic Acid (HNCO) [2]	0.11	0.02	0.05	0.20	0.11	0.07
Carbonyl Sulphide (COS) [2]	0.18	0.11	0.00	0.31	0.16	0.11
Sulphur Dioxide (SO_2) [2]	0.17	0.12	0.40	0.00	0.00	0.00
Formaldehyde (CH_2O) [2]	3.66	2.93	3.15	1.12	1.12	1.16
Acetaldehyde (CH_3CHO) [2]	9.28	8.11	3.97	10.2	9.02	7.70
Methanol (CH_3OH) [2]	0.98	0.74	0.64	1.42	1.13	1.07
Formic Acid (HCOOH) [2]	2.68	1.73	0.82	2.12	1.66	0.37
Acetic Acid (CH_3COOH) [2]	2.78	2.37	2.28	4.24	3.59	3.35
Phenol (C_6H_5OH) [2]	1.46	0.65	0.73	2.31	1.83	1.99
Acetone (CH_3OCH_3) [2]	2.05	2.15	1.63	1.80	1.86	2.10

[1] Weight percent of initial sample, on as received basis

[2] Weight percent of initial sample, on dry and ash free basis

Table 2: Kinetic parameters and precursor pool size (yields) for the main and N-species determined by TG-FTIR analysis of *Miscanthus giganteus* for use in the FG-DVC model.*

Species	A (s^{-1})	E/R (K)	σ/R	Yield (wt. % daf)
1-CO_2	5.20E+11	16300	900	0.98
2-CO_2	5.20E+11	19500	500	7.50
3-CO_2	2.80E+12	24830	900	0.93
1-CO	5.20E+11	19550	400	3.90
2-CO	6.50E+12	26300	3200	1.85
3-CO	2.80E+12	35930	3300	1.40
1-H_2O	5.20E+11	19100	400	14.14
2-H_2O	4.40E+12	24000	2500	4.50
1-CH_4	6.50E+11	19500	600	0.15
2-CH_4	6.50E+11	22300	500	0.23
3-CH_4	3.10E+12	26900	1600	0.76
1-Tars	5.20E+11	19700	500	21.34
1-HCN	2.80E+12	20530	1900	0.07
2-HCN	2.80E+12	28000	3100	0.07
1-HNCO	2.80E+12	16930	900	0.02
2-HNCO	2.80E+12	25930	2800	0.09
1-NH_3	5.20E+11	18600	800	0.07
2-NH_3	5.20E+11	25000	2500	0.03
1-C_2H_4	2.80E+12	24930	800	0.21
1-CH_3OH	5.20E+11	17500	500	0.20
2-CH_3OH	2.30E+11	19930	400	0.95

Species	A (s^{-1})	E/R (K)	σ/R	Yield (wt. % daf)
1-CH_2O	5.20E+11	17100	500	0.33
2-CH_2O	5.20E+11	19500	200	0.82
1- CH_3CHO	5.20E+11	19500	300	9.02
1- HCOOH	5.20E+12	20500	500	1.66
1- CH_3COOH	5.20E+11	18700	900	3.59
1- C_6H_5OH	7.40E+18	29100	2100	2.03
1- CH_3OCH_3	5.20E+12	20500	800	1.86

*Numbers before species names indicate precursor-material sources (pools). Some species evolve as a single peak (one source), the evolution of other species is bimodal or trimodal (two or three sources).

EXPERIMENTAL EQUIPMENT

The experiments were carried out in an electrically heated grid reactor in a nitrogen environment. The evolution of light gas species was analysed in-situ by means of infrared absorption spectroscopy with an infrared tuneable diode laser. This technique makes analysis frequencies up to 20 kHz possible; high enough to monitor the evolution of gas species during flash pyrolysis at heating rates up to 10^4-10^5 K/s. The method uses a tuneable diode laser in the deep infrared spectrum, which makes scans in a narrow spectrum window of about 2-4 cm^{-1} of specific absorption spectrum of the gas to be detected. Using this narrow window makes advanced mathematical analysis techniques on the detector site, as used in the FTIR method, unnecessary.

The grid reactor, shown in Figure 1, is a closed cylindrical reactor chamber of stainless steel with a length of 227 mm and an inner diameter of 15 mm. In the middle of the cylinder, a platinum-rhodium grid is placed between 2 copper electrodes. The 0.31 mm mesh grid (0.08 mm wire diameter) is a U-shape of 6 mm long and 2 mm wide, with the opening to the side. The pieces of biomass or coal are sandwiched between the folded grid and heated from top and bottom. An external power supply heats up the grid. Manipulating the electrical current controls the temperature of the grid and the heating rate. Heating rates up to 10^3-10^4 K/s are possible, the temperature is maximized at 2045K, the melting point of platinum.

The reactor has been designed to withstand an internal pressure of 30 bar. Gas enters and leaves the reactor at the two sides of the cylinder. Sintered porous material in the gas inlets reduces the gas velocity in order to prevent the biomass sample is blown away. Before a pyrolysis experiment is started, the reactor is evacuated with a vacuum pump and refilled with nitrogen in order to create inert conditions.

In the middle of the reactor two windows are positioned, making it possible to observe the sample on the grid during the experiment. The intensity of the radiation is measured with a photodiode through one of the windows. The temperature of the grid is measured with a thermocouple. At temperatures above ca. 1100K the final grid temperature was measured with a manual colour pyrometer as well. This was done before an experiment was started. Assumed is that the particles on the grid are small enough to have the same temperature of the grid. Using the steady-state end temperature of the pyrometer and the signal of the photodiode, the heating rate of the grid (time-temperature profile during experiment) was reconstructed. The accuracy of the temperature measurements was estimated at ± 30K. At heating rates above 500 K/s difficulties were found in reconstructing the time-temperature profile at this accuracy.

On both sides of the reactor, ceramic CaF_2 windows are mounted. The beam of the tuneable diode laser enters and leaves through these windows, scanning the full cross-sectional area of the cylinder for absorbing gas molecules. This is shown in Figure 2. The IR diode laser was tuned by changing the diode temperature and current to the wavelength of an absorption peak of the gas to be detected. The laser was set to a saw tooth signal; covering 2-4 cm^{-1} in each scan. The laser is adjustable in a range of 1500-3000 cm^{-1}. The laser beam was split in two. A reference cell was used to make sure the right compound was detected. The signals from the IR detectors were recorded with an oscilloscope.

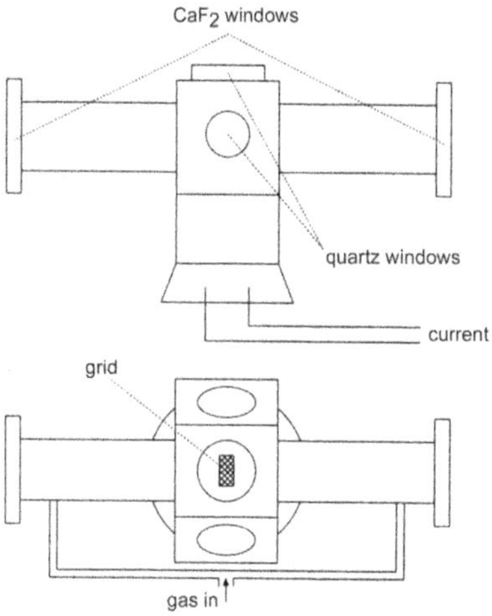

Figure 1: Schematic representation of the grid reactor. Top: side view, bottom: bird view.

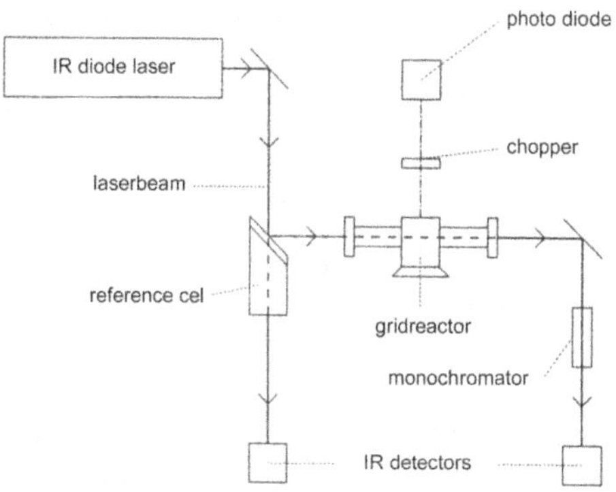

Figure 2: Set up of the grid reactor and diagnostic tools.

MATERIAL AND EXPERIMENTAL PROCEDURE

The experiments were carried out with *Miscanthus giganteus* grown in the Netherlands for research and other applications. The *Miscanthus* pellets were milled and sieved. The fraction with sieve size 38-63 μm was conditioned for 2 weeks at 20°C at 60% relative humidity. During the experiments, the fraction was stored at the same conditions. Batches of about 50 mg of the conditioned *Miscanthus* powder were pressed at 1130 MPa under vacuum conditions to tablets of 13.0 mm diameter and about 0.6 mm thick. The tablets were cut into small pieces with a mass between 0.30 and 1.00 mg, with a characteristic diameter between 0.6 and 1.0 mm. The mass of the pieces was measured with a balance of an accuracy of 10^{-6} g. The experiments were carried out with these samples of a known mass in order to be able to calculate the yield of a specific gas species as part of the initial sample mass. The samples were sandwiched in the grid. The reactor was closed, evacuated and filled to atmospheric pressure with nitrogen gas before an experiment. The grid was heated electrically up to the desired final temperature, and then the temperature was kept constant. The experiments lasted 6 seconds in total. A stable yield of the measured gasses was reached within this period in all experiments.

Absorption was defined at the central wavelength of the absorption peak as shown in Figure 3 by means of equation (1.4). Applying equation (1.4) to experimental data of a typical 6 seconds pyrolysis experiment with a *Miscanthus* sample in the heated grid reactor results in an absorption curve as shown in Figure 4. The data in Figure 4 is fit with function (1.5).

$$Absorption = \frac{I_h - I_g}{I_h - I_{a,b}} \tag{1.4}$$

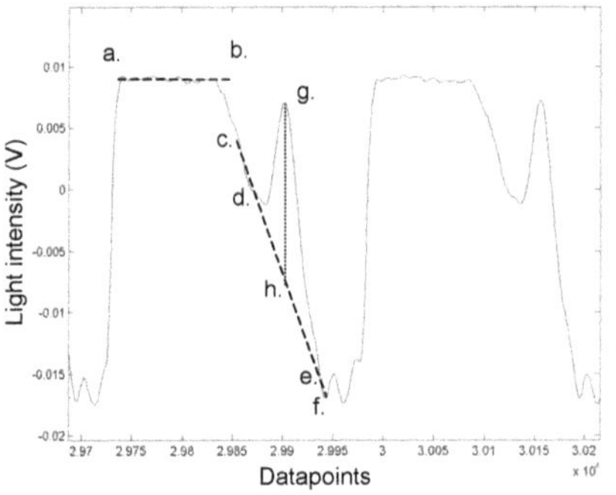

Figure 3: Saw tooth laser signal with a typical IR absorption peak for CO_2. Letters refer to equation (1.4).

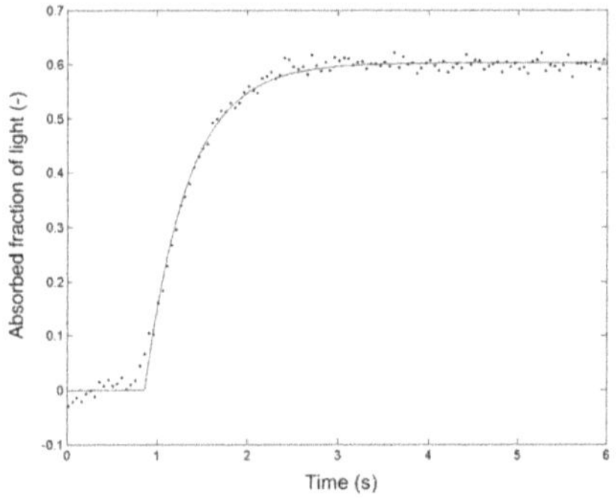

Figure 4: Typical IR absorption pattern of CO_2 evolution during pyrolysis, with curve fit.

In equation (1.4) I_h is the light Intensity at point h, I_g is the light Intensity at point g, and $I_{a,b}$ is the average light Intensity between points a an b. In equation (1.5) the absorbed fraction of light, I_{abs}, is a function of time t and constants c_1, c_2, and c_3.

$$I_{abs} = c_1 + c_2 \cdot \left(1 - e^{-\frac{t}{c_3}}\right) \tag{1.5}$$

By using a computer program with a database of the absorption and transmittance spectra of the gasses, or by performing experiments (done in this research) with a reference gas containing a known concentration CO, CO_2 or NH_3 in Nitrogen, the gas concentration as function of time can be calculated from the absorption curve of Figure 4. For the last method, ideal gas and a linear relation between the concentration gas and the absorption of light was assumed. This assumption is correct for constant temperature and pressure. Both the volume of the reactor and the initial mass of the biomass sample are known very accurately, making it possible to calculate the yield of a specific gas as function of time during pyrolysis, using equation (1.6) and equation (1.7). The parameters of both equations can be found in the Nomenclature.

$$m_{gas} = \frac{pV_cM}{RT_g} \cdot f_{gas} \tag{1.6}$$

$$Y\left(wt\%\ daf\right) = \frac{m_{gas}}{m_{sample} \times \left(1 - f_{ash,moisture}\right)} \times 100\% \tag{1.7}$$

where

$f_{ash,moisture}$ = Fraction ash and moisture in sample, from TGA $(-)$
f_{gas} = Mol fraction gas in reactor $(-)$
M = Molar mass gas (kg/mol)
m_{gas} = Total mass of a specific gas released in reactor (kg)
m_{sample} = Initial sample mass (kg)
p = Final pressure in reactor (Pa)
R = Universal gas constant $(= 8.3145\ J/mol.K)$
T_g = Gas temperature in the reactor $(= 298K)$
V_c = Corrected volume of the reactor $\left(= 4.548 \cdot 10^{-5}\ m^3\right)$
Y = Final gas yield $(wt\%$, dry and ash free$)$

RESULTS

The experiments were focused on the pyrolysis product yield of CO, CO_2 and NH_3 as a function of temperature at a high heating rate. Attempts to detect NH_3 have failed since no absorption peak could be found. The reason for this could be that the concentration of NH_3 in the reference cell was too low or that incorrect laser settings were used. Figure 5 shows that the yield of CO is strongly correlated with the final temperature in the range 1050-1400K, for a heating rate of 250-320 K/s. The weight percent CO of the initial *Miscanthus* sample varies from 4.0% at 1050K to 13.0% at 1350K. The estimated error made in the experiments is within -16% and +21%. Opposite to CO, Figure 6 shows that the yield of CO_2 is not correlated with temperature in the same range and for the same experimental conditions. In average the CO_2 yield was 0.59wt%.

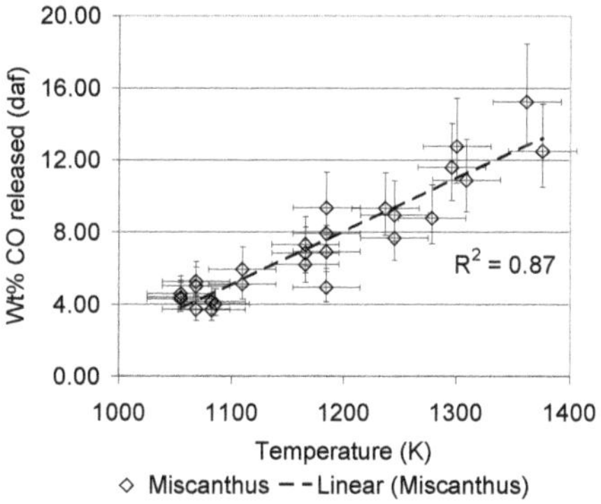

Figure 5: CO yield (wt%) of pyrolysed *Miscanthus*: strongly correlated to the final temperature in the range 1050-1400K, ambient pressure, heating rate 250-320K/s.

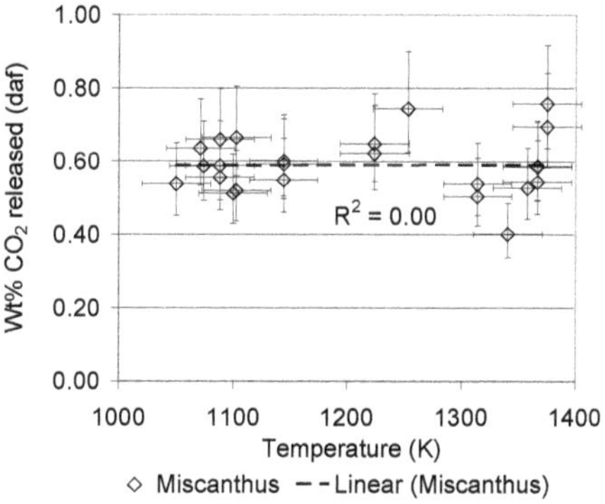

Figure 6: CO_2 yield (wt%) of pyrolysed *Miscanthus*: not correlated to the final temperature in the range 1050-1400K, ambient pressure, heating rate 250-320K/s.

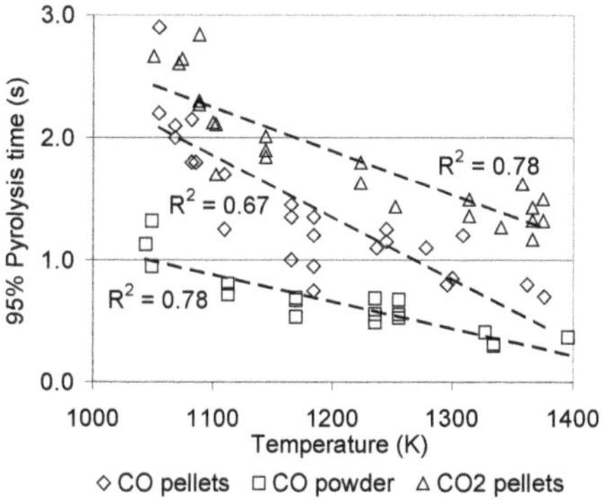

Figure 7: The 95% pyrolysis-time correlated with temperature. Deviations are caused by variation in sample mass.

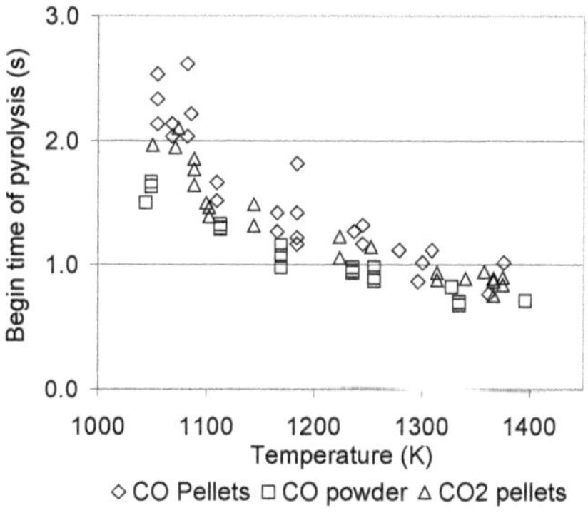

Figure 8: Start moment of devolatilisation of CO and CO_2 from *Miscanthus* as a function of temperature.

The 95% pyrolysis time is defined as the time required for release of 95% of the CO or CO_2 final yield. An Arrhenius plot is shown in Figure 9 to determine the activation energies of the reactions.

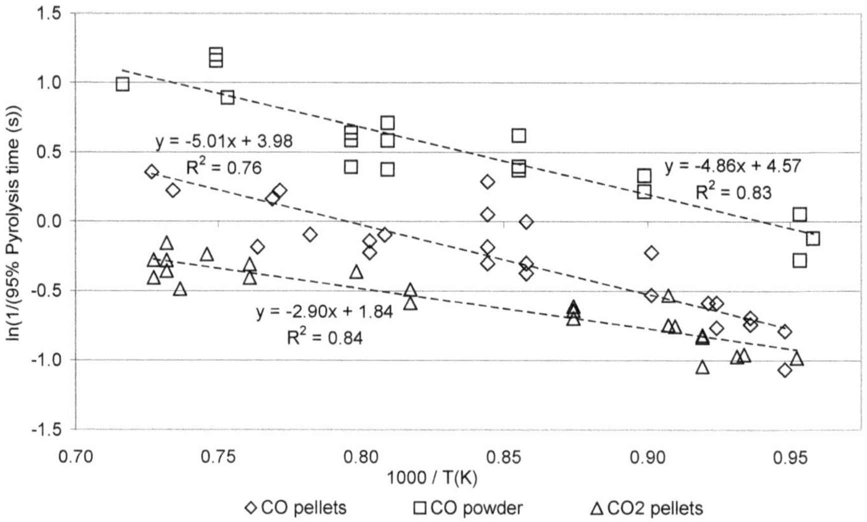

Figure 9: Arrhenius plot for the devolatilisation of CO from *Miscanthus* pellets and powder and from CO_2 pellets. The dashed lines are the least square linear fits.

The slope of the dashed lines in Figure 9 represents the activation energy of the decarbonylation (release of CO) and decarboxylation (release of CO_2) reactions. Obviously, the slope of the two lines for CO release during pyrolysis from *Miscanthus* pellets and powder are practically equal to each other since the reaction mechanism -decarbonylation- is similar. The interception of the line in Figure 9 with the y-axis represents the value of the pre-exponential factor in equation (1.2). Heat transfer effects are reflected in this factor also. The observed activation energy for both reactions is shown in Table 3.

Table 3: Activation energy from Arrhenius plot for decarbonylation and decarboxylation reactions.

Species	Reaction	E_a/R (K)	Activation energy, E_a (kJ/mole)
CO	Decarbonylation	$4.94\cdot10^3$	41.0
CO_2	Decarboxylation	$2.90\cdot10^3$	24.1

These values for the activation energies determined for a heating rate of 300 K/s, are significantly lower than the values found with the TG-FTIR pyrolysis experiments (at 10-100 K/min) used to obtain the kinetic for the FG-DVC model. The E_a/R value of the TG-FTIR experiments found in Table 2 are $19.6\cdot10^3$ up to $35.9\cdot10^3$ K for CO and $16.3\cdot10^3$ up to $24.8\cdot10^3$ K for CO_2. This would imply that decarbonylation and decarboxylation reactions at high heating rates are initiated at lower temperatures.

The experimental results obtained in this research are compared to simulations with the FG-DVC biomass pyrolysis model at a heating rate of 300 K/s and to results of the limited number of comparable experiments published in accessible literature. The comparison is shown in Figure 10 and Figure 11. Most experimental research focuses on pyrolysis at low heating rates (up to 100K/min) and/or low temperature (up to 900 K), e.g. liquefaction. A series of systematic flash pyrolysis experiments were carried out with sweetgum hardwood, wood lignin en cellulose at high heating rates (100-15000 K/s) in a heated grid reactor in the period 1980-1985 at the Massachusetts Institute of Technology (MIT) in Cambridge, USA [5,9,10]. Chen *et al.* (1998) [2] have used these experimental results in a first attempt to validate the FG-DVC biomass model at high heating rates.

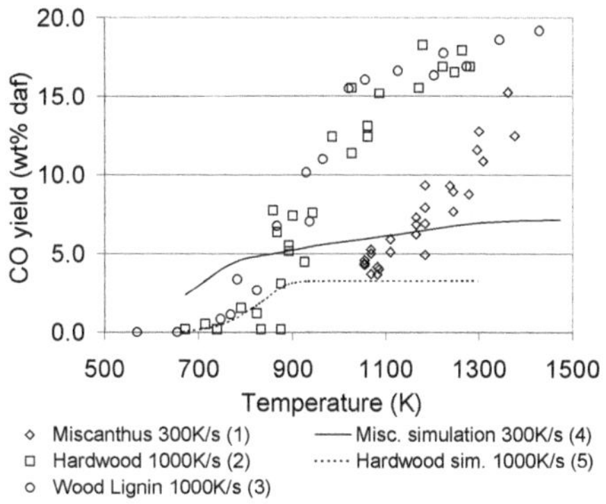

Figure 10: Comparison of heated grid pyrolysis experiments and FG-DVC model simulations. CO yield in wt% (dry and ash free) of the initial biomass sample as function of final (or peak) pyrolysis temperature.

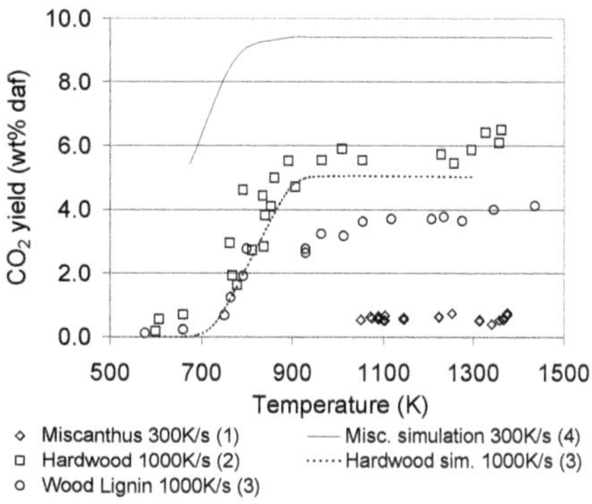

Figure 11: Comparison of heated grid pyrolysis experiments and FG-DVC model simulations. CO_2 yield in wt% (dry and ash free) of the initial biomass sample as function of final (or peak) pyrolysis temperature.

Remarks at Figure 10 and Figure 11:
(1) Experimental results of this work.
(2) Milled Sweetgum hardwood pyrolysis experiments from [9].
(3) Milled wood lignin pyrolysis experiments from [10].
(4) FG-DVC model simulations of pyrolysis of *Miscanthus* (this work).
(5) FG-DVC model simulations of pyrolysis of hardwood [2]. Note that the type of hardwood used in (2) differs slightly from (5).

At a final pyrolysis temperature of 900°C the main product yield is compared at different heating rates and biomass feedstocks, as shown in Table 4.

Table 4: Comparison of the pyrolysis product yields of the main species at 900°C as a function of heating rate (HR), for different biomass materials.

Material	Wood pellets (1)			*Miscanthus* pellets (1)				Lignin (2)	SG Hard wood (2)	Cellulose (3)
HR (K/s)	0.17(4)	0.5(4)	1.7(4)	0.17(4)	0.5(4)	1.7(4)	300(5)	1000(5)	1000(5)	1000(5)
CO (daf)	7.5	6.8	5.1	8.6	7.2	5.7	7.2	16.0	16.0	20.0
CO_2 (daf)	5.9	5.1	4.9	10.4	9.4	9.1	0.6	3.8	5.5	3.3
H_2O (daf)	10.6	16.0	13.6	17.7	18.5	20.6	-	3.8	5.0	9.2
CH_4 (daf)	1.3	1.3	1.2	1.2	1.1	1.2	-	2.8	1.6	2.6
Tar (daf)	37.2	38.8	48.4	18.1	21.9	28.3	-	46.0	45.0	50.0
Char (daf)	13.8	13.8	13.8	19.8	20.6	18.1	-	15.0	8.0	4.0
Volatiles (daf)	86.2	86.2	86.2	80.2	79.4	81.9	-	-	-	-
Ash (ar)	0.0	0.0	0.0	1.5	3.8	-	-	-	-	-
Moisture (ar)	5.6	6.3	6.6	5.0	2.5	6.0	-	-	-	-

(1) This research: TG-FTIR analysis of *Miscanthus giganteus* pellets by AFR Inc. [1] and heated grid reactor experiments. (2) Wood lignin and Sweetgum Hardwood from [9,10]. (3) Cellulose from [5]. (4) TG-FTIR analysis. (5) Heated grid reactor experiments.

FORMATION OF METAL CARBONYLS

Most likely formation of transition metal carbonyls on the stainless steel reactor wall has been observed in all CO absorption curves as shown in Figure 12. In average 21 wt% of the released CO disappears with in 1-2 seconds after being released. The fraction of CO that disappears is correlated to the initial mass of the *Miscanthus* samples.

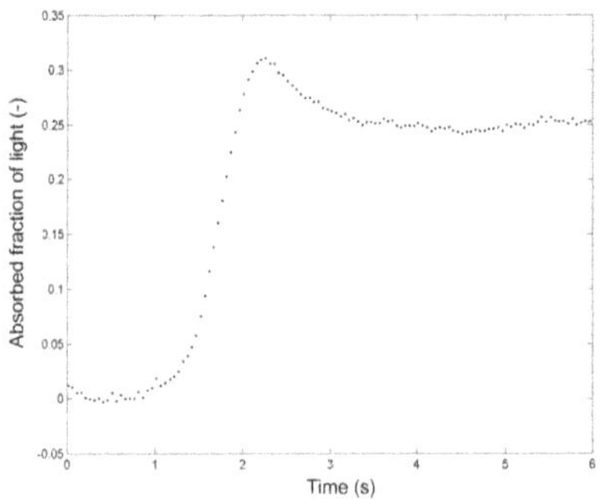

Figure 12: Typical absorption curve for CO release from *Miscanthus* pellets. Final temperature 1240 K, heating rate 280 K/s, atmospheric.

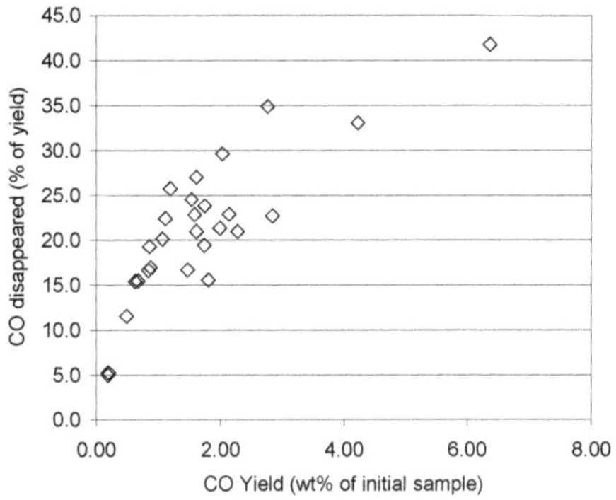

Figure 13: Disappeared part of CO yield (in % of maximum yield) versus total CO yield. *Miscanthus*, 250-320 K/s, 1050-1400 K, atmospheric.

Figure 13 shows that at a low partial pressure of CO in the reactor, the absorbed amount of CO is correlated with the partial pressure of CO in the reactor. This curve suggests that a maximum amount of CO molecules can be absorbed at higher partial pressure of CO in the reactor. This supports the hypothesis that the reactor wall absorbs parts of the CO yield. An additional set of experiments with a larger sample could clarify whether the data fits a Langmuir absorption isotherm.

Concerning the CO yield results shown in Figure 5 and Figure 10 it is assumed that the results are hardly affected by the absorption phenomenon, since the absorption peak value of the experiments, shown in Figure 12, was used to construct Figure 5 and Figure 10. The absorption phenomenon is just slow enough to make detection of the final CO yield possible. The experiments with the CO reference gas are most likely not affected by the absorption phenomenon since the grid reactor was flushed with the reference gas several times before measurements were done.

DISCUSSION OF THE RESULTS

The CO_2 yield of flash pyrolysis (300 K/s) of *Miscanthus* is a factor 16 lower than the yield of slow pyrolysis (10-100 K/min) of *Miscanthus* at a temperature of 900°C, as shown in Table 4. The CO_2 yield is a factor 6-10 lower than found in Nunn *et al.* (1985) and Hajaligol *et al.* (1982) [5, 6, 9, 10] using lignin, hard wood and cellulose. Assuming that differences in biomass materials, like wood and *Miscanthus*, will result in a maximum CO_2 yield difference of a realistic factor 2, questions are raised on the correctness of the results. Experiments with a stainless steel grid, chemical equilibrium calculations and thorough re-examination of the experimental method learns that it is highly unlikely that the set-up or method affects the results. Then somehow the pyrolysis conditions in the grid reactor of this research must have been different from the conditions in the grid reactor used by [5, 6, 9, 10], causing a relatively low CO_2 yield.

Reviewing the TG-FTIR results with wood and *Miscanthus*, shown in Table 4 and Table 2, could lead to the conclusion that the tar yield increases at the expense of the light gas species CO, CO_2 and acetaldehyde if the heating rate increases from 10 to 100 K/min. This could be explained by differences in kinetic effects associated with the release of tar and light species. Tar fragments leave the reaction zone carrying with them precursor material for formation of light gases. Hence, the tar molecules are larger and heavier. The heating rate does not significantly affect the yield of char, methane, formaldehyde and acetone, according to the TG-FTIR results alone. However, the results of [5, 6, 9, 10] at higher heating rates experiments at about 1000 K/s show in Table 4 no significant decrease in CO_2 yield, and a factor two increase in CO yield. The tar yield however is almost doubled. Apparently the increasing amount of tar fragments, or heavier tar fragments, do not carry with them precursor material for CO_2, the carboxyl groups. The specific precursor material for CO_2 containing carboxyl groups must have another origin in the pyrolysis process.

Considering abovementioned, the specific precursor groups for CO_2 could be formed during the primary stage of flash pyrolysis. In the experiments of [5, 6, 9, 10] these precursor groups further decomposed to form CO_2. In the experiments in this research they didn't. These specific groups are likely to be Carboxylic Acids, like HCOOH and CH_3COOH, released in the primary stage of flash pyrolysis from lignin and hemi-cellulose compounds of the biomass. In [5, 6, 9, 10] large samples of 100 mg fine powder (45-88μm) were placed on a 325-mesh folded grid of 60x20mm. This corresponds with 0.083 mg/mm^2. In this research small samples of 0.3-0.8 mg were folded in a 76-mesh grid of 6x2 mm, corresponding to 0.042 mg/mm^2. The fine structure the grid in [5, 6, 9, 10] (325 wires per linear inch, 12.8 per mm) combined with a large sample could explain the high CO_2 yield of their experiments. The residence time of the Carboxylic Acids in the hot reaction zone is just long enough to decompose into CO_2. In the experiments of this research the coarse-meshed grid allows the Acetic Acids to leave the reaction zone faster from the small samples. The cold nitrogen environment ensures immediate quenching, making secondary reactions to form CO_2 impossible. More grid reactor experiments at high heating rates focussing on a broader range of pyrolysis products, including Acetic Acids, are necessary to further establish this hypothesis.

A similar hypothesis could explain the higher CO yield of [5, 6, 9, 10] compared to the results of this work. During the pyrolysis process in this research the primary tar fragments are formed and removed quickly from the reaction zone, carrying precursor material, for example ether-links, carbonyl and carboxyl groups, for CO with them. This results in a lower CO yield. The transition process of large primary tar fragments (e.g. levoglucosan) into smaller secondary tars (more aromatic compounds) involves cracking of ether-links at higher temperatures accompanied by formation of CO, is expected to have contributed significantly to the high CO yield in the experiments in [5, 6, 9, 10]. The relatively small difference in CO yield (factor 1.5-2.5 at 900°C) could be ascribed to differences in material properties as well.

VALIDATION OF THE FG-DVC MODEL

In Figure 10 and Figure 11 the results of a FG-DVC flash pyrolysis simulation at a heating rate of 300K/s, a total residence time of 6 seconds and a varied final temperature, are compared to the experimental results obtained in this research. The figures show also a FG-DVC simulation of [2] with experimental data of [9, 10]. The relation between final pyrolysis temperature and yield is predicted correctly in all cases. However, difficulties are found to predict correctly the yield quantitatively. The straightforward extrapolation that the model uses to predict the evolution of gases high heating rates based on low heating rate kinetic constants is found to be risky. It is believed that the difficulties are mainly caused by the occurrence of secondary pyrolysis reactions. It is impossible to exclude the occurrence of these reactions during the TG-FTIR experiments at low heating rates (10-100 K/min) to determine the kinetic constants, as it's impossibility to ensure immediate quenching of the pyrolysis products using hot gas to heat up the sample in a TGA. The large differences between the CO and CO_2 yields (factor 1.5-2.5 and 6-10 respectively) measured during flash pyrolysis experiments at high heating rates of 300 and 1000 K/s using different experimental set ups are explained by secondary reactions as well. This shows the sensitivity of the pyrolysis reaction mechanisms to the experimental equipment in terms of a not perfectly controlled time-temperature profile of the sample and pyrolysis products during the pyrolysis experiment.

The implications for a typical one-stage multi-reaction model as the FG-DVC biomass model are obvious. The tars and Carboxylic Acids are modelled in the basic FG-DVC model as one functional group for each, as shown in Table 2. No relations are included between formation or deformation of these groups and the formation of light gases (CO, CO_2, H_2O, CH_4, etc.) The statistical network submodel for Depolymerization, Vaporization and Cross-linking that was applied in the coal version of the FG-DVC model is de-emphasised in the biomass version because of the different chemical structure of biomass. Currently new subroutines are being developed at AFR Inc. for the FG-DVC biomass model taking into account secondary reactions and the relation between formation of primary tars and light gases, in order to improve the performance of the model.

CONCLUSIONS

The biomass version of the FG-DVC pyrolysis model was validated by flash pyrolysis experiments with *Miscanthus giganteus* in a heated grid reactor. The pyrolysis gas yield was analysed by means of in-situ infrared absorption spectroscopy with an infrared tuneable diode laser. The research was focused on the yield of CO, CO_2 and NH_3 at a heating rate of 280-320K/s and a final temperature of 1050-1400K. Reliable experimental results were obtained for the release of CO and CO_2 during pyrolysis. The attempts to detect formation of NH_3 during pyrolysis have failed so far, since no absorption peak was found in wavelength window of the diode laser.

The experiments show exceptionally low CO_2 and CO yields compared to similar experiments carried out by [5, 6, 9, 10], indicating considerable yields of primary products like primary tars and acetic acids. The evolution of these primary products into light gasses by secondary reactions most likely didn't occur in the grid reactor experiments in this research due to quicker quenching and different pyrolysis conditions as a result of a coarse grid construction.

The FG-DVC model predicts well the qualitative trends of the CO and CO_2 formation. However, the yield prediction is not quite accurate at high heating rates. The straightforward extrapolation that the model uses to predict the pyrolysis product yield at high heating rates, based on kinetic input data that was determined by Thermo-Gravimetric experiments with Fourier Transform Infrared analysis (TG-FTIR) at low heating rates (10-100 K/min), is found to be precarious. The heating rate dependent character of secondary reactions and occurrence of these reactions during TGA pyrolysis experiments is believed to be the reason for this. The new version of the FG-DVC biomass pyrolysis model will be released in 2002 and is expected to have a significantly improved yield prediction since tar will be considered as a pool and a pyrolysis product, rather than a pyrolysis product alone.

ACKNOWLEDGEMENTS

The Dutch Agency for Energy and Environment, NOVEM, is acknowledged for funding the research in the framework of biomass gasification modelling via the project Biomassavergassing-CFB-STEG (contract EWAB 355298/2010).

REFERENCES

1. Bassilakis R., Carangelo R.M., Wójtowicz M.A. (2001) TG-FTIR analysis of biomass pyrolysis. *Fuel*, **80**, 1765-86.
2. Chen Y., Charpenay S., Jensen A., Wójtowicz M.A., Serio M.A. (1989) *Modelling of biomass pyrolysis kinetics*. 27th Symposium (International) on Combustion, The Combustion Institute, 1327-43.
3. Di Blasi C. (1993) Modeling and simulation of combustion processes of charring and non-charring solid fuels. *Prog. Energy Combust. Sci.*, **19**, 71-104.
4. Di Blasi C. (1998) Comparison of semi-global mechanisms for primary pyrolysis of lignocellulosic fuels. *Journal of Analytical and Applied Pyrolysis*, **47**, 43-64.
5. Hajaligol M.R., Howard J.B., Longwell J.P., Peters W.A. (1982) Product Compositions and Kinetics for Rapid Pyrolysis of Cellulose. *Ind. Eng. Chem. Process Des. Dev.*, **21**, 457-465.
6. Hajaligol M.R., Howard J.B., Peters W.A. (1993) An Experimental and Modeling Study of Pressure Effects on Tar Release by Rapid Pyrolysis of Cellulose Sheets in a Screen Heater. *Combustion and Flame*, **95**, 47-60.
7. Miller R.S. & Bellan J. (1997) A generalized biomass pyrolysis model based on superimposed cellulose, hemicellulose and lignin kinetics. *Combust. Sci. and Tech.*, **126**, 97-137.
8. Moors J.H.J. (1998) *Pulverised char combustion and gasification at high temperatures and pressures*. DPhil thesis, Eindhoven University of Technology, Department of Applied Physics, The Netherlands.
9. Nunn T.R., Howard J.B., Longwell J.P., Peters W.A. (1985) Product Compositions and Kinetics in the Rapid Pyrolysis of Sweet Gum Hardwood. *Ind. Eng. Chem. Process Des. Dev.*, **24**, 836-44.
10. Nunn T.R., Howard J.B., Longwell J.P., Peters W.A. (1985) Product Compositions and Kinetics in the Rapid Pyrolysis of Milled Wood Lignin. *Ind. Eng. Chem. Process Des. Dev.*, **24**, 844-52.
11. Shafizadeh F. (1984) *The chemistry of Pyrolysis and Combustion* Wood. Chemistry laboratory, University of Montana, Missoula, USA, American Chemical Society, 0065-2393/84/0207-0489, In: Rowell R. M., The chemistry of Solid Wood, American Chemical Society, ISBN 0065-2393.
12. Solomon P.R., Hamblen D.G., Carangelo R.M., Serio M.A., Deshpande G.V. (1988) General Model of Coal Devolatilization. *Energy and Fuels*, **2**, 405.
13. Solomon P.R., Hamblen D.G., Serio M.A., Yu Z.Z., Charpenay S. (1993) A characterization method and model for predicting coal conversion behaviour. *Fuel*, **72**, 4.

Neural Networks as a Free Model Approach for Waste and Biomass Pyrolysis Products Forecasting

F. Fantozzi

Università di Perugia - Dipartimento di Ingegneria Industriale – Via G. Duranti 1A/4, 06125 Perugia, Italy

ABSTRACT Pyrolysis of waste and biomass may represent an interesting solution for energy conversion in a highly regenerative Gas Turbine (GT) based power cycle. The combined production of a medium LHV gas that fuels the GT, and of a high LHV byproduct (tar and/or char) that may contribute to maintain the pyrolysis process, makes pyrolysis highly competitive when compared to gasification. Nevertheless few simulations of such integrated plants are available in literature also because of the lack of general and robust modeling tools for the pyrolysis process.

The difficulty in modeling the pyrolysis process is a consequence of the strong dependence of the equilibrium constants from experimental data being experimental data strongly dependent from the equipment and procedure utilized.

On the other hand, it is well known that the behavior of a generic particle that undergoes a pyrolysis process is dependent from physical quantities such as temperature, pressure, size, heat and time. These dependencies, that may be more interesting than precise data when simulating complex power plants, may be extrapolated from data available in literature.

This paper presents a new approach to modeling pyrolysis of biomass and waste based on a neural network model trained and tested on data available in literature, and taking into account only general inputs, such as fuel composition and temperature, while not referring to specific equipment and procedure. The model outputs pyrolysis products rates of production and their energy content with a sufficient precision making it an interesting and general prevision tool that can be easily integrated and utilized in any power plant simulation software.

INTRODUCTION

Highly regenerated gas turbine based power plants can be easily coupled to gasification devices to obtain high efficiency IGCC plants. This makes it possible to utilize renewable energy sources, such as biomass and waste, at a considerably higher energy conversion efficiency, with respect to traditional incineration methods [1].

Pyrolysis, in particular, is experiencing a favorable conjuncture also because the byproducts of the reaction have an interesting energy content that can be reused in the cycle, for example in providing heat to the reactor, to ameliorate the overall efficiency. This makes it also possible to obtain small and micro scale Integrated Pyrolysis Combined Plants (IPCP) that may satisfy distributed generation demand with residual fuels [2-4].

When evaluating plant performance, however, many problems arise because no general and easily accessible simulation model is available for the pyrolysis process. Actually there are many simulation models for biomass pyrolysis in literature [5-10], but they are based on chemical reactions and kinetics, that provide the yield of pyrolysis products as a function of time, heat rate, biomass size, thermodynamic parameters etc. These information may not be necessary when evaluating pyrolysis as a part of a combined cycle while mass and energy balances in steady state conditions are much more useful. Moreover these models (i.e. the kinetic constants) are strongly dependent from experimental data that are characteristic of the particular procedure, equipment and fuel utilized, therefore may not be extendable to other situations.

On the other hand, many experimental studies have been carried out, and are available in literature, [11-13] to describe the process by evaluating yield, energy content and composition of pyrolysis products on test-benches. This provides scientists with real data to determine, for example, equilibrium and kinetic constant of chemical reactions that are necessary to model the process.

The aim of this work is to demonstrate the possibility of using directly these experimental data in a model-free approach that builds a simple and practical tool to forecast the yield of pyrolysis products and their energy content, given only some descriptive elements of the fuel in input, and of the thermodynamic process.

Such a tool would be extremely useful for technical feasibility studies and thermodynamic optimization of IPCP where no reference to reactor type and dimension, fuel size, heat rate or general process characteristics should restrain the results to particular pyrolysis technologies and methodologies. Moreover, plant performance evaluation and optimization procedures are often referred to the steady state, therefore, the tool will not consider time dependencies.

METHODOLOGY

Traditional and available pyrolysis simulation models are strongly dependent from the above describes variables, therefore a model free approach was followed in which there is no pre-defined knowledge of the phenomenon under study, and where the above mentioned dependencies can become negligible. This can be obtained by means of auto-learning methods, such as Artificial Neural Networks (ANN), which extrapolate dependencies between variables, by numerical regressions on experimental data [14-16].

Forcing inputs and outputs of the ANN model to those quantities that the user is interested on, the regression will determine the configuration that best fits the data. When coming to pyrolysis, this turns out into an ANN that outputs the percentage in weight of pyrolysis products and their energy content, when inputted with the fuel composition and thermodynamic variables.

The possibility of obtaining an efficient modeling tool, while neglecting technological peculiarities and other variable dependencies, should be guaranteed by the choice of experimental data obtained with different technologies and at different conditions. The ANN will therefore extrapolate those macro dependencies that are actually common to every pyrolytic process regardless the technology and procedure used.

Data available in literature from three different laboratory plants were considered to train and test the model as briefly described in Table 1 [11-13]. They were chosen because of the different pyrolysis processes and different residues that were utilized though covering similar temperature ranges. Moreover it is noticeable that data provided are always incomplete and do not allow a mass and energy balance of the process, however missing data are not always the same therefore it is possible that a missing information in a data set can appear in another data set and vice versa. This may be explained as if the whole set of data contains the knowledge of the phenomenon, which is partially described in each experiment. Therefore ANN ability to reproduce a reality even with partial information or missing data will be proven to simulate pyrolysis.

Considering data available in Table 1, and the above mentioned considerations, input quantities to the model were assumed to be descriptive of the residual fuel, in terms of percentage in weight of Carbon (% C), Oxygen (% O), Hydrogen (% H) and Nitrogen (% N) obtained from the ultimate analysis, and of the ash content (% ash), obtained from the proximate analysis. The chemical reaction is described with the pyrolysis temperature (Tpyr) which is another input. The model will output the percentage in weight of gas, tar and char yields and also char HHV and gas LHV. Tar LHV was not simulated because there are too few examples available in the selected literature of Table 1.

Figure 1 shows the architecture of the ANN which has 1 input and 1 output layer, respectively with 6 neurons and 5 neurons as many as the input/output quantities, and two hidden layers of neuron that are responsible for internal modeling.

Table 1: Data description.

	Di Blasi C., et al [12]	Demirbas, A [11]	Li, A. M. et al. [13]
Pyrolysis Test Bench	Stationary packed vertical bed (Φ40 mm) in cylindrical steel reactor and infrared lamps; vacuum pump;	Stationary cylindrical reactor (245 x Φ21 mm) in electric muffle furnace; nitrogen flow;	Rotary kiln(450 x Φ205 mm) horizontal in electrical furnace; friction type seal
Temperature	650-1050 K	550-1150 K	550-850 K
Residue	Wood chips Wheat straw Olive husks Grape residues Rice husks	Hazelnut kernel husk, Olive husk Hazelnut shell Spruce wood Beech wood Corncob	Wood PE plastic Waste tire
Proximate Analysis	yes	yes	yes
Ultimate Composition	yes	yes	yes
Gas Composition/Lhv	yes	yes	yes
Gas Yield	yes	yes	yes
Tar Composition/Lhv	no	no	yes
Tar Yield	yes	yes	yes
Char Composition Hhv	HHV @ 850 K	yes	no
Char Yield	yes	yes	yes
Data Availability	7 cases x 5 species incomplete	7 cases x 6 species incomplete	4 cases x 3 species incomplete
Total Examples	35	42	12
Total	89		
Training Set	75		
Testing Set	14		

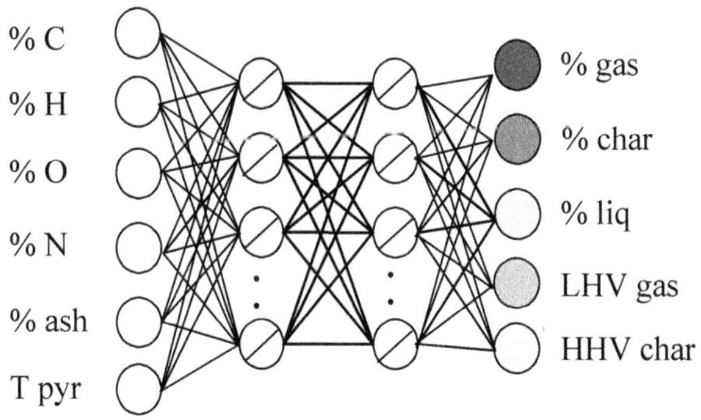

Figure 1: Example of figure caption.

Data from Table 1 were then organized in 89 pairs of input-output vectors that represent the examples that will describe the phenomenon to the ANN, and containing 6 input variables and 5 output variables respectively. The set of pairs was then randomly divided in two subsets. The bigger set (training set) consists of 75 pairs of vectors and it was shown to the ANN in a training phase, in which the values of the connection weights between neurons are continuously varied, in an iterative mode, to minimize the error committed in forecasting the output. A homemade software takes care of missing data in the training set, not varying the weight when it either goes to a missing output or comes from a missing input.

The smaller set (testing set) consists of 14 pairs of vectors and was used to test the training process to avoid curve over-fitting and eventually to determine the overall performance of the ANN (Table 1). The training was suspended either when the SSE reached the goal or when the overfitting appears, that is to say when the NN looses the skill of generalization not improving the performance on the test set while still diminishing the error on the training one and making thus useless any further prosecution of the training.

The ANN used are of the back-propagation type [16] with dynamic variation of the learning rate [17], momentum [18] and choice of the initial weights with the Nguyen-Widrow method [19]. Different ANN architectures were obtained by changing the number of neurons in the hidden layers. They were trained and tested to obtain the best performing one.

Before discussing the results it should be remembered that NN always work with normalized data thus, for this operation, we have used Eq. [1] for each quantity. The performance of the NN has been evaluated with respect to the RMS error defined in Eq. [2].

$$x_{norm}(j) = \frac{x_{real}(j) - \min(x_{real})}{\max(x_{real}) - \min(x_{real})} \qquad [1]$$

$$RMS = \sqrt{\frac{\sum_e \sum_j \left(t_{je} - a_{je}\right)^2}{n_n n_e}} = \sqrt{\frac{SSE}{n_n n_e}} \qquad [2]$$

MAIN RESULTS

Figure 2 shows the results of the training procedure in terms of the RMS committed on the testing set for each output variable, and the RMS committed on global training and testing sets, for every architecture tested. Different architectures are identified with two numbers corresponding to the number of neurons respectively in the first and second hidden layer. The arrow shows the best architecture: it was chosen because it produces the smallest error on the testing set. To be noted that the 10-5 architecture has a higher RMS on the training set, with respect, for example, to the 20-20 architecture that performs also well, but the first has a lower RMS on the testing set, which is considered more representative of the ANN performance.

The overall performance of the ANN is good with a global RMS below 5%, while the error committed on the single variable is around 4% for gas yield and LHV and around 4,5% for char yield and HHV. To be noted that the error committed on LHV and HHV is comparable with the others even though the data set were incomplete for these variables. The error committed on Liquid yield forecasting is slightly above 6% and this is coherent with data availability and completeness for liquid shown in Table 1.

Figure 2 shows the performance of the best ANN (10-5) on single examples and variables together with real data from the testing set. Data provided by the neural model are in good agreement with real data both in trends and in numerical values. The performance on liquid yield is somehow poorer but still sufficiently reliable.

The ANN is performing very well also on the variables with missing data, i.e. char HHV and gas LHV. The results for those examples where there is no possibility to verify the performance, show trends that are however coherent with the chemical reaction giving low LHV of gas for high HHV of char (examples 10-11) and viceversa (examples 8-9). This demonstrates the ability of the ANN model to understand and reproduce the pyrolysis process.

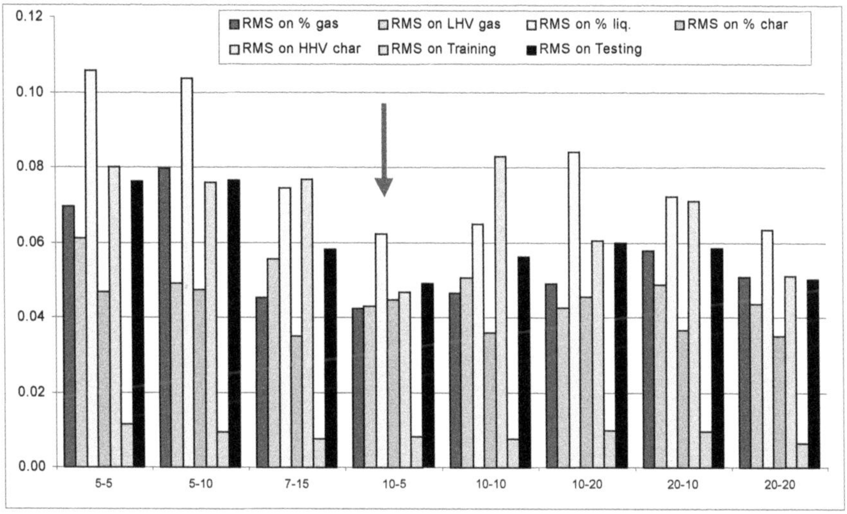

	A 5-5	A 5-10	A 7-15	A 10-5	A 10-10	A 10-20	A 20-10	A 20-20
RMS on % gas	0.06945	0.07981	0.04517	0.04225	0.04647	0.04919	0.05770	0.05084
RMS on LHV gas	0.06097	0.04911	0.05535	0.04290	0.05058	0.04248	0.04885	0.04349
RMS on % liq	0.10576	0.10370	0.07441	0.06227	0.06472	0.08407	0.07213	0.06335
RMS on % char	0.04662	0.04729	0.03490	0.04473	0.03587	0.04559	0.03659	0.03499
RMS on HHV char	0.08000	0.07581	0.07667	0.04658	0.08278	0.06030	0.07082	0.05111
RMS Training set	0.01126	0.00947	0.00753	0.00821	0.00763	0.00978	0.00966	0.00628
RMS Testing set	0.07629	0.07653	0.05798	0.04912	0.05614	0.05977	0.05849	0.05020

Figure 2: Training and testing results.

CONCLUSIONS

This works demonstrates the possibility of using directly experimental data, in a model-free approach, to build a simple and practical tool to forecast the yield of pyrolysis products and their energy content, given only chemical composition of the residual fuel in input, and of the pyrolysis temperature. To this aim different architectures of ANN were built and trained on data available in literature.

Experimental studies, on different residual fuel using different equipments and procedures, however not complete with all the necessary data, were utilized to provide the ANN with a global view of the phenomenon. The results show a good agreement between real and forecasted data also in those cases were missing data were present. In this case the energy content of the products forecasted by the ANN are coherent with a mass and energy balance. The error on real data is always below 6% and often below 5% therefore demonstrating the possibility of extracting general rules that reproduce pyrolysis main outputs even with incomplete or missing information. Future work will refine the model by training the ANN on a wider set of examples obtained from literature and from experimental plants, focusing more precisely on waste derived fuels.

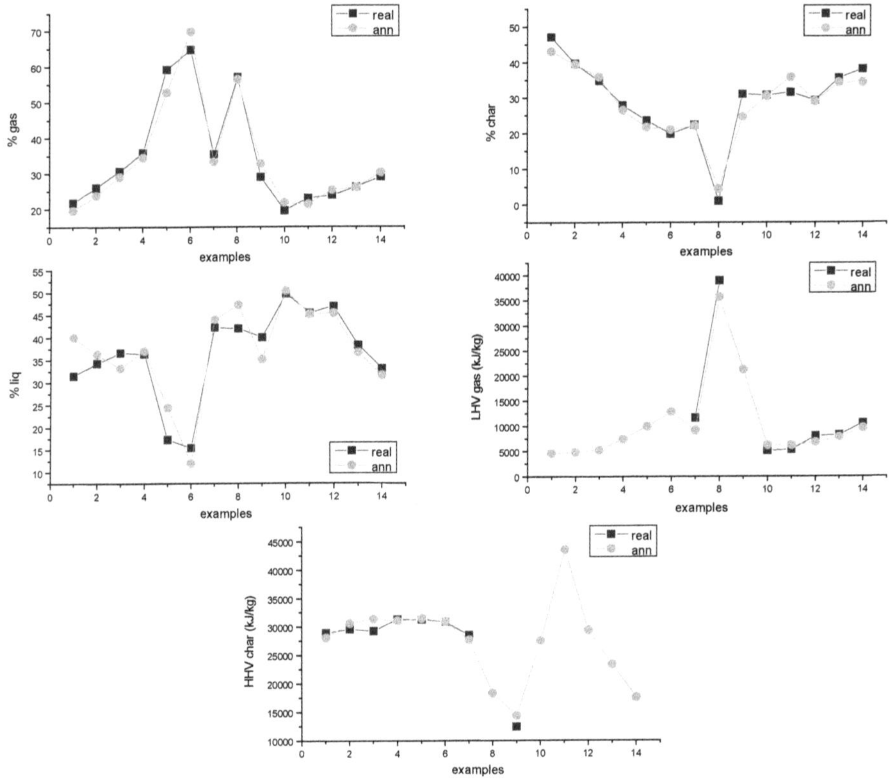

Figure 3: Performance of the ANN in reproducing real data.

ACKNOWLEDGEMENTS

The author would like to acknowledge the contribution of Mr. Bruno D'Alessandro for literature survey, Mr. Daniele Mosca for training the ANN models and Dr. Cinzia Buratti for useful discussions.

NOMENCLATURE

ANN	Artificial Neural Network
C	Carbon
H	Hydrogen
pyr	pyrolysis
IGCC	Integrated Gasification Combined Cycles
IPCP	Integrated Pyrolysis Combined Plants
GT	Gas Turbine
N	Nitrogen
O	Oxygen
RMS	Root Mean Square Error
SSE	Sum Squared Error

T	temperature
a	output value
x	generic variable
j	generic value
n	number
t	target value

SUBSCRIPTS

e	example
j	neuron index
n	neurons
norm	normalized
real	to be normalized

REFERENCES

1. Bridgwater A.V. (1995) *"The technical and economic feasibility of biomass gasification for power generation"*. Fuel, 74 p.631-653
2. Fantozzi, F. and D'alessandro, B., (2002), "IPCP – Integrated Pyrolysis Combined Plant – Gas Turbine and externally heated Rotary Kiln as a biomass and waste to energy conversion system. Influence of thermodynamic parameters", presented *to Proceedings of the Institution of Mechanical Engineers, Part A: Journal of Power and Energy,*
3. Fantozzi F., Di Maria F., Desideri U., (2001) "Micro-Turbine Fuelled By Pyrolysis Gas Thermodynamic Analysis", proceedings of the POWERGEN EUROPE conference held in Helsinky June 2001;
4. Fantozzi F., Di Maria F., Desideri U., (2002), "Integrated Micro-Turbine and Rotary Kiln Pyrolysis as a waste to energy solution for a small town in central Italy - Cost Positioning and global warming assessment ", ASME paper GT-2002-30652
5. Orfao, J.J.M., Antunes, F.J.A. and Figueiredo, J.L., (1999), "Pyrolysis kinetics of lignocellulosic materials – three independent reactions model", Fuel 78, p. 349-358
6. Liang, X. H. and Kozinsky, J. A., (2000), "Numerical modeling of combustion and pyrolysis of cellulosic biomass in thermogravimetric systems" Fuel 78, p . 1447-1486
7. Pletka, R., Brown, R. C. and Smeenk, J., (2001), "Indirectly heated biomass gasification using a latent heat ballast. Part 2: modeling", Biomass and Bioenergy 20 p. 307-315
8. Kastanaki, E., Vamvuka, D., Grammelis, P., and Kakaras, E., (2002), "Thermogravimetric studies of the behavior of lignite-biomass blends during devolatization", Fuel Processing Technology 77-78, p. 159-166
9. Bryden, K. M., Ragland, K. W and Rutland, C. J., (2002), "Modeling thermally thick pyrolysis of wood", Biomass and Bioenergy 22 41-53
10. Liliedhal, T. and Sjostrom, K., (1996), "Modelling of char-gas reaction", Fuel 76, p. 29-37
11. Demirbas, A. (2001)"Carbonisation ranking of selected biomass for charcoal, liquid and gaseous products". Energy Conversion and Management, 42 p.1229-1238
12. Di Blasi C., Signorelli G., Di Russo C., Rea G. (1999)*"Product distribution from pyrolysis of wood and agricultural residues"*. Ind. Eng. Chem. Res, 38 p.2216-2224.
13. Li, A.M. Li, X.D. Ren, Y. Chi Y., Yan J.H, Cen K.F. (1999) "*Pyrolysis of solid waste in a rotary kiln: influence of final pyrolysis temperature on the pyrolysis products*". Journal of Analytical and Applied Pyrolysis, 50 p.149-162.
14. Guo, B., Shen, Y., Li, D., Cui, L., (1997), "Modeling coal gasification with a hybrid neural network", Fuel 76, 1159-1164

15. Leib, T. M., Mills, P. L., Lerou, J. J., (1996), "Fast response distributed parameter fluidized bed reactor model for propylene partial oxidation using feed-forward neural network methods", Chem. Engng. Sci. 51, 2189-2198
16. Guo, B., Dingkai, Li., Congming, C., Zi-an, L., and Youting, S., (2001) "Simulation of biomass gasification with a hybrid neural network model", Bioresource Technology 76 (2001) 77-83
17. Rumelhart, D., E. and McClelland, J., L., (1990), *"Parallel Distributed Processing,"* MIT Press, Cambridge, USA.
18. Vogl, T., P., et al, 1988, "*Accelerating the Convergence of the Back-Propagation Method,*" Biological Cybernetics, Vol. 59, pp257-263.
19. Dayoff, J., 1990, *"Neural Network Architectures. An Introduction,*" Van Nostrand Reinhold.
20. Nguyen, D. and Widrow, B., (1990), "Improving the Learning Speed of a 2-layer Neural Network by Choosing Initial Values of the Adaptive Weights," *Proceedings, International Joint Conference of Neural Networks*, Vol. 3 pp. 21-26.

The Influence of a Preoxidation Stage on Pyrolysis of Alkaline Black Liquor from Straw

S. García; G. Gea; M.B. Murillo; J. Arauzo
Chemical and Environmental Engineering Department, University of Zaragoza
Centro Politécnico Superior, María de Luna, 3, 50018 Zaragoza, Spain

ABSTRACT: Black liquor is an industrial biomass waste, which is considered as an important fuel in pulp and paper manufacturing countries. Nowadays, black liquor is burned in Tomlinson-type recovery boilers, which have several weaknesses: relatively low energy efficiency and some safety and environmental problems. Therefore, new alternative processes, such gasification at low temperature, are currently being investigated. Some studies were carried out at the University of Zaragoza in order to evaluate the feasibility of black liquor gasification at low temperature in a fluidized bed reactor. Two main problems were found during the tests: agglomeration of the bed as consequence of swelling and plastic properties of black liquor, and low gas yield because of low char reactivity at low temperature.
Based on some studies about the influence of a low temperature oxidation stage (preoxidation) on the coal plastic properties and char reactivity, a previous work was carried out in order to evaluate the influence of a preoxidation stage on the black liquor at the University of Zaragoza. The results seemed to be important improvements for the development of new fluidized bed gasifiers for black liquor gasification.
The aim of the present work is to analyse experimentally the effect of a preoxidation stage on the pyrolysis rate of soda straw black liquor by means of a thermo gravimetric system.

INTRODUCTION

Black liquor is a biomass waste generated as a by-product in the production of pulp and paper. Black liquor contain organic components and a large amount of inorganic chemicals dissolved during the digestion of wood, straw, or other biomass fibre source. The organic solids are composed mainly of degraded alkali lignin and polysaccharide degradation products (aliphatic carboxylic acids), together with a minor fraction of extractives.

Due to its high organic matter content black liquor is an industrial residue which is considered as an important fuel in pulp and paper manufacturing countries[1]. Nowadays, black liquor is burned in Tomlinson-type recovery boilers to generate energy from the organics and to recover the cooking chemicals. Although this technology has been used almost exclusively in alkaline pulp mills for nearly 70 years, it has several weaknesses: relatively low energy efficiency and some safety and environmental problems.

Therefore, new alternative processes, based on gasification, are currently being investigated. These could be more energy efficient, cheaper, safer and easier to control than conventional recovery boilers. One of the new alternative processes could be gasification at low temperature, below inorganics melting point (around 750 °C). Fluidized bed systems seem to be appropriate for that gasification technology[2]. Some studies were carried out at the University of Zaragoza in order to evaluate the feasibility of soda straw black liquor gasification in a fluidized bed reactor[3, 4]. Two main problems were found during the tests: agglomeration of the bed as consequence of swelling and plastic properties of black liquor, and low gas yield because of low char reactivity at temperature below melting point.

There are some studies in the literature about other materials, like bituminous coals, which are focused in the influence of a previous oxidation stage on the coal plastic properties and char reactivity. In all of them, the preoxidation stage results in lower swelling capacity and enhancement of the char gasification reactivity[5-9]. Based on that literature, a previous work was carried out at the University of Zaragoza in order to evaluate the influence of a preoxidation stage at low temperature (180 °C) on the black liquor swelling and on specific surface area of the black liquor char[10]. The resulting black liquor char showed a clear decrease in swelling level, as well as, an increase on their specific surface area, which would enhance the black liquor char reactivity. Both results seem to be important improvements for the development of new fluidized bed gasifiers for black liquor gasification.

The aim of the present work is to analyse the effect of a preoxidation stage on black liquor pyrolysis rate. It was demonstrated that the preoxidation stage increases the specific surface area of black liquor char[10], here the expected variation in the black liquor pyrolysis rate, which could be related to the enhancement in the specific surface area, has been checked by thermogravimetric experiments. It is also known that the preoxidation treatment decreased the swelling level of black liquor[10], in this work, scanning electron microscope (SEM) analysis has been performed to confirm if the black liquor thermoplastic properties have disappeared completely.

The two main variables that determined the preoxidation level of black liquor are time and temperature of preoxidation. Several runs have been performed in a thermobalance varying time (4-20 h) and temperature (160-200 °C) of preoxidation in order to evaluate the influence of both variables on the black liquor pyrolysis rate. The statistical study of the data obtained will indicate which of both variables influences more on the thermochemical behaviour of black liquor.

The results showed that the preoxidation pre-treatment influences the pyrolysis rate of the black liquor. Thermal degradation rate of the inorganic fraction of the non-preoxidized black liquor is lower than the preoxidized ones.

EXPERIMENTAL

MATERIAL

The raw material used in this work was black liquor produced from the soda pulping of straw. The ultimate and proximate analyses for this alkaline black liquor are listed in Table 1. The elemental analysis was obtained in a CHNS Carlo Erba elemental analyser (mod. EA1108). The proximate analysis was carried out following the standard norms, ISO 1171 for ashes and ISO 5623 for volatiles. The metal analysis was performed by atomic absorption after previous alkaline fusion in accordance with norm ASTM D368296.

Table 1: Ultimate and Proximate analysis (dry basis) for the soda straw black liquor.

Elements	wt %
C	39.05
H	4.54
N	1.00
S	0.78
Cl	3.50
K	4.10
Na	8.83
Si	0.23
Others	37.97

Components	wt %
Ashes	20.63
Volatiles	65.98
Fixed Carbon	13.39

For the present work, the black liquor, which is 90 % water in its original state, was dried to 100 % solid content. First, the black liquor was concentrated to 45 % in a heating plate with continuous stirring, and afterwards, it was completely dried in an oven at 105 °C. After drying, the black liquor solids were grounded and sieved in a 53 μm particle size.

EXPERIMENTAL PROCEDURE

The runs performed in the present study were carried out in a Cahn TG-131 (TGA) thermobalance (schematic shown in Figure 1).

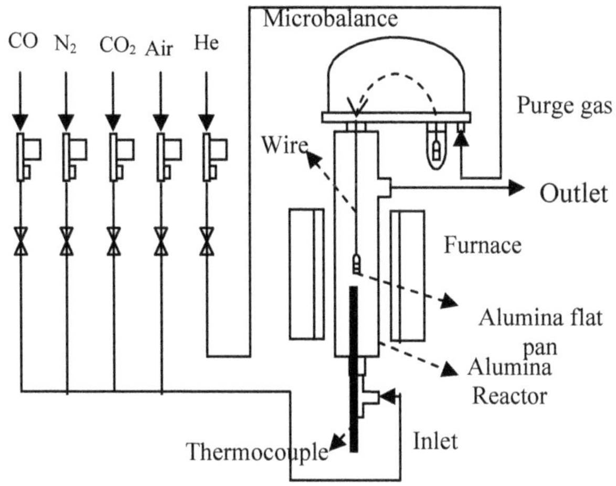

Figure 1: TGA set up with gas supply system.

The runs consisted of two different stages, both performed consecutively in the thermobalance. The first stage was a preoxidation with air at low temperature, the second one was the pyrolysis in a N_2 atmosphere. Figure 2 illustrates a scheme of these two stages followed during the experiments.

The sample used for all the experiments was approximately 5 mg of dried black liquor. Once the sample was added to an alumina pan, the thermobalance was purged with 150 cm^3NTP/min of N_2 during 30 minutes. 150 cm^3 NTP/min was the chosen flowrate for the study, it was maintained constant for the whole run. After purging, the furnace temperature was raised from 25 °C to the pre-chosen preoxidation temperature in a N_2 atmosphere, maintaining the inert atmosphere for 30 minutes to allow the sample weight stabilize. The preoxidation temperature (T_p) is one of the studied variables, it was varied from 160 °C to 200 °C. Then, the preoxidation stage started by changing N_2 by air flowrate maintaining the preoxidation temperature for the preoxidation time. The preoxidation time (t_p) is the second studied variable, it was varied from 4 h to 20 h. Once the preoxidation finished, the air was changing by N_2 and the pyrolysis stage began. The furnace temperature was raised from the preoxidation temperature to 750 °C at a heating rate of 5 °C/min during the pyrolysis stage. Once the final pyrolysis temperature (750 °C) was reached, it was maintained for 2 hours to allow the sample weight stabilize. The chosen final pyrolysis temperature was 750 °C for all the runs in order to work below the melting point of the inorganic compounds of the black liquor.

Figure 3 shows a typical loss of weight curve for a preoxidation-pyrolysis run.

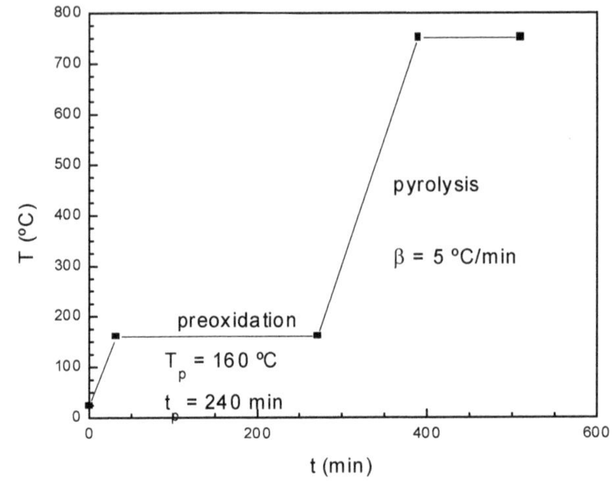

Figure 2: Scheme of the two stages followed during the experiment performances.

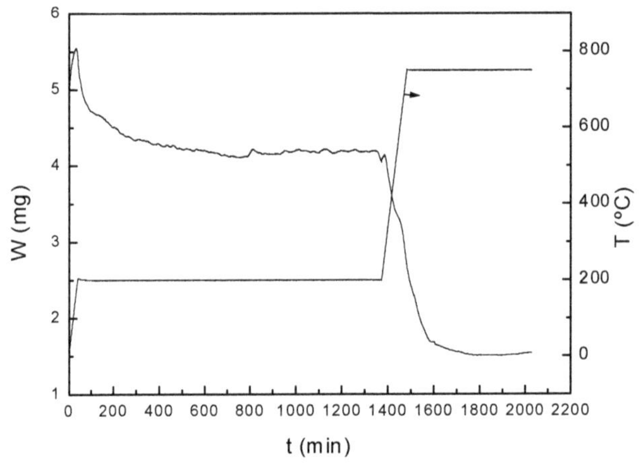

Figure 3: Loss of weight curve for a typical preoxidation-pyrolysis run. T_p = 200 °C, t_p=20 h

A factorial experimental design was used in the present work in order to perform the minimum number of runs required to evaluate if the two main variables, which determine the preoxidation level (T_p and t_p), influenced on the black liquor pyrolysis rate. The binomial design used (2 variables and 2 levels) is represented in Figure 4. Table 2 shows the summary of the experiment performed. For statistical purposes each experiment was duplicated.

Table 2: List of the resulting experiments from the factorial design.

Experiment	A	B	C	D
T_p (°C)	160	200	160	200
t_p (h)	20	20	4	4

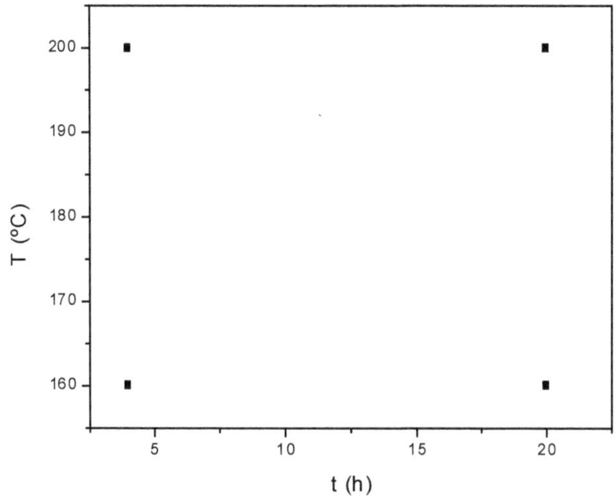

Figure 4: Scheme of the experimental design.

RESULTS

Previous works related to coal have shown that an air preoxidation stage of some types of coals increased both, the specific surface area of the final active coals and the gasification and combustion reactivity of the preoxidized chars[8, 9]. Due to the similarities between coal activation and black liquor gasification the same enhancement is expected for the black liquor. Puértolas et al.[10] observed that the specific area of the char produced from preoxidized black liquor was larger that non-preoxidized ones. However, there are not data about the effect of the preoxidation stage on the pyrolysis rate of the black liquor. The main objective of the present work was to analyse the influence of a preoxidation step at low temperature on the black liquor pyrolysis rate. For that reason the variation of the pyrolysis rate with time for the non-preoxidized black liquor pyrolysis is compared with the preoxidized black liquors in Figure 5. The preoxidation stage of the run showed in Figure 5 was performed at T_p = 160 °C for 4 h.

Pyrolysis rate (r) was defined as the variation of the solid conversion with time during the pyrolysis stage, according to the equation 1,

$$r = \frac{dX_s}{dt} \qquad 1$$

solid conversion (X_s) during the pyrolysis stage was defined by equation 2:

$$X_s = \frac{W_0 - W}{W_0} \qquad 2$$

where W_0 was the initial solid weight, and W the solid weight remaining at time t. In figure 5, the time t = 0 represents the beginning of the pyrolysis stage, without considering the preoxidation step in order to get clarity in the illustration. The decomposition rate was negligible during the preoxidation step.

By observing Figure 5 it can be detected 2 maximum values of the pyrolysis rate for the black liquor. The first maximum belongs to the pyrolysis of the organic matter of the black liquor, which takes place at temperatures below 550 °C and the second maximum belongs to the thermal decomposition of the inorganic matter of the black liquor, which takes place at temperatures above 600 °C[11].

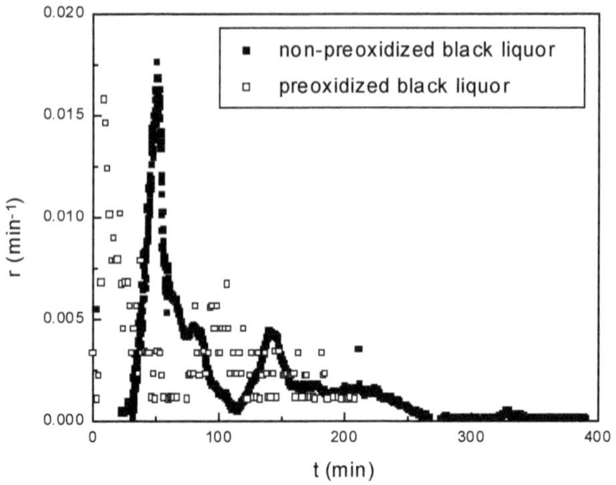

Figure 5: Variation of the pyrolysis rate with time for a) non-preoxidized black liquor and b) preoxidized black liquor at $T_p = 160°C$ and $t_p = 4$ h.

The maximum value of the pyrolysis rate for the organic matter fraction of the non-preoxidized black liquor ($r = 0.017$ min^{-1}) was approximately 5% higher than the rate for the organic matter of the preoxidized one ($r = 0.016$ min^{-1}). However, the maximum value of the thermal decomposition rate for the inorganic fraction was 35 % higher for the preoxidized black liquor ($r = 0.007 min^{-1}$), which indicated that the preoxidation step could provoke some structural changes in the black liquor matrix, which increased the degradation of its inorganic matter. For the sake of clarity, in Figure 5, only one of the runs listed in Table 2 is showed, but the maximum value of the decomposition rate of the inorganic fraction for the preoxidized black liquor was higher than non-preoxidized one for all the performed experiments. However, the pyrolysis rate maximum value for the organic fraction was lower for the preoxidized samples. The variation of solid conversion during the pyrolysis stage for non-preoxidized and preoxidized black liquors ($T_p = 160$ °C and $t_p =$ 4h) is presented in Figure 6. As can be observed, the preoxidation step did not influence the solid conversion. The final solid conversion did not practically vary with the preoxidation step.

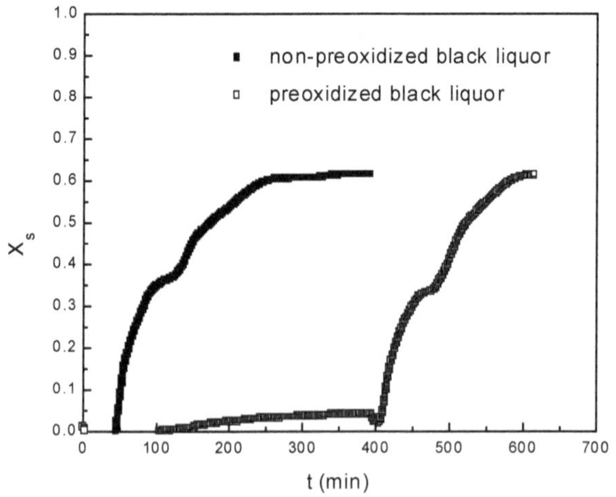

Figure 6: Variation of the solid conversion during pyrolysis step with time for a) non-preoxidized black liquor and b) preoxidized black liquor at $T_p = 160°C$ and $t_p = 4$ h.

The preoxidation level is mainly determined by two variables: preoxidation temperature and time. The second objective in this work was to figure out if both variables influenced the pyrolysis rate, and which one more. For that, the experiments listed in Table 2, which are based on a statistical experimental design, were performed. The variable defined as the response or independent variable in the design was the maximum value of the pyrolysis rate reached in every experiment. As it has been pointed out before, the black liquor pyrolysis consisted of two different thermal degradations (organic and inorganic matter fractions), therefore there are two maximum rate values in every experiment, one for the organic matter degradation and the second one for the inorganic matter. In the present work, the influence of preoxidation temperature and time on both thermal degradation steps was analysed separately.

Table 3 shows the maximum pyrolysis rate values (response variables) obtained for all the experiments.

Table 3: Maximum pyrolysis rate values obtained during the thermal degradation of the organic and inorganic matter fractions of the black liquor.

Run	Average maximum pyrolysis rate for the organic matter fraction (min^{-1})	Average maximum pyrolysis rate for the inorganic matter fraction (min^{-1})
A	0.01628	0.0071
B	0.01132	0.0077
C	0.00978	0.0067
D	0.01674	0.0067

A variance analysis was applied to the data presented in Table 3 to determine if the variables, which are preoxidation temperature and time, influenced both thermal degradation rates, as well as if there was any interaction between both variables. The results obtained from statistical F-test are listed in Table 4. These results showed that, preoxidation temperature and time affect the maximum pyrolysis rate value of the organic matter fraction at the 95 % confidence level (F-value tabulated is lower than F_0-value calculated). Besides, statistically there was an interaction effect between both variables at the 95 % confidence level, which means that the influence of preoxidation temperature and time on the pyrolysis rate of the organic matter cannot be studied separately. However, there was not effect of the variables on the maximum decomposition rate value of the inorganic fraction. This decomposition rate increased with a preoxidation stage, but this increase does not vary with the preoxidation conditions in the range of preoxidation temperature and time studied.

A linear regression model was calculated to show the effect of both variables on the maximum pyrolysis rate value for the organic matter fraction. This model is expressed by the equation 3:

$$\hat{y} = 0.010505 + \left(\frac{-0.00563}{2}\right)x_1 + \left(\frac{0.000825}{2}\right)x_2 + \left(\frac{0.000955}{2}\right)x_1x_2 \qquad 3$$

where,

y = response variable = maximum pyrolysis rate value for the organic fraction.

$$x_1 = \frac{T_p - (160+200)/2}{(200-160)/2} \qquad 4$$

$$x_2 = \frac{t_p - (4+20)/2}{(20-4)/2} \qquad 5$$

x_1 and x_2 are respectively are related to T_p (°C) and t_p (h) by the limits chosen for both variables. The superior limit for T_p was 200 °C and the inferior 160 °C, for t_p the superior limit was 20 h and the inferior 4h.

Table 4: F-test values calculated.

Factors	F_0 value for the maximum pyrolysis rate for the organic matter fraction	F_0 value for maximum pyrolysis rate for the inorganic matter fraction
Preoxidation temperature	970.44	2.47
Preoxidation time	20.84	4.18
Temperature-time interaction	27.92	4.18
F value tabulated at the 95 % confidence level	7.71	7.71

It is important to note in the equation 3 that the effect of the preoxidation temperature and time on the pyrolysis rate of the organic fraction was substantially different. An increase in the preoxidation time always supposed an enhancement in the pyrolysis rate, however an increase in the preoxidation temperature could result in decreasing the pyrolysis rate for low preoxidation time values. According to the coefficients values in the equation 3 and the limits of the variables chosen in this work, it can be said that the preoxidation time influenced the pyrolysis rate of the organic matter fraction of the black liquor more than the preoxidation temperature did.

The preoxidation step affects the surface area of the black liquor char formed after pyrolysis and the swelling behaviour of the black liquor as was showed in a previous work[10]. Here, in the present study, it was observed an influence on the black liquor pyrolysis rate. Therefore, an effect of the preoxidation step on the morphology of the char formed after pyrolysis can be expected. A scanning electron micrograph of the original black liquor is shown in Figure 7A, and in 7B a micrograph of the char produced after the pyrolysis of the non-preoxidized black liquor. An agglomerate of the particles was formed upon pyrolysis. The preoxidation of the black liquor at 180 °C for 4 h slightly decreased the agglomeration of the char, but did not destroy it completely (Figure 7C). By comparing Figures 7D and 7E it can be appreciated how the preoxidation step changed the morphology of the black liquor char. Figure 7D illustrates a char particle produced from the non-preoxidized liquor and Figure 7E a char particle from the preoxidized black liquor.

CONCLUSIONS

A preoxidation stage at low temperature of the soda straw black liquor affected its pyrolysis rate. The black liquor pyrolysis can be divided in 2 steps. The first one, which consists of thermal degradation of the organic fraction of the black liquor, at temperatures below 550 °C, and the second one, which consists of the thermal degradation of the inorganic fraction, at temperatures above 600 °C. Both steps were influenced by the preoxidation stage.

The pyrolysis rate of the organic fraction decreased with the preoxidation, although the percentage of decreasing depended on the preoxidation conditions. According to the statistical analysis performed in the study, both variables, which are preoxidation temperature and time, affected the pyrolysis rate of the organic matter. This pyrolysis rate increased when the preoxidation time increased. However, the preoxidation temperature could enhance or reduce the pyrolysis rate depending on the preoxidation time. This fact is due to the interaction effect between preoxidation temperature and time, as can be observed in the regression model showed in the equation 3. The effect of variables, T_p and t_p, cannot be studied separately. By comparing the effect of both variables, it can be said that the effect of the preoxidation time on the pyrolysis rate of the organic fraction was higher than the effect of the preoxidation temperature.

Regards the thermal degradation rate of the inorganic fraction is of interest to note that increased with the preoxidation stage, however this increase was not influenced by the preoxidation temperature nor time.

By means of scanning electron micrographs it can be observed that the agglomeration of the black liquor char formed after pyrolysis was slightly reduced by a preoxidation stage but it did not completely destroy. Besides, the morphology of the char particles was noticeably changed by a preoxidation step.

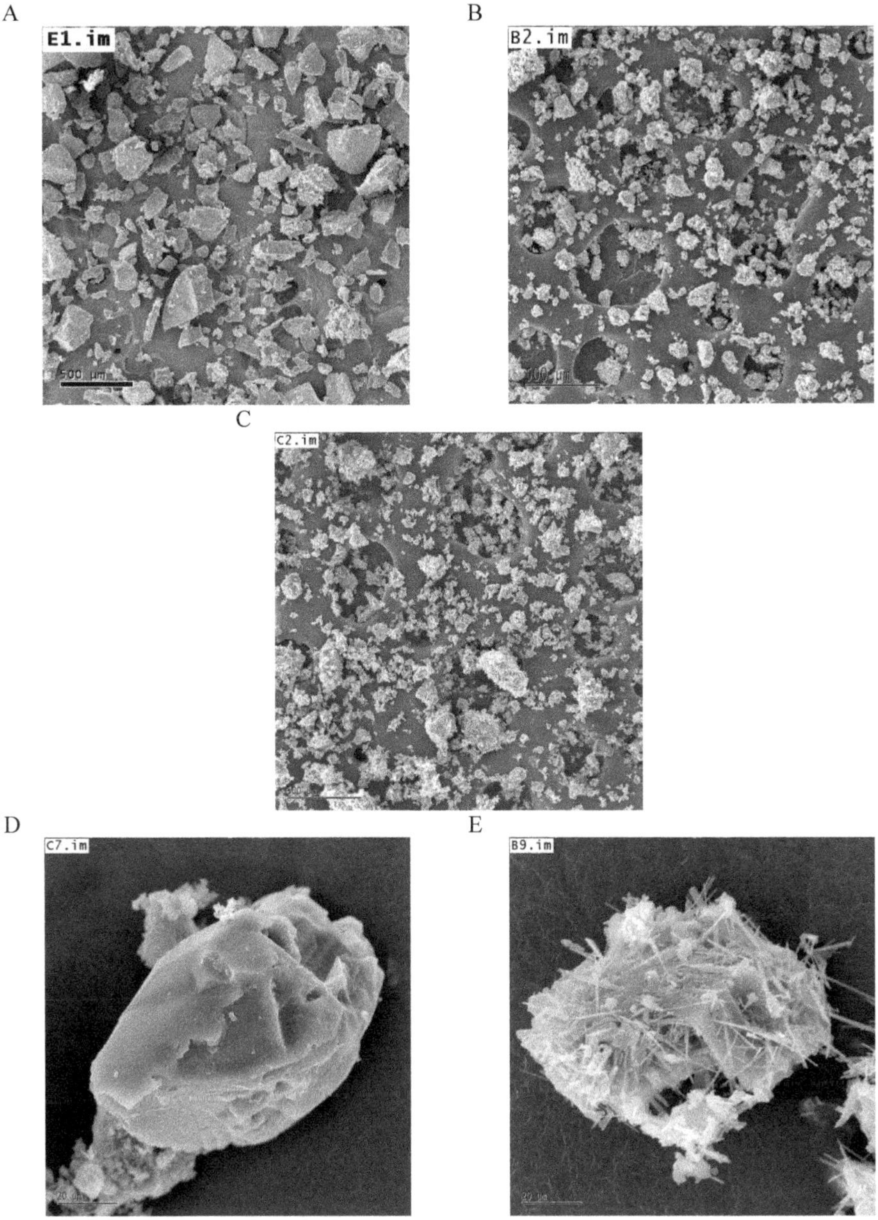

Figure 7: Scanning electron micrographs of black liquor and black liquor chars after pyrolysis (at 5°C/min at 750 °C in N_2). A) Original black liquor sample (scale 500 μm); B) char from non-preoxidized black liquor (scale 500 μm); C) char from preoxidized black liquor at 180 °C for 4h. (scale 500 μm); D) char from non-preoxidized black liquor (scale 20 μm); E) char from preoxidized black liquor at 180 °C for 4h (scale 20 μm).

NOMENCLATURE

r: variation of the solid conversion with time during the pyrolysis stage (min^{-1}).
T_p: preoxidation temperature (°C).
t_p: preoxidation time (h).
X_s: solid conversion.
W_0: initial solid weight (mg).
W: solid weight remaining a time t (mg).

REFERENCES

1. Schricharoenchaikul, V.; Phimolmas, V.; Frederick, W.J.; Grace, T.M. (1998) Pyrolysis of Kraft Black Liquor: Formation and Thermal Conversion of Volatile Products and Char. J. Pulp Paper Sci. 24(2), J43.
2. Berglin, N.; Persson, L.; Berntsson, T. (1996) Energy System Options with Black Liquor Gasification. Pulp & Paper Canada. 97(5), 19.
3. Sánchez, J.L. (1999) Thermochemical Processing at low temperature of Black Liquors from Alkaline Pulping of Straw. Ph. D. Dissertation, University of Zaragoza, Zaragoza, Spain.
4. Gea, G. (2000) Thermochemical Behaviour of Black Liquors from Alkaline Pulping of Straw. Ph. D. Dissertation, University of Zaragoza, Zaragoza, Spain.
5. Mahajan O.P., Komatsu, M.; Walker, P.L. (1980) Low-Temperature air oxidation of caking coals. Fuel. 59, 3.
6. Maloney, D.J.; Jenkins, R.G. (1985). Influence of coal preoxidation and the relation between char structure and gasification potential. Fuel. 64, 1415.
7. Tromp, P.J.J.; Kapteijn, F.; Boon, J.J.; Moulijn, J.A. (1987) Low-temperature pre-oxidation of coals. The effect on the char reactivity. International Conference on Coal Science. Amsterdam, Netherlands. 537.
8. Serrano-Talavera, B.; Muñoz-Guillena, M.J.; Linares-Solano, A.; Salinas Martinez de Lecea, C. (1997) Activated Carbons from Spanish coals. 3. Preoxidation effect on Anthracite activation. Energy & Fuel. 11, 785.
9. Pis, J.J.; Parra, J.B.; De la Puente, G.; Rubiera, F.; Pajares, J.A. (1998) Development of macroporisty in activated carbons by effect of coal preoxidation and burn-off. Fuel. 6, 625.
10. Puértolas, R.; Gea, G.; Murillo, M. B.; Arauzo, J. (2001) Pyrolysis of Black Liquors from Alkaline Pulping of Straw. Influence of a Preoxidation stage on the Char Characteristics. J. of Analytical and Applied Pyrolysis. 58-59, 955.
11. Gea, G.; Murillo, M.B.; Arauzo, J. (2002) Thermal degradation of alkaline black liquor from straw. Thermogravimetric study. Ind. Eng. Chem. Res. 41, 4714.

Scale Effects And Distribution Problems In Fluid Bed Fast Pyrolysis Using CFD Models Integrated With Reaction Kinetics

H. Gerhauser and A.V. Bridgwater
Bio-Energy Research Group, Aston University, Birmingham, B4 7ET, United Kingdom

ABSTRACT: Scale-up of fluidised beds used for bio-oil production via fast pyrolysis is a complex task, which has received only limited attention in the literature. This paper describes the integration of pyrolysis reaction kinetics into a CFD (computational fluid dynamics) model of the reactor's hydrodynamics. The reaction kinetics are based on a four step reaction scheme that includes primary and secondary pyrolysis reactions. Promising results are obtained, which, however, could not yet be subjected to validation. Notably, the model suggests that the interaction of the hydrodynamics with the reaction kinetics can be of considerable importance. Recirculation patterns within the fluidised bed are found to have the potential to significantly lengthen pyrolysis vapour residence times and therefore deleteriously affect yields due to enhanced secondary cracking. The pyrolysis reaction is found to be substantially completed within about 10 cm of the feed-point. This leads to the potential for mal-distribution of vapours, which is considerably enhanced by larger ratios of bed diameter to bed height and is also affected by scale.

INTRODUCTION

Renewable fuels in a liquid form are widely considered to be particularly desirable. Bio-oil obtained from fast pyrolysis qualifies for this need. It still suffers, however, from some perceived and some real quality problems. Furthermore, the process needs to have a certain scale, in order to be economical. On the other hand, transportation costs often tend to put an upper limit on plant size, which is generally determined by the amount of biomass available within about 50 km of the plant[1].

Particularly promising sites are those where copious amounts of biomass are produced as a waste product and currently burned to produce process heat and electricity through the low efficiency Rankine cycle, for example paper mills, ethanol plants or similar operations. In addition to raising the efficiency of electricity production in such co-generation schemes, surplus bio-oil could be stored to provide electricity and/or heat in the off-season on site or near to the site, or alternatively could be transported to demand centres further away. Transportation would be considerably easier than for virgin biomass, as the density of bio-oil is around 1.2 kg/l and therefore much higher than for solid biomass, the density of which is commonly around 0.1 to 0.4 kg/l.

To meet these opportunities, pyrolysis processes need to be scaled up to around 100 metric tonnes per day; or more, if feedstock availability and technical barriers to scale-up allow greater throughputs. At the same time, quality concerns need to be met to be able to take advantage of the higher efficiencies offered by Diesel engines or combined cycle. Scale up is inherently fraught with risks, as the hydrodynamics of the scaled up system may be difficult to predict. This is well illustrated with the failure of the Hydrocol natural gas to synthetic gasoline process, where the inability to replicate laboratory yields of around 90% in the scaled up reactor led to the eventual abandonment of the project after considerable amounts of capital had been lost[2].

The traditional way industrial fluidised beds are in general designed is by a combination of correlations and laboratory experimentation with the latter commonly relying on rules of similarity to study the hydrodynamics in cold flow models[2].

A much more recent tool is CFD, computational fluid dynamics. Here, the reactor is subdivided into a large number of small computational cells / control volumes. Mass, momentum and energy balances are then performed for each cell. With thousands of cells (typically between 5000 and 20000 for the work here reported) and interaction between them this is of course a monumental computing task. It does, however, give values for pressure, temperature and velocity in particular locations, and as such it gives a very visual picture. It enables the designer to look at the impact of local differences and flow patterns. Correlations, on the other hand, tend to give estimates of the most important variables averaged over the whole reactor; they also tend to have a very limited range of applicability, which is in stark contrast to CFD, which relies on first principles and is therefore capable of dealing with a wide variety of geometries and materials.

While modelling of the hydrodynamics is required to obtain the necessary reactor dimensions and fluidising gas requirements for a given throughput, a knowledge of the reaction kinetics helps with optimising reactor temperatures to maximise liquid yields and with assessing the impact of using a different feedstock.

MODELLING STRATEGY EMPLOYED

The hydro-dynamics of the fluidised bed are modelled with the help of computational fluid dynamics. A more detailed explanation of the underlying theory has been given in an earlier paper[3], which will only be summarised very briefly here.

Both phases are modelled as continua, which is known as a Eulerian-Eulerian as opposed to a Eulerian-Lagrangian approach. Instead of considering individual particles in other words, the particles are treated as if they were a fluid. Interaction between the two phases, that is the solid particles and the mixture of pyrolysis vapours and inert fluidising gases surrounding them, is taken care of via inter-phase transfer terms.

In the earlier paper[3] the temperature of the reactor was not considered. The relevant transport equation for energy is:

$$\frac{\partial}{\partial t}\left(r_\alpha \rho_\alpha H_\alpha\right)+\nabla \bullet \left(r_\alpha\left(\rho_\alpha \mathbf{U}_\alpha H_\alpha-\lambda_\alpha \nabla T_\alpha\right)\right)$$
$$=\sum_{\beta=1}^{N_P} c_{\alpha\beta}{}^{(h)}\left(T_\beta-T_\alpha\right)+\sum_{\beta=1}^{N_P}\left(\dot{m}_{\alpha\beta} H_\beta-\dot{m}_{\beta\alpha} H_\alpha\right)$$

The reaction scheme employed in the modelling (see Figure 1) is loosely based on Broido-Shafizadeh[4]. This is a widely employed type of scheme, which is a useful first approximation to biomass fast pyrolysis kinetics, even though it is originally based on data obtained in relatively low temperature slow pyrolysis of cellulose[5].

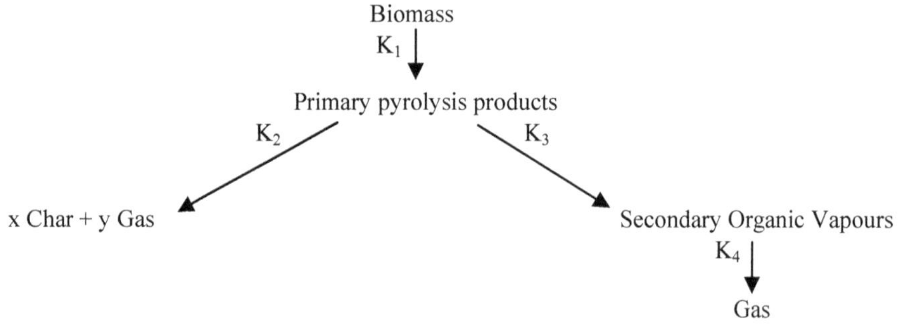

Figure 1: Illustration of reaction scheme.

As discussed in more detail at the end of the paper, the scheme needs to be capable of accounting for the time distribution and temperature dependence of both vapour evolution from the pyrolysing particles and of secondary cracking reactions of the formed vapours. Ideally the model should be based on parameters determined experimentally at bench scale.

The values of the kinetic parameters used in the code are based on work by Bradbury[4] and Liden[6] and are summarised in Table 1.

Table 1: Kinetic parameters.

Parameter	Value adopted
E_1	$242.8*10^6$ J kmol^{-1}
E_2	$150.7*10^6$ J kmol^{-1}
E_3	$196.8*10^6$ J kmol^{-1}
E_4	$108.0*10^6$ J kmol^{-1}
A_1	$2.8*10^{19}$ s^{-1}
A_2	$1.3*10^{10}$ s^{-1}
A_3	$3.2*10^{14}$ s^{-1}
A_4	$2.6*10^{6}$ s^{-1}
x (see Figure 1)	0.35
y (see Figure 1)	0.65

Biomass, active cellulose and char were implemented as scalars within the solid phase, while organic vapours and gas were taken to be components of the gas/vapour phase. Scalars are equivalent to concentrations or mass fractions. They share the same temperature, velocity and pressure, but can diffuse along a concentration gradient. Sand was not implemented as a particular scalar. Instead it is the assumed background fluid for the solid phase.

The changes in mass fractions resulting from reaction in a particular control volume during the course of a particular time step were calculated as follows for the solid phase components:

$$X_{t+\Delta t}^{biomass} = X_t^{biomass} - X_t^{biomass}\left(A_1 \exp\left(\frac{E_1}{RT}\right)\right)\Delta t$$

$$X_{t+\Delta t}^{active} = X_t^{active} + X_t^{biomass}\left(A_1 \exp\left(\frac{E_1}{RT}\right)\right)\Delta t - X_t^{active}\left(A_2 \exp\left(\frac{E_2}{RT}\right)\right)\Delta t$$
$$- X_t^{active}\left(A_3 \exp\left(\frac{E_3}{RT}\right)\right)\Delta t$$

$$X_{t+\Delta t}^{char} = X_t^{char} + xX_t^{active}\left(A_2 \exp\left(\frac{E_2}{RT}\right)\right)\Delta t$$

The evolution of vapours and gas resulting from the decomposition of active cellulose is modelled with the help of a mass source term in the vapour phase. No sink was implemented in the solid phase (sinks and sources are terms used in CFD to describe additions or subtractions in the mass balance that are not due to standard hydro-dynamic considerations). This is permissible, as the amount of mass leaving the solid phase during the relatively short simulations that are feasible with the available computing hardware is a negligible fraction of the total mass of the solid phase. The mass source term for the vapour/gas phase on the other hand is highly significant. The difference is explained by the fact that the density of the solid phase is three orders of magnitude higher than for the vapour/gas phase.

The mass source term complicates the calculation of the mass fractions of the components of the vapour/gas phase and makes it necessary to first compute the absolute amounts of vapour and gas in each control volume, as well as the absolute amounts of vapour and gas added in each control volume over the course of a time step from the pyrolysis reaction:

$$m_t^{gas} = V\rho_\alpha r_\alpha X_t^{gas}$$

$$m_{\Delta t}^{gas} = V\rho_\beta r_\beta \; yX_t^{active}\left(A_2 \exp\left(\frac{E_2}{RT}\right)\right)\Delta t$$

$$m_t^{vapours} = V\rho_\alpha r_\alpha X_t^{vapours}$$

$$m_{\Delta t}^{vapours} = V\rho_\beta r_\beta \; X_t^{active}\left(A_3 \exp\left(\frac{E_3}{RT}\right)\right)\Delta t$$

$$X_{t+\Delta t}^{gas} = \frac{\left(m_t^{gas} + m_{\Delta t}^{gas}\right)}{(m_t^{gas} + m_{\Delta t}^{gas} + m_t^{vapours} + m_t^{vapours})}$$

$$X_{t+\Delta t}^{gas} = \frac{\left(m_t^{vapours} + m_{\Delta t}^{vapours}\right)}{(m_t^{gas} + m_{\Delta t}^{gas} + m_t^{vapours} + m_t^{vapours})}$$

Then the fourth reaction in the extended Broido-Shafizadeh scheme is applied in each control volume:

$$X_{t+\Delta t*}^{gas} = X_{t+\Delta t}^{gas} + X_{t+\Delta t}^{gas}\left(A_4 \exp\left(\frac{E_4}{RT}\right)\right)\Delta t$$

$$X_{t+\Delta t*}^{vapours} = X_{t+\Delta t}^{vapours} - X_{t+\Delta t}^{vapours}\left(A_4 \exp\left(\frac{E_4}{RT}\right)\right)\Delta t$$

Entrainment of the char is estimated in a separate procedure and does therefore not interact directly with the rest of the modelling effort. The pyrolysis reaction is assumed to be controlled by the kinetics, rather than by mass and heat transfer limitations to the biomass particles.

Yields were calculated once the simulation had approached a steady state. The yield of organic vapours was obtained by comparing the biomass feed rate with the flow rate of organic vapours through the outlet patch. The yield of gas was calculated by subtracting the flow rates of organic vapours through the outlet patch, of gas added with the biomass and of fluidising gases from the mass flow rate through the outlet. Finally, the yield of char was calculated by difference.

All calculations were performed on single processor PCs equipped with a 1.2 GHz AMD Athlon processor running Windows NT. The CFD software, CFX, was supplied by AEA Technology. The Fortran 77 compiler required to implement the above equations in User Fortran was Compaq Digital Fortran.

RESULTS

EFFECTS OF SCALE-UP ON BUBBLE DIAMETER AND HOLD-UP

A number of simulations were performed to study the effect on the hydro-dynamics of scaling up to bed diameters of 5 m without taking into account reaction kinetics. Some of the most important variables in the hydro-dynamic design of fluidised beds are the bed expansion and bubble diameter, as they are required inputs for estimating vapour residence time, entrainment, heat transfer and fluidising gas requirements. Figure 2 shows a plot of bubble diameter against bed diameter derived from these simulations. It indicates that the bubble diameter increases with scale, but also that this increase flattens out beyond a bed diameter of roughly 1 m. The results are based on air at room temperature as the fluidising gas and a fluidising velocity of 0.5 m/s. In visual terms this effect is illustrated in Figure 3. The maximum bubble diameter to bed diameter ratio for the 0.5m bed is typically 0.4 under the conditions just described while for the 2.5 m bed, the ratio is 0.2. Thus the bubble size and bed expansion do not scale proportionally when the bed diameter and other bed dimensions are raised by a factor of 5. It should be noted that the bubble diameter of irregularly shaped bubbles is calculated from the bubble cross-sectional area based on an equivalent circle with the same area.

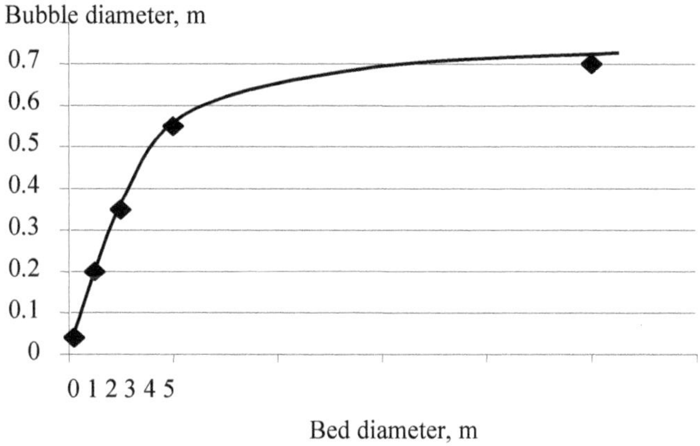

Figure 2: Plot of bed diameter against bubble diameter.

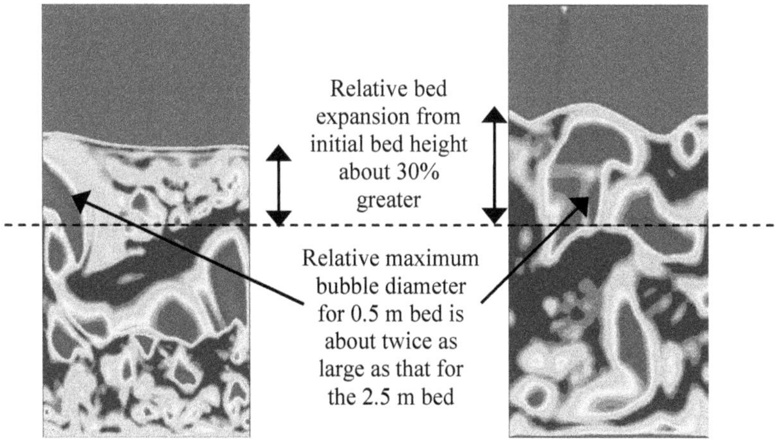

Figure 3: Same conditions (fluidising velocity 0.5 m/s, ambient temperature and pressure, average particle diameter 0.503 mm) except for scale – Left 2.5 m bed diameter, Right 0.5 m bed diameter Plots show volume fractions of the two phases.

Relatively few simulations were performed for diameters beyond 1 m. There is therefore still some uncertainty attached to these novel simulation results. They are, however, in broad agreement with the results of simulations done for cracking catalyst particles by Krishna and van Baten [7], who find that the bubble hold-up declines with increased scale, while the bubble velocity tails off in a similar manner as shown for bubble diameter in Figure 2. Lower bubble hold-up would imply a decrease in bed expansion, which is the variable considered in this work. It should be noted, however, as Geldart points out, that [8] *"As the technique [fluidisation] was applied to processes far removed from oil cracking, different solids had to be used, but there was a tendency to assume that published conclusions drawn from using cracking catalyst were also applicable to these other powders, which often had quite different particle sizes and densities."*

The experience from cracking catalyst is not necessarily a good guide to the behaviour of sand. Furthermore, different results may be obtained, when changing other variables, such as the fluidising velocity, the particle diameter or the fluidising medium.

Kunii and Levenspiel[2] give graphs for bubble diameter against bed diameter that extend to Geldart class B and D particles. They are in broad agreement with Figure 2 in giving a nearly linear correlation between bed and bubble diameter for bed diameters below 1 m. They do not, however, cover values greater than 1 m, which is the novel aspect considered in this work.

IMPACT OF FLOW PATTERNS AND SCALE ON YIELDS

The temperature in the reactor is effectively constant and the same everywhere in the fluidised bed. The heat capacity of the sand is several orders of magnitude larger than that of the fluidising gases and organic vapours. This gives the fluidised bed reactor a significant thermal inertia and means that the reactor temperature will not change much during a simulation. In practical operation, this property of fluidised beds is actually very desirable giving good temperature control that allows operation close to the optimum temperature that maximises liquid yields. Heat transfer within the bed is also excellent, both in terms of heat transfer between the two phases and in terms of heat transfer, mostly through particle convection, between different parts of the bed.

The average residence time of the organic vapours can change much more rapidly than the temperature of the bed and is consequently the variable responsible for a degree of fluctuation in product yields during short simulation runs.

Yields have been calculated for a scaled up reactor, illustrated in Figure 4, with a diameter of 50 cm and a height from bottom to top of nearly 2.8 m.

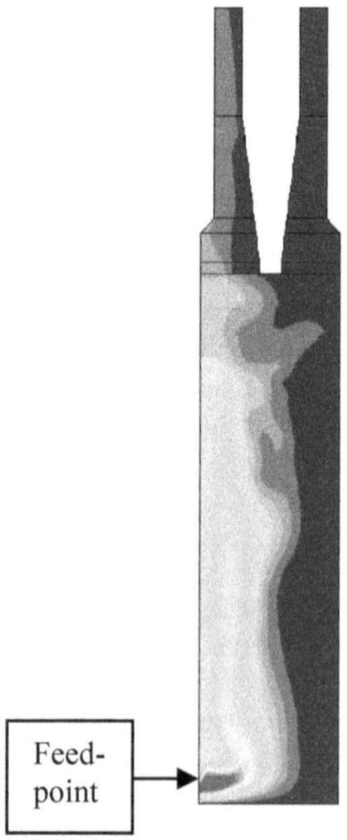

Figure 4: Illustration of modelled reactor (0.5 m diameter, 2.8 m height), plot shows concentration of organic vapours after about 2 seconds of simulation time.

The yield as detailed above mainly depends on the residence time, which in turn depends on the superficial velocity. The conversion itself determines the amount of gas and organic vapours added and therefore influences the superficial velocity. With a superficial velocity of the fluidising gases of just over 0.2 m/s and a total superficial velocity (which includes the gas and organic vapours added via pyrolysis) fluctuating around 0.7 m/s, the yields (moisture free basis, wood) shown in Table 2 were obtained.

Table 2: Calculated product yields.

Product	Yield
Char	13%
Bio-oil	57%
Gas	30%

These low yields can be explained by the long residence time experienced by the organic vapours, which allows significant secondary cracking; in smaller laboratory reactors with residence times below 2 seconds yields of 75% are routinely obtained experimentally, which is in agreement with the predictions of reaction kinetics schemes of the Broido-Shafizadeh type also used in this work. The residence time of the vapours is primarily determined by the path they take and the overall superficial velocity of both vapours and fluidising gases.

A further scale-up by a factor 10 in each dimension (from a diameter of 0.5 m to a diameter of 5 m) was performed as a test. This translates into a factor 1000 by volume giving 250 tonnes per hour or of the order of 20000 barrels per day of oil equivalent, which compares to fluid catalytic cracking of oil, but would be exceptionally high for any conceivable commercial bio-oil plant. To calculate yields, the run would have had to be continued to the point where the vapours reached the exit and steady operation was achieved. This was not done, as it would have required excessive computing power.

However, the initial 12 seconds of simulation are quite instructive. The resulting flow patterns are illustrated in Figure 5. Rather than moving significantly upwards, the biomass vapours move towards the centre of the bed, as they are caught in a very large recirculation pattern.

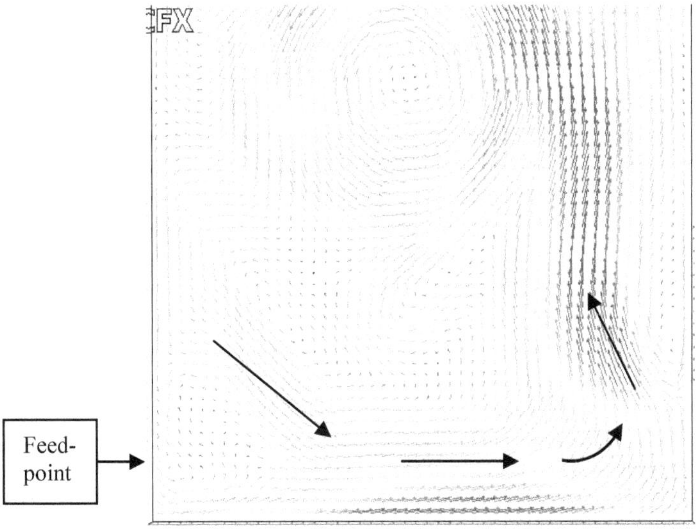

Figure 5: Flow patterns in the bottom section of a scaled up reactor (5 m diameter). Darker areas show higher gas velocities.

As can be seen in the vector plot illustrated in Figure 5, the initial movement of the vapours is largely sideways, not up. It is quite noticeable that very high speeds are obtained on the side opposite to the feed-point. Therefore the vapours have to travel an additional 5 metres in this simulation before they join the upward movement illustrating the potential for flow patterns to lengthen the residence time of the vapours, which may deleteriously affect bio-oil yields.

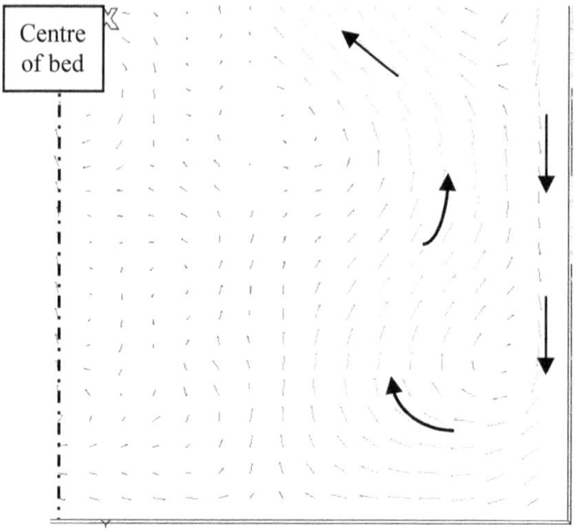

Figure 6: Illustration of common flow patterns of the solid phase observed in larger fluidised beds (section shown has dimensions of 0.25 by 0.25 m and is located within a reactor with a diameter of 0.5 m).

For larger scales, that is for diameters of roughly 0.5 m or greater, it is frequently observed in the simulations that down movement at the walls is combined with upward movement towards the centre of the fluidised bed (see Figure *6*). This kind of solids circulation has also been noted in the literature[2,8] and is therefore not entirely unsuspected. As the biomass is commonly fed through an opening in the wall, it could tend to be caught up in this downward movement. The flow patterns of the solid and vapour/gas phase do differ markedly, as there is no net upward movement for the solid phase, while the movement of the vapour/gas phase must average out to be equivalent to the superficial velocity. It is therefore not surprising that the downward movement at the walls is driven by the solid phase. The gas/vapour phase is either slowed down there, or even forced to move down together with the solid phase.

POTENTIAL FOR MALDISTRIBUTION

There is a close interdependency between the hydro-dynamics and the reaction kinetics in that the path taken by the organic vapours influences the residence time and therefore the bio-oil yield, while the location where the pyrolysis reaction liberates organic vapours and gas impacts the flow in the fluidised bed. Because the speed of the solid phase tends to be low close to the feed-point and as the pyrolysis reaction proceeds quickly, the bulk of the pyrolysis reaction takes place quite close to the feed-point, generally within on the order of 10 cm. As the vapours and gas released by the pyrolysis reaction make up a sizeable proportion of the overall gas/vapour phase flow rate, this very localised evolution holds the potential for mal-distribution, as is shown in Figure 7 for a bed with a diameter of 50 cm and a height of 2 m. Figure 7 also illustrates that mal-distribution is less of an issue for thin and tall beds than it is for shallow beds with a very large diameter.

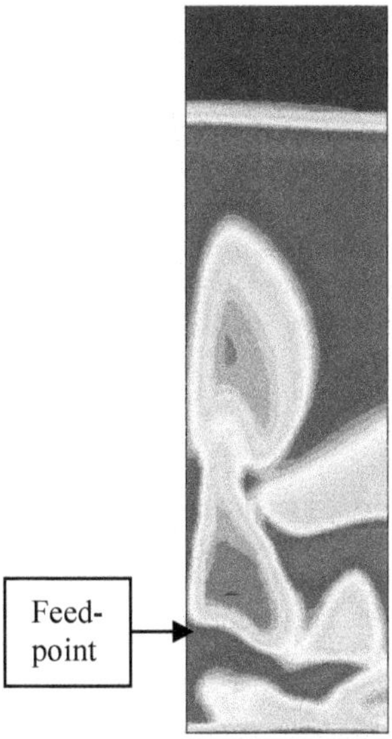

Figure 7: Illustration of potential for mal-distribution, plot shows volume fractions of the two phases (reactor diameter 0.5 m).

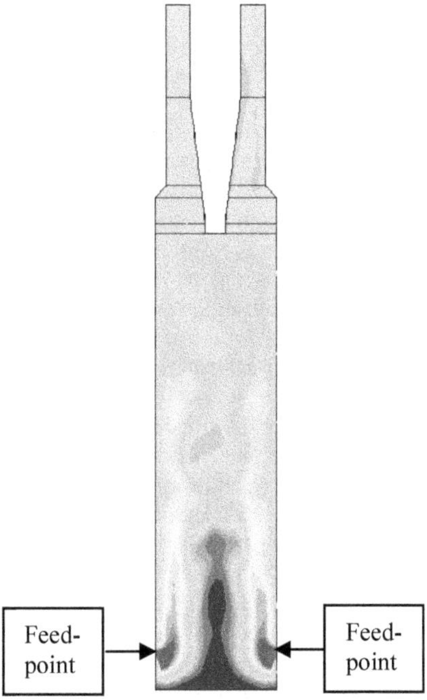

Figure 8: Pyrolysis reactor with two feed points, plot shows concentration of organic vapours (diameter of 0.5 m).

Whether mal-distribution actually represents a problem depends on the individual case considered. Its effect on particular reactor geometries will differ. A few potential problems are: localised temperatures close to the feed-point that are lower than the average in spite of the excellent heat transfer customarily afforded by fluidised bed reactors, lower heat transfer to the bed via the bed walls and higher char concentrations in the bed leading to more secondary cracking reactions.

The velocity of the bubbles is in general higher than the superficial velocity. Consequently, mal-distribution may have a positive impact, if it leads to a greater fraction of the organic vapours in the bubbles than in the rest of the fluidised bed, as this would shorten the average residence time of the organic vapours. In the case of the simulations performed to obtain the yields given in Table 2, however, mal-distribution was not significant and the average bubble actually contained a slightly lower fraction of organic vapours than the bed as a whole. Therefore this factor contributed to the long average residence time of the organic vapours.

To avoid mal-distribution it may be necessary to use several feed-points. This option has been explored with two feed-points in a simulation of a reactor with a diameter of 0.5 m and a height of 2.8 m, which is illustrated in Figure 8. When comparing with Figure 4 it becomes clear that the objective of an even distribution of the organic vapours has been achieved in this case.

REQUIREMENTS FOR THE REACTION KINETICS MODELLING

As has been pointed out above, the main purpose of combining the CFD modelling of the hydro-dynamics of fluidised beds with pyrolysis reaction kinetics is to consider their interdependency and the resulting impact on the design of scaled up reactors. In particular, the location where vapours evolve significantly impacts the hydro-dynamics, while the residence time, and to a lesser extent the local temperature, of the organic vapours and biomass, and therefore product yields, are determined by the hydro-dynamics.

The reaction kinetics themselves are not scale dependent. The decomposition of a given biomass particle only directly depends on the local conditions, such as the temperature, of the medium in its immediate vicinity. The size of the fluidised bed has an indirect impact, namely in the way it influences the local conditions experienced by individual biomass particles and by the organic vapours.

There are a number of reaction kinetics models available that try to predict the effect of these local conditions on the decomposition of the biomass particles and on the secondary reactions of the organic vapours e.g. 9, 10, 11. It is, however, currently feasible to obtain more reliable and accurate results from experimental work on the bench-scale, which is furthermore easily affordable when compared to the use of pilot plants for scale-up purposes. It should therefore be stressed that design of commercial plant should primarily rely on experimental data for obtaining reaction kinetics data.

The choice of a modified Broido-Shafizadeh type model in this work is not to be understood as an endorsement of that particular scheme as a method for obtaining reaction kinetics data for new conditions. Nor is it the aim of this work to predict those reaction kinetics data from first principles, as they are clearly much more easily and accurately obtained from experimental data. The use of this particular scheme should rather be regarded as a first attempt to codify experimental data into an input for the CFD model that then provides a prediction for the effect of the interaction between the known or assumed to be known, and not predicted, reaction kinetics with the hydro-dynamics. The most important points that need to be verified, be it against fresh experiments or against published data for similar feedstocks, are the composition of the primary pyrolyis products and the way in which secondary reactions are impacted by the local temperature and the residence time of the vapours. Another important consideration is the time taken by the primary gases and vapours to evolve, which could be estimated from experimentation[5] or from simplified single particle models such as those proposed by Di Blasi[12].

CONCLUSIONS

Reaction kinetics have been successfully integrated into an existing CFD model.

Mal-distribution and longer residence times caused by the nature of the flow in the fluidised bed have been identified as potential problems and the use of multiple feed-points has been shown to be one potential solution, particularly to the problem of mal-distribution.

It has also been shown that bubble diameter, which is intimately related to hold-up and therefore bed expansion, only scales linearly up to a bed diameter of around 0.5 m.

NOMENCLATURE

A	frequency factor	1/s
$c_{\alpha\beta}^{(h)}$	interphase heat transfer term	$J\, m^{-3} s^{-1}\, K^{-1}$
E	activation energy	J/kmol
H	enthalpy	J/kg
K	reaction rate	1/s
m	mass	kg
$\dot{m}$	mass flow between phases	$kg\, m^{-3} s^{-1}$
N_P	number of phases, here 2	
r	volume fraction	
R	universal gas constant	$J\, kmol^{-1}\, K^{-1}$
t	time	s
T	temperature	K

U	velocity	m/s
V	volume of computational cell	m^3
x	fraction of char produced in reaction 2	
y	fraction of gas produced in reaction 2	
α	subscript for the vapour/gas phase	
β	subscript for the solid phase	
Δt	time step	s
λ	thermal conductivity	$W\ m^{-1}\ K^{-1}$
ρ	density	kg/m^3
X	mass fraction	

REFERENCES

1. Bridgwater A.V. et al. (2001) An Overview of Fast Pyrolysis. In: *Progress in Thermochemical Biomass Conversion*, (Ed. by A.V. Bridgwater), pp. 977-997. Blackwell Science Ltd, Oxford.
2. Kunii D. and Levenspiel O. (1991) *Fluidization Engineering,* 2nd edn. Butterworth-Heinemann, Newton.
3. Gerhauser H. et al. (2001) CFD for the Modelling of Entrainment in Fluidised Bed Fast Pyrolysis of Biomass. In: *Progress in Thermochemical Biomass Conversion*, (Ed. by A.V. Bridgwater), pp. 1281-1295. Blackwell Science Ltd, Oxford.
4. Bradbury A.G.W. et al. (1979) A Kinetic Model for Pyrolysis of Cellulose. *Journal of Applied Polymer Science*, **23**, 3271-3280
5. Lédé J. et al. (1999) The nature and properties of intermediate and unvaporized biomass pyrolysis materials. In: *Fast Pyrolysis of Biomass, A Handbook*, published by CPL Press for the Bio Energy Research Group of the University of Aston in Birmingham
6. Liden A.G. et al. (1988) A kinetic model for the production of liquids from the flash pyrolysis of biomass. *Chem. Eng. Comm.*, **65**, 207-221
7. Krishna R., van Baten J.M. (2001) Using CFD for scaling up gas-solid fluidised bed reactors with Geldart A powders, *Chemical Engineering Journal*, **82**, 247-257
8. Geldart D. (1986) *Gas Fluidization Technology*, A Wiley-Interscience Publication
9. Maa P.S. and Baillie R.C. (1971) Influence of particle sizes and environmental conditions on high temperature pyrolysis of the cellulosic material. Phd Dissertation. West Virginia University, Morgan Town, USA, 1971
10. Tran D.Q. and Rai C. (1978) A kinetic model for pyrolysis of Douglas fir bark. *Fuel*, **57**, 293-298
11. Diebold J (1985) The cracking kinetics of depolymerised biomass vapours in a continuous tubular reactor. MSc Thesis, Colorado School of Mines, Golden, Colorado, USA
12. di Blasi C. (1996) Heat, momentum and mass transport through a shrinking biomass particle exposed to thermal radiation. *Chemical Engineering Science*, **51**, 1121-1132.

Environment, Health And Safety PyNe Working Group : Assessment of Bio-oil Toxicity for Safe Handling and Transportation

P Girard & J Blin

Centre de Coopération Internationale de Recherche Agronomique pour le Développement (CIRAD), TA 10/16, 34398 Montpellier, Cedex 5, France

ABSTRACT: Pyrolysis is one of the three main thermochemical routes to convert biomass into useful primary energy products. Fast pyrolysis has benefited from active research programme since 1980's in order to obtain bio-oils which can be used in engines for the generation of electricity or after refining in transport. Today, several demonstration plants are operating in Europe and North America where significant quantities of bio-oils are produced for research and development purposes and several commercial plants are at an advanced stage of planning. Thus, for a commercial development the question of safety procedures for human health and environment preservation is raised. The aim of this programme supported by the European Union is to comprehensively assess toxicity and biodegradability of a representative bio-oil after preliminary screening of a wide range of bio-oils from different processes and temperatures. In the project, the relation between process parameters on one hand and chemical composition and toxicity for human health and environment on the other hand will be investigated so as to recommend the operating conditions to produce bio-oils with the lowest impacts. Then, the optimised compositions of bio-oils will be submitted to the mandatory tests required by the EU legal authority, the objective being the definition of secure handling and storage procedures, in order to control the risks related to the product for the population and the environment. A MSDS containing safety procedures and a report on transport requirements, will be published in order to obtain the bio-oils inscription in the ELINCS and give authorisation for marketing with the corresponding preventive accompanying measures. An appropriate assessment of the risk involved will contribute to reduce the production costs and make pyrolysis bio oil more competitive. In addition the knowledge of the parameters potentially responsible for higher toxicity will limit production losses. The structure of the consortium is defined to insure dissemination and utilisation of the results, as producers and users are involved in the project through PyNe. During the different meetings of the steering committee, they will be asked to give their feedback on their own experience, give their opinions on the tests performed and participate to the definition of the safety procedures to be applied. This will enhance awareness among the producers and end-users, and the procedures defined will be directly applicable on industrial sites.

This paper presents the foreseen work-programme and discusses the expected results.

OBJECTIVES

Diebold [1] has made a review of available toxicological data on individual components that can be found in pyrolysis liquids and smokes. Primary vapours are mainly oxygenated compounds from the partial decomposition of biomass macropolymers (lignin, cellulose, hemicellulose).

Secondary vapours are formed by cracking of the primary vapours, at higher temperature or higher residence times, and are phenolic compounds. Tertiary vapours are mainly polyaromatic hydrocarbons. In the whole, the pyrolysis oils and smokes are a mixture of primary, secondary and tertiary vapours, one type being predominant depending of process conditions. 74 chemicals have been reported to be either in oils, smokes or both, with a concentration of at least 0.1 wt %.

The toxicity of each chemical has been examined. The conclusion is that aldehyde and furan families can lead to acute toxicity of pyrolysis vapours, by ingestion or inhalation. Chronic health effects could also be related to aldehydes unless the severity of the pyrolysis is so high as to create significant quantities of benzene and PAH, potentially leading to tumour development.

It has been assumed until now that similar properties can be more or less expected for fast pyrolysis bio-oils. However, fast pyrolysis bio-oils contain less condensed compounds, which should decrease long term toxicology. On the other hand, they contain different compounds compared to smokes which can give synergetic effects from a toxicological point of view (leading to either increase or decrease the health effects). One study has been reported concerning tests on bio-oils, [2][3] but contradictory results have been obtained on chronic toxicity, one indicating no tumour promoting ability and the other showing some mutagenicity. Furthermore, some processes are "known" to be at the origin of a higher PAH content into bio-oil than others.

Up to now, the high cost of the necessary analysis to potential toxicity and biodegradability study prevents their financing by generally SMEs fast pyrolysis companies. The transition from R&D plant to demonstration and commercial unit as well as the transportation of large quantities of bio oil for test purpose make this work necessary. Such programme by grouping producers and users will have a wide impact and allow a complete evaluation of bio-oil potential toxicity and biodegradability. That is why a program is needed at EU level. The aims of this project are:

- to determine toxicological and eco-toxicological data on bio-oils. The work will concern both acute and chronic effects on human health, carcinogenic and mutagenic effects. The environmental impacts of bio-oils liquids will also be measured in water and soil. This will allow a comprehensive and definitive MSDS to be produced with the proper preventative and remedial procedures to adopt during production, transport and use of bio-oils,
- to identify the best operating conditions to avoid or minimise the formation of toxic products from the composition of the bio-oils,
- to produce fast pyrolysis bio-oils with low impact on human health and environment by avoiding bio-oils production presenting potential toxic characteristics,
- to encourage the active involvement of industry in the development and exploitation of the results through an existing network, which includes developers of bio-oils production and application technologies.
- to disseminate information on this activity from the database and results of meetings by newsletters, reports, conferences, specific workshops, web site and other publications, to those of the network and other interested parties in regional, national and international programmes.

PROJECT WORK PLAN

METHODOLOGY AND APPROACH

The necessary steps to achieve the different goals of the projects are :

- Collect oils from the different process manufacturers as well as producers and produce representative fast pyrolysis oils from a selected feedstock and process conditions on fluidised bed reactor available on lab scale so as to obtain a variety of products from which the relation with toxicity and biodegradability can be identified.
- Analyse physically and chemically bio oil as well as to carry out preliminary toxicological and eco-toxicological tests. This will provide a good description of the product and able to relate composition with process parameters, toxicity and biodegradability.
- Carry out toxicity, eco- toxicity and biodegradability tests. This is the heaviest and the more crucial part of the project. Indeed, these tests are essential if bio-oils are to be freely traded and transported. This knowledge is of public interest if large volumes of bio-oils are to be produced, transported and stored all over Europe. The risks related to the handling of bio-oils have to be assessed for prevention and cure.
- Produce recommendations for handling, storage and transport
- Disseminate results.

After this work has been done, the necessary data for product notification will be available. The data will be presented to the legal authority of one of the European countries. This legal authority will be different from the institution chosen in the project for impartiality reason. The legal authority will then statute on the product classification, labelling and inscription in the ELINCS and give authorisation for marketing with the corresponding preventive accompanying measures. This is a mandatory procedure for any companies aiming at commercialising a new chemical substance [4]. In this case, the owners of industrial pilot plants are responsible for initiating the procedure, with the available data, to obtain trading authorisation over Europe.

The project will work in close collaboration with PyNe network. This close relationship will allow the opinions and feedback of industry on practical experience already available on oil handling, brought through research projects. It will also facilitate the dissemination of the results of this project to the main actors concerned.

WORK PROGRAMME

Bio-oils production and procurement

It is known that oil composition strongly depends on feedstock, pyrolysis technology and process conditions. The aim is to produce bio-oils from different reactors and temperature in order to relate those parameters to oil composition, toxicological characteristics and biodegradability. The different processes typically will be involved:

- fluidised bed reactors,
- circulating fluidised bed reactor
- vacuum reactor,
- ablative reactor
- rotating cone reactor.

These different technologies induce differences in process parameters, leading to varying oil properties and composition. The necessary bio-oils for the investigation of the influence of temperature on toxicity and biodegradability will be produced by Aston and the others obviously by industries.

Aston University will manufacture several oils with a fluidised bed reactor varying the temperature: three different levels of temperature will be choose (in the range 450 - 600 °C). Eleven bio-oils will be produced. The samples will then be sent to the different laboratories for analysis.

Bio-oils analysis and screening tests

Bio-oils are complex mixtures of the thermal degradation products from the three main constituents of biomass: cellulose, hemicellulose, and lignin plus extractives such as resin acids, fatty acids and terpenes. Thus, the chemical analysis requires sophisticated methods in order to identify and quantify the single compounds. The main tools for analysis will be high resolution capillary gas chromatography (HRGC) and high performance liquid chromatography (HPLC).

Gas chromatographs are equipped either with flame ionization detector (FID) or mass selective detector (MSD) for identification and quantification. The liquid chromatograph (LC) is equipped with UV/VIS detector and an ion trap mass spectrometer (MS). These methods allows to separate, identify and quantify both the polar and non polar fractions of the bio-oils. LC/MS belongs to the latest development in liquid chromatography and will be applied for the first time in the analysis of pyrolysis liquids. Detection sensitivity will be further increased by use of a mass spectrometer operating in the single ion monitoring (SIM) mode. It is expected that also dibenzofurans and dibenzodioxins will be measured if present in the bio-oil. In order to lower the detection limit special pre-separation techniques will be applied in order to facilitate and improve the chromatographic separation by GC and LC. One key technique will be solid phase extraction

(SPE) on cartridges packed with different sorbents. SPE allows both the separation of complex mixtures into fractions of compounds with a similar chemical nature such as phenols, catechols, aldehydes, furans, polyaromatic hydrocarbons etc. It is also expected that especially polyaromatic hydrocarbons can be analysed more easily as they are not detected in bio-oils with standard GC techniques.

BFH and Cirad have a long tradition and experience in the analysis of pyrolysis products from biomass [5][6]. Apart from the chemical analysis, also physical properties will be determined including stability by viscosity index, water content by Karl-Fischer titration, initial viscosity, char content, char particle size and range, and acid number, freezing point/freezing range, boiling point/boiling range, relative density, vapour pressure, static method, surface tension in aqueous solution, water solubility, flask method, partition coefficient (n-octanol/water), flash point, flammability, explosive properties and auto-flammability. BFH will be in charge of those tests.

The screening tests consist in the evaluation of irritating effects, as well as the oxidising properties. They will be performed on the 11 bio-oils. They concern acute oral and dermal toxicity, dermal and ocular irritation, sensitisation, genotoxicity (Ames) and ecotoxicity to *Daphnia*.

One of these 11 bio-oils will then be selected for complete toxicological tests. This bio-oil will be selected based on their representativity and the results in screening tests.

Complete toxicological and eco-toxicological test

Complete analyses will be carried out on a selected oil, representative of the market and based on previous results, to test its comportment:

The complete tests follow the European regulation. The test will be done by an European accredited laboratory. The tests to be performed include the basic set of annex VII A, corresponding to a production of 1ton/year/producer. In this first study on the evaluation of bio-oils toxicity, we will not perform the tests of annex VIII, which are required in case of very harmful products. These tests consist of:

For toxicological tests:

- Acute toxicity: oral, by inhalation, by skin contact, by eye contact : two of these three tests have to be performed for notification. This will be discussed at the beginning of the project on the basis of the results of the screening test.
- Irritation test (under GLP)
- Skin sensitisation
- Toxicity with repeated doses (oral, by inhalation, by skin contact)
- Mutagenicity : « in vitro » mammalian chromosome aberration test with CHO cells and reversed mutation test on *Salmonella typhimurium*
- Sub acute toxicity: either oral, dermal or inhalation tests must be performed for notification. This will be discussed at the beginning of the project.

For eco-toxicological tests:

- Acute toxicity towards fishes and *Daphnia*
- Algal growth inhibition tests,…

For biodegradability tests:

- Bacterial growth inhibition, activated sludge test
- Biotic/Abiotic degradation ,…..

Recommendations for safety procedures and dissemination of the results

This will include the preparation and presentation of the “dossier” for substance notification (MSDS elaboration, redaction of handling, storage and spillage procedures), elaboration of recommendations on the best operating conditions to be used to obtain friendly products. This activity will particularly involve the industrial partners (through the pyrolysis network), which already have some practical experience in the handling of bio-oils and bio-oils processes.

The safety procedures will be defined by comparison with existing MSDS of substances presenting analogous risks. The MSDS will contain the following information :

- Product identification : generic name
- Composition (range of concentration for family compounds, results of chemical analyses)
- Hazards Identification : brief description of acute effects
- First Aid Measures : what to do in case of exposure
- Fire Fighting Measures : what to do and to avoid in case of fire
- Accidental Release Measures : what to do in case of release of small and large quantities.
- Storage and Handling : type of containers, type of clothes , temperature
- Physical and Chemical properties : as measured in WP 2 : density, viscosity, pH, vapor pressure, ...
- Stability and Reactivity : incompatibilities with current solvent or operation (heating ...)
- Toxicological and Ecological Information
- Transport Information

Specific recommendations for the transport of fast pyrolysis liquids ensuring compliance with UN recommendations relating to the transport of dangerous goods will also be derived :

- appropriate transportation code for pyrolysis liquids which would then be internationally recognised,
- procedures for the treatment of spills.

A guide for shipment from a few grams to tonnes, with protocols for labelling, packaging and shipment will be prepare. All these information will be arranged in the form required to present a dossier to a European legal authority to apply for a notification of the bio-oils. The project progress will be monitored & feedback provided to all relevant persons and organisations and those expressing an interest in the results of the project.

On completion of the work, a workshop will be organised to disseminate the results and recommendations. The results will widely published on the existing Web site, as well as in the newsletter. A comprehensive programme of dissemination of the results and their implications for the industry will be carried out as specifications, protocols and guidelines.

The Pyne working group added value is guaranteed by the strong involvement of industrialists. The results will benefit directly to European bio oil producers and users at the end of the project as they are all invited to supply samples and discussed the results during a workshop to be held at the end of the programme (at least : BTG, FORTUM, ENEL, Dynamotive, Wellman, ...).

PROJECT EXPECTED OUTCOMES

The expected results are:

- a comprehensive assessment of the toxicity and biodegradability of bio-oil,
- a definitive Materials Safety Data Sheet (MSDS)
- definition of the best practice for the production of the most benign bio-oil in terms of health and the environment
- an official notification to the authorities: the product will be recorded in the ELINCS (European List of Notified Chemical Substances) data base, so that all relevant information on health and environment care is available for anyone interested in the production or use of bio-oils such as producers, users, social and medical services, transporters, trade associations etc.
- widespread dissemination of the results to all production and utilisation organisations,
- recommendations for a safe and acceptable distribution system for bio-oils. This will encourage larger scale applications of both production processes and end-use processing of bio-oils.

The structure of the consortium is defined to insure dissemination and utilisation of the results, as producers and users are involved in the project through PyNe. During the different meetings of the steering committee, they will be asked to give their feedback on their own experience, give their opinions on the tests performed and participate to the definition of the safety procedures to be applied. This will enhance awareness among the producers and end-users, and the procedures defined will be directly applicable on industrial sites. The results of the laboratory works will be directly usable by the producers for the optimisation of their own equipment to produce low impact oils as well as to notify their products to the legal authorities in their country.

To enlarge the diffusion of the results, the information will be disseminated by reports, PyNe newsletter, a specific page on the web-site of PyNe and other publication. This will be directed to the interested parties in regional, national and international programmes, which have been identified within PyNe, though members of the network.

THE PARTNERSHIP

The project partnership includes a co-ordinator also involved in chemical analyses (CIRAD), a partner involved in the production of bio-oils at lab scale and also responsible for the relations with the PyNe network and for dissemination of the results (ASTON), a laboratory performing chemical and physical analysis (BFH). ASTON is also responsible for the sub-contractor CIT performing the toxicological and eco-toxicalagical tests. CARE as subcontractor to CIRAD will have in charge transport requirements.

ACKNOWLEDGEMENT

These activities are supported by the European Union DG TREN programme.

REFERENCES

1. Diebold J., (1999) *A review of the toxicity of biomass pyrolysis liquids formed at low temperatures.* In: Fast Pyrolysis Of Biomass - A Handbook, (Edited by A Bridgwater, S Czernik, J Diebold, D Meier, A Oasmaa, C Peacocke, J Piskorz, D Radlein), pp 135-164 CPL Press, ISBN 1-872-691-072.
2. Elliott D.C. (1988) *Relation of Reaction Time and Temperature to Chemical Composition of Pyrolysis Oils.* In: Pyrolysis Oils from Biomass, (Edited by E.J. Soltes and T.A. Milne) pp 55-65, ACS Symposium series 376.
3. Scott D.S. (1987) *Technical Evaluation of the Waterloo Fast Pyrolysis Process,* Energy, Mines and Resources Canada, Ottawa, Ontario, pp104-123, DSS Contract File N° 20ST.31926-3-5026, ISBN 0-662-15614-5.
4. Girard P., Diez C. (2002) *Environment, health and safety aspects related to fast pyrolysis oils.* In: Fast Pyrolysis Of Biomass - A Handbook, volume 2. pp 69-87, CPL Press.
5. Meier D., Oasmaa A., Peacocke G.V.C. (1999) *Properties of fast pyrolysis liquids : status of test methods,* In: Fast Pyrolysis Of Biomass - A Handbook, (Edited by A Bridgwater, S Czernik, J Diebold, D Meier, A Oasmaa, C Peacocke, J Piskorz, D Radlein), pp 75-92 CPL Press, ISBN 1-872-691-072.
6. Meier D., (1999) *New methods for chemical and physical characterisation and round robin testing*, In: Fast Pyrolysis Of Biomass - A Handbook, (Edited by A Bridgwater, S Czernik, J Diebold, D Meier, A Oasmaa, C Peacocke, J Piskorz, D Radlein), pp 92-102 CPL Press, ISBN 1-872-691-072.

Determination of Basic Fuel Quality Standards for Biomass-Derived Pyrolysis Liquids

S. Gust
Fortum Oy, Porvoo, Finland
R.J. Mclellan
Wellman Process Engineering Ltd., Oldbury, UK
D. Meier
Institute of Wood Chemistry, Hamburg, Germany
A. Oasmaa
VTT Processes, Espoo, Finland
G.V.C. Peacocke
Conversion And Resource Evaluation Ltd., Birmingham, UK

ABSTRACT: The uptake of biomass-derived pyrolysis liquids for use in heat and power applications depends on the ease of use and acceptability of the fuel by the end user and equipment providers. As part of an EU-funded ALTENER II contract, one of the aims is to derive standards for biomass derived pyrolysis liquids, based on a consensus between equipment providers (boilers, engines and turbines) and the liquids producers. Five basic properties (homogeneity, water content, solids content, stability, flash point) for the liquids are used as the primary criteria for pyrolysis liquid evaluation. Specific values are proposed to ensure that pyrolysis liquids meet a minimum grade acceptable for use as a fuel oil ("premium" grade) in boilers and engines. Data on emissions from boilers, engines and turbines are presented. At present, there is not enough long duration data available to allow further more detailed specifications on secondary properties to be made, or define standards for liquids in turbines. The aim is to ensure that a realistic set of specifications are determined to allow the introduction into existing fuel infrastructures and markets, initially perceived as the domestic heating fuel oil markets in Scandinavia.

INTRODUCTION

The work presented here is part of an EU funded ALTENER II project [contract no. 4.1030/C/00-015/2000], looking at the determination of norms and standards for biomass derived fast pyrolysis liquids. Pyrolysis liquids offer some distinct advantages over gasification and combustion for power production:

- High energy density [>2.5-3.5 times the energy density of softwood],
- Liquids production can be decoupled from power generation,
- More easily transported and handled than the raw biomass or producer gas.

The five tasks within the project are:

1. Review fast pyrolysis technologies and describe processes at pre-commercial and commercial scale, suitable for heat and power production in the near [1-4 years] to medium term [5-10 years]. Review incentives to develop pyrolysis technologies at national and EU level.
2. Derive norms and standards for biomass fast pyrolysis liquids. The project will review end user requirements and specifications for biomass fast pyrolysis liquids to obtain specifications and standards in liquid fuel quality.
3. Sector and market strategies for the production of power from pyrolysis liquids.

4. Long-term cost/benefit analyses comparing biomass fast pyrolysis to traditional forms of energy and other alternative renewable energy sources and comparing the overall conversion efficiencies to electricity.
5. Quantification of benefits obtained in improving the producer-converter-user interface and improvement of the energy/environmental balance in pyrolysis liquids production.

The results of Task 2 are discussed in this paper. Over the past 10 years, developments in fast pyrolysis technologies has led to the planning, construction and operation of plants ranging from laboratory scale to commercial/demonstration scale. These are summarised in Table 1. One other plant, which is now closed down was the Union Fenosa plant, with a design capacity of 250 kg/h, which provided some of the first liquids for testing in the EU (1, 2).

Table 1: Biomass Pyrolysis companies active in the EU for the production of liquids.

Organisation	Plant Capacity, kg/h	Status	Products and Use
BTG, the Netherlands	250 2000	Operational Planned	Liquids for fuels & chemicals
Border Biofuels & Dynamotive., UK	1000	Planned	Liquids for fuels & chemicals
ENEL, Italy	650	Re-design	Liquid fuel
Ensyn Technologies, UK	Up to 10000	Available	Liquids for fuels & chemicals
Fortum Oy, Finland	500	Operational	Liquid fuel
PyTec, Germany	20	Ongoing	Liquid fuel
VTT Processes, Finland	20	Operational	Liquid fuel
Wellman Process Engineering Ltd., UK	250	Commissioning	Liquid fuel

Analysis of liquids from some of these technologies is given in Table 2. Additional data has been included to address the wide spectrum of properties that may be required in different applications and to assist in the design of process equipment and power generation systems.

It can be seen that there is some variation in all of the properties, but general agreement in the ranges. From these typical properties and other work, a consensus was reached on what properties were important should be used to classify the liquids

BOILER, ENGINE AND TURBINE EMISSIONS

There have been several small-scale attempts to combust pyrolysis liquids in flame tunnels, boilers and engines, but no experience yet on turbines in the EU. Limited data is available on boiler emissions, but some data is summarised in Table 3 for a range of different liquids, typically fired in small boilers. Comprehensive test data may be found in the work of Oasmaa et al.(8) for example, on a range of liquids. Also, there has been limited engine work on engines and turbines, primarily due to a lack of liquids in large quantities and of consistent properties. Boiler work has therefore attracted more attention, usually in dual fuel mode, but with the aim of single fuel combustion in a boiler.

There has been limited development on engines, as noted, due primarily to a lack of a consistent supply of pyrolysis liquids, the perceived high cost of the technology and the inherent unusual characteristics of the liquid. Some work has been carried out by Ormrod Diesels on their dual fuel diesel engine and by VTT using Ensyn liquids (3). The emission results are given in Table 4, in conjunction with trst work on a small Stirling engine burner. Work on testing of pyrolysis liquids in a rotary engine has also been noted (4). Data for the early VTT test work is presented in Table 5.

Table 2: Pyrolysis liquids properties from different feedstocks.

			Source			
Physical property	Ensyn	NREL HVF	NREL	BTG	VTT	VTT
Feedstock	Mixed hardwood	Poplar	Switch grass	Pine	Straw	Forest residue*
Water content, wt%	22	18.9	n.d.	28.2	19.9	28.5
Ph	2.5	2.8	n.d.	n.d.	3.7	3.2
Specific gravity (at 15°C) kg/dm^3	1.18	1.2	n.d	1.2	1.19	1.20
Elemental analysis (moisture free), wt%						
- C	56.4	57.3	55.8	52.9	55.3	55.0
- H	6.2	6.3	6.9	6.1	6.6	7.6
- O (by difference	37.1	36.2	36.3	39.6	37.7	36.9
- N						
- S	0.2	0.18	0.79	1.2	0.4	0.3
- Ash	<0.01	0.02	0.03	0	0.05	0.03
- K+Na, ppm	0.01	<0.01	n.d.	0.2	0.14	0.11
- Cl, ppm	460	9.9	128	n.d.	2	60
HHV (moisture	3	7.9	1 900	n.d.	330	<100
free basis), MJ/kg	23.1	22.3	23.8	n.d.	23.1	21.9
HHV (as produced), MJ/kg	17	18.7	n.d.	n.d.	18.5	15.7
Viscosity (at 20°C), cSt						
Viscosity (at 40°C), cSt	n.d.	128	n.d.	22.3	55	n.d.
Flash point, °C	50 (50°C)	13.5 (50°C)	n.d.	n.d.	11 (50°C)	21
Pour point, °C	55	64	n.d.	n.d.	56	42
Solubility, wt%	-25	36	n.d.	n.d.	-36	-12
insolubles in						
- ethanol						
- methanol	0.015	0.045	n.d.	0.68	0.3	0.10
- methylene chloride	n.d.	n.d.	n.d.	n.d.	n.d.	0.05

*Bottom phase (80 wt% of total liquid)

Several potential problems were identified in the preliminary tests on the Valmet engine (see Table 5). Wear and corrosion were detected in injection needle and pump elements during both bench tests and engine operation. Relatively high CO emissions were analysed in the engine exhaust gas during the test. However, the commercial catalyst installed in the unit removed most of CO and hydrocarbons.

Table 3: Boiler emissions - some examples.

Liquid Source	Ensyn	Union Fenosa [1]	Ensyn	VTT	VTT	CFB [2]	BFB [2]	BFB [2]
Feedstock	Hardwood	Eucalyptus	Hardwood	Pine Sawdust	Forest Residues	Hardwood	Pine	Pine
Solids	0.5	0.7		0.1	0.1	~0.4	0.17	0.03
Boiler	Arimax Eetta 200 kW boiler	Arimax Eetta 200 kW boiler	Water-wall boiler 10 MWth	Arimax Eetta 200 kW boiler	Arimax Eetta 500 kW boiler	Oilon Lenox GRT-5L	Oilon Lenox GRT-5L	Oilon Lenox GRT-5L
O_2 [vol%]	4	5-6	6	8	4-5	3.4-3.5	3.3-3.6	3.4-3.7
CO [ppm]	32	28-40	20	50-70	40-50	2-5	1-2	2-4
CO_2 [vol%]	--	--	--	--	--			
NOx [ppm]	142	137-170	150	170	250-500	193-205	159-164	88-105
SOx [ppm]	--	--	--	--	--	--	--	--
THC	--	--	--	--	--			
Particulate [g/MJ]	--	--	--	--	--	15	15	98
Bacharach No.	5	2.5-5	2.8	3-4	2	1.3	2	5.9
Reference	5	6	7			8	8	8

1 with additional 3wt% ethanol and 3wt% water, modified refractory in boiler to ensure complete combustion

2 CFB - circulating fluidised bed, BFB - bubbling fluidised bed

Table 4: Engine and burner emissions.

Liquid Source	BTG			BTG
Feedstock	Mixed hardwood			
Solids [wt%]				0.35
Pilot fuel	Diesel			None
Engine & User	250 kWe Mirrlees 6 cylinder. Ormrod Diesels, UK			Stirling Engine FLOX® burner. ZSW Germany
Pilot fuel [wt%]	7	17	100	
O_2	15	15	15.8	6-10
CO [ppm]	3475	2057	271	<35-125
CO_2 [vol%]	4.4	4.6	3.9	
NO [ppm]	240	313	510	
NO_2 [ppm]	41	77	76	
NOx [ppm]	384	266	586	20-95
SOx [ppm]	0	33	86	
THC [mg/m^3]	--	--	--	20-40
Particulate [mg/Nm^3]	--	--	--	
Reference	9			10

Table 5: Valmet 420 DS 84kW diesel engine emissions. Diesel as pilot fuel (11).

Main fuel	NOx g/kWh	CO before catalyst g/kWh	CO after catalyst g/kWh	THC before catalyst g/kWh	THC after cat. g/kWh	Smoke number[1]
Diesel	7.1	1.8	0.1	1.7	< 0.1	2.7
Ethanol	4.6		0.1		< 0.1	0.5
Pyrolysis oil[2]	3.2	9.9	0.1	1.3	0.4	0.7

[1] Bosch-units NOx = Nitrogen oxides, CO = Carbon monoxide, THC = Total hydrocarbons
2 Ensyn Mixed hardwood. Solids content 0.45 wt%.

TURBINE EMISSIONS

There has been work performed only on one turbine in Canada (12) and a small turbine combustor in Germany (13, 14). Only 6 continuous hours have been obtained on the Orenda Aerospace gas turbine using Ensyn liquids in 1997. Work by the University of Rostock on a 300 kWth turbine combustor test facility and a 75 kWe gas turbine give relative values for CO_2, CO, NOx and hydrocarbons of 105, 780, 45 and 981% respectively compared to operation of the turbine on 100% diesel (14). These values are given in Table 66 below. Other work on a small turbine combustor [100 kg/h 20:80 ethanol pyrolysis liquids mixture] with some data on COx and NOx emissions is available (15).

Table 6: Turbine combustor emissions using pyrolysis liquids.

Liquid Source	BTG	NO_2 [ppm]	42
Feedstock	Mixed hardwoods	NOx [ppm]	72
Turbine	80 kWe MAN	SOx [ppm]	n.a.
O_2	5	THC [mg/m^3]	---
CO [ppm]	3434	Particulate [mg/Nm^3]	---
CO_2 [vol%]	10.5	Bacharach No.	---
NO [ppm]	30	Reference	16

PROPOSED STANDARDS FOR BIOMASS DERIVED PYROLYSIS LIQUIDS

Pyrolysis technologies are slowly emerging from pilot and demonstration scale to commercial scale for heat and power applications. One of the difficulties facing the industry is that pyrolysis liquids will have to compete with conventional fuel oils, which are well established and known to the end user [utility, local grid, on-site use for heat and power, etc.].

Some of the main problems in the uptake of pyrolysis liquids have been:
1. Irregular supply,
2. Variable moisture content [and heating value],
3. High solids [and method for accurate determination],
4. Chemical instability [inherent in all liquids],
5. Inconsistent physical properties,
6. Poor fuel quality [relative to conventional fuel oils],
7. Lack of co-ordination between producers and end-users/developers [see above],
8. High cost of production [compared to other alternative fuels].

To define a "standard" pyrolysis liquid is therefore very difficult, if not impossible, as the characteristics of the liquid change with feedstock to be pyrolysed and the processing conditions.

Recent surveys carried out among end users, who participate in this project and fuel oil providers have highlighted that the following properties are viewed [in priority] of significance in the pyrolysis liquids for boiler, engine and turbine testing. Based on the diesel engine and boiler tests performed, an engine manufacturer, Wärtsilä NSD Oy, and a potential bio-oil user, Birka Energi (former Stockholm Energi AB), set the requirements for heating values and water, solids, and ash content that are presented in Table 7. This formed the basis for further discussions with Oilon Oy and Fortum Oy, who also input their boiler experiences to the liquid specifications. As fuel oil specifications cover a range of applications from boiler to turbines, the grade of fuel required also increases, for a wider range of physical and chemical properties. This means that properties important for one application may not be as important for another. The size of unit will also influence the properties of the liquids, as smaller domestic boilers require liquids with low viscosities, while larger commercial boilers may accept a liquid with a higher viscosity.

Table 7: Specifications for biomass pyrolysis oil for diesel engines and boilers.

	Diesel engine tests by Wärtsilä		Tests at Birka
	Specifications	Allowable variation (10%) for a container (surface/bottom)	Combustion Specifications
Homogeneity	7-day storage	no phase-separation	-
Water, wt%	max 26	max 10% difference, max 26	max 25
HHV, MJ/kg	min 18	max 10% difference,	min 19
LHV, MJ/kg	min 16	min 18 and 16	min 17
Ash, wt%	max 0.1	0 - 0.1	max 0.1
Solids, wt%	max 1	0 - 1	max 1
<50 μm	100	-	-
<25 μm	min 90	-	-

Several activities have resulted in some generalised standards for biomass derived pyrolysis liquids, but with limited scope, as proposed by Diebold et al. (17). Within the aims of the work, based on discussions with the equipment providers, liquid producers and end users noted, five properties were identified. Values for these five basic properties are given in Table 8. HHV is not included, as it is a function of the water content, therefore with the accepted maximum of 27 wt% water, the minimum HHV of the liquids should be above 18 MJ/kg.

Table 8: Key properties and specifications for "premium" pyrolysis liquids.

Property	Specification
Homogeneity	Single phase after 6 months
Water, wt%	Maximum 27
Solids, wt%	< 0.1
< 50 μm	100%
< 25 μm	>90%
Stability index	viscosity increase max 50% in 6 months
Flash point, °C	> 40

CONCLUSIONS

There is limited operational experience on fast pyrolysis liquids in boilers, engines or turbines, which can instil sufficient confidence to implementation of heat or CHP plants. Liquids quality is still inconsistent and there is no long term testing of liquids in engines or turbines. Technical difficulties concerning the use of the liquids remain and there are insufficient commercial pyrolysis plants producing liquids for long duration testing.

ACKNOWLEDGEMENTS

The authors would like to thank the European Commission ALTENER II programme which has part-funded the work which has been carried out.

REFERENCES

1. Cuevas A., Reinoso C. and Lamas P. (1995) Advances and Developments at the Union Fenosa Pyrolysis Plant. In: *8th Biomass Conference for Energy and the Environment*, (Ed. by P. Chartier, A.A.C.M. Beenackers and G. Grassi), Vol. 2 pp. 1506-1512. Pergamon.
2. Cuevas A, Reinoso C. and Scott D.S. (1994) The production and handling of WFPP bio-oil and its implications for Combustion. In: *Proceedings Biomass Pyrolysis Oil Properties and Combustion Meeting*, (Ed. by T.A. Milne), pp. 151-156. NREL.
3. Sipilä K., Oasmaa A., Arpiainen V., Westerholm M., Solantausta Y., Angher A., Gros S., Nyrönen T. and Gust S. (1996) Pyrolysis oils for power plants and boilers. In: *Biomass for energy and the environment. 9th European Conference on Bioenergy*, (Ed. by Chartier, P. et al. (eds.). Vol 1., pp. 302–307, Elsevier.
4. See http://www.rotarymicropower.com/Biooilsubpages/Rotary%20Engine%20Biofuel.htm for details of preliminary test work.
5. Gust S. (1997) Combustion experiences of flash pyrolysis fuel in intermediate size boilers", Developments. In: *Developments in Thermochemical Biomass Conversion*, (Ed. by A.V. Bridgwater & D.G.B. Boocock), Vol. 1, p 481-488, Blackie Academic and Professional.
6. Gust S. (1997) Combustion of Pyrolysis Liquids. In: *Biomass Gasification and pyrolysis - state of the art and future prospects,* (Ed. by M. Kaltschmitt & A.V. Bridgwater), pp. 498-503, CPL Press.

7. Huffman D.R. and Freel B. A. RTP™ Biocrude: a combustion/emissions review. In: *Developments in Thermochemical Biomass Conversion*, (Ed. by A.V. Bridgwater & D.G.B. Boocock), Vol. 1, p 489-494, Blackie Academic and Professional.
8. Oasmaa A., Kytö M. and Sipilä K. (2001) Pyrolysis oil combustion tests in an industrial boiler. In: *Progress in Thermochemical Biomass Conversion*, (Ed. by A.V. Bridgwater), Vol. 2, pp. 1468-1481, Blackwell Science Ltd..
9. JOR3-CT97-0197 Contractors' Report, Aston University [co-ordinator], Birmingham, UK, 2001.
10. Bandi A. and Baumgart F. (2001) Stirling Engine with FLOX® Burner fuelled with fast pyrolysis liquid In: *Progress in Thermochemical Biomass Conversion*, (Ed. by A.V. Bridgwater), Vol. 2, pp. 1459-1467, Blackwell Science Ltd..
11. Solantausta Y., Nylund N-O., Oasmaa A., Westerholm M. and Sipilä K. (1994) Preliminary tests with wood derived pyrolysis oil as fuel in a stationary diesel engine In: *Proceedings Biomass Pyrolysis Oil Properties and Combustion Meeting*, (Ed. by T.A. Milne), pp. 355-361, NREL.
12. Andrews R.G., Zukowski S. and Patnaik P.C. (1997) Feasibility of firing an industrial gas turbine using a biomass derived fuel. In: *Developments in Thermochemical Biomass Conversion*, (Ed. by A.V. Bridgwater & D.G.B. Boocock), Vol. 1, pp. 495-506, Blackie Academic and Professional.
13. Strenziok R., Hansen U. and Wickboldt P. (2000) Investigation of flame characteristics and emissions of pyrolysis oil in a modified flame tunnel. In: *Biomass: a growth opportunity in green and value-added products*, (Ed. by R.P. Overend & E. Chornet), Vol. 2, pp. 1241-1246, Elsevier Science.
14. Strenziok R., Hansen U. and Wickboldt P. (1999) Combustion of pyrolysis oil in a gas turbine. In: *Combustion and Incineration*, Vol. 1492, Chapter 113, pp. 417-421.
15. López Juste, G. and Salvá Monfort, J.J. (2000) Preliminary Test on combustion of wood derived fast pyrolysis oils in a gas turbine combustor. In: *Biomass and Bioenergy*, 19, pp. 119-128.
16. Wagenaar B.M., Venderbosch R.H., Carrasco J., Strenziok R. and van der Aa, B.J. (2001) Rotating cone bio-oil production and applications. In: *Progress in Thermochemical Biomass Conversion*, (Ed. by A.V. Bridgwater), Vol. 2, pp. 1268-1280, Blackwell Science Ltd..
17. Diebold J.P., Milne T.A., Czernik S., Oasmaa A., Bridgwater A.V., Cuevas A., Gust S. Huffman D. and Piskorz J., (1996) Proposed specifications for various grades of pyrolysis oils. In: *Developments in Thermochemical Biomass Conversion*, (Ed. by A.V. Bridgwater & D.G.B. Boocock), Vol. 1, pp. 1-26. Blackie Academic and Professional.

Forestera™ - Liquefied Wood Fuel Pilot Plant

Steven Gust
Fortum Oil and Gas Oy, POB 310, 06101, Porvoo, Finland
Jukka-Pekka Nieminen
Fortum Oil and Gas Oy, POB 310, 06101, Porvoo, Finland
Timo Nyrönen
Vapo Oy, POB 22, 40101 Jyväskylä, Finland

ABSTRACT: Fortum and Vapo of Finland have constructed a pilot plant for process development and test production of liquefied wood fuels. Commissioning has been completed and the plant was officially opened May 14, 2002. Both the fuel and process have been registered under the tradename Forestera™. The fuel is produced by a pyrolysis process, e.g. heating of feed up to approximately 500 °C followed by condensing of vapours. Plant capacity is 0.5 dried tonnes per hour of feed. The pilot plant is fully automated and integrated including wood chips receiving, crushing, drying using process heat, Forestera™ cracker, condensers, fraction and additive mixing and transfer to storage tanks. In this presentation, experiences from the planning, constructing and commissioning of the plant together with early operations are described.

BACKGROUND

PROJECT DRIVERS

As EU member states are preparing to increase their use of renewable energy and reduce their greenhouse gas emissions, various incentives are being offered for wind, solar and biomass systems in the form of fuel tax reductions and investment support. In the area of biomass, a variety of processes are available to convert biomass into an energy carrier such as combustion, co-combustion, gasification, pyrolysis, fermentation, pressing into pellets or vegetable oils. The heating of wood to form tars has a long history in Finland of more than 300 years and was Finland's first major export product. But it wasn't until the oil crises of the 1970s and the search for alternative liquid fuels that it was discovered that high liquid yields could be obtained from lignocellulosic feedstocks by a fast cracking or pyrolysis process. Both the advantages: high liquid yields, good energy and CO_2 balance, reasonable costs and disadvantages: similarity to wood tar, acidity, low heating value, poor stability and lack applications of pyrolysis processes have been documented.

Due to the physical and chemical properties of pyrolysis liquids and the high cost for refining the liquid into a traffic fuel, the main applications that are currently being pursued for pyrolysis liquids are as a stationary engine fuel or as a boiler fuel. In Scandinavia, good market opportunities exist for heating fuels while in central Europe, power applications are emphasised. Since consumer prices of light fuel oil (LFO) is 9-10 €/GJ and heavy fuel oil 4-6 €/GJ in most EU countries except for Sweden and Denmark where prices are higher, the replacement of LFO, although more technically demanding, is more economically attractive. On the EU level, this has the added advantage that the replacement of LFO will free up refinery middle distillates, which can be used as a diesel fuel.

In order to achieve customer and authority approval for this technology, large-scale demonstration will be necessary. This will also give added incentive to application developers.

FORTUM AND VAPO PROJECT

Both Fortum (formerly Neste and IVO, now Fortum Oil & Gas and Fortum Power & Heat) and Vapo have been assessing pyrolysis technologies and fuel properties for over 10 years together with various partners. In 1998, these companies entered into an agreement where they would jointly develop their own pyrolysis process and related production technologies. Vapo with it's expertise in solid fuels handling and logistics, Fortum Oil and Gas in process development and Fortum Power and Heat in heat and power plant operations cover the entire process chain. Together they have registered Forestera™ for both their process and the products. The pilot plant that has been built will provide the needed liquids for further application testing development and allow the companies to assess the technology required to achieve the fuel quality that will be required in the marketplace.

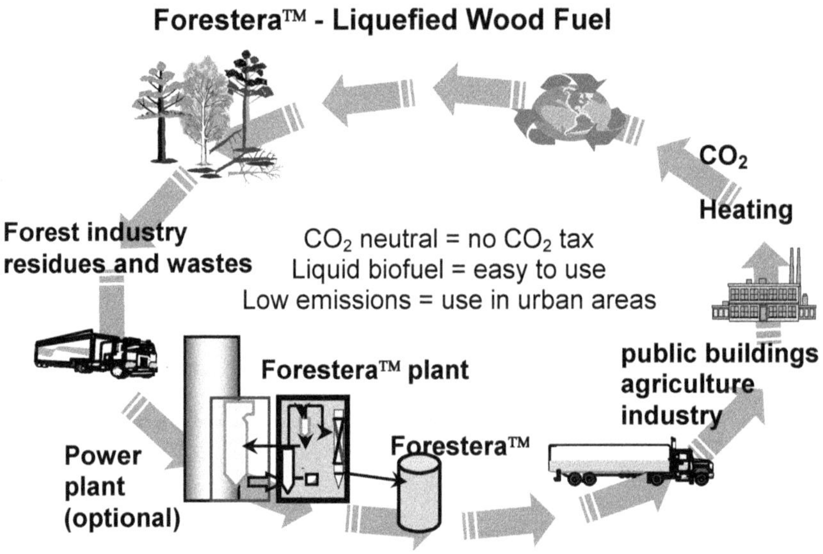

Figure 1: Business concept for Forestera™.

FORESTERA™ PROCESS AND PRODUCT

The fast cracking or thermal degradation of biomass in an essentially oxygen free atmosphere produces a combination of solids, condensable vapours and gases. The ratio of these three can be controlled by reaction conditions. The patented Forestera™ fast cracking process transfers heat to the feed material by a unique and novel combination of gases, bed materials and hot surfaces as shown in Figure 2. A special feature of the process is the solids separation unit where the solids content is reduced to under 0.1 weight percent, which has been found to improve stability, reduce particulate emissions during combustion and eliminate sludge formation during storage. A number of Forestera™ product grades will be produced as work with applications progresses.

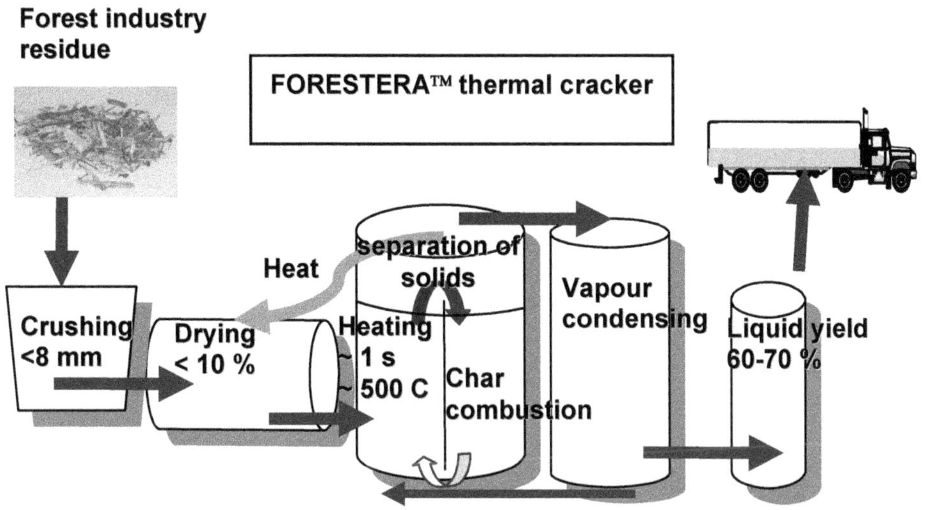

Figure 2: Forestera™ fast thermal cracking process.

DESIGN AND CONSTRUCTION OF PILOT PLANT

BASIS FOR THE DESIGN

The basis for the design of the pilot was the Forestera™ fast cracking process. Special attention was paid to the flexibility of the plant to enable us to test and modify different process conditions and solutions. The size of the pilot plant was a compromise based a number of criteria. First it was to be large enough to allow rapid scale up to commercial sizes. This meant that commercially available. normal industrial equipment that could also be used also in larger plants of commercial size are used. We selected the size of the process units in order to minimise risks of process scale up to commercial sizes especially taking into account residence time and wall effects. But at the same time, the pilot plant size was carefully chosen that process changes and modifications would not become prohibitively expensive or that they would not fit into our existing infrastructure on the chosen building site and availability of utilities. The optimum size for our pilot plant was chosen as 500 kilograms dried feed input per hour with a minimum feed rate of 100 kg/hour.

The development of our patented Forestera™ fast cracking process evolved from a combination of various in-house related process development projects together with our critical assessment of public domain knowledge and our experiences from various pyrolysis projects over the past 10 years. Special features of the plant relate to the safety aspects such as multiple inertisation systems: nitrogen, steam and water spray to prevent dust explosion; spark detection in drier loop, constant oxygen, CO, temperature and pressure monitoring.

The pilot plant has been constructed in an existing building in the Technology Centre at Fortum's Porvoo Refinery. The site was chosen as the 12 meter high building came equipped with a crane and all required utilities of steam, cooling water, natural gas, instrument air etc.

The plant is fully automated: it has separate systems for operation and safety with manual override. Feedstock preparation equipment for the plant was delivered as two larger packages by subcontractors. The other package includes receiving station with walking floor, crusher with different sieve sizes transport systems and silos for solid feedstock. The other package covers flue gas drum drier with associated equipment e.g. spark detection and flame extinguishing system.

PILOT CONSTRUCTION AND COSTS

After the decision was taken to construct the pilot plant in the Combustion Laboratory at the Fortum Porvoo Refinery Technology Centre Construction time one year. Mechanically ready early 2002. Investment costs 3,5 M€. Located at Fortum Porvoo Technology Centre.

Figure 3: Construction of the pilot plant showing receiving station, crusher and drier.

COMMISSIONING AND OPERATION OF THE PILOT PLANT

COMMISSIONING

Upon completion of plant construction, the pilot plant has undergone a thorough and extensive commissioning. Commissioning of the plant took place during the spring of 2002. The Minister of Trade and Industry Sinikka Mönkäre officially opened the plant 14.5.2002.

The commissioning consisted of the usual testing of unit operations, warranty runs of purchased equipment, verification of safety systems and plant automation. Much of the time during commissioning has been spent modifying the biomass preparation subsystems in order to be able to deal with different feedstocks under various feed rates since although biomass systems are commercial on larger scales, it has proven difficult to find appropriate commercial equipment to cover the scale of our operations.

An example of one of the particular interesting features that was noted during the commissioning was concerning the manner in which the plant reached normal operating conditions. During the planning of the plant, the alarm levels and safety shutdown levels for gases, temperatures and pressures were set for normal operations. During start up it was noted that a number of these levels could not be attained immediately and so it has been necessary to either modify the levels or in some cases by-pass the forced shutdown. For example the maximum oxygen level in the flue gas before entering our drier is normally 10% but during warming of the drier and before biomass has been fed in, higher levels can be accepted.

Figure 4: Opening of the pilot plant by Sinikka Mönkäre, Minister of Trade and Industry:

Figure 5: The Forestera™ Pilot Plant showing Reactor (centre) and Vapour Recovery (left).

OPERATION

The process development work will concentrate on reliability and durability of equipment and components, defining of the operating window for the various process steps, setting the specification and quality requirements of the process, determining the requirement of various utilities etc.

The initial operational experiences from the plant are from summer 2002. A number of different feedstocks have been tested and based on these experiences, modifications have been made to the feed systems and to the reactor conditions. Therefore the total number of hours in which the entire plant has been operating are limited and the observations presented here are preliminary.

Experiences and conclusions from the first test runs includes following observations:

(a) the pyrolysis part of the plant as a whole operates as planned
(b) the process is stable, reactor temperatures can be held to $\pm 2^{\circ}C$
(c) plant has good turndown ratio from 500 to 100 kg/hour feed

FUTURE PLANS

Forestera™ will be developed and commercialised as both a heating fuel (one or two grades) and as a technology. Both will have separate business plans and time schedules. Because neither fuel nor the processing technology is yet fully developed, business plans are preliminary.

The consortium will carry out technical development utilising the pilot plant and as this work has just commenced, the time required for this development is uncertain but will most likely take between 18 months and three years.

In parallel to this process and product development we are developing together with Oilon Oy required combustion system for Forestera™. This development includes field testes with fuel from pilot plant, which we are hoping to be able to start for next heating season.

Subsequent to this technology development we will design and construct the first commercial size plant. Engineering and construction will take one to two years. When this phase is finished, we are then introduce the products to commercial markets.

Low-temperature Pyrolysis of CCA Treated Wood Waste: Mechanism Responsible for Arsenic Release

L. Helsen and E. Van den Bulck
Department of Mechanical Engineering, Katholieke Universiteit Leuven, Celestijnenlaan 300 A, B-3001 Leuven (Heverlee), Belgium

ABSTRACT: Low-temperature pyrolysis may be a good solution to the disposal problem of CCA treated wood waste. In this paper, different experimental observations with respect to the behaviour of As during the pyrolysis process are brought together in order to identify the mechanism responsible for As release. Lab-scale pyrolysis experiments, performed at different working conditions, show that As releases are substantial (2.5 % up to 17.5 %) at a pyrolysis temperature of 320 °C. Thermogravimetric (TG) analysis of hydrated chromium arsenate ($CrAsO_4.6H_2O$, the major As compound in CCA treated wood) reveals that the decomposition of this compound probably results in the formation of solid Cr_2O_3 and gaseous H_2O, O_2 and As_4O_6, indicating that As(V) is reduced to As(III) and arsenic is released in the trivalent form. TG studies carried out with arsenic trioxide (As_2O_3) show that As_2O_3 is already released at temperatures as low as 200 °C. This release is driven by temperature dependent vapour pressures. Pure arsenic pentoxide (As_2O_5.aq) does not decompose (or reduce), nor volatilise at temperatures lower than 500 °C, while in the pyrolysis residue, resulting from pyrolysis of CCA treated wood at 350 °C for 20 minutes, trivalent arsenic is found and arsenic is already released at temperatures as low as 320 °C. An explanation could be that the reduction of As(V) to As(III) during pyrolysis is favoured by the reducing environment formed by char and pyrolysis vapours. Another explanation could be that part of the arsenic in the CCA treated wood is already present in the trivalent form. The most important conclusion is that limiting the As release during the pyrolysis of CCA treated wood implies that arsenic has to be kept in pentavalent form and thus the reduction reaction has to be controlled and avoided.

INTRODUCTION

Waterborne salts have been used to preserve wood from insects, fungi and water damage for many years. One of the more common formulations contains copper, chromium and arsenic salts and is known as chromated copper arsenate or CCA. After impregnation of the wood with a CCA solution the metal compounds will be fixed to the cell walls of the wood matrix. Substantial amounts of CCA remain in the wood for many years and the disposal of scrap wood is a growing problem in Europe and in the United States. Telephone poles, railway sleepers, timber from landscape and cooling towers, wooden silos, hop-poles, cable drums and wooden playground equipment generate wood waste for which environmental benign disposal technologies need to be developed.

The recent consciousness of the need for abatement of air pollution leads to further interest and investigation of pyrolysis as a major process for the disposal of enormous quantities of cellulosic wastes and residual materials. Low-temperature pyrolysis may be a good solution to the growing disposal problem of CCA treated wood waste. A study has been set up to design a low-temperature pyrolysis facility for the CCA treated wood waste such that the metals are concentrated in the solid product stream, and the pyrolysis gases and liquid are used to their maximum potential with respect to energy recuperation [1,2].

A set of experimental and theoretical studies has been carried out in order to gain more insight in the metal (Cr, Cu and As) behaviour during the pyrolysis process. Among these studies are: lab-scale pyrolysis experiments using different temperatures and residence times [1,3], determination of mass balances over the total system using inductively coupled plasma mass spectrometry (ICP-

MS) [4,11], sequential extraction of the pyrolysis residue [4], speciation study of As in the pyrolysis residue using hydride generation coupled to ICP-MS [4], study of metal microdistribution in wood and pyrolysis residue by energy dispersive X-ray analysis coupled to scanning electron microscopy (SEM-EDX) [5], thermogravimetry (TG) and evolved gas analysis (EGA) of CCA treated wood [6], TG analysis of the major arsenic compound in CCA treated wood ($CrAsO_4$) [7], development of a sampling train for arsenic in pyrolysis vapours, derivation of a kinetic scheme for the pyrolysis of CCA treated wood [7], modelling of arsenic release in the temperature range from 350 °C to 450 °C [8] and TG analysis of arsenic model compounds (pure arsenic oxides).

Since arsenic is identified as the problem compound with respect to metal release during pyrolysis, the aim is to identify the mechanism of arsenic release. Knowledge of this mechanism may help to control and prevent arsenic release. In this paper, all experimental observations that contribute to the insight in the mechanism of arsenic release are brought together. In the first part observations concerning the thermal behaviour of CCA treated wood are outlined, the second part focuses on the thermal behaviour of some arsenic model compounds (chromium arsenate and arsenic oxides). The paper is organised as follows. In the "Methods and materials" section the different samples and experiments are briefly described. In the "Results and discussion" section the results of the two parts (1) thermal study of CCA treated wood and (2) thermal study of arsenic model compounds are discussed. Finally, the most important conclusions are summarised in the "Conclusions" section.

METHODS AND MATERIALS

SAMPLES AND SAMPLE PREPARATION

CCA treated wood

The type of wood used in the lab-scale pyrolysis study is *Pinus sylvestris* peeling, that is impregnated with a 3.3 % solution of type CCA 1-C salt: 32.5 % $CuSO_4.5H_2O$, 41.1 % $Na_2Cr_2O_7.7H_2O$ and 26.4 % $As_2O_5.2H_2O$. The peeling wood chips are between 20 and 80 mm long, between 2 and 12 mm wide and between 0.5 and 1.5 mm thick. Double impregnation of the peeling is carried out in order to simulate a CCA level that is higher than the levels found in CCA treated wood waste (limiting case for industrial scale processes or problems). The peeling is treated twice for class 4 impregnation (wood may be in contact with groundwater and sweet water, requiring a salt retention of 9 kg CCA per m^3 of wood). A full-cell treatment (according to the procedure "Bréant-Bethell") is applied to the wood chips, which have a moisture content of less than 22 %, in an industrial autoclave in five stages. The impregnation process is described in detail elsewhere [8]. The treatment procedure as well as the post preservation conditioning is carried out according to the French norm 7925-NF C67-100-1500-3/82 [9], which is developed for the treatment of electricity and telephone poles. The samples have been prepared by the company Beaumartin (France).

The CCA treated wood samples used in the lab-scale pyrolysis experiments are oven-dry wood samples. The drying of the wood was accomplished by putting the wood in a conventional oven at a pre-set temperature of 105 °C, until the weight of the wood sample is constant (drying time is approximately 3 hours). The percentage of moisture, calculated from the difference in mass before and after drying, is between 11 and 12 %. The exact CCA retention of these double-treated samples is not known, but the metal (Cr, Cu and As) content is determined using inductively coupled plasma mass spectrometry (ICP-MS), resulting in mean concentrations of 9875 ± 138 µg/g for Cr, 3746 ± 60 µg/g for Cu and 9204 ± 156 µg/g for As in the CCA treated wood [3].

Chromium arsenate

Since chromium arsenate can not be bought as such, it has to be synthesised [7]. The following procedure is used. Concentrated aqueous solutions of CrO_3, As_2O_5.aq and D(+)-glucose are prepared separately. These three solutions are mixed and water is added to obtain a mixture of 1 M D(+)-glucose, 0.028 M As_2O_5.aq and 0.05 M CrO_3. All reagents used are of analytical-reagent grade and ultra-pure Milli-Q water (Millipore, Bedford, MA, USA) is used for the dissolutions. The mixture is stirred using a magnetic stirrer and heated to approximately 60 °C for 4 hours to accelerate the reactions. A dark brown solution is formed, but after 6 hours a precipitate can not yet be observed. After 50 hours the whole solution is turned into green and a fine precipitate is visible. The solution is filtered (after 70 hours) through a 0.4 μm membrane-filter using a vacuum pump. During this filtration process, some of the green precipitate is collected on the filter, whereafter it is dried in a vacuumoven (-30 inch Hg) for 90 minutes, resulting in a very viscous gel-like paste. This compound should be pure hydrated chromium arsenate ($CrAsO_4.6H_2O$), if temperatures higher than ambient temperature and enough reducing agents, such as glucose, are used, whereby enough time is given [10].

Arsenic oxides

Arsenic pentoxide hydrate (As_2O_5.aq, aq being 2 up to 3 mol H_2O) from purum quality is purchased from Fluka Chemika. Arsenic trioxide (As_2O_3) from pro analyse quality is purchased from Riedel-de-Haen.

LABSCALE PYROLYSIS EXPERIMENTS

The influence of pyrolysis temperature and residence time of the wood chips on the metal content of the resulting pyrolysis residue is determined by use of a lab-scale downdraft fixed-bed pyrolysis reactor. The experimental set-up is schematically shown in Figure 1.

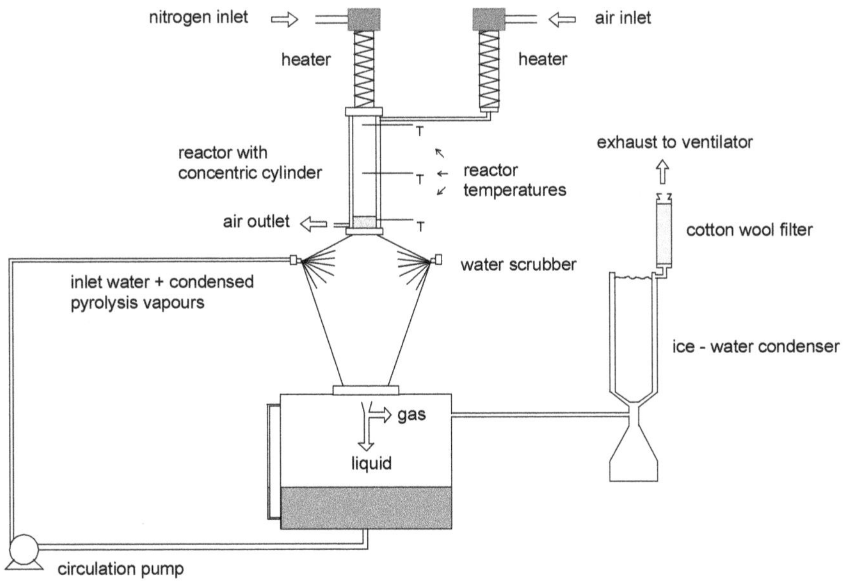

Figure 1: Experimental facility for the pyrolysis of wood chips: downdraft fixed bed reactor with gas cleaning system.

The reactor that holds the wood is a stainless steel circular tube with a diameter of 50 mm and a length of 200 mm. Reactor temperatures are measured at three positions (bottom, middle and top) in the reactor. In this study, it is very important to control the reactor temperature very accurately and to reach this reactor temperature as fast as possible with a high repeatability. In a typical experiment, 2 g of wood is loaded into the reactor. As a consequence, the top of the wood column in the reactor reaches just above the thermocouple that measures the bottom temperature. The bottom reactor temperature is taken as the pyrolysis temperature that characterises the process.

Temperature control is performed as follows. A flow of heated nitrogen (2 Nm^3/h) is forced through the reactor. The gas heater is a very compact 3 kW electrical resistance heater, which is capable of producing exit gas temperatures up to 650 °C. The heater outlet temperature is regulated by use of an auto-tune PID temperature controller. To avoid radiation influences of the red hot electrical heating element on the reactor-top-thermocouple, a radiation shield is provided by putting a ball of stainless steel gauze above the reactor-top-thermocouple. The reactor tube is thermally guarded with an outer concentric cylinder, through which a flow of heated air is forced. At the outer side of this concentric cylinder a thick layer (thickness of 60 mm) of rockwool isolation limits the heat losses to the environment. This isolation layer is not drawn in Figure 1 in order to keep the figure clear enough. The amount of heat transported from the hot air flow through the reactor wall depends on the airflow rate, which is set to 50 Nm^3/h, and the temperature of the air. The airflow is heated by use of a very compact 8 kW electrical resistance heater, which is capable of producing exit gas temperatures up to 650 °C. The air temperature at the inlet of the outer cylinder is regulated by use of an auto-tune PID temperature controller. A nitrogen flow rate of 2 Nm^3/h, combined with a rise of the (bottom) reactor temperature from room temperature to the pre-set pyrolysis temperature in 5 minutes (actually it takes 6.5 minutes before the pre-set temperature is reached) results in a fast temperature increase that is well controllable and repeatable. The nitrogen flow rate is set by a mass flow controller, while for the air flow rate a rotameter is used.

At the outlet of the reactor the nitrogen flow contains non-condensable pyrolysis gases as well as microscopic tar droplets. The gas flow that exits the reactor is immediately quenched in a water scrubber. The cold wet gas flow at the outlet of the scrubber forms an aerosol of microscopic tar and water droplets. In the bottom vessel the liquid (water and condensed pyrolysis vapours) and gaseous (nitrogen and non-condensable pyrolysis gas) streams are separated. The liquid is recirculated with a circulation pump and is re-used to feed the water scrubber. The gas is forced through an extra ice/water condenser in which the condensate is collected and finally through a tube filled with cotton wool that acts as a filter. Each experiment is carried out in duplicate in order to check the repeatability. To determine the metal retainment in the pyrolysis residues, the total amount of Cr, Cu and As in the CCA treated wood and in the pyrolysis residues is measured by ICP-MS. Therefore, the metals have to be brought in solution. During previous research work, several leaching and dissolution methods were tested and compared to find the optimal procedure [4,11]. To determine the total amount of Cr, Cu and As in the dried CCA treated wood, a leaching procedure using H_2O_2 and H_2SO_4 (BSI method) is used. To determine the total amount of Cr, Cu and As in the pyrolysis residues, the sample is dissolved in concentrated HNO_3. The dissolution procedure for the pyrolysis residues is adjusted in order to be able to dissolve the total mass of the pyrolysis residue for each experiment instead of only 0.1 g [3]. More details about the ICP-MS analyses are given elsewhere [4].

The working conditions studied here, are the combinations of the pyrolysis temperatures 300 °C and 320 °C and the residence times 20 minutes, 40 minutes and 60 minutes. Higher temperatures (350 °C, 400 °C and 450 °C) have been studied in earlier research work [1,12], where they give rise to substantial metal releases.

SPECIATION STUDY

A pyrolysis residue, resulting from pyrolysis at 350 °C for 20 minutes, is subjected to two successive extraction steps: (1) using dilute NH_4Ac (pH = 7) and (2) using dilute NaAc + HAc (pH

= 5). A hydride generator (HG) is coupled online with the ICP-MS system to determine the oxidation state of As in these two extraction steps (neutral environments).

THERMOGRAVIMETRIC ANALYSIS

Thermogravimetric (TG) analysis is carried out using a DuPont Instruments 951 Thermogravimetric Analyser, supported by a PC and software for control and data handling. To analyse the gas stream, resulting from the decomposition of the sample, the TG apparatus is coupled to a VG gas analysis system (FISONS Instruments - Thermolab) which is able to measure components up to mass 300. In this study the sample, approximately 20 mg in weight, is introduced into a quartz sample pan and heated to a pre-set temperature following a pre-set temperature profile using nitrogen (to have an inert atmosphere) or dry air as purge gas. Only when gas analysis is carried out, helium is used as purge gas. In order to ensure inert working conditions, the TG apparatus has been modified as described elsewhere [13]. Moreover, the installation is flushed with nitrogen for 1 hour before starting the temperature rise. The apparatus provides for the continuous measurement of sample weight as a function of temperature and provision is made for an electronic differentiation of the weight signal to give the rate of weight loss.

When TG experiments are carried out with arsenic model compounds (such as chromium arsenate or arsenic oxides), a sampling train for arsenic is coupled to the outlet of the TG apparatus in order to keep the laboratory environment safe to work in. The sampling train consists of a cooling tube, a filter section containing an alkali impregnated cellulose ester membrane filter with backup pad and an impinger containing 50 ml NaOH 0.1 N. In order to have substantial bubbling in the impinger, a higher purge gas flow rate (220 ml/min) is used in these TG experiments.

RESULTS AND DISCUSSION

THERMAL BEHAVIOUR OF CCA TREATED WOOD

Metal retainment in the pyrolysis residue

The measured metal concentrations in the CCA treated wood and in the pyrolysis residues are recalculated to obtain the percentages of metals that are retained in the pyrolysis residues for the different working conditions. These percentages are calculated as:

$$\text{metal retainment (wt \%)} = \frac{[\text{metal}]_{\text{res}} \cdot m_{\text{res}}}{[\text{metal}]_{\text{dw}} \cdot m_{\text{dw}}} \cdot 100$$

with $[\text{metal}]_{res}$ and $[\text{metal}]_{dw}$ the metal concentrations (in µg/g) in residue and dried wood, respectively, and m_{res} and m_{dw} the masses (in g) of residue and dried wood, respectively. These calculated values are presented in Figure 2 for the three metals as a function of temperature with the residence time as parameter. Each working condition (combination of temperature and residence time) is represented by two points. Error bars are not shown for the sake of clearness. The possible causes of errors are:

(1) Determination of metal concentrations in the pyrolysis residues: errors can come from dissolving the sample by the Reflux method (weighing, dissolution, filtration), dilution and ICP-MS analysis. The corresponding standard deviation on the metal concentrations in the pyrolysis residues is estimated to be 1.7 %.
(2) Determination of mass of the pyrolysis residues: a balance with an accuracy of 0.03 g is used, resulting in an estimated standard deviation of 3.5 %.

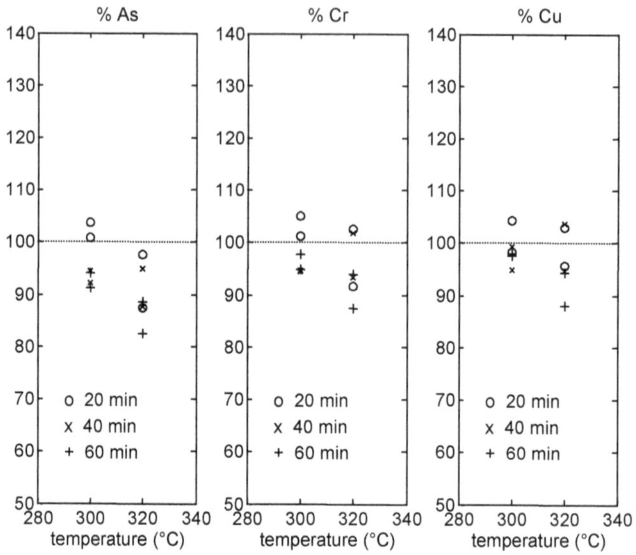

Figure 2: Percentages of metals (As, Cr and Cu) found in the solid residues after pyrolysis at different temperatures (300 °C and 320 °C) for different residence times (20 minutes, 40 minutes and 60 minutes).

(3) Determination of metal concentrations in the dried wood samples: errors can come from leaching the metals by the BSI method (weighing, leaching, filtration), dilution and ICP-MS analysis. The standard deviations corresponding to the calculation of the mean value of six concentrations are 1.7 %, 1.4 % and 1.6 % for As, Cr and Cu respectively.

(4) Determination of mass of the dried wood samples: a balance with an accuracy of 0.03 g is used, resulting in an estimated standard deviation of 1.5 %.

These errors are quantified based on estimates, resulting in global standard deviations of 4.3 %, and assuming a normal distribution of the metal retainment values, 95 % confidence intervals of 8.4 % are calculated.

Figure 2 clearly shows the influence of temperature and residence time on the metal retainment: higher retainment at lower temperatures and smaller residence times, with the influence of temperature being more pronounced. For all working conditions, except the combination of 320 °C and 60 minutes, the mean values (mean value of the duplicate runs) of both Cr and Cu retainment are around 100 %. According to these results, pyrolysis at 320 °C for 60 minutes would result in Cr and Cu releases of almost 10 %. The corresponding As release at these conditions is 14 %. This working point should thus be avoided. In the range of working conditions studied here, metal release seems to be zero, but is inconclusive (because of the high experimental uncertainty) for the three metals at a temperature of 300 °C which is held for 20 minutes. Arsenic is the problematic compound for which retainment values lower than 100 % are already found for temperatures as low as 300 °C, combined with residence times longer than 40 minutes.

For future industrial applications, this study has already shown that very careful control of process parameters will be required to ensure a "zero" arsenic emission. Furthermore, it is very difficult to draw conclusions with respect to metal release based on the calculated retainment values. Standard deviations of 4.3 % are too high to determine working conditions corresponding to "zero" release. Therefore, a procedure to measure arsenic emissions in the pyrolysis gas is set up and validated for the specific case of pyrolysis of CCA treated wood (presence of sticky tar compounds and aerosols).

Speciation study

The amount of As(III) found in the leachates of the first and second extraction steps are 52 % and 82 %, respectively. This speciation study shows that the major part of As in the pyrolysis residue is present in the trivalent state. The presence of As(III) in the pyrolysis residue indicates that arsenic, present as As(V) compound ($CrAsO_4$) in the CCA treated wood, is partly reduced to As(III) during the pyrolysis process. If the pyrolysis residue would be disposed on a landfill, mainly As(III), which is the more toxic form, will be liberated into the environment. Hence, the pyrolysis residue of CCA treated wood waste can not be landfilled without pre-treatment.

THERMAL BEHAVIOUR OF ARSENIC MODEL COMPOUNDS

Thermogravimetric study of chromium arsenate

The thermal behaviour of hydrated chromium arsenate ($CrAsO_4.6H_2O$) in a nitrogen atmosphere is presented in Figure 3 for a sample predried for 120 minutes and subjected to the following temperature program: heating from room temperature to 850 °C at 10 °C/min with an isothermal part of 20 minutes at 100 °C. The presentation of the DTG curve in Figure 3 is limited to the ordinate range [0,0.30] to obtain a better view of the peaks at temperatures higher than 100 °C. In Figure 3 the height of the DTG peak at 100 °C has no physical meaning since an isothermal part was programmed at 100 °C, resulting in high values of the derivative with respect to temperature.

Figure 3 shows besides the peak at 100 °C, four other peaks: at 174 °C, 241 °C, 328 °C and 456 °C:

(1) The peak at 100 °C corresponds to a weight loss of 16 % and is attributed to the release of free and crystal water.

(2) The release of crystal water (dehydration) may continue up to temperatures of 210 °C [14], which means that the peak at 174 °C is attributed to water release too. The total water release results in a mass reduction of 35 %.

(3) TG analysis of glucose reveals that two successive DTG peaks describe the decomposition of glucose: the first occurring between 194 °C and 261 °C with a peak temperature of 231 °C and the second occurring between 261 °C and 370 °C with a peak temperature of 320 °C. The DTG peak, observed for hydrated chromium arsenate at 241 °C, can thus be ascribed to the decomposition of traces of glucose.

(4) The fourth DTG peak is a broad peak located around a temperature of 328 °C. This is close to the values found for the decomposition temperature of arsenic pentoxide (327 °C [15,16,17,18] and 315 °C [17,19,20]) according to the reaction:

$As_2O_5\,(s) \rightarrow As_2O_3\,(s) + O_2\,(g)$

However, the fourth DTG peak results in a mass reduction of 12 %, which can not be attributed to this oxygen release only. Part of this mass reduction is probably due to the further decomposition of traces of glucose, since the second DTG peak for glucose exhibits its maximum at 320 °C. Moreover, part of the arsenic trioxide (As_2O_3) formed at this stage may evaporate due to the relatively high vapour pressure (60 mmHg at 310 °C [21]).

(5) The fifth peak is neither well defined at one particular temperature, but located around a temperature of 456 °C, which coincides with the volatilisation temperature of arsenic trioxide (boiling point at atmospheric pressure, being 457.2 °C [19,21]) according to:

$2\,As_2O_3\,(s) \rightarrow As_4O_6\,(g)$.

Since the fourth and fifth DTG peaks are not clearly defined, they could alternatively be interpreted as one broad peak, representing the release of arsenic trioxide. The glucose decomposition products may promote the reduction of arsenic pentoxide, resulting in arsenic trioxide release as soon as it is formed.

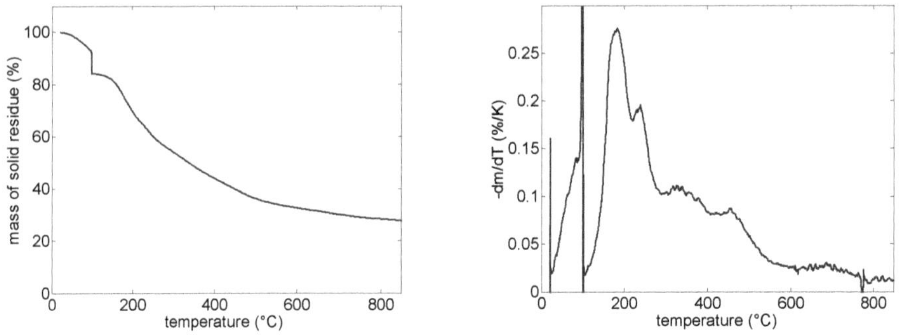

Figure 3: Thermal behaviour of hydrated chromium(III) arsenate in nitrogen atmosphere (temperature profile: heating at 10 °C/min to 850 °C with an isothermal part of 20 minutes at 100 °C). Left: TG curve, right: DTG curve.

The DTG peak temperatures may change if the thermal decomposition of a mixture, containing chromium arsenate besides some other compounds, is studied instead of the pure chromium arsenate compound. The total weight reduction is 73.8 %. About 56 %, which is about 76 % of the total amount evolved, is released at temperatures lower than 400 °C. This means that the major part of chromium arsenate is decomposed and volatilised before a temperature of 400 °C is reached.

Based on these observations and the results of equilibrium calculations carried out by other researchers [18,22,23], it is concluded that the decomposition of hydrated chromium arsenate results in the formation of solid Cr_2O_3 and gaseous H_2O, O_2 and As_4O_6 (together with the decomposition products of the traces of glucose) according to the scheme [12]:

$$4\ CrAsO_4.6H_2O\ (s) \rightarrow 24\ H_2O\ (g) + 2\ Cr_2O_3\ (s) + 2\ As_2O_5\ (s)$$
$$2\ As_2O_5\ (s) \rightarrow 2\ As_2O_3\ (s) + 2\ O_2\ (g)$$
$$2\ As_2O_3\ (s) \rightarrow As_4O_6\ (g)$$

The combination of this TG study and equilibrium studies shows that starting from chromium(III) arsenate, arsenic may already be released at temperatures between 300 and 400 °C, while Cr is not released for temperatures up to 850 °C. Furthermore, assuming that the decomposition/reduction of As_2O_5 to As_2O_3 and O_2 is responsible for the As release during pyrolysis, the experimental results of the lab-scale pyrolysis study (metal retainment in pyrolysis residue) suggests that the decomposition temperature associated with pyrolysis is lower than 320 C.

Thermogravimetric study of arsenic oxides

Arsenic pentoxide

TG analysis of arsenic pentoxide (As_2O_5.aq) results in the TG curve, DTG curve and MS spectra presented in Figure 4 (respectively the top, middle and bottom plot). The sample is heated to 500°C with an isothermal part of 10 minutes at 220 °C in a helium atmosphere. The outlet of the TG apparatus is coupled to the mass spectrometer (MS) for gas analysis. The TG curve in Figure 4 shows that the total mass loss at 500 °C is less than 12 %. Furthermore, comparison of the DTG curve and the MS spectra reveals that both DTG peaks (at 231 °C and at 348 °C) coincide with the MS peaks for H_2O. The mass loss of almost 12 % can thus be attributed to the release of H_2O (As_2O_5.aq contains between 2 and 3 mol of H_2O per mol of As_2O_5, corresponding to 13-19 % H_2O). For O_2, on the other hand, an MS peak is not visible. It can thus be concluded that no O_2 is released at temperatures lower than 500 °C, which means that the reduction reaction $As_2O_5 \rightarrow$

$As_2O_3 + O_2$ does not take place and trivalent As is not formed when pure As_2O_5.aq is heated to 500 °C. This result does not agree with the decomposition temperatures of arsenic pentoxide, as reported in the literature [15,16,17,18,19,20].

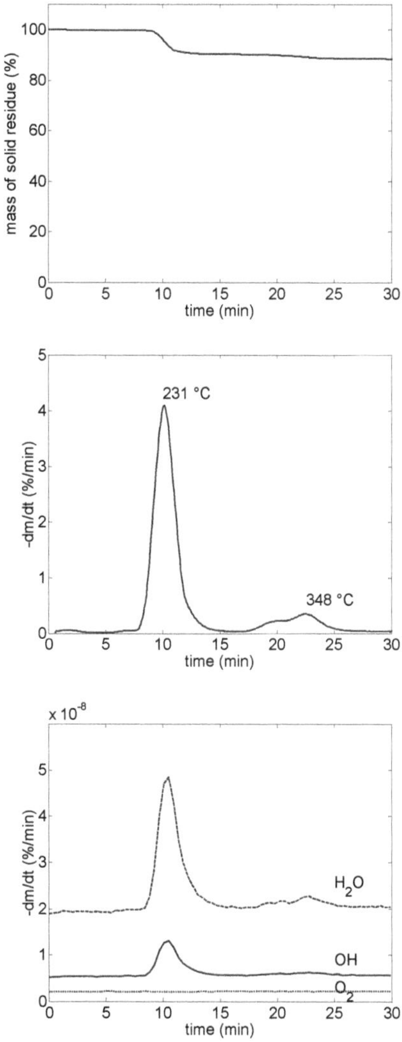

Figure 4: Thermal behaviour of arsenic pentoxide (As_2O_5.aq) in helium atmosphere (temperature profile: heating at 20 °C/min to 500 °C with an isothermal part of 10 minutes at 220 °C). Top: TG curve, middle: DTG curve, bottom: MS spectra for OH (solid line), H_2O (dashed line), O_2 (dotted line).

To examine whether As_2O_5 decomposes at higher temperatures, TG analysis of As_2O_5.aq is performed in a nitrogen atmosphere while heating from room temperature to 850 °C at 10 °C/min. The resultant TG and DTG curves are presented in Figure 5. From these curves it is clear that decomposition takes place at higher temperatures, leading to arsenic release.

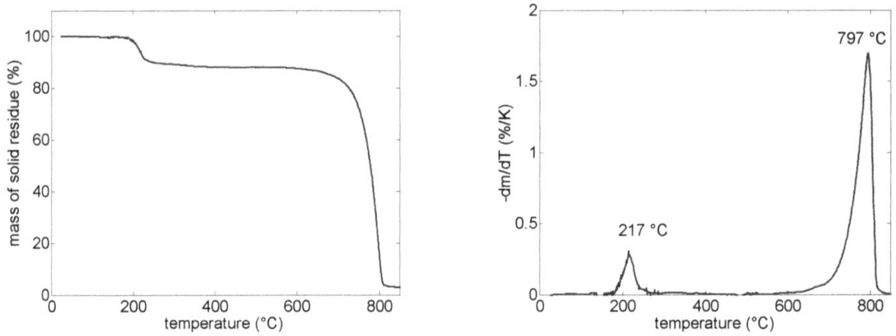

Figure 5: Thermal behaviour of arsenic pentoxide (As_2O_5.aq) in nitrogen atmosphere (temperature profile: heating at 10 °C/min to 850 °C). Left: TG curve, right: DTG curve.

Arsenic trioxide

Besides the thermal behaviour of arsenic pentoxide (As_2O_5.aq) the thermal behaviour of arsenic trioxide (As_2O_3) is studied through TG analysis. First, TG analysis of As_2O_3 is performed in a nitrogen atmosphere (the same results are obtained in a dry air atmosphere) by heating the sample at 5 °C/min to 500 °C with an isothermal part of 10 minutes at 200 °C. The resultant TG and DTG curves are presented in Figure 6. These curves show that arsenic is already released at temperatures as low as 200 °C.

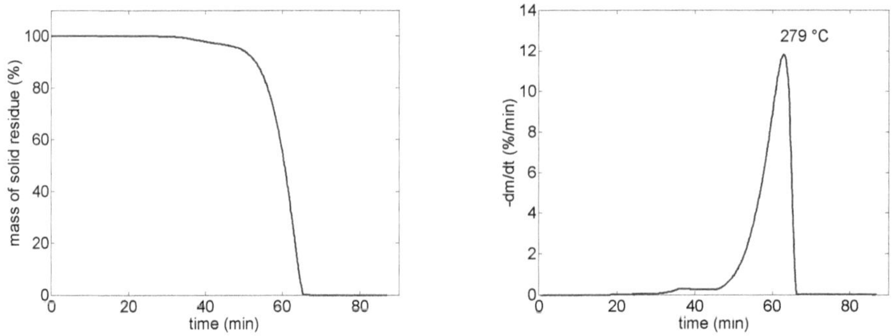

Figure 6: Thermal behaviour of arsenic trioxide (As_2O_3) in nitrogen atmosphere (temperature profile: heating at 5 °C/min to 500 °C with an isothermal of 10 minutes at 200 °C). Left: TG curve, right: DTG curve.

To check whether this release is caused by high values of the vapour pressure, TG analysis of As_2O_3 is carried out at different constant temperatures (170 °C, 180 °C and 190 °C) in a dry air atmosphere (the same results are obtained in a nitrogen atmosphere). The sample is heated to the pre-set temperature at 10 °C/min and held at that temperature for 60 minutes. The corresponding TG and DTG curves are presented in Figure 7. A linear decrease of mass at constant temperature with the slope increasing at higher temperatures is observed (see TG curve). The DTG curve gives rise to the same observation: an almost constant -dm/dt for the isothermal parts with a higher value for higher temperatures. Therefore, it can be concluded that the release of As_2O_3 is driven by temperature dependent vapour pressures; the higher the temperature, the faster the release.

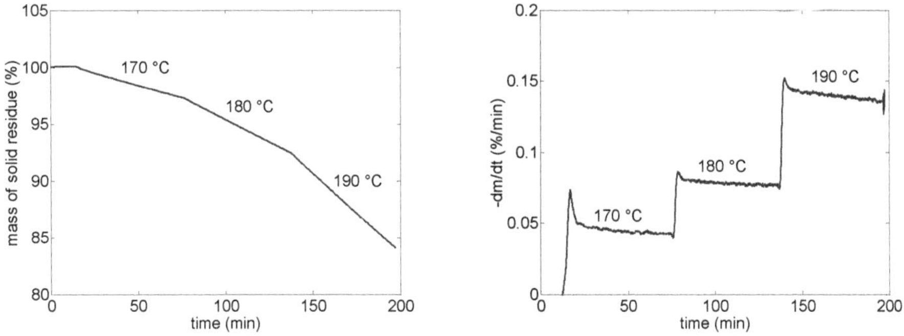

Figure 7: Thermal behaviour of arsenic trioxide (As_2O_3) in dry air atmosphere (temperature profile: heating at 10 °C/min to 170 °C, isothermal for 60 minutes, heating at 10 °C/min to 180 °C, isothermal for 60 minutes, heating at 10 °C/min to 190°C, isothermal for 60 minutes). Left: TG curve, right: DTG curve.

CONCLUSIONS

Low-temperature pyrolysis is suggested as a possible technology for the disposal of CCA treated wood waste. With respect to metal release copper and chromium do not pose any problem, but arsenic is identified as the problematic compound. Temperature and residence time highly influence the amount of arsenic released during pyrolysis. All studies performed to get more insight in the mechanism of arsenic release lead to the same hypothesis: the reduction of pentavalent arsenic to trivalent arsenic is responsible for the release of arsenic.

Pure As_2O_5.aq does not decompose (or reduce) nor volatilise at temperatures lower than 500°C. On the other hand, during pyrolysis of CCA treated wood arsenic is already released at a temperature of 320 °C. Moreover, in the pyrolysis residue resulting from pyrolysis at 350 °C for 20 minutes trivalent arsenic is found.

Originally, arsenic is present in pentavalent state in the CCA solution (As_2O_5), as well as in the CCA treated wood ($CrAsO_4$). This means that arsenic is partly reduced to the trivalent state during pyrolysis at temperatures lower than 400 °C. Since pure As_2O_5.aq does not decompose (or reduce) nor volatilise at temperatures lower than 500 °C, it can be concluded that the presence of wood, char and pyrolysis vapours may influence the thermal behaviour of arsenic oxides. Another explanation for the difference in thermal behaviour could be that part of the arsenic in the CCA treated wood is already present in trivalent form. This is a subject for further research.

The most important conclusion is that limiting the arsenic release during the pyrolysis of CCA treated wood implies that arsenic has to be kept in pentavalent form and thus the reduction reaction has to be controlled and avoided. Once the trivalent arsenic oxide is formed, it is released, obeying a temperature controlled solid-vapour equilibrium.

ACKNOWLEDGEMENTS

The financial support, provided by the Fund for Scientific Research – Flanders (Fonds voor Wetenschappelijk Onderzoek – Vlaanderen) is acknowledged. The authors would like to thank the Department of Chemical Engineering of the K.U.Leuven for their valuable help in conducting the quantitative analysis of the samples. We are especially grateful to Prof. C. Vandecasteele and H. Cooreman from the Chemical Department for their comments through the course of the study. Also the assistance of dr. M. Van Bael and Prof. J. Mullens in the TG analysis at the Laboratory of Inorganic and Physical Chemistry of the "Limburgs Universitair Centrum" are acknowledged. Furthermore, we are grateful to Beaumartin S.A. and Mr. J.S. Hery in particular for the financial support, the helpful discussions and the wood samples.

REFERENCES

1. Helsen L. and Van den Bulck E. (1997) Release of metals during the pyrolysis of preservative impregnated wood. In: Developments in Thermochemical Biomass Conversion, (Ed. by A.V. Bridgwater), pp. 220-228. Chapman & Hall, London.
2. Helsen L., Van den Bulck E. and Hery J.S. (1998) Total recycling of CCA treated wood waste by low-temperature pyrolysis. *Waste Manage.*, **18**, 571-578.
3. Helsen L. and Van den Bulck E. (2002) Metal retainment in the pyrolysis residue after low-temperature pyrolysis of chromated copper arsenate (CCA) treated wood. *Submitted for publication in Environ. Eng. Sci.*.
4. Van den broeck K., Helsen L., Van den Bulck E. and Vandecasteele C. (1997) Determination and characterisation of copper, chromium and arsenic in chromated copper arsenate (CCA) treated wood and its pyrolysis residues by inductively coupled plasma mass spectrometry. *Analyst*, **122**, 695-700.
5. Helsen L. and Van den Bulck E. (1998) The microdistribution of copper, chromium and arsenic in CCA treated wood and its pyrolysis residue using energy dispersive X-ray analysis in scanning electron microscopy. *Holzforschung*, **52**(6), 607-614.
6. Helsen L., Van den Bulck E., Mullens S. and Mullens J. (1999) Low-temperature of CCA treated wood: thermogravimetric analysis. *J. Anal. Appl. Pyrolysis*, **52**, 65-86.
7. Helsen L. and Van den Bulck E. (2000) Kinetics of the low-temperature pyrolysis of chromated copper arsenate treated wood. *J. Anal. Appl. Pyrolysis*, **53**, 51-79.
8. Helsen L. and Van den Bulck E. (2000) Metal behaviour during the low-temperature pyrolysis of chromated copper arsenate treated wood waste. *Environ. Sci. & Technol.*, **34**, 2931-2938.
9. Norme Française enregistrée NF C67-100 (1982) l' Union technique de l'Electricité, 12 place des Etats-Unis, 75783 Paris Cedex 16, France.
10. Pizzi A. (January 1999) Personel communication.
11. Helsen L., Van den Bulck E., Van den Broeck K. and Vandecasteele C. (1997) Low-temperature pyrolysis of CCA treated wood waste: chemical determination and statistical analysis of metal input and output; mass balances. *Waste Manage.*, **17**(1), 79-86.
12. Helsen L. (2000) *Low-temperature pyrolysis of CCA treated wood waste*. Ph.D. Thesis, Katholieke Universiteit Leuven, Belgium.
13. Mullens J., Vos A., Carleer R., Yperman J. and Van Poucke L.C. (1992) The decomposition of copper oxalate to metallic copper is well suited for checking the inert working conditions of thermal analysis equipment. *Thermochim. Acta*, **207**, 337-339.
14. Dahlgren S.E. (1971) *Thermobalance investigation of the preservative pastes boliden K33 and Celcure AP*. Collaborative Programme Report 71/17.
15. Barin I. (1989) *Thermochemical data of pure substances*. VCH Verlagsgesellschaft mbH, Weinheim.
16. Chesworth S., Yang G., Chang D.P.Y., Jones A.D., Kelly P.B. and Kennedy L.M. (1994) The fate of arsenic in a laminar diffusion flame. *Combust. Flame*, **98**, 259-266.
17. Kirk R.E. and Othmer D.F. (1978) Encyclopedia of chemical technology. Vol. 3, pp. 255, Wiley, New York.
18. Wu C.Y. and Biswas P. (1993) An equilibrium analysis to determine the speciation of metals in an incinerator. *Combust. Flame*, **93**, 31-40.
19. The Chemical Rubber Co. (1970) *CRC Handboek of Chemistry and Physics*, (Ed. by R.C. Weast), Ohio.
20. Webelements - The periodic Table on WWW: Professional Edition: Arsenic: compound data [arsenic ({V}) oxide] (December 2001) http://www.webelements.com/webelements/compounds/text/As/As2O5-1303282.html
21. Perry R.H. and Chilton C.H. (1973) *Chemical Engineers' Handbook*, 5th edn. McGraw-Hill, New York.
22. Ebbinghaus B.B. (1993) Thermodynamics of gas phase chromium species: the chromium oxides, the chromium oxyhydroxides, and volatility calculations in waste incineration processes. *Combust. Flame*, **93**, 119-137.
23. Frandsen F.J., Helble J.J., Erickson T.A. and Kuhnel V. (1997) Prediction of trace element partitioning in utility boilers. *presented at the EPA/DOE/EPRI Symposium*.

Development of Combustors for Pyrolysis Liquids

Matti Kytö
Oilon Oy, POB 5, 15801, Lahti, Finland
Pasi Martin, Steven Gust
Fortum Oil and Gas Oy, POB 310, 06101, Porvoo, Finland

ABSTRACT: Flash pyrolysis has been identified as an economic method to convert solid biomass into a liquid fuel which can be shipped, stored and combusted more economically than solid biofuels. This liquid differs significantly from existing liquid fuels and a combustion system specifically designed for this fuel is required. The long term use of this fuel in practical systems has not yet been demonstrated and there remain questions concerning the type of combustion technology that will be required, the level of emissions and long term storage. Oilon, a burner manufacturer, together with Fortum, an energy company have embarked on a development programme where the liquids produced in the Fortum/Vapo 500 kg/hour pilot plant will be test combusted in a Oilon prototype combustion system during the fall of 2002. Results of these tests will be used to design and construct commercial systems.

BACKGROUND

Although flash pyrolysis liquids have been combusted in a large number of laboratories and test programmes over the past ten years and the special features of combustion have been documented (1-7) there does still not yet exist commercial combustors. The main reasons for this are the lack of commercial production and variability of fuel qualities produced to date.

With the decision taken by Fortum and Vapo of Finland in 1999 to develop and scale up pyrolysis production technology to commercial sizes, Oilon, a burner manufacturer in Finland, has embarked on the task of developing commercial combustion systems. As the properties and quality of pyrolysis liquids have been known to vary in the past and since there does not yet exist a fuel specification, it was further agreed that the pyrolysis liquid producers would take an active role in the development work.

The main goals of the development work were to find clean and economic ways to burn pyrolysis liquids. This report presents results gained with in this work. As the development work and the required technology have not yet been decided, the technical solutions are not presented here. This will be verified with further measurements and combustion tests were performed in Porvoo and Lahti and at selected customer sites.

MODIFICATIONS REQUIRED TO COMMERCIAL BURNERS

Existing burner equipment that is normally used with light fuel oils has been optimized with a much higher heating value fuel than pyrolysis liquid (roughly double). This means that to achieve an adequate fuel / air mixtures and velocities with pyrolysis liquids, much less air is required for combustion. This causes the flame to expand in size and to extend to the end of the combustion chamber. It was observed that this type of flame is such that internal heating is not adequate to combust particulates and high molecular tars which increases emissions. This required that a new burner retention head be developed. Also it was observed that since pyrolysis viscosity is higher than light fuel oils, a higher pressure for atomisation is required to achieve the small drop size required for efficient combustion. One drawback of this high pressure is large droplet blow through which is caused by a small fraction of the drops being larger than the average size and leaving the flame centre before they are combusted. This leads to high emissions. The low pH due to the high concentration of formic and acetic acids has meant also that normal carbon steel cannot be used in pumps and fuel lines. Stainless steel pumps are available but they are considerably more expensive than normal fuel pumps. Since pyrolysis liquids contain significant amounts of water and have a low heating value, it has also been found that they are difficult to ignite. This has meant that the combustion chamber must first be warmed up to ignition temperatures by a separate fuel. This temperature was found to depend on the type of burner head solution that was adopted. In practise this is possible in less than 30 seconds. Finally, it was found that after burner shut-down, pyrolysis liquids must be rinsed out of nozzles to prevent clogging.

RESULTS

EMISSIONS

The results of combustion tests are shown in Tables 1 and 2. Good combustion, as seen by carbon monoxide levels around 60-70 ppm, can be obtained with oxygen content in the flue gas at around 5% levels. Clean combustion was also determined by measuring tar and PAH emissions as shown in Table 1. Here, the influence of improvements in the burner retention head are illustrated. Especially the PAH emissions have been reduced to very low levels. This was done while at the same time reducing the amount of excess air required for good combustion. The amount of excess air has very strong influence on the type of boiler heat exchange surface that is required. We started with 10-8 vol % and now we are at 4 vol %.

Table 1: Tar and polyaromatic emissions for pyrolysis liquid combustion.

Emission	Unit	1998	2000	2002
Tar	mg/MJ	-	3	*
PAH	mg/MJ	1000	50	15
CO	mg/MJ	200	100	20
NO_x	mg/MJ	50-200	50-200	50-200
O_2	%	8	6	4

* difference between content and background is too small. Method used SP-1686

Finally in Table 2, the combustion emissions of pyrolysis liquids using our improved burner retention head are compared to those of typical mineral oils and to those of pellets and woodchips. Tempera 15 is a typical light fuel oil used in intermediate size boilers while POR L 180 is a typical heavy fuel oil containing 1% sulfur. Here it is seen that combusting wood in the form of this pyrolysis liquid is much cleaner than that in a solid form but still not as clean as for light fuel oil.

Other emissions of interest are sulfur dioxide, nitrogen oxides and particulates. Sulphur dioxide is not measured because wood contains very small amounts of sulphur. Nitrogen oxides come primarily from the nitrogen content of the fuel and so attempts to reduce NOx emissions by changing combustion conditions have very little effect. Particulate emissions continue however to be one of the problems with this fuel. This is seen by a higher smoke number or Bacharach number than for light fuel oils. Whereas LFO has a Bacharach number less than 1, pyrolysis liquids have smoke numbers in the range of 2 to 4 depending on the fuel quality. With our emission testing results taken from Table 2 it can be seen that the emissions are mainly inorganic. They are a combination of ash that is found in the residual char and some sand from the process. The reduction of these emissions will be determined by producing pyrolysis liquids with very low levels of solids, at or below the 0.1 weight percent range.

Table 2: Pyrolysis emissions compared to other fuels.

	Tempera 15	Pyrolysis Oils	POR L 180	Pellet	Wood Chips	Fire Wood
HHV (MJ/kg)	42	15	39	17	5	13
Water wt-%	-	28	-	8	50	25
Typical Boiler Size	~ 500 kW	~ 300 kW	> 1 MW	~ 20 kW	> 1 MW	~ 20 kW
Excess air %	4	4	4	8 - 12	5 - 10	10 - 20
CO (mg/MJ)	10	15	15	50-1000	100-5000	1000 - 10000
NO_X (mg/MJ)	30	***	100-150	***	***	***
SO_X (mg/MJ)						
Tar mg/MJ	-	*	-	< 10**	< 20 **	1000**
PAH mg/MJ	-	8****	1	1000*****	2000*****	8000*****

* Method used SP-1686. Difference between measured value and background is too small. Background ~1 mg / MJ

** Measuring method used, Not known

*** NO_X Content comes from the biomass nitrogen, typically 50 - 300 mg/MJ. In bigger boiler (10-300 MW) one can use NO_X reduction technology

**** Method used SP-1686, Analysis EPA 610

***** Information from 1986

PROTOTYPE DEVELOPMENT

With the more than 5 years of experiences of Fortum and Oilon from various combustion tests and the scale up of Fortum / Vapo Forestera technology to 500 kg/hour, a prototype combustion system utilising these experiences has been developed. The platform for the prototype is a standard burner that has been modified to combust pyrolysis liquids. This is a fully automated combustion system that will be installed at a location in Finland in the fall of 2002 to begin long term testing of both the burner, possible deposit formation in the boiler and test qualities of different fuels.

Figure 1: Oilon KP 50H.

REFERENCES

1. Shihdah, A., et al. Combustion characterisation of wood-derived flash pyrolysis oils in industrial scale turbulent diffusion flames. Paper presented at the International Workshop "Properties and Applications for Pyrolysis Oils" Estes Park, CO, USA, September 1994.
2. Baxter, L. Jenkins, B. Baseline NOx emissions during combustion of wood-derived pyrolysis oils. Paper presented at the International Workshop "Properties and Applications for Pyrolysis Oils" Estes Park, CO, USA, September 1994.
3. Gust, S., "Combustion experiences of flash pyrolysis fuel in intermediate size boilers", Developments in Thermochemical Biomass Conversion (A.V. Bridgwater and D.G.B. Boocock, Eds.), Blackie Academic and Professional, London,1997, pp. 481–488
4. Huffman, D.R., Vogiatzis, A.J. and Clarke, D.A. Combustion of Bio-Oil in Bio-Oil Production and Utilisation, Proceedings of the 2nd EU-Canada Workshop on Thermal Biomass Processing eds. Bridgewater & Hogan.
5. Solantausta, Y., Nylund, N-O., Gust, S. Use of Pyrolysis Oil in a Test Diesel Engine to Study the feasibility of a Diesel Power Plant Concept. Biomass and Bioenergy 7 (1994) 1-6, 397-306.
6. Shaddix, C.R., and Huey, S.P. "Combustion characteristics of fast pyrolysis oils derived from hybrid poplar", Developments in Thermochemical Biomass Conversion (A.V. Bridgwater and D.G.B. Boocock, Eds.), Blackie Academic and Professional, London, 1997, pp. 465–480.
7. Andrews, R.G., Zukowski, S., and Patnaik, P.C. (1997) "Feasibility of firing an industrial gas turbine using a bio-mass derived fuel", Developments in Thermochemical Biomass Conversion (A.V. Bridgwater and D.G.B. Boocock, Eds.), Blackie Academic and Professional, London, pp. 495–506.

Fast Pyrogasification and/or Pyroliquefaction of Biomass in a Cyclone Reactor

J. Lédé, M. Ferrer, F. Broust
Laboratoire des Sciences du Génie Chimique, CNRS-ENSIC, 1, rue Grandville - BP 451 - 54001 NANCY Cedex (France)

ABSTRACT: The concept of biomass fast pyrolysis has been introduced as soon as the 70's, with the aim to produce high yields of gases. Later on, the technique of ablative pyrolysis evidenced the formation of high yields of liquids. Nowadays, fast pyrolysis is mainly carried out for the preparation of bio-oils (fast pyroliquefaction). The possibility of gas production by fast pyrolysis (fast pyrogasification) is no more considered in spite of the promising potentialities of such gases. Resulting from thermal cracking reactions, they contain ethylene and acetylene, molar fractions of H_2 and CO reaching 80 % and have heating values up to 3 times higher than in usual air gasification processes.

The paper describes the specific case of the cyclone vortex reactor inside which fast pyrolysis and cracking occur simultaneously. The experiments show that it is possible to control the fractions of oils and gases according to the experimental parameters. Typically, the yields of oils can reach 75 % (10 % gases) and those of gases 85 % (10 % oils) for reactor's walls temperatures ranging from 880 to 1220 K. Under partial steam atmosphere the low fractions of char may be considerably reduced by in situ gasification. A model describing the possible locations of the cracking reactions is presented. One of the conclusions of the paper is to recommend the study of multifunctional type reactors with which it is possible to prepare high yields of either bio-oils or of gases by simply changing the operating conditions.

INTRODUCTION

The thermal conversion of biomass is probably at the basis of the oldest processes carried out by mankind. Since a great number of centuries, they include mainly combustion (for direct heat generation) and pyrolysis. Carried out in the absence of oxygen, pyrolysis refers to the thermal transformation of biomass under inert atmosphere. It gives rise to three phases : solid charcoal, liquid products and permanent gases. Their proportions may considerably vary according to the operating conditions. It is why pyrolysis has been and continues to be considered for the preparation of various products and for various objectives that may vary according to the economical situations. Charcoal production (carbonisation of biomass) is a recurrent application since the pre-historical ages. Until the middle of the 20th century, low temperature pyrolysis of biomass has been also carried out for the production and valorisation of tars and liquids (acetic acid, methanol, acetone, ...). However, the gases produced by pyrolysis were very little exploited for their own qualities.

It is only around the 1940th (last world war) that the conditions for a maximum production of gases have been searched, through the optimisation of gasification processes (thermal conversion in the presence of controlled pressure of air, 0_2 and/or steam). About 30 years later, appeared the new notion of fast pyrolysis of biomass. At that time, it was supposed that the needed conditions were to operate with fine particles and high temperatures, in order to reach high heating rates of the feedstock. The reaction produced only little quantities of char and the objective was maximum gas production. The gases had much higher heating values than those produced in usual air gasification processes. This new technology became rapidly promising for possible energy and chemicals productions, as underlined in 1982 by R. Dumon and M. Gelus [1] : "Ainsi à travers le monde, mais en particulier en France, la pyrolyse éclair doit se révéler comme un concurrent certain pour la gazéification à l'oxygène tant pour la production de gaz combustibles que de gaz de synthèse. Mais pour atteindre l'industrialisation, bien des recherches restent à effectuer, en particulier dans le devenir des cendres et des résidus carbonés". In the same times, a key event has been the Copper Mountain Workshop [2] that gathered the main actors working all over the world on fast pyrolysis. Besides the gases, high quantities of liquids were also formed and, rapidly, the objectives of fast pyrolysis became the maximum production of oils. This tendency increased also after the invention of ablative pyrolysis [3], in which the source of heat was a hot surface placed in contact with biomass. Contrary to usual fast pyrolysis, ablative pyrolysis could be operated under lower temperatures and with big pieces of biomass. Ablative pyrolysis proved also that the primary products formed by fast pyrolysis were mainly liquids [4]. Thus, since the mid 85th, most of the works and published papers in the field of biomass fast pyrolysis deal with the preparation of liquids – the so-called bio-oils – produced for their promising potentialities (energy and/or chemistry) [5].

Summarizing the present situation, air, O_2 and steam gasification processes are considered for gas production, while fast pyrolysis processes are only considered as a source of bio-oils.

The first purpose of this paper is to underline the possibilities of biomass fast pyrolysis to produce high yields of either oils (fast pyroliquefaction) or of permanent gases (fast pyrogasification). The second purpose is to report experimental and theoretical results of biomass fast pyrolysis performed in a cyclone. Such a reactor offers the two main following potential advantages :

- It is a "multifunctional reactor" inside which several steps of a process can be simultaneously performed (heating, pyrolysis, quenching, separation). In more usual processes, these functions are fulfilled in separate stages.
- It is possible to steer the reaction to the maximum production of oils or of gases by only adjusting the operating conditions.

GENERAL PROPERTIES OF THE CYCLONE CONSIDERED AS A CHEMICAL REACTOR

THE CYCLONE REACTOR

Cyclone devices have been known and used since several decades for dust removal from gases, particularly in wood industries such as in sawmills plants. Even if perfect separations are not achieved, cyclones are among the simplest and easiest fabricated processing equipments. In principle, they rely on the separation of solid particles under the influence of centrifugal forces, the dispersed matter being separated at the bottom, while the cleaned fluids escape at the top. Such a working suggests its use as a chemical reactor for biomass fast pyrolysis for the potential following reasons:

- The particles flow behaviour against the walls probably induces efficient heat exchanges with the heated walls.
- The motion of biomass particles pressed against the hot walls (by centrifugal forces) induce similar phenomena as in ablative pyrolysis.
- The cyclones are basically good separators and are hence attractive for the automatic separation of charcoal, and unreacted biomass, from the gaseous fluid products (gases and vapours).

MAIN PROPERTIES

Most of the works devoted to the cyclone separators deal with the optimization of collection efficiencies of the particles to be separated as a function of their size distribution, total loading, pressure drop and cyclone configuration [6]. Many papers report also the results of the modelling of computarized particles trajectories and of the carrier gas field of velocities.

The main characteristics of the cyclone considered as a chemical reactor have been determined through global approaches. Scaling up laws have been derived for cyclones diameters ranging from 2.8 to 40 x 10^{-2} m, particles sizes from 10 x 10^{-6} to 3.2 x 10^{-2} m, particles mass densities from 720 to 8800 kg m^{-3} and several natures of carrier gases, flowing under Reynolds numbers from 400 to 250 000. These laws include the hydrodynamics of the carrier gas [7] and of the particles [8], as well as their heat transfer efficiencies with the walls [9,10]. It has been shown for example that the hydrodynamics of the particles is quite close to that of a plug flow, leading to much better reaction selectivities than, for example, in a fluidized bed reactor. In addition, the solid residence time can be easily and quantitatively adjusted by simply changing the carrier gas flow rate. It increases as the gas residence time decreases. The solid-walls heat transfer efficiency is excellent and the particles may be heated at higher temperatures than the carrier gas. These general properties suggest two possible opposite situations that can be achieved in a cyclone reactor:

- Under high gas flow rate (long particle residence time) the solid may be completely converted, while, in the same time, the vapours products are rapidly cooled (quenched) because of small gas residence times and low temperatures. This is a favourable case for preventing secondary gas phase reactions.
- Under small gas flow rate (short particle residence times), the solid may be only partially converted, but the vapours may undergo further gas phase reactions such as thermal crackings because of the longer gas residence time.

In the case of biomass fast pyrolysis, we can hence expect:

- In the first case, high yields of bio-oils after condensation of the unreacted vapours
- In the second case, high yields of gases.

Associated to the possibility of changing, at will, the walls temperatures, these potential properties show that the cyclone could be used for producing high fractions of either bio-oils or gases.

Finally, it is clear that the efficiency and occurrence of these extreme situations are intimately connected to the natures of the basic controlling factors of particles consumptions.

To these favourable potentialities are of course added the basic advantages of self separation of the unreacted solids or of the solids formed during the reaction.

Finally, the cyclone appears to be mostly recommended to perform reactions of the types:

Solid 1 $\rightarrow$ (Liquids) $\rightarrow$ Gases

or Solid 1 $\rightarrow$ (Solid 2 + Liquids) $\rightarrow$ Solids 2 + Gases.

A FEW EXAMPLES OF APPLICATIONS

The feasibility of carrying out thermochemical reactions of solids in hot walls cyclones has been demonstrated at the laboratory scale in the cases of $NaHCO_3$ [11] and $CaCO_3$ decarbonations and the pyrolysis of coal [12]. An extensive study has been also made with the thermal reaction of isocyanuric acid giving rise to isocyanic acid [13]. This example proved that the cyclone could be also efficiently used for processing sublimation reactions [14]. The most advanced works have been performed in the case of biomass fast pyrolysis.

FAST PYROGASIFICATION AND/OR PYROLIQUEFACTION OF BIOMASS IN A CYCLONE REACTOR

FORMER RESULTS

They have been obtained nearly 20 years ago in the framework of the preparation of two PhD thesis. The first one (Soulignac, July 1984 [15]) demonstrated that it was possible to gasify biomass in a 4 x 10^{-2} m diameter standard cyclone reactor, with gas yields reaching 95 %. The second one (Diebold, July 1985 [16]) operating with a 14 x 10^{-2} m vortex cylindrical reactor showed that up to 75 % of water soluble vapours could be obtained and further cracked into gases.

Two years after Soulignac's works other experiments made with a smaller 2.8 x 10^{-2} m cyclone suggested that the gases/condensable vapours mass fractions could mainly depend on the cyclone walls temperature [17].

The experimental results obtained in the 2.8 and 4 x 10^{-2} m cyclone reactors are discussed in a more recent paper [18]. The main results are :

- The feasibility of processing fast pyrolysis of biomass in a cyclone reactor and in steady state has been demonstrated.
- The mass balances derived from the analysis of gases, liquids and solids products are close to 100 %. The char fractions are of the order of 3 % when operating with walls temperatures of the order of 1273 K.
- It is possible to process up to 0.5 kg h^{-1} with the 4 x 10^{-2} m cyclone corresponding to capacities of the order of 9 x 10^{-3} kg s^{-1} per square meter of heated surface, in good agreement with the values of Diebold [16] in his vortex reactor.
- It is possible to reach, in steady state operation, mass yields of gases of up to 80 % when operating with steam as carrier gas. They can be increased to nearly 100 % by recycling the fluid products of the reaction : because of the relatively long residence times in the recycling loop, the vapours entirely crack and the resulting gases can be used as carrier gas.
- The gas compositions in the one way and the recycling processes are rather similar for cyclones walls temperatures between 1166 and 1321 K. The extreme values of the fractions of the different species are reported in table 1. The total fraction of H_2 + CO can be very high, offering interesting potentialities for chemical synthesis. It has been shown that the fractions of H_2, CO and of unsaturated hydrocarbons increase and CO_2 decrease when the temperature increases.
- The heating value of the gases logically increases also with the temperature and may reach 19 MJ/Nm^3 mainly because of the high mass fractions of hydrocarbons. These values are significantly higher than for the gases obtained in more usual air slow gasification processes (a factor close to 3). In addition, their productivity is much higher because of the rapidity of the reaction.
- First experiments made in a 2.8 x 10^{-2} m reactor showed that the yield of condensable vapours increased with lower walls temperatures confirming the well known properties of the vapours to give rise to gases by thermal cracking.
- The basis of a mathematical model describing the possible locations of these cracking reactions has been suggested. The model assumes that the reaction occurs near the close vicinity of the walls of the reactor, preferably than in its main volume, because of the too low values of residence times and temperatures reached by the gases in the cyclone.

Table 1: Extreme values of the fractions (mol %) of the species contained in the gases produced in the fast pyrogasification of biomass performed in 2.8 and 4 x 10^{-2} m cyclone reactors. The results are the same for experiments made under pure steam atmosphere or after the recycling of the fluid species. The walls temperatures range between 1166 and 1321 K.

H_2	CO	CO_2	CH_4	C_2H_4	C_2H_6	C_2H_2	Total hydrocarbons
18-28	37-50	15-28	6-10	2-6	1-4	1-5	10-25

NEW DEVELOPMENTS

New experimental campaigns are being performed in order to confirm the multifunctional properties of the cyclone. The objectives are to demonstrate the possibility of carrying out the fast pyrogasification or pyroliquefaction of biomass in the same vessel by simply changing operating parameters such as the walls temperature.

Description of the experiments [19]

The reactor is a 6 x 10^{-2} m diameter standard Lapple type cyclone built in refractory stainless steel. Its internal volume and surface are respectively 0.461 x 10^{-3} m^3 and 3.7 x 10^{-2} m^2. It is heated by induction, a technique offering the advantage of a very high flexibility for laboratory uses. The walls temperature, measured by pyrometry, is varied between 880 and 1220 K. The reactants tangentially arrive at the top and near the vertical axis of the reactor according to an original device already described [15]. Biomass particles are obtained from wood pellets ground and classified into several classes of sizes between 0.1 and 1.2 x 10^{-3} m. A calibrated screw feeder placed at the bottom of an air tight hopper delivers known biomass flow rates ranging from 0.03 to 1.3 kg h^{-1}. At the exit of the screw feeder, the particles are transported by a controlled flow of N_2 or of N_2 containing up to 25 % of steam. The inlet gas Reynolds numbers range from 3000 to 10 000.

The solid by-products (char and possible non converted biomass particles) are automatically separated and collected in the hopper placed at the bottom of the cyclone. Their mass is weighted. The fluid products (carrier gas, gas of the reaction, vapours, aerosols) leave the cyclone at its top. Vapours are then condensed in three tubular water cooled heat exchangers, followed by an ice cooled condenser and a glass wool filter. All these condensed matters are weighted. Cleaned and dry gases are then sampled and analysed by gas chromatography from which it is possible to derive the total mass of produced gases.

From all the measurements, the mass fractions of each gaseous, liquid and char products are calculated, leading to the determination of global mass balances. These fractions are expressed with respect to the total amount of crude (not dried biomass), which initial water content is close to 8 % wt.

Results

For all the experiments performed under N_2 and/or N_2 + steam atmosphere, the mass balances closures range between 85 and 108 %.

Table 2: Orders of magnitudes of the solids, liquids and gases mass fractions (%, calculated on the basis of crude - not dried – biomass) obtained in the fast pyrolysis of biomass performed in a 6 x 10^{-2} m cyclone reactor, and for two extreme values of walls temperatures. The results are given for experiments made under N_2 and N_2 + 25% steam atmosphere.

T_w (K)		Solids	Liquids	Gases
880	N_2	15	75	10
	N_2 + steam	15	65	20
1220	N_2	10	15	75
	N_2 + steam	5	10	85

The mean values (about 30 experiments) of the products fractions obtained for the two extreme walls temperatures of 880 and 1220 K are reported in Table 2. The gases and liquids fractions range inside confidence intervals of about ± 5 %. The solids contain charcoal but also a small fraction of biomass particles that have not been completely converted, mainly at low walls temperatures. The solids fractions only slightly depend on the nature of the carrier gas. The liquids fractions are about 10 % lower and the gases 10 % higher in the presence of steam. The gases/liquids ratios depend considerably on the walls temperature, from 0.13 at 880 K under N_2, until 8.5 at 1220 K under partial steam atmosphere. This result clearly shows the possibility of the cyclone reactor to process the fast pyrolysis of biomass for the production of either bio-oils (fast pyroliquefaction around 880 K) or gases (fast pyrogasification around 1220 K). The variations of the liquids (and gas fractions) as a function of T_W are similar with other sizes of cyclones [20], as shown by Fig. 1.

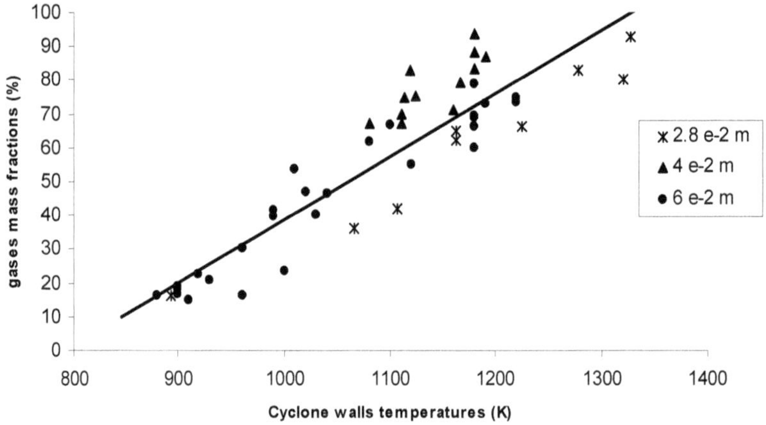

Figure 1: Variations of the gases mass fractions (%) as a function of the temperature (K) of the walls of cyclone reactors of three different sizes.

Table 3 reports the extreme values of the compositions of the gases obtained for walls temperatures in the range 1160 – 1220 K. The results are the same when operating with pure N_2 or with N_2 + 25 % steam. The H_2 + CO fractions range from 74 to 82 %. Compared to the values reported in table 1, the proportions of H_2 are slightly lower and those of CO are higher. The total fractions of hydrocarbons are similar in both cases. The gas heating values range from 14 to 18 MJ/Nm^3. As already noticed, these properties confer to the gas, interesting potential applications for chemical synthesis or for energy generation.

Table 3: Extreme values of the fractions (mol %) of the species contained in the gases produced in the fast pyrogasification of biomass performed in a 6 $\times 10^{-2}$ m cyclone reactor with walls heated at temperatures of 1180 ± 20K.

H_2	CO	CO_2	CH_4	C_2H_4	C_2H_6	C_2H_2	C_3H_8	Total hydrocarbons
15	59	7	5	5	0.3	0.9	0.2	11.4
20	62	9	8	6	0.6	1.3	0.6	16.5

For the same walls temperatures of 1160 – 1220 K and particles sizes between 500 and 800 μm, the maximum loading allowing a steady state operation (i.e. longer than about 900 s) is of about 1 kg h^{-1}. It is interesting to notice that this limit corresponds to 7.5 x 10^{-3} kg s^{-1} m^{-2} which is quite similar to the value obtained with smaller cyclones [18] and also in Diebold's experiments. We can hence expect that in a vortex type reactor it is possible to process, in steady state conditions, up to a maximum of about 10^{-2} kg s^{-1} m^{-2}. For higher biomass loadings two phenomena are observed. The first one is a clogging at the bottom of the cyclone. The second one is a deposition of charcoal on the hot walls of the reactor (mainly inside its conical part). The first phenomenon can be probably prevented by changing the cyclone geometry (greater angles of the cone and/or increase of the cross section of the bottom of the cyclone). It could be also recommended to increase regularly the biomass flow rate at the beginning of the process preferably than operating immediately with the nominal loading value. The second phenomenon is always observed under pure N_2 atmosphere even if it is less important at high temperature. For walls temperatures greater than about 1080 K, no wall deposition is observed if the cyclone is operated with N_2 to which about 25 % of steam is added.

As a conclusion, the 6 x 10^{-2} m diameter cyclone can be driven in steady state conditions with up to 1 kg h^{-1} of biomass flow rate, under partial steam atmosphere and for walls temperatures above about 1100 K, for processing the fast pyrogasification of biomass. In these conditions, the solids (char) formed in the process are completely separated at the bottom of the cyclone and their fraction does not exceed 5 %. When operating the cyclone with the objectives of fast pyroliquefaction of biomass at temperatures lower than 900 K, about 50 % of the total solids formed during the process (about 5 % of introduced biomass) stick on the hot walls of the reactor, thus preventing a too long steady state operation. In that case, a very satisfactory solution consists in eliminating this deposition by steam gasification. For that purpose, the study of the possibility of alternating periods of biomass fast pyrolysis (producing high yields of bio-oils) and periods of steam gasification of char (producing high yields of CO and H_2) is now in progress.

MODELLING OF THE CYCLONE REACTOR

GENERAL CONSIDERATIONS

The extrapolation of the results to different cyclone and biomass sizes, and for different objectives such as gases or bio-oils productions needs the complete modelling of a great number and interdependent phenomena (Fig. 2). Some are already known, others are still in progress. They include:

- Hydrodynamics of the particles. They behave as in a plug flow. Their residence time linearly increases with the gas inlet Reynolds number [8].
- Hydrodynamics of the gases. They behave as in the association of a plug flow followed by a more or less by passed stirred part. The importance of each of these zones is also a function of the gas inlet Reynolds number [7].
- Heating of the particles. In usual conditions, the solids are mainly heated by direct exchanges with the hot walls. The corresponding heat transfer coefficient depends, in particular, on the solid loading [10].
- Heating of the carrier gas. The Nusselt number defined at the inlet conditions is, in first approximation, proportional to the inlet Reynolds number [9]. An empirical linear relationship has been also established between the gas temperature at the exit of the reactor and the walls temperature [20].
- Elementary processes controlling the consumption of the particles. The mode of external heat transfer is similar as in contact ablative pyrolysis. However, because of the usual small particle sizes, they preferably decomposes in chemically controlled regimes [21].
- Biomass fast pyrolysis reaction scheme. It is supposed that the Broido Shafizadeh model applies in our conditions, with the formation of intermediate short life time liquid compounds (ILC) [22,23].
- Localization and modelling of the formation of gaseous products. The study of these phenomena is now in progress. The main results will be summarized in the following section.

The solid residence time (directly connected to the carrier gas hydrodynamics) should be such that every particle is completely consumed at the exit of the reactor and hence as close as possible as its reaction time. This latter depends on solid-wall heat transfer efficiencies, of the model of fast pyrolysis reaction and of the internal controlling consumption factors. Finally, the reaction gives rise to primary liquids producing vapours that can be either quenched (fast pyroliquefaction) or further cracked into gases (fast pyrogasification). The relative importance of each of these last phenomena depends on their localizations inside the cyclone and also on the field of gas temperature. This latter, in turn, depends on gas-walls heat exchanges and hence on the carrier gas hydrodynamics.

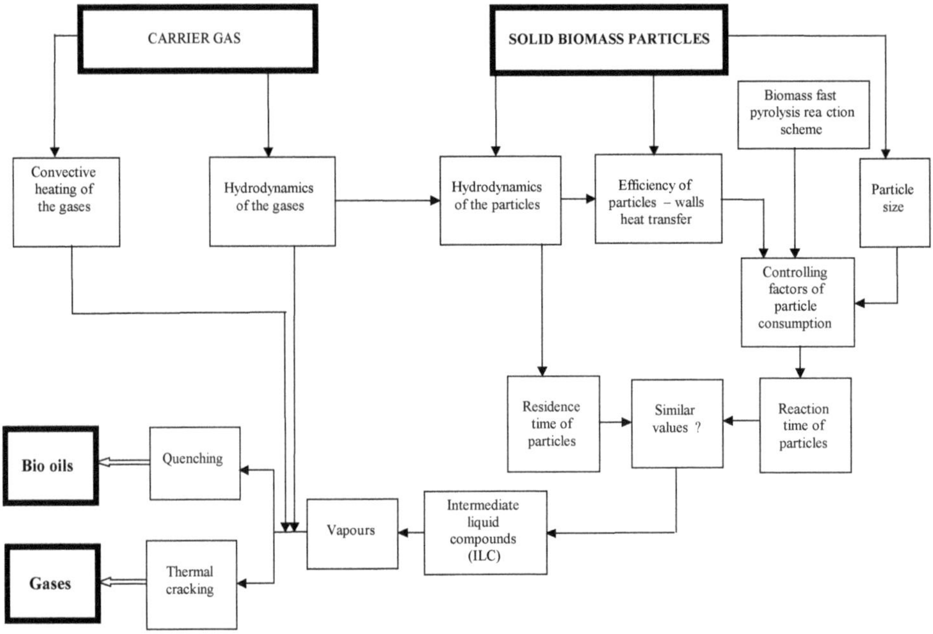

Figure 2: Main connections existing between all the phenomena that should be taken into account in the modelling of pyrolysis in a cyclone reactor.

MODELLING OF THE CRACKING REACTIONS

This last part of the modelling work is important for predicting the conditions required for maximising the production of either bio-oils or gases.

It is supposed that the gases are formed by the thermal cracking of vapours produced from the intermediate short life time liquid compounds (ILC) primary formed during biomass fast pyrolysis. At the present time, the model does not take into consideration the possibility of formation of gases directly and primarily from biomass or from ILC. It is important to locate, at first, these elementary cracking processes inside the cyclone reactor.

A first assumption is to suppose that they occur during the crossing of the vapours through the volume of the cyclone. It is possible to show that the temperature reached by the gas inside the cyclone is, most of times, much lower than T_W. The differences may reach several 100 K. In addition, the global gas mean residence time may be as low as 0.1 – 0.2 s. Considering the usual values of the gas phase cracking kinetic constants [18,24], it can be easily shown that only low conversion yields could be reached in these conditions (specially for small cyclone sizes). This result demonstrates that the products of reaction formed at moderate temperatures have chances to stop reacting once liberated in the heart of the reactor. This property proves that the multifunctional aspect of the cyclone includes also the possibility of quenching.

The second assumption is based upon the Fig. 1 showing that the gasification yields can be satisfactorily correlated with the walls temperatures for several sizes of cyclone reactors. Thus, T_W seems to be of a major importance. The model supposes that the cracking reactions occur inside a thin boundary layer close to the walls of the cyclone. This small volume is supposed to be at T_W and made of a great number of small elementary stirred cells arranged side by side and covering the whole internal surface of the cyclone. Each cell is a small cylinder having a cross section of πR^2 and a thickness equal to biomass particle diameter. Each of these cells is fed only by the vapours formed from the vaporization of ILC (no carrier gas enters the cells). The mixture of cracked (into gases) and uncracked vapours enters the cooler bulk of the reactor inside which they mix with the carrier gas and undergo a rapid cooling (quenching). Assuming a mean molar mass of the vapours species and a number of mole increases during vaporization, it is possible to derive an expression giving the first order cracking kinetic constant as a function of measured gasification yields, walls temperatures, particles sizes, biomass flow rates and cyclone surface. In the simplified assumption of a cyclone behaving as a Continuous Stirred Tank Reactor, the values of the rate constants, thus calculated on the basis of experimental results, are in a quite good agreement with the constants reported in the literature. To be noticed that such a model can be used for any reaction of the type Solid → (liquids) → vapours → gases carried out in a vortex reactor. For example, the same cyclone has been used to perform the reaction of production of isocyanic acid from isocyanuric acid. The derived cracking kinetic constants, thus derived, are also perfectly comparable with the values obtained in other and more usual devices [13]. This model is actually improved in order to take into account:

- The fact that the small elementary stirred cells do not cover the whole internal surface of the cyclone but only the surface in contact with the particles that are present at any time in the reactor.
- Imperfect quenching and thus, possible additional cracking inside the volume of the cyclone. In this more realistic model, the overall gasification yield would result from two processes in series: cracking at T_W near the walls and also inside the reactor at the lower carrier gas temperature.
- The fact that the hydrodynamics of the gas is more complicated than in a simple Continuous Stirred Tank Reactor.
- The intimate interactions between the solids and gases flows.
- The fact that biomass particles have not the same diameters but have a size distribution.
- The possibility of an additional pathway of gas formation resulting from cracking reactions occurring directly inside ILC [25].

CONCLUSIONS

- Since a great number of centuries, pyrolysis is probably one of the most often considered route of biomass upgrading. Its most frequent applications concern carbonisation and various liquids preparations. Nowadays, maximum gas production is supposed to be obtained only in the air, O_2 and/or steam gasification processes. Unfortunately, the direct gasification of biomass by fast pyrolysis is no more mentioned, in spite of the very promising results published in the 1970th.
- The present paper has reminded that according to the experimental conditions fast pyrolysis can give rise to high fractions of bio-oils (fast pyroliquefaction) or of gases (fast pyrogasification), both having very interesting potentialities for energy generation or chemicals preparations.
- The cyclone, heated at its walls, is a very promising multifunctional chemical reactor inside which can be simultaneously performed : heating and pyrolysis of solid particles, quenching and/or cracking of primary produced species, separation of the solid products.
- The cyclone reactor is particularly well adapted to the case of biomass flash pyrolysis. By the only change of the walls temperature it is possible to control the process for producing up to 75-85 % of bio-oils, or of gases. These latter have heating values reaching 19 MJ/Nm^3, i.e. 3 times higher than those produced in usual air gasification processes.
- The optimisation of a model representing the localizations and efficiencies of cracking reactions in cyclone type reactors is now in progress. A major result is that these gases can be

competitively produced in the close vicinity of the hot walls, and in the heart of the cyclone according to the experimental conditions.

- At low temperatures, a layer of char develops on the hot walls. It disappears if the process is carried out in the presence of steam. The optimization of a steam gasification process relying on the succession of pyrolysis and steam gasification stages is now in progress.

ACKNOWLEDGMENTS

The authors want to thank ADEME (Agence de l'Environnement et de la Maîtrise de l'Energie) and CNRS (Centre National de la Recherche Scientifique) for supporting this work.

REFERENCES

1. Dumon R. and Gelus M. (1982) *La valorisation chimique du bois.* Masson, Paris
2. Diebold J.P. (1980) *Proceedings Specialists' Workshop on Fast Pyrolysis of Biomass* , SERI/CP-622-1096.
3. Lédé J., Panagopoulos J., Li H.Z. and Villermaux J. (1985) Fast Pyrolysis of Wood : Direct Measurement and Study of Ablation Rate. *Fuel*, **64**, 1514-1520.
4. Lédé J., Li H.Z. and Villermaux J. (1988) Pyrolysis of Biomass : Evidence for a Fusionlike Phenomenon. In: *Pyrolysis Oils from Biomass, Producing, Analyzing and Upgrading* (Ed. by J.Soltes and T.A. Milne), pp. 66-78, chapter 7, ACS Symposium Series 376.
5. Bridgwater A.V., Czernik S. and Piskorz J. (2001) An Overview of Fast Pyrolysis. In: *Progress in Thermochemical Biomass Conversion* (Ed. by A.V. Bridgwater), pp. 977-997, IEA Bioenergy, Blackwell Science Ltd, Oxford.
6. Zenz F.A. (1999) Cyclone Design. In: *Fluidization Solids Handling and Processing* (Ed. by Wen-Ching Yang) pp. 773-816, NP Nayes Publications.
7. Lédé J., Li H.Z. and Villermaux J. (1989) Le cyclone réacteur. Partie I : Mesure directe de la distribution des temps de séjour de la phase gazeuse. Lois d'extrapolation, *Chem. Eng. J.*, **42**, 37-55.
8. Lédé J., Li H.Z., Soulignac F. and Villermaux J. (1989) Le cyclone réacteur. Partie II : Mesure directe de la distribution des temps de séjour de la phase solide. Lois d'extrapolation, *Chem. Eng. J.*, **42**, 103-117.
9. Lédé J., Li H.Z., Soulignac F. and Villermaux J. (1990) Le cyclone réacteur. Partie III : Mesure de l'efficacité des transferts de chaleur et matière entre les parois et un gaz circulant seul, *Chem. Eng. J.*, **45,** 9-24.
10. Lédé J., Li H.Z., Soulignac F. and Villermaux J. (1992) Le cyclone réacteur. Partie IV : Mesure de l'efficacité des transferts de chaleur entre les parois et les phases gazeuse et solide, *Chem. Eng. J.*, **48**, 83-99.
11. Lédé J., Li H.Z. and Villermaux J. (1992) Le cyclone : un réacteur chimique. Application à la réaction de décarbonatation de $NaHCO_3$, *Can. J. Chem. Eng.*, **70**, 1132-1142.
12. Houzelot J.L., Hilaire P., Lebrun-Cabon A., Villermaux J., Berton R. and Isler D. (1991) Etude d'un procédé de pyrolyse rapide du charbon dans un réacteur cyclone, *Can. J. Chem. Eng.*, **69**, 876-882.
13. Lédé J. and Mercadier J. (2002) Simulation of the Thermal Cracking of Biomass Derived Vapours, by the Model Reaction of Decomposition of Isocyanuric Acid, *J. Anal. Appl. Pyrolysis,* In press.
14. Lédé J., Mercadier J. and Villermaux J. (1995) Le cyclone, un réacteur de sublimation : application à la sublimation de l'acide isocyanurique, *Chem. Eng. J.*, **62**, 13-21.
15. Soulignac F. (1984) *Le cyclone : un nouveau réacteur chimique. Application à la pyrolyse éclair de bois divisé.* Thèse INPL, LSGC, Nancy, France (2 juillet).

16. Diebold J.P. (1985) *The Cracking Kinetics of Depolymerized Biomass Vapours in a Continuous Reactor*. Thesis T-3007, Colorado School of Mines, Golden Co, USA.
17. Lédé J., Verzaro F., Antoine B. and Villermaux J. (1986) Flash Pyrolysis of Wood in a Cyclone Reactor, *Chem. Eng. Process.*, **20**, 309-317.
18. Lédé J. (2000) The cyclone : a Multifunctional Reactor for the Fast Pyrolysis of Biomass. *Ind. Eng.Chem. Res.*, **39**, n° 4, 893-903.
19. Broust F., Ferrer M. and Lédé J. (2002) Fast Pyrolysis of Biomass in a Cyclone Reactor for the Production of either Liquids and Gases. *Proceedings of the 12th European Conference and Technology Exhibition on Biomass for Energy, Industry and Climate Protection.* Amsterdam (Nl), 17-21 June.
20. Broust F., Ferrer M. and Lédé J. (2001) Procédé de pyrolyse de la biomasse dans un réacteur multifonctionnel et multi-usages de type cyclone. In : *Récents Progrès en Génie des Procédés : Génie de la Réaction Chimique et des Réacteurs* (Ed. by J.L. Houzelot, J. Boudrant et D. Tondeur), pp. 377-384, vol. **15**, n° 80, Lavoisier, Technique et Documentation, Paris.
21. Lédé J. (1994) Reaction Temperature of Solid Particles undergoing an Endothermal Volatilization. Application to the Fast Pyrolysis of Biomass. *Biomass and Bioenergy*, **7**, n° 1-6, 49-60.
22. Lédé J., Diebold J.P., Peacocke G.V.C. and Piskorz J. (1999) The Nature and Properties of Intermediate and Unvaporized Biomass Pyrolysis Materials. In: *Fast Pyrolysis of Biomass : a Handbook* (Ed. By A.V. Bridgwater et al.), pp. 51-65, CPL Press, Newbury, UK.
23. Boutin O., Ferrer M. and Lédé J. (2002) Flash Pyrolysis of Cellulose Pellets Submitted to a Concentrated Radiation : Experiments and Modelling, *Chem. Eng. Sci.*, **57**, 15-25.
24. Baumlin S., Broust F., Ferrer M., Meunier N. and Lédé J. (2002) Measurement of the Cracking Kinetics of Biomass Pyrolysis Vapours inside a Continuous Stirred Tank Reactor. *Proceedings of the 12th European Conference and Technology Exhibition on Biomass for Energy, Industry and Climate Protection*, Amsterdam (Nl), 17-21 June.
25. Lédé J., Blanchard F. and Boutin O. (2002) Radiant Flash Pyrolysis of Cellulose Pellets : Products and Mechanisms involved in Transient and Steady State Conditions, *Fuel*, **81**, 1269-1279.

An Evaporation Model for Bio-Oil Fuel

T. Lederlin and J.E.D. Gauthier
Department of Mechanical and Aerospace Engineering. Carleton University, 1125 Colonel By Drive, Ottawa, ON K1S 5B6 Canada

ABSTRACT: A Canadian aerospace company has modified a 2.85-MW industrial gas turbine engine to burn biomass pyrolysis oil (bio-oil). Tests show that the engine runs properly on bio-oil but has to be modified to account for corrosive and erosive properties of bio-oil and the products of its combustion. To assess the effectiveness of various coatings, to protect the gas turbine hot components from erosion and corrosion when burning bio-oil, requires knowledge of the combustion emissions. A critical aspect of emission prediction in liquid spray combustion is the correct modelling of evaporation. A model of bio-oil droplet evaporation is proposed here. It is based on the quasi-steady evaporation theory and is applicable to multi-component fuels such as bio-oil. The thermodynamic properties of the bio-oil, enthalpy of vaporization and saturation pressure, are obtained from a semi-empirical method for multi-component fuels, based on distillation data. The chemical composition of the bio-oil is determined from experimental values found in the literature. This computationally low-cost model is suitable to implement in CFD codes to simulate bio-oil combustion in combustion devices such as furnaces and gas turbine combustors.

INTRODUCTION

The combustion of biomass-derived liquid fuels in a gas turbine combustor generates emission products such as alkali metals that may cause damages to the combustion chamber and thus need to be predicted accurately. Modelling the evaporation of fuel droplets is of major importance in liquid spray combustion studies since this process is a controlling step of combustion emissions.

For a single-component fuel, the classical approach of droplet evaporation has been first achieved by Spalding [1] and is based on the simultaneous integration of the energy and species conservation equations with a Lewis number equal to one. The droplet temperature is assumed to be uniform and constant throughout its lifetime. A law for the temperature dependence of the vapour-pressure at the droplet surface (e.g. Clausius-Clapeyron equation) allows closing of the system and finding the fuel vapour mass flux as well as the droplet radius law, known as the d-square law. This method does not account for the liquid phase heat and mass transfer and is basically a gas-phase model. When the droplet consists of more than one liquid component, the analysis becomes more complicated: the interior of the droplet must be modelled to account for the variation in temperature and composition due to transport process.

For spherically symmetric vaporization, three models with different internal heat and mass transport description exist. In the diffusion-limit model [2], no internal motion is assumed; heat and mass transfers are achieved through conduction and diffusion. This model represents the lowest limit for transport process and best applies to viscous fuels. It is solved through numerical integration and thus requires a large computational effort. Although this model has proved to be the most accurate for multicomponent evaporation, it depends on the binary diffusivity correlations and therefore becomes inadequate for a large number of components, especially in non-classical fuels. The rapid-mixing model [3] assumes equilibrium evaporation of the droplet, which is similar to a batch distillation. It is the highest limit for transport process in the droplet: convective and diffusive transfers are assumed to be so fast that the droplet temperature is uniform and temporally varying. It applies well to low-viscosity fuels and slow evaporation problems; therefore it was mostly applied to non-combustion problems [4]. The thin-skin model [5] assumes a high liquid Lewis number, no diffusion inside the droplet, a composition and temperature unchanged and only the surface layer of the droplet being heated. It proved to be valid at high temperature only and the final solution highly depends on the thermo-physical properties so that this model becomes inaccurate for a fuel made of radically different components. For a more precise description of the processes inside the droplet, theories have been developed, which account for internal mixing due to vorticity induced by the shear stress at the drop surface [6]. The problem becomes non-spherically symmetric and mathematically complex.

A simpler way to calculate multicomponent droplet evaporation was found by Bardon and Rao [7]. The authors recognized the importance of saturation vapour pressure, which determines the temperature and fuel concentration at the droplet surface in the d-square approach. Based on distillation data (ASTM distillation curves), a numerical technique allows computing of the important variables as the evaporation proceeds. Thus, the fuel is treated as a single component with varying properties and this procedure can be easily implemented in the d-square model to account for the multicomponent nature of the fuel.

A newer method must also be quoted here: Hallett [8] uses continuous thermodynamics to calculate the mixture properties. The composition of the fuel is described through a probability density function (pdf), which enables to solve the problem for only three distribution parameters. This method proved to be fast and accurate. It requires though to have specific data (pdf) about the fuel studied and is more suitable for fuels with a very large number of components since a statistical method is employed.

In the present paper, due to the lack of information about the components of bio-oil and their thermo-physical properties, the simple d-square model is used. The multicomponent nature of bio-oil is treated with time-varying droplet variables, as for a rapid-mixing model, and varying properties as the droplet evaporates, following the idea of Bardon and Rao [7].

THE ANALYTICAL D-SQUARE MODEL

HYPOTHESES

- The evaporation process is quasi-steady: at any instant, the system is described as if it was in a steady state.
- The droplet temperature is uniform and constant for each pure component; it is determined through an iterative process.
- The geometry of the problem is strictly spherically symmetric, thus only radial gradients appear in the equations.
- The flow is not reacting, pure evaporation is modelled.
- The velocities are low, the pressure is constant, there is no need for a momentum equation in the analysis.
- There is no radiation
- The droplet fully evaporates; there is no dissolved gas (microexplosions) or solid in the fuel ($Y_{FR} = 1$).
- There is no fuel vapour in the surrounding air ($Y_{F\infty} = 1$).
- The surface of the droplet is in liquid-vapour equilibrium: the gas is at its saturation vapour pressure.
- The Lewis number ($Le = Sc/Pr = \alpha/D$) equals 1.
- The liquid components of the fuel fully evaporate one after another, according to their boiling point. There are no coupling effects; at each instant the process is similar to single-component evaporation.

This last assumption does not appear in the classical theory; in the present model, it is the basic rule for multicomponent evaporation.

GAS PHASE ANALYSIS

The following analysis, largely inspired from [9] and [10], summarizes the approach of the d-square law model. The geometry of the problem is represented in Fig.1.

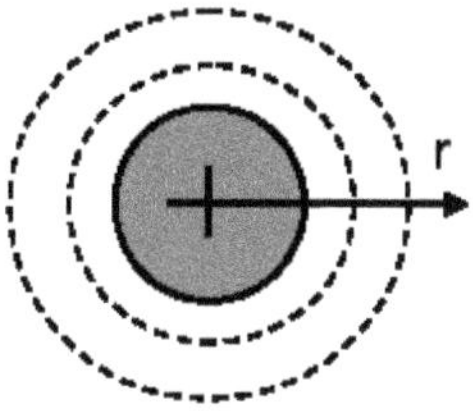

Figure 1: Geometry of the problem.

The continuity states:

$$\dot{m}_F = 4\pi * r^2 \rho_g V \qquad (1)$$

and $\dot{m}_F$ is spatially constant. In particular,

$$\dot{m}_F = 4\pi * R^2 \dot{m}''_F \qquad (2)$$

where $\dot{m}''_F$ is the fuel mass flow rate per unit area at the droplet surface.

The conservation equations for energy (using Fourier's law) and species (using Fick's diffusion law), with constant density, heat capacity, conductivity and diffusivity are as follows

$$\rho_g Cp_g \frac{DT}{Dt} = k_g \nabla^2 T \qquad (3) \qquad \text{and}$$

$$\rho_g \frac{DY_F}{Dt} = \nabla[\rho_g D_F \nabla Y_F] \qquad (3')$$

The profiles of temperature and fuel mass fraction are shown on Fig.2.

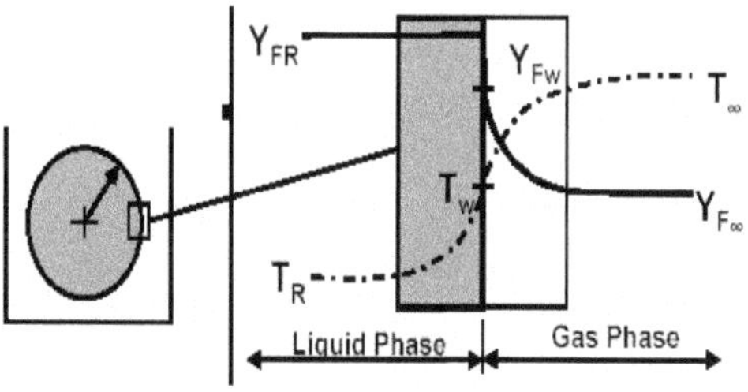

Figure 2: Temperature and mass fraction profiles at the droplet surface.

For a steady-state case, with spherically symmetric geometry, the conservation equations become:

$$k_g \left[\frac{1}{r^2} \frac{\partial}{\partial r} \left(r^2 \frac{\partial T}{\partial r} \right) \right] - \rho_g Cp_g V \frac{\partial T}{\partial r} = 0 \qquad (4) \qquad \text{and}$$

$$\rho_g D_F \left[\frac{1}{r^2} \frac{\partial}{\partial r} \left(r^2 \frac{\partial Y_F}{\partial r} \right) \right] - \rho_g V \frac{\partial Y_F}{\partial r} = 0 \qquad (4')$$

using the continuity equation

$$k_g \frac{\partial}{\partial r}\left(r^2 \frac{\partial T}{\partial r}\right) - \frac{\dot{m}_F}{4\pi} Cp_g \frac{\partial T}{\partial r} = 0 \qquad (5) \qquad \text{and}$$

$$\rho_g D_F \frac{\partial}{\partial r}\left(r^2 \frac{\partial Y_F}{\partial r}\right) - \frac{\dot{m}_F}{4\pi} \frac{\partial Y_F}{\partial r} = 0 \qquad (5')$$

Finally

$$k_g \frac{\partial}{\partial r}\left(r^2 \frac{\partial T}{\partial r}\right) - \dot{m}''_{FW} R^2 Cp_g \frac{\partial T}{\partial r} = 0 \qquad (6) \qquad \text{and}$$

$$\rho_g D_F \frac{\partial}{\partial r}\left(r^2 \frac{\partial Y_F}{\partial r}\right) - \dot{m}''_{FW} R^2 \frac{\partial Y_F}{\partial r} = 0 \qquad (6')$$

An analogy between heat and mass transfer appears in equations (6) and (6'). A non-dimensional temperature b_T and normalized mass fraction b_D are defined as

$$b_T = \frac{Cp_g(T - T_\infty)}{L_H + Cp_l(T_W - T_R)} \qquad \text{and} \qquad b_D = \frac{Y_{F\infty} - Y_F}{Y_{FW} - Y_{FR}}$$

so that equations (6) and (6') become:

$$\rho_g \alpha_g \frac{\partial}{\partial r}\left(r^2 \frac{\partial b_T}{\partial r}\right) - \dot{m}''_{FW} R^2 \frac{\partial b_T}{\partial r} = 0 \qquad (7) \qquad \text{and}$$

$$\rho_g D_F \frac{\partial}{\partial r}\left(r^2 \frac{\partial b_D}{\partial r}\right) - \dot{m}''_{FW} R^2 \frac{\partial b_D}{\partial r} = 0 \qquad (7')$$

With the assumption of a Lewis number equal to one, equations (7) and (7') become identical. Equations (6) and (6') are solved directly through a double integration; the boundary conditions for temperature are:

at $r = r_W$, $T = T_W$ and at $r = r_\infty$, $T = T_\infty$

and for mass fraction:

at $r = r_W$, $Y_F = Y_{FW}$ and at $r = r_\infty$, $Y_F = Y_{F\infty}$

The final solution is the expression of the fuel vapour mass flow rate at the droplet surface, under two different forms:

$$\dot{m}''_{FW} = \frac{\rho_g \alpha_g}{R} \ln\left(\frac{Cp_g(T_\infty - T_W)}{L_H + Cp_l(T_W - T_R)} + 1\right) \qquad \text{or}$$

$$\dot{m}''_{FW} = \frac{\rho_g \alpha_g}{R} \ln(b_{TW} + 1) \qquad (8) \qquad \text{and}$$

$$\dot{m}''_{FW} = \frac{\rho_g D_F}{R} \ln\left(\frac{Y_{F\infty} - Y_{FW}}{Y_{FW} - Y_{FR}} + 1\right) \qquad \text{or}$$

$$\dot{m}''_{FW} = \frac{\rho_g D_F}{R} \ln(b_{DW} + 1) \qquad (8')$$

At this point, the last two unknowns are the temperature and the fuel mass fraction at the droplet surface. A liquid-vapour equilibrium is assumed at the evaporating droplet surface, thus:

$$\chi_{FW} = \frac{P_{F\,sat}}{P}$$

and since $\chi_{FW} = \frac{Y_{FW} MW_g}{MW_F}$, with $MW_g = (1 - \chi_{FW}) MW_a + \chi_{FW} MW_F$,

the mass fraction at the droplet surface is

$$Y_{FW} = \frac{1}{\left(\frac{P}{P_{sat}(T_W)} - 1\right)\frac{MW_a}{MW_F} + 1} \qquad (9)$$

Finally, the comparison of (8) and (8'), with a Lewis number equal to unity, gives the equation for T_W:

$$b_{TW} = b_{DW} \qquad (10)$$

The solution procedure is an iterative one since P_{sat} and L_H are calculated at T_W and k_g and Cp_g at T_{ev}, which is determined by the "one-third rule":

$$T_{ev} = T_W + \frac{T_\infty - T_W}{3}$$

To obtain the d-square law, the droplet mass change is written as

$$\frac{dm_d}{dt} = -\dot{m} \qquad (11)$$

or $$\rho_l \pi \frac{D_d^{\,2}}{2} \frac{d(D_d)}{dt} = -2\pi D_d \frac{k_g}{Cp_g} \ln(b_{TW} + 1) \qquad (12)$$

which gives $$\frac{dD_d^{\,2}}{dt} = -\frac{8k_g}{\rho_l Cp_g} \ln(b_{TW} + 1) \qquad (13)$$

The right-hand-side term in (13) represents the evaporation constant λ_{ev}:

$$\lambda_{ev} = \frac{8k_g}{\rho_l Cp_g} \ln(b_{TW} + 1),$$ so that $D_d^{\,2}(t) = D_{d0}^{\,2} - \lambda_{ev} t$,

which is the final expression of the d-square law.

Throughout this analysis, the effect of the ambient air velocity was not accounted. This can be easily achieved through the use of a correlation for the Nusselt number around the droplet [11].

APPLICATION TO BIO-OIL

In the case of a multicomponent fuel such as bio-oil, the simple procedure developed above is successively applied to all components, sorted in the order of growing boiling temperature: after one component has fully evaporated following its own d-square law, the next component evaporation is calculated and applied to the droplet, whose size has been reduced by the volume of liquid already evaporated. The result is a piecewise d-square evaporation law. As in the rapid-mixing model, the evaporation occurs at equilibrium and the droplet temperature is varying in time.

COMPOSITION AND COMBUSTION BEHAVIOUR OF BIO-OIL

Bio-oils, or pyrolysis oils, are complex mixtures of the degradation products of ligno-cellulose. Compared to other liquid fuels, their main characteristics are: a high viscosity and acidity, a high water and solid content. More than four-hundred organic molecules have been reported in pyrolysis liquids [12], which can be viewed as poorly defined mixtures of acids, alcohols, aldehydes, esters, ketones, sugars, phenols, guaiacols, syringols, furans and multi-functionnal compounds, in widely varying proportions, depending on the raw material and the pyrolysis technique employed. The water content is most often around 25%. The solids in bio-oils are tars and chars; tars are more abundant but chars contain the inorganic compounds (calcium, lithium, iron, magnesium, manganese, zinc) that are more likely to corrode the components of the combustor exposed to the hot products. Again, the solid content varies with the type of biomass and the filtering technique.

COMBUSTION BEHAVIOUR

Experimental analyses of bio-oil droplets have been carried out in the last years and can be found in the literature [13, 14]. It appears from these publications that the combustion process of a bio-oil droplet is roughly divided in three successive steps as the temperature increases:

1) The droplet temperature rises up to 100°C, it starts burning in a quiescent blue flame; this corresponds to the evaporation of the lighter compounds, the most abundant of them being water.
2) As the temperature increases until around 600°C, an intense chemical activity develops: high molecular mass compounds evaporate and burn simultaneously.
3) Finally, the droplet expands due to internal boiling, microexplosions occur, the solid particles are expelled and burn, producing glassy-like solid residues.

From this information, a qualitative profile of bio-oil distillation curve can be drawn (see Fig.3).

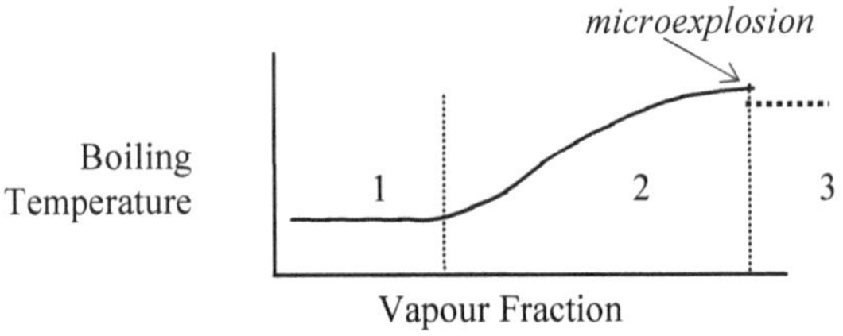

Figure 3: Qualitative profile of a bio-oil distillation curve.

Some data in reference [15] prove that this deduced profile is correct. In the present paper, the analysis focuses on parts 1 and 2 of Fig.3: only the evaporation of the liquid part of bio-oil is modelled. It must be noted that the prediction and modelling of microexplosions has been studied by various authors [16,17,18,19] and could be added to the present model for a more complete combustion analysis.

CHEMICAL COMPOSITION OF THE LIQUID PART OF BIO-OIL

Many chemical analyses of bio-oils [12, 20, 21, 22, 23] have been used to produce a realistic composition of a typical bio-oil. Nine chemicals are retained (see Table 1) so that their boiling points cover the whole range of temperature found in the distillation curves.

This artificial composition is obviously simple regarding the complexity of bio-oil but it can be modified and refined without any change in the method used here.

Table 1: Chemical composition of the bio-oil.

Name	Formula (molecular weight in g/mol)	Boiling Temperature (K)	Mass Fraction in Bio-Oil (%)
Formaldehyde	O=CH2 (30)	253.5	5.2
Glyoxal	CHO-CHO (58)	324	3.4
Ethanol	CH3-CH2-O-H (46)	351	5.3
Water	H2O (18)	373	22.9
Formic Acid	O=CH-OH (46)	374	9.2
Hydroxyacetaldehyde	O=CH-CH2-O-H (60)	383	22.5
Acetic Acid	O=COH-CH3 (60)	391	10.8
Acetol	CH3-CO-CH2-OH (74)	419	9.7
Levoglucosan	1,6-anhydro-β-D-glucopyranose (233)	659	11

PHYSICAL AND CHEMICAL PROPERTIES

The thermo-physical properties of the nine bio-oil components could not be found in the literature, they are estimated with semi-empirical methods. Below is a summary of the methods used and their references:

- Cp of gaseous compounds: Joback group contribution method [24]
- Cp of liquids: Chueh-Swanson group contribution method [24]
- Cp of air: the curve-fit coefficients found in [25] give a polynomial expression for two different temperature ranges.
- Enthalpy of vaporization at boiling point: formula of Kistyatowski [26]

  $\Delta H = Tb * (36.61 + 19.13 Log Tb)$ (J/mol)

- Variation of enthalpy of vaporization with temperature: Watson's relation [24]

$$\Delta H_2 = \Delta H_1 \left(\frac{1 - T_2}{1 - T_1} \right)^{0.38}$$

- Saturation vapour pressure: Clausius-Clapeyron equation

$$\frac{dP_{sat}}{dT} = \frac{P_{sat} L_H}{RT^2}$$

- Thermal conductivity: for most of the components (except ethanol and water), no relation between the thermal conductivity and the temperature was found. The thermal conductivity of air is used instead. The validity of this assumption has been checked with the results of ethanol and water, the resulting error in the evaporation constant never exceeds 5%.

RESULTS

For any operating conditions (ambient temperature, initial droplet size), the model provides an evaporation law. Fig.4 shows the d-square curve of a 100 μm droplet at two different ambient temperatures.

As the ambient temperature is increased, the evaporation curve tends to look like the one of a single-component d-square law. At 500 K, the levoglucosan has not reached its boiling point, its low evaporation rate clearly appears on Fig.4. At 800 K, all components have reached their boiling point, the boiling occurs immediately, the differences between the physical properties of the components and the corresponding evaporation constants (mainly P_{sat} and Cp) are damped.

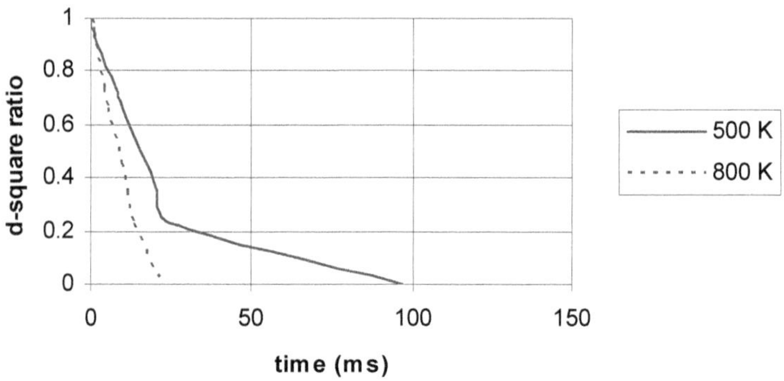

Figure 4: D^2-law of a 100 μm drop at 500 K and 800 K.

In order to easily implement a CFD code with this evaporation model, a numerical regression of the evaporation data is carried out and provides a power law for the diameter. The regression power law is more accurate at higher temperatures since the evaporation curve itself is smoother (see Fig.5).

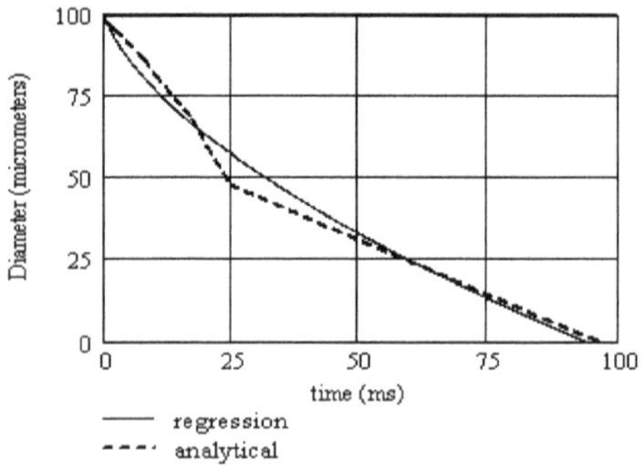

Figure 5a: Diameter Power law at 500 K

$$D(t) = -6t^{0.625} + 102$$

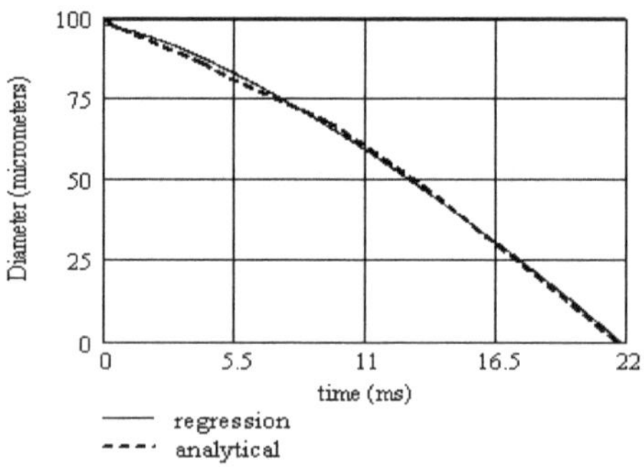

Figure 5b: Diameter Power law at 800 K

$$D(t) = -1.46t^{1.366} + 98$$

Each power law describes the evaporation process for one set of operating conditions. A further improvement of the model will be to achieve a three-dimensional mapping of the diameter law against temperature and initial droplet size with a view to obtaining a tool applicable to any bio-oil spray.

DISCUSSION

As a first check of the model, it is interesting to compare the evaporation law of bio-oil with the single-component evaporation laws obtained for the pure components that are present in the bio-oil. The graph on Fig.6 shows the evaporation at 500 K of a 100 μm doplet of bio-oil, formaldehyde or levoglucosan.

Formaldehyde and levoglucosan respectively are the fastest and the slowest evaporating constituents of bio-oil. As expected, the bio-oil evaporation time is between those of formaldehyde and levoglucosan, which is consistent with the composition of bio-oil defined here.

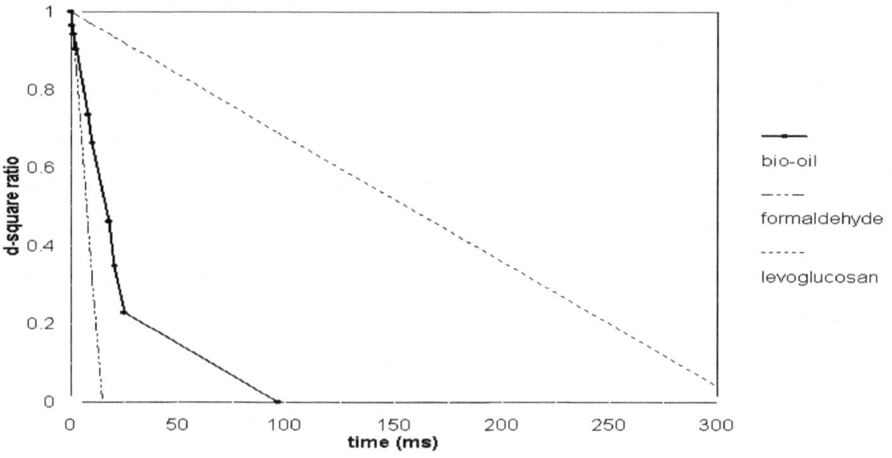

Figure 6: Comparison of the evaporation of bio-oil and its single components.

In the experimental work of Wornat et al. [13], the combustion of a 320-μm-diameter bio-oil droplet is investigated. The burning time obtained is 135 ms at an ambient temperature of 1600 K. However, the flame temperature of bio-oil is lower than this value since the high water content in bio-oil has a cooling effect on the flame. The evaporation model should give a reasonable approximation of the burning time since, for combustion, the same d-square law model can be applied with the flame temperature as ambient temperature. The evaporation of a 320-μm-diameter bio-oil droplet was therefore computed with the present model at different ambient temperatures (see Fig.7).

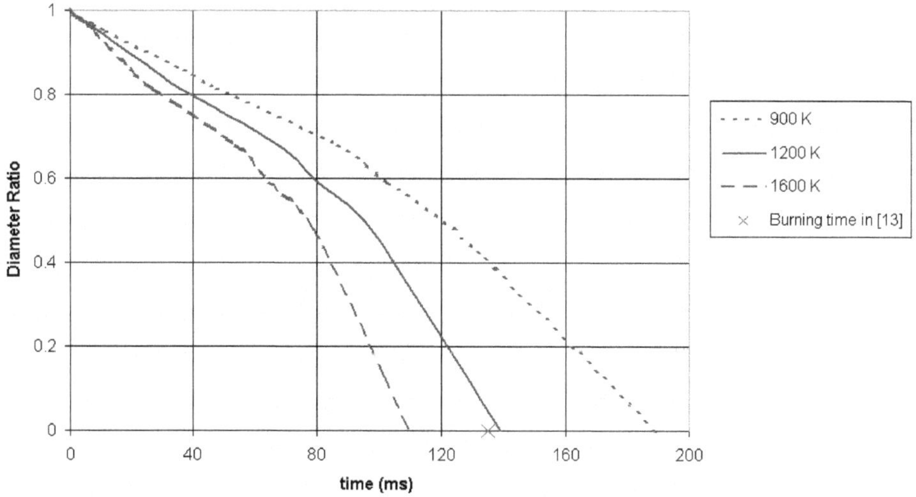

Figure 7: Burning time of a 320-μm-diameter droplet at different flame temperatures.

The value of 135 ms is reached for a flame temperature around 1200 K, which is in the range (1100 K-1300 K) of bio-oil flame temperatures found in the literature [27,28].

CONCLUSIONS AND PERSPECTIVES

An evaporation model for bio-oil droplets is proposed. It is based on the quasi-steady theory of vaporization with a spherically symmetric geometry. Only the liquid fraction of bio-oil is accounted and the components are assumed to fully evaporate one after another, according to their boiling temperature. The composition of bio-oil is estimated from different experimental studies found in the literature and can be easily refined. The thermo-physical properties of the chemicals are estimated through the use of semi-empirical correlations. The result is an evaporation curve for any droplet size and ambient temperature, which can be analyzed with numerical methods to give a diameter power law, easy to implement in a CFD code. The results show a good qualitative agreement with the experimental results published and an encouraging quantitative accuracy. Among the many possible improvements to this model, the most necessary would be a detailed analysis of the chemical processes occurring in the liquid fuel during heat-up and vaporization: at high temperature polymerization and pyrolysis of species that have not evaporated are known to affect the composition of the fuel and thus to modify the evaporation characteristics. Also, specific values for the important properties of the components of bio-oil would improve the accuracy of the results.

ACKNOWLEDGEMENTS

This work was funded by Materials and Manufacturing Ontario (MMO) and Orenda Aerospace, whose support is gratefully acknowledged.

NOMENCLATURE

Dimensional parameters

Cp	heat capacity at constant pressure ($J.kg^{-1}.K^{-1}$)
D_d	drop diameter (m)
D	mass diffusivity ($m^2.s^{-1}$)
k	thermal conductivity ($W.m^{-1}.K^{-1}$)
L_H	enthalpy of vaporization ($kJ.kg^{-1}$)
m	mass (kg)
$\dot{m}$	mass flow rate ($kg.s^{-1}$)
$\dot{m}''$	mass flow rate per unit area ($kg.m^{-2}.s^{-1}$)
MW	molecular weight ($kg.kmol^{-1}$)
P	pressure (Pa)
P_{sat}	saturation vapour pressure (Pa)
r	radial coordinate (m)
R	droplet radius (m)
t	time (s)
T	temperature (K)
Tb	boiling temperature (K)
V	velocity ($m.s^{-1}$)

Non-Dimensional parameters

Le	Lewis number
Pr	Prandtl number
Sc	Schmidt number
vf	vapour fraction
Y	mass fraction

Greek symbols

α	thermal diffusivity ($m^2.s^{-1}$)
λ_{ev}	evaporation constant ($m^2.s^{-1}$)
χ	mole fraction
ρ	density ($kg.m^{-3}$)

Subscripts

a	air
d	droplet
F	fuel
g	gas
l	liquid
R	interior of the droplet (reservoir)
W	droplet surface (wall)
0	initial condition
∞	surroundings

REFERENCES

1. Spalding D.B., Fourth Symposium (Int.) On Combustion, The Combustion Institute: Pittsburgh, PA, 1954, p847
2. Landis R.B. and Mills A.F., *Effects of Internal Resistance on the Vaporization of Binary Droplets*, Fifth Int. Heat Transfer Conf., Tokyo, Japan, Paper B7-9, 1974
3. Newbold F.R. and Amundson N.R., *A Model for Evaporation of a Multicomponent Droplet*, AIChE Journal, 1973, V19, N1, p22
4. Lehtinen K.E., Kulmala M., Vesala T., Jokiniemi J.K., *Analytical Methods to Calculate Condensation Rates of a Multicomponent Droplet*, Journal of Aerosol Science, 1998, V29, N9, p1035
5. Chen G. and Aggarwal S.K., *Experimental Study of Pure and Multicomponent Fuel Droplet Evaporation in a Heated Air Flow*, Atomization and Sprays, V7, p317, 1997
6. Sirignano W.A. and Law C.K., *Transient Heating and Liquid-Phase Mass Diffusion in Fuel Droplet Vaporization*, Evaporation-Combustion of Fuels, American Chemical Society, 1978, p3
7. Bardon M.F. and Rao V.K., *Brief Communication*, Combustion and Flame, 1991, V84, p427
8. Hallett W.L.H., *A Simple Model for the Vaporization of Droplets with Large Numbers of Components*, Combustion and Flame, 2000, V121, N1-2, p334
9. Gauthier J.E.D., *Droplet Evaporation and Combustion*, Lecture Notes of the *Gas Turbine Combustion* Graduate Course Given at Carleton University in Winter 2002
10. Catoire F., Gauthier J.E.D., Bardon M.F., Benaissa A., *Quasi-Steady-State Evaporation Model for Real Multi-Component Fuel Droplets*, Journal of the Institute of Energy, 1999, V72, N493, p134
11. Kanury A.H., *Introduction to Combustion Phenomena*, Gordon and Breach, New York, 1975
12. Diebold, J.P., *A Review of the Chemical and Physical Mechanisms of the Storage Stability of Fast Pyrolysis Bio-Oils*, Aug.26, 1999, http://www.eren.doe.gov/biopower/bplib/library/reviewchemicalphysical.pdf
13. Wornat M.J., Porter B.G., Yang N.Y.C., *Single Droplet Combustion of Biomass Pyrolysis Oils*, Preprint from Energy and Fuels 8, p1131 (1994)
14. D'Alessio J., Lazzaro M., Massoli P., Moccia V., *Thermo-Optical Investigation of Burning Biomass Pyrolysis Oil Droplets*, 27th Symposium (Int.) on Combustion, The Combustion Institute, 1998, p1915
15. Boucher M.E., Chaala A., Roy C., *Bio-oils Obtained by Vacuum Pyrolysis of Softwood Bark as a Liquid Fuel for Gas Turbines. Part1: Properties of Bio-Oils and its Blends with Methanol and a Pyrolytic Aqueous Phase*, Biomass and Bioenergy 19 (2000), p337
16. Yamasaki Hiroshi, Tsue Mitsuhiro, Kadota Toshikazu, *Evaporation and Combustion of Emulsified Fuel (Onset of Microexplosion)*, JSME International Journal-Series B, V36, N4, 1993
17. Leite L.F.T. and Lage P.L.C., *Modeling of Emulsion Droplet Vaporization and Combustion Including Microexplosion Analysis*, Combustion Science and Technology, 2000, V157, p213
18. Li T., Zhu D.L., Law C.K., *Droplet Combustion, Microexplosion, and Sooting Characteristics of Several Energetic Liquid Propellants*, Journal of Propulsion and Power, V14, N1, 1998, p45
19. Chang J., Kailasanath K., *Dynamics and Microexplosion of High Energy Fuels Injected into a Combustor*, Combustion Science and Technology, V137, N1, 1998, p217
20. Oasmaa A., Leppamaki E., Koponen P., Levander J., Tapola E., *Physical Characterization of Biomass-Based Pyrolysis Liquids, Application of Standard Fuel Oil Analyses*, VTT Publications, Technical Research Centre of Finland, ESPOO 1997
21. Bakhshi N.N., Adjaye J.D., *Properties and Characteristics of ENSYN Bio-Oil*, Catalysis & Chemical Reaction Engineering Laboratory, University of Saskatchewan, Sept. 1994
22. Radlein D., *Chemicals and Materials from Bio-Oil-Part 1*, Transforms International Ltd., http://www.pyne.co.uk/pdf/apps1a.pdf

23. Dynamotive Energy Systems Corporation, *Fast Pyrolysis of Bagasse to Produce Biooil Fuel for Power Generation*, 2001 http://www.dynamotive.com/biooil/technicalpapers/2001SugarPaper.pdf
24. Reid R.C., Prausnitz J.M., Poiling B.E., *The Properties of Gases and Liquids*, 4th Edition, McGraw Hill, New York, 1987
25. Chappell M.S. and Cockshutt E.P., *Gas Turbine Cycle Calculations: Thermodynamic Data Tables for Air and Combustion Products for Three Systems of Units*, Division of Mechanical Engineering, National Research Council Canada, Ottawa, August 1974
26. Bardon M.F., and Rao V.K., *Calculation of Gasoline Volatility*, Journal of the Institute of Energy, Sept.1984, p343
27. *Preliminary Pilot-Scale Combustion Evaluation of ENSYN RTP Oils*, CANMET, Canada Centre for Mineral and Energy Technology, Energy, Mines and Resources Canada, June 1992
28. Buhre B.J.P., Andries J., *Biomass-Based, Small-Scale, Distributed Generation of Electricity and Heat Using Integrated Gas Turbine-Fuel Cell Systems*, Paper Presented at the International Gas Turbine & Aeroengine Congress & Exhibition, Munich, May 8-11, 2000

Commercial Scale Carbonization of Briquetted Tropical Wood Sawdust

K.O. Lim
School of Physics, Universiti Sains Malaysia, 11800 Penang, Malaysia
P.B. Hii
Cipta Briquettes Sdn. Bhd., Lot 2604, Block 32, Kemena Industrial Estate, 97000 Bintulu, Sarawak, Malaysia

***ABSTRACT*:** This paper outlines a commercial scale carbonization system that rapidly carbonizes briquetted sawdust from tropical wood species at 850° - 875°C. The various issues involved in the carbonization process and the steps taken to overcome problems encountered are discussed. To date the plant is able to produce briquetted charcoal for export. The qualities of the product are as follows: Moisture content 5.0 - 9.5%, Volatile Matter 6.3 - 9.0%, Ash 1.1 - 2.5%, Fixed Carbon 82 - 92%, Calorific Value 7200 - 8500 kcal/kg.

INTRODUCTION

Tropical rainforests are home to hundreds of tree species. In Malaysia some of these species are harvested for their timber. The harvested trees are processed into sawn timber or plywood. In the process large quantities of wastes such as bark and sawdust are produced. Lim et al (1) estimated that in 1995 the timber industry in Malaysia generated roughly 0.226 million dry tonnes of sawdust. Cipta Briquettes Sdn. Bhd., located in Bintulu, Sarawak was set up 5 years ago to capitalise on the large quantities of sawdust produced by the sawmills in and around Bintulu. The sawdust gathered by Cipta Briquettes come from about 30 - 40 wood species and are used for the production of charcoal briquettes, most of which are then exported.

THE CHARCOAL PRODUCTION PROCESS

As mentioned, the raw material used in Cipta's factory is sawdust from various species of tropical wood. The moisture content of these green or fresh sawdust varies from about 40 - 50%.

The green sawdust are first screened to remove contaminants and oversized particles. They are then dried to a moisture content of about 8 - 10% in an impulse dryer with hot air that is generated using wood wastes. The fairly dry sawdust are then fed into a screw type briquetter (Shida, Taiwan) for compaction into briquettes. The density of the briquettes formed is around 1200 kg m^{-3}. The briquettes that leave the briquetting machine are hot and slightly roasted on the outside. As such they have to be left in the open to cool. The cooling process further enhances the binding between the sawdust particles. In Cipta's factory, briquettes with 2 types of cross-sections are produced, namely hexagonal and square with a circular hole in the centre (see Fig 1).

Before carbonization, the dimensions of these 2 shapes of briquettes are as follows in *Table 1*.

Table 1:

	Hexagon (mm)	Square (mm)
Length	410	410
Width	52	50
Diameter of central hole	22	15

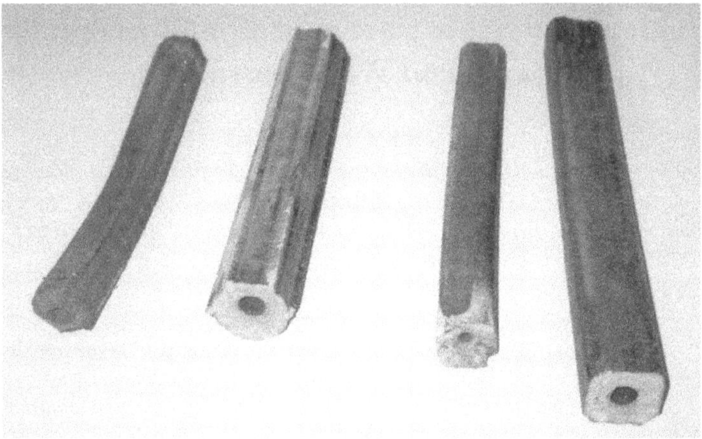

Figure 1: Sawdust briquettes before (brownish) and after carbonization (dark grey).

The sawdust briquettes are then neatly arranged on a bugee which is an open cart with only 2 end walls, with air flow holes at the lower portion and rollers at its undersurface (see Fig. 2). Each bugee which is made of steel and refractory cement can hold about 3000 pieces of briquettes. The loaded bugees are then pushed into the carbonization kilns. This motion is guided by 2 parallel steel rails embedded on the floor of the kiln.

Figure 2: Sawdust briquettes loaded onto bugees.

Fig. 3 is a picture of one kiln before the bugee is pushed in. The internal dimensions of the kiln are roughly 1.5 m wide, 2.4 m deep and 1.5 m high and the height of the curved roof is about 0.2 m. The walls of the kiln are constructed from refractory fire bricks sourced locally while the kiln floor is of concrete of high grade cement. The roof of the kiln on the other hand is of mild steel cast with refractory castable cement. The kilns are constructed in rows with outlets for the escape of emissions built into one side - wall. This arrangement allows 2 adjacent kilns to share one emission outlet while 1 chimney is shared by 4 kilns (see Figs. 4 and 5).

Figure 3: Picture showing the kiln used for carbonization.

Figure 4: Emissions from 2 adjacent kilns are connected to a common outlet which is then connected to a chimney that serves 2 other kilns. Thus one common chimney serves 4 kilns. Note the flame that burns off most of the combustible emissions.

After the bugee is pushed into place inside the kiln, the kiln is then sealed. The carbonization process is started by putting a few pieces of lighted briquettes on top of the pile before the bugee is pushed into the kiln. The sawdust briquettes are carbonized at a temperature of 850 - 875°C for 108 hours with air flow that is controlled at various stages of the process. A good control of the air-flow is essential in ensuring that good quality charcoals are produced.

Some of the emissions from the carbonization process are combustible and these are burnt as they exit the kilns as shown in Figs. 4 and 5. Rejected briquettes are used for this purpose. As most of the combustible emissions from the carbonization process are burnt before they reach the chimney, the system adopted by Cipta is therefore fairly free of pollutants when the final emissions are allowed to escape into the atmosphere.

When the carbonization process is complete which is indicated by the emissions from the chimneys becoming invisible, the bugees with the red hot charcoal inside are removed from the kilns and immediately covered with steel boxes which are then completely sealed with sand to avoid any entry of air (see Fig. 6). The steel boxes with their contents are then allowed to cool for 24 hours after which the charcoals are ready for packaging (Fig. 7).

Figure 5: A row of kilns, showing some empty, some with bugees inside but unsealed and some already sealed.

Figure 6: A row of cooling boxes.

THE CHARCOAL PRODUCT

Figure 7: The charcoal product.

After undergoing carbonization, the sawdust briquettes were observed to have shrunk in size as can be seen in Fig. 1 and Table 2 below:

Table 2

	Hexagonal cross-section Briquettes (in mm)		Square cross-section Briquettes (in mm)	
	Before carbonization	After carbonization	Before carbonization	After carbonization
Length	410	350	410	350
Width	52	40-42	50	37-39
Diameter of central hole	22	15	15	11

When the product was targeted for the Korean market, charcoal briquettes of hexagonal cross-section and having the following qualities were produced.

Moisture content	8.8 - 9.5%
Volatile matter	6.3 - 9.0%
Ash	2.0 - 2.5%
Fixed carbon	82 - 89%
Calorific values	>7200 kcal/kg

The Japanese consumers however prefer charcoal briquettes with a square cross-section and having the following qualities:

Moisture content	5.0 - 8.0%
Volatile matter	6.3 - 9.0%
Ash	1.1 - 2.0%
Fixed carbon	85 - 92%
Calorific value	>8200 kcal/kg

Cipta is able to produce both the above market requirements (2). With 28 briquetting machines and 90 kilns, Cipta processes 100 tonnes of green sawdust per day or 2800 tonnes per month. This amount of green sawdust results in the production of about 558 tonnes of charcoal a month. Taking into consideration that about 5% of the raw material are lost during the screening process, the carbonization yield works out to be roughly 35% of the weight of sawdust briquettes used.

Of the 558 tonnes of charcoal produced a month, roughly 15% cannot be exported as they do not meet the desired specifications. These then end up as charcoal scraps which are then sold to local consumers. Presently Cipta exports monthly about 474 tonnes of charcoal briquettes to Japan.

CONCLUDING REMARKS

When Cipta started operations about 5 years ago, not all were smooth sailing. The company encountered numerous problems that prevented export quality charcoal to be produced. Among the numerous obstacles encountered, the following were the more major ones:

(1) Some species of wood sawdust were found to be unsuitable for quality charcoal production. These were therefore not further used as raw material.

(2) The variation of moisture in fresh sawdust resulted in uneven drying and this in turn affected the quality of the sawdust briquettes produced in that more cracks were observed. As a result the drying temperature needed more constant adjustments.

(3) The length of screw and profile of screw fins in the briquetter were observed to affect the density of the sawdust briquettes produced. After some R&D work, the problems were overcome.

(4) It was found that natural draught was not the most suitable for the carbonization process as it was affected by chimney height as well as weather conditions. Forced draught, which would increase cost, would be preferred. However by a suitable choice of chimney height, Cipta continued using natural draught.

(5) Air input control was found to be essential and this was done at various stages during the carbonization process.

Even though there were problems in Cipta's initial stages of operation but through R&D activities as well as trial and error operations, most of the problems were overcome and now Cipta is proud to say that the company currently has a long term supply contract with the Japanese. This is indeed an achievement as we are all well aware of the demand for quality by the Japanese consumer.

Besides Cipta, there are 9 other similar operations in the country and together Malaysia exports some 2000 tonnes of charcoal from sawdust briquettes a month.

Besides Malaysia, as far as the authors are aware, charcoal from sawdust briquettes are also produced by China, Indonesia, Japan, Korea, Myanmar and Vietnam.

REFERENCES

1. Lim K.O., Zainal Z.A., Quadir G.A. & Abdullah M.Z. (2000) Plant Based Energy Potential and Biomass Utilization in Malaysia. *International Energy Journal*, **1**, 77-88.
2. Analysis Reports (2002), Department of Mineral and Geoscience Malaysia, Kuching, Sarawak.

Chemical Stability of Wood Fast Pyrolysis Liquids

D. Meier, G. Jesussek, S. Radtke
Federal Research Centre for Forestry and Forest Products, Institute for Wood Chemistry and Chemical Technology of Wood, Hamburg
Leuschnerstr. 91, D-21031 Hamburg, Germany

ABSTRACT: The chemical changes of a fast pyrolysis liquid from beech wood have been followed over 32 weeks by monitoring 70 calibrated compound with gas chromatography. The fast pyrolysis liquid (bio-oil) was freshly made in a laboratory fluid bed reactor system according to the Waterloo Fast Pyrolysis Process. Immediately after production a GC analysis was made taking 70 compounds into account for quantitative analysis. Then the sample was divided into three parts and stored under 3 different conditions for 32 weeks: room temperature with daylight, room temperature in the dark, and cool and dark (refrigerator). Analyses were performed in the beginning once a week then the interval was increased to 4 weeks. All analysed peaks were grouped into categories such as acids, phenols, furans, sugars etc. Surprisingly, only minor changes were observed over the monitoring period with the exception of some compounds such as vinyl-guaiacol, vinyl-syringol, and 3-hydroxy-5,6-dihydro-(4H)-pyran-4-one. Obviously, they contribute by reaction with pyrolytic lignin to sludge formation observable after storage. Cooling of the sample delays but does not prevent polymerization.

INTRODUCTION

Pyrolysis liquids (bio-oil) are currently being generated by several processes aimed at using them at first as substitute for petroleum-based fuel oils (1). Secondly, applications are being investigated to use the oils as source for chemicals (2). To achieve high yields of condensates rapid heating of the biomass in the reactor, short residence times of the vapours and rapid cooling or quenching of the vapours are essential features of any of the existing fast pyrolysis processes. However, these fast chemical processes are the reason why the resulting bio-oil is an thermodynamically unstable product (3). Bio-oil has a high oxygen content and contains many reactive components that can form higher molecular weight species. In general, these reactions result in an increase in viscosity and a decrease in volatility. Some attempts were made to evaluate the effects of storage conditions on physical and chemical properties of biomass fast pyrolysis oils. They were exposed to elevated temperatures over extended periods of time. Several physicochemical properties such as pH, water content, viscosity were measured. Chemical changes were monitored using molecular weight determination by GPC and FTIR (4). It was reported that the water content, viscosity and molecular weight of the oil increased with time and temperature of storage. Furthermore, FTIR provided evidence that etherification or esterification are mechanisms for condensation.

A comprehensive review of chemical and physical mechanisms of the storage stability of fast pyrolysis bio-oils is given by Diebold (3). A list of probable chemical reactions which might occur during storage are presented based on functional groups of the many compounds present in bio-oil. The author presents calculated equilibrium compositions of a pseudo bio-oil using water, acetaldehyde, methanol, and acetic acid as components. However, there are no real data available in literature on the fate of components in bio-oil after storage at different conditions. Therefore, this paper presents results of an investigation into the detailed qualitative and quantitative determination of single oil compounds detectable by gas chromatography using three different storage conditions.

EXPERIMENTAL

Beech wood (*Fagus sylvatica L.*) with a particle size range of 0.15-0.50 mm and a moisture content of 8.6 % was selected for fast pyrolysis.

Generation of bio-oil

Bio-oil was generated in a fluidized bed reactor system (*Figure 1*) at 475 °C and a vapour residence time of 1.5 s.

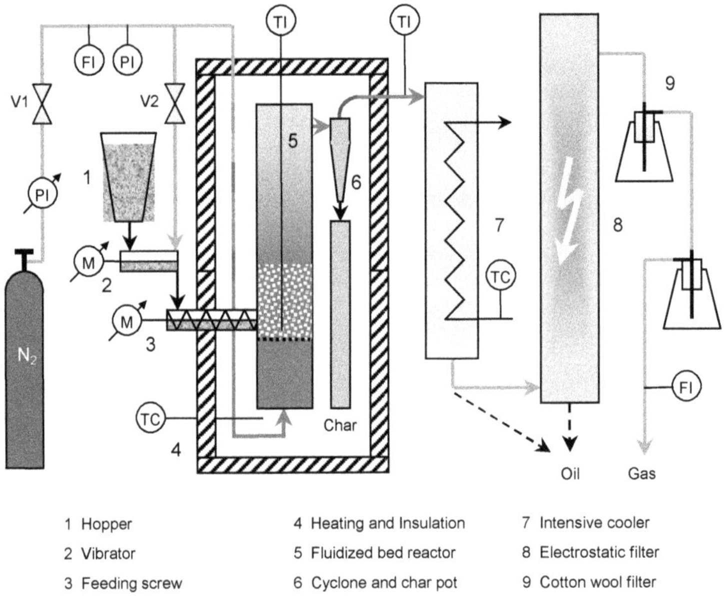

Figure 1: Laboratory fluidized-bed reactor system.

Oil is collected in a flask which is connected both to the intensive cooler and the electrostatic filter. Immediately after production a sample was taken for GC analysis.

Storage and sampling of bio-oil

The whole oil was divided immediately after production into three samples which were stored under the following conditions in sealed glass vials: (a) room temperature with daylight (occasionally direct sunlight), (b) dark at room temperature, and (c) in a refrigerator (dark at 8 °C). Sampling frequency was once a week over a period of 7 weeks, after that frequency was decreased to monthly sampling up to week 32.

GC analysis

Oil samples were taken from the sealed glass vials after intensive shaking. To 60 mg of oil sample, 1 ml of acetone containing 0.2008 mg of internal standard (fluoranthene) was added. This solution was injected into a Chrompack CP 9001, NL, gas chromatograph using the following conditions: column Zebron ZB-1701 (Phenomenex) 60m L x 0.25mm ID x 0.25µm, split injection 1:30, carrier gas He, injector pressure 200 kPa, oven programme 45 °C for 4 min, 3 °C/min to 280 °C, 15 min at 280 °C, FID 280 °C. Prior to analysis the GC system was calibrated with known compounds which are presented in Table 1.

Table 1: Calibrated compounds for quantitative GC analysis.

Hydroxyacetaldehyde	p-Cresol	Acetoguaiacone
Crotonaldehyde	m-Cresol	4-Vinyl-syringol
Acetic acid	4-Methyl guaiacol	Guaiacyl-acetone
Hydroxypropanone	2,4- and 2,5-Dimethyl phenol	4-Allyl-syringol
1,2 Ethandiol	3-Ethyl-phenol	4-Propyl-syringol
1-Hydroxy-2-butan-1-one	4-Ethyl-phenol	Isomer of Coniferyl-alcohol
3-Hydroxypropanal	4-Ethyl-guaiacol	4-Propenyl-syringol (cis)
2-Furaldehyde	1,4:3,6-Dianhydro-manno-pyranose	Levoglucosan
α-Angelicalactone	3-Methoxy-catechol	4-Propenyl-syringol (trans)
2-Hydroxy-2-cyclopenten-1-one	4-Vinyl-guaiacol	Syringaldehyde
Benzaldehyde	Eugenol	Homo-syringaldehyde
Dihydro-methyl-furanone	4-Propyl-guaiacol	Aceto-syringone
3-Methyl-2-cyclopentene-1-one	5-Hydroxymethyl-2-furaldehyde	Coniferylalcohol (trans)
Dihydro-2(3H)-furanone (γ-Butyrolactone)	Catechol	Coniferylaldehyde
(5H)-Furan-2-one	Syringol	Syringylacetone
3-Methyl-(5H)-furan-2-one	Isoeugenol (cis)	Propiosyringone
4-Hydroxy-5,6-dihydro-(2H)-pyran-2-one	Isoeugenol (trans)	Isomer of Sinapyl alcohol
2-Hydroxy-1-methyl-1-cycolpentene-3-one	4-Methyl-syringol	Dihydrosinapyl alcohol
Phenol	Vanillin	Sinapyl alcohol (cis)
Guaiacol	Resorcin	Sinapyl alcohol (trans)
o-Cresol	Methyl-resorcin	Sinapaldehyde
Maltol	Homo-vanillin	4.4-Dimethyl-2-cyclopenten-1-one
3-Ethyl-2-hydroxy-2-cyclopenten-1-one	4-Ethyl-syringol	Acetonylacetone
		Cyclopentanone

RESULTS AND DISCUSSION

The results are summarized in the following Figures 3-5. Not all 70 monitored compounds can be presented here. To facilitate interpretation components were grouped into chemically similar categories. The main compounds from each group are presented.

The behaviour of the main bio-oil compounds is presented in

Figure 3. The contents of acetic acid, hydroxypropanone and levoglucosan remain more or less constant under the three storage conditions. There is some fluctuation in the determinations which could result from inhomogeneous samples. But the overall trend shows little variation.

The next three main components hydroxyacetaldehyde, 3-hydroxypropanal, and 2-hydroxy-3-oxobutanal already exhibit a more distinct behaviour. Two aldehydes, hydroxyacetaldehyde and 3-hydroxypropanal remain rather stable under cool storage conditions. Room temperature, regardless the effect of light, leads to a decrease for all components. 2-hydroxy-3-oxobutanal has almost disappeared after the storage period.

Figure 4 shows pyrans with a keto-group (pyranones) and guaiacols representing a smaller fraction of the bio-oil but exhibiting a much higher reactivity. From the pyranones 3-hydroxy-5,6-dihydro-(4H) pyran 4-one is the most reactive one. Cooling helps in slowing down reactivity. From the guaiacols, vinyl-guaiacol and iso-eugenol are the most reactive species. Their unsaturated double bond is very prone to polymerization reactions. Just after production reaction is delayed slightly by cooling. Remarkably, the content of coniferylaldehyde with a propenyl side chain remains constant.

Figure 5 shows the behaviour of phenols and furans with an aldehyde or keto-function. Phenols are rather stable components in the bio-oil. The storage conditions do not affect their content significantly. The cyclic aldehydes and ketones show the tendency to decrease slowly, despite the storage conditions.

Surprisingly, the overall changes in the course of the storage period of 32 weeks are not very big. The most significant effects can be observed with pyranones derived from cellulose degradation and lignin derived compounds which have an unsaturated bond in their side chain of the aromatic ring from the vinyl and propenyl type. Various reactions of the carbonyl and alkenyl groups are possible. It would be beyond the scope of this paper to discuss them all. Furthermore, it is impossible to give a clear picture of reactions due to the presence of several hundreds of compounds present in the pyrolysis liquid. However, it is worth to mention reactions which could take place under acid conditions as is the case with bio-oils. Recently it was reported that alkylation of carbonyls and lactonization of alkenes are possible (5). Moreover, two principle reaction mechanisms are possible for the unsaturated bonds: firstly radical polymerization with each other. This was also mentioned by Diebold (3), and secondly cationic polymerization through addition of other molecules. A probably candidate for this type of reaction could be pyrolytic lignin which would explain the increase in molecular weight and viscosity as reported by Czernik et al. (4). Pyrolytic lignin was comprehensively characterized by Scholze and Meier (6) and Scholze et al. (7). Based on their results a possible structural formula of pyrolytic lignin can be established which represents the current knowledge of its composition (Figure 2). Possible reactive sites are marked with an arrow.

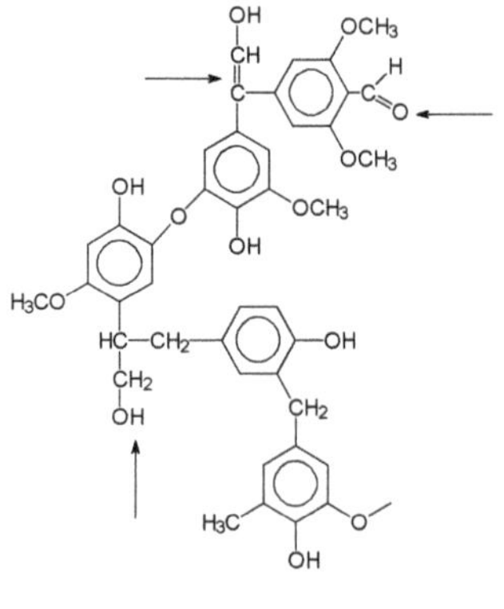

Figure 2: Proposed configuration of pyrolytic lignin.

Based on the findings that the most reactive compounds in bio-oils have unsaturated double bonds of the aryl-alkenyl-type, the determination of their amounts in crude bio-oils could be used as a measure for instability. Therefore, it would be desirable to inhibit polymerization for example through addition of stabilizers or inhibitors. Commonly used inhibitors are phenolic derivatives such as dibutyl-phenyl-phenol (DBT). As already proposed earlier (3), mild hydrogenation of the double bonds could also be an alternative route. The addition of small amounts of solvents is also beneficial (8).

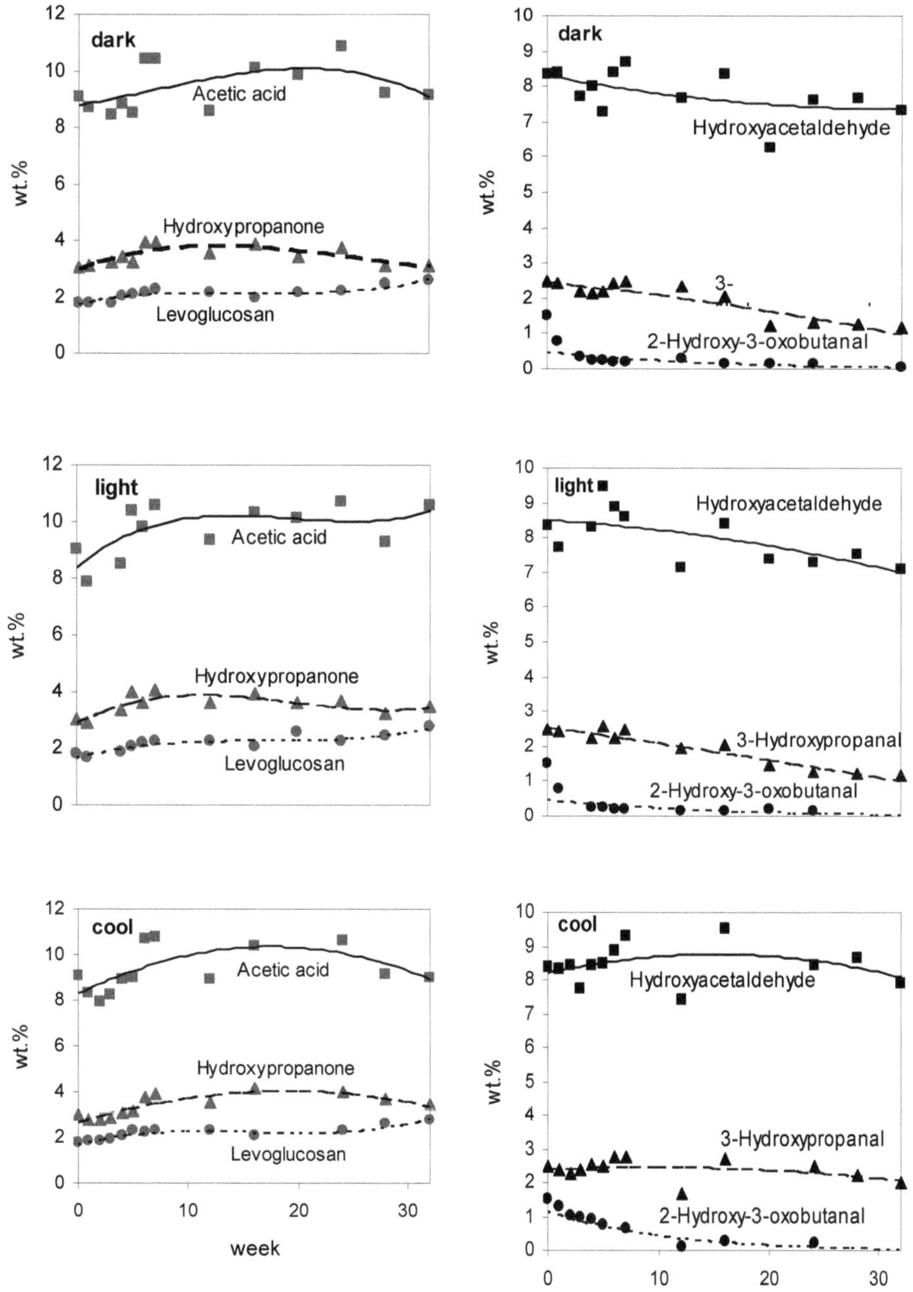

Figure 3: Behaviour of the main bio-oil components.

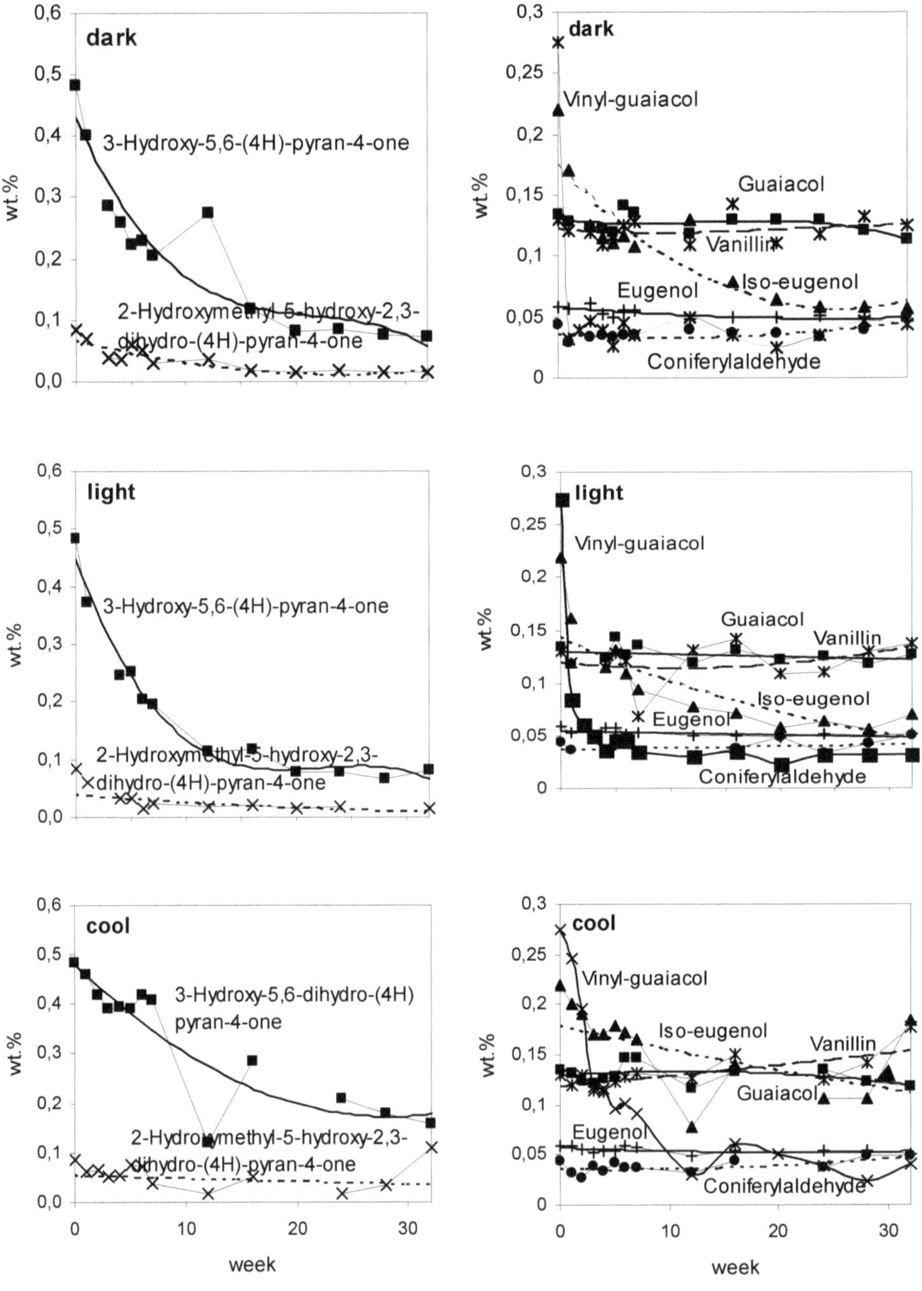

Figure 4: Behaviour of pyrans and guaiacols.

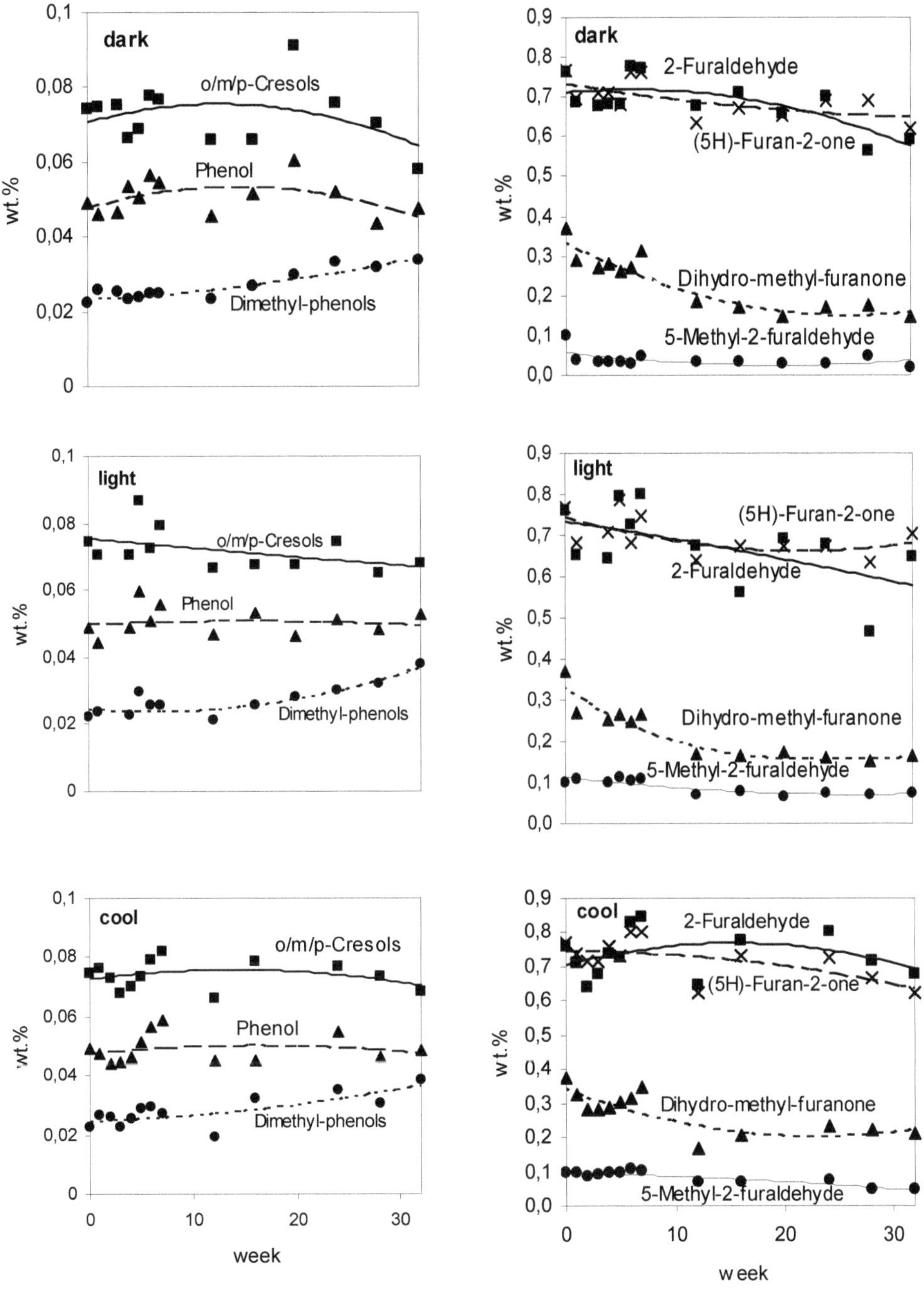

Figure 5: Behaviour of phenols and furans.

CONCLUSIONS

Fast pyrolysis liquids (bio-oils) could substitute light or heavy fuel oil in special applications. Bio-oil production is made from renewable inhomogeneous feedstocks using an unspecific thermal treatment which results in a product which is not in an thermodynamic equilibrium. Therefore, reactions in the bio-oil occur after production which cause some oils to separate into two phases or to form sludge. Many of the GC detectable compounds which represent ca. 40 % of the whole liquid (9) show a rather stable behaviour under different storage conditions. Most reactive are compounds with aryl-vinyl structure and pyranones. Cooling of the sample helps to delay reactions. It is proposed to test either the effect of stabilizers or inhibiters or mild hydrogenation to minimize the ageing effects of fast pyrolysis liquids.

REFERENCES

1. Bridgwater A. V., Czernik, S & Piskorz J. (2002) The status of biomass fast pyrolysis, In: *Fast Pyrolysis of Biomass: A Handbook Vol. 2* (Ed. by A. V. Bridgwater), pp 1-22. CPL Press, Newbury.
2. Radlein D. The production of chemicals from fast pyrolysis liquids, In: *Fast Pyrolysis of Biomass: A Handbook* (Ed. by A. Bridgwater et al.), pp 164-188, CPL Press, Newbury.
3. Diebold J. P. (2002) A review of the chemical and physical mechanisms of the storage stability of fast pyrolysis bio-oils, In: *Fast Pyrolysis of Biomass: A Handbook Vol. 2* (Ed. by A. V. Bridgwater), pp 243-292, CPL Press, Newbury.
4. Czernik S., Johnson D. K. & Black, S. (1994) Stability of wood fast pyrolysis oil. *Biomass & Bioenergy*, **7**, 187-192.
5. Melikyan G. G. (1998) Manganese-based organic and bio-organic transformations. *Aldrichimica Acta*, **31**, 50-64.
6. Scholze B. & Meier D. (2001) Characterization of the water-insoluble fraction from fast pyrolysis liquids (pyrolytic lignin). Part I. Py-GC/MS, FTIR, and functional groups. *J. Anal. Appl. Pyrolysis*, **60**, 41-54.
7. Scholze B., Hanser Ch. & Meier D. (2001) Characterization of the water-insoluble fraction from fast pyrolysis liquids (pyrolytic lignin). Part II. GPC, carbonyl groups, and 13C-NMR. *J. Anal. Appl. Pyrolysis*, **58-59**, 387-400.
8. Diebold J. P. & Czernik S. (1997) Additives to lower and stabilize the viscosity of pyrolysis oils during storage. *Energy & Fuels*, 11, 1081-1091.
9. Oasmaa A. & Meier D. (2002) Analysis, characterization and test methods of fast pyrolysis liquids, In: *Fast Pyrolysis of Biomass: A Handbook Vol. 2* (Ed. by A. V. Bridgwater), pp 23-40, CPL Press, Newbury.

Pyrolysis of Plastics and Tyres Wastes

M Miranda, P Costa, F Pinto, I Gulyurtlu and I Cabrita
INETI-DEECA, Estrada Paço do Lumiar, 22, 1649-038, Lisboa, Portugal

ABSTRACT: The main aim of the work is to characterize the nature of liquids produced by pyrolysing various blends of wastes were several mixtures of tyres and plastics wastes were pyrolysed. The plastics wastes used were polyethylene (PE), polypropylene (PP) and polystyrene (PS). The tyre waste also included natural rubber (NR), styrene butadiene rubber (SBR) and butadiene rubber (BR).

The influence of the following operating conditions as well as different mixture of wastes on nature of the products was investigated; run temperature: 380-450 °C, run pressure: 0.2-8.0 MPa, reaction time: 10 to 60 minutes, waste composition: several mixtures of plastics and tyres in different quantities. The final objective was to optimize operating condition to maximize the production of liquids and to achieve selectivity in the nature of the liquids obtained. The results obtained showed that the composition of the end products was highly dependent of the components of the blend of wastes used.

Several solvents with different H-donor capacity were also added to the reaction mixture to investigate the nature of liquids formed to verify if a higher degree of selectivity could be achieved with regard to the nature of liquid products. Solvents used included 1,2,3,4-Tetrahydronaphthalene and 1,2,3,4-tetrahydroquinoline. Liquids yields obtained so far were found to be very promising, reaching values around 70% (v/v), while gas and residual solid amounts were around 10% (v/v) and 20% (v/v).

INTRODUCTION

Petroleum is the fossil fuel with the greatest contribution to the worldwide energy production and also is the natural raw material for plastics and tyres production. As petroleum reserves have a limited lifetime [1, 2], it is advisable to have a better management of the remaining petroleum resources.

Both plastics and tyres have been used in modern societies in almost everything, which have created increasing amounts of wastes with negative impact on environment. Due to the fact that the main solutions, used so far, to deal with these wastes present several problems and do not seem to be the right ones, to decrease, in an effective way, the increasing amount of wastes produced all over the world. Landfilling of plastic and tyres residues is not a solution, because it has been increasingly difficult to find suitable places for building technically adequate landfills, due to the danger of leaching and soil impregnation, with the subsequent contamination of underground waters. Apart from this, landfilling does not allow the recovery of the organic content of these wastes that should be part of the organic lifetime cycle. Incineration has the advantage of taking profit of wastes energetic content, but produces pollutants like light hydrocarbons, nitrous and sulphur oxides, dusts and dioxins, which have highly negative bearing on the environment [1]. On the other hand, these wastes combustion destroys completely an important resource, as wastes organic content are converted only into CO_2 and H_2O.

Pyrolysis technology applied to these hazardous wastes may have an important role in future [3], as it may allow the conversion of these residues into economical valuable products. The process is achieved through thermal decomposition under moderate conditions of pressure and temperature were polymeric structure is broken down, producing smaller intermediate species (radicals or ions), which can further react and produce a mixture of smaller hydrocarbon molecules, being liquid and gas in nature [3]. These hydrocarbons could be used as fuel or as feedstock in petrochemical industry [4].

Several mixtures of plastics and tyres wastes were pyrolysed with the aim of producing liquids with specific characteristics that could be re-utilized in different end-use applications. Some mixtures of plastics wastes chosen to study present a similar composition to the one existing in Portuguese municipal solid waste (MSW); these plastics contained polyethylene (PE), polypropylene (PP) and polystyrene (PS). The wastes also included tyres witch main components are natural rubber (NR), styrene butadiene rubber (SBR) and butadiene rubber (BR).

EXPERIMENTAL PART

In Figure 1 is shown a schematic representation of the experimental installation used for waste pyrolysis studies. The main component is a 1 liter autoclave built of Hastelloy C276, by Parr Instruments.

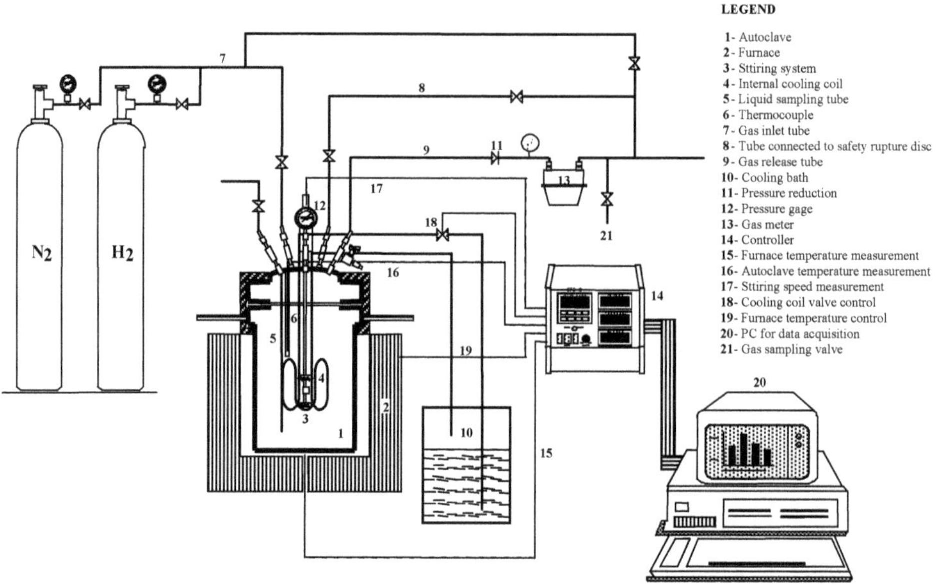

Figure 1: Schematic representation of the waste pyrolysis installation.

After the autoclave was loaded and closed, it was pressurised to a pre-set value with nitrogen. Afterwards, it was heated till the desired temperature value was reached, at which it was maintained during the settled reaction time. The autoclave was then cooled down to room temperature and gases were subsequently collected, measured and analysed by GC (gas chromatography) [1, 2]. Finally, the autoclave was opened and its content was weighted.

Liquid hydrocarbons were then collected and separated from solid fraction and distillated to obtain three fractions using the ASTM standard D 86 [6]. The lighter fraction distilled to a value lower than 150°C; whilst the other one presented a distillation range between 150°C and 270°C; the third fraction had a distillation range higher than 270°C. Liquid fractions were analyzed by GC-MS (gas chromatography-mass spectrometry) with capillary columns in order to identify their main compounds [7], and they were also characterized by using ASTM standards [6] regarding theirs chemical and physical properties. The values of the chemical analysis allowed the theoretical calculation of research octane number using the method developed by Protic-Lovasic et al. [2, 8]. Solid fraction was solvent extracted, first with dichloromethane and then with tetrahydrofuran to recover liquids hydrocarbons impregnated in initial solid fraction.

DISCUSSION OF RESULTS

In Figure 2 is shown the effect of tyre composition on products yields. The highest liquid yield (more than 60% w/w) was archived when no tyre amount was included in the plastic mixture witch is similar composition to the one existing in Portuguese municipal solid waste (MSW); in this conditions it was also archived the lowest value on solid yield reaching values around 12 % (w/w). As the amount of tyre is included in plastic mixture the solid yield increases in 77% (w/w) wile liquid yield decrease in 45% (w/w). The release of gases seam to show a constant behavior varied between 25 and 15% (w/w).

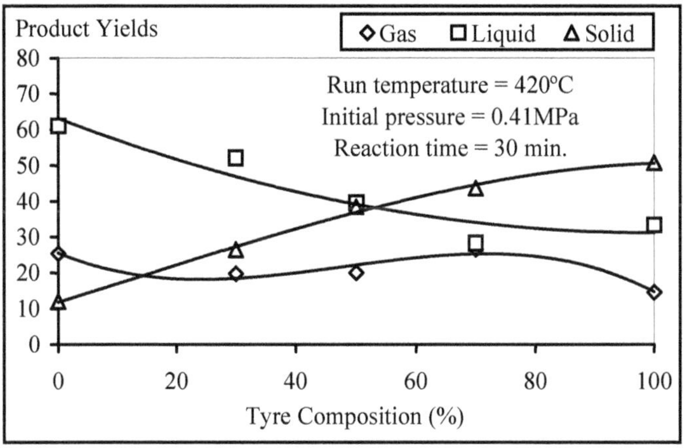

Figure 2: Effect of tyre composition on products yields.

In Figure 3 and 4 the relative distribution of liquid hydrocarbons was highly dependent on input plastic waste composition.

The highest value of alkanes, around 72% was obtained with 100% of PE. The rise of alkanes production with the increase of PE content in the input mixture suggests that the presence of this kind of plastic favoured the formation of alkanes. Higher values of PP favoured the formation of alkenes, being the highest value of alkenes obtained when 100% of this plastic was used. Probably, because the presence of the methyl group on the PP polymeric structure favoured the stabilisation of the intermediate radicals by the formation of a double bond.

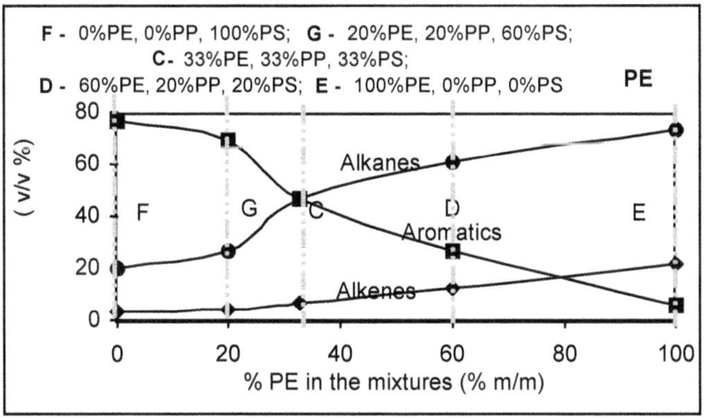

Figure 3: Plastic waste composition effect on liquids hydrocarbons relative distribution for polyethylene.

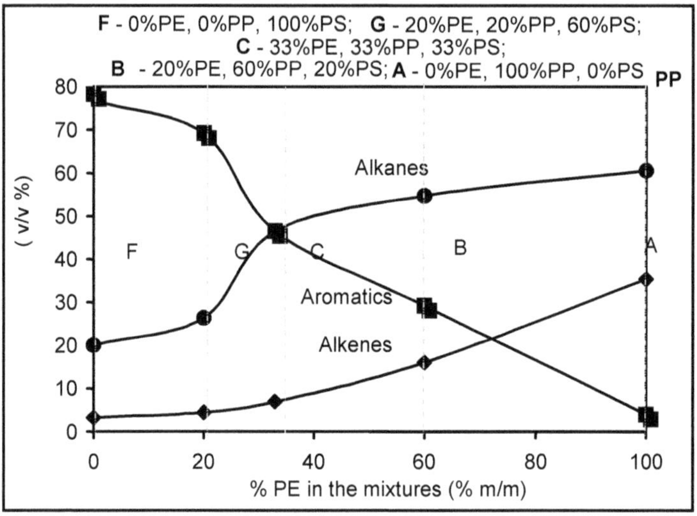

Figure 4: Plastic waste composition effect on liquids hydrocarbons relative distribution for polypropylene

In Figure 5 PS increased dramatically, the aromatic content of the end product being more than 76% while the lowest alkanes and alkenes percentage were obtained. This probably means that in the presence of PS the experimental conditions used did not favour the cracking of the aromatic ring. Only the C – C bonds were weak enough to be broken, which was followed for some rearrangement. Therefore, both PS and PP increased octane number and calorific value of end products.

For all runs carried out the results showed that the alkenes fraction were mostly hexene in all runs, reaching the highest values when more PP was used. The aromatic fraction was mainly constituted ethylbenzene, being the highest values obtained when a higher percentage of PS was used. As regarding alkanes, a more complex mixture of end products was produced, ranging from C_5 to C_{20}, containing mostly pentane, hexane, heptane and 2-methylpentane (the last one only when higher percentages of PP was used).

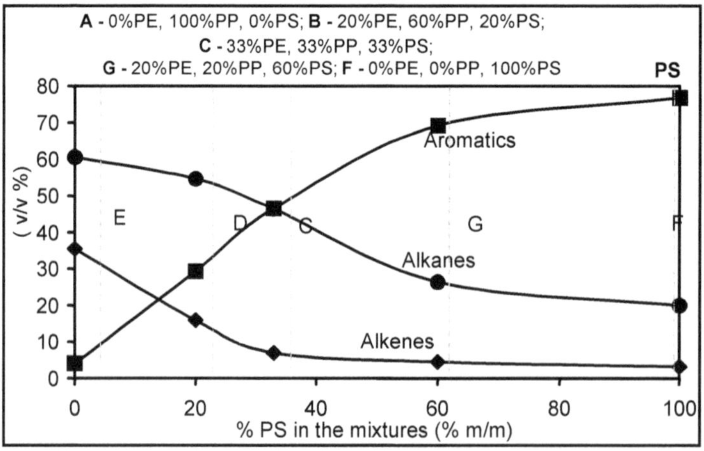

Figure 5: Plastic waste composition effect on liquids hydrocarbons relative distribution for polystyrene

Figure 6 shows that for all temperature runs the alkane fraction is a complex mixture of end products, ranging from C_5 to C_{17}, being mostly pentane and compounds with carbons atoms higher than 12.

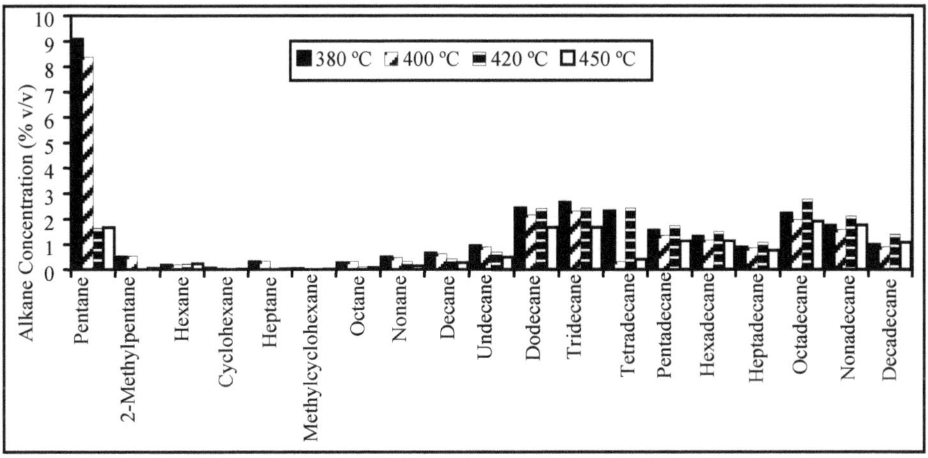

Figure 6: Effect of run temperature on alkane concentration.

Besides the fact that that several alkenes with carbons atoms from C_5 to C_{15} were found, most of the compounds reach values lower than 1% (v/v) except hexene and tridecene who reach values around 3% (v/v) at 380 and 400°C and dodecene and tridecene witch present similar values in all temperature runs as we can see in Figure 7.

The small values obtained in alkenes show that the stabilization of intermediate species by hydrogen seems to be more probable than the formation of double bonds between carbon atoms.

In Figure 8 the two compounds who present higher concentrations are toluene and ethylbenzene reaching values near 20 and 30% (v/v) respectively. The other aromatic compound should be red in the left scale because they present small concentrations. In that group of compounds who present lower concentrations, n-propylbenzene is the highest one reaching values near 0.5% (v/v). The use of temperatures higher than 400°C clearly favored the formation of three aromatic compounds respectively n-propylbenzene, toluene and ethylbenzene.

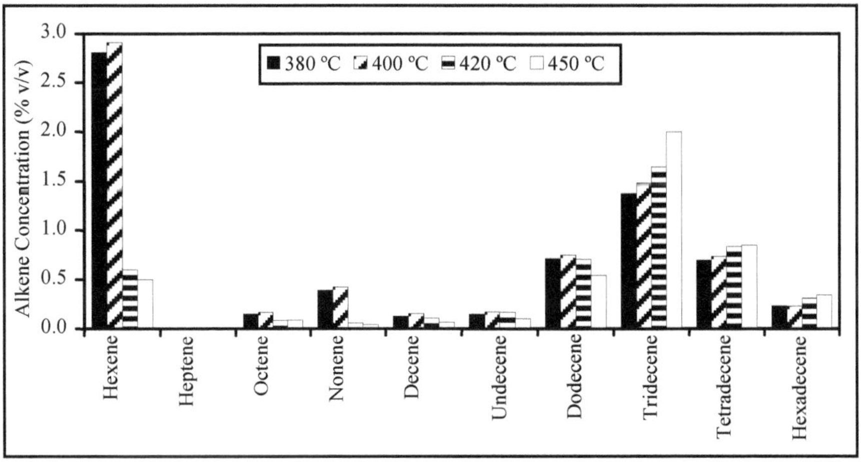

Figure 7: Effect of run temperature on alkenes concentration.

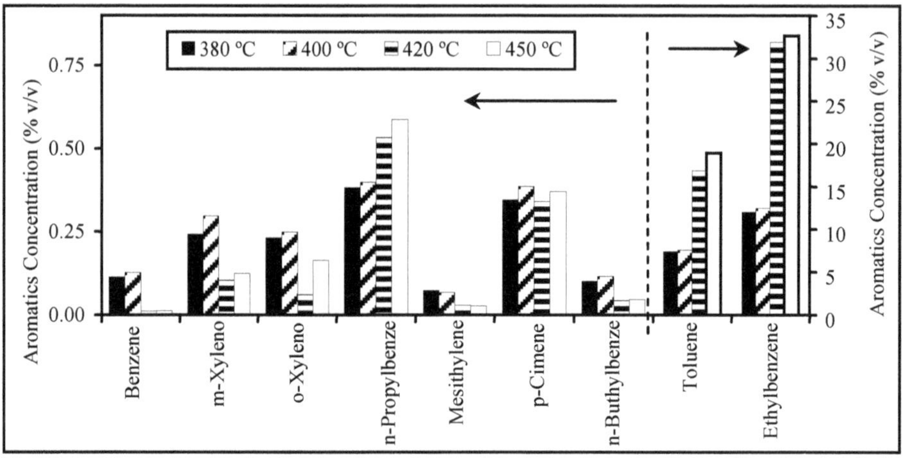

Figure 8: Effect of run temperature on aromatics concentration.

Table 1 presents the solvents and quantities used in the tyre wastes pyrolysis.

Table 1: Study of different solvents type.

Solvents	Quantity [g]	
1,2,3,4-Tetrahydronaphthalene	50.0	100.0
1,2,3,4-Tetrahydroquinoline	50.0	100.0
Mixture of PS, PP and PE	50.0	100.0
PE	50.0	
PP	50.0	
PS	50.0	

Figure 9 shows the effect of solvent type on product yields were the highest liquid yields was obtained in presence of either tetrahydronaphthalene or tetrahydroquinoline. Also with the use of higher solvent content (100g) it was accomplish an increase in liquid yields of around 14%. The presence of mixture PS, PP and PE (in equal amounts) led to gas and liquid yields similar to the ones obtained when each of these plastics were used separately. However, the presence of PP and PS favored the amount of solid residue obtained.

Figure 10 shows the effect of tyre waste composition on liquid fraction distillated. The increase of tyre waste content in pyrolysis (more than 50% v/v) led to a significant decrease in all distillation fractions; for the runs were tyre composition was higher than 70% (v/v), the volume of liquid produced was not enough to distillate. The fraction with a distillation range lower than 150°C led to the highest quantity of product distillate reaching values around 50% (v/v).

Pyrolysis liquids were distilled according to ASTM D 86 [9] and more than 70% (v/v) of pyrolysis liquids had a distillation point lower than 270°C. Distillation curves of the liquid products obtained at different experimental conditions were always found to lie between the curves of standard gasolines and gas oils, which may mean that the composition of the pyrolysis liquids was probably similar to these fuels. Although, the distillation of gasoline and gas oil reach amounts of residue lower then 5% the residue obtained in pyrolysis liquids increase to values near 30% (v/v) depending on experimental parameters tested. For temperatures higher than 270°C, the remaining liquid residue started to burn and to release dark fumes, due to the decomposition of the sample, which prevented its further distillation.

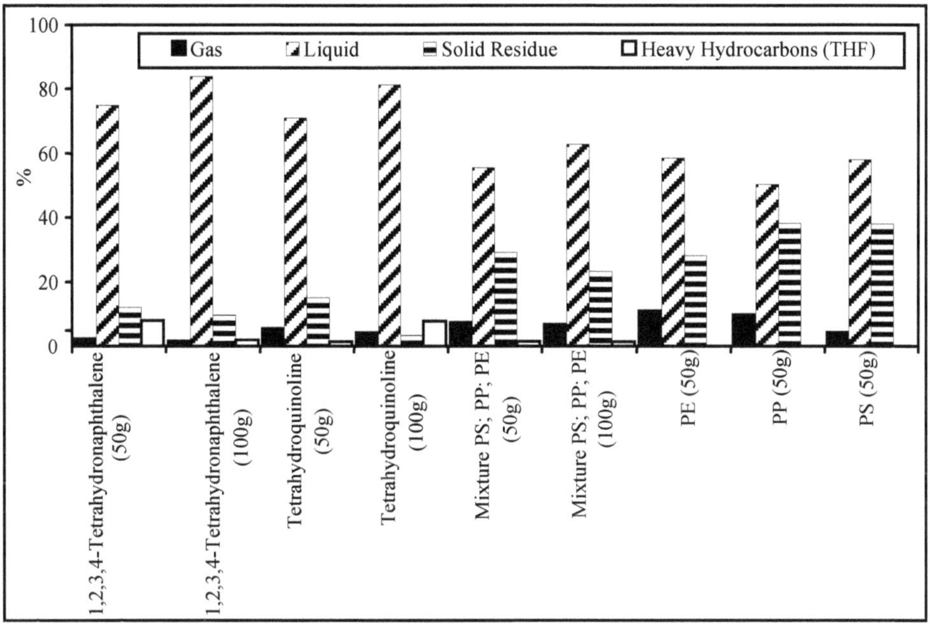

Figure 9: Effect of solvent type on products yields.

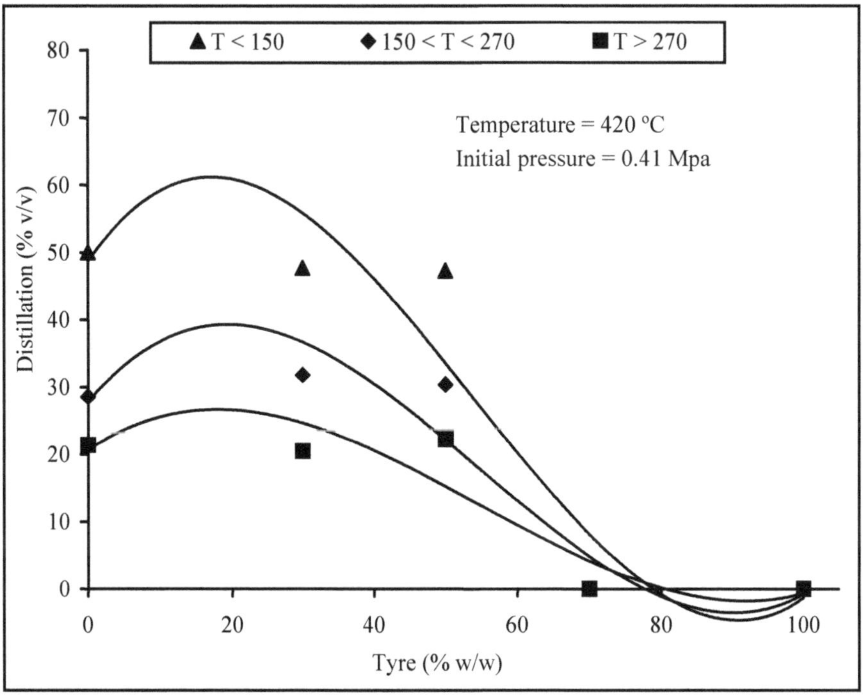

Figure 10: Effect of tyre waste composition on liquid fraction distillation.

CONCLUSIONS

Pyrolysis has proven to be a possible way of converting plastics and tyres wastes, into economical valuable products, which can be used as fuels or as feedstock in petrochemical industry. Total conversion was always around 80%, being mainly due to the liquids either for plastics mixtures or for all the plastics used alone. As the amount of tyre is included in plastic mixture the solid yield increases in 77% (w/w) wile liquid yield decrease in 45% (w/w). The release of gases seam to show a constant behavior varied between 25 and 15% (w/w). The main influence in products yields is the type of waste used.

The lighter liquid fraction, (boiling point lower than 270°C) was approximately 70% (v/v) of the total liquid obtained. The presence of PS increased solids formation and decreases gases. The results obtained showed that total conversion and products yields were not much affected by the range of experimental conditions used although some variations were observed.

The gaseous compounds formed in higher concentrations for all the runs were: methane, ethane and propane. The use of PP increased the gas alkenes content in expense of the alkanes. The amount of alkanes decreased with the increase of carbon number, being methane production always the highest.

Relatively to the liquid fraction, it was observed that the relative distribution of liquid hydrocarbons was highly dependent on input plastic waste composition. The production of alkanes rises with the increase of PE content in the input mixture. Higher values of PP favoured the formation of alkenes. When PS was used, the aromatic content of the end product increased dramatically while the lowest alkanes and alkenes percentage was obtained. Therefore, both PS and PP increased octane number and calorific value of end products. Also, the increase of tyre waste content to values higher than 50% (v/v) led to a significant decrease in all distillation fractions. All the runs were tyre composition was higher than 70% (w/w), the volume of liquid produced was not enough to distillate.

The use of solvents like tetrahydronaphthalene or tetrahydroquinoline allow us to obtained higher liquids yields wile the presence of mixture PS, PP and PE (in equal amounts) led to gas and liquid yields similar to the ones obtained when each of these plastics were used separately. Also, the presence of PP and PS favored the amount of solid residue obtained.

REFERENCES

1. F Pinto, I. Gulyurtlu, P Costa, I. Cabrita, "Journal of Analytical and Applied Pyrolysis", 1999, 51, 39-55.
2. F Pinto, M Miranda, I. Gulyurtlu, I. Cabrita, "Study of Pyrolysis Process to Recycle Tyres and Plastic Wastes", "International Symposium - Recycling and Reuse of Tyres", Dundee, Scotland, p. 27-38, 2001.
3. M Miranda, F Pinto, I. Gulyurtlu, I. Cabrita, "The Economical Valuable Products of Pyrolysis Conversion on Tyres and Plastic Wastes", "R'02 – Integrated Resources Management, 6th World Congress", Palexpo-Geneva, Switzerland, 2002.
4. M Miranda, F Pinto, I. Gulyurtlu, I. Cabrita, "Pyrolysis Process to Convert Plastics and Tyres", "4th International Symposium on High Pressure Process Technology and Chemical Engineering", Venice, Italy, 2002.
5. Annual Book of ASTM Standards, 1994, Volumes 05.01 and 05.02, ASTM D 86-93, ASTM D 240-92, ASTM D 396-92, ASTM D 910-93, D 975-93, ASTM D 1655-93, ASTM D 2069-91, ASTM D 2699-89, ASTM D 2880-92, ASTM D 3699-92, ASTM D 4485-92b.
6. Annual Book of ASTM Standards, 1994, Volumes 05.01 and 05.02.
7. F Pinto, I. Gulyurtlu, P Costa, I. Cabrita, "Journal of Analytical and Applied Pyrolysis", 1999, 51, 57-71.
8. G. Protic-Lovasic, N. Jambrec, D. Deur-Siftar, M. V. Prostenik, Fuel, 69 (1990) 525.
9. Annual Book of ASTM Standards, 1994, Volumes 05.01 and 05.02.

Pyrolysis of Wastes and By-Products in a Multistage Pyrolyser

J-P Mauchien & P-D Oudenne
NESA Product of Umicore Engineering, Rue du Bosquet, 4
1348 Louvain-la-Neuve, Belgium

INTRODUCTION

For many years, NESA has been gaining considerable experience in a range of alternative thermal processes, through which the destruction of raw wastes generates useful by-products.

This paper describes the development work carried out at the pilot stage over the last twenty years as well as some of the industrial realisations, applied across the world, including much experimental and commercial data. Recent projects will be presented such as the recovery of aluminium from Tetrapack residues as well as treatment of "Stabilaat-fluff-RDF".

MAIN CHARACTERISTICS OF THE NESA PYROLYSIS PROCESS

Pyrolysis can be defined as the thermal conversion of solid and liquid material into gaseous compounds and a solid residue that consists of fixed carbon and mineral matter.

Chemical reactions which occur during the pyrolysis step are complex and include thermal decomposition, atomic or molecular rearrangements and polymerisations.

The overall pyrolysis process followed by the combustion of the fixed carbon proceeds via several distinct steps.

The multiple hearth furnace is particularly well suited to pyrolysis where the individual requirements of the stage combustion are:

- pyrolytic gasification of the volatile matter which can be further oxidised in a separate chamber (afterburner) with an adequate excess of air, and
- separate combustion zone for the fixed carbon which requires a higher excess air.

To achieve these requirements, the classical NESA incineration furnace is modified by adding:

- two ducts equipped with slide gates that can selectively extract a volume of flue gas which is oxygen-starved compared with its counterpart in the classical incineration process;
- several inlets for the combustion air which has been preheated in an air/flue gas exchanger located downstream in the gas train.

GENERAL DESCRIPTION OF THE MULTISTAGE PYROLYSIS PROCESS

The combustion process entailing a pyrolysis stage has been developed by NESA to meet the following requirements:

1) To obtain, at the thermal treatment level, the best separation of solids and gas, with a view to obtain waste gas with very low dust content. The benefits are:
    - good operating conditions for the recovery boiler or heat exchanger installed in the waste as circuit
    - in the case of wet dedusting, the washwater, often sent back to the plant head for economic reasons, will not be heavily laden with dust and so no additional load is to be feared for the water purification circuits

- obtaining high dedusting efficiency.

2) To improve the heat balance by the reduction to a minimum value of the excess air required for the thermal treatment.
3) To obtain a stable thermal treatment easy to regulate in spite of the wide variations of the supply characteristics, whether they are variations of flow, calorific value or grading size distribution.
4) To be able to provide the treatment of materials whose ash has a fairly low melting temperature.
5) To control the process temperature of materials with high calorific value.

It has been possible to develop this pyrolysis process thanks to the capabilities of the Multiple Stage Pyrolysis Reactor.

In this reactor the materials to be treated circulate from hearth to hearth, from the top downwards, whilst generally the gas is on reverse flow circulation, that is to say from the bottom upwards.

On the higher hearths, the matter is dried by the action of the gas from the lower hearths. In this zone, temperature and sensible heat of the gas drop, to the benefit of the matter to be treated.

If the organic matter of the materials is partly volatile at the temperature of free water vaporization (which is sometimes the case for sludge of industrial origin), the gas leaving the reactor contains non-burned matter. In this event the use of a post-combustion chamber is required so as to provide for the complete oxidation of the non-burned volatile matter.

Whilst conventional incineration provides for the complete oxidation of organic matter with a high air excess (of the order of 75 %), the pyrolysis process only requires a low global air excess, which is matched to the characteristics of the materials or to the legal standards.

This reduction of excess air and of the resulting gas volume, entails a decrease of the thermal losses by the waste gas sensible heat.

In this thermal process by pyrolysis, the materials are first dried and then heated in an oxygen-poor atmosphere. Under these conditions, the organic matter is partly volatilized to produce a combustible gas. The heat required for the process is produced by the complete or partial combustion of the volatilized matter and of the make-up fuel, if any.

After pyrolysis of the volatile matter, the materials only contain mineral matter and fixed carbon. The combustion of this fixed carbon takes place on the lower hearths of the furnace and requires high excess air.

One part of the gas from the combustion of the fixed carbon rises towards the upper zones of the furnace, whilst the other part is taken off and sent straight to the post-combustion chamber with the gas drawn in from the top of the furnace.

The gas is very finely mixed with the air by its tangential inlet in the post-combustion chamber designed to comply with the legal standards.

In the event of the gas being too poor in combustible matter, a burner is provided at the top of the post-combustion chamber, so as to maintain the temperature set point.

Thanks to the low excess air, high waste gas temperatures can be reached economically. It is then possible to envisage a heat recovery in a boiler or heat exchanger, and taking into consideration the low dust content of the waste gas, both can be of simple construction.

In conclusion, the process gives the best performances from the standpoints of environment, flexibility, ease of operation and energy consumption. It also allows a simple and economic recovery of energy.

Figure 1 presents a cross section of a multistage pyrolysis reactor. Figure 2 presents a flowsheet of the pyrolysis followed by the combustion of the fixed carbon.

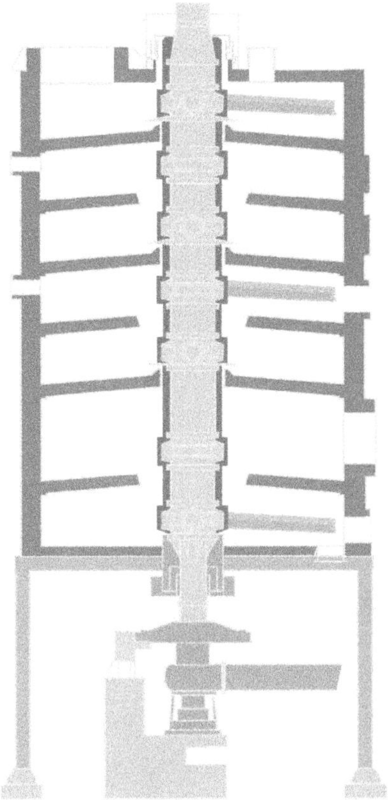

Figure 1: Cross section of a multiple stage pyrolyser.

ADVANTAGES OF THE PYROLYSIS PROCESS

The pyrolysis process presents many advantages over classical incineration, namely:

- *Combustion Performance*: the combustion conditions for the solids and the gases to be closer to its ideal.
- *Low dust carry-over*: the flue gas volume and the dust carry-over levels are significantly reduced.
- *Thermal Efficiency* : this is appreciably higher
- *Furnace control* : more flexible and very stable operation
- *Feed Calorific Value*: feed materials with high calorific contents do not cause problems since the furnace temperatures are readily controllable.
- *Ash Sintering*: ashes with low melting temperature do not agglomerate.
- *Low NOx*: as the kinetics of reaction is slow.
- *Low Operating Cost*: very robust equipment and low power consumption.

PHYSICAL AND CHEMICAL COMPOSITION OF THE PRODUCTS

The determination of the physical and chemical characterisation of the products to be treated is essential before undertaking pyrolysis tests on whatever products. Table 1 indicates the main data to be defined or characterised for pyrolysis tests.

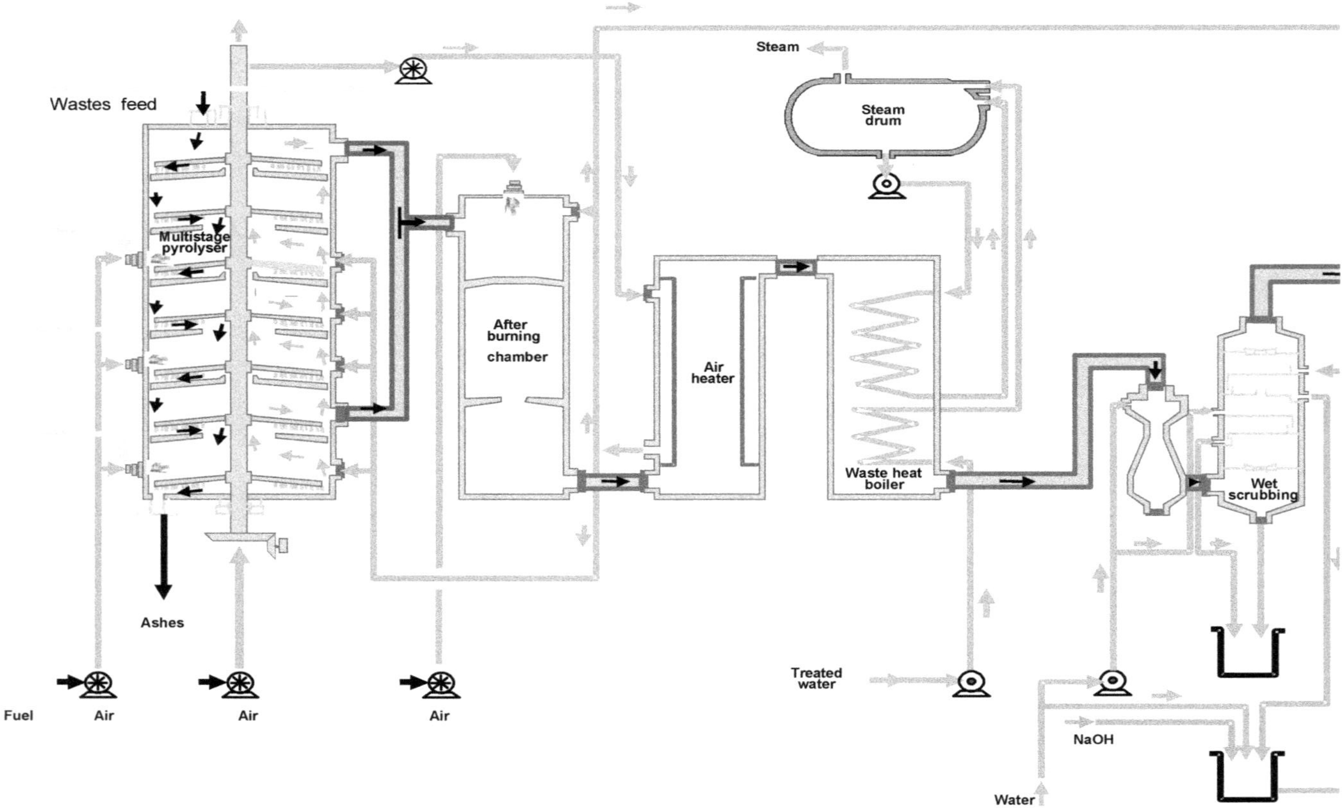

Figure 2: Pyrolysis unit.

Table 1:

- % moisture content
- % dry matters / total matters
- on dry matters
  - % organic matters / dry solid (% OM/DS)
  - % mineral matters / dry solid (% MM/DS)
- composition of mineral matters (M Ca, %K, % Na)
- elemental composition of organic matters
  - % C
  - % H
  - % O
  - % N
  - % S
  - % Cl
- fixed carbon content of organic matters
- LHV (low heating value) of organic matters (kJ/kg OM)
- ash temperature of mineral matters
- visual aspect of products, etc.

EXAMPLES OF INDUSTRIAL APPLICATIONS

ZANDERS PAPER MILL PLANT AT BERGLISH GLADBACH (GERMANY)

Two pyrolysis plants :

- in 1980: 4000 kg/h of paper mill sludge at 34 % DS
- in 1990: 5476 kg/h of paper mill sludge at 45 % DS

Each unit includes :

- a pyrolysis furnace
- an after-burning chamber
- a waste heat boiler (WHB)
- a venturi followed by a scrubber with neutralisation stage
- an I.D. fan
- a stack

The off-gases of each unit comply with the stringent German Bimsch 17 emissions standard. Figure 3 presents a general view of the unit built in 1990.

CIBA GEIGY PLANT AT MONTHEY, SWITZERLAND

The feed is combined municipal and industrial bio-solids sludge. The plant includes:

- a multistage pyrolysis reactor.
- an after-burning chamber
- an air/flue gas heat exchanger
- a waste heat boiler

It has two optional operating modes:

- a direct feeding from a centripress (i.e. without drying);
- a feeding after a dryer

Figure 3: Zanders paper mill plant at Berglish Gladbach (Germany).

The temperature profile for each of the operating cases is given in Fig. 4 below:

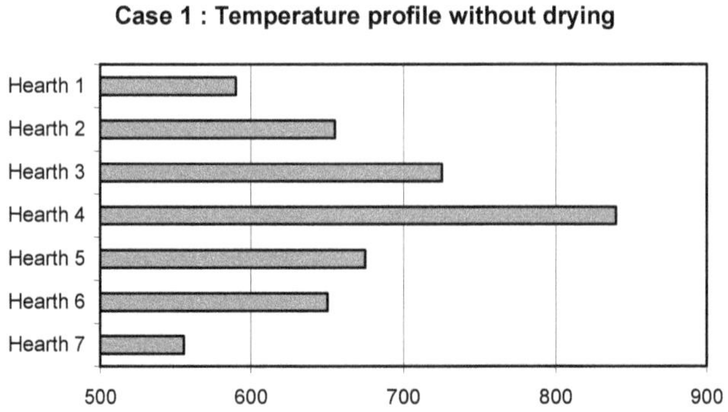

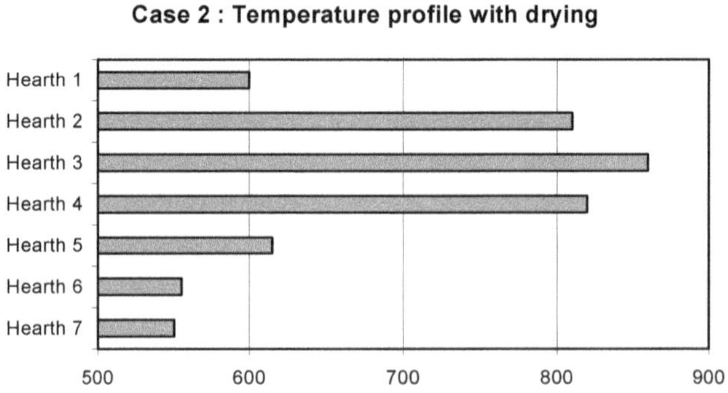

Figure 4: Furnace temperature profile.

ORIGNY AT ROCHEFORT, FRANCE: MACHINING WASTE (PYROLYSIS ONLY)

This is a unique process which uses a pyrolysis system to separate machining waste into both fuel and raw materials. When metal is machined or lathed, the resulting oil soaked swarfs and fines are difficult to dispose of. The machining waste is sticky and prone to spontaneous combustion through the oxidation of metal and ignition of the hydrocarbons. The ash produced by this pyrolysis unit is clean metal oxide and can be recycled to the clinker kiln (Fig. 5).

The pyrolysis furnace associated with a traditionnal cement kiln allows the destruction of the rectification sludge and at the same time provides supply of iron for the fabrication of clinker.

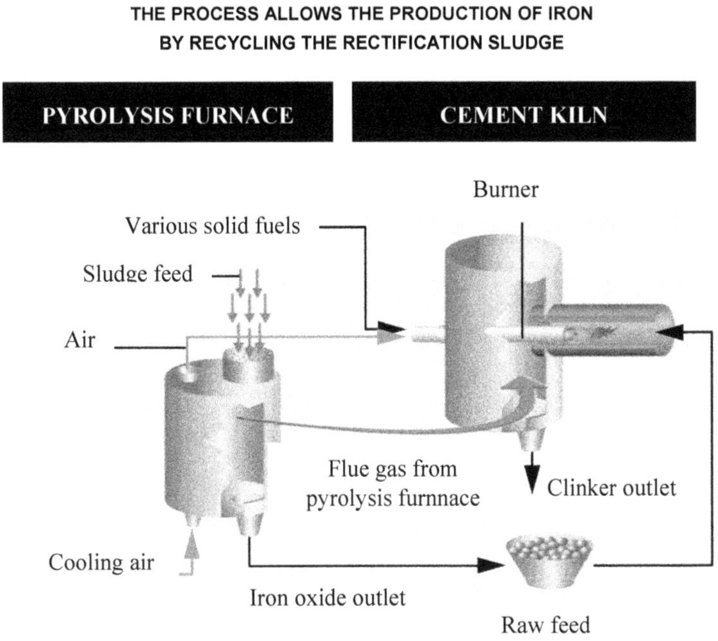

Figure 5: Origny at Rochefort, France.

TREATMENT OF RESIDUES FROM TETRAPAK (PYROLYSIS)

Some paper mills used waste paper and especially liquid board packaging (LBP) as raw materials.

The application of pyrolysis at low temperature to the remaining waste of cellulose fibre recovery process from LBP allows the complete valuation of the residues.

Indeed, the organic fraction of the waste (mostly polyethylene and fibre) is not burned but decomposed into molecules of lower molecular weight, generating a mixture of gases (CH4, C_xH_y, H_2, CO). This gas is burned in a post-combustion chamber where all the heat from the combustion gases is taken advantage in a waste heat recovery boiler. All the non oxidised aluminium and fixed carbon is recovered in the ash at the pyrolyser outlet. A block diagram of the industrial unit is shown in Figure 6.

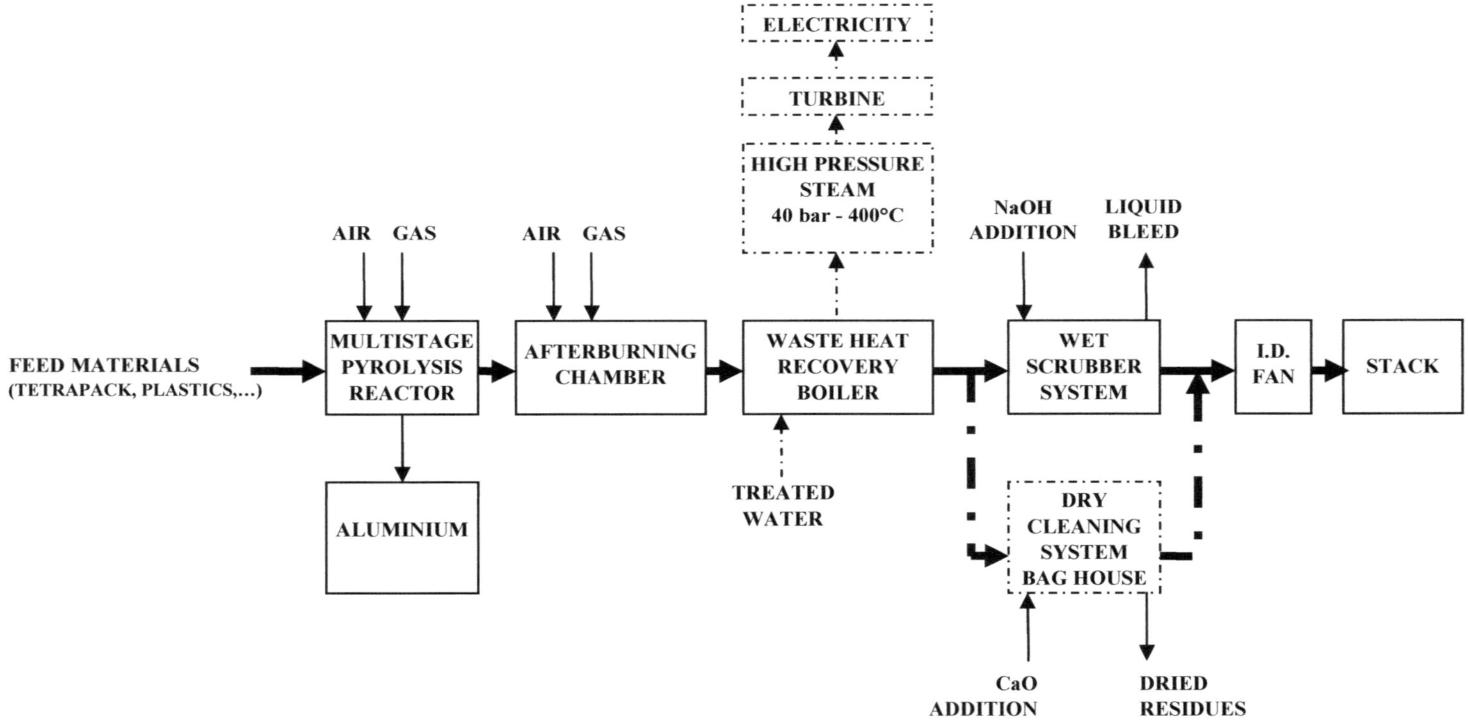

Figure 6: Block diagram of industrial unit.

PILOT TESTS ON RDF FROM MUNICIPAL GARBAGE

Several European companies are treating household wastes in order to produce after aerobic decomposition, densimetric separation and classification an energetically usable product called "stabilaat" or fluff or RDF. Some of the characteristics of this RDF are presented in Table 2.

Table 2: Characteristics of the RDF.

~ bulk density	:	0.20 kg/dm^3
~ moisture content	:	10 – 20 %
~ organic content on dry basis	:	65 %
~ mineral matters on dry basis	:	35 %
~ L.H.V. of dry fluff	:	15 MJ/kg (3 582 kcal/kg)

This RDF has been successfully testes on batch and continuous scale in NESA pilot station in a multistage reactor during several days.

Pictures below present the products before and after treatment.

Picture 1: Product before treatment.

Picture 2: Product after treatment.

CONCLUSIONS

The pyrolysis process, sometimes called the Starved Air Combustion process, developed by NESA and applied commercially for more than twenty years presents several advantages not only from an operational point of view but also in terms of improved energy and environmental impact.

The low level of dust carry-over in the flue gases allows the use of dedusting equipment downstream which is more robust and reliable, and also the maintenance of the air/flue gas heat exchanger is considerably easier. Moreover, the process described above, as well as the associated equipment incorporated within the process, provides for compliance with the most stringent emissions regulations without the need to select expensive and sensitive equipment for the flue gas cleaning.

Changes on the Hydrodynamic Properties of Scrap-Wood Particles during Pyrolysis Processes

J. Reina
Department of Energy and Waste Technologies, HERA Holding
Numáncia 185. 08034 Barcelona (Spain)
E. Velo, L. Puigjaner.
Department of Chemical Engineering, Universidad Politécnica de Cataluña. Diagonal 647. 08028 Barcelona (Spain)

ABSTRACT: The objective of this study is to carry out the analysis of the hydrodynamic properties of scrap-wood particles (from buildings demolition, old furniture and scrap palettes) when submitted to the same thermal decomposition that takes place in a fluidised-bed pyrolysis reactor.

Results show that the nature of particles, as well as the treatment temperature has a fundamental role in the system hydrodynamics. The main conclusions on the effect of the thermal treatment on the particle properties are as follows. Changes observed in the apparent minimum fluidisation velocity, the apparent density, and the terminal velocity of particles with the treatment temperature can bee approximated by a second order equation.

Particles shape factor increases upon increasing the thermal treatment temperature. Due to this change, very irregular particles become slightly irregular particles. Furthermore, it is shown that the temperature range in which a larger change of the particles is observed, coincides with the temperature range where the main thermal degradation of these types of materials takes place (below 500°C).

INTRODUCTION

Gasification or pyrolysis processes constitute an economically attractive route for residue utilization (lignocellulosic materials or urban and industrial wastes), allowing either the synthesis of high-value chemical products or power generation. Among others, scrap wood is one of the wastes generated in a larger quantity and diversity. According to its origin, wood can be classified in wood from conifers or softwood (pine, red fir, and larch), hardwood from leafy trees (oak, beech, ash-tree, etc.), and softwood from leafy trees (birch, poplar, linden, etc.).

The fluidized-bed technology, widely used for the gasification or pyrolysis of wood residue, requires the knowledge of both the process kinetics and particle hydrodynamics for its development and operation. In such processes, wood undergoes a progressive loss of weight, due to its thermal degradation, that modifies the basic hydrodynamic properties of its particles inside the reactor.

The hydrodynamics of a fluidized-bed depends on particle parameters such as size, apparent density, sphericity, minimum fluidization and terminal velocity. The possible formation of networks between particles is also a key factor when studying the hydrodynamics of fibrous materials. The system under study is formed by residual wood particles from different origins. These particles are different in shape, as well as in size; therefore constituting a polydisperse system. For such systems, it is not advisable to use the classic definition for particle intrinsic and hydrodynamic parameters. In this case, it is better to treat them as apparent values. Although various studies on wood fluidization have been published, they do not take into account the change that the matter undergoes with the temperature and, by so much the change in its characteristic hydrodynamics.

In a fluidized bed feedstock, the shape of wood particles depends on the method used for their size reduction. If these particles are submitted to thermal decomposition at high temperature, the loss of weight and carbonization produces a change in shape that not only depends on the temperature of treatment, but also on the original structure of the particle system.

The shape factor is commonly used to quantify the shape of particles. It is defined as the ratio between the surface of a sphere and the actual surface of the particle both having the same volume. This value is comprised between 0 and 1. Using this parameter, Lucas et al.[1] classified solid particles in three different groups. Group I, with a shape factor between 0.8 and 1; particles with a shape factor between 0.5 and 0.8 are in Group II, and Group III is characterized by a shape factor between 0.1 and 0.5. Particles belonging to Group I were identified by the same authors as spherical particles; those in Group II as irregular particles, and those in Group III as difficult-to-fluidize particles.

EXPERIMENTAL

Three different types of scrap wood particles (forest clearing, demolition of buildings, scrap slot machines, old furniture and scrap palettes) were used in this study. Wood was milled and sieved.

The study on properties of the scrap wood particles was carried out submitting the material to thermal conversion at different temperature levels (250, 500, 700, 900 °C). It was thus possible to analyze how the properties of the materials change due to a thermal decomposition as occurs in gasification or pyrolysis processes.

EXPERIMENTAL SET-UP FOR THE CARBONIZATION OF THE SAMPLE WOOD PARTICLES

Thermal treatment of the scrap-wood samples was carried out in a laboratory oven (W.C. Heraeaus Hanau, type KR 170) specially adapted to allow a continuous flow of nitrogen over the sample. A rotameter was used to measure the nitrogen flowrate, which was always set at 100 l/h. Samples were placed in a stainless steel (ANSI 316) cylindrical crucible (70 mm I.D. and 71 mm high). A K-type thermocouple was placed inside the crucible to measure the temperature of the sample.

Three samples of each type of wood were conveniently selected. One of them was kept without thermal treatment. The remaining samples were treated at 500, and 900 °C, respectively. Experiments started by preheating the oven to the set temperature while nitrogen was added. The sample was thereafter introduced and kept inside the oven for 5 minutes. The oven was then turned off and let to cool overnight while passing nitrogen. The thick layer of ash formed on the top of the sample was always removed by injecting air over the crucible after extracting it from the oven at room temperature.

EXPERIMENTAL SET-UP FOR HYDRODYNAMIC EXPERIMENTS

The equipment (see Fig. 1) is similar to that used by Geldart[2]. The main part is a glass column (51 mm I.D. and 1000 mm high) where particles are supported by a metallic mesh with 2304 orifices by cm^2. A Lapple type cyclone is placed at the top end of the column in order to remove fine particles from the air stream. The pressure drop through the fluidized-bed is measured using a water manometer. The air pressure is also measured at the column inlet by a Bourdon gauge. Air is supplied from two calibrated rotameters (0 to 2 m^3/h and 0 to 11 m^3/h).

Experiments were conduced with air flowrates between 0.4 and 6 m^3/h at room temperature and atmospheric pressure. In each experiment, the gas flowrate was increased stepwise from zero to its maximum value (fluidization stage). Then, the gas flowrate was gradually decreased (defluidization stage). At every step, the bed depth and the pressure drop were registered.

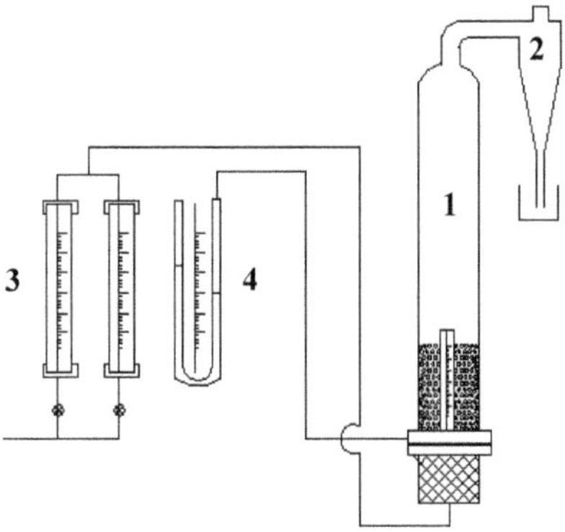

Figure 1: Diagram of the experimental facility for fluidization experiments
(1) Fluidized bed column; (2) cyclone; (3) air flow meter; (4) pressure drop meter

PARTICLE SIZE CALCULATION

Methods to determine the particle size are well described in the literature[3, 4]. Sieving, elutriation, microscopy, sedimentation and permeability can be mentioned among others.

Wood is a fibrous material with low density. Due to particle shape irregularities, sieving alone was discarded as the way to calculate the particle size. Instead, sieving coupled with microscopy was used to obtain more precise measurements.

The wood particles can be considered as small cylinders whose diameter (d) is smaller than their length (l). The ratio d / l depends on the wood source. A graduated lens was used to measure the mean cylinder length for every sieved fraction of particles. In this way, the basic dimensions of the particles could be defined. For each particle fraction a different shape was observed by microscopy. Fine particles tended to be more spherical while big particles were more elongated.

For each sieved fraction, the equivalent diameter was calculated as the diameter of a sphere with the same volume according to equation 1. These results are illustrated in Table 1. The mean diameter for each type of wood (for different treatment temperatures) is then calculated from equation (2), with the results shown in Table 2.

$$de_{eq} = \sqrt[3]{3/2 * d^2_m * l_m} \tag{1}$$

$$d_{ms} = \frac{1}{\sum X_i / d_i} \tag{2}$$

Table 1: Characteristic dimensions of particles for each sieving fraction (all dimensions in mm).

		X_1 (2 to 1.20 mm)			X_2 (1.20 to 0.71)			X_3 (0.71 to 0.315)			X_4 (0.315 to 0.177)		
		d_m	l_m	de_{eq}	d_m	l_m	de_{eq}	d_m	l_m	de_{eq}	d_m	l_m	de_{eq}
Raw particles	Demolition	1.6	5.0	2.67	0.95	3.0	1.60	0.51	2.00	0.92	0.25	1.00	0.45
	Furniture	1.6	5.0	2.67	0.95	3.5	1.68	0.51	2.50	0.99	0.25	1.50	0.51
	Palettes	1.6	5.0	2.67	0.95	3.5	1.68	0.51	2.00	0.92			
		X_1 (1.20 to 0.89)			X_2 (0.89 to 0.514)			X_3 (0.514 to 0.25)			X_4 (0.25 to 0.042)		
Treated at 500°C	Demolition	1.0	3.5	1.74	0.7	2.25	1.18	0.38	1.0	0.69	0.15	0.75	0.32
	Furniture	1.0	3.0	1.65	0.7	2.25	1.18	0.38	2.0	0.76	0.15	0.75	0.28
	Palettes	1.0	3.25	1.69	0.7	2.25	1.20	0.38	1.75	0.73	0.15	0.75	0.28
Treated at 900°C	Demolition	1.0	2.25	1.50	0.7	1.75	1.10	0.38	1.5	0.69	0.15	0.75	0.32
	Furniture	1.0	3.0	1.65	0.7	2.00	1.14	0.38	1.5	0.69	0.15	0.75	0.28
	Palettes	1.0	3.0	1.65	0.7	2.00	1.14	0.38	1.5	0.69	0.15	0.75	0.28

Table 2: Mean diameter of particles (in mm) at different treatment temperature.

Treatment T(°C)	Demolition	Furniture	Palettes
20	1.48	1.57	1.69
500	0.91	0.96	0.94
900	0.84	0.80	0.80

PARTICLE DENSITY

The method proposed by Casal et al.[5] was used for the measurement of the density and the shape factor of particles. This method was selected in view of its simplicity and accuracy and also because it is very easy to reproduce the experiments. According to this method, the density is determined from the plot of $\Delta p/H$ vs. u. Two set of values were found, one corresponding to the bed before its expansion ($\Delta p/H_1$), and the other to the expanded bed ($\Delta p/H_2$) (see Fig. 2).

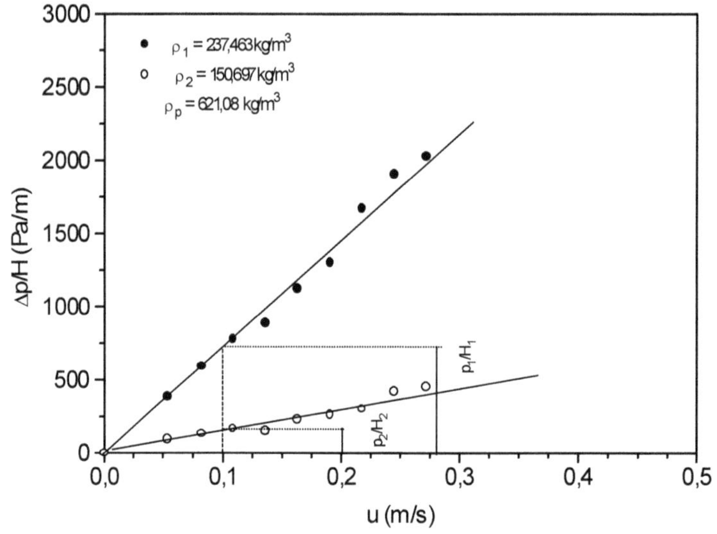

Figure 2: Plot for calculating the apparent density. Furniture wood.

The apparent density of the bed for each case (before and after its expansion) was calculated from equation 3, and the particle density was thereafter calculated from equation 4. Results are given in Table 3.

$$\rho_b = \frac{M}{V} \qquad (3)$$

$$\rho_{app} = \frac{\left[\rho_{b2} - \left\{\frac{\Delta p_1/H_1}{\Delta p_2/H_2}\left(\frac{\rho_{b2}}{\rho_{b1}}\right)^2\right\}^{1/3} \rho_{b1}\right]}{\left[1 - \left\{\frac{\Delta p_1/H_1}{\Delta p_2/H_2}\left(\frac{\rho_{b2}}{\rho_{b1}}\right)^2\right\}^{1/3}\right]} \qquad (4)$$

Table 3: Properties and hydrodynamic parameters of particles.

	ρ_{app} (kg/m^3)	ϕ	S (cm^2/cm^3)	u_f (m/s)	u_t (m/s)
Raw particles					
Demolition	759	0.35	116	0.82	5.26
Furniture	621	0.32	119	0.82	4.33
Palettes	505	0.33	108	plug	3.80
Treated at 500°C					
Demolition	475	0.79	83	0.41	2.20
Furniture	345	0.63	99	0.41	1.67
Palettes	335	0.82	78	plug	1.71
Treated at 900°C					
Demolition	436	0.78	92	0.33	1.35
Furniture	322	0.80	94	0.33	1.47
Palettes	315	0.79	95	0.39	1.22

SHAPE FACTOR

Let us define the parameter A as:

$$A = [\, K*\mu*u*\,(1-\varepsilon)^2 / (\varepsilon 3*d_p^{\ 2}\,)\,] \quad (5)$$

where K = 180 for particles of irregular shape. The bed porosity, ε, takes two different values: ε_1 for the bed before expansion, and ε_2 for the expanded bed. Both can be calculated from the bed density as

$$\varepsilon = 1 - \rho_b / \rho_{app} \quad (6)$$

Using both values of the bed porosity and plotting experimental values of Δp/H vs. A, a single straight line can be found (see Fig. 3) of slope $1/\phi^2$.

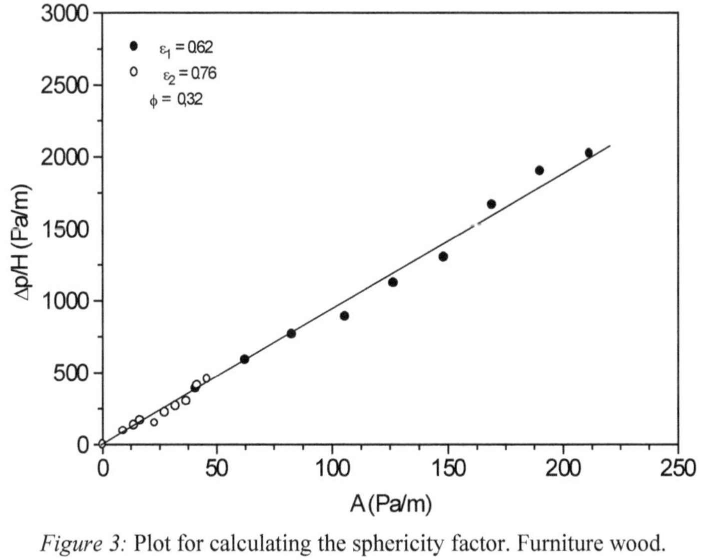

Figure 3: Plot for calculating the sphericity factor. Furniture wood.

SPECIFIC SURFACE

Once the shape factor and the particle diameter are known, the specific surface can be calculated by:

$$S = 6/\phi * dp \tag{7}$$

TERMINAL VELOCITY

The graphic correlation suggested by Heywood was used to measure the terminal velocity of particles. After calculating the Heywood number, defined as N_H = (2/3) Ar, the graphical correlation gives the value of Re_t, from which the terminal velocity of every kind of particle can be calculated. The values found for all these properties are summarized in Table 3 according to the nature of the particle and the treatment temperatures.

RESULTS AND DISCUSSION

MEAN DIAMETER

For wood samples treated at temperatures 500 and 900°C, it has been observed than particles can be described as cylinders with well-defined diameter and length. For raw wood samples, dimensions are not easily defined due to the irregular form of the particles.

Comparing the effect of the thermal decomposition on the particle size, it can be seen (Table 1) that this effect is different depending on the particle original size.

It has been observed that the thermal decomposition has more influence on the particle diameter than on its length. This is reflected by analyzing the dimensions shown in Table 1 or by the visualization of micrographs. In the scrap-wood samples tested, it has been observed (see Table 2) that the larger percentage of reduction in the mean particle diameter occurs for the samples from slot machines and scrap palettes.

PARTICLE DENSITY

Results obtained for the density of the five types of scrap wood suggest that one of them can be classified as softwood (wood from scrap palettes), while the rest (demolition and furniture) can be classified as hardwood. Nevertheless, because of the evolution shown by the furniture wood with temperature, it would also be classified as softwood. This classification also agrees with the reduction of size observed in Table 2, where softwoods (palettes) show a higher decrease in size compared to hardwoods (demolition), being that of furniture-wood in between.

The variation of the apparent density (ρ_{app}) with the treatment temperature can be well correlated, for all the types of residual wood, by a second-degree equation:

$$\rho = B_0 + B_1 T + B_2 T^2 \tag{8}$$

Values for constants B_0, B_1, and B_2 for each type of scrap wood are given in Table 4.

Table 4: Parameters for the variation of the particle density with the treatment temperature (eq. 8).

	B_0	B_1	$B_2 \times 10^4$	r
Demolition	770	- 0.828	5.20	0.99
Furniture	646	- 0.770	2.60	0.98
Palettes	514	- 0.550	3.14	0.99

SHAPE FACTOR

Results obtained in this study for the shape factor show that scrap wood particles are very irregular. Raw particles, without thermal treatment, have shape factor values in the range of non-spherical particles (i.e.: fibers, rashig rings or spirals). It has also been observed that the thermal decomposition has a strong influence on the shape factor. Table 3 shows that the shape factor increases with temperature. Values near 0.8 are obtained for most of the wood types for temperatures higher than 500 °C. These values are, according to the classification suggested by Lucas et al.[1], in the limit between irregular shaped and spherical particles. This was confirmed by visualization through a microscope. It is clear that irregular particles become well-defined cylinders after thermal treatment.

The increase in the shape factor with the treatment temperature produces a change in the fluidization characteristics of wood particles. According to the classification of Lucas et al.[1], particles belonging to group III, can be reclassified as group II particles after thermal decomposition and hence difficult-to-fluidize particles become easy-to-fluidize particles. In effect, scrap palette wood particles fluidize for a treatment temperature of 900 °C. The complete fluidization velocity is shown in Table 3.

VELOCITY OF COMPLETE FLUIDIZATION

This velocity defined as the gas velocity at which the bed is visually well-fluidized, could only be defined for scrap woods classified as leafy tree wood and decreases upon increasing the thermal treatment temperature. Therefore the nature of the particles has also in this case a remarkable influence and stable fluidization for most of the particles tested is doubtful or highly unlikely.

SPECIFIC SURFACE

The smaller diameter particles possess larger specific surface. However, results obtained in this work show that the particles with larger specific surface are those that have a smaller shape factor and a larger diameter, i.e., those that present a higher irregularity (fibers). It has been observed that upon increasing the treatment temperature, the specific surface of the particles decreases, due precisely to the fact that the thermal decomposition increases the shape factor, making the particles less irregular.

TERMINAL VELOCITY

The influence of the thermal decomposition temperature on the terminal velocity can be calculated from the Heywood number when taking measured wood particle properties into account. In doing so, the calculated values can be correlated by a second order equation:

$$u_t = C_0 + C_1\,T + C_2\,T^2 \qquad (9)$$

Values for constants C_0, C_1, and C_2 for each type of scrap wood are given in Table 5.

Table 5: Parameters for the variation of the terminal velocity with the treatment temperature (eq. 9).

	C_0	C_1	$C_2 \times 10^6$	r
Demolition	5.27	- 0.009	5.48	0.98
Furniture	4.41	- 0.008	5.80	0.99
Palettes	3.80	- 0.006	3.77	0.98

CONCLUSIONS

The experimental hydrodynamic study carried out on three different types of scrap-wood particles (from buildings demolition, old furniture and scrap palettes) treated at different temperatures, allows to conclude the following:

(1) The hydrodynamic behavior and properties of the residual wood particles are different according to the origin of the wood and the thermal treatment at which it is submitted.

(2) All types of particles studied can be classified as Geldart C type powders. Nevertheless, some differences in fluidization properties have been observed between scrap wood particles classified as softwood and hardwood.

(3) Changes observed for the apparent density, and the terminal velocity of particles with the temperature of treatment, can be approximated by a second order equation.

(4) The particles shape factor increases upon increasing the thermal treatment temperature. Due to this change, very irregular particles become rounded particles. Some particles that are difficult-to-fluidize in their raw form can be fluidized after carbonization.

(5) It is also shown that the temperature range in which a greater change of the particle properties is observed coincides with the temperature range where the main thermal decomposition of these types of materials is below 500 °C.

LIST OF SYMBOLS

A	constant used to calculate the shape factor, -
Ar	Archimedes number, $d_p^3\, g\, \rho_f (\rho_{app} - \rho_f)/\mu^2$
B_i	i^{th} coefficient in eqn. 8
C_i	i^{th} coefficient in eqn. 9
d_i	size of particles in i^{th} fraction, mm
d_{ms}	mean diameter of a powder containing a mixture of sizes, mm
d_m	mean diameter of particles by sieving, mm
d_p	particle diameter, mm
de_{eq}	equivalent diameter of particles, mm
g	acceleration due to gravity, 9.81 m/s^2
H	bed depth, m
K	Constant, -
l_m	mean length of particles, mm
M	mass of wood in the bed, kg
N_H	number of Heywood, -
Δp	pressure loss across the bed, Pa
Re_t	Reynolds number at terminal velocity, -
S	specific surface, cm^2/ cm^3
u_f	velocity of complete fluidization, m/s
u_t	terminal velocity, m/s
V	volume of the bed, m^3
X_i	fraction held in the ith sieve, -

Greek symbols

ε	voidage in the bed, -
ϕ	shape factor, -
μ	gas viscosity, Pa·s
ρ_{app}	apparent density of particles, kg/m^3
ρ_b	bed density, kg/m^3
ρ_f	gas density of particles, kg/m^3

REFERENCES

1. Lucas, A.; Arnaldos, J.; Casal J.; Puigjaner, L. (1986) Improved Equation for Calculation of Minimum Fluidization Velocity. *Ind. Eng. Process. Des. Dev.*, **25**, 426.
2. Geldart, D. (1990) Estimation of Basic Particle Properties for Use in Fluid-Particle Process Calculations. *Powder Technol.*, **60**, 1.
3. Arnaldos, J.; Coll, T.; Llop M.F.; Perales, J.F. (1990) Caracterización de Partículas Sólidas. *Ing. Química,* July.
4. Kaye B.H.; Trottier, R. (1995) The Many Meassures of Fine Particles. *Chem. Eng.*, April.
5. Casal, J.; Lucas A.; Arnaldos, J. A (1985) New Method for Determination of Shape Factor and Particle Density. *The Chem.Eng. J.*, **30**, 155.

Low Temperature Thermal Decomposition of Straw Black Liquor. Thermogravimetric Study

J.L. Sánchez; R. Bilbao; J. Arauzo
Chemical and Environmental Engineering Department, University of Zaragoza. María de Luna, 3, 50018 Zaragoza, Spain

ABSTRACT: Thermal decomposition at low temperature of a straw Black Liquor has been studied by thermogravimetric techniques. Isothermal and dynamic experiments have been performed in nitrogen and air atmospheres in order to know the effect of oxygen on the thermal decomposition. At the low temperatures used (under 575°C) oxygen slightly increases the final conversion reached. Its main effect is to increase the conversion of some compounds at a lower temperature range. This trend is observed for all the β studied in the dynamic experiments, not being influenced by the value of the heating rate.

INTRODUCTION

Black liquor (BL) is the residue of cooking wood or straw in the production of pulp and paper. Lignin, hemicellulose, other carbohydrates and, in the case of straw, inorganic compounds dissolve and form part of the black liquor. Recovery of the chemicals used, as well as of the energy contained in the organics, is important for the economics of the paper mill.

Nowadays the recovery cycle is based on the performance of the recovery boiler, in which BL is burned producing steam and a smelt. Pulping chemicals are recovered from the smelt in a suitable form to be reused in the pulp making process. Alternative processes which in terms of energy can be more efficient, cheaper, safer and environmentally friendlier than the conventional recovery boiler are being investigated, and gasification is probably the one which is generating more expectation.

Even though there hasn't been any process based on BL pyrolysis proposed to a commercial scale to give a solution for the BL treatment problem, the thermal decomposition is a common step for combustion and gasification processes, and is recognized as an important one, because the properties of the char which will be oxidized depend on the pyrolysis step. Anyhow few papers are found on this subject in the literature [1, 2, 3], and are based on kraft BL, not on straw BL.

The thermal decomposition of lignocellulosic materials, as BL, takes place through a complex series of chemical reactions, which makes a extremely difficult task to establish a clear mechanism for the pyrolysis. Moreover, BL is an heterogeneous material, so during its pyrolysis compounds of different reactivity can decompose at different times, doing the interpretation of experimental results more complex. Because of this, it is considered interesting to know the BL weight loss under different experimental conditions. The experimental procedure used for the study also influences the results obtained [4].

In this work thermogravimetric techniques, which have been widely used to research into the thermal behavior of different organic materials, have been applied to deepen into the knowledge of Black Liquor Solids (BLS) thermal decomposition. Temperatures used in the isothermal experiments range from 200 to 300°C, and in dynamic experiments from ambient temperature to 575°C. The reason to use those low temperatures is to assure that the weight loss only can be assigned to organic material decomposition, and at temperatures above 575°C, carbonate decomposition can happen [5]. As neither in the combustion nor in the gasification processes, the thermal decomposition of BL happens in an inert atmosphere, but usually in an air (oxygen) one, the weight loss of BL during its pyrolysis in nitrogen atmosphere is compared in this study with the weight loss in air atmosphere in the same conditions. The presence of oxygen can cause an oxidative degradation of the material and the subsequent oxidation of the volatile released during this degradation and of the char residue. These reactions add further complexity to the already complex process of thermal decomposition.

EXPERIMENTAL

The black liquor used in this study was supplied by a Spanish paper mill where cereal straw is cooked with soda as the only chemical.

The straw black liquor, as produced, has a solid content ranging between 5 and 10% in weight. Before using it in the thermogravimetric system, it is carefully dried in an oven at 105 °C until completely dry. The black liquor solids (BLS) obtained are smashed and sieved to a particle diameter smaller than 150 μm. The ultimate analysis of the BLS, obtained in a CHNS Carlo Erba elemental analyzer (mod. EA 1108), is: N 1.29%, C 45.85%, H 4.49 % and S 0.55 %.

The ash weight content of the BLS is 22%, with the main components being 33,6% carbonate, 8% chloride, 28% sodium, 7,2% potassium, and 8,8% silica. The chloride comes from the water used for cooking while K, N, SiO_2 and sulfur come from the straw, and were dissolved while cooking.

The BLS organic fraction is composed by hemicellulose (18.8 % in weight of the organic content), aliphatic carboxylic acids (49.5 %) and lignin (31.7 %). Extractives were not detected, probably they were released during the drying step.

The thermogravimetric study was performed in a thermobalance Cahn Instruments TG-131. A gas flow rate of 150 Ncm^3/min was set with a mass flow controller in order to purge the reaction chamber. Two kinds of experiments were performed: isothermal and dynamic. For both around 15 mg of BLS were evenly placed in the sample pan. In the isothermal experiments they were preheated at a constant heating rate of 5 °C/min up to the final temperature, which was kept time enough (48 hours) to assure complete decomposition. Experiments were performed from 200°C to 300°C, at higher temperatures the sample is almost completely decomposed when the final temperature is reached.

In dynamic experiments, BLS was heated at different heating rates, ranging from 1 to 75°C/min, up to a temperature of 575°C, as higher temperatures can lead to a volatilization of part of the inorganic content of the BLS.

RESULTS

ISOTHERMAL EXPERIMENTS

The results obtained are shown here as solid conversion values, X_S, which has been defined as the ratio between the solid weight loss at a given time (ΔW) and the initial organic weight (W_0), equation 1. This definition has been used in order to be able to compare the weight loss obtained in nitrogen atmosphere with the one obtained in air atmosphere (a term of sample weight at a very long time has been avoided which usually is included to take into account the fixed carbon remaining after pyrolysis).

$$X_S = \frac{\Delta W}{W_0} \qquad (1)$$

The X_s values obtained in nitrogen and air experiments are shown in Figures 1 (200, 250 and 275 °C) and 2 (275 and 300 °C).

It can be observed that the different conversion curves show two sections, in the first one X_s increases sharply, which is due to the initial heating and initial minutes of the isothermal period. During the second section, X_s increases smoothly to reach a final value and happens during the rest of the isothermal period.

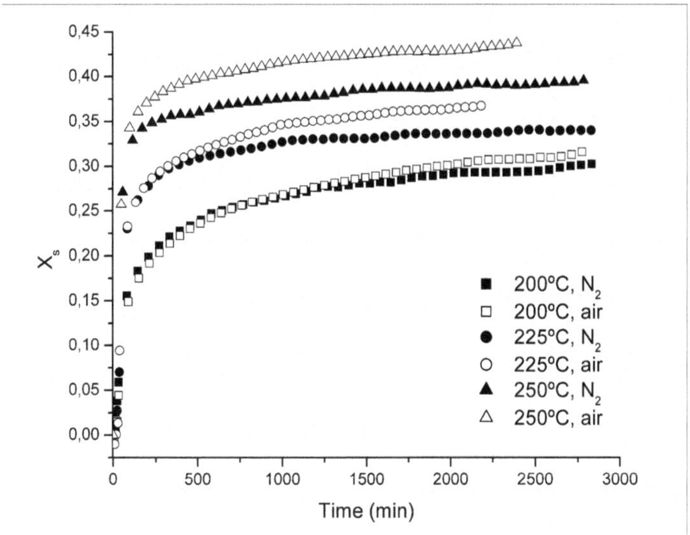

Figure 1: 200, 225 and 250°C isothermal experiments in N_2 and air atmospheres.

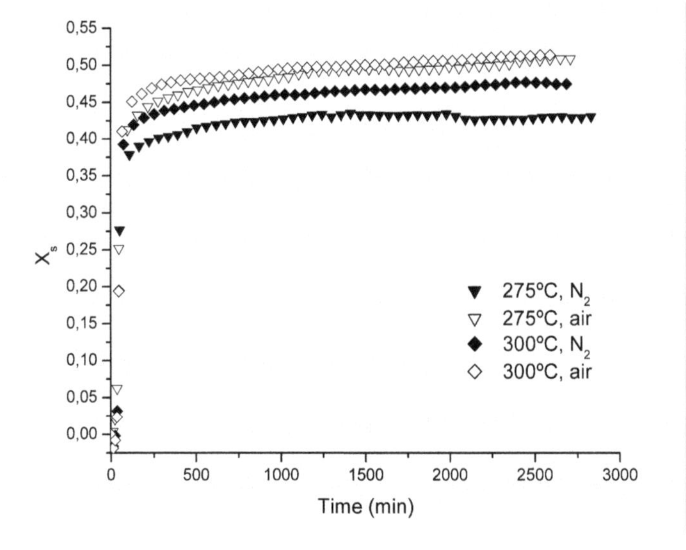

Figure 2: 275 and 300°C isothermal experiments in N_2 and air atmospheres.

In figures 1 and 2, it can be observed that, at the same isothermal temperature, the conversion obtained in air atmosphere is slightly higher than in nitrogen atmosphere. In figure 3 the derivative of conversion respect time, which represents the solid conversion rate, is shown for the isothermal experiments carried out at 200 and 225°C. Isothermal experiments performed at higher temperatures are not represented, as they show the same trend as the 225°C experiment. The dX_s/dt.curves show a maximun, which is reached at the end of the heating period, once the temperature of the isothermal period has been reached, the dX_s/dt smoothly diminishes as the decomposition of the remaining material takes place, reaching a value of 0. It can be observed that at 200°C, the values of dX_s/dt are similar, suggesting that oxygen effect is very small, if any, at this temperature. At 225°C, the values of dX_s/dt obtained in air atmosphere are higher than in nitrogen atmosphere, which could be due to the effect of oxygen, and obviously higher than the values obtained at 200°C.

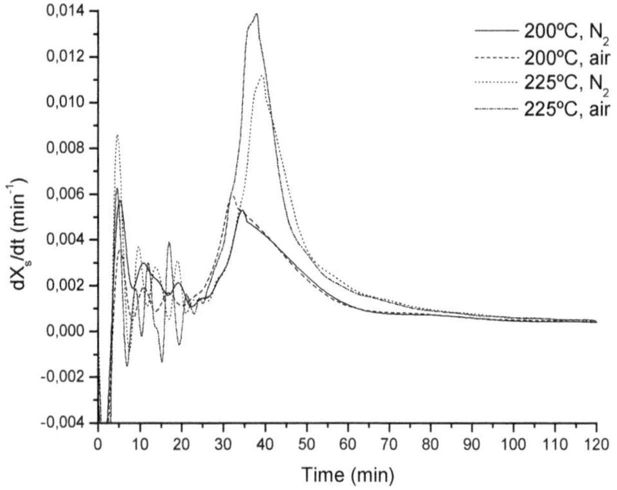

Figure 3: Derivative of X_s respect time for the different isothermal experiments.

DYNAMIC EXPERIMENTS

Isothermal experiments are used to determine weight loss rate at a given temperature. Its main disadvantage is that the BL sample is almost completely decomposed when temperatures higher than 300°C are set for the isothermal experiment. Consequently in order to study BL thermal decomposition over 300°C, dynamic experiments must be carried out. Figure 4 shows solid conversion for a heating rate of 1°C/min in N_2 and air atmospheres,

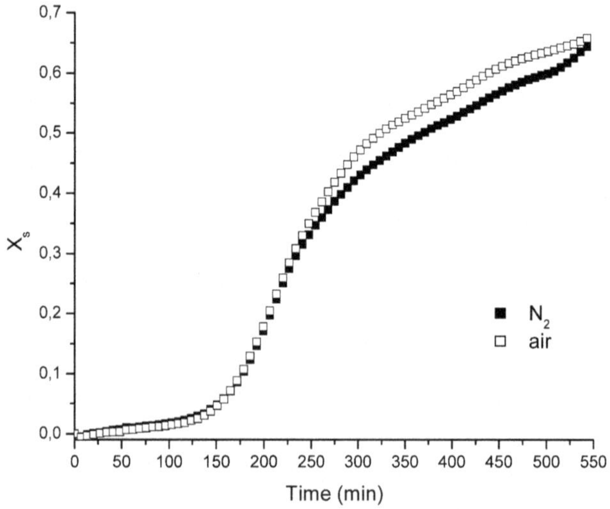

Figure 4: Dynamic experiment at 1°C/min.

In figures 4, 5 and 6 it can be observed that as heating rate increases, conversion curves move to shorter times, and that final conversion values tend to be slightly higher in air atmosphere than in N_2.

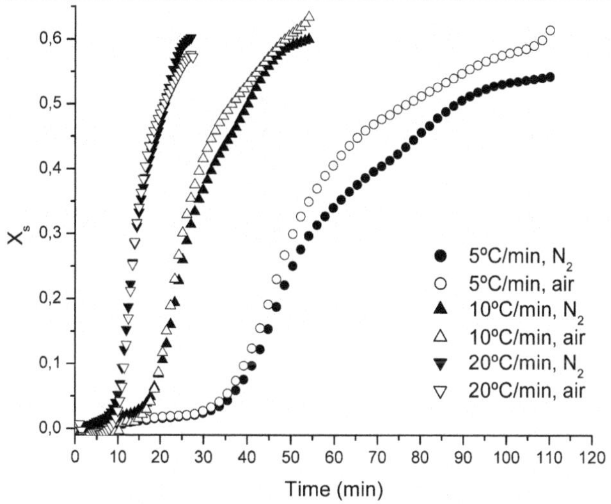

Figure 5: Dynamic experiments at 5, 10 and 20°C/min, X_s vs time.

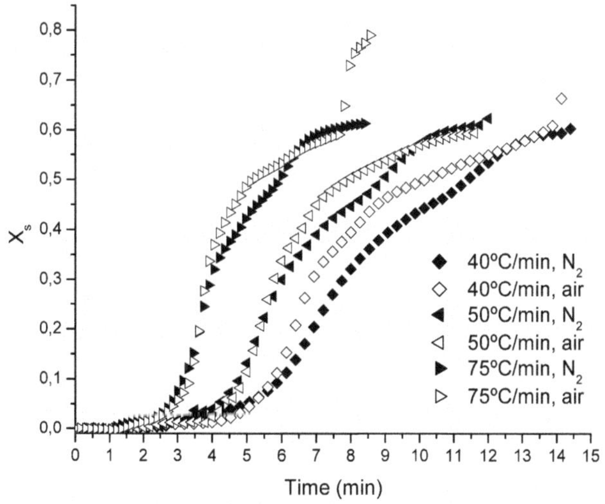

Figure 6: Dynamic experiments at 40, 50 and 75°C/min, X_s vs time.

Table 1 shows the values of X_s and temperature at which conversion in air atmosphere gets higher than in air atmosphere. Although with some dispersion, temperature is between 200 and 225 °C and the conversion around 0.1. It also can be observed that solid conversion shows two different plateaus in all the dynamic experiments performed in nitrogen atmosphere, where as this is not so evident in the air atmosphere ones.

Table 1: Temperature and X_s, separation of conversion in nitrogen and air atmosphere.

β (°C/min)	T (°C)	X_s
1	211	0.10
5	190	0.04
10	220	0.09
20	215	0.10
40	212	0.08
50	220	0.13
75	220	0.18

In figures 7, 8 and 9, dX_s/dt is represented versus time for the experiments of 1, 5 and 50 °C/min, the other heating rates giving similar trends, but displaced in time and with higher values of dXs/dt as heating rate was higher. Except for the experiment at 1°C/min (in which, due to the fact that as heating rate is very low, noise in the weight recording is more important), in all the dynamic experiments performed can be observed that, in nitrogen atmosphere, two maximum values of dXs/dt can be observed, the first one being higher than the second one. This could be due to the fact that BLS is a heterogeneous material and during its thermal decomposition, more reactive compounds decompose earlier, giving the first maximum, and less reactive compounds decompose giving the second maximum.

In air atmosphere only the first maximum appears and it is bigger than in nitrogen atmosphere. This could be due to some compounds, which presented a lower reactivity to decompose in nitrogen atmosphere, behave in the presence of oxygen as the more reactive compounds, decomposing at the same time and temperature.

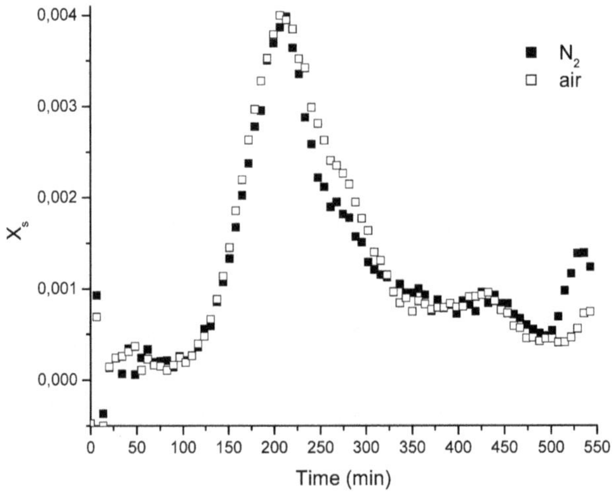

Figure 7: Dynamic experiment at 1°C/min, dX_s/dt vs time.

The effect of the oxygen in the thermal decomposition of BLS in the plateaus observed can be more easily seen in Tables 2 and 3, where the starting temperature, T_1, of the first plateau, the intermediate temperature, T_2, and the final temperature of the second plateau, T_3 are shown with the solid conversion increase of each plateau, ΔX_{s1} and ΔX_{s2}. T_1, T_2 and T_3 were selected graphically as is shown in figure 8.

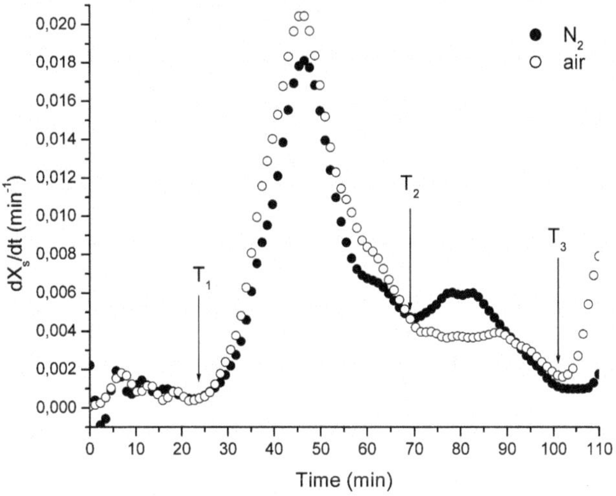

Figure 8: Dynamic experiment at 5°C/min, dX_s/dt vs time.

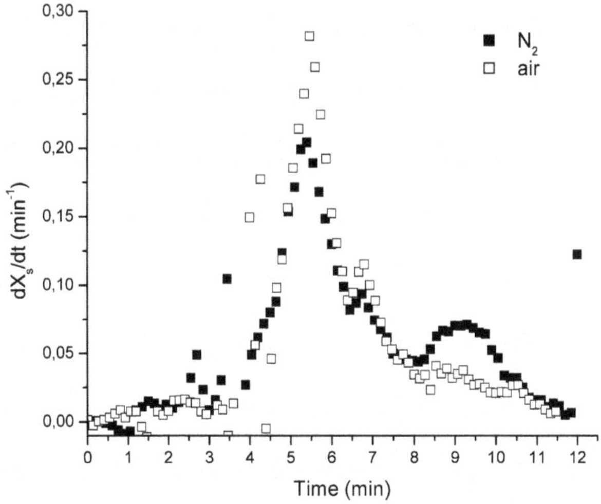

Fig. 9 Dynamic experiment at 50°C/min, dX_s/dt vs time

Table 2: T_1, T_2, T_3 and ΔX_s in N_2 atmosphere.

β (°C/min)	T_1 (°C)	T_2 (°C)	T_3 (°C)	ΔX_{s1} (%)	ΔX_{s2} (%)
1	154	396	540	47.3	10.9
5	155	377	555	37.7	13.3
10	171	387	559	41.9	15.0
20	152	392	566	43.0	15.6
40	154	316	562	41.6	13.8
50	164	386	552	40.9	15.9
75	115	352	561	42.0	16.9

Table 3: T_1, T_2, T_3 and ΔXsi in air atmosphere.

β (°C/min)	T_1 (°C)	T_2 (°C)	T_3 (°C)	ΔX_{s1} (%)	ΔX_{s2} (%)
1	141	377	558	50.5	12.7
5	151	388	555	46.3	10.7
10	149	390	554	48.8	12.3
20	149	402	563	48.0	9.9
40	141	356	548	46.0	13.0
50	175	355	564	45.9	12.3
75	150	355	555	47.0	9.2

From Tables 2 and 3 can be observed that T_1 is approximately 150°C, T_2 370°C and T_3 550 °C. Solid conversion increase for the first plateau is about 40% in N_2 atmosphere, where as in air is represents almost 50%. On the other hand, the second plateau slightly diminishes in air atmosphere (11%) compared to nitrogen atmosphere (14%).

CONCLUSIONS

The thermal decomposition of Black Liquor Solids has been studied using thermogravimetric techniques at low temperature (under 575°C) in nitrogen and air atmospheres. From isothermal experiments it can be concluded that the presence of oxygen does not increase the solid conversion rate at 200°C, but its effect is noticeable already at 225°C.

In dynamic experiments two different plateaus can be observed in nitrogen atmosphere, which can be due to the separetely decomposition of compounds with a higher reactivity (as can be hemicelluloses and carboxylic acids) between 150 and 370°C and compounds with a lower reactivity (as can be lignin) between 370 and 550 °C. The presence of oxygen increases the conversion of the first plateau and diminishes the conversion which happens in the second one.

REFERENCES

1. R. Alén, S. Rytkönen and P. McKeough (1995) Thermogravimetric behavior of black liquors and their organic constituents. *Journal of Analytical and Applied Pyrolysis*, **31**, 1.
2. P.K. Bhattacharya, A.S. Shrinath and D. Kunzru (1985) Pyrolysis of black liquor. *Journal of Chemical Technology and Biotechnology*, **35A**, 223.
3. M.M Nassar (1984) Thermal studies on kraft black liquor. *Journal of Chemical Technology and Biotechnology*, **34A**, 21.
4. W.J. Frederick and M. Hupa (1991) Evidence of sodium fuming during pyrolysis of black liquor. *Tappi J.*, **74(11)**, 192.
5. R. Bilbao, A. Millera., and J. Arauzo (1987) Kinetics of thermal decomposition of cellulose. Part I. Influence of experimental conditions. *Thermochimica Acta*, **120**, 121.

An Alternative to Incineration: the Japanese Experience of New Thermal Processes

E. Sèveno and S. Degrange
Research and Development Division, Gaz de France, 361 Avenue du Président Wilson, B.P.33, 93211 Saint-Denis La Plaine Cedex, France.

ABSTRACT: Incineration is currently one of the most common ways to reduce the amount of household waste. However, for the last decade, interests have been focused on dioxin and furan emissions; this has lead partly to the creation of the NIMBY (Not In My BackYard) syndrome. To cope with the pressure of the public, and in an approach of sustainable development, local authorities now look for alternatives of traditional incineration. Japan, affected by the NIMBY phenomenon and also by the end of landfilling in its territory, has decided to use emerging thermal technologies like pyrolysis, gasification and vitrification. This document deals with the Japanese experience, which appears as an innovative technological showcase. The particular context, the applied technologies and the main actors are outlined. In conclusion, a global analysis with future tendencies completes this paper.

INTRODUCTION

After World War II, in 1945, Japan's priority was given to the economical growth, and at that time environmental laws did not exist. However, natural and industrial disasters contributed to the importance of environmental awareness within the population. The first law on environmental protection was draught in 1963. Today, the creation of new environmental industrial projects and the action of non-governmental environmental organizations, among with some particular engagements of the government, demonstrate the current environmental commitment.

With its huge population (125 million of inhabitants), Japan produces about 51 million tons of municipal solid waste (MSW) and 500 million tons of industrial waste per year. Penalized by its lack of space (370 000 km^2 among which 70% are hilly), Japan has to adapt the existing waste treatment technologies to its own case. Consequently, the market of environmental equipment and services is growing quickly.

JAPANESE CONTEXT

Since 1991, 9 new laws concerning all the environmental domains (waste, air, water…) have been established. The development of clean energies is adopted, and above all, the government has defined clear purposes for waste treatment issues:

- Decrease of 50% of waste volume by 2010;
- Decrease of the incinerated waste amount by 15% for MSW and 22% for industrial waste, in 2010.

According to these decisions, regulations, though the Basic Environment Law, aim to increase the recycling rate by 24% concerning MSW, and by 48% for industrial waste in 2010.

The credo of authorities is summarized by the "3R" rule: Reduce (reducing used resources), Reuse (reusing parts) and Recycle (material recycling).

Many research and development programs have also been created (e.g.: "Waste to Energy" program).

Recently, a ministry of Environment has been created, and the NEDO (New Energy and Industrial Technology Development Organization) was founded to lead common actions between industrial and public partners.

Since 1992, selective waste collecting is compulsory, and the amendment of 2000 extends responsibility to the waste producer, who is in charge of an appropriate disposal for the waste.

Moreover, some specific laws concerning air pollution have been established (CF Table 1, Table 2 and Table 3):

Table 1: Emission limit values of pollutants (Source: Ministry of Environment, 1998).

Component	Emission level
Cadmium	1,0 mg/Nm^3
Chloride	30 mg/Nm^3
Hydrogen chloride	700 mg/Nm^3
Fluorine	1,0 - 20 mg/Nm^3
Hydrogen fluoride	1,0 - 20 mg/Nm^3
Lead and its components	10 - 30 mg/Nm^3

Table 2: Emission limit values of Dioxins (Source: Ministry of Environment, 2001).

Capacity	New plants	Existing plants
> 4 t/d	0,1 ng/m^3	1 ng/m^3
2 to 4 t/d	1 ng/m^3	5 ng/m^3
< 2 t/d	5 ng/m^3	10 ng/m^3

In comparison with European Directive, the emission limit value for Dioxins is the same for Japanese regulations in case of high capacities of waste treatment.

Table 3: Emission limit values of pollutants (Source: Ministry of Environment, 1998).

	Estimated emission		Objective for 2001 (g– TEQ/year)
	Emitted quantity 1997 (g-TEQ/year)	Emitted quantity 1999 (g-TEQ/year)	
Municipal waste incinerators	5 000	1 350	310
Industrial waste incinerators	1 500	690	200
Small capacity incinerators	340 ~ 591	279 ~ 481	66 ~ 112
TOTAL	6 841 ~ 7 092	2 320 ~ 2 522	576 ~ 622

ACTUAL TREATMENT PROCESSES AND TENDENCIES

The 550 Million tons of waste produced are mostly incinerated or recycled, depending on the nature of waste (see Fig. 1).

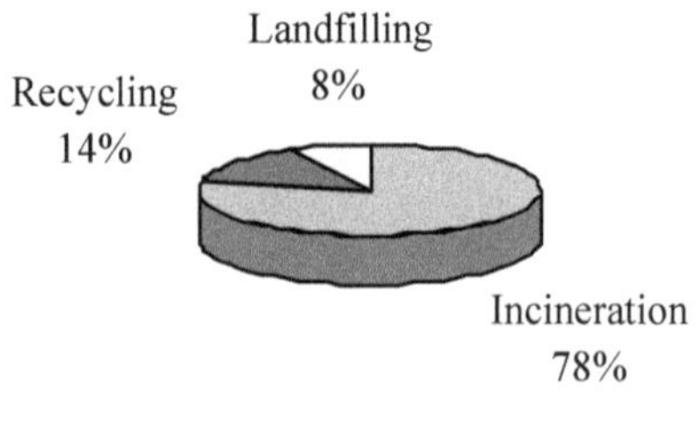

Figure 1: MSW: Municipal Solid Waste.

LANDFILLING AND RECYCLING

Landfilling comes up against a space issue: because of the hilly landscape, at least 67% of the country area is unusable for landfilling. The average cost of landfill treatment is 9150 Yen per ton (84€ per ton).

- Recycling is well developed in Japan: 37% of the waste are currently recycled.
- Waste composting is also commonly used but preferably for agricultural products (e.g.: manure) and wastewater sludge. Cost for composting is about 7.700 Yens per ton (71 € per ton), including compost sells (4.000 to 5.500 Yens per ton - the population is interested in natural fertilizers).

INCINERATION

In 1996, Japan had about 15 000 incinerators, among which 1870 process MSW, each with an average capacity of 37 000 tons per year. According to the government, most of the existing incinerators have to be modernized in order to respect the new environmental policy. The closing of 2050 incinerators has followed the implementation of the policy. This is equivalent to the third of the plants into service.

Today, subsidies are available for the construction of large plants (> 300 tons per day), and the government encourages communities to build melting furnaces instead of incineration ones, in order to reduce the final amount of solid residue. Preference is also given to the systems, which allows the energetic yield to be improved.

Therefore, new opportunities in the waste treatment market have been created. The Japanese companies involved search out new processes, like pyrolysis and gasification. They are very interested by foreign emerging technologies.

JAPANESE WASTE TREATMENT MARKET

GENERAL POINTS

The Japanese market of environmental activities represents around 25 000 billions of Yen (220 billions of Euros), and should reach 40 000 billions of yens by 2010. Some big companies dominate the market of specific equipments for waste treatment (Figure 2).

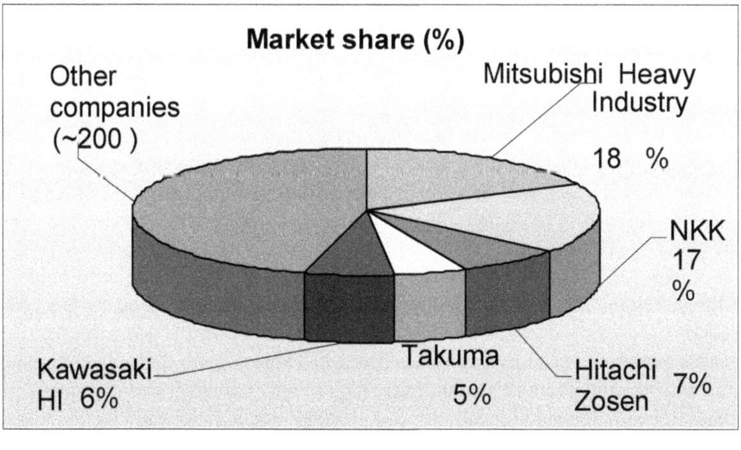

Figure 2: The Japanese waste treatment market.

For the past years, there has been a tendancy for importation of technologies, especially from Germany, Switzerland and USA. The main end is to shorten the duration of Research and development programs in oder to respect the new regulation on time.

MUNICIPAL SOLID WASTE

MSW treatment costs have increased a lot since the last 10 years (e.g. 2 284 million yens in 1996, and 1 846 million yens in 1992). The fact that 75% of waste treatment costs are financed locally, has led to a raise of local taxes. The population has not appreciated this inflation. This factor has to be taken into account for the new waste treatment processes. The cost of municipal waste management averages currently 59000 yens per ton (544 euros per ton).

INDUSTRIAL WASTE

Industries are totally in charge of their waste treatment. Therefore, 70% of them are on the equipped spot; others contract specialized external companies. Treatment processes are diversified, but 29% of plants into service use incineration (especially concerning plastic waste).

STATE OF THE ART OF JAPANESE WASTE TREATMENT

The following review only concerns the alternative systems to incineration. By definition, pyrolysis corresponds to a molecular thermal cracking without oxygen whereas gasification is a combustion in shortage of oxygen.

PYROLYSIS

Pyrolysis is an endothermic reaction, which occurs at an average temperature of 400-600 °C.
Two kinds of pyrolysis can be found:

- Flash pyrolysis : thanks to a short residence time of material, it leads to an oil production that represents 30 to 80% of the feed mass.
- Slow pyrolysis : it is more usual than flash pyrolysis. The residence time averages from 15 to 120 minutes. Rotating or fixed furnaces can be used. The transformation of the feed results more or less in 40% of char, 20% of water vapor, 34% of synthesis gas and 6% of tar.

Metals can directly be recycled because they are not oxidized (lack of oxygen). Moreover, pyrolysis is well adapted to polluted waste because of its improvement in atmospheric emissions, in comparison to traditional combustion. Therefore, its success depends on the effective valorization of by-products (char and syngas).

The char can be used in the cement industry. It can also be burned or gasified to produce heat or synthesis gas, on the plant site : the process is then called "integrated pyrolysis". In Japan, most of the plants use the char as a fuel in a direct combustion or vitrification, in order to provide the necessary heat to furnace. The syngas has to be treated before any valorization (except direct combustion).

At the moment, Japan has few pyrolysis plants into service, but the country holds a favorable context for the development of the technology. Therefore, the number of plants has gown up since the last 2 years.

The existing plants of waste pyrolysis are listed in Table 4, with their characteristics (process manufacturer, information about the technology, type of waste treated, and capacity).

GASIFICATION

The reaction of gasification occurs for a temperature of 800-1000°C; it is slightly exothermic. Downdraft or updraft processes can be used, depending on the unit capacity (and the proportional amount of gas then produced). There are two types of gasification furnaces: the fixed bed and the fluidized bed. The latter needs a restrictive and expensive waste preparation (especially in terms of size). The synthesis gas obtained by the gasification of waste has a low calorific value (2–3 kWh/Nm^3).

The current evolution of treatment capacity (Figure 3) shows the fact that gasification is in a recent stage of development.

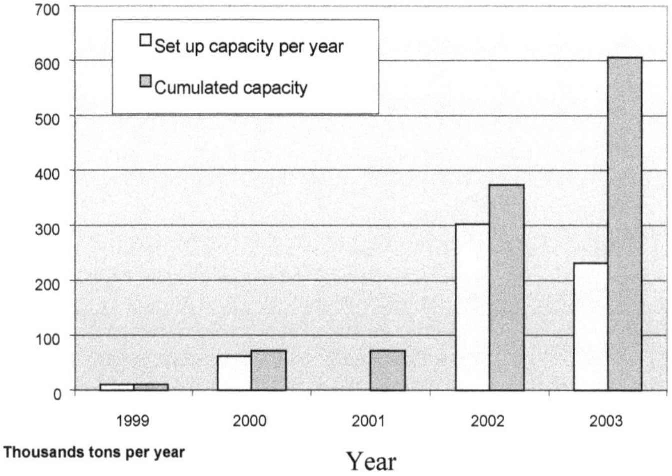

Figure 3: Evolution of capacity for pyrolysis treatment (based on existing and in construction plants listed by Gaz de France; demonstration plants are not included).

Table 4: List of Japanese pyrolysis plants into service (Ref: Gaz de France).

Plant Name *prefecture*	Process manufacturer	Technology	Type of waste	Capacity
Pyrolysis Gas Slagging Type Incineration Power Plant *Kumamoto*	Takuma Co. (Licence Siemens)	Rotating pyrolysis furnace and vitrification of "coke" and ashes	Automobile shredder residue	90 t/day
Toyohashi Shigenka Center *Aichi*	Mitsui Engineering & Shipbuilding Co (Licence Siemens)	Rotating pyrolysis furnace and vitrification of "coke" and ashes	Municipal waste	2 x 200 t/day
Yame Seibu Clean Center *Fukuoka*	Mitsui Engineering & Shipbuilding Co (Licence Siemens)	Rotating pyrolysis furnace and vitrification of "coke" and ashes	Municipal waste	2 x 110 t/day
Itoïgawa	Hitachi (Licence Thide Environnement)	Rotating furnace External "coke" valorization	Municipal waste	50 000 t/year

The increase is thus fast since the last years: in 2003, the capacity of gasified waste will reach 2 million tons. This amount is nevertheless low, since it only represents 1% of the incinerated waste amount.

Gasification processes used in Japan are listed below in Table 5.

Table 5 : List of Japanese gasification plants into service (Ref: Gaz de France)

Plant name *Prefecture*	Process manufacturer	Technology	Type of waste	Capacity
Aomori plant *Aomori*	Ebara Corporation	Gasification in fluidized bed and gas combustion (vitrification of flying ashes)	Automobile shredder residue	2 x 225 t/day
Chubu Kamikita Seiso Center *Aomori*	Kobe Steel	Gasification in fluidized bed and gas combustion (vitrification of flying ashes)	Municipal waste and sludge	2 x 30 t/day
Hokubu Incineration Plant(*) *Chiba*	Tsukishima Kikai Co.	Vapor gasification in fluidized bed and coke + sand combustion	Municipal and industrial waste	450 t/day (Dismantled)
Iizuka Clean Center *Fukuoka*	Nippon Steel	Updraft gasification and vitrification (injection of O_2 enriched air)	Glass, papers, plastics, bulky things, dangerous waste, sludge	2 x 90 t/day
Kamaishi Seiso Jigyosho *Iwate*	Nippon Steel	Updraft gasification and vitrification (injection of O_2 enriched air)	Municipal and hospital waste, ashes, freon	2 x 50 t/day
Kazusa Clean System *Chiba*	Nippon Steel	Updraft gasification and vitrification (injection of O_2 enriched air)	Municipal and industrial waste, sludge	200 t/day
KSC Chiba *Chiba*	Kawasaki Steel Corporation / Thermoselect	Pyrolysis tunnel and gasification/ vitrification furnace	Industrial waste (plastics, sludge, papers, wood, etc.)	2 x 150 t/day
Niihama Works *Ehime*	Sumitomo Heavy Industries (Licence Krupp Uhde)	Gasification in fluidized bed and vitrification	Municipal waste	20 t/day

(*) Built in 1983, the Hokubu incineration plant has been replaced by a traditional incineration unit in 1989. The technical problems were too numerous, especially due to the tar accumulation in the gas washing system and by the valves blocking by metal pieces. This failure had a bad influence on population, which had to pay for the replacement of the unit.

Nippon Steel is the leader in the Japanese waste gasification market. NKK (demonstration unit into service at Tsurimi) and Nippon Steel both use very similar technologies. This is the same for the processes of Ebara and Kobe Steel.

The use of fluidized bed furnace implies an effective pre-treatment (shredding in particular), which is difficult to operate for municipal solid waste.

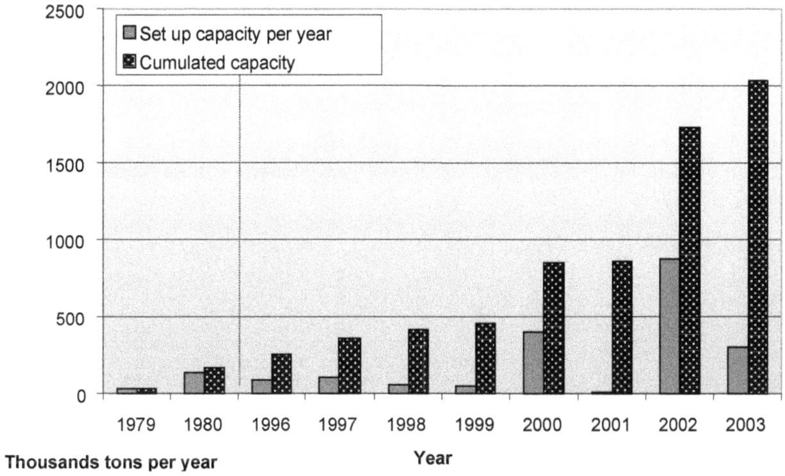

Figure 4: Evolution of capacity for gasification treatment (based on existing and in construction plants listed by Gaz de France; demonstration plants are not included).

VITRIFICATION

Vitrification aims to valorize the by-products as stabilized material that can be used in construction. It is an energy consuming process: 600 to 3000 kWh are spent per ton. In general, vitrification is applied as a post-treatment of other thermal processes, like pyrolysis, gasification, in order to transform carbon residues in vitrified slag. Vitrification is divided in two stages:

- The carbon residue is heat at high temperature (1200 to 1700 °C). It melts and is transformed in a liquid phase.
- The melting mixture obtained is then suddenly cooled down by passing through a cold-water bath. The mixing is thus solidified and becomes the vitrified slag.

Different technologies of melting furnace are available: some of them use electricity as energy sourcing (electric resistor, electric arc, plasma flare or induction) and others use the combustion heat from another surrounding process.

In Japan, vitrification is a systematic post-treatment, which is often realized directly into the gasification/incineration furnace, by injection of additional air or oxygen. The plants of gasification or pyrolysis have been previously listed (Table 4 and 5). Concerning the vitrification of incineration ashes, this does not belong to the study, and thus it is not detailed.

CONCLUSION

Japan benefits from a particular and favorable context for the emerging technologies in the waste treatment field. In spite of a number of existing plants of pyrolysis and gasification that is still low, there is a great potential for alternative thermal processes to incineration.

These thermal treatments are systematically followed by a vitrification post-treatment, which is applied to the carbon residue or ashes. They also have to be coordinated with a recycling disposal.

It is interesting to note that European countries tend to have a similar waste management issue.

The Japanese pattern could be then partly reproduced in other countries if:

- The main actors, like government and specialized companies, support the action;
- Regulations do not impede the alternative thermal treatments, especially concerning the status of the by-products.

In the end, due to some differences relating to the organization of the waste management market and to the economical factor, we can conclude that only tendencies can be reproduced from the Japanese model.

Fast Pyrolysis of Forestry Residues

Y.Solantausta, A. Oasmaa, K. Sipilä
VTT Processes, P.O.Box 1601, 02044 VTT, Finland

ABSTRACT: Liquid biofuels have several advantages over solid or gaseous fuels. Pyrolysis liquids have been estimated as low cost, but their fuel quality needs improvement. In Finland the aim is to use pyrolysis liquid to replace light fuel oil in boilers in heating of institutional buildings. Forestry residue, an industrial biofuel, is estimated as a feasible biomass feedstocks for liquid fuel production. Production of pyrolysis liquid from this biofuel is described, and liquid fuel properties are discussed.

RATIONALE FOR THE USE OF LIQUID BIOFUELS

Biofuels are typically solid, in-homogeneous materials. Although they are rather CO_2 neutral, use of solid fuels is expensive compared to fluid fuels. The reason for this is the high investment costs of solid handling equipment. Utilisation of solid fuels results often also in lower efficiency than use of fluids.

Biofuel properties may be improved, for example by pelletization, but converting them either to gases or liquids may further expand their utilisation. Liquid biofuels have following advantages:
(1) Possibility to decouple solid handling and utilisation, for example in heat and power production;
(2) Liquid may be stored;
(3) Transportation of liquids more practical than solids;
(4) Use of liquids is less expensive than solids in existing boilers;
(5) Emissions in boilers are less than when firing with solid fuels;
(6) Potential for intermittent operation.

Fast pyrolysis liquids have been estimated less expensive than other liquid biofuels, when biofuel production costs are compared to the corresponding mineral oil product value. However, it is well known that pyrolysis liquids are not suitable as transportation fuels. It has also been reported (1) that it is not feasible or viable to upgrade pyrolysis liquids to be used in transportation. They may be used as boiler fuels or diesel engine fuels in power production. The first alternative is technically more developed. The alternative is also economically more attractive in Scandinavia (2). The aim is to use pyrolysis liquid to replace light fuel oil in boilers in the current development effort by Fortum and Vapo (2). An additional advantage in this approach is that light fuel oil is available to be used in production of diesel fuel, making more middle distillates available for the transportation sector.

FEEDSTOCK FOR PYROLYSIS

Forestry residue, besides pine sawdust, is one of the most feasible biomass feedstocks for liquid fuel production in Finland. The majority of forestry residue in the Finnish market comprises spruce and pine from end-cuttings. The composition of forestry residue is affected by wood species, quantity of trees, sturdines of growing timber, knottiness and extent of decay. Forestry residue can be logged either as "green" (moisture about 55%) with needles or as "brown" after seasoning (moisture about 35%) when around half of the needles and a small amount of thin branches remain at logging area. Green residue of spruce contains 40% of wood, 23% of bark and 37% of needles on the average. The corresponding figures for brown residue are: over 60% of wood, below 30% of bark and below 10% of needles.

Compared to bark-free pine sawdust, forestry residue is high in extractives, ash, alkali metals, and nitrogen due to presence of needles and bark (Table 1). The amount and composition of extractives depends on wood species, wood fraction and storage conditions.

Forestry residues are used at present industrially as power plant fuel. Total amount of wood fuel used in Finland in 2001 was 23 TWh/a, which accounts for about 6% of the total energy consumption. Out of the 23 TWh about 6% was forestry residues (Figure 1). Costs and amounts for wood fuels available in Finland are shown in Figure 2. It can be seen that forestry residues are typically more expensive than industrial by-products (sawmill residues etc). However, they are less expensive than wood from forest thinnings.

Table 1: Feedstock analysis.

		Pinus sylvestris	Forestry residue brown[a]	Forestry residue green[b]
Moisture as received	wt-%	3.3	4.9	8.1
Ash, d.m.[c]	wt-%	0.2	3.8	2.1
Volatiles, d.m.	wt-%	82.5	73.2	76.7
Carbon, d.m.	wt-%	50.3	51.1	51.4
Hydrogen, d.m.	wt-%	6.0	5.9	6.0
Nitrogen, d.m.	wt-%	<0.1	0.5	0.5
Oxygen,d.m.	wt-%	43.5	38.7	40
Sodium	mg/kg	-	80	250
Potassium	mg/kg	600	1800	1523
Calcium	mg/kg	1200	4800	5332
Magnesium	mg/kg	200	660	642
Chlorine, d.m.	mg/kg	<50	-	220
Sulphur, d.m.	wt-%	-	-	0.04
HHV, d.m.	MJ/kg	20.6	20.5	20.8
LHV, d.m.	MJ/kg	19.2	19.3	19.5
Extractives, d.m.[d]	wt-%	5.9	5.4	10.5

[a] 80 % spruce, 10 % pine, 10 % birch
[b] 86 % spruce, 9 % pine, 5 % birch
[c] d.m. = dry matter
[d] Solubles in acetone and dichloromethane

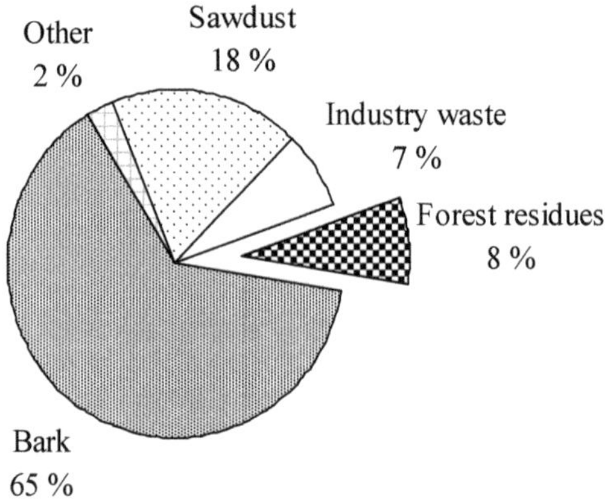

Figure 1: Wood fuels used in Finland 2001.

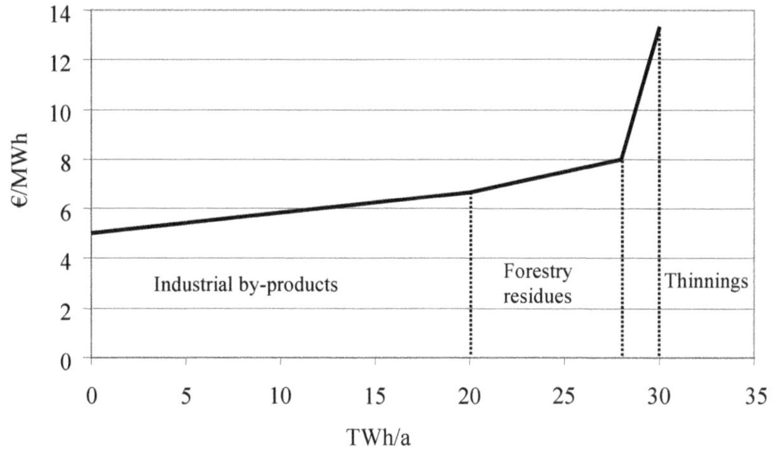

Figure 2: Cost of wood fuels as a function of amount available, Finland 2001.

PYROLYSIS OF FORESTRY RESIDUES

Pyrolysis liquid of forestry residues were produced with a 20 kg/h capacity Process Development Unit (PDU) at VTT (Figure 3). The fast pyrolysis unit is described in more detail in ref. (3). The ground, sieved and dried (moisture 3-10 wt-%) feedstock was fed to the reactor. The residence time for pyrolysis vapours at 520 °C was 1 – 2 s. The product vapours were condensed using pyrolysis liquid condensers. Condensed liquid product was transferred to phase separation towers equipped with warm water-circulation, by which the temperature was adjusted. After a certain storage time, typically 24 hours at 35 °C, the liquid was recovered from bottom of the tower as ”bottom” and ”top” phases.

Figure 3: Pyrolysis PDU at VTT.

Pyrolysis liquid yields from three wood fuels are compared in Figure 4. It should be noted that the value presented for each feed correspond to an average of three to four days of continuous operation. During this time 520 - 1,180 kg of pyrolysis liquid was produced.

Fast pyrolysis of forestry residue yields less liquid product than bark-free wood. While organic liquid yield from dry pine feed is around 67 wt%, organic yields from green and brown forestry residues are 53 and 48 wt%, respectively. Pyrolysis water yields are between 10 and 11 wt% based on dry wood. Using 8 wt% moisture feed, the overall final moisture of pyrolysis liquids is around 25 and 29 wt% for pine and forestry residues, respectively.

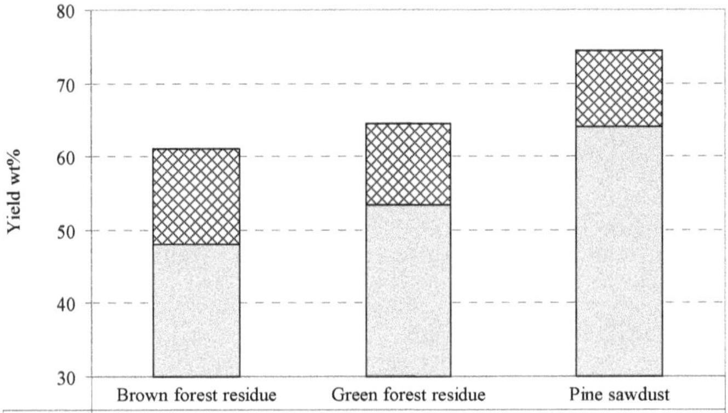

	Brown forest residue	Green forest residue	Pine sawdust
⊠ Water	13.1	11.2	10.3
□ Organic	48.0	53.3	64.1

Figure 4: Organic liquid and pyrolysis water yields from wood fuels based on dry feed.

PYROLYSIS LIQUID QUALITY

Production of even quality pyrolysis liquid is a necessary requirement for industrial operation. Primary objectives related to the operation of the PDU at VTT have been to:

(1) To define pyrolysis liquid fuel quality;
(2) To use (and develop new where methods unavailable) appropriate methods for quality determination;
(3) To verify quality variation, and
(4) To demonstrate the production of even quality liquid in the PDU.

Among others, water and solid content of pyrolysis liquid have earlier been determined as critical properties regarding homogeneity of the product. The criteria developed were used, for example in an EU project, when large quantities of liquids were handled and tested (4). Other properties analysed and methods tested have been presented in references (5) and (6).

An example of product liquid water content during a three-day PDU-operation is illustrated in Figure 5. During initial operation liquid quality changes because of the nature of the process (seed oil will be replaced during the first day). After the initial stage, a relatively constant water content is achieved. Forestry residue pyrolysis liquid separates immediately after condensation to polar, liquid bottom phase (80-90 wt-%) and to hydrophobic viscous top phase (10-20 wt-%). The amount and composition of top phase are affected by the feedstock (wood species, amount of needles, feedstock storage time), pyrolysis and recovery of product.

The reason for this type of phase separation is not connected to poor quality of the liquid (or better known phase separation caused by water). Pyrolysis liquid is a solution of organic compounds, in which water is dissolved up to approximately 30%. Large, high molecular weight lignin compounds are thought to exist as micro-emulsions. If the water content exceeds this level, the acids and other solvents required to keep the lignin in solution are thought to be removed from around the lignin, the solubility balance of the pyrolysis liquid matrix changes and the least hydrophilic lignin-derived material sediments out of the liquid. This type of liquid is of poor quality and cannot be used as fuel. Extractives are various neutral compounds of low oxygen content and hence have a tendency of separating out of the polar matrix.

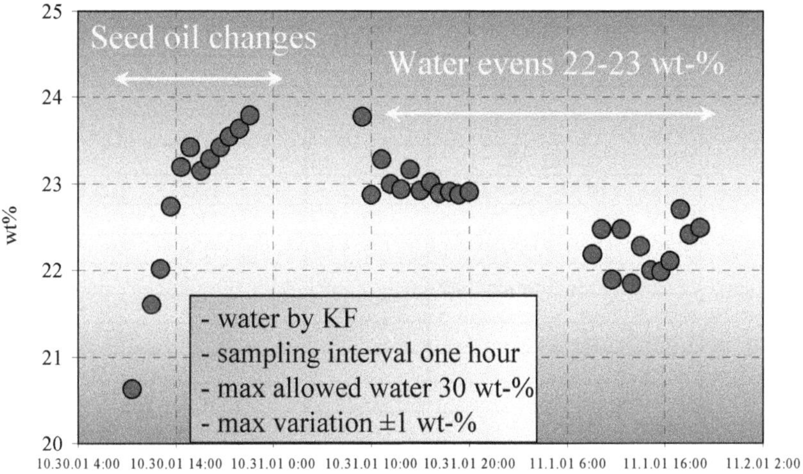

Figure 5: Water content of pyrolysis liquid during three days time of operation.

The composition of bottom phase (Figure 5) is similar to bark free wood pyrolysis liquids. Bottom phase contains 75-80 wt-% water-soluble material: water (25-30 wt-%), carboxylic acids (5-12 wt-%) and alcohols (1 wt%), aldehydes and ketones (7-15 wt-%), carbohydrates "sugars" (30-39 wt-%), and 20-25 wt-% water-insoluble material: "HMW (High Molecular Weight) Lignin fraction" (3-15 wt-%), "LMW (Low Molecular Weight) Lignin fraction" (6-12 wt-%), and extractives (3-6 wt-%). The top phase (Figure 6) differs significantly from bottom phase containing high amount of water-insoluble material, especially extractives (35-50 wt-%). Extractives in forestry residue liquids consist mainly of fatty acids, resin acids, triglycerides, and hydroxy acids, the proportion of which depends on the feedstock. Green forestry residue produces liquid high in resin and fatty acids and triglycerides. Brown forestry residue liquid is high in resin and fatty acids but low in triglycerides.

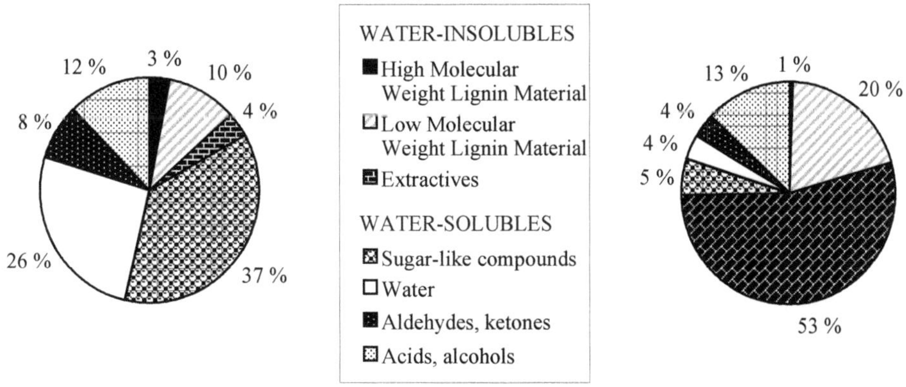

Figure 6: Chemical composition of bottom and top phase of a green forestry residue liquid.

Forestry residue liquid from brown feedstock contains less extractives than that from green feedstock. Compared to bottom phase top phase contains relatively more fatty acids, and less hydroxy fatty acids, resin acids, and dicarboxylic acids than the top phase.

Physical properties of forestry residue and pine liquids are shown in Table 2.

Table 2: Physical properties of pine and forestry residue liquids.

Feedstock	Green FR	Brown FR	Pine
Solids. wt-%	0.03*	0.02*	0.20**
PH	2.97	2.9	2.6
Water,wt-%	28.9	24.1	21.4
Viscosity @40C, cSt	13	29	35
Density (15°C), kg/dm3	1.21	1.22	1.22
HHV, MJ/kg	15.7	17.4	17.8
LHV, MJ/kg	14.1	15.8	16.3
LHV, MJ/kg, dry	20	21	21
Ash, wt-%	0.16	0.08	0.02
C, wt-%	38.7	43	43.5
H, wt-%	7.4	7.4	7.1
N, wt-%	0.3	0.1	< 0.1
O (diff.)	53	49	49

* Determined as methanol-dichloromethane insolubles

** Determined as ethanol insolubles

ACKNOWLEDGEMENTS

The work was funded by Tekes under agreement 40921/99, Fortum Oyj, Vapo Oyj, Oilon Oy, EU Contract ERK5-CT1999-00011, and by VTT. At VTT Eeva Kuoppala, Sirkka-Liisa Huru, and Sirpa Lehtinen are acknowledged for the analytical work. Steven Gust, Jukka-Pekka Nieminen and Johan Selin from Fortum are acknowledged for valuable discussions.

REFERENCES

1. Beckman, D., Elliott D.C., Gevert, B., Hörnell, K., Kjellström, B., Östman, A., Solantausta Y., Tulenheimo, V., (1990). Techno-ecomonic Assessment of Selected Biomass Liquefaction Processes. Espoo: Technical Research Centre of Finland. 161 p. + app. 95 p. (VTT Research Reports 697).
2. Gust, S., Nieminen, J., Nyrönen, T., Forestera - Liquefied wood fuel pilot plant. In these proceedings.
3. Oasmaa, A., Kuoppala, E., Gust, S., Solantausta, Y. Fast pyrolysis of Forestry Residue. I. Effect of Extractives on Phase Separation of Pyrolysis Liquids. Accepted for Energy & Fuels 2002.
4. Oasmaa, A.; Kytö, M.; Sipilä, K., Pyrolysis oil combustion tests in an industrial boiler. Progress in Thermochemical Biomass Conversion. Bridgwater, A. V. (ed.). Vol. 2. Blackwell Science. Oxford (2001), 1468 - 1481.
5. Oasmaa, A., et al., Physical characterisation of biomass-based pyrolysis liquids. Application of standard fuel oil analyses. Espoo 1997. VTT Research Notes 306.
6. Oasmaa, A., Peacocke, C. A guide to physical property characterisation of biomass-derived fast pyrolysis liquids. Espoo: VTT Energy, 2001. 65 p. + app. 34 p. (VTT Publications 450)

Behaviour of Corn Stalk in an Indirectly Heated Pyrolysis Reactor

L. Sun, M. Xu and R. F. Sun
Energy Research Institute of Shandong Academy of Sciences, Jinan, China

ABSTRACT: This paper presents the research on biomass pyrolysis process. The research explored the behavior of biomass in an indirectly heated pyrolysis reactor, which uses agricultural residues such as corn stalk as feedstock. The influences of temperature and reaction time on parameters such as gas composition, heat value of gas, and characteristics of residual charcoal were studied. The paper described the experimental apparatus, experimental data, and analysis of main influence factors of the process. This research paves the way to two directions. One combines with a gasifier to produce gas with very low tar content, which can be used to generate electricity. The other one, in association with secondary decomposition process, generates hydrogen-rich gas.

INTRODUCTION

In Chinese rural area, corn stalk is the main biomass resource available as energy. Since 1990, the authors have developed a kind of down draft gasifier using corn stalk as feedstock and supplying cooking gas for farmers. Hundreds of systems have been built up so far. While the gasifier produced the gas with considerable tar content when use agricultural residues like corn stalk, wheat straw etc. For solving the problem, a so-called "two-stage gasifier" has been studied and developed under the project supported by MOST (Ministry of Science and Technology) since 1995. The gasifier comprises an indirectly heated pyrolysis reactor and a combustion/gasification reactor. By thermal cracking of pyrolysis gas in second stage, a gas with very low tar content was generated.

This paper focuses on the behavior of corn stalk in the indirectly heated pyrolysis reactor. It was just a part of research work of whole project. The experimental results, analysis and conclusion have used to guide the development of two-stage gasifier. An interesting thing was the hydrogen concentration increasing obviously along with the rising of pyrolysis temperature. It showed a possibility that a hydrogen-rich gas could be generated by decomposing of the pyrolysis gas at a higher temperature. A new project has been started to explore the process of converting biomass to hydrogen by thermal chemical conversion.

EXPERIMENTAL APPARATUS

The main experimental apparatus was shown in Fig. 1. A 100mm diameter tube shape pyrolysis reactor heated by an electric heater surrounding the tube. In the tube, a screw feeder droved by a motor was used to make the biomass material moving through the tunnel at a constant speed. By adjusting the speed of the motor, the biomass can be kept inside the reactor for a predetermined time. It was kept a micro-positive pressure inside the reactor to keep the process away from outside air. Constant temperature was achieved by adjusting power input of the heater. A charcoal collector located at the end of the reactor. The produced gas was sampled after removing the tar by passing a filter connected with the reactor outlet. The filter worked also as a cooler to decrease the gas temperature to normal level.

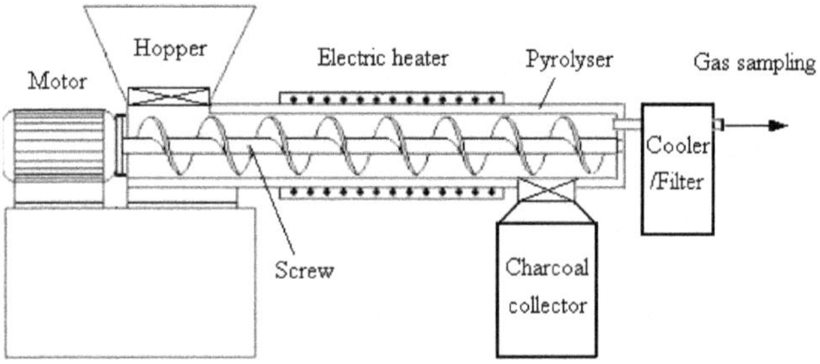

Figure 1: Experimental indirect heated pyrolysis reactor.

Other experimental apparatus include:

(1) A automatic control system for adjusting the temperature in the reactor and the rotate speed of the screw by a PC;
(2) A gas chromatogram analyzer for gas components analysis;
(3) A thermogravimetry and differential thermal analyzer to analyze the characteristics of charcoal;
(4) A balance for weighing the raw material and residual charcoal.

EXPERIMENTAL RESULT AND DISCUSS

The corn stalk was crushed into 0~10mm length, and desiccated naturally to moisture content within 8% weight ratio. Before experiment, the pyrolysis reactor was preheated to predetermined temperature. During the experiment the temperature and the rotate speed of the screw were kept constantly.

By choosing different reaction temperatures and the time of biomass staying in the reactor, a series of tests have been made. The effects of temperature and reaction time on weight ratio of residual charcoal, gas compositions, and characteristics of charcoal have been analyzed.

WEIGHT RATIO OF RESIDUAL CHARCOAL

During the pyrolysis process large molecules of carbohydrates were cracked to smaller molecules of volatile matters along with temperature rising. These volatile matters were released from the solid material, only charcoal remained. Commonly saying, the charcoal should consist of fixed carbon and ash. While in an uncompleted pyrolysis reaction, there is always part of hydrocarbon remained in the charcoal. The weight ratio of residual charcoal to raw material could represent the deepness level of pyrolysis process. By weighing the raw material and remained charcoal, the weight ratio of charcoal can be obtained and convert to dry basis of raw material. Corresponding different conditions of corn stalk tests, the results were shown in Table 1.

Table 1: Weight ratio of residual charcoal.

Temperature, ℃	Reaction time, min		
	18	9	6
350	41.21		
400	39.06	38.82	38.98
450	36.05	37.98	38.02
500	35.03	35.00	37.48
550	31.78	32.42	31.35
600	27.05	24.33	24.49
650	25.28		

Effects of temperature on charcoal remaining

Figure 2 shows the weight ratio of residual charcoal versus temperature when reaction time kept at 18 minutes. It was saw that the reaction temperature has a significant effect on deepness level of pyrolysis process. When temperature was in a lower range of 350~400℃, around 60% of corn stalk was converted to volatile. Along with the temperature rise, weight ratio of residual charcoal dropped dramatically. When the temperature reached 650℃, only 25% of charcoal remained. This means that more carbohydrates were cracked at higher temperature.

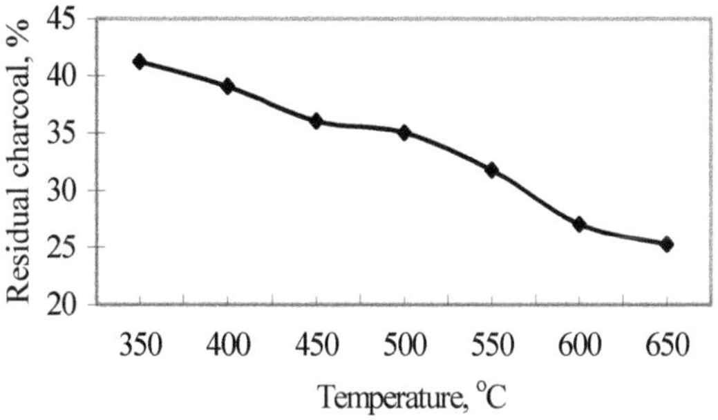

Figure 2: Weight ratio of residual charcoal (reaction time= 18 min).

Effect of reaction time on charcoal remaining

Figure 3 showed the effect of reaction time on weight ratio of residual charcoal at different temperatures. From common sense, when temperature was kept constant, the longer the reaction time was, the smaller the charcoal remaining would be. However no evident conclusion can be drew from the test results. Thus it was reasonable to assume when the raw material was heated to a certain temperature, pyrolysis process happened in a very short time interval. Prolonged reaction time had little effect on helping cracking the volatile remained in charcoal.

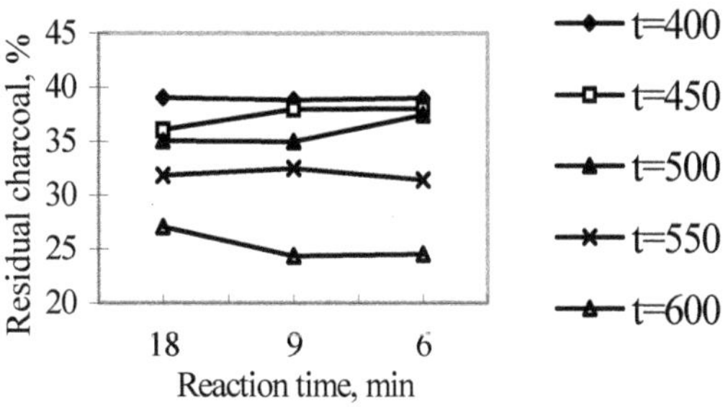

Figure 3: Weight ratio of residual charcoal at different temperature.

GAS COMPOSITION

After cooling to normal temperature and removing the tar, the pyrolysis gas was sampled for analysis in a gas chromatogram analyzer. Compositions and heat values were collected and tabulated as table 2.

From table 2, it was seen that gas composition varies with both temperature and reaction time. However, the heat value of the gas mainly followed the temperature. The higher the temperature, the more total combustible contents there were, and the higher the heat value was.

Table 2: Composition and heat value of pyrolysis gas.

Temp., ℃	Reaction time, min	Composition, %									Heat value, MJ/Nm³
		H_2	O_2	N_2	CO_2	CO	CH_4	C_2H_4	C_2H_6	C_3+	
400	18	2.12	1.09	5.38	43.96	32.02	8.33	0.85	1.33	3.06	10.475
	9	2.99	0.51	3.26	41.81	35.32	10.26	1.08	1.48	3.14	11.968
450	36	6.83	0.67	3.32	39	30.96	12.95	1.21	1.76	3.56	13.304
	18	4.94	0.32	1.69	40.36	32.77	13.29	1.23	1.91	3.58	13.565
	9	4.43	0.4	1.91	37.79	35.92	13.37	1.38	1.75	3.26	13.745
500	18	13.72	0.21	1.36	29.74	32.17	14.81	1.78	1.97	3.96	15.593
	9	11.07	0.75	1.24	29.99	33.21	16.92	1.43	1.92	4.09	16.043
	6	9.78	0.32	0.61	29.79	34.73	17.07	1.84	2.34	4.38	16.816
550	18	22.87	0.78	4.76	25.58	25.62	14.09	1.49	1.8	3.47	14.933
	9	20.04	0.29	2.99	28.23	27.87	14.59	1.72	1.54	3.71	15.217
	6	11.05	0.21	0.63	29.19	32.84	17.41	2.14	2.08	4.81	17.117
600	18	32.47	0.11	0.71	23.79	20.09	16.97	2.12	1.29	2.47	15.805
	9	26.73	0.78	1.41	22.37	25.13	16.68	2.28	1.64	3.5	16.629
	6	24.18	0.05	0.51	22.59	28.42	16.53	2.64	1.52	4.01	17.161
650	18	30.26	0.18	3.25	21.67	20.55	19.24	1.84	1.27	1.72	14.802

In any case, the temperature has the most important effect on indirectly heated pyrolysis process. Fig. 4 shows the trends of main gas compositions versus different reaction temperatures at 18 minutes reaction time. The intuitionistic concepts can be got from the trends as follow.

(1) H_2 content rises along with increasing of the temperature obviously;
(2) CH_4 content rises along with increasing of the temperature;
(3) CO content dropped along with increasing of the temperature;

Apparently, raising temperature was an effective way to convert the biomass material to fuel gas.

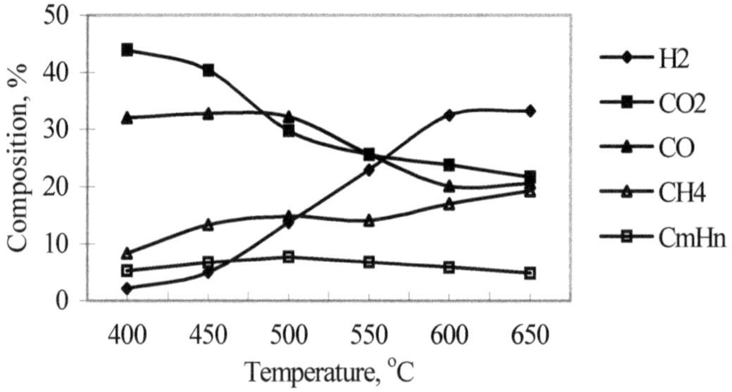

Figure 4: Composition of pyrolysis gas (reaction time=18 min)

CO_2 content dropped dramatically along with increasing of the temperature.

CHARACTERISTICS OF RESIDUAL CHARCOAL

Completeness level of the pyrolysis process

To evaluate the completeness level of the pyrolysis process, the residual charcoal samples were analyzed in a thermogravimetry and differential thermal analyzer. During the tests the samples were protected under N_2 with 10ml/min flux. The samples were heated at a constant rate of 100℃ /min. Fig 5-6 show the curves for samples respectively from 450℃ and 650℃ pyrolysis tests.

In the figures, two of larger weight losing were found. The first one (occurred around 100℃) was due to the evaporation of moisture absorbed by samples when stored at normal condition. The second one (occurred at 800~900℃) was the volatilization of residual volatile matter.

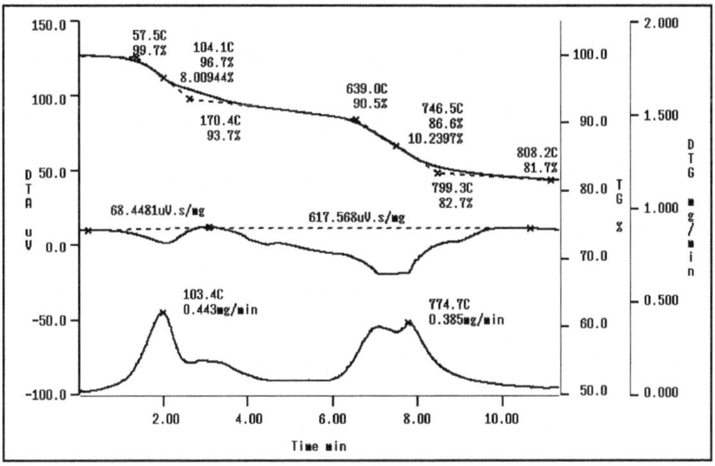

Figure 5: TG, DTG and DTA analysis of charcoal under N_2 protection (Sample from 450℃ pyrolysis).

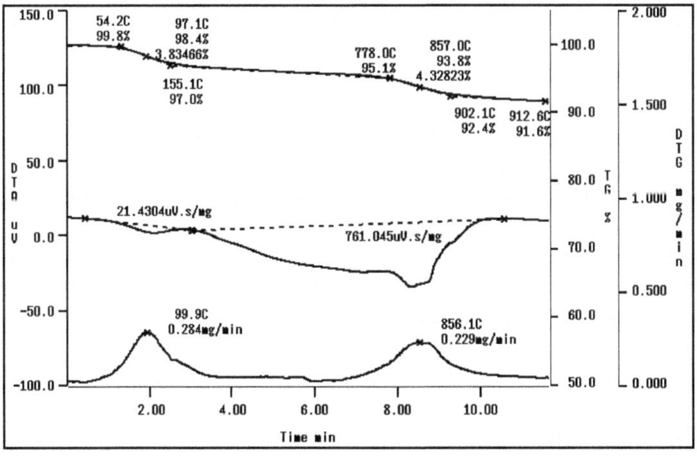

Figure 6: TG, DTG and DTA analysis of charcoal under N_2 protection (Sample from 650℃ pyrolysis).

Comparing these curves one could conclude: in the pyrolysis reaction, the higher the temperature was, the lesser the volatile remained in the charcoal. When pyrolysis temperature was 450℃, the charcoal contained about 12% of volatile matter. If the temperature rose to 650℃, less than 5% of volatile matter still remained in the charcoal. In fact, it could be seen from Fig. 6, if the temperature rises to as high as 900℃, a completely pyrolysis can be reached. However it required special material to make the apparatus and the cost would rise accordingly.

Combustion characteristics of the charcoal

To explore the charcoal action in combustion/gasification reactor that is the second stage of two-stage gasifier, the combustion characteristics of the charcoal was analyzed and the results was shown as Fig 7. In the test without any protection medium was used, the sample touched with air.

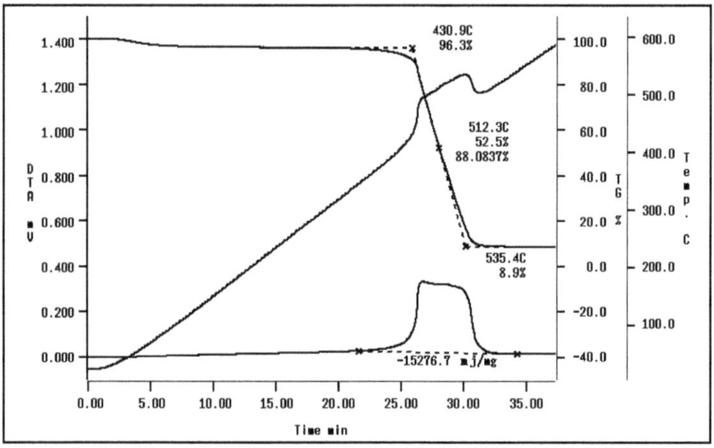

Figure 7: TG, DTG and DTA analysis of charcoal.

Fig 7 showed a drastic combustion occurred in temperature range of 430-540℃. The charcoal lost its weight rapidly and an amount of heat was released. The whole heat value of sample was 15279 kJ/kg. The results means that the residual charcoal produced from corn stalk pyrolysis is a kind of good fuel with higher activity and heat value.

CONCLUSIONS

(1) In any case, the temperature has the most important effect on biomass pyrolysis process;
(2) Higher temperature would raise the completeness level of biomass pyrolysis, and reduce the weight ratio of residual charcoal and the amount of volatile matter remained in the charcoal;
(3) H_2 concentration rises along with increasing of temperature obviously, that shown a possibility to produce a hydrogen-rich gas at a higher temperature;
(4) The residual charcoal produced from corn stalk pyrolysis is a kind of good fuel with higher activity and heat value.

REFERENCES

1. Gaur S. and Reed T. B (1995) *An atlas of thermal data for biomass and other fuels.* NREL/TP-433-7965.
2. Sun L. (1998) *The properties of biomass materials relation to pyrolysis and gasification in a solid bed gasifier. Shandong Science*, **11**, No. 4.

Secondary Decomposition of Pyrolysis Gas for Hydrogen-Rich Gas Production

L. Sun, M. Xu and R. F. Sun
Energy Research Institute of Shandong Academy of Sciences, Jinan, China

ABSTRACT: Biomass is the carrier of both hydrogen and energy. It is thus possible to extract hydrogen from biomass using the energy itself contained. A thermal chemical conversion process for hydrogen-rich gas production was proposed in this paper. The process consists of a pyrolysis process of raw material and a secondary decomposition process of the produced pyrolysis gas. By means of secondary decomposition, almost all of tar in the pyrolysis gas was cracked down at high temperature and a hydrogen-rich gas was generated. This paper presents the conversion process, experimental data, and evaluation of influences of temperature and reaction time on hydrogen content. All these showed that it is technically feasible to convert biomass to hydrogen-rich gas by a relatively simple method. It was also forecasted that the hydrogen content could be further increased by means of optimizing operation parameters and further reform of the gas.

INTRODUCTION

Hydrogen is well known as a kind of cleanest fuel. Therefore it will play an important role in world's energy structure of 21st century. With the successful development of fuel cell technology, hydrogen's share in the energy market will increase. In recent years, more and more attention has been paid to develop different hydrogen production technologies in many countries. The authors believe that, for an energy system with high efficiency and friend environment impact, the hydrogen should come from renewable energy resources instead of fossil fuels. Advantages of hydrogen production from biomass are:

1. Biomass is the carrier of both hydrogen and energy. The energy itself contained can be used to extract hydrogen from it. Optimized process can even decompose extra water to produce more hydrogen.
2. Biomass energy comes from solar energy by foliage's photosynthesis and it is renewable and sustainable.
3. Biomass energy is the zero CO_2 emission energy.

Commonly saying biomass materials contain average 6% weight percentage of hydrogen element, or equal to 0.672 Nm^3 of gaseous hydrogen can be produced from 1kg of biomass. If all of hydrogen is extracted from biomass, over 55% of biomass energy will convert as chemical energy of hydrogen.

This paper describes the processes of hydrogen-rich gas production by second decomposition of pyrolysis gas. The objective of this research is to establish a technique that converts biomass to hydrogen using its own energy. Primary experiment has been carried out and the results showed a bright prospect.

TECHNICAL ROUTE

The processes proposed to produce hydrogen from biomass are shown in Fig. 1.

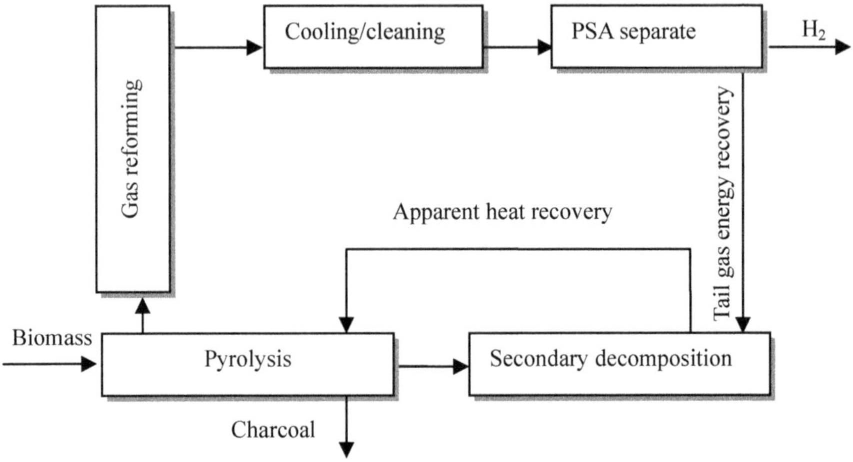

Figure 1: Processes for hydrogen production from biomass.

The key points of the flowchart are as follow:

1. Through an indirectly heated pyrolysis (primary decomposition) process in absence of air, around 70~75% of biomass material can be converted into gaseous volatile. The volatile include some uncondensable components, such as H_2, CO, CO_2, CH_4 etc., and some large molecule hydrocarbons that condense into so-called "tar" at normal temperature. The residual charcoal is removed out at the end of pyrolysis.
2. By means of a secondary decomposition process, in which the pyrolysis gas underwent a further thermal cracking at a temperature higher than one of the pyrolysis, the tar will be decomposed into smaller molecule gas. The process raises not only the hydrogen content but also the volume of uncondensable gas. After the conversion, a hydrogen-rich gas should be generated.
3. For further increasing hydrogen content, a gas reforming process was proposed. Depending on the final experimental results, either carbon monoxide shift reaction or methane reforming reactions can be chosen.
4. A pressure swing adsorption (PSA) system should be adopted to separate pure gaseous hydrogen from the gas.
5. In order to realize the self-energy balance, a reasonable energy recovery program has to be designed carefully. It is necessary to leave at least one of combustible components, carbon monoxide or methane, in the tail gas after separating hydrogen. According to the temperature levels needed for each process, the tail gas could supply energy to secondary decomposition, and then the apparent heat of the hydrogen-rich gas and the flue gas should be recovered to pyrolysis process.

PRIMARY EXPERIMENTAL RESULTS

During this research, primary experiments on pyrolysis and secondary decomposition were carried out under different temperature and reaction time. The experimental data was collected and tendency of hydrogen concentration variety was analyzed.

PYROLYSIS TESTS

Biomass pyrolysis and its intermediate reaction are very complicated. While in our research what we just concerned was the relationship between reaction condition and hydrogen content. In an indirectly heated pyrolyzer, a series of tests within temperature range from 400~650°C were done to find the tendency of hydrogen concentration variety with temperature. Then an 800°C pyrolysis test was carried out to check the extending trend.

The tendency of hydrogen concentration in pyrolysis process

Fig 2 shows the gas compositions tested in an indirectly heated reactor using corn stalk as feedstock. 4 curves represent respectively H_2, CO, CH_4, and C_mH_n , which are able to convert as hydrogen finally, concentration variety corresponding reaction temperature.

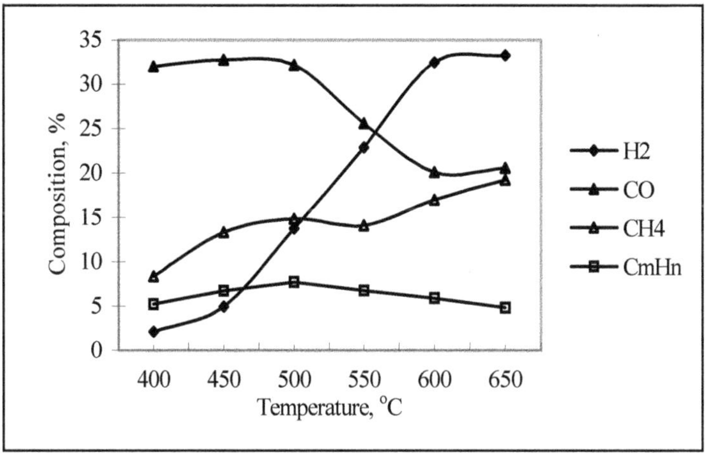

Figure 2: Volatile gas composition variety with temperature.

From Fig 2 an evident tendency could be seen that the hydrogen content rose observably along with the pyrolysis temperature and the methane content rose also. When temperature rose from 400°C to 650°C, the hydrogen content moved up from 2.12% to 33.26%, and the methane content also changed from 8.33% to 19.24%.

It was easy to understand the pyrolysis is a thermal process that cracks the large molecule of carbohydrates in biomass materials to small molecule volatile matters. The higher the temperature was, the more thorough the reaction went along, and the more the small molecule components could be produced.

It should be noticed that the rising of temperature caused not only the increasing of hydrogen and methane content but also the increasing of total volume of volatile gas. Corresponding the temperature range from 400°C to 650°C, the amount of residual charcoal was dropped from 40% to 25% after pyrolysis. This means more matters in feedstock were converted to volatile gas.

Pyrolysis tests under 800℃

In order to prove the extending variety tendency on hydrogen content, pyrolysis tests under 800°C, the highest temperature bearable for the experimental apparatus, were carried out. In these tests wood dust was used as feedstock. The main gas compositions involved with hydrogen production are shown in Table 1.

Comparing with the data of 650°C pyrolysis in Fig 2, it could be found that the hydrogen content increased further. This is accordant with expected trend appeared in above pyrolysis tests. In the tests, the weight ratio of charcoal reduced to around 18%. This is a satisfactory result to increase the converting efficiency.

Table 1: Results in 800°C pyrolysis tests.

	Gas composition, %			
	H_2	CO	CH_4	C_mH_n
Test No. 1	36.73	22.45	18.12	4.96
Test No. 2	36.82	21.46	17.52	4.03

TESTS ON SECONDARY DECOMPOSITION REACTION

Secondary decomposition reaction (SDR) is a gaseous phase reaction because the charcoal has been removed out of the pyrolysis gas. Main purpose of SDR was to convert the tar into simple gas by thermal cracking. In the research of biomass gasification, tar removing was always a difficult problem. Many scientists have put a great effort to remove it from the gas or destroy it. The thermal cracking was considered as an effective method to destroy the tar without any pollution.

The tar comprised of several hundred complicated chemicals. But from point of view of energy converting, all of these substances are the heavy hydrocarbons that condense into mucous liquid when temperature drops. By means of thermal cracking, these heavy hydrocarbons could be decomposed into light ones like CH_4, C_2H_4, C_2H_6, etc. or H_2. These uncondensable gases are easy to convert to gaseous hydrogen.

By a rough estimating, in the pyrolysis tests described above, around 10~12% of biomass was became the tar kept in the volatile gas. This means a 30% of energy loss without SDR process.

An experimental SDR reactor was designed for primary test. The apparatus can work at a temperature as high as 1100°C. By controlling gas flux the reaction time could be adjusted.

Presented in Fig 3 are the tests data corresponding respectively to the pyrolysis tests under 800°C. In test No. 1, the gas produced in pyrolyzer was sent into the SDR reactor and stayed there for 3 seconds. And in test No. 2 it stayed there for 10 seconds.

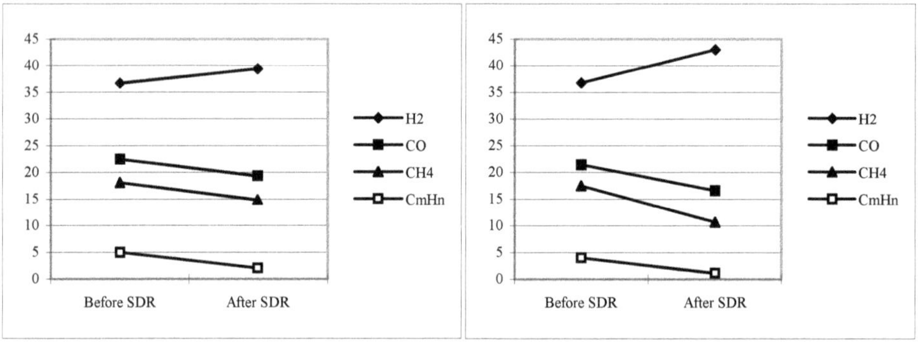

(a) Reaction time =3 s (b) Reaction time =10 s

Figure 3: Gas concentration variety in SDR process.

From Fig 3 some important concepts could be seen clearly, as follows:

1. After the SDR, hydrogen concentration increased observably, while all the other compositions decreased. It showed that the SDR was an effective method to decompose the hydrocarbons even for the small molecules hydrocarbon like methane. The final product was hydrogen.
2. Compare Fig 3 (b) with (a), the influence of reaction time could be found. For test No. 1 with 3 seconds reaction, the hydrogen content increased from 36.74% to 39.43%; while for test No. 2

with 10 seconds reaction, the hydrogen content increased from 36.83% to 43.01%. The influence also was found from the curves of other compositions although they sloped down. To ensure completeness of SDR, enough reaction time was necessary.
3. In test No. 2 the concentration of uncondensable C_mH_n dropped from 4.03% to 1.11%. Therefore it was presumed that almost all of tar should be decomposed at high temperature. Actually in the test the tar could find hardly any.

The gas chromatogram analysis result of the gas was shown as Fig 4. The upper one is TCD curve, and the lower one is FID curve. It could be seen clearly that the main contents in the gas were H_2, CH_4 and CO. The experimental data proved the technical feasibility of producing hydrogen from biomass by secondary decomposition method.

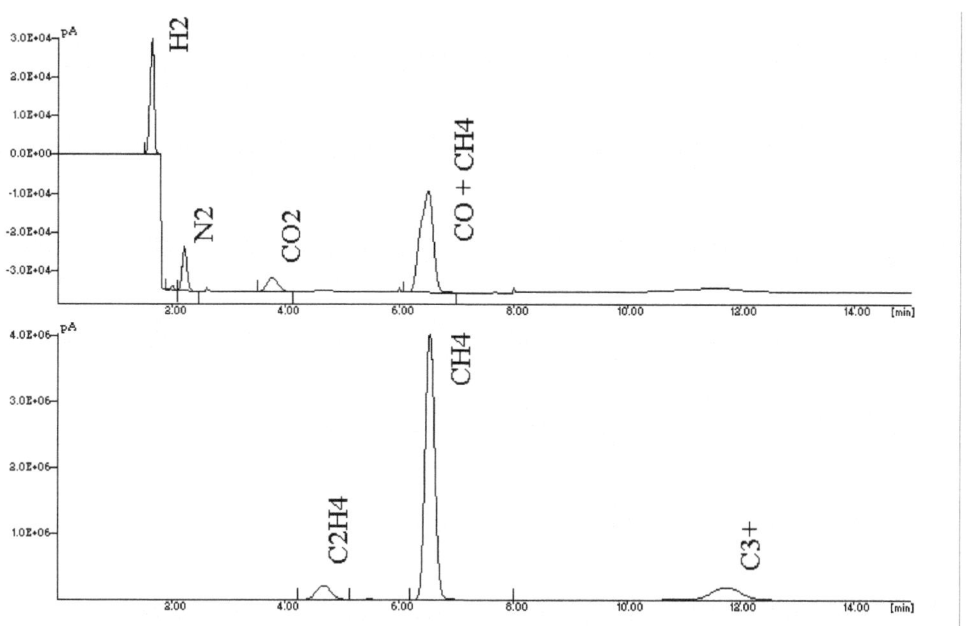

Figure 4: The gas chromatogram analysis of SDR gas

FURTHER RESEACH WORK

Comparing with other ongoing studied technologies of hydrogen production from biomass, the technical route described in this paper has the following advantages:
1. All of apparatus worked at normal pressure to avoid rigorous process conditions;
2. None of additional energy was required;
3. Broad adaptability for raw biomass material;
4. Air was kept away from the process to increase gas energy density and reduce apparatus cost.

Primary experiment has shown a bright prospect. While for developing the technology a lot of further research work needed to do, including:
1. Thorough study on the properties of pyrolysis and secondary decomposition processes for determine processes parameter;

2. Function of steam in the reactions to increasing the yield of hydrogen;
3. Selection of gas reforming process;
4. Measurement of energy consumed for each processes and study on the energy balance method.
5. Optimization of the processes for designing of system.

REFERENCES

1. Thomas A. M., Carolyn C. E. & Robert J. E. (2001) *Hydrogen from biomass.* IEA/H2/TR-02/001.
2. Sun L. (1998) *The properties of biomass materials relation to pyrolysis and gasification in a solid bed gasifier. Shandong Science*, **11**, No. 4.

BioSlurry^{TM} for Co-firing and Gasification

R.H. Venderbosch, B.M. Wagenaar, E. Gansekoele, J. Florijn
BTG Biomass technology Group B.V., Drienerlolaan 5, 7522 NB Enschede

ABSTRACT: Fast pyrolysis is a process to convert solid biomass into 70 wt% bio-oil, 20 wt% char and 10 wt% gas. Pulverising the char followed by homogenisation with bio-oil yields BioSlurry^{TM} with a caloric efficiency of 90%. Advantages of the product are its fluid like properties which renders it pumpable, atomiseable and storable and to be used for co-firing in power plants or as feedstock for gasification. The caloric loss to produce the slurry is about 10 to 15%, which has an energy density of about 21 MJ/m^3.

The physical characteristics of BioSlurry^{TM} are largely unknown yet. This paper addresses some of these aspects in relation to its application (density, heating value, atomisation and sedimentation behaviour).

It contains a prognosis of the anticipated slurry delivery cost at the gate of a Dutch power station for the case that it is produced in biomass countries and transported by ship to the Netherlands.

INTRODUCTION

Despite the widespread availability of biomass, its non-uniform appearance and generally low energy density hinders its commercial application for the production of sustainable energy or products. Conversion of biomass in a versatile liquid can enhance its use significantly. A process enabling the conversion of biomass into a liquid is fast pyrolysis, yet approaching commercial application. The rationale is that (cheap) biomass can be transformed in a liquid on a scale corresponding with its availability by specialised production companies, while customers are acquainted with the use of a (systematically specified) liquid.

Fast pyrolysis of bio-oil is characterised by high heating rates of the biomass particles in the absence of an oxidising agent, and a rapid quenching of the produced vapours. For this reason, conversion technologies are based upon the use of a heat carrier material, in or through which small biomass particles are rapidly heated. Several techniques and methods have been suggested and tested during the last 10 years. The group of Bridgwater presented numerous overviews and updates, of which the most recent is published in 2002 [1].

In all of the review papers on pyrolysis the emphasis was on the production of bio-oil: continuous pyrolysis in a 250 kg/hr pilot plant revealed that self-sustaining operation could be achieved (no additional fuel), but moreover there is a surplus of charcoal. Approx. 60 to 70 % of the char can be removed, and this increases the efficiency considerably. In the present paper a concept of an integrated system is proposed, in which part of the char is removed. This concept is evaluated for both the Rotating Cone Reactor (RCR) and for the Integrated Fluid Bed (IFB). Actual data are presented only for the latter case. The char can be sold as a separate product, apart from the bio-oil, but also as a mixture of char and bio-oil, in the present paper referred to as BioSlurry^{TM}.

PYROLYSIS PROCESSES

The oldest pyrolysis processes are based upon fluid beds. In laboratory small-scale units, the required inert fluid bed carrier gas is usually taken from bottles. In commercial large-scale units the re-circulated inert pyrolysis gases will be used. The latter requires extensive cleaning of the gas to avoid deposition of tar and dust on the compressors and (if required) heat exchangers and blocking of the distributor plate of the fluid bed. In the gasification processes referred to above, cleaning of the gas is considered a major bottleneck for implementation, and much research is related to this particular issue.

In particular North American organisations were and are involved in commercialising fluid bed technology (Resource Transforms International, DynaMotive Technologies Corporation, and Ensyn Technologies Inc.). Latest developments took place in the UK (Wellman Process Engineering) and Finland (Fortrum Oy).

It has been demonstrated that also mechanical mixing devices can do the mixing of the biomass into a hot sand bed, as is the case in the rotating cone reactor or a screw reactor. This avoids the use of any inert carrier gas, because the mixing is now taken over by a mechanical device. The latest development is a fully automated continuous operating unit at a scale of 250 kg/hr biomass [3]. In this unit more than 50 tonnes has been produced since 2000, in several continuous runs of over 100 hours. The installation showed self-sustaining characteristics, as the produced charcoal is combusted in a separate fluid bed. A 2 t/hr demonstration unit is emphasised within the next two year, with support from the EC.

Some aspects of fast pyrolysis are worthwhile to summarise for the purpose of the present paper:

a) Almost all technologies developed for the production of bio-oil show liquid mass and energy yields above 70%. In comparison, gasification of biomass yields cold gas efficiencies of near 70%. The difference is due to the relatively strong exothermicity in gasification versus a small endothermicity in pyrolysis.
b) A series of experiments in continuous pyrolysis systems showed that the overall efficiency of pyrolysis can be almost fifteen percent points higher if char is removed as a saleable (by)product.
c) The liquids derived from fast pyrolysis are generally similar, and the basic appearance is almost independent on the pyrolysis process. The differences observed can me mainly attributed to feedstock (preparation) and operating conditions (pressure and temperature). Nevertheless, characterisation of the liquid is difficult, and analysis techniques used during the last decades are shown to be inappropriate to analyse the complete bio-oil.
d) It is considered troublesome to address the distinct effects of the various parameters on the process. This is due to several reasons, amongst other the variable feedstock characteristics (ranging from type of feedstock to its preparation), the large variation in operating conditions, the semi-batch wise operation, etc. As a typical example, detailed mass and energy balances for commercially aimed processes are seldom presented and considered proprietary information.

PRODUCTION OF BIOSLURRY™ IN THE IFB CONCEPT

The basic difference in pyrolysis systems aimed at for commercial purposes is whether fluid bed systems or mechanically driven reactors should be applied. For the present case, the system discussed is the Integrated Fluid Bed approach (IFB: Figure 1) which shows a basic similarity to the Rotating Cone Reactor [3].

Biomass logs are chunked (70 x 70 x 70mm, 30 wt.% moisture), dried (10 wt.%) and further sized down to 10 mm, and subsequently injected through a series of screws in the fluid bed, operated at 530°C. The fluid bed agent is derived from the non-condensables from pyrolysis, scrubbed and cleaned to meet the requirements of the compressor, heat exchangers and the fluid bed distributor design. The ratio between the inert carrier gas flow rate and the biomass is taken as 1 : 2.2 (kg fluidizing gas : kg biomass).

The pyrolysis vapours are produced by rapid heating of the biomass, and fed to one or more condensing tower (one is shown). The non-condensable vapours, mainly comprising of CO, CO_2 and CH_4 but also some aerosols are directed through an electrostatic precipitator, where after part of these gases are re-cycled through a compressor. The char produced upon pyrolysis is separated in two fractions: one part is separated and collected as such, while the remaining is transported from the fluid bed pyrolysis reactor to a second fluid bed where air is used to burn the char (570°C). The ratio between the sand circulation flow and the biomass flow rate is taken as 30 : 1 (kg sand : kg biomass) The heat of combustion from the char is used to re-heat the sand required for the pyrolysis reactions, and to heat the biomass to the required pyrolysis temperature.

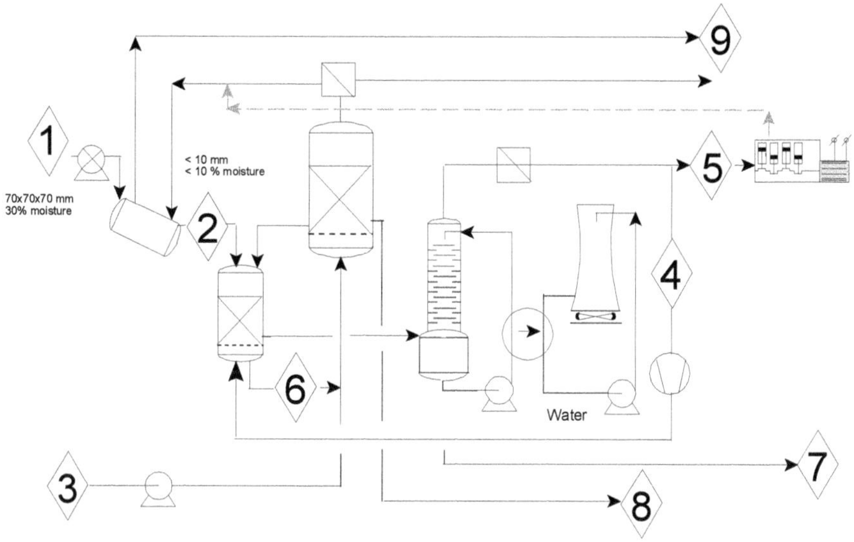

Figure 1: Flowsheet for the Integrated Fluid Bed.

Table 1: Mass and energy balance balance for 2000 kg/hr (daf) Integrated Fluid Bed pyrolysis.

	Input	Reactor feed	Air	Inert carrier gas	Pyrolysis vapours	Heat carrier	Bio-oil	Surplus char	Flue gas
Stream no.	1	2	3	4	5	6	7	8	9
Organics	2,000	2,000					1200		
Water	900	200					400		740
O_2			600						250
N_2			2,400						2,400
CO_2									470
Pyrolysis gas				1,000	200				
Char						155		245	
Ash	20	20						12	8
Sand						66,600			
Total	2,920	2,220	3,000	1,000			1,200	257	3,860
Temp. (K)	291	291	291	326	304	804	317	804	400
Pressure (bar)	1	1	1	1.2	1	1	1	1	1
Chem. heat (MW)	10.4	10.4		2.1	0.42	1.3	6.6	2.1	
Heat (MW)						7.41		0.05	0.33

For the reference case of the IFB a mass and energy balance is prepared for a pyrolysis plant. The results are summarised in Figure 1. The overall reaction equation for the pyrolysis system (excluding the gas engine) can be given as:

$$CH_{1.60}O_{0.68} + 0.136\ H_2O + 0.230\ O_2 \quad \Psi$$

(daf wood) (water) (air)

$$0.624\ CH_{2.59}O_{1.05} + (0.0250\ H_2O + 0.095\ O_2 + 0.131\ CO_2) + 0.185\ CH_{0.716}O_{0.219} + 0.0735\ CH_{0.543}O_{1.300}$$

(Bio-oil incl. water) (Flue gases) (Char) (Pyrolysis gas)

The main difference for the RCR is that a recycle gas flow is absent. This leads to a higher thermal efficiency since heating the recycle gas requires a thermal duty. As a result, for the RCR 65% of the char can be removed, whereas 60% can be removed for the IFB. No expensive compressors for the recycle gas, additional heaters and extensive gas cleaning is required. For the RCR concept it appeared that a single vessel can be used to condense more than 98 % of the liquid product, in contrast to literature data on fluid bed pyrolysis where two or more vessels are recommended.

In summary, 2.92 t/hr of biomass (as received, 30 wt.% moisture) yields as main products 1.6 t/hr bio-oil (6.6 MW_{th}), 200 kg/hr of pyrolysis gas (0.42 MW_{th}) and 400 kg/hr of charcoal. Part of the charcoal is combusted to supply the heat required for pyrolysis (heating the biomass and supply heat for the pyrolysis reaction). About 40% of the charcoal is required. In principal, there are two possibilities to retain the char, viz. separate from the bio-oil in the form of mm-sized char particles, or collected and emulsified in the bio-oil in a specified weight content. On average, the weight fraction of the char in the oil (on oil basis) is approx. 15 wt.%.

The pyrolysis gas (7.5 MJ/kg) can be used to fuel a gas engine: the electrical power output will be 0.16 MW_e, with additional heat production of 0.21 MW_{th}. The flue gases derived from the combustion of the char contain approx. 0.5 MW_{th} sensible heat. Both, the heat from the flue gases and from the gas engine can be used to dry the feedstock from over 30 wt.% down to the 10 wt.% required for pyrolysis (air-dried wood contains approx. 25 wt.% moisture).

Proprietary data indicate that for the pyrolysis process roughly 0.015 to 0.03 MW_e is required per $MW_{th,in}$. The exact value depends amongst others on whether the feedstock should be chunked, milled and dried upon reception: 0.016 $MW_e/MW_{th,in}$ is produced on basis of the pyrolysis gases, which is sufficient if the feedstock is delivered at certain specifications.

The general mass and energy balances for the RCR are almost similar, except for the recycle flow of pyrolysis gases. Only a small difference can be calculated in the overall efficiency (86 % for the IFB versus over 88 % for the RCR).

USE OF FAST PYROLYSIS PRODUCTS

In contrast with the usual papers on pyrolysis reporting 70% energy efficiency on bio-oil, it is possible to increase this value to over 90% on 'cold' products, if the char is entrained from the system and the gas can be used for another purpose. On basis of oil and char alone, the theoretical efficiency is over 86%. As stated above the pyrolysis gas can be used to generate 'green' electricity to drive the relevant process equipment.

The overall potential of all pyrolysis products is shown in Figure 2. Oil can be used for turbines and engines (most likely only if the oil is filtered), combustion, syngas and chemicals. In the latter three options in which some char is allowed, BioSlurry™ can be considered.

PROPERTIES OF BIO-OIL – CHAR SLURRY

A slurry of around 1200 to 1250 kg/m^3 can be produced by suspending charcoal in bio-oil. The slurry can contain up to 25 wt.% of charcoal, but it is still easy to transport, to store and to handle. However, one special issue to consider is the possible separation of the char. The viscosity of such a slurry can be a factor two higher than for the 'pure' bio-oil, but even then no preheating before pumping seems to be required [6]:

$$\eta_{sl} = \eta_L \frac{1+4.5\varepsilon_s}{1+\left(\dfrac{\rho_s}{\rho_L}-1\right)\varepsilon_s} \cong \eta_L \cdot (1+4.5\varepsilon_s)$$

in which η_L and η_{sl} are the dynamic viscosity of the bio-oil and the BioSlurry™ respectively (kg m^{-1} s^{-1}) with a volume fraction of solids of ε_s (-) and ρ_s and ρ_l the densities of char and bio-oil (kg/m^3).

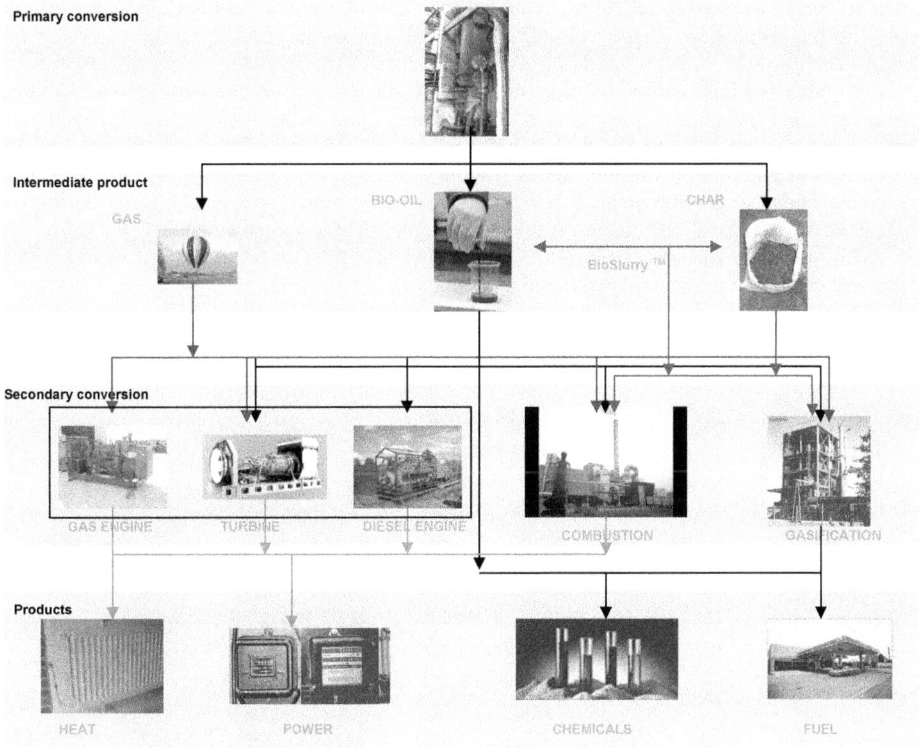

Figure 2: Potential application of fast pyrolysis derived products

The energy content of the slurry is increased due to the presence of the higher energetic value of the char: values of around 21 MJ/kg are reasonable.

Issues related to the use of BioSlurry™ are its stability, in the sense of phase separation due to polymerisation and / or due to the solid separation. This factor is of particular importance as char is reported to be a catalyst for polymerisation reactions.

(CO-)COMBUSTION

For the combustion of bio-oil or BioSlurry™, the most straightforward approach is to co-combust it in existing power stations. It is expected that co-combustion in a coal-fired installation is the more convenient way to produce green electricity from a BioSlurry™. The straightforward possibility is to co-fire the bio-oil using oil lances (usually already foreseen in the power station) and the charcoal together with the fossil coal. The other option is to co-fire the BioSlurry™. As an intensive gas cleaning section is available to meet the environmental requirements, it is not foreseen that there is an upper limit for co-firing bio-oil or BioSlurry™. The efficiency of the bio-oil combustion will be similar to that of coal combustion. Ensyn was the first company who demonstrated the commercial use of bio-oil in co-firing a coal-fired boiler in a 20 MW_e boiler. Bio-oil was co-fired for about 370 hours. Co-firing stopped however when no surplus was available anymore [2].

The challenge is to co-combust the bio-oil or the BioSlurry™ in a natural gas (NG) fired installation. For just bio-oil co-firing is possible just up to a level of 5 to 10 energy% of an existing power plant. The upper limit is determined by the maximum NO_x-level and dust allowed in the flue gases, as no intensive gas cleaning system is foreseen in NG fired power plants. Some efficiency loss compared to NG is foreseen, because the NG gas will be fired in the gas turbine section of the installation, whereas bio-oil will be fired in the heat recovery boiler section of the power plant.

Electrabel and BTG carried out experiments concerning bio-oil co-combustion in a 350 MW_e natural gas fired installation in the period of September 2002 in the Netherlands. A preliminary paper on this issue was already presented in Amsterdam [3]. During a campaign of several hours, several tons of pure bio-oil was combusted without problems. The average substitution of natural gas corresponded to a value up to 5 MW_e. Figure 3 shows some photographs taken during the combustion experiments. The detailed results will be published in a separate paper [7].

With respect to char or BioSlurry™, with similar nitrogen content as the bio-oil but with a higher ash content, it is yet unknown if and how much can be co-fired.

Figure 3: Co-firing bio-oil in the 350 MW_e NG-fired power station in Harculo. On the top is the bio-oil truck connected to the supply pump, the middle shows the back of the nozzle, and the lower picture is a visualisation of bio-oil combustion. The lance at the front and at the back is fuelled on natural gas, the lance in the middle is fired on pure bio-oil (up to 5 MW_e). On the left side the stack during bio-oil combustion is shown.

(CO-)GASIFICATION

In contrast to green transportation fuels, for the production of 'green' electricity various sustainable resources can be considered, like wind, water, solar and biomass. Only biomass however is expected to play a major role for the production of green hydrocarbons; it is impossible to produce them from the other sustainable sources, which do not contain any carbon. Currently green fuels can be produced directly (e.g. bio-diesel), biologically (bio-ethanol), or via thermochemical routes on basis of biomass 'syngas' production. 'Syngas' is a mixture of CO and H_2. It is the feedstock of various processes for methanol, dimethylether, or Fischer-Tropsch fuels. In case of solid biomass gasification, large amounts of biomass must be collected, because syngas production and Fischer-Tropsch processing are only economically feasible on a very large scale, in the order of over 200 tons wood per hour. Apart from such scale-related considerations, another problem of solid biomass gasification is that the gasification conditions usually do not match properly with synthesis gas requirements. Direct biomass gasification is carried out at relatively low temperatures with air or with steam / oxygen mixtures to avoid ash melting in dedicated biomass gasifiers. The syngas produced therefore contains large amounts of nitrogen (when air is used), and high 'tar' and methane concentrations. Pressurization of biomass is expensive, and biomass gasifiers are yet developed for atmospheric conditions only. High-pressure syngas processing requires expensive syngas compression to 60 bar. The use of bio-oil instead of solid biomass offers a simple way to diminish transportation costs, facilitate the gasification process and making use of existing gasifiers. Conventional entrained flow gasifiers are common in large-scale oil gasification plants. Its scaling-up rules are simple, it has a high conversion efficiency and the gas quality is excellent (low tar concentrations and little methane) due to the high temperatures applied (1200 to 1500°C, depending on whether air or oxygen is used). Several projects are focussing on the use of bio-oil as a feedstock for entrained flow gasifiers. Worthwhile to mention are a) a report presented in Amsterdam by BTG, b) work to be carried out by a partnership of the Forschungszentrum Karlsruhe (FZH) and the Institute for Wood Chemistry (IWC), and c) future work scheduled by a partnership of Choren Industries GmbH.

Figure 4: UET's entrained flow gasifier in Freiberg.

The first paper mentioned presents the results for lab-scale bio-oil gasification with air in an entrained flow reactor at temperatures up to 1100°C. The producer gas composition showed a clear similarity with gas derived from direct biomass gasification. Ratios of the concentrations of H_2 over CO in the syngas are in agreement with the shift equilibrium. Gasification temperatures over 1250°C and longer residence times are required to reduce the methane concentrations. Experimental results using oxygen (to be reported later) confirmed these observations. Current work is focussing on these aspects, and preliminary results have been obtained. Later this year, experiments together with Shell are planned in a 1 MW_{th} entrained flow gasifier (Figure 4). This gasifier is owned by UET Umwelt- und Energieteknik located in Freiberg (Germany), which is associated with Choren Industries GmbH. Choren is involved in an 11 M€ research project together with DaimlerChrysler AG. This project, scheduled to end in mid-2003, is to clarify at what prices and conditions these renewable fuels can be produced. Researchers will also evaluate the fuel quality and quantities available and compile a comprehensive energy and material balance, covering collection and processing of biomass and syngas, and fuel distribution.

The work of FZH-IWC was also aimed at the use of bio-oil in entrained flow gasifiers, but now fed together with the char in a slurry similar to BioSlurry™. The authors reported the principal feasibility of gasification of the slurry in a 5 MW_{th} 26 bar slagging entrained flow gasifier. It is reported that about 8 tonnes of slurries up to 26 wt.% char were gasified, yielding a practically tar-free syngas. The results of the experiments are expectedly reported later in these conference proceedings.

ECONOMICS

The main factors affecting the price of the biomass delivered at the gate of say a power plant are feedstock costs, transportation costs, and additional pretreatment costs. Although the biomass costs are relatively low in biomass countries, transportation costs can be high. Conversion of biomass into a liquid or BioSlurry^{TM} reduces the transportation costs with a factor over 2, but this is at the expense of the efficiency of the pyrolysis process and the investment costs at the site of the pyrolysis plant.

Roughly, if biomass is transported from biomass countries (Brazil, Indonesia, Baltic States etc.) the additional costs at the gate in the Netherlands for the slurry compared to the wood logs is in the order of 0.2 €/GJ to 1.5 €/GJ. The exact value depends on:

a) the feedstock costs: due to the efficiency of pyrolysis the additional costs are higher for higher feedstock costs (taken from zero to 4 €/GJ);
b) the transportation distance: the larger the distance between the country from where the biomass is taken and the Netherlands, the smaller the difference because the transportation costs for the slurry are more than half of the value for the solids (distance taken from 2500 to 5000 km)
c) The scale of the pyrolysis system, due to the economy of scale. The investment costs for the pyrolysis plants are taken from the present 2 t/hr design, scaled to 10 t/hr.

Although the slurry may be a bit more expensive, compared to the original biomass the handling of the slurry is much less complicated at the site of its utilization. Most likely conventional conversion systems (as explained before) can be applied, in combination with existing expertise of operators to handle an oil or slurry. For comparison, additional costs for seizing and drying of wood logs alone at the site of a biomass conversion plant in West-Europe is in the range of € 0.2 /GJ and 1 €/GJ, depending on the end-requirements. Additional cost are also required for further processing (f.i. the biomass gasifier or a modified combustion system).

For the case that biomass is imported from countries where it is abundantly available, almost for free and in a shape and texture due to which it can not be easily transported, pyrolysis seems the only economic possibility. Examples of these biomass resources are numerous, and include bagasse, rice husk, residues, etc (bio-cascade). In this way it can promote BioTrade between European and less industrialized countries.

CONCLUSIONS

This paper discusses the pyrolysis of a solid biomass into a slurry of bio-oil and pulverised char, yielding a BioSlurry^{TM}, with a potential overall caloric efficiency of over 90 %. Similar to bio-oil, the slurry can be pumped, atomised and stored. It can easily be fed for co-firing in power plants or to produce syngas in existing pressurised gasifiers.

Experiments in a continuous Rotating Cone pyrolysis system showed that producing such a slurry is possible, as the heat required for pyrolysis is much less than the heat than can be produced from combustion of char. Calculations for an Integrated Fluid Bed system showed similar results. The slurry's characteristics are in accordance with potential end-applications, such as co-combustion and gasification.

If biomass is transported from so-called biomass countries, such as Brazil, Indonesia and the Baltic States, the anticipated additional price at the Dutch gate of the slurry compared to wood logs is 0.2 to 1.5 €/GJ. The exact value depends on the local feedstock cost, the transportation distance, and the scale of the pyrolysis system. These costs can well compensate cost to be made at the site of the biomass conversion plant (f.i. seizing and drying).

REFERENCES

1. Bridgwater A.V., 2002, Fast Pyrolysis of Biomass: A Handbook Volume 2, CPL Press, ISBN 1-872691-47-1
2. Sturzl R., The commercial co-firing of RTP™ bio-oil at the Manitowoc Public Utilities power generating station, available at http://www.ensyn.com
3. Wagenaar B.M., Venderbosch R.H., Prins W., Penninks F.W.M., Bio-oil as a coal substitute in a 600 MW_e Power Station, 12th European Conference and technology Exhibition on Biomass for Energy, Industry and Climate Protection, 17-21 June 2002, Amsterdam, The Netherlands
4. Venderbosch, L. van de Beld, and W. Prins, Entrained Flow Gasification of Bio-Oil for Synthesis Gas, 12th European Conference and technology Exhibition on Biomass for Energy, Industry and Climate Protection, 17-21 June 2002, Amsterdam, The Netherlands
5. Henrich E., Dinjus E., Weirich F., Meier D., 2002, A new Concept for biomass gasification at high pressure, 12th European Conference and technology Exhibition on Biomass for Energy, Industry and Climate Protection, 17-21 June 2002, Amsterdam, The Netherlands
6. Deckwer W.D., Louisi Y., Zaidi A., Ralek M., 1980, Hydrodynamic properties of the Fischer-Tropch Slurry process, Ind. Eng. Chem. Proc. Des. Dev., 19, p.699-708
7. Wagenaar B.M., 2002, Co-firing bio-oil in a natural gas fired power station, to be published

Pyrolysis as a Method for Utilization of Solid Wastes in Greece (Agricultural Wastes and Used Tyres)

A. Zabaniotou and A.J. Karabelas
Chemical Engineering Department, Aristotle University of Thessaloniki, Un. Box 455, 54124 Thessaloniki, Greece.

ABSTRACT: This study evaluates agricultural wastes and used tyres as an energy resource, by studying the effect of temperature on the pyrolysis product yields. Experiments were carried out in a captive sample batch reactor at 1 bar, in helium atmosphere. The heating rate was 80 – 100 °C/s and the temperature range studied was 350 – 850 °C. The pyrolysis yielded char, liquid HC and tar and gaseous products.

The purpose of the present work is to establish a record for the products yields from pyrolysis of various agricultural wastes such as olive cuttings, kernels and cotton ginning wastes and other wastes such as used tyres. The data obtained, can contribute to the development of new bioenergy conversion methods for those wastes consisting an environmental problem but also alternatives sources of energy or other high added value products.

INTRODUCTION

The potential of the available biomass in Greece in the form of agricultural wastes, is sufficient for the supply of a big percentage of the energy demands regarding production. Although the above wastes represent a considerable amount they are hardly used as renewable energy source, except the traditional burning of olive kernels. Proper management and exploitation of this potential could lead to economically profitable approximations and solutions for the agricultural industry. Pyrolysis of the above residues under a non-oxidizing atmosphere or gasification in-situ with energy or alternative fuels production could be a solution to the environmental problem that landfilling or combustion could create and furthermore, can establish a new approach for a more efficient utilization of biomass.

In Mediterranean basin countries, Italy, Greece, Spain and Portugal, olive cultivation is a typical activity. However, residues from olive plantations are hardly used as a renewable energy source. As pointed out in the literature (Della Rocca et al., 1997; Zanzi et al, 1996; Buttiglieri, 1996; Buttiglieri and Nicoletti, 1995; Bosio et al. 1992; Figueiredo, 1990), only few studies have dealt with olive residues, although olive residues could be a source for fuels, energy generation and activated carbon production. Amongst thermochemical transformations, pyrolysis has received much attention, since the process conditions can be optimized to produce liquids, which may be upgraded to fuels, and solid. Char may be useful as a fuel, either directly (briquettes) or as char-oil and char-water slurries and also in activated carbon production (Kaltschmitt and Dinkelbach, 1997; Bridgwater 1997)

Cotton gin waste valorization has a two-fold aim. The first is environmental, relating to disposal of the waste. The second is to provide an alternative energy source. Cotton gin waste as an energy feedstock has a major advantage over other biomass resources because it does not need to be collected; it is concentrated around the ginning house.

Used tyres represent a considerable quantity of solid wastes. Traditional disposal methods such as open-air incineration and landfilling are no longer adequate to increasing environmental concerns. Because used tyres contain an appreciable energy content, thermal treatment is considered an interesting disposal solution. Among the thermal treatment processes, pyrolysis has received increasing interest, as an alternative method to obtain raw materials and fuels as remarked in the literature (Williams et al, 1999; Sharma et al., 2000; Kim et al., 1995). Pyrolysis might be used for char production. The solid char may be used either as smokeless fuel, carbon black or for preparation of activated carbon. It could be also burned both in pulverized fuel or fluidized bed combustors, alone or in co-combustion with other fuels. The derived oils may be used directly as fuel or added to petroleum feedstock. They may also be an important source of refined chemicals. The gases are also useful as fuel, Fig. 7.

EXPERIMENTAL

WASTES CHARACTERISTICS

Samples were prepared by cutting and crushing pieces of the raw material to fractions of the desired particle size (18-80 U.S standard mesh). The elemental analysis of waste was carried out using a LECO COR analyser CHN-800. Oxygen was calculated on wt. % basis after elimination of C, H percentages. Determination of ash and moisture was performed according to ASTM Standards (D-1102-84 for ash in wood and D-2016-74 for moisture).

APPARATUS AND PROCEDURE

Pyrolysis of *wastes* was carried out in a laboratory captive sample reactor. The apparatus employed for the pyrolysis of cotton gin waste was based on a captive sample technique. The advantages of this apparatus are:

a) Independent control of parameters such as final temperature, heating rate and residence time.
b) Effective heat transfer.
c) Nearly zero residence time at the final temperature and rapid cooling of the gaseous products

Table 1: Elemental analysis of waste.

Raw material characteristics (maf : moisture and ash free)	Olive wood	Olive kernel	Cotton gin	Used tyres
C *(wt. % maf)	45.6	44.30	41.23	79.89
H (wt.% maf)	6.19	5.85	5.03	6.74
N (wt. % maf)	0.00	0.00	2.63	
O (wt.% maf)	40.11	49.85	34.0	
Ash (wt. % mf)	1.06	3.90	13.3	
Moisture (wt. %)	15.45	21.50	6.0	0.51
Heating value (kcal/kg)	4658	4916	3775	8210.5

The schematic diagram of pyrolysis reactor is presented in Figure 1. The experimental set up is shown in Figure 2.

The reactor is a cylindrical Plexiglass vessel, of an inner diameter of 7cm and is 12 cm high. The reactor vessel closes with two pairs of flanges (top and bottom). The upper pair of flanges is made of stainless steel SS316. Between the flanges, elastic o-ring is fitted, in order to achieve isolation. The stainless steel flange has a diameter of 12 cm and a thickness of 1cm. In the centre there a 0.3125 mm diameter hole serving for the gas exit, while on it are fitted a filter for tar collection and traps (T_1, T_2) for liquids collection. On the flange, a second similar hole serves for the fitting of a manometer, in order to measure the pressure in the reactor.

About 200 mg of the raw material is spread in a layer on a screen of stainless steel, which is inserted between the electrodes. Helium is passed through the reactor at the rate of 30 cm^3/min. The sample temperature is raised at the peak temperature.

The reaction effluent includes fine charcoal, gas and volatile compounds. Their quenching occurs by natural cooling. The charcoal remains on the screen and is determined gravimetrically.

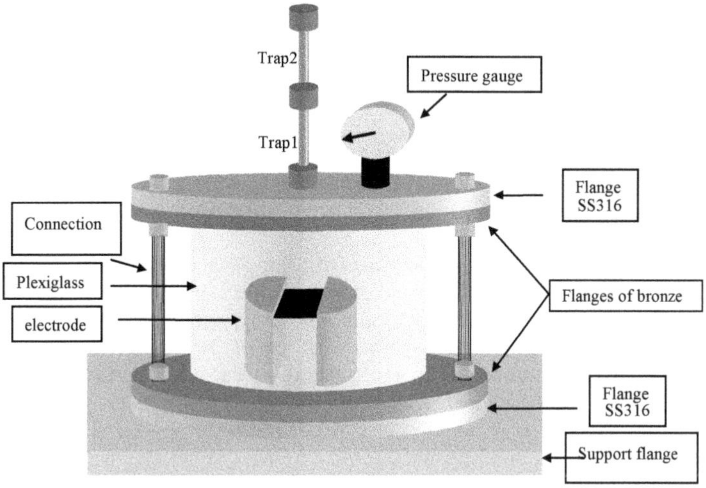

Figure 1: Schematic diagram of pyrolysis reactor.

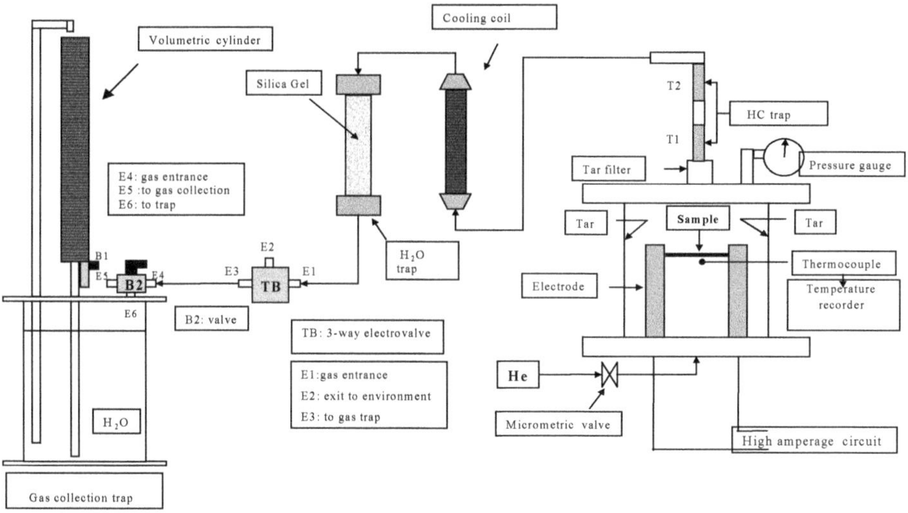

Figure 2: The experimental set up.

Tar is defined as the material condensed within the reactor vessel, on the wall, flanges and on a paper filter at the exit of the reactor, at room temperature. Tar condensed inside the reactor is removed by washing with CH_2Cl_2 soaked filter paper and is measured gravimetrically. Hydrocarbons (in the vapour phase at room temperature) are collected in two lipophilic traps placed at the exit of the reactor and containing 80/100 mesh Porapaq Q chromatographic packing and then measured gravimetrically.

Gaseous products pass through two traps (T_1 and T_2), where the liquid HC are collected. Then, they pass through a water-cooling coil to be cooled to ambient temperature and then through a moisture collection trap, which is a cylindrical plexiglass tube containing silica gel.

Finally they reach the gas collection trap through the electro valve TB. The volume of the removed water determines the gas volume. The gaseous products are selected in plastic sacs and they follow gas chromatography analysis. Gaseous products were analysed in a Perkin-Elmer model Sigma 300 gas chromatograph, with thermal conductivity detector connected to a PC applied for the integration of the exit signal and using a cart type HP 35900D A/D and a softword HP 3365 Series II ChemStations.

RESULTS

OLIVE WOOD

Data on the effect of temperature on the yields of char, gases and tar from pyrolysis of dry olive cuttings are presented in Figure 3. The yield of charcoal decreases, increasing the temperature. The charcoal yield reaches the value of 25 wt.%, at 550°C. Tar yield goes through a maximum of 35 wt.%, at about 550°C. Above this temperature tar yields decrease due to secondary reactions, producing more gaseous products. Tar and gaseous products are initially evolved at the same temperature (270°C).

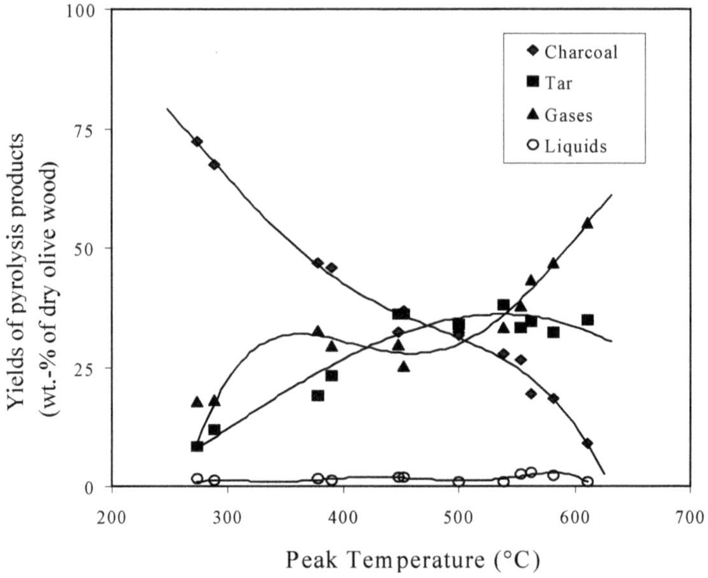

Figure 3: Effect of temperature on the yields of char, gases and tar from pyrolysis of dry olive cuttings.

OLIVE KERNEL

Data of the temperature effects on the yields of char, gases and tar from pyrolysis of *dry olive kernels*, are presented in Figure 4. The yield of charcoal decreases to a value of 33 wt.% of dry sample, at 500°C. Above 500°C, this yield tends to be stabilized to a value of 30 (wt.%). Tar yield goes through a maximum of ~30 wt.% at about 450-500°C.

COTTON GIN

Figure 5 depicts the product yield (char, liquid and gas) in relation to temperature. Pyrolysis of cotton gin waste apparently starts before reaching the temperature of 350°C. The yield of char decreased with temperature approaching the asymptotic value of 35-40 wt.%, at nearly 700°C (Fig. 5).

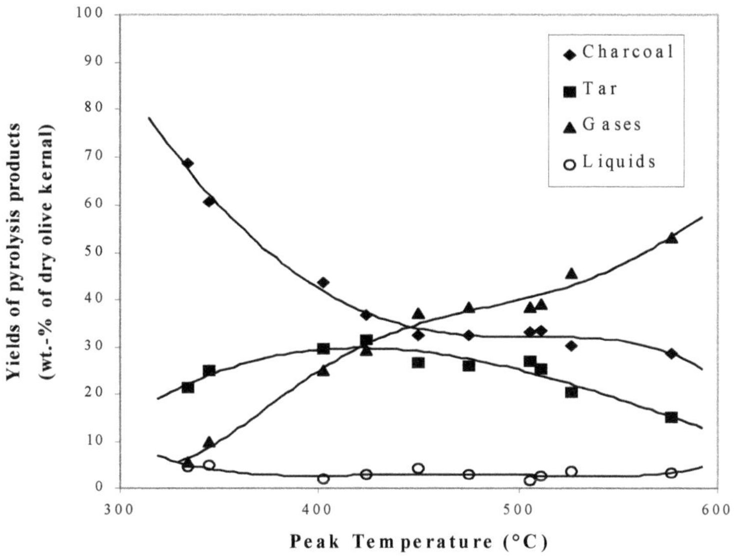

Figure 4: Effect of temperature on the yields of char, gases and tar from pyrolysis of dry olive kernel.

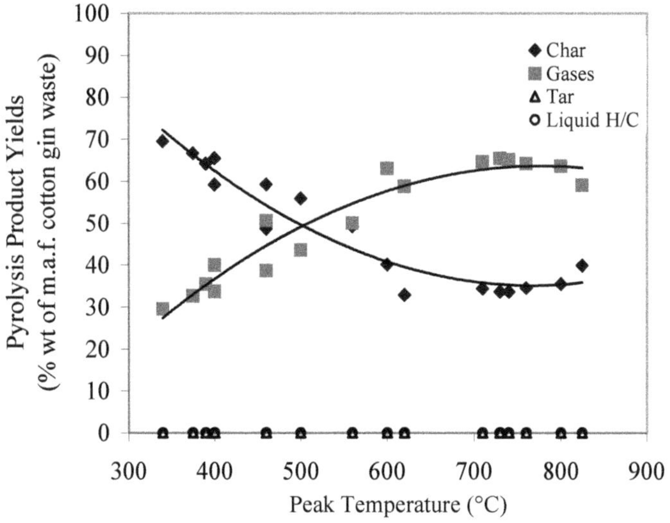

Figure 5: Effect of temperature on the yields of char, gases and tar from pyrolysis of dry cotton gin.

USED TYRES

Figure 6 shows the products yields (char, tar, liquid and gaseous) of used tyres in relation to pyrolysis temperature. As illustrated, char yield decreased with temperature, reaching a value of 20 wt.% of the raw material, at about 830 °C. In contrary, gas yield increased with temperature reaching a value of 73 wt.% of raw material, at the same temperature (830 °C). Concerning liquid and tar yields, these were in small amounts and showed a slightly increasing trend with temperature.

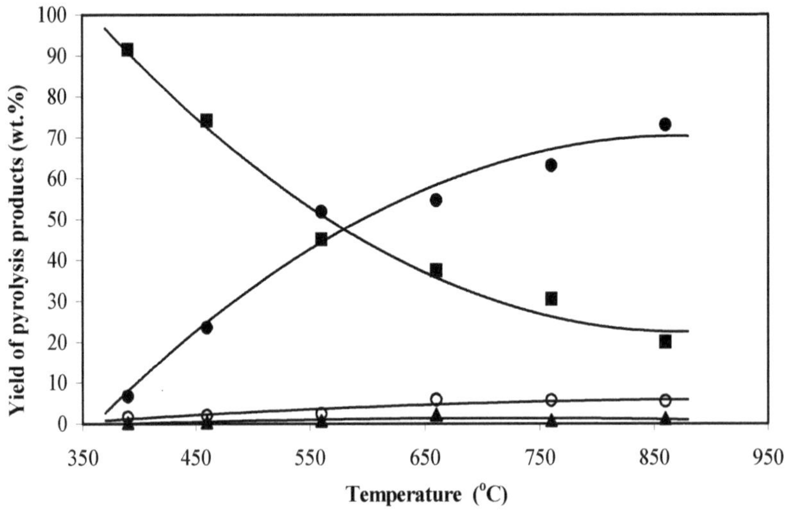

Figure 6: Effect of temperature on the yields of char, gases and tar from pyrolysis of used tyres.

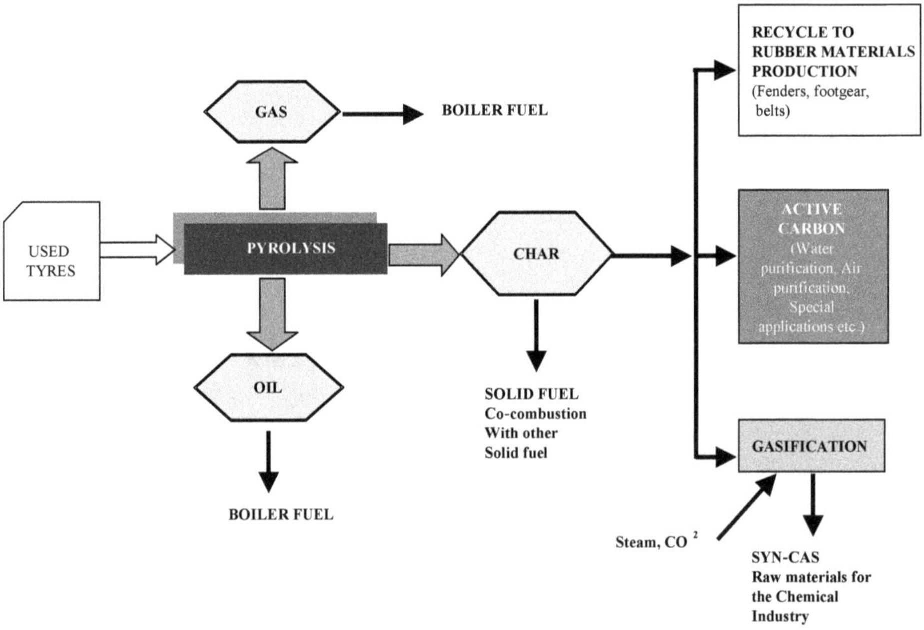

Figure 7: Products from tyres pyrolysis.

CONCLUSIONS

- With pyrolysis of solid wastes various products might be obtained. The solid char may be used either as smokeless fuel, carbon black or for preparation of activated carbon. It could be also burned both in pulverized fuel or fluidized bed combustors, alone or in co-combustion with other fuels. The derived oils may be used directly as fuel or added to petroleum feedstock. They may also be an important source of refined chemicals. The gases could be also useful as fuel.
- Pyrolysis could be a good utilization method for agricultural waste of negligible cost
- Pyrolysis is simple and suitable for plants located near a raw material source.

REFERENCES

1. Bosio A., Cioli R., Desideri U.. Marchi L. (1992). Alternative uses of olive husks: A technical-economic evaluation. In: *Biomass for Energy, Industry and Environment, 6th E.C. Conference.(Ed. by* G. Grassi, A. Collina, H. Zibetta), pp. 1116-1120.Elsevier Applied Science, London.
2. Bridgwater A.V . (1977).Fast pyrolysis of Biomass in Europe. In: in *Biomass Gasification and Pyrolysis – State of the Art and Future Prospects. (Ed. By* M. Kaltschmitt and A.V. Bridgwater), pp53-67 , cpl Press, Newbury, UK.
3. Buttiglieri M., Nicoletti G. (1995). Cheap fuel from olive husks, production and distribution for heating in residential buildings. In *Biomass for Energy, environment agriculture and industry, 8th Biomass Conference*.(Ed. by Ph. Chartier, A.A.C.M. Beenackers, G. Grass),pp. 740-747 Pergamon, Elsevier Science Ltd, Oxford, UK.
4. Buttiglieri M. (1966). The energy use of the olive-husks in southern Italy - A feasibility study of low-cost production and distribution. In: *Biomass for Energy and the Environment, 9th European Bioenergy Conference. (Ed. by* P. Chartier, G.L. Ferrero, U.M. Henius, S. Hultberg, J. Sachau, M. Wiinblad), pp 1894-1899. Pergamon, Elsevier Science Limited, Oxford.
5. Della Rocca P.A., Horowitz G.I., Bonelli P., Cassanello M.C. and Cukierman A.L (1997). Olive stones pyrolysis: chemicaal, textural and kinetics characterization. In: *Developments in thermochemical biomass conversion,(Ed. by* A.V. Bridgwater and D.G.B Boocock), pp 176-189. Blackie Academic & Professional, London.
6. Figueiredo J.L. (1990). Pyrolysis of Holm-oak and olive wood. In: *Biomass for Energy and Industry, 5th European Community Conference. (Ed. by,* G. Grassi, G. Gosse, G. dos Santos), pp 2.585-2.589. Elsevier Applied Science, London.
7. Kaltschmitt M, Dinkelbach L. (1977). Biomass for energy in Europe – Status and Prospects.In: *Biomass Gasification and Pyrolysis – State of the Art and Future Prospects,(Ed. By* M. Kaltschmitt and A.V. Bridgwater), pp7-23 , cpl Press, Newbury, UK.
8. Sharma V.K., Fortuna F., Mincarini M., Berillo M. and Cornacchia. Applied Energy (2000) 65: p. 381.
9. Zanzi R., Sjostrom K. andBjornbom E. (1996). Rapid pyrolysis of olive waste at high temperature. In: *Biomass for Energy and the Environment, 9th European Bioenergy Conference. (Ed. by* P. Chartier, G.L. Ferrero, U.M. Henius, S. Hultberg, J. Sachau, M. Wiinblad), pp.1584-1588. Pergamon, Elsevier Science Limited, Oxford.

Lump Charcoal Production in Latvia: State of the Art and Prospects

A. Zhurinsh and J. Zandersons
Latvian State Institute of Wood Chemistry, 27 Dzerbenes Str., LV 1006, Riga, Latvia

ABSTRACT: The annual production of the Latvian charcoal industry is about 10000 t, mainly exported to Scandinavian countries and West Europe. Barbecue lump charcoal and briquettes are produced from deciduous tree wood, and their market has a seasonal character. The recent growing demands for sustainable forestry and FSC certificates, as well as the tropical forests destruction control, hamper the export of barbecue charcoal from Asia and South America. Undoubtedly, charcoal production in direct heating kilns is much more economically attractive. However, the European ecological demands are the reason for the absence or restricted operation of such kilns.

Latvia's equipment with the annual capacity from several tens of t to even 2000 t includes indirectly heated stationary or interchangeable retorts, whose number and dimensions depend on the production scale. A common furnace provides the environmentally friendly combustion of the pyrolysis gas-vapour mixture. The content of fixed carbon in the obtained charcoal exceeds 80% (on the average, 82-85%), while the ash and moisture contents are 1.5-3% and 3-5%, respectively. The problem of charcoal production for industrial applications is the high price of barbecue lump charcoal, to be 205-220 EURO/t in pile or 250-280 EURO/t of fixed carbon. The high price is governed by the wood price, which is 8.5 EURO, on the average, for 1 m^3 of freshly cut low-value deciduous tree wood, and up to 15 EURO for birch wood and plywood industry wastes, while the average consumption of wood for production of 1 t of charcoal is 8-10 m^3 and 6.5-7.5 m^3, respectively.

INTRODUCTION

In Latvia, the charcoal production industry, after a 50-year interruption, has developed only in the recent decade, and the current annual output of charcoal is about 10000 t. The prerequisite for the development of this field has been the demand for barbecue lump charcoal in West Europe, where it is a popular prestigious fuel widely used in grilling trips, household and restaurants as well as fireplaces, an is produced only from deciduous trees. More than 90% from the charcoal output is exported mainly to Germany, France and Scandinavian countries. In this case, the attraction of Latvia is its proximity to potential clients, since the product can be supplied by motor transport, flexibly and promptly reacting to variations in the demand and nomenclature. The consumption of barbecue charcoal and its briquettes is commonly assumed to be about 1000 t per 1 million of inhabitants. In Sweden alone, where the annual volume of output of domestic charcoal is above 2000 t, its consumption in the three summer months exceeds 7500 t [1], while 60000 t of charcoal is imported to England each year [2]. In its turn, the total market of charcoal in Germany, applicable in household, exceeds 80000 t a year. Recently, the increasing demands of the customers for the green certificate, which testifies the sustainable forestry, as well as the combat against the destruction of tropical forests, hamper the export of charcoal from South Asia, South Africa and South America. Therefore, the inscription "Charcoal is not made of tropical timber", and the logos indicating the certified production with the assessed range of raw materials appear on the charcoal bags. This promotes the development of lump charcoal production in Europe, although, at the same time, despite the stable market volume, the developed countries cannot supply themselves with home-produced charcoal. Also France, with several "Lambiotte"-type continuous installations, cannot ensure the mean consumption. The existing plants, where small ecologically unreliable direct heating kilns are mainly used, cannot ensure the local consumption, which results in a high demand for export.

Charcoal consumption grows not only for household, but also for industrial applications (e.g. in Norway [3], 60000 t of charcoal is consumed for silicon production, and 300000 t would be required). Hence, with the decrease in export of charcoal and banning of small direct heating kilns in countries such as France and Poland, the deficit of lump charcoal can increase.

Figure 1: Metallic kilns with partial interior combustion (Slovakia, 1998).

MAIN PROBLEMS OF LUMP CHARCOAL PRODUCTION

RAW MATERIALS

The basis of the raw materials for production of lump charcoal, which is based on deciduous tree wood and can vary, is wide and tends to decline as a result of the application of modern sawing technologies (for example, the minimum saw-out diameter for wood materials has decreased now already to 8 cm, also the plywood peeling cores diameter is decreased to 6 cm). It applies especially to restaurant lump charcoal, whose diameter can exceed even 20 cm. With the decrease in the dimensions of the raw materials, the dimensions of the obtained charcoal also tend to diminish, since shrinkage in the carbonisation process exceeds 50%. The full utilisation of oak and beech wood, as well as the use of birch wood for pulp production (minimum butt-end diameter 6 cm) makes to take up also logging debris and low-value deciduous wood (black and grey alder, aspen, etc.) for carbonisation. The utilisation of aspen is limited, and its admixture in barbecue lump charcoal does not commonly exceed 10%. Softwood charcoal admixture in briquettes cannot exceed 10%, too. The promotion of utilising wood as a local raw material basis for energy production makes a mark on the accessibility of resources, which, in the far future, especially as the fossil fuel accessibility falls and its prices rise, is not promising. Hence, one should reckon with the growing fine charcoal raw material basis (including that from different types of wood waste, also demolition wood), which will intensify the production of barbecue charcoal briquettes that have come into increased demand. This applies especially to pillow-shaped briquettes, since their production and especially drying are complicated in comparison with the case of egg-shaped or cylindrical charcoal briquettes. In Latvian forests which are located more than 50 km from the ports (the export of fuel wood reached 749300 t in 2001), a great amount of unused wood, including low-value one, currently remains, which is formed by deciduous trees such as alder, bird-cherry tree, willow, rowan-tree, logging debris and decayed wood. Such wood is utilised mainly as a fuel, for which the demand is currently limited, and the firewood, because of its low price, is often left in the felling areas and burned there by the woodsmen. Similar problems are faced also for edgings, mainly slabs and sawdust, although waste wood is the cheapest raw material for charcoal production in comparison with stem wood.

PRODUCT AND ITS QUALITY

An essential indicator of charcoal quality is the content of fixed carbon. As the content of fixed carbon in charcoal increases, the charcoal yield from the bulk wood as well as its piling volume decrease (as the fixed carbon content grows by 1%, the yield decreases even to 2%) correspondingly. At the same time, the final temperature of the process should be also increased, which increases the process duration, since the speed-up of heating, which leads to the formation of voids and cracks in lump charcoal, decreasing its mechanical strength and dimensions, and increases the yield of charcoal fines, is undesirable. Hence, when producing lump charcoal, the dramatic increase in temperature is not advisable, especially in the case of freshly cut wood, since the obtained charcoal sometimes can be pressed into powder by hand.

It is clear that the high content of fixed carbon, i.e. above 80%, can be ensured only by "Lambiotte" installations (even above 90%) and indirect heating retorts. It is readily demonstrated by Brazil's experience of employing Missouri kilns [4], where the content of fixed carbon in charcoal utilised for production of crystalline silicon had grown from 65% in 1979 to only 72% after 1991. Data on mound kilns [5] also show the content of fixed carbon 75%. However, in Poland, in metallic furnaces using half-dry firewood, it was possible to obtain charcoal in compliance with the DIN requirements. The experience of Latvia, Belarus and Russia also shows that this is possible in the summer or in the case of dry wood, but problematic in other seasons. However, in this case, it is difficult to observe the technological conditions precisely; the charcoal is often moist and contains admixtures of sand. This is possibly the reason why the European barbecue charcoal standards, which envisage the increase in the fixed carbon content to 82%, are not put into force.

The carbonisation of the wood of different tree species with different density leads to different piling volumes, which is important when packing up the produce. For example [6], the apparent density of charcoal obtained at 530°C was 345 g/cm^3 for birch wood, and 290 and 245 g/cm^3 for grey alder and aspen, respectively, which clearly demonstrates that the difference reaches 20-30%. Barbecue charcoal is commonly packed up in craft paper bags with the weight from 1.6 even to 20 kg, which are stacked on pallets. The volume of a standard 2.5 kg barbecue charcoal bag is 15 litres, and it is intended for valuable deciduous wood (e.g. birch) charcoal. Therefore, problems arise when packing, for example, alder charcoal. Owing to the low piling volume of charcoal, standard trailers, (depending on the packing) are able to transport only 9.9-13 t of packed-up barbecue charcoal, which restricts the application of motor transport to great distances, since the transportation costs make up even 30-40% from the price. Hence, it is more profitable to use raw materials from hard deciduous wood (oak, birch).

In fact, all European charcoal producers and sellers are aware of and act in compliance with the German Standard DIN 51 749, which applies to fixed carbon, ash and moisture content. This does not mean that one cannot act according to other regulations. For example, in the Belgian Standard, fixed carbon is calculated based on non-dried charcoal, and not dried one. At the same time, the charcoal size composition can vary in a wide range depending on the customer's requirements.

The consumption and sales of barbecue charcoal have a seasonal character. Apart from this, the climatic conditions have a strong effect (in the sunny summer, barbecue charcoal consumption exceeds dramatically the mean one, while in rainy summers the warehouse remainder reaches even 30%), which affects the production character. The demand is maximal from April to July, when 80-90% from the annual charcoal output can be sold out. Of great importance in this field are warehousing costs, since, on the average, not more than 0.3 t of packed-up charcoal (up to 0.9 t in many-storeyed modern warehouses) can be stored per 1 m^2 of the warehouse area, and a wide area is required for the conditioning freshly obtaining charcoal. Therefore, finished products are also stored in the open air, preliminary packing in stands in a thermofilm.

ENVIRONMENT PROTECTION PROBLEMS

Undoubtedly, the production of barbecue charcoal in simple direct heating kilns (for example, Missouri-type) or mounds and pits yields considerably cheaper products (Fig. 2). However in

Europe, as a result of ecological requirements, such kilns are not available at all in some countries or their operation is limited and is beyond question. It is well known that the products of incomplete combustion from such installations are harmful for the environment, increasing the greenhouse effect, and also are not indifferent against the soil and humans. This activity is justified by the fact that this is a traditional production way, which ensures the jobs for people in rural regions. It is known that the temperature 750-800°C and the time minimum 2 sec are required for full combustion of pyrolysis products [7], which is difficult to reach in direct heating kilns. Different measures for decreasing the emissions, such as "afterburning" in Missouri-type furnaces (which are applied in U.S.A.), make the production more expensive and does not fully solve the emissions problem – it applies especially to particulates. Catalytic "afterburning" is low-efficient and expensive. In "Lambiotte"-type retorts, as the non-condensable gases are being cooled with water, wastewater is formed. In any case, in the combustion process, owing to the reaction equilibrium, it is practically impossible to get rid of nitrogen oxide emissions, and, when burning the tar products, it is difficult to get rid of even minor emissions of carbon monoxide and particulates (the latter are formed also when reloading and packing charcoal). Hence, in fact, environmentally friendly charcoal production installations with the emission level that does not exceed the corresponding legislation would be dealt with. At the same time, the level of emissions could approximate the one permissible for a boiler with the corresponding capacity, where fuel wood is used.

At the same time, it should be mentioned that, for example, in Germany, Sweden and Finland, direct heating kilns (built in bricks or mound) have become a tourism object, where, for example, charcoal production and tar distillation begin with a solemn ceremony and proceed for several days. Besides, tar and products with its admixture are sold there as souvenirs.

Figure 2: Direct heating kiln (Belarus, 2000).

CHARCOAL PRODUCTION EQUIPMENT APPLIED IN LATVIA

To satisfy the potential charcoal producers, technologies and equipment (Flow diagram see Fig. 3) for obtaining lump charcoal [8] have been developed at the Latvian State Institute of Wood Chemistry (LS IWC), whose development is based on the following requirements:

(1) The content of fixed carbon in charcoal should at least exceed 80% (i.e. to conform to the requirements of the German Standard DIN 51749), but, at best, should be capable of exceeding the requirements set up by European Standards, namely, 82%.

(2) The emissions from the equipment should conform to the requirements of the Latvian environment protection legislation, which are similar to the EU requirements.

(3) The technology should be cheap, easy-to-maintain, with easily changeable capacity (from hundred to a couple of tons a year, as well as a chance of extending or cutting the production promptly).

(4) The heat energy necessary for the carbonisation process should be ensured by the equipment itself or by additional burning of wood, excluding the use of fossil fuel.
(5) Equipment must be capable of operating also with moist wood.

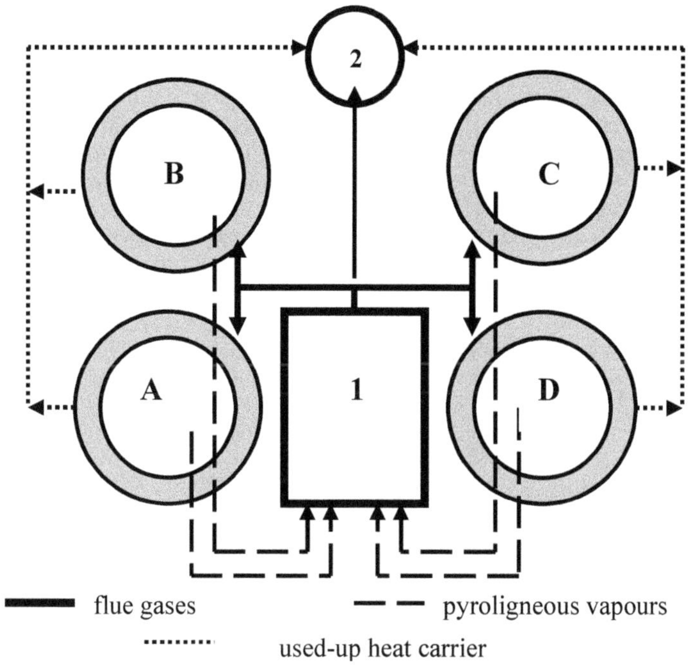

Figure 3: Flow diagram of the equipment of LS IWC: 1 - furnace; 2 - stack; A, B, C, D – retorts.

Based on the above-mentioned, the following equipment falls off:
(1) The metallic furnace for partial interior combustion, whose attraction is the cheapness and simplicity, but which cannot ensure the conformity to ecological requirements (Poland, Slovak Republic, Russia, etc.) and hardly ensures the necessary content of fixed carbon.
(2) Continuous "Lambiotte"-type retorts which, owing to large capital investments and also for economic reasons, are not accessible to all producers and cannot ensure a production with the relative moisture content of wood 20-25 %.

Therefore, the version with batch exterior heating metallic retorts has been chosen, since the release of the pyrolysis gas-vapour mixture in such type of apparatus is uneven, as 70% from the total amount of heat is obtained when burning the gas-vapour mixture that is released at temperatures of 280-420°C. This makes up only 30-40% from the total time of the apparatus' turnover and exceeds its heat energy amount that is necessary for maintaining the process.

In its turn, the energy produced in the glowing period is insufficient. Therefore, separate retorts were combined in blocks, which operate in different periods, thereby levelling-off the total gas-vapour mixture composition that is burned in the joint furnace, and the total heat flow working time does not vary by more than 10%. In the case of wood with the moisture content up to 30%, heat energy is generated when burning up in the joint retort the pyrolysis gas-vapour mixture from the retorts. In the case of wood with a higher moisture content, burning of firewood (commonly softwood slabs) in the same furnace is required to cover the amount of the heat necessary for wood drying. At the same time, by ensuring constant burning conditions, it is possible to achieve an environmentally friendly operation of the equipment. The temperature in the heating system in front of the retort is 700-900°C. However, if several retorts reach the maximal release of the pyrolysis gas-vapour mixture, then the temperature can reach 1200°C. Therefore, of great importance is the personnel's experience, which does not allow the situations leading to the burning-out of retorts' walls. In the case of wood with the moisture content < 20%, a problem arises that the amount of the burning product heat exceeds the necessary one, and, to diminish the temperature in the heating system, it is necessary to burn-up the excess gas-vapour mixture in flare-stack or use the heat energy for other purposes (warming-up of water, drying of charcoal briquettes, etc.). At present, two types of installations are distributed in Latvia:

(1) Stationary vertical retorts (Fig. 4) with the capacity 5-10 m^3 and diameter 1.60-2.0 m, in which wood sawn-up in chunks (length 80- 150 mm, diameter not more than 150 mm, the thickest logs are chopped) is used.

(2) Interchangeable retorts (Fig. 5) with the capacity 2-8 m^3 and diameter 0.6-1.8 m, in which logs are charged depending on the retort length, which is 2.0 –2.75m.

Figure 4: Installation with four vertical stationary retorts (capacity up to 800 t /year).

Figure 5: Loaded interchangeable retorts.

Charcoal production installations in Latvia comprise as minimum 4 up to 16 exterior heating batch retorts, a common furnace and a stack, whose height ensures the self-draft in the heating system. The retort furnace and heating chambers are lined with fire-resistant bricks. The retorts lean on the brickwork. The vertical stationary retorts with the brickwork rise above the ground and lean on steal load-carrying structures to drive the charcoal cooler, which serves for cooling of charcoal to ambient temperature, below the discharge hatch. The cooling of the charcoal in the interchangeable retorts proceeds exactly therein, after their removal from the installation. The pyrolysis gas-vapour mixture from the retorts is inlet directly to the furnace through the gas pipeline. An access door in the furnace is designed for charging the firewood. The heating intensity in the retorts is controlled with gates that are located in front of and behind the flue gas inlet and outlet, following the readings of the thermocouple connected with a display or a self-recorder.

The technology is environmentally friendly, since the amount and composition of the emissions are the same as in a boiler with the corresponding capacity, which operates on wood. The amount of harmful emissions per 1 t of the produced charcoal comprises, on the average, 7 kg of particulates, 1-6.5 kg of carbon monoxide (which can grow to 40 kg in the autumn-winter period in the case of freshly cut frozen wood, when the air temperature is below –10°C), while the amount of nitrogen oxide does not exceed 2 kg.

In the case of moist birch wood (w_{rel} = 40-50%), the mean turnover time for a retort is 16-28 h depending on the season. On the average, a plant with eight interchangeable retorts (diameter 1.8 m, volume 8 m^3) yields 5 t of charcoal per day at positive air temperatures.

Both types of installations, in which export products are successfully produced, have their own attractions and drawbacks, although the constant improvement of the design and materials is under way as the experience is gathered.

Interchangeable retort's attractions:
(1) Especially suitable for processing of plywood cores, since, owing to the ideal form, a high stacking factor, namely, up to 0.76 is reached, as well as for sawmill wastes (slabs).
(2) Sawing is practically unnecessary.
(3) Retort heating is realised well, especially in the lower part.

Interchangeable retort's drawbacks:
(1) Processing of low-value deciduous wood becomes more difficult, since logs are not ideally straight, and this diminishes the stacking factor; therefore, they should be sawn up to logs 1 m in length.
(2) Three retort sets are necessary, which increases metal consumption.
(3) When reloading, retort's experience dramatic temperature fluctuations (in winter periods, even from +500°C to -20°C).
(4) Replenishment of retorts should be done manually, since it is difficult to mechanise the process.
(5) It is difficult to perform drying, as logs have great longitudinal dimensions in the case of retorts with a great capacity.
(6) A powerful hoist is necessary (more than 5 t).

Vertical stationary retort's attractions:
(1) In the case of chunks, wood of different form and length can be processed.
(2) Easy mechanisation the felling, loading (belt conveyer can be also used) and drying.

Vertical stationary retort's drawbacks:
(1) In the case of chunks, about 5% of sawdust is formed, and electric energy is consumed.
(2) Hardly suitable for processing of slabs.
(3) It is difficult to warm up the retort's lower part; as a result, in the case of moist wood, half-burnt charcoal remains, and condensation of pyrolysis vapours in the lower part of the retorts at the discharge hatch occurs.
(4) A high difference in temperatures throughout the retort height.

The largest plant in Latvia, operating according to the technology developed at LS IWC, produces annually 2000 t of lump charcoal with the fixed carbon content 85%, and 800 t of charcoal briquettes. At the same time, there are also small plants with the maximum annual output up to 100 t.

It should be noted that horizontal or vertical VMR retorts (the Netherlands), which successfully operate with furniture industry waste at the maximal moisture content 25% at the production capacity that is the same as for the equipment developed by LS IWC, require twice as much capital investments. Besides, in the case of moist wood in the coldest seasons, it is necessary to use additionally the diesel fuel.

For charcoal obtained in Latvia, the content of fixed carbon exceeds 80% (82-85% on the average), ash content is 1.5-3% and moisture content 3-4%, which would be a prerequisite for using the obtained charcoal for production of active carbon or crystalline silicon. However, a crucial problem is the high price of barbecue charcoal in Latvia, which is 205-220 EURO/t in terms of lump charcoal in bulk or 250-270 EURO/t in terms of fixed carbon. At the same time, in Norway, charcoal for production of crystalline silicon is purchased at 311 USD/t in terms of fixed carbon [3]. A similar situation is for utilising charcoal as a fuel, as the expenses for 1 GJ of energy are considerably higher than for fossil fuel. The high charcoal price is determined by the wood price, which, on the average, is 8.5 EURO for freshly cut low-value deciduous wood (average consumption 9-10 stack m^3 per 1 t of charcoal). In the case of birch wood and plywood industry wastes, whose average consumption is 6.5-7 m^3 per 1 t of charcoal, it reaches 15 EURO per stack m^3. Apart from this, on the average, 1 m^3 of softwood sawmill waste should be added, which is used for additional heating per 1 t of charcoal, as the moisture content of the processed wood exceeds 30%. Charcoal fines, which are a waste of the lump charcoal production, are currently realised in Latvia at 90 EURO/t for charcoal briquette production, using corn starch as a binder. The prices for charcoal briquettes are similar to those for barbecue lump charcoal, to be 350 EURO/t on the average.

Owing to the above-mentioned economic conditions, charcoal production for industrial aims is not so profitable in comparison with barbecue charcoal production. The release of separate chemical products from liquid products of slow pyrolysis is not envisaged in Latvia, since ensuring of production with heat energy is primary in this field. Charcoal production for industrial purposes is possible only in the case of the further processing of charcoal here on the spot, in Latvia. The major effect is exerted by the wood price, which is directly influenced by the local consumption and export of the fuel wood.

REFERENCES

1. Dyremark A. & Westerholm R. (1995) Polycyclic aromatic hydrocarbon (PAH) emissions from charcoal grilling. *Atmospheric Environment*, **29**, 1553-1558.
2. http://www.bioregional.com.
3. Grønli M. & Eikeland I.J. (2002) The use of charcoal in Norwegian Ferrosilicon Production. *PyNe* **13**, 6-7.
4. Morales R.C., Mathieux F., Godoy P.H. & Suzuki C.K. (1998) Environmental implications of charcoal production in Brazil. In: *Abstr. and Proc. of the First Workshop on Quartz Industrial Trade System: Materials Life-Cycle and Sustainable Development,* pp. 23-30. Campinas, Tokyo.
5. Schenkel Y., Bertaux P., Vanwijnbserghe S. & Carre J. (1998) An evolution of the mound kiln carbonization technique. *Biomass and Bioenergy*, **14**, 505-516.
6. Zhurinsh A. (1997) A feasibility to utilize wood of low-value deciduous species in charcoal production. *Baltic Forestry*, **2**, 53-57.
7. Mezerette C. & Girard P. (1991) Environmental aspects of gaseous emissions from wood carbonization and pyrolysis processes. In: *Biomass PyrolysisLliquids Upgreading and Uses*, (Ed. by Bridgwater A.V., Grassi G.), pp. 263-287. Elsevier Applied Science, London and New York.
8. Zandersons J., Zhurinsh A. & Gravitis J. (1999) Carbonization of wood residues by utilising volatile thermolysis products combustion heat. *Cellulose Chem. Technol.*, **33**, 1-2, 147-160.

Overview of Small Scale Biomass Gasification

H.A.M. Knoef
BTG biomass technology group, P.O. Box 217, 7500 AE Enschede, The Netherlands

ABSTRACT: The production of a combustible gas from carbon containing materials is already an old technology. The first commercial updraught gasifier for continuous gasification of solid fuels with air was installed in 1839. In anticipation of unreliable petroleum supply, between 1920 and 1940 compact tar-free down draught gasifier systems for automotive application were developed in Europe and widely used during the 2" World War. The energy crisis of the 1970's brought a renewed interest in biomass gasification, both at small and large scale. The technology was perceived as a relatively cheap indigenous alternative for industrial and utility power generation in those countries that suffered from high world market petroleum prices and had sufficient sustainable biomass resources. In the beginning of the 1980's several biomass gasification systems were installed through donor-financed projects and local entrepreneurs in a large number of countries. Most of those demonstration projects have been abandoned due to technical, financial/economic and institutional problems. The renewed interest in biomass gasification at the beginning of the 1990's was due to the climate change discussion and the commitments of the Kyoto protocol. Gasification is seen as an important route for power production and meet the objectives laid down in (inter)national bodies. The technology is now close to commercialization and therefore it is required to inform the international community in detail on the status, including the current manufacturers and installations, the problems and solutions as well as non-technical issues during implementation. The paper will focus on the state-of-the-art, the achievements, the problems that have been solved and the problems that remain to be solved.

INTRODUCTION

The production of a combustible gas from carbon containing materials is already an old technology. Dry distillation was first practiced on a commercial scale in 1812, when a towngas company in London started production. The first commercial updraught gasifier for continuous gasification of solid fuels with air was installed in 1839. Updraught gasifiers were subsequently further developed for different fuels and were in widespread use in specific industrial power and heat applications up to the 1920's, when their function was gradually taken over by oil fuelled engines and furnaces.

In anticipation of unreliable petroleum supply, between 1920 and 1940 compact tar-free down draught gasifier systems for automotive application, were developed in Europe. During the 2" World War tens of thousands of those gasifiers were used in Europe and elsewhere. Shortly after the War most gasifiers were decommissioned because of widespread availability of inexpensive liquid fuels.

The energy crisis of the 1970's brought a renewed interest in biomass gasification. The technology was perceived as a relatively cheap indigenous alternative for small-scale industrial and utility power generation in those developing countries that suffered from high world market petroleum prices and had sufficient sustainable biomass resources. In the beginning of the 1980's at least 10 (mainly European) manufacturer's were offering small-scale wood and charcoal fired power plants (upto approximately 250 kW_{el}). At least four developing countries (Philippines, Indonesia, Brazil and India) started gasifier implementation programmes based on locally developed technologies. Several biomass gasification systems were installed through donor financed projects and local entrepreneurs in a large number of developing countries. Almost none of these projects have been successful. Most of the demonstration projects have been abandoned due to technical, financial/economic and institutional problems. Nevertheless, gasification is considered as one of the most promising routes for power production because of several potential benefits (higher efficiency, lower emissions and lower production costs).

TECHNOLOGY DESCRIPTION

Biomass gasification is an endothermic thermal conversion technology where a solid fuel is converted into a combustible gas. A limited supply of oxygen, air, steam or a combination serves as the oxidizing agent. The product gas consists of carbon monoxide, carbon dioxide, hydrogen, methane, trace amounts of higher hydrocarbons (ethene, ethane), water, nitrogen (with air as oxidant) and various contaminants, such as small char particles, ash, tars, higher hydrocarbons and oils. Figure 1 illustrates the gasification process (v. Swaay, 2002).

A biomass gasification system consists primarily of a reactor into which fuel is fed along with a limited (less than stoichiometric or that required for complete combustion) supply of air. Heat for gasification is generated through partial combustion of the feed material. The resulting chemical breakdown of the fuel and internal reactions result in a combustible gas usually called "producer gas". The heating value of this gas varies between 4.0 and 6.0 MJ/Nm^3. Figure 2 shows different reactor types with their main characteristics.

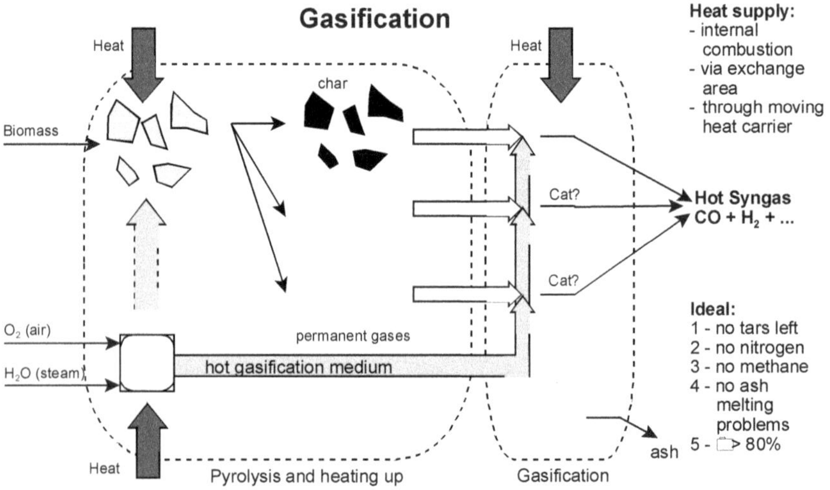

Figure 1: The principle of gasification.

Table 1: Main technical characteristics of small scale gasifiers.

Gasifier type	down draught	up draught	'open core'	cross draught
max. proven commercial capacity (kW_{el})	350	4,000	appr. 200	appr. 150
sensitivity to fuel characteristics	high	low	Very high	high
tar production full load (g/NM^3 gas)	< 0.5	1 - 15	10 - 15	< 0.1†
size & volume gas cleaning section	small	big	big	small
quantity residual tars	small	big	big	very small†
sensitivity to load fluctuations	high	low	low	high
turn down ratio	3 - 4	5 - 10	5 - 10	2 - 3
HG_{eff} full load (%)	85 - 90	90 - 95	70 - 80	80 - 90
CG_{eff} full load (%)	65 - 75	40 - 60	35 - 50	60 - 70
cold gas heating value full load (MJ/Nm^3)	4.5 - 5.0	5.0 - 6.0	5.5 - 6.0	4.0 - 4.5

* kW_{th} † low volatile content (< 10 %wgt) charcoal

E only rice husk n.a. not applicable

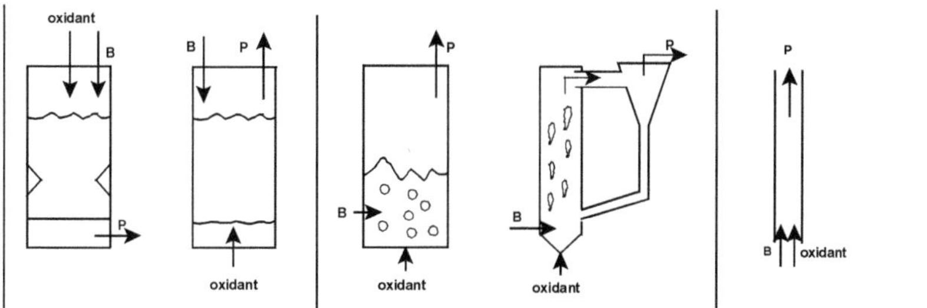

	Moving beds		Fluid beds		Entrained beds
	Co-current	Counter current	dense	circulating	
T°C	700-1200	700-900	< 900	< 900	[illegible]1500
tars	low	very high	intermediate	intermediate	absent
control	easy	very easy	intermediate	intermediate	very complex
scale	< 5 MW_t	< 20 M_t	10<MW_t<100	20<MW_t<?	> 100 MW_t
feedstock	very critical	critical	less critical	less critical	very fine particles

Figure 2: Different gasifier types.

Main characteristics of small scale gasifiers are listed in Table 1 (Quaak, 1998).

STATE OF THE ART

Since the beginning 90's there has been an increasing awareness that advanced gasification concepts can play an important role in stimulating bioenergy and reducing greenhouse gas emissions. The technology is now close to commercialization and therefore it is required to inform the international community in detail about the status of biomass gasification. Information on fixed bed gasifiers is available on several websites like www.gasifiers.org, www.woodgas.com or www.iitb.ac.in/~mech/. Only gasification for heat production has achieved commercial status, meaning that guarantees are supplied and the technology is competitive. An overview for different applications is given below.

GASIFICATION FOR HEAT PRODUCTION

Heat gasifiers are commercially available. The most well-known technologies are those of **Bioneer** (fixed-bed, updraft), **PRM Energy** (fixed-bed, updraft), Ahlstrom (now **Foster Wheeler**) and **Lurgi Umwelt** (both CFB). Less well-known are the small scale heat gasifiers installed in several developing countries supplying heat for lime kilns, tea drying (Jayah, 2001).

- About ten Bioneer gasifiers have been in operation successfully for a number of years in Finland and Sweden. In most cases the gas is used for combustion in boilers and district heating purposes. Bioneer is now market by Condens Oy, Finland.
- PRM Energy Systems has eighteen (18) units operating on four (soon to be five) continents gasifying 500,000 tons annually of biomass, mostly rice husk. Usually, the heat is used for industrial drying applications or low pressure saturated steam for an industrial process.
- The first commercial fluid bed gasifiers were installed by Ahlstrom, Finland in the mid 1980's. Those 'Pyroflow' circulating fluid bed gasifiers had a capacity of 15 – 35 MW_t. The product gas from these gasifiers is used for fueling lime-reburning kilns of pulp factories. Four are still in operation.

- The first Lurgi circulating fluidised bed (CFB) gasifier was built in 1987 in Pöls, Austria by a large paper mill; heat was used in the limekiln. Since 1996 a CFB gasifier plant of Lurgi Umwelt is in operation at the Rüderdorfer Zement GmbH, Germany.
- Hundreds of small-scale heat gasifiers are installed in several developing countries (India, China) supplying heat for cooking, lime kilns, tea drying, etc. Figure 3 shows a mobile charcoal gasifier, which was used successfully for water irrigation purposes.

Figure 3: A mobile charcoal gasifier.

FIXED BED GASIFIERS FOR POWER OR CHP PRODUCTION

Stimulated by favouring policy measures a great number of small-scale, fixed bed gasifiers are either in operation or under development. Below, a selection is made of installations which are in operation or which are nearly completing its construction. It must be noted that this list is not exhaustive and must be considered as an indication of the current situation (March 2002).

Small-scale gasification for CHP in distributed generation and village power applications in developing countries is a field that has expanded very rapidly. In India and China over 100 rice husk gasifiers of around 200 kWe are installed (Overend, 2000). Over the last decade, India has implemented about 500 units of 100-200 kWe in rural industries and villages. There is hardly any information on the technical performance available and its viability. Distributed generation is considered to have a higher value than grid-distributed electricity, especially when it is in a CHP configuration. This is a niche that is well suited to biomass.

ACHIEVEMENTS

Table 2 shows that several developments have resulted in successful pilot plants and working demonstrations. However, only a very few projects have achieved commercial status where the technology has a competitive advantage in the market. Compared to direct combustion, the gasification technology has several advantages such as higher conversion efficiency and reduced environmental impact because gas can be cleaned prior to combustion. Moreover, combustion of a gaseous fuel can be more accurately controlled than combustion of the solid biomass. In terms of capital costs advanced gasification have lower gas clean up equipment costs as a result of reduced volumes of gas to be handled.

Due to the RD&D efforts the following achievements can be reported:

- Over 100 small scale fixed bed gasifiers are installed in Europe and the USA
- Hundreds are installed in developing countries, mainly for rural electrification

Table 2: Selection of existing fixed bed gasifiers.

Location	System, supplier	Power MWe	Status
Harboore, Denmark	Babcock & Wilcox Volund, updraft, chips	1.5	> 1000 hours since 2000
Viking, Denmark	DTU 2-stage for wood chips	0.017	In operation since 2002
Regal, Belgium	Xylowatt sa, downdraft, fuelled with SRC chips	0.3	Operating since nov. 2002 at farm, > 3000 hrs/yr
Gazel, Belgium	Xylowatt sa, downdraft fuelled with SRC chips	0.15	Operating since 2000
Eckenförde, Germany	EVN, two-zone downdraft, wood chips	0.18	Operating since 2001
Civitas Nova, Austria	Downdraft, wood chips	0.75	Under construction
SK Industrietechnik, Austria	Downdraft, wood chips	0.3	Commissioning
Austria	Grübl, downdraft wood gasifiers	0.05	Several in operation at farms
Siebenlehn, Germany	Pipeline Systems, downdraft, combined cycle	2.3	Shut-down for unknown reasons
Brook Hall Estate Londonderry, N-Ireland	Rural Generation Ltd. Downdraft, SRC	0.1	15,000 operating hours since three years
Blackwater Valley Museum, N-Ireland	B9 Energy Downdraft, wood	0.2	> 1000 operating hours
Herning, Denmark	Hollenson, downdraft gasifier, wood chips	0.4	> 7000 hrs, now being tested after modifications
Legnano, Italy	Downdraft, CCT	1	Under construction
Spiez, Switzerland	Pyroforce, downdraft	0.2	Operational since 2002 > 1400 hours
Bulle, Switzerland	Xylowatt A.S, open-top gasifier	0.2	Operational since spring 2002, > 600 hours
Camp Lejeune, USA	Thermal Technologies, downdraft	0.7	Operational
Tervola, Finland	Entimos Oy, combined updraft/downdraft	0.45	Commissioning
Bladel, Netherlands	Chicken manure	0.06	Commissioning
Rossano, Italy	PRM Energy Systems, updraft	3.8	Nearing completion
Hengelo, Netherlands	HoSt, downdraft, wood chips	0.2	Operational since 2001
Petten, Netherlands	ECN, downdraft	0.1	Operational as test unit
Warrington, UK	Biomass Engineering	0.1	Operational as test unit

- Innovative concepts have been developed as an alternative to conventional moving-bed gasifiers like 2 and 3-stage process based on the ideas of De la Cotte and Kiener.
- Gasifiers have been developed for "difficult" fuels like poultry litter, straw and wood chips.
- New gas cleaning devices have been developed like electrostatic filters, tar crackers, rotating particle separators, scrubbers, solid bed filters, etc. including cleaning of condensate water.
- Detailed characterisation of fuels for biomass gasification, like the phyllis database at www.ecn.nl/phyllis

- Fundamental research has improved the understanding of mass and heat transfer mechanisms of the complex gasification technology
- Complete automatic control devices have been developed for continuous and automated operation.
- There is a consensus on the need for standard methods for monitoring the gasification process as indicated by the development of a standard tar measurement method.
- Scale-up and demonstration of many different fixed bed gasifiers up to 10 MW_t input.
- Exchange of knowledge and experience by the use of internet (like www.gasifiers.org, www.woodgas.com and www.gasnet.uk.net) and networks (like IEA Task 33 and GasNet).
- Commercial operation of several community fixed bed gasifiers with low emissions like the Babcock & Wilcox Volund updraft gasifier for wood chips in Denmark or the Brook Hall Estate gasifier in Londonderry, N-Ireland with > 15,000 operating hours.
- Gasifiers are in operation, which do not produce any condensate since the gas temperature is kept above the dewpoint.
- In several countries policy measures are in place to stimulate biomass gasification. The most important market introduction measure is the governmental support in the form of subsidies on investments and guaranteed feed-in tariffs for green electricity.

TECHNICAL AND NON TECHNICAL BARRIERS

TECHNICAL BARRIERS

Despite all R,D&D efforts several problems remain to be solved. The main technical problems or barriers with gasification systems are listed below.

Tar contamination

Because wood has a higher hydrogen/carbon ratio and oxygen/carbon ratio than fossil fuels, a higher yield of both gases and hydrocarbons such as tars is the result. The composition of tars is dependent on the fuel, pyrolysis conditions and secondary gas phase reactions. Tars are a problem if they polymerise or condense when the fuel gas is cooled down (condensation temperature of tar from woody biomass = 200 to 500°C): considerable problems with contamination of downstream gas cooling and gas cleaning equipment (e.g. filter clogging) can occur (Abatzoglou, 2000). Many research and development work has been initiated on tar removal (like scrubbers) and tars conversion (thermal and catalytic), see Figure 4 (Stassen, 2002).

Ash melting

The amount of ash between different type of feedstocks differs widely (0.1% for wood up to 15% for some agricultural products) and influences the design of the reactor, particularly the ash removal system. The chemical composition of the ash is also important because it affects the melting behaviour of the ash. Ash melting can cause slagging and channel formation in the reactor. Slags can ultimately block the entire reactor.

Too low gas heating value

Biomass gasification with air results in a gas heating value of about 4-5 MJ/Nm^3 with the following problems:

- Peak power output derating in IC engines
- Flame stability in gas turbines
- Gas pipeline transport becomes less economical
- Too diluted for syngas

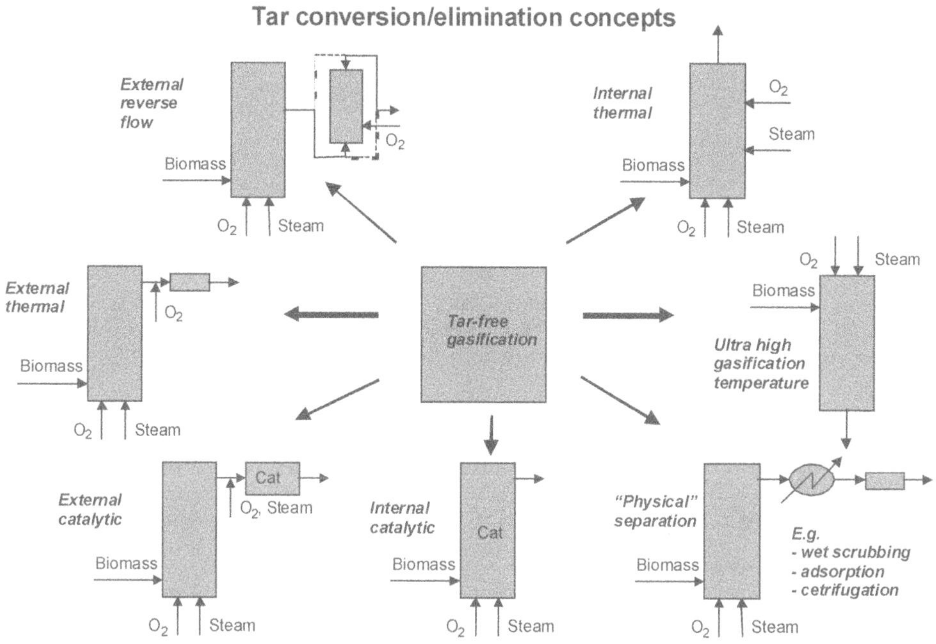

Figure 4: Tar removal/conversion options.

Feedstock

Variations in biomass morphology, composition and moisture content give rise to problems with reactor feed and operation. For most gasifier types a certain degree of standardisation is required for which pre-treatment steps like sieving, drying, densifying, chipping are available. Contaminations like sand, stones, plastic, paper and the like can cause operational and emission problems for which they should be removed. Feedstock availability, logistics, costs are other areas of concern. To justify the high initial investment costs of biomass gasifiers, long-term contracts of low-prized, quality controlled feedstock is needed (sustainable fuel supply). In practise this appears to be difficult; feeding of low-dense materials and mixed residues with varying characteristics is still problematic. There is a need for standardised fuel like sawdust, woodchips, rice husk, etc.

Gas cleaning

Different gas cleaning devices have been developed with varying success, like wet/dry cleaning, tar removal/cracking systems. There is a need for both technical and economical evaluation of different processes.

Prime mover

Regarding prime movers, gas engines are most frequenting used. R&D effort on fuel cell, Stirling engines and gas turbines is on-going. In general there is insufficient operational experience to determine the specifications of producer gas and therefore very strict demands are requested. On the other hand, very strict limits on allowable emissions require an adequate clean gas. There is hardly any R&D work on adaptions of prime movers to producer gas.

Reliability

In general the technology is still considered as being unreliable; in most cases guarantees are lacking or very limited. This is mainly due to the lack of operational experience, unfamiliarity with the technology and the tendence to closedown projects.

NON-TECHNICAL BARRIERS

Financial

Biomass gasifier installations are cost intensive. Most demonstration project receives a certain amount of subsidy, but in most cases this is not significant and therefore not really stimulating the technology. It appears to be difficult to attract private investment to share the risk of building first-of-a-kind gasification plant. Scale-up and demonstration of reliable systems, followed by commercial replications and optimisation seems the best approach to increase the production volume and reduce the capital costs.

The financial viability depends highly on the remuneration for the electricity and heat as well as the feedstock price. The high initial investment demands for favourable prices for both the electricity and heat. On the other hand there is a need for a sustainable supply of quality-controlled feedstock at a guaranteed price. If all conditions are favourable large-scale deployment of biomass gasifiers is justified.

Permitting procedures and emission limits

In general permitting procedures for biomass gasifiers are complicated and time consuming. This is mainly caused by a lack of knowledge on the technology by the permitting authorities. The implementation and the promotion of biomass gasification technology is hampered by the very strict emission levels applied in many projects. In most cases emission limits of combustion technology are applied without taking the difference of both technologies into consideration. In particularly the very strict CO emission limits applied are detrimental to gasificaton. In combustion processes the CO emission is an indication for the production and emission of hydrocarbons. This is not valid for gasification; CO is one of main constituents in the fuelgas and due to the well-known slip in engines a high CO emission in the exhaust can not be avoided. In this case the CO is no indication for the hydrocarbons. Furthermore, emission limits vary from country to country.

Safety, health and environmental aspects

During gasification a hot combustible gas is produced which has several impacts on health and environment. Figure 5 illustrates different HSE impacts of a biomass gasifier plant. Most critical is the toxicity of the gas and the risk of explosions. There is an increasing awareness about the importance of safety aspects of biomass gasification plants.

Many European directives exist on HSE aspects of thermal conversion of biomass. However, is not very clear which directives are valid to gasification. This aspect is therefore one of the items discussed within the European Gasification Network, GasNet.

CONDITIONS FOR COMMERCIALIZATION

Before efforts are made to commercialize small-scale power gasifiers, it is essential that the following conditions be fulfilled:

- The technology should be mature and proven in demonstration models. This includes long-duration testing of prototype installations. Further development is needed for the production of high-efficient biomass gasifiers. Installed plants should be reliable having a availability of 90% with low maintenance. Due to the limited operational experience, maintenance requirements are not well-known. Improvements should also come from learning by doing

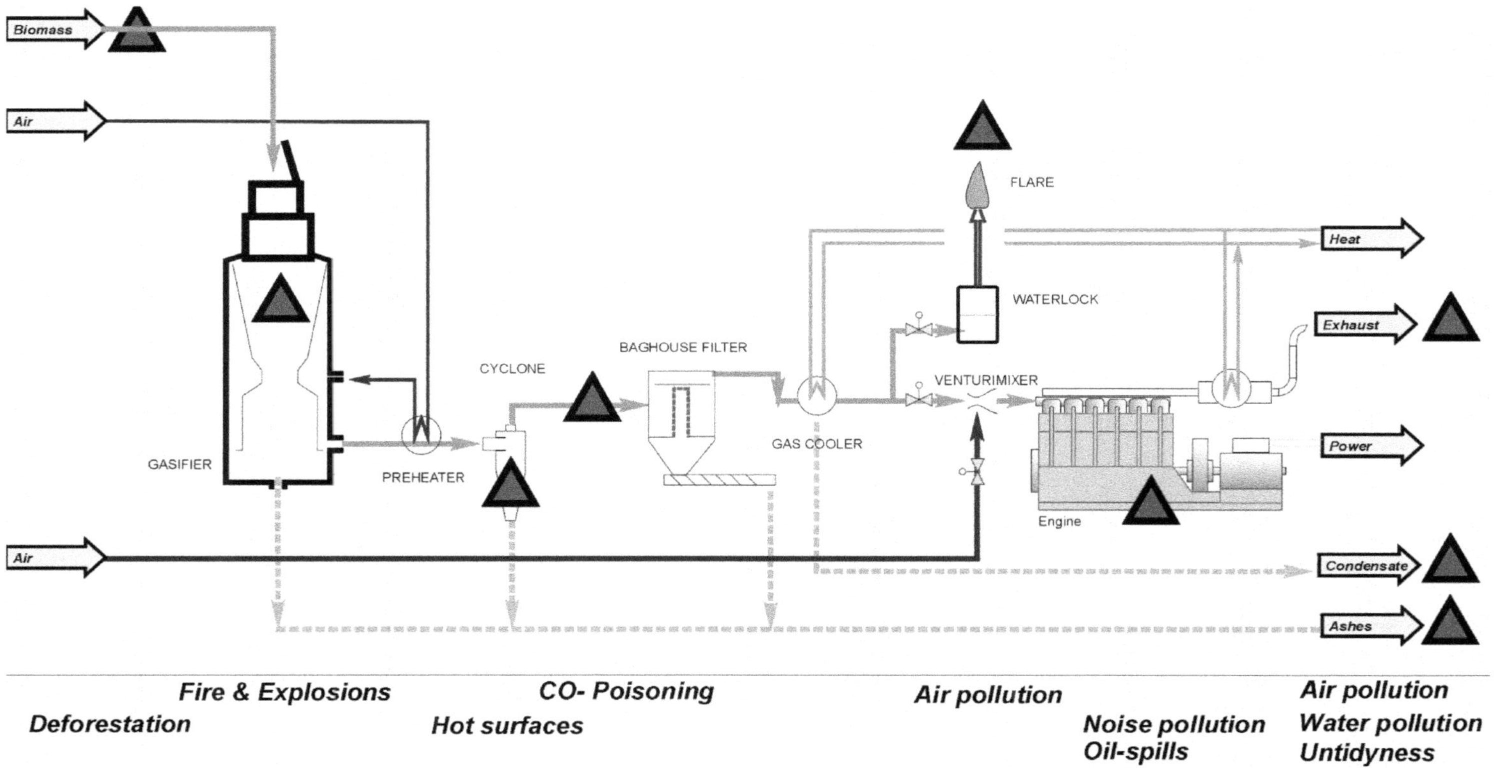

Figure 5: Major HSE impacts of biomass gasifiers.

- The quality of the product should meet the client's specification. During the UNDP/World Bank monitoring program it appeated that the technology does not meet the expectations regarding technical, financia;/economic and operational performance (Knoef, 2000)
- A local capacity for manufacturing the gasifier system must exist. Manufacturers must provide adequate after-sales service, guarantees and training of operators
- Labour, with the skills and motivation required for operation of gasifiers, should be available. Incentives to operators will stimulate optimal operation of a gasifier plant
- Accurate and objective information on the technology, the gasification system and its capabilities, its limitations, and how it compares with competing systems should be available to potential customers
- Loan finance should be available for purchasers of approved systems and for the reduction of risk with 'first-of-its-kind' installations. Government policies and incentives will reduce risk for private investments
- Scaling-up and demonstration of proven concepts followed by replication and optimalisation is the best approach to commercialise biomass gasification; economy of numbers will lead to reduced capital costs and knowledge on the pros and cons of the technology
- Regulations covering the safety, quality of gas produced, permitted level of noxious emissions and discharges, and other aspects of gasifier operation should be in force. In turn, the installation should meet the legislation on liability, health and safety. Proper design and documentation with instructions on HSE are essential
- Certification is important in promoting and gaining confidence in the technology. When a product meets certain and defined quality standards, munufacturers can market their product with full operarational guarantees
- Standardisation of the technology (components) as well as for acceptance tests and guarantee measurements. In many cases with new technologies, ultimately one dominant design remains becoming the market leader
- Any legal obstacles to the sale of surplus electricity should be removed. Long-term fixed prices are needed for feed-in of electricity, sale of heat and purchase of quality controlled biomass

REFERENCES

1. v. Swaay W. (June 2002) Presented at the 12th EU Conference on Biomass and Waste, Amsterdam, Netherlands
2. Quaak P., Knoef H., Stassen H. (1998) *Energy from Biomass, A review of Combustion and Gasification Technologies*, World Bank Technical Paper No. 422
3. Jayah T.H., Fuller R.J., Lu Aye, Steward D.F. (Dec. 2001) *The Potential for Wood Gasifiers for Tea Drying in Sri Lanka*. International Energy Journal, Vol. 2, No. 2
4. Abatzoglou N., Barker N., Hasler P. & Knoef H. (2000) *The development of a draft protocol for the sampling and analysis of particulate and organic contaminants in the gas from small biomass gasifiers*. Biomass and Bioenergy **18** (1) 5-17
5. Stassen, H., v. Swaay, W., & Prins W. (2002) *Thermal Conversion of Biomass into Secondary Products, the case of gasification and pyrolysis* Presented at the 12th EU Conference on Biomass and Waste, Amsterdam, Netherlands
6. Overend R. (2000) *Biomass Gasification: the enabling technology*. In Renewable Energy World
7. Knoef H. (2000) *Impediments to successful comercialisation of thermal biomass gasification*. Presented at the Conference "Gasification-4, The Future". Institution of Chemical Engineers

The Biosyngas-Estrie Pilot Project for the Gasification of Sorted Municipal Solid Waste

N. Abatzoglou[1], M. Gagnon, B. Valsecchi, R. Toupin, H. Cabana, L. Sarbaek & E. Chornet[1]
Enerkem Technologies Inc., Sherbrooke (Québec), Canada

ABSTRACT: The Biosyngas-Estrie Pilot Project aims at determining the technical, environmental and economic feasibility of sorting urban waste, recycling a fraction of the waste, and gasifying (at a maximum rate of 150 kg/h) the residue of the sorting operation. The Pilot Project has been co-financed by a public/private partnership and the effort is being monitored by the City of Sherbrooke. The latter is evaluating the possibility of implementing this technology within its Integrated Waste Management Plan.

The Pilot Project comprises the following steps:
(1) Urban waste reception, analysis, sorting, conditioning and preparation of a residual feedstock for gasification.
(2) Feeding (via lock hoppers, a rotary valve and a transfer screw) and gasification using a BIOSYN bubbling fluid bed reactor.
(3) Gas conditioning, via cyclones and scrubbing, to produce a synthetic gas ready to be valorized energetically.
(4) Lean gas burner design and operation and off-gases analysis for the evaluation of the burner energetic and environmental performances.
(5) Wastewater cooling/recycling loop. A coagulation/flocculation/decantation /filtration treatment has also been studied. Wet oxidation tests are underway.

The results obtained indicate that, technically, it is possible to conduct a sorting operation, which will enhance recycling levels in an urban context. The residue from the sorting operation can be adequately prepared for gasification. The synthetic gas produced can be conditioned and combusted to satisfy emissions regulations.

The key technical aspect of the gasification process with sorted urban waste is the choice and management of the conditions and configurations that result in a) stability of the synthesis gas and b) no agglomeration of the materials present in the bed.

Data covering a) gas sampling, b) gasification solid residues, c) the behaviour of the fluid bed material as function of the characteristics of the material used, is reported and discussed. A preliminary techno-economic analysis, evaluating the market conditions under which the commercialization of this technology might be successful, is available at Enerkem Technologies inc.

INTRODUCTION – THE CONTEXT OF THIS PROJECT

The Biosyngas-Estrie project aims at finding a solution to the problem of the Urban Solid Residues (USR). USR is used from now on, instead of Urban Waste, because it defines more precisely the nature of this material and avoids its association with positive or negative financial value. The solution sought must be socially acceptable, which means that a large majority of the citizens must accept it.

---

[1] Also affiliated with: Université de Sherbrooke, Dept of Chemical Engineering, Sherbrooke(Que), J1K 2R1, Canada

The technological strategies for disposing the USR (i.e. household refuse) are the following:
(1) Landfill with, in some cases, recovery of biogas.
(2) Incineration with recovery of energy (this approach can be preceded by the sorting and recycling of a part of the USR).
(3) Sorting to recover a) a fraction that can be recycled; b) a fraction which contains a significant proportion of the putrescible materials, and which will be digested biologically to produce biogas and/or compost; c) a fraction, which constitutes typically around 50% of the USR (dry basis), and which is landfilled.
(4) New approaches, compatible with (3) and which aim at valorizing the non-recyclable and non-compostable residue. This valorization is, in the majority of the cases, energetic.

Biosyngas-Estrie falls within the last category and is the result of consultations, having preceded the project, among the developers, representatives of the authorities and the citizens of Sherbrooke. These consultations clearly showed that the citizens seek socially acceptable solutions for the growing problem of "urban solid residues" management. The consultations indicate that the citizens are ready to support projects which:
(1) Are technologically and environmentally reliable.
(2) Show a favourable cost/benefit ratio.
(3) Are compatible with the new societal paradigms, namely Sustainable Development.

The citizens specified that to prove the technological and environmental reliability of BIOSYN gasification technology a pilot demonstration unit is required using "real USR of the City of Sherbrooke". As this was presented by the citizens as a "must" condition, Biosyngas-Estrie had been conceived as a multiphase project whose Phase 1 was the demonstration at pilot level.

OBJECTIVES

The objectives of Phase 1 of the project were specified as follows:
- Obtain a data base, at a reasonable scale, which allows to evaluate if ENERKEM's approach is feasible from the technical and environmental standpoints
- Establish, with the obtained data, the mass and energy balances and a Process Flow Diagram (PFD). This information would lead to the basic engineering and a preliminary P&ID allowing a relatively detailed cost calculation.

METHODOLOGY

Enerkem's approach consisted of two steps which were closely linked between them, as the second one depends on the success of the first one:
- Sorting and separation of the USR as well as the preparation of the fraction which cannot be recycled; this fraction ought to have appropriate physico-chemical and biological characteristics in order to be transported, handled, stored and fed to the gasifier safely and efficiently.
- The conversion of this fraction into a synthetic gas of typically low calorific value (5 to 6 MJ/Nm^3), which, after appropriate conditioning, will be clean enough to produce thermal energy in an appropriately designed burner. Proofs that the burner can be operated safely and efficiently while respecting the atmospheric emissions norms had also to be presented at the end of this phase of the project.

To reach the targets of Phase 1of the project, ENERKEM developed a methodology to separate the fraction which can be recycled from the fraction to be valorized energetically by gasification. This effort gave place to the implementation of the following sequential operations:

- Opening the green bags.
- Manual and Mechanical Sorting to separate the potentially recyclable material (a description of this operation with a tentative material balance is presented in the next sections of this paper)
- Shredding the residual fraction with recovery, by magnetic separation, of ferrous metals.
- Separation of the shredded material, by a trommel, in two sub-fractions:
  - Fine material (< 15 mm) and
  - Gross material (> 15 mm)
- Granulation of the fines. Dolomite was added in small quantities as a possible tar reforming catalyst during gasification.
- Drying to reach a moisture < 5%, which ensures odourless storage with negligible biological activity.
- Storage

GENERAL DESCRIPTION OF THE BIOSYNGAS-ESTRIE UNIT

The gasification methodology is based on BIOSYN design. A unit having a nominal capacity of 100 kg/h (able to operate between 50 and 150 kg/h) was conceived and built for this project. The granules are conveyed into a lock hopper system, composed of two cylindro-conical pressure vessels from where they are gravity-driven, through a calibrated rotary valve, into an Archimedean lateral screw which feeds the bubbling bed gasifier. The gasification takes place inside the fluid bed composed of alumina or synthetic olivine as fluidizing media. Air was used in all our tests. Oxygen-enriched air tests are programmed in Phase 2 of this project.

The temperature of the gasifier is set by adjusting the % of the stoichiometric air (λ) and the humidity of the gasified solids. Typically, a desirable operation temperature is 800°C. The particles entrained by the hot synthetic gas exiting the gasifier are captured by two cyclones in series. The hot gas exiting the cyclones is driven into a scrubbing train where the tar is condensed and the particles having average dimension higher than 1 μm are collected. A droplets coalescing unit allows the removal from the gas stream of the fine droplets produced by the scrubbing tower and the venturi. The cooled and clean gas is then ready for its use in a combustion chamber. ENERKEM designed, in the course of project, a special burner for this type of producer gas.

Sampling trains are located at the exit of the gasifier and at the entrance of the burner; the composition of the synthetic gas and the efficiency of the conditioning system is thus determined experimentally. Finally an external laboratory has installed at the outlet of the burner (chimney) a sampling train to evaluate the composition of the process off-gases and determine the environmental balance of the system.

The scrubbing water is recirculated, after being cooled in a water-water plate heat exchanger. Detailed wastewater analyses were performed in order to determine the contaminants present and define in the second stage of the project, along with external wastewater treatment companies, the most appropriate treatment for the eventual commercial unit. The part of the water, which has to be treated prior to its disposal to the sewage system of the area, is the purge; its quantity is approximately equal to the humidity of the solids entering the gasifier.

SORTING AND CONDITIONING PRIOR OF GASIFICATION

The following diagram present the sorting steps taken for the preparation of the feedstock prior of Sherbrooke's urban solid residues (USR) gasification. The mass balance and all steps involved in the preparation process are presented.

The USR sorting was made manually at the recycling sorting centre located in Victoriaville. All inorganic or recyclable materials were removed to respect the 3R-V-D hierarchy (V stands for valorization and D for disposal). The organic, non-recyclable fraction was compacted in a baler for transportation.

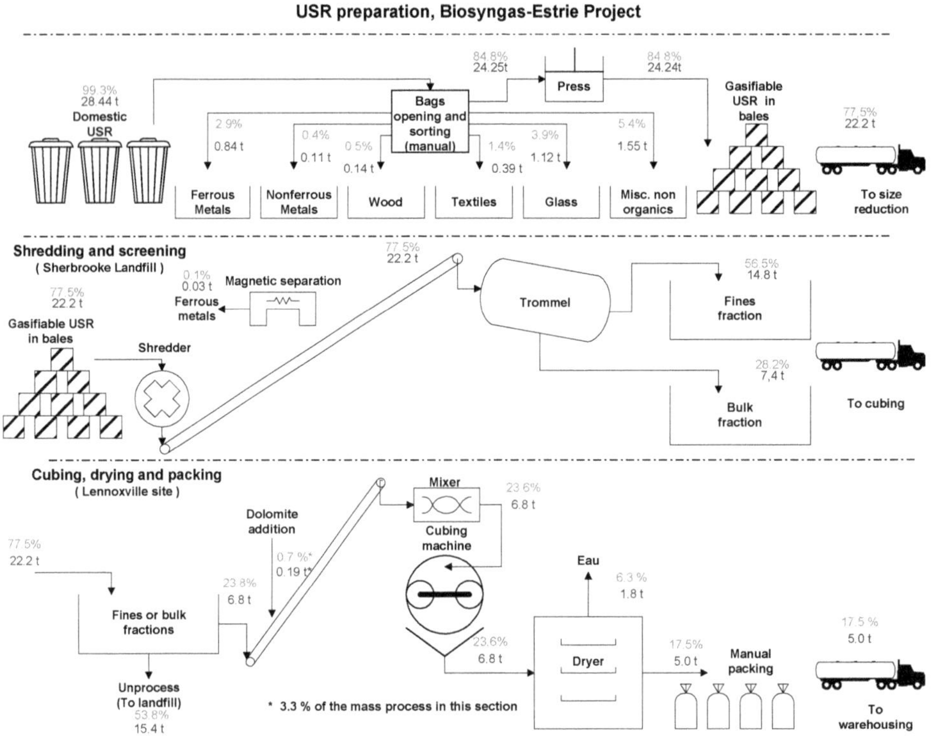

Figure 1: Sorting and pre-treatment steps prior to gasification.

The shredding and screening were accomplished at the municipal landfill in Sherbrooke. The size reduction equipment used was a tub grinder, a mobile hammer mill shredder. The tub grinder was equipped with a magnetic conveyor for the separation of ferrous metals. A trommel was attached to the tub grinder for separation into fines and a gross bulk fraction. Both fractions were stored and transported separately using conventional USR containers.

In the next step, the granulation and drying were performed at the Agriculture Canada site in Lennoxville. The fines fraction was taken from the containers and transported by a conveyor to the granulator . 2 to 3 wt% dolomite was added. The moisture of the granules was adjusted to a level of 5% using Enerkem's tray dryer. Finally, the dry granules were put in bags, transported and stored at Enerkem's pilot plant.

DETAILED DESCRIPTION OF THE BIOSYN GASIFICATION UNIT

The Biosyn technology can be adapted to different types of residues: wood, plastics, USR, sludges, meat and bone meal and other biomass materials of several shapes (pellets, briquettes, granules, chips or simply ground material) having different physical characteristics (humidity, density and sphericity) can be fed to the Biosyn system. Air or oxygen-enriched air can be used as partial oxidation agents for solid gasification. The partial pressure of water (vapour) inside the reactor can be adjusted so that the reforming reactions are appropriately controlled. Gas conditioning can be accomplished either under hot-dry or cold-wet (using water) options. The first conditioning option includes a novel moving granular bed filter to separate the entrained particles from the Syngas stream (Canadian Patent Application No. 2,268,376 1999 and Spanish Patent Application No 9901711 1999) and a catalytic tar-reforming reactor. The second conditioning option is composed of a scrubbing tower, a scrubbing venturi and a demister. The solid and the aqueous gasification residues can be re-injected in the gasifier for a better carbon conversion efficiency or can be disposed or valorized. Finally, the clean Syngas can be energetically valorized using the following devices: boiler, gas turbine, diesel engine, fuel cell. With an appropriate $CO/CO_2/H_2$ ratio liquid bio-fuels can also be produced via catalytic synthesis. Figure 2 presents a general mass and energy balance of the three basic modules of the technology.

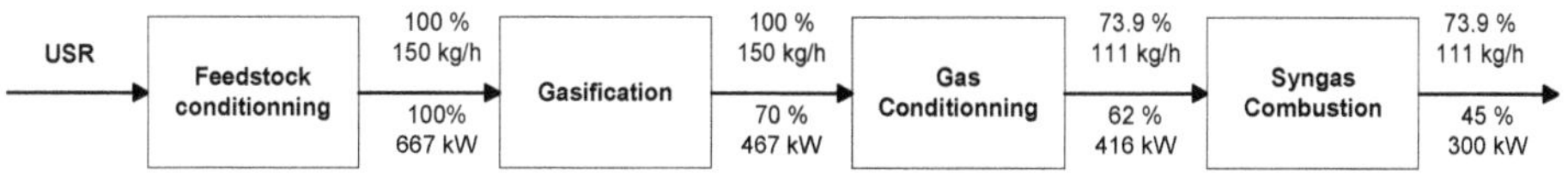

Figure 2: Mass and Energy Balances on sorted USR.

In the BioSynGas-Estrie project the feeding system is composed of a conveying screw that transports the feedstock from the storage area into the first lock hopper. Two sliding gate valves are needed to operate the feeding system in conjunction with the gasifier which is slightly pressurized at 1.5atm. The feedstock is transferred by gravity from the first lock hopper to the second lock hopper and then to a rotary valve that controls the solids flow. A screw feeder transfers the feedstock directly into the hot zone of the gasifier. Figure 3 gives a detailed description of this Biosyn unit.

The gasifier is a bubbling fluidized bed reactor with a nominal biomass capacity of 100 kg/h. The system operating pressure is 1.5 Atm even though certified for operation up to 4atm. The bed temperature is comprised between 750 and 850°C. The gasifier is designed to obtain and maintain the desired bed temperature, to optimize the turbulence and residence time (between 10 and 20s) and minimize the particle entrainment. An enlarged freeboard is used for the particle disengagement. Enough quantities of alumina or olivine are used as the fluidized bed for a proper heat and mass transfer and for minimizing the bed pressure drop.

Two cyclones in series collect the coarse particles (over 10 μm) at an efficiency comprised between 90 and 95 %. The particles collected fall by gravity into two bins and through a series of valves they are removed intermittently. It has been found that the production of solid gasification residues (SGR) is comprised between 15 to 25% of the USR feedstock; 20-28% of these SGR produced from USR is unconverted carbon while the other 80-72% is the inorganic content of the USR itself plus abraded alumina/olivine fines.

The Biosyn cold-wet conditioning is composed of:

a) A scrubbing tower to separate the fines and to cool the synthetic gas. The average inlet and outlet gas temperatures are 750°C and 80-55°C (depending upon water rate). The scrubbing water to gas ratio is 10-20 L/m^3.
b) A venturi, which captures the finer particles. The scrubbing water to gas ratio is 4 L/m^3.
c) A demister equipped with a mesh pad to coalesce the entrained fine droplets using an inertial impact mechanism.

The scrubbing water is continuously recirculated in a close loop at a rate of 4.25 m^3/h. Only a certain amount of this scrubbing water is withdrawn and replaced by clean water to insure an optimum absorption and scrubbing capacity. Finally, the recirculating water is cooled by a water-water heat exchanger. Table 1 presents typical operating conditions of the Biosyn gasification system used in the Biosyngas-Estrie project.

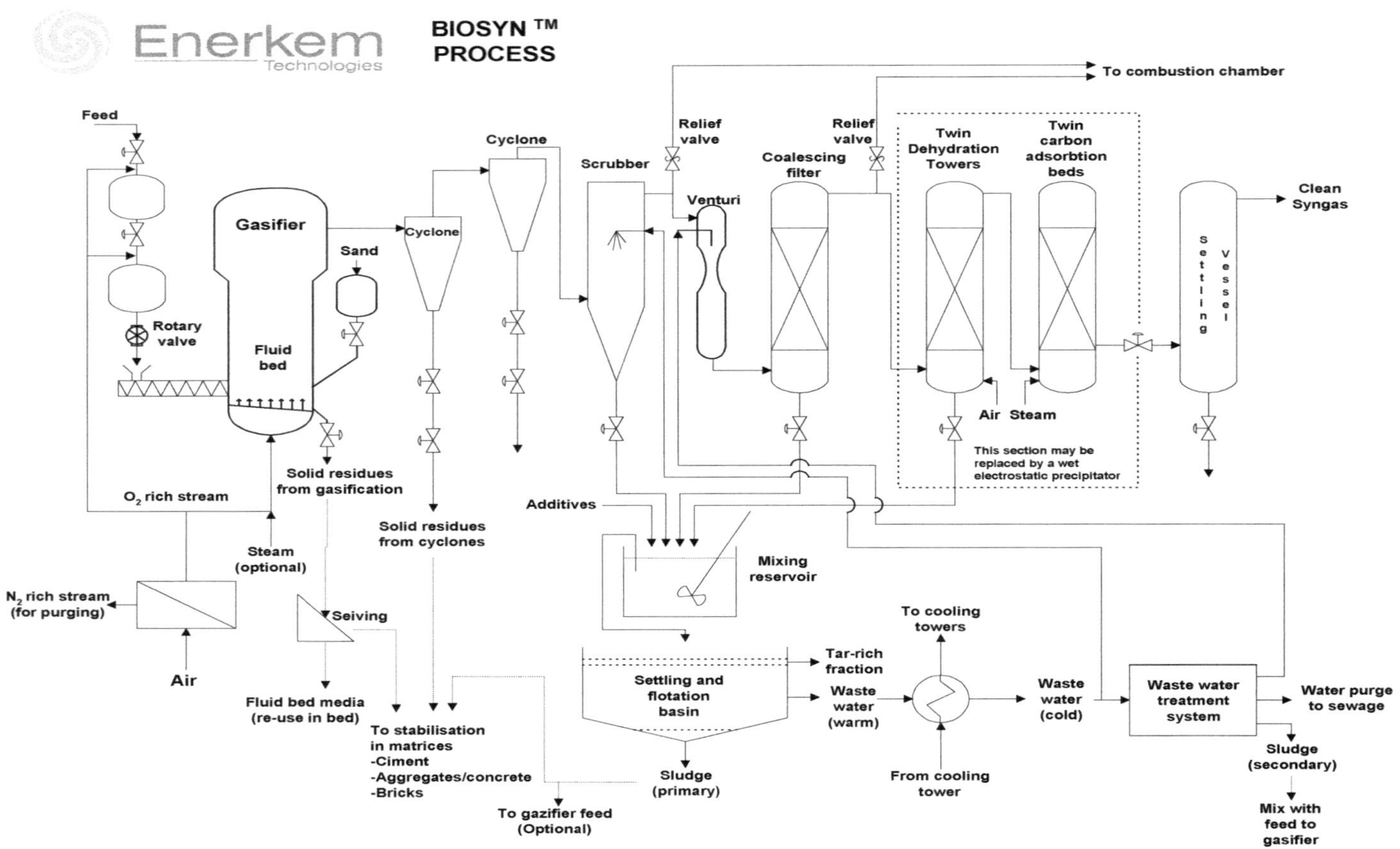

Figure 3: Basic PFD of Biosyn gasification plant.

Table 1: Typical BiosynGas-Estrie pilot plant operating conditions.

No	Systems	Operating Conditions	Values	Units
1	Feeding system	Biomass feeding rate	100 to 150	Kg/h
2	Gasifier	Temperature	750 to 850	°C
		Pressure	1.5 to 3	Atm.
		Gas residence time	8 to 15	s
		Fluidizing agent	Alumina / Olivine	***
3	Cyclones	Number of cyclones	2	***
		SRG production rate from USR	20 to 25	%
		Unconverted carbon in SRG	20 to 28	%
4	Scrubbing tower	Average inlet and outlet temperature	750 and 55	°C
		Scrubbing water to gas ratio	20	L/m^3
5	Scrubbing Venturi	Gas throat velocity	54	m/s
		Scrubbing water to gas ratio	4	L/m^3
		Type of demister	Mesh pad	***
7	Scrubbing water loop	Recirculating rate	4.25	m^3/h
8	Combustion chamber	Average temperature	960	°C
		External diameter	1	m

THE BURNER

The energetic valorization of the Syngas was tested in a burner designed by Enerkem. Even though the Syngas has a low calorific value, the burner operated efficiently. A commercial burner using propane is used to preheat the combustion chamber. The syngas, the primary and the secondary air are injected in the burner at an appropriate velocity in such a way to obtain adequate turbulence and mixing as well as residence time. The Syngas burner is currently under disclosure for patent protection.

Some basic specifications of Enerkem's burner design are summarized in the Table 2.

Table 2: Burner specifications.

Design	Enerkem Technologies Inc.
Construction material	SS-304
Average Power (kW)	293 - 440
Fuel	Syngas
Oxidant	Air
Syngas/Air mixture	External
Safety control module	Fireye

RESULTS

The Biosyngas-Estrie pilot plant has operated so far just over 200 hours. The bed preheating for the start-up is made using a propane preheater and lasts 2 hours. When an appropriate bed temperature is reached (around 400°C), solids are fed into the reactor, under an intermediate combustion/gasification regime ($0.6<\lambda<0.7$), until the bed temperature reaches 800°C (duration of 3h). The process simulator developed by Enerkem Technologies predicts accurately the mass and energy balances as function of the system operating conditions. It also allows a rapid and efficient process intervention as a function of the data obtained by the system probes during the experiment.

During the gasification regime, isokinetic sampling was made to determine particles and tar present in the syngas at the exit of the gasifier and after wet conditioning (sampling was made just before the combustion chamber). A gaseous phase chromatograph (GC) is used to analyze the syngas. An external laboratory has installed the appropriate probes to characterize the flue gas emissions at steady state operation. Lambda varied from 0.3 to 0.4 during the various steady states tested. The dolomite added to the sorted USR aimed at increasing the tar conversion to simple molecules. No results are yet available. Typical data obtained from the USR gasification during steady state operation are shown in Table 3 for a feed rate of 105kg/h of sorted USR granules at 6 wt% moisture content.

Table 3: Typical Results obtained for the USR gasification during the BioSyngas–Estrie project.

USR granules composition	**Values**	**Units**
Putrescibles	59	%
Paper	20	%
Plastics	10.5	%
Carboard	10.5	%
Inorganic (including dolomite)	21.2	%
Inorganic (without dolomite)	17.9	%
GSR	**Values**	**Units**
Inorganic - Cyclone 1	71.4	%
Carbon - Cyclone 1	28.6	%
Inorganic - Cyclone 2	72.9	%
Carbon - Cyclone 2	27.1	%
Synthetic gas	**Values**	**Units**
Flowrate produced	220	Nm^3/h
Lambda ratio	0.38	*****
Mean Syngas HHV	5.2	MJ/m^3
Average installation power	0.3	MW
Hydrogen	5.9	%
Argon / Oxygen	1.4	%
Nitrogen	63.3	%
Methane	3.3	%
Carbon monoxyde	10.1	%
Carbon dioxyde	12.8	%
Ethylene	2.2	%
Ethane	0.2	%
Propylene	0.6	%
Propane	0.1	%
Flue gas composition	**Values**	**Units**
Oxygen	5.5	%
Carbon dioxide	8.7	%
Excess air	36.1	%
Carbon monoxide	123.9	ppm
NO_x	119.2	ppm
SO_2	47.2	ppm
Average stack temperature	960	C

A comparison between results obtained using two different lambda values is presented in Table 4.

Table 4: Results for two different λ values.

	For λ=0,38	For λ=0,30
SG Flowrate	220 Nm^3/h	205 Nm^3/h
-SG HHV	5.2 MJ/Nm^3	6.1 MJ/Nm^3
Composition of SG (% v/v)		
•H2	5.9	12.8
•CO	10.1	12.3
•CH4	3.3	3.5
•N2	63.3	53.4
•Ar	1.4	1.2
•CO2	12.8	14.7
•CxHy	3.1	2.2

The pH, conductivity, total organic carbon (TOC), total carbon (TC) and suspended solids (SS) were among the variables monitored in the recirculating scrubbing water to define the subsequent treatment prior to their final sewage disposal. The recirculating water pH stabilizes at 8.5 for the USR gasification (Figure 4). The scrubbing water conductivity increased following a constant slope revealing salts accumulation in the system (Figure 4).

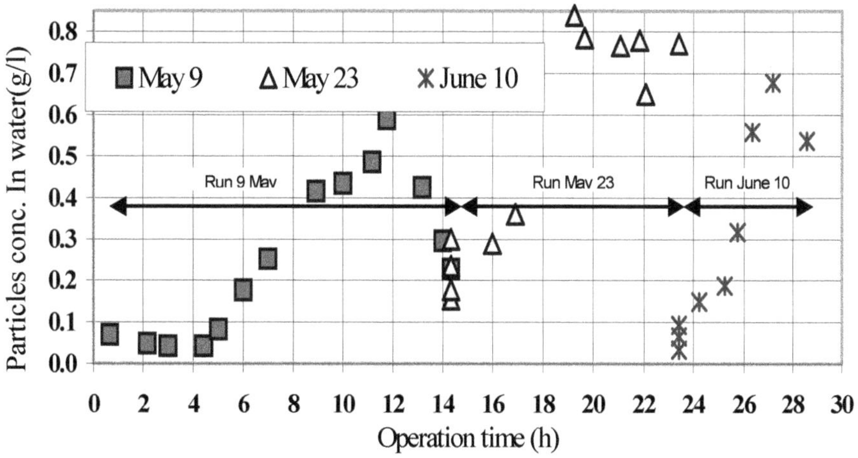

Figure 4: pH and conductivity as function of gasification time, May 9 & 23, June 10.

Figure 5 demonstrates the increasing difference between the TOC and the TC, indicating that the inorganic forms of carbon (basically carbonates) accumulate in the system. During the testing period the scrubbing water did not reach its TOC and TC saturation levels. The increasing and decreasing trends in Figure 6 show the accumulation of particles of 20μm and over in the scrubbing water and their decantation between runs.

TECHNICAL PROBLEMS AND SOLUTIONS

Even if the feeding system works well with various heterogeneous type of solids, there were still some bridging observed at the throat of the feeding system. This problem was encountered for materials having a specific gravity less than 0.2. It may thus be imperative to have a densification step in order to render the feeding as versatile and trouble-free as possible. Eutectics can be formed at a certain level of temperature and inorganic material content in the fluid bed, when different types of inorganics, contained in the USR, remain at the bottom of the bubbling fluidised bed. These inorganics must be intermittently removed from the bed during operation to avoid accumulation. The eventual commercial unit should be equipped with automatic evacuation port(s) to avoid accumulation of inorganics in the fluid bed.

In the cold-wet gas conditioning, level monitoring and control at the scrubbing tower and demister have been initially problematic. This was due to the stickiness and the foamy tendency of the condensing tar. A solution was found by taking the level in a parallel, relatively clean side pipe. A challenge in the control of the scrubbing efficiency is how to adjust the absorption capacity of the scrubbing water recirculating loop when cold-wet syngas conditioning is chosen. Finally another challenge is to find a use for the Solid Residues from Gasification (SRG); work is being undertaken on this topic to insure a better social acceptability for any gasification technology.

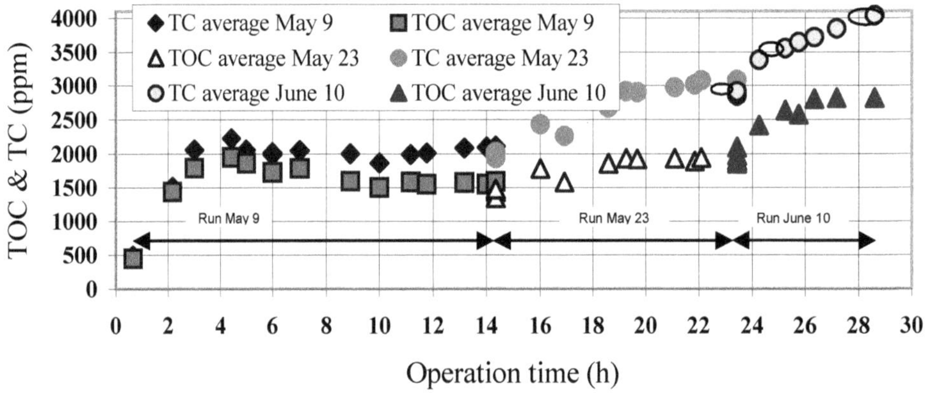

Figure 5: TOC & TC as function of gasification time; May 9, 23 and June 10.

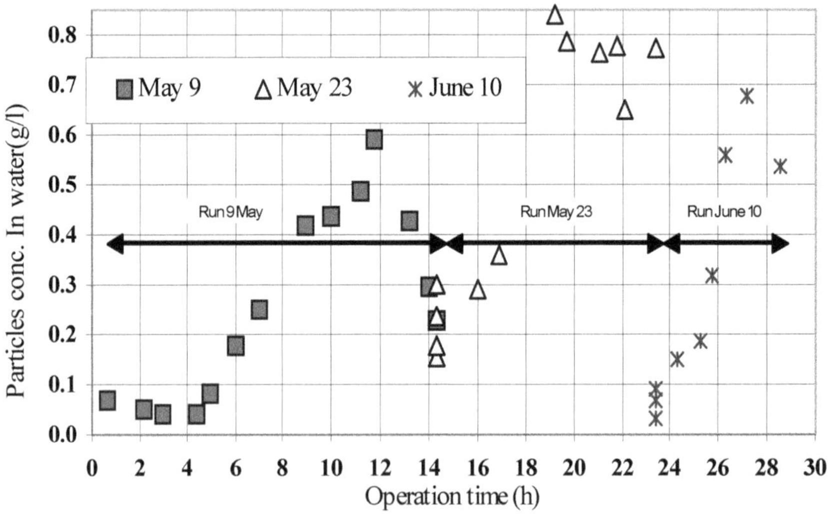

Figure 6: Suspended solids as function of gasification time; May 9, 23 and June 10.

DISCUSSION ON THE RESULTS

The results from the sorting step have shown that, on the basis of 28 tones of delivered and sorted USR:

- 15.4% can be recycled.
- 84.6% consists of putrescible matter (more half of the total mass at certain times), plastics, cardboard and papers
- The percentage of inorganics after shredding and granulation is 18% (on a dry basis). This includes the small amount of dolomite added in the granules.

The results of the gasification of the granulated feedstock indicate that the synthetic gas has a calorific value ranging between 5.2 and 6.2 MJ/Nm^3 and that its average vol/vol composition is depending mainly on the "λ" used, as shown in Table 4.

The results of the flue gas emissions following the combustion of the synthetic gas have shown that CO, NO_x and SO_2 are within environmental regulations.

The results obtained from Phase 1 of the work show that the BIOSYN gasification technology can handle USR. Nevertheless, its reliability requires the preparation of the feedstock as granules and the management of the inorganics entering (as part of the USR) the fluid bed. In the eventual commercial unit a bed removal port must be added to allow for a regular renewal of the bed+inorganics material in the gasification reactor. If this is not done the probability of bed agglomeration is real depending upon the nature of the USR inorganics and its content.

CONCLUSIONS

- The BIOSYN Gasification technology has proven to be feasible for the valorization into energy of non-recyclable urban residues. It is an alternative to incineration or mass burning.It is technically feasible to sort and condition the urban residues before gasification. Such conditioning implies an increase in the recycling efficiency of the process
- The conditioning steps are: mechanical sorting, shredding, granulation and drying, if the humidity is higher typically than 25% w/w.
- The gasifier operated and performed according to the design specifications.The operation of the cyclones was very successful and performed according to the design specifications.
- The water recirculation system demonstrated a trouble-free and efficient operation. Some problems with the monitoring of the water level were encountered and solved by changing the instrumentation.
- The burner, although fed with a lean synthetic gas (5,2-6,2 MJ/Nm3), performed efficiently both energetically and environmentally.
- An amount of water equal to typically 20% w/w of the feed solids is withdrawn continuously from the unit as a purge.

ACKNOWLEDGEMENTS

The authors are indebted to:

- All Enerkem team members, Jacques Bureau, Vincent Chornet, Lyne Champagne and Alex Mincic for their work and contributions to Phase 1 of this project.
- National Resources Canada, Ministère des ressources naturelles du Québec and Ministère de l'industrie et du commerce du Québec for financial support.
- The City of Sherbrooke for their time spent to follow the various stages of the project.
- Innovatech du sud du Québec, SOQUIP/SGF, and FSTQ, as Enerkem's financial partners.

REFERENCES

1. "The development of a draft protocol for the sampling and analysis of particulate and organic contaminants in the gas from small biomass gasifiers", N. Abatzoglou, N. Barker, P. Hassler and H. Knoef", Biomass and Bioenergy, Vol. 18, No 1, 2000, pp. 5-18.
2. "PROFILE-Citizen Involvement in Waste Management: An Application of The STOPER Model via an Informed Consensus Approach", P. DeConinck, M. Seguin, E. Chornet, L. Laramée, A. Twizeyemariya, N. Abatzoglou and L. Racine, Environmental Management-Berlin, Vol. 23, No 1, 1999, pp. 87-94.
3. "Gas Conditioning Technologies for Biomass and Waste Gasification", N. Abatzoglou, P. Legast, P. Delvaux, D. Bangala and E. Chornet, 3rd Biomass Conference of the Americas, (Ed. By R.P. Overend and E. Chornet, Pergamon), ISBN 0 08 0429 963, Montreal, Que., August 24-29, 1997, pp. 599-606.
4. (4) "The Improved Biosyn Waste Gasification Technology. Environmental Performances", N. Abatzoglou, J.-C. Fernandez, L. Laramée, S. Gasso and E. Chornet, published in the International Conference in Gasification and Pyrolysis of Biomass, State of the art and Future Prospects, (Ed. by M. Kaltschmitt and A.V. Bridgewater, CPL Press), ISBN 1 872691 71 4, Stuttgart, Germany, April 9-11, 1997, pp. 259-268.
5. "Application of Gasification to the Conversion of Wood, Urban and Industrial Wastes", N. Abatzoglou, J.-C. Fernandez, P. Jollez, and E. Chornet, in "Developments in Thermochemical Biomass Conversion", Banff, Canada, 20-24 May, 1996, Edited by A. V. Bridgewater and D.G.B. Boocock, Blackie Academic & Professional, London, 1997, ISBN 0 7514 0350 4, pp. 960-972
6. "Performance of Different Dolomites on Hot Raw Gas Cleaning from Biomass Gasification with Air", A. Orió, J. Corella and I. Narváez, Industrial & Engineering Chemistry Research, Vol. 36, No 9, 1997, pp . 3800-3808.
7. "Biomass Gasification with Air in an Atmospheric Bubbling Fluidized Bed. Effect of six Operational Variables on the Quality of the Produced Raw Gas", I. Narváez, A. Orió, M. P. Aznar and J. Corella, Industrial & Engineering Chemistry Research, Vol. 35, No 7, 1996, pp. 2110-2120.
8. "Destabilization of Fluidized Beds Due to Agglomeration. Part II: Experimental Verification", G. Tardos, D. Mazzone and R. Pfeffer, The Canadian Journal of Chemical Engineering, Vol. 63, June 1985, pp. 384-389.
9. "Correlations for Bubble Characteristics in Fluidized Beds", K. Viswanathan, The Canadian Journal of Chemical Engineering, Vol. 63, June 1985, pp. 332-335.
10. "Gasification of Urban Waste Wood (Altholz) ", Ph. Hasler and R. Bühler, Report submitted to the IEA, Biomass Gasification Group, September 1994.
11. "Evaluating Gasification and Advanced Thermal Technologies for MSW", Walter R. Niessen, Solid Waste Technologies, Vol. 10, No 5, 1996, pp. 28-35.
12. ''Pyrolysis and Gasification of Waste: a Worldwide Technology and Business Review'', Technology and business review, Juniper Consultancy Services Ltd volume 1, UK, ISBN 0-9534305-6-1, (2000).

Tar Removal in an Entrained Flow Cracker (EFC) with Application to Biomass Gasification

Z. Abu El-Rub[*a], E. A. Bramer[a], G. Brem[a,b]
[a]*University of Twente, Laboratory of Thermal Engineering, P. O. Box 217, 7500 AE Enschede, The Netherlands*
[b]*TNO, P. O. Box 342, 7300 AH Apeldoorn, The Netherlands*

ABSTRACT: A project is carried out to develop a so called Entrained Flow Cracker (EFC) for tar removal downstream of a gasifier making use of cheap and active materials as catalysts. These catalysts convert tars in the producer gas to useful gases. In this reactor the fine catalyst is mixed intensively with the producer gas and then separated and recycled to the reactor and or to the gasifier. In the first stage of the project screening of catalysts was carried out in a fixed bed tubular reactor. Calcined dolomites and olivine are considered as such active and inexpensive catalysts for hot gas cleaning of a gas produced by biomass gasifiers. In addition to them, other cheap catalysts were investigated such as FCC catalyst, char and ashes. The efficiency of tar removal for these catalysts was compared with that of an inert material (silica sand) and with that of high activity catalyst (commercial nickel catalyst). Tar was modeled using naphthalene and its decomposition over the different catalysts was modeled by naphthalene reaction with CO_2 and steam. Tests were carried out in a fixed bed tubular reactor at a temperature of 900 °C under atmospheric pressure and a gas residence time in the catalyst bed of 0.3-0.4 s. For the selected catalysts, kinetic studies were carried out in the temperature range of 700 to 900 °C and atmospheric pressure. A simple first order kinetic model was used to describe the tar conversion. The gas composition and tar content in the gas were analyzed downstream the reactor using solid-phase adsorption (SPA) sampling method in combination with gas chromatography and mass spectrometry (GC/MS). The results of the fixed bed experiments will be used in the design of the entrained flow reactor.

INTRODUCTION

Biomass can be converted to energy carriers by biological or thermochemical processes. Among the various thermochemical processes gasification has attracted significant interest. This is due to the higher efficiencies produced by this technology either for small or large-scale systems. One of the most important technical barriers in biomass gasification development is the removal of tars. These tars are always present in producer gas from the gasifier as a side product. If tar content in the exit gas is high, then it will condense in the gas transfer lines and cause severe plugging problems resulting in serious operational interruptions. For engines and turbines, high concentration of tar can damage them or lead to unacceptable level of maintenance. For most people working on gasification, high temperature catalytic tar removal is technically and economically interesting approach for gas cleaning. Such approach is intuitively interesting because it has the potential to increase conversion efficiencies while simultaneously eliminating the need for collection and disposal of tars.

CATALYSTS

Calcined dolomites and olivine are considered as active and inexpensive catalysts used for hot gas cleaning of a gas produced by biomass gasifiers. In addition to these catalysts, other catalysts are investigated such as FCC catalyst, char and ashes.

Tar removal over calcined dolomites (MgO.CaO) is mostly due to steam and dry (CO_2) reforming reactions [1]. The reacting network for the overall tar elimination over calcined dolomite is not known yet, but at least includes: reactions of steam and CO_2, thermal cracking and hydrocracking of tars [2]. Simell *et al.*, related the activity of dolomite to the large pore size, CaO surface area of the corresponding calcinates presence of active metal such as iron and may be due to its relatively high alkaline (K, Na) content [3]. The main problem with dolomite is its fragility [4]. It has a lower mechanical strength than olivine [5,6]. Delgado *et al.*, (1996), presented the factors that influence the deactivation of dolomite: space time or space velocity, Particle diameter and temperature [7]. Simell *et al.*, (1999), concluded that thermal reactions at a temperature of 900 °C did not have a measurable effect on the decomposition rates of tar model component [8].

Olivine consists mainly of a silicate mineral in which magnesium and iron cations are embedded in the silicate tetrahedral. Natural olivine is ($(Mg,Fe)_2SiO_4$) [8]. Its mechanical strength is comparable to that of silica sand, even at high temperatures, it presents a higher attrition resistance than dolomite [4,5]. Olivine is available in the market at about the same price as that for dolomite (120 Euro per metric ton) [4].

FCC catalysts used nowadays are composite catalysts, made of zeolite component and a matrix component consisting mostly of amorphous silica alumina. FCC catalysts are readily poisoned by substances whose molecules react with the catalyst acidic sites. Basic nitrogen compounds and alkaline metals present in the feedstock show such poisoning effect. Corella *et al.*, (1998) tested an "in equilibrium" spent catalyst in a fluidized bed and found that FCC catalyst was quickly elutriated from the bed [9]. Herguido *et al*., (1992) tested an "in equilibrium" spent FCC catalyst in a 15 cm i.d. rider-gasifier with a stable fluidized bed of sand at its bottom. Tar was reduced from 78 to 9 g/Nm^3 with recirculation and continuous regeneration of the catalyst [10].

Charcoal is produced by heating wood under limited access of oxygen. When wood is heated slowly to ~280°C, an exothermic reaction occurs. In the usual carbonization procedure, heating is prolonged to 400 to 500°C in the absence of air. Sjöström *et al*., (1999) [11] run co-gasification of birch wood with both Daw Mill coal and Polish bituminous coal in a pressurized fluidized bed reactor. They found that both tar and ammonia yields seemed lower in the co-gasification experiments compared with the yields from gasification of the individual fuel. Joseph et al. (1996) [12] note that an entrained-flow vortex gasifier, with re-injection of char ensures that a larger proportion of tars and higher molecular weight hydrocarbons are cracked.

Ash is a waste product from the gasification process. The chemical composition of ash determines the physical properties of the material such as softening, melting points or vaporization points [13]. The autocatalytic effect of ash is due to its minerals content [14]. The analysis of wood ash after gasification showed that there was a high content of alkali metals. Hauserman [14] investigated the use of wood ash as a possible catalyst for the production of hydrogen by steam gasification of wood and coals. The catalyst was directly added to the feed by dry mixing. He found that the addition of 20 wt.% wood ash increased the reactivity of the steam gasification reaction of the bituminous coal at 700 °C by a factor of almost 9 and that of wood by a factor of 32 relative to the uncatlyzed results.

MODEL TAR COMPONENT

Model tar compounds are single molecules chosen as substitutes for real tar mixture. Since tar is a complex mixture of hydrocarbons, it is difficult to prepare and to study its conversion. Researchers used several model tar components to represent the real tar in their studies for tar reduction. However a new study show that tar can be divided into five different lumped groups based on their problematic nature, i.e., solubility and condensability [ref?]. The commonly used model tar compounds in literature are naphthalene [15], benzene [8], toluene [16], cyclohexane and n-heptane [17]. Naphthalene represents a light poly aromatic hydrocarbon or a tertiary tar and at a temperature of 900 ^{0}C it is the major component in tars. [18]. The order of thermal reactivity of tar model compounds as found by Jess [15] is toluene >> naphthalene > benzene. Based on information from literature, it seems that it is more accurate to use naphthalene as a model tar compound than other model tar compounds.

This paper presents a comparative study of tar decomposing catalytic activity for several potential low cost materials suitable for bulk use in biomass gasification processes. The activity of these materials was compared with that of an inert material (silica sand) and with that of high activity material (commercial nickel catalyst). In addition, a kinetic study is done to catalyst that shows high activity.

EXPERIMENTAL

This is the first stage of developing a so called Entrained Flow Cracker (EFC) for hot gas cleaning downstream of a biomass gasifier. In this stage, catalysts screening tests were done using a fixed bed tubular reactor at a mean temperature of 900 °C, an operating pressure of 1 atm. and a residence time of 0.3-0.4 s.

APPARATUS

The experiments were performed in a tubular reactor at atmospheric pressure and a temperature of 900 °C. The feed gas is typically 70 $L.h^{-1}$ (STP). The gas (N_2) is metered from a cylinder controlled by critical nozzles. The nitrogen is used as the carrier gas, which flows through naphthalene and water saturation units made from glass. The desired mole fractions are obtained by controlling the temperatures of the saturators. CO_2 is added to the feed gas and the mixture is passed through the reactor. The reactor is an empty quartz tube with an internal diameter of 4 cm and a length of 90 cm, and heated by a tubular electric furnace. The reactor has two thermocouples, the first is in the center (considered as the reference temperature) and the second is close to the wall. These two thermocouples can be moved along the bed height in small quartz pipes to measure longitudinal temperature profiles. Temperatures are measured by K-type thermocouples. The bed material is supported by a porous quartz disc, which allows the gas to pass through its bottom. The catalyst is placed above a bed of an inert silica sand of equal height to the catalyst bed. Fig. 1 shows the scheme of the setup.

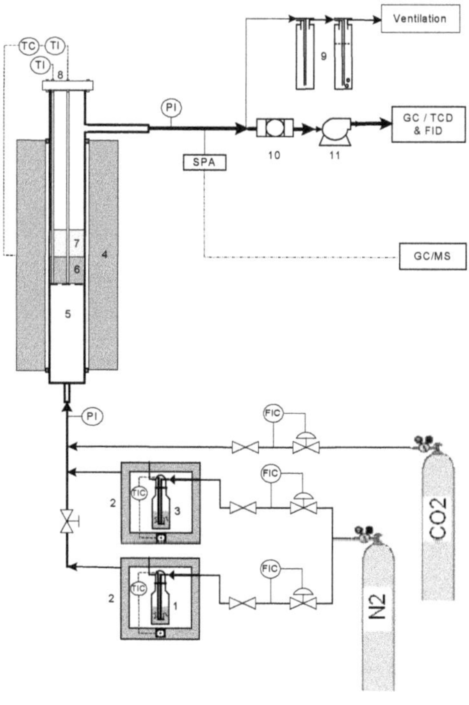

Figure 1: Experimental setup.
1. water saturator; 2. oven; 3. naphthalene saturator; 4. tubular furnace; 5. quartz tubular reactor; 6. inert bed; 7. catalyst bed; 8. quartz tubes for thermocouples; 9. water condenser; 10. filter; 11. heated pump; GC: gas chromatography; FID: flame ionization detector; TCD: thermal conductivity detector; MS: mass spectrometry; SPA: solid phase adsorption

TEST PROCEDURE

The experimental runs were started by pouring a weighed sample of the bed material (catalyst) on the reactor grid. The feed gases flow rates were regulated to give the desired space time, which is 0.3-0.4 s. dolomite calcination was carried out in situ at 900 °C and atmospheric pressure for 1 h at constant nitrogen flow. After calcination, all the gaseous reagents were fed and the reactor is heated to the required temperature. The catalysts were then stabilized for at least 15 minutes [12] before tests were started.

TAR ANALYSIS

The gas composition and tar content in the gas are analyzed downstream the reactor using solid-phase adsorption (SPA) sampling method in combination with gas chromatography and mass spectrometry (GC/MS).

CATALYSTS

Dolomite, olivine, nickel and "in-equilibrium" (once used) fluid catalytic cracking (FCC) catalyst were obtained from commercial suppliers. Char and ash were produced from wood biomass in our laboratory. Char is produced by pyrolysing the biomass wood at 500 °C and ash is produced by burning the produced char at 600 °C. FCC catalyst was obtained at an average particle size of 57 μm. The particles of dolomite, olivine and char were sieved to a particle size range of 1.40-1.70 mm. Nickel catalyst particles were crushed and sieved to particle size range of 1.4-1.7 mm. Ash produced from biomass wood was very fine, of particle size less than 0.3 mm.

Table 1, 2 and 3 show some chemical characteristics of the catalysts.

Table 1: Chemical composition of the catalysts (wt.%).

Element	Olivine	Dolomite	Nickel	Ash
MgO	48.5-50.0	21.5	-	15
CaO	0.05-0.10	30.5	-	44.3
SiO_2	41.5-42.5	0.15	7	-
Fe_2O_3	6.8-7.3	0.20	-	-
Al_2O_3	0.4-0.5	0.061	12	-
NiO	0.3-0.35	-	70	-
MnO	0.05-0.10	-	-	-
Cr_2O_3	0.2-0.3	-	-	-
$NiCO_3$	-	-	5	-
K_2O	-	-	-	14.5

Table 2: Chemical characteristics of FCC catalyst.

APS μ	SA m^2/g	MSA m^2/g	ABD g/cc	Fe wt.%	Na wt.%
57	172	79	0.86	0.3	0.15

C wt.%	Ti wt.%	ReO wt.%	SiO2 wt.%	Mg wt.%	Al_2O_3 wt.%
0.08	0.97	3.73	50.58	0.21	44.3

Table 3: Ultimate char analysis (wt. %).

Element	Dry	daf
C	92	93
H	2.45	2.5
O	3	3
N	0.53	0.54
S	1	1.01

EXPERIMENTAL DATA EVALUATION

TAR CONVERSION

The conversion of the tar model compound (naphthalene) was calculated from the inlet and outlet naphthalene concentrations of the gas. This tar elimination includes catalytic conversion as well as thermal cracking. Tar conversion reactions are related to steam and dry (CO_2) reforming reactions with the model tar component. The concentrations of steam and CO_2 in the feed gas are maintained at 10 vol.% for each component. The conversion is defined as:

$$X_n = \frac{(C_{n,in} - C_{n,out})}{C_{n,in}} \quad (1)$$

RESIDENCE TIME

The residence time (τ) in the catalyst bed with respect to the empty reactor is defined as:

$$\tau = \frac{V_{R,cat}}{Q(T, P_{tot})} \quad (2)$$

KINETIC MODEL

A first order kinetic model is used for making a kinetic study for the selected catalyst. This model is easy for data evaluation and comparison of results with others [14].

$$(-r_n) = k_{app,n} C_n \quad (3)$$

Under plug flow conditions, the apparent kinetic constant can be integrated as:

$$k_{app,n} = \frac{[-\ln(1 - X_n)]}{\tau} \quad (4)$$

The apparent rate constant is temperature-dependant variable according to Arrhenius' law:

$$k_{app,n} = k_{o,app} e^{-E_{app}/RT} \quad (5)$$

RESULTS AND DISCUSSION

Catalysts screening experiments were performed at a high temperature (900 °C) in order to get high tar conversion or low tar content in the exit gas. Fig. 2 illustrates the temperature profile inside the reactor.

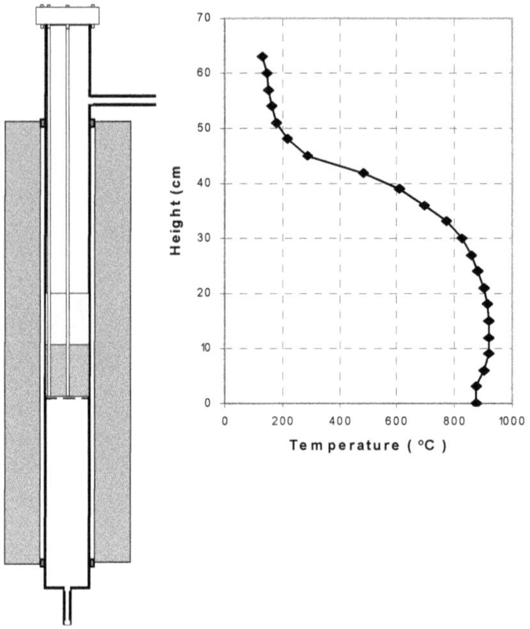

Figure 2: Reactor temperature profile.

The effect of the first four catalysts on conversion of the model tar component at 40 g/Nm^3 is illustrated in Fig. 3. The order of reactivity obtained is: commercial nickel > dolomite > olivine > silica sand.

The relatively low activity of the used dolomite and olivine could be related to the low active iron content, especially in dolomite and the large particle size used in the experiments, which caused internal mass transfer limitations. Commercial nickel based catalyst is, as expected, much more active than dolomite and olivine, but they are more expensive and can be also deactivated.

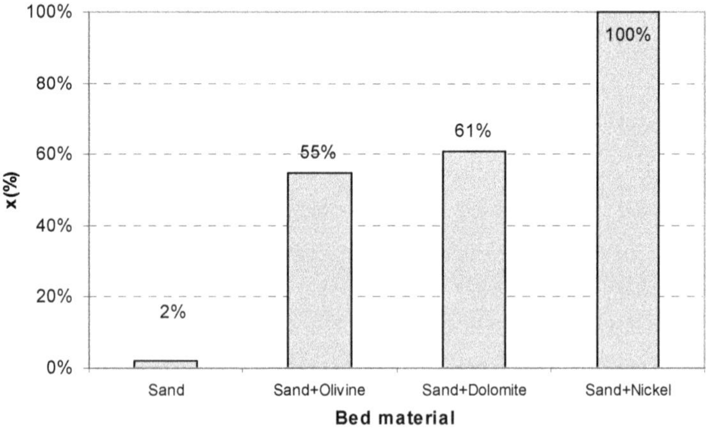

Figure 3: Effect of catalyst bed material on conversion of the model tar component. Bed temperature 900 °C, 40g/Nm^3 initial tar concentration,1 atm and 0.3 s residence time.

FCC, biomass char, commercial char and ash were tested at a bed temperature of 900 °C, 1 atm, 90 g/Nm^3 initial tar concentration and 0.3-0.4 s residence time. All catalysts were tested at 0.3 s residence time, where FCC was tested at 0.3 and 0.4 s residence times. Fig. 4 shows the decomposing tar activity of these materials. The order of reactivity obtained is: commercial char > biomass char > FCC (0.4 s) > ash > FCC (0.3 s). Conversion using FCC catalyst was increased by increasing the residence time from 0.3 to 0.4 s.

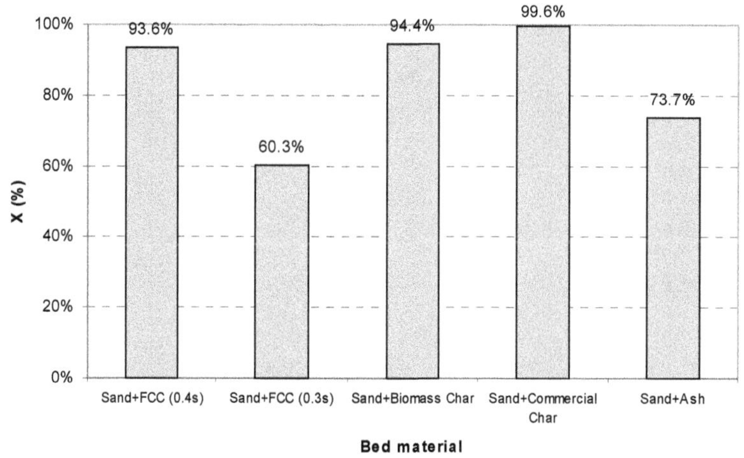

Figure 4: Effect of catalyst bed material on conversion of the model tar component. Bed temperature 900 °C, 90 g/Nm^3 initial tar concentration, 1 atm and 0.3-0.4 s residence time.

The temperature effect on tar conversion was studied for biomass char and FCC catalyst in the temperature range of 700 to 900 °C., as shown in Fig. 5. Tar removal was increased by increasing the bed temperature. As bed temperature decreased the amount of coke formation increased. The coke covered the catalyst active sites and blocked the pores, which leads to deactivation of the catalyst.

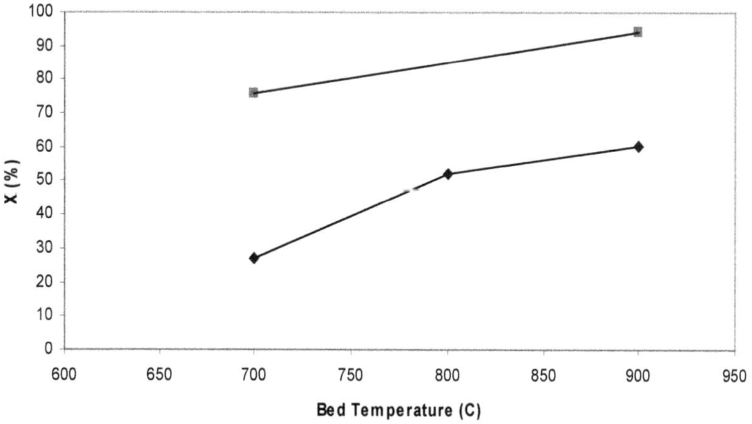

Figure 5: Effect of temperature on tar conversion with biomass char and FCC catalysts at 1 atm and 0.3s.♦, FCC; ■, biomass char.

A kinetic study was done for FCC catalyst at a temperature range of 700 to 900 °C. The apparent rate constant of FCC was varied with temperature by an Arrhenius-type relationship, as shown in Fig. 6. The apparent activation energy (E_{app}) and the apparent preexponential factor ($k_{o,app}$) are found to be 50 kJ/mol and 650 s^{-1} respectively.

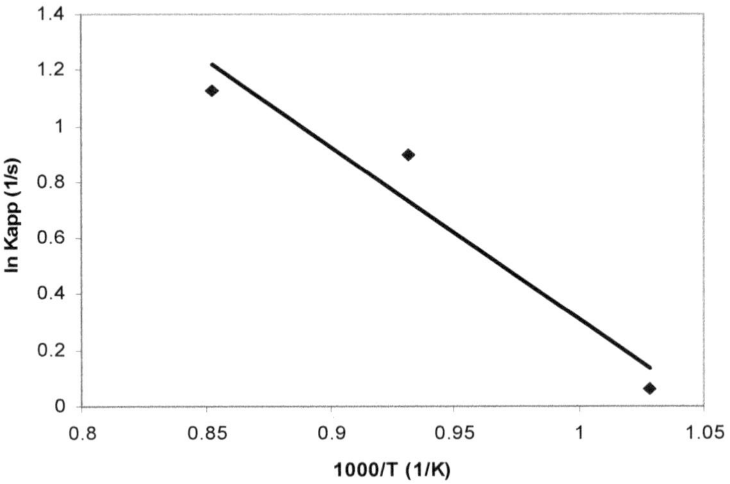

Figure 6: Temperature dependency of the apparent reaction rate constant of FCC according to Arrhenius' low.

CONCLUSIONS

(1) As a first stage of developing a so called entrained flow cracker downstream a biomass gasifier, the catalytic conversion of tarry fuel gas was studied with different bed materials.
(2) The tar removal of naphthalene with CO_2 and water over different solid catalysts was studied at 900 °C, atmospheric pressure and 0.3-0.4 s residence time.
(3) The order of reactivity of these catalysts was found to be: commercial nickel > commercial char > biomass char > commercial FCC (0.4 s residence time) > Ash > dolomite > olivine > silica sand.
(4) Tar removal increased when increasing the bed temperature for both FCC and biomass char in the temperature range of 700 to 900 °C.
(5) For tar removal over FCC catalyst, the activation energy and the preexponential factor for the first order rate coefficient between 700 and 900 °C were found to be 50 kJ/mol and 650 s^{-1}, respectively.

ACKNOWLEDGEMENT

This work is carried out within the framework of the Center of Separation Technology of University of Twente. This center is financed by the Netherlands organization for Applied Scientific Research (TNO) and the Process Technology Institute Twente (PiT).

NOMENCLATURE

$C_{n,in}$	Inlet naphthalene concentration to the reactor, $g.m^{-3}$
$C_{n,out}$	Outlet naphthalene concentration to the reactor, $g.m^{-3}$
E_{app}	Apparent activation energy for the overall tar removal reaction, kJ/mol
$k_{app,n}$	Apparent kinetic constant for naphthalene removal, s^{-1}
$K_{app,o}$	Preexponential factor in the Arrhenius' law, s^{-1}
P_{tot}	Total pressure, Pa
Q	Volume flow rate, $m^3.s^{-1}$
r_{in}	Reaction rate with respect to the volume of empty reactor, m^3
T	Reaction temperature, °K
$V_{R,cat}$	Volume of catalyst bed with respect to the volume of empty reactor, m^3
X_n	Conversion of naphthalene, $g.g^{-1}$
τ	Residence time with respect to the empty reactor, s

REFERNCES

1. Narváez, I., Corella, J. and Orío, A. (1997) Fresh Tar (from a Biomass Gasifier) Elimination over a Commercial Steam Reforming Catalyst. Kinetics and Effect of Different Variables of Operation. *Ind. Eng. Chem. Res.*, **36**, 317-327.
2. Narváez, I., Orío, A., Anzar, P.M. and Corella, J. (1996) Biomass Gasification with Air in an Atmospheric Bubbling Fluidized Bed. Effect of six Operational Variables on the Quality of the Produced Raw Gas. *Ind. Eng. Chem. Res.*, **35**, 2110-2120.
3. Simell, P.A., Leppälahti, J.K. and Kurkela, E.A. (1995) Tar-Decomposing Activity of Carbonate Rocks under High CO_2 Pressure. *Fuel*, **74**(6), 938-945.
4. David Sutton, Brian Kelleher and Julian R.H. Ross (2001) Review of Literature on Catalysts for Biomass Gasification. *Fuel Processing Technology*, **73**, 155-173.
5. Rapagna, S., Jand, N., Kiennemann, A. and Foscolo, P.U. (2000) Steam Gasification of Biomass in a Fluidized-Bed of Olivine Particles. *Biomass and Bioenergy*, **19**, 187-197.
6. Olivares, A., Anzar, M.P., Caballero, M.A., Gil, J., Franés, E. and Corella, J. (1997) Biomass Gasification: Produced Gas Upgrading by In-Bed Use of Dolomite. *Ind. Eng. Chem. Res.*, **36**, 5220-5226.
7. Delgado, J., Anzar, M.P. and Corella, J. (1996) Calcined Dolomite, Magnesite, and Calcite for Cleaning Hot Gas from a Fluidized Bed Biomass Gasifier with Ateam: Life and Usefulness. *Ind. Eng. Chem. Res.*, **37**, 3637-3643.
8. Simell, P., Kurkela, E., Ståhlberg, P. and Hepola, J. (1996) Catalytic Hot Gas Cleaning of Gasification Gas. *Catalyst Today*, **27**, pp. 55-62.
9. Gil, J., Caballero, M.A., Martín, J.A., Anzar, M.P. and Corella, J. (1999) Biomass Gasification with Air in a Fluidized Bed: Effect of the In-Bed Use of Dolomite under Different Operating Conditions. *Ind. Eng. Chem. Res.*, **38**, 4226-4235.
10. Herguido, J., Corella, J., Antal, G., Garcia-Bordeje, J.F. (1992) Results with a Multisolid Circulating Fluid Bed Pilot Plant for the Improved Steam Gasification of Biomass. In: *Biomassfor Energy, Industry and Environmen*t, 6th E.C. Conference. London and New York, pp. 792–796.
11. Sjöström, K., Chen, G., Yu, Q., Brage, C. and Rosén, C. (1999) Promoted Reactivity of Char in Co-Gasification of Biomass and Coal: Synergies in the Thermochemical process. *Fuel*, **78**, 1189-1194.
12. Joseph, S.; Denniss, R.; Lipscombe, R.; Errey, S.; Fletcher, D.F.; Haynes, B.S. (1996) The Development and Testing of an Air/Steam Blown Entrained Flow Gasifier Fueled with CottonWaste and Sawdust. In: *Bioenergy'96-The Seventh National Bioenergy Conference*, pp. 304–311.

13. Don J. Stevens (2001) Hot Gas Conditioning: Recent Progress with Larger-Scale Biomass Gasification Systems, Update and Summary of Recent Progress. NREL/SR-510-29952, Pacific Northwest National Laboratory, Richland, Washington.
14. David Sutton, Brian Kelleher and Julian R.H. Ross (2001) Review of Literature on Catalysts for Biomass Gasification. *Fuel Processing Technology*, **73**, 155-173, ...(ref 147)
15. Jess, A. (1996) Mechanisms and Kinetics of Thermal Reactions of Aromatic Hydrocarbons from Pyrolysis of Solid Fuels. *Fuel*, **75**(12), 1441-1448.
16. Simell, P.A., Hakala, N.A.K. and Haario, H.E. (1997) Catalytic Decomposition of Gasification Tar Benzene as the Model Compound. *Ind. Eng. Res.*, **36**, 42-51.
17. Carnö, J. Berg, M. and Järås, S. (1996) Catalytic Abatement of Emissions from Small-Scale Combustion of Wood – A Comparison of the Catalytic Effect in Model and Real Flue Gases. *Fuel*, **75**(8), 959-965.
18. Evans, R.J. and Milne, T.S. (1997) In: Developments in Thermochemical Biomass Conversion. (Ed by Bridgewater, A.V. and Boocock, D.G.B), pp. 803-8. Blackie Academic & Professional, London, Great Britain.

The Novel "OLGA" Technology for Complete Tar Removal from Biomass Producer Gas

Patrick C.A. Bergman, Sander V.B. van Paasen, and Harold Boerrigter
Energy research Centre of the Netherlands (ECN), Unit ECN Biomass
P.O. Box 1, 1755 ZG Petten, The Netherlands

ABSTRACT: A new process for the removal of tar from biosyngas, OLGA, is presented. This gas cleaning technology is based on physical principles and is designed to eliminate tar problems that relate to fouling and wastewater pollution. The applied principles are confirmed by experimental validation. OLGA can be operated without suffering from harmful effects on important gasification characteristics as cold gas efficiency, gasifier simplicity, and feedstock flexibility.

Some fundamental considerations in the development OLGA are discussed. Amongst these is a different approach to the tar problem. This approach does not consider the total tar content as decisive in the evaluation of gas cleaning equipment. Instead, evaluation is focussed on whether tar properties as condensation behaviour and water solubility can still occur. A biosyngas free of tar should therefore be synonymous to a biosyngas that is free of tar related problems.

INTRODUCTION

THE TAR PROBLEM

Tar can be considered as the Achilles heel of biomass gasification as tars are the major technical obstacle in the implementation this technology. Condensing tars dramatically foul gas-cleaning equipment and liquid tar droplets that enter prime movers hamper the operation of these end-use applications of the biosyngas [3]. In the last few years, it has become clear that tar also plays an important – but negative – role in wastewater management. In conventional water-based gas cleaning systems tars and condensed water are mixed, creating an often costly and difficult water treatment problem.

Regarding the presence of tars in biosyngas, it may be stated that "tars" are equivalent to a major economic penalty in biomass gasification. Tar aerosols and deposits lead to more frequent maintenance and repair of especially gas cleaning equipment and resultantly lower plant capacity factors. This leads to a decrease of revenues or to higher investments, as some equipment will be installed in duplicate to overcome standstills. Furthermore, removal of tar components from the process wastewater requires considerable investments that can even be dramatic as some tar components show poisoning behaviour in biologic wastewater treatment systems (*e.g.* phenol).

Several measures for tar removal have been studied or are under investigation. These measures can be divided in primary measures, measures inside the gasifier, or secondary measures, measures downstream of the gasifier. With the prospect of operating an integrated biomass gasification installation without struggling with tar anywhere downstream, many researcher focus on effective primary measures. This should make complex and expensive gas cleaning equipment obsolete. Although measures inside the gasifier may be fundamentally more ideal, they have not yet resulted in satisfactorily solutions. Some of the primary measures do result in low tar emissions, but suffer from disadvantages related to, for instance, limits in feedstock flexibility and scale-up, the production of waste streams, a decrease in cold gas efficiency, complex gasifier constructions, and/or a narrow operating windows. Although primary measures can reduce the tar content considerably, it is foreseen that complete removal is not feasible without applying secondary measures (see Figure 1).

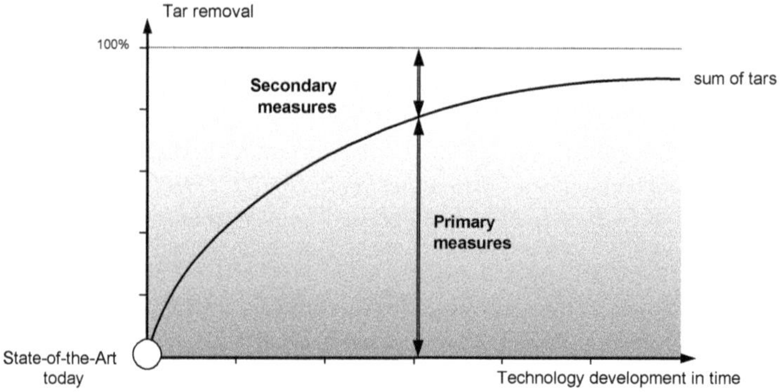

Figure 1: Illustration of the need of primary- and secondary measures versus technology development in time.

Secondary measures that have been investigated till now, exhibit similar deficiencies. The measures are either not effective enough, too expensive, or the tar problem is shifted to the treatment of wastewater. However, a secondary measure can be feasible without needing primary measures. This becomes even stronger when the problems with wastewater treatment can be eliminated as well. A secondary measure should therefore form the basis for tar removal from biosyngas and primary measures could possibly be used for its optimisation.

CURRENT SITUATION (STATE OF MIND)

During the 12th European Biomass Conference held in Amsterdam in 2002, some attendants ventilated great disbelieve in the future prospects of biomass gasification. Tar was tagged to be the major reason for this lack of confidence. If the tar problem could not be solved within the last few decades, after trying so many alternatives for tar removal, why should one believe in future systems wherein tar would not be the major obstacle?

The authors believe that there are arguments enough to disprove this pessimistic view. ECN started mid 2001 with the development of a new technology called "OLGA". The starting point of the development of the patented OLGA technology was to establish tar removal from syngas on the basis of a scrubbing liquid other than water (a secondary measure). This paper will outline some of the features of OLGA, especially with respect to the design approach and performance characteristics.

TAR DEFINITION AND PROPERTIES

Tar comprises a wide spectrum of organic components, generally consisting of several aromatic rings. Reported tar concentrations are strongly dependent on the tar definition used and the measurement method applied. The large amount of tar definitions and measurement methods, as well as the wide spectrum of organic compounds, makes it almost impossible to capture "tars" with a clear definition. This already has resulted in unrealistic tar specifications for end-use applications, as such, that in many cases practically complete removal of tar is demanded. For the development of OLGA, a totally different approach was chosen. This approach does not concentrate on the tar content, but on the behaviour (*i.e.* the properties) of the tar. A feasible tar removal system is not a system that totally *removes* tars, but is a system assuring that tar related problems do not *occur* anymore. Hence, a biosyngas free of tar is in this approach synonymous to a biosyngas that is free of tar related problems

CLASSIFICATION SYSTEM

According to the ECN definition, tar comprises all organic components having a higher molecular weight than to benzene. Benzene is not considered to be a tar. ECN uses a tar classification system comprising six classes (see Table 1). This classification system is in particular developed to provide 'easy' insight in the general composition of tar. Trends are easier recognised on the basis of these classes. However, for more specific problems or issues the detailed data will remain necessary.

Table 1: Tar classification system.

Class	Type	Examples
1	GC undetectable tars.	biomass fragments, heaviest tars (pitch)
2	Heterocyclic compounds. These are components that generally exhibit high water solubility.	phenol, cresol, quinoline, pyridine
3	Aromatic components. Light hydrocarbons, which are important from the point view of tar reaction pathways, but not in particular towards condensation and solubility.	toluene, xylenes, ethylbenzene (excluding benzene)
4	Light polyaromatic hydrocarbons (2-3 rings PAHs). These components condense at relatively high concentrations and intermediate temperatures.	naphthalene, indene, biphenyl, antracene
5	Heavy polyaromatic hydrocarbons (≥4-rings PAHs). These components condense at relatively high temperature at low concentrations.	fluoranthene, pyrene, crysene
6	GC detectable, not identified compounds.	unknowns

From the practical viewpoint, the classification comprises only tar components that can be measured. Classes 2 to 6 are sampled using the solid phase adsorption (SPA) method and measured by gas chromatography (GC). Although class 6 tars are sampled and measured (a peak is found in the chromatogram), it is unknown what the individual components are. In principle components in this class belong to the other classes, but are here lumped to a single concentration representing the 'unknowns'. Class 1 represents the heavy tar fraction (roughly ≥7-ring PAHs). These components cannot be determined by the combination of SPA and GC. The components are measured by weight and thus represent the gravimetric tars.

TAR CONDENSATION: THE TAR DEWPOINT

Tar leads to fouling once the gas becomes (over) saturated with it. This leads to aerosol formation and depositions inside the installation. These fouling phenomena are not of concern as long as all the tar is present in the gas phase. It is therefore believed that the tar problem is fundamentally not concerned with the tar quantity, but is with the properties and the composition of the tar.

The condensation behaviour of tar is an integral effect of all tar components that are present in the syngas. The components their individual contribution to the total tar vapour pressure is therein decisive. When the tar vapour pressure exceeds the saturation pressure of the tar, the gas becomes (over)saturated according Raoult's Law [1]. Thermodynamically, this state leads to condensation of the saturated vapour. The tar dewpoint is the temperature at which the real total partial pressure of tar equals the saturation pressure of tar. Hence, in condensation related issues, the tar dewpoint is a powerful parameter to evaluate the performance of gas cleaning systems. It is believed that, when the dewpoint of tar is reduced to levels below the lowest expected temperature, fouling related problems by condensation or tar aerosols are solved.

To use this approach in design issues, a calculation tool has been developed to predict the tar dewpoint on basis of the concentration of the individual tar components in the syngas. An illustration of the relation between the tar dewpoint and tar concentrations is provided by Figure 2. Condensation curves are given for the individual tar classes (as defined in Table 1), *e.g.* the dewpoint curve for class 5 is calculated including only class 5 tars. Furthermore, each tar component is contributes equal to the total concentration on mass basis. The dewpoint calculation excludes tar class 1, as the components are not known. For a CFB gasifier it is believed that tars that belong to class 1 start to condense around 300-350°C.

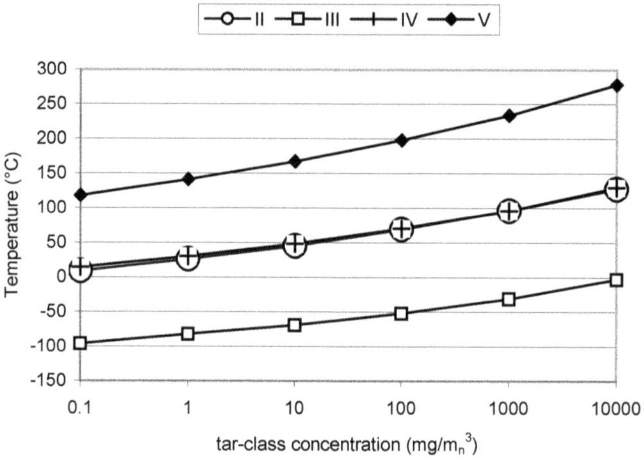

Figure 2: The tar dewpoint of the different tar classes in relation to the concentration.

Leaving out class 1 in this discussion, it can be derived from Figure 2 that class 5 tars dominate the dewpoint of tar. Even for very low concentrations of class 5 tars (*e.g.* <1 mg/m$_n^3$) a dewpoint below 100°C cannot be obtained. The graph clearly points out that, dependent on the concentration in the syngas, classes 2 and 4 need to be partially removed for a proper tar dewpoint of about 25°C. The class 3 tars play an unimportant role in this matter.

WATER SOLUBILITY

The pollution of wastewater is in strong relation to the type of tar components being present in the biosyngas. The (poly)aromatic non-polar components will practically not dissolve, however, small non-polar components may still form a problem as they dissolve in small amounts that can exceed allowable concentrations. In general this will not cause a problem as due to the volatility of these components they are easily removed from water. Polar components on the other hand, in particular phenol, dissolve in large quantities and are very difficult to remove. Waste process water from biomass gasification must be clean, which is much easier to accomplish when pollution of tar can be avoided.

A similar tool as for the calculation of the tar-dewpoint is in development for the calculation of the water solubility of the tar classes. This tool was not available for this work. The class 2 tars are most important with respect to water solubility. This class comprises the oxygen, nitrogen, and sulphur hetero-atoms containing components that dissolve well. Also the class 3 (and benzene) can be important with respect to wastewater treatment. These components may dissolve in large quantities but they are readily removed. Other classes are typically insoluble and form a two-phase liquid/liquid system of tar and water with a rather low mutual solubility.

DESIGN OF OLGA

PROCESS OBJECTIVES

The (ambitious) objective in the design of OLGA was the creation of a new process that eliminates issues involved with tar condensation and water solubility. The process to develop should be competitive with competing technologies.

PROCESS TASKS

The approach to design a process for complete and selective tar removal in a controlled way, started with a definition of the required tasks for such a process. Primary and secondary tasks are distinguished (see Table 2). Primary tasks deal directly with the objective. Secondary tasks are additional and need to be accomplished to obtain a system that also meets specifications from, for instance, the economic and legislation points of view. In contrast to the primary tasks, the secondary tasks are only indirectly responsible for the technical feasibility of the system.

Table 1: Primary and secondary tasks of OLGA.

Primary tasks	
1.	Selective tar removal (*viz.* no water removal).
2.	Deep removal of tar components resulting in a syngas quality for which no tar condensation or tar desublimation occurs, and simultaneously absence of tar aerosols, while applying the desired operating conditions.
3.	Specific removal of heterocyclic tar components (in particularly phenol), to avoid water contamination in the wet syngas cleaning that is necessary to remove contaminants like NH_3 and HCl.
Secondary tasks	
4.	Avoiding waste streams
5.	Avoiding a (too) high scrubbing liquid consumption. This is in particularly important with respect to process economics, but surely also with respect to de sustainable image of biomass gasification.
6.	Removing dust and/or fines that have not been removed by dust separators upstream of OLGA.
7.	Preventing a high gas-side pressure drop over de gas cleaning system.

BASIC PROCESS STRUCTURE

The biggest challenge is to remove tar selective from the syngas (task **1**). In particular water must not be absorbed by the applied scrubbing liquid, as that would still lead to the pollution of process water. Similarly, the permanent gas components in the biosyngas (*e.g.* CO, H_2, and CO_2) should not dissolve in, or be absorbed by, the scrubbing liquid. This would not contribute to the simplicity of the process.

Deep removal of classes 1 and 5 tars is desired in order to decrease the tar dewpoint and to eliminate condensation problems (task **2**). Complete collection of these tar classes yields a dewpoint below 100°C (*cf.* Figure 2). Furthermore, to operate end-use applications that require syngas temperature below 50°C, without the risk of tar condensation, classes 2 and 4 tars need to be removed partly. The required collection efficiency depends on the actual amount and composition of the tars in the syngas (*cf.* Figure 2). Although the collection efficiency of class 2 tars (the heterocyclic tars) needs not to be complete from the condensation point of view, essentially quantitative removal is required to avoid of pollution of process water (task **3**).

The elimination of all condensation-related issues means also that no tar condensation may occur upstream OLGA. Hence, the syngas inlet temperature of OLGA must be higher than the tar dewpoint of the raw syngas. As a consequence of task 1, water present in the syngas may not condense simultaneously with tar. Therefore, the exit temperature of OLGA must remain above the water dewpoint of the syngas. Figure 3 illustrates this further by positioning OLGA with respect to both the tar and the water dewpoint.

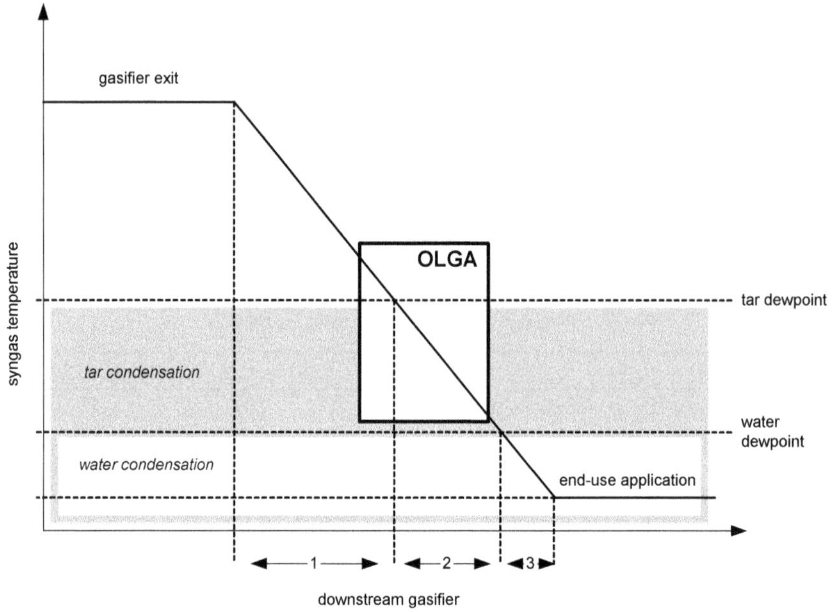

Figure 3: The position of OLGA with respect to the syngas temperature and to the dewpoints of water and tar. Explanation on zones 1, 2, and 3 downstream the gasifier: (tar phase/ water phase) 1: (G/G), 2: (L/G), 3: (L/L)

Upon cooling of the biosyngas, the temperature decreases below the tar dewpoint and tar condensation gradually takes place until syngas is not cooled further. At the resulting temperature, between the dewpoint of tars and water, a liquid/gas (L/G) phase system is obtained (L represents liquid tar). The scrubbing liquid acts as the medium to collect these liquid tars. The remaining tars are collected into the scrubbing liquid by absorption (*i.e.* the scrubbing liquid acts as absorption medium). This is illustrated by Figure 4.

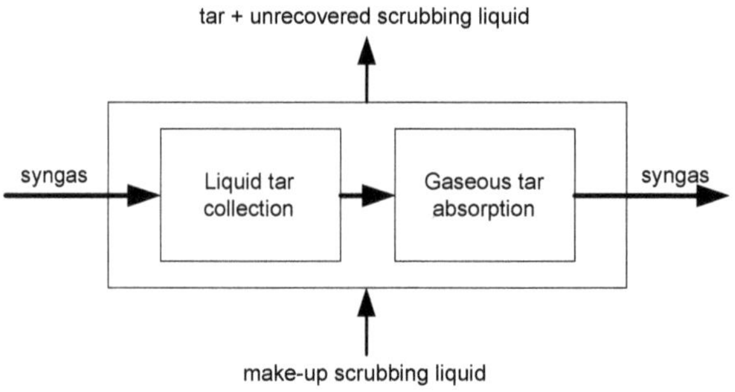

Figure 4: Basic structure of OLGA.

The degree of absorption latter can be controlled by changing the operation conditions and will be determined by the desired tar dewpoint of the outlet syngas. In the regeneration of the scrubbing liquid the tar is removed upon which some scrubbing liquid may evaporate.

CONCEPTUAL PROCESS STRUCTURE

Figure 5 depicts the general concept of the air-blown biomass gasification process using OLGA. The produced syngas is first cooled and de-dusted upstream of OLGA. Downstream OLGA, the main (inorganic) non-tars impurities NH_3 and HCl are removed by wet scrubbing and water is condensed out due to further cooling of the gas. The syngas is then suitable for most of end-use applications as it is free of condensable tars, tar aerosols, as well as inorganic impurities. Tar and unrecovered scrubbing liquid are fed to the gasification process and also the separated NH_3 is recycled to the gasification process. Similarly to the tar recycling, separate experiments at ECN have shown that NH_3 is converted to nitrogen in the gasifier [2].

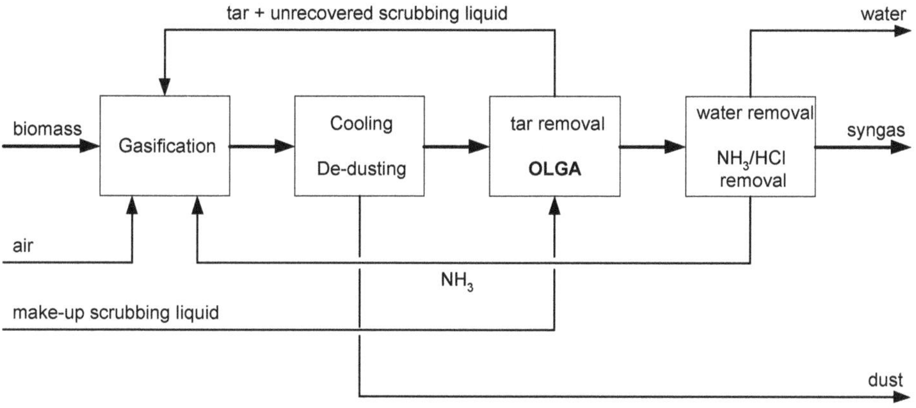

Figure 5: General concept of air-blown biomass gasification using OLGA for tar removal.

Although the major part of dust and/or fines will be collected upstream OLGA, it is inherent to the use of a scrubbing liquid as medium that fines will collected in the scrubbing liquid. The removal of small particles in OLGA is considered as a secondary task (task **6**) that is optimised as much as possible so that small particles do not have to be dealt with further downstream. It is considered as one of the interesting and economical attractive optimisation options of OLGA to remove the full dust load from the biosyngas, making the separated dust removal step upstream obsolete.

Intrinsically related to the use of a scrubbing liquid as process utility, is the consumption of liquid due to bleed streams and volatilisation upon stripping the tar. Even if the scrubbing liquid is very effective and the losses minimal, the process may not create another waste stream (task **4**) as this causes an economic penalty for the waste handling.

It is furthermore undesirable as the consumption of scrubbing liquid creates an image for the process, which is in contradiction with the 'green' nature of the biomass gasification. The necessity to minimise scrubbing liquid losses motivated the inclusion and design of a regeneration step and has been a major selection criterion for the scrubbing liquid to be applied (task **5**). The collected tar and the unrecovered scrubbing liquid from the regeneration step are recycled to the gasifier, preventing a waste stream. Separate experiments carried out at ECN have shown that tars can be recycled and gasified without accumulation [2].

APPLIED EQUIPMENT

A simplified flow sheet of OLGA is provided by Figure 6. OLGA consists of scrubbing towers interacting with each other in a classical absorption-regeneration mode. Syngas is fed to the tar collector in which tars are removed from the gas to the desired level. The scrubbing liquid with the dissolved tars is regenerated in the stripper. Part of the scrubbing liquid exiting is purged and charged to the gasifier. In case of air-blown gasification, air is used to strip the tar. Subsequently, the air with the stripped tars is used as gasifying medium.

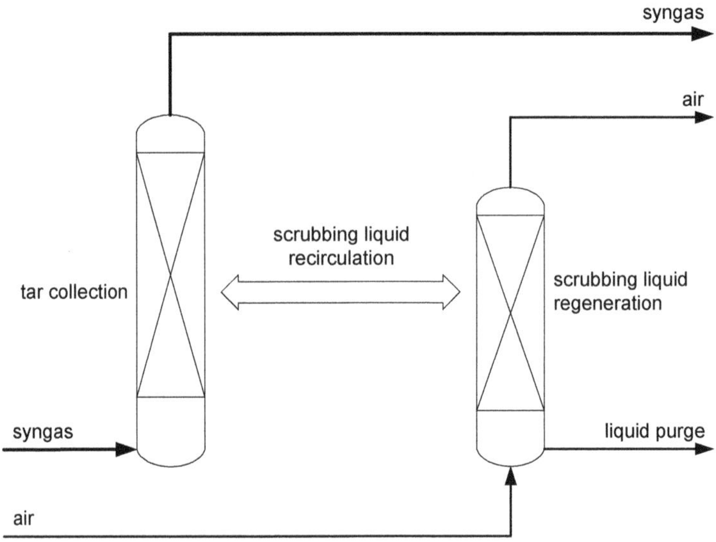

Figure 6: Simplified flow sheet of OLGA.

A practical reason for the selection of a scrubbing tower for the removal of tar is that the gas-side pressure drop can be limited (task **7**). The first application of OLGA is foreseen for atmospheric gasification processes and it is important to limit the total gas side pressure drop over the whole installation from gasifier to end-user of the syngas. The typically used simple solid feeding systems, which encounter the highest absolute pressure, used can generally only function at small pressure drops. Therefore, equipment in OLGA with inherent high-pressure drop, such as venturi scrubbers, is avoided. The selected equipment for OLGA is mature, a lot of operational experience is available, and moreover, it is well known how to scale-up this type of equipment.

EXPERIMENTAL VALIDATION

OLGA was built on lab-scale to demonstrate its feasibility (Proof-of-Principle). The lab-scale OLGA was designed for this capacity and was built according general scale-up rules valid for the selected equipment. This ensured that the generated data could be used for scale-up of OLGA. The experimental program consisted of two series of experiments. The first series dealt mainly with the primary tasks (*cf.* Table 1). The second series dealt with the generation of design data for scale-up of the OLGA to bench-scale. The ECN atmospheric air-blown bubbling fluidised bed (BFB) gasifier "WOB" was used to generate biomass-derived syngas. The WOB has a capacity of 1 kg/h of fuel producing 2 m_n^3/h wet biosyngas. The syngas typically contains 18 g/m_n^3 of tar (including toluene and xylenes; on dry basis). Dust was removed from the biosyngas with a high efficiency filter before the syngas was fed to OLGA.

Figure 7 shows a typical result obtained from the proof-of-principle experiments. The outlet gas of OLGA is free of heavy tars (class 1 and 5), also free of class 4 tars and nearly free of tars that are classified as unknown (class 6). The collection efficiency of the class 3 tars is about two-third and nearly all tar present in the syngas exiting OLGA are class 3 tars (*viz.* toluene and xylenes with a concentration of 1,534 mg/m_n^3). A few milligrams of class 2 tars are still present in the cleaned biosyngas (removal efficiency 97%). Permanent gas components (*e.g.* CO and H_2) do not dissolve in the scrubbing liquids and also water is not absorbed.

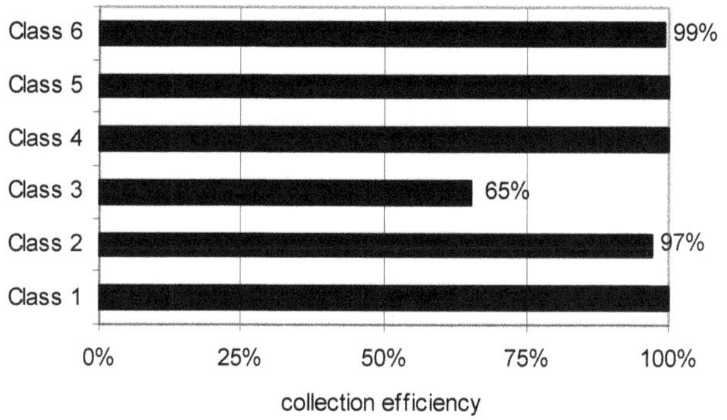

Inlet tar concentration: 18,000 mg/m_n^3

Outlet tar concentration: 1,550 mg/m_n^3

Tar dewpoint: -16°C

Figure 7: Typical results obtained from the lab-scale OLGA facility.

The dewpoint corresponding to the presented tar collection efficiencies is -16°C. In considering water solubility, the class 2 tars were not completely removed. However, the generated design data clearly indicate that the system can be easily optimised towards deep removal of class 2 tars, especially phenol. This was confirmed in an additional experiment in which the optimal conditions for phenol removal were applied. This experiment revealed that 99% of the phenol can be removed. The design data that was generated from the experimental results reveal that OLGA is also capable of complete tar removal if that would be necessary.

The second series of experiments showed that the applied scrubbing liquid can be easily regenerated without significant losses by evaporation. The tar collection and stripping of the scrubbing liquid proved to be very effective and without unacceptable losses of scrubbing liquid. The syngas conditioned by OLGA did also not contain any particulate matter anymore. Although the concentration of particulate matter before OLGA was not measured, small amounts of dust have been found in the scrubbing liquid. This indicates that fines are removed as well.

CONCLUSIONS

Based on the experimental results it can be concluded that with the OLGA technology a new and effective secondary measure is introduced for tar removal in biomass gasification. The approach of eliminating the problems due to tar (condensation and water solubility) instead of eliminating tar itself has resulted in a structural solution to the major tar problems. Despite the high tar collection efficiency of OLGA, no concessions need to be made to other important gasification characteristics, for instance, the cold gas efficiency remains unaffected. Also complex gasifier constructions as a consequence of primary measures are not required. Consequently, a simple and robust gasifier can be deployed. This assures the desired fuel flexibility and flexibility in operation.

OUTLOOK

The future for biomass gasification is prosperous now the tar issue has been resolved and this major (technical) obstacle for implementation is removed: *Achilles is no longer vulnerable*. Together with the Dutch company Dahlman Industrial Group, ECN Biomass is working on the further development of the OLGA technology and demonstration on bench-scale. Engineering has already started, construction is foreseen to start in December 2002, and the first tests are scheduled for June 2003. After this Proof-of-Concept phase the OLGA technology is ready for commercialisation.

ACKNOWLEDGEMENTS

The work described in this paper was partly sponsored by Novem-NECST. Furthermore, the authors acknowledge Herman Bodenstaff, Ruud Wilberink, and Johan Kuipers for carrying out the experimental work and gas analyses.

REFERENCES

1. Reid, R.C.; Prausnitz, J.M.; Polling, B.E. (1988) *The properties of gases & liquids*, McGraw-Hill, 4th edition.
2. ECN Biomass, *Publication in preparation.*
3. Swaaij, W.P.M. van; Aarsen, F.G. van den; Bridgwater, A.V.; Heesink, A.B.M. (1994), "A review of Biomass Gasification", A report to the European Community DGXII Joule Programme,

Reforming of Biomass Wastes into Fuel Gas with High Temperature Air and Steam

W. Blasiak, D. Szewczyk, C. Lucas
Royal Institute of Technology (KTH), Division of Energy and Furnace Technology, S-100 44 Stockholm, Sweden

ABSTRACT: The presented work aims to provide information on solid biomass conversion into fuel gas as a result of air and steam gasification process. In contrary to the conventional gasification in this work highly preheated air and steam is used as a gasifying agent. Preheat of air and steam is realised by means of the high-cycle regenerative air/steam preheater. Use of highly preheated gasifying media provides additional energy into the gasification process that enhances the thermal decomposition of the gasified solids. The objective of this work is to provide information on the process of biomass wastes reforming to fuel gas using high temperature air and steam gasification.

High Temperature Air/Steam Gasification of biomass wastes has very clear economical and environmental benefits. It will increase consumption of biomass thus decreases CO_2 emissions. Apart from CO_2 reduction possibility, the new process, High Temperature Gasification has the following advantages:

- Clean fuel gas for production of heat or electricity,
- No need to treat ashes from gasification since there is no ash or at least no carbon in the ash produced,
- No need to landfill since the produced slag can be used for example as a building material.

In this work high temperature air and steam gasification of charcoal and wood pellets in a fixed bed updraft gasifier is tested. The following conclusions were found out:

- For both charcoal and wood pellets gasification cases it was seen that an increase of steam molar fraction in the feed gas decreases the temperature of the gasification, the gasification rate, the mass yield of the fuel and molar fraction of carbon monoxide but increases molar fraction of hydrogen,
- An increase of the feed gas temperature reduces production of tars, soot and char residue as well as increases calorific value of the fuel gas produced,
- Preheating of the feed gas obtains a higher gasification efficiency and a higher gasification rate,
- High temperature air and mixture of air and steam used as feed gas in a fixed bed gasifier shows the capability of this technology of maximising the gaseous product yield,
- High lower heating value of fuel gas and high molar fraction of hydrogen and hydrocarbons obtained by using high temperature air and steam gasification technology increase the technical possibility of the application of fuel gas produced.

INTRODUCTION

Research and Development carried out in Japan under leadership of Ryoichi Tanaka [1] on High Temperature Air Combustion (HiTAC) of gas and liquid fuels was followed by development of a new solid fuel gasification technology [2]. Particularly progress has been made in the area of High Temperature Air Combustion [3] and on designing of highly efficient regenerative air and steam preheaters [4] made this development faster. This progress has created the possibility to begin a systematic research on gasification processes with high temperature air and steam called High Temperature Air/Steam Gasification (HTAG) [5,6,7].

Gasification involves partial oxidation of the feedstock to obtain a mixture of hydrogen, carbon oxides, water, nitrogen if air is used and small amounts of methane and higher hydrocarbons. The product gas also contains variable amounts of ash particles, volatile alkali metals and tar, which is a complex mixture of aromatics that includes a significant fraction of polycyclic aromatic hydrocarbons [8]. Currently, methods to remove tars from fuel gas fall into one of the three categories [9]: physical removal, thermal conversion of tars, and catalytic destruction of tars. Removal of tars via physical processes is a separation process. Thermal conversion, also called "cracking", is the use of extremely high temperatures to disintegrate complex organic compounds into more benign form. Catalytic destruction of tars involves the use of catalysis to promote disintegration of tars.

A limited supply of oxygen, air, steam or a combination serves as the oxidizing agent. Using steam on the composition of the feed gas, water gas reaction ($C + H_2O \leftrightarrow CO + H_2$) and water gas shift reaction ($CO + H_2O \leftrightarrow CO_2 + H_2$) are assumed to take place during gasification process. The primary water gas reaction (endothermic) becomes significant at temperatures from 1000 to 1100°C and upward. The secondary water gas reaction (exothermic) begins and predominates between 500 and 600°C while the water gas shift reaction is purely a gaseous reaction, which takes place in the presence of undecomposed steam [10,11,12]. In total, both reactions are net endothermic and supplemental heat must be supplied to carry the reaction to completion [13]. Therefore, an increase or decrease of the temperature displaces either of the both reactions in opposite directions [13]. Consequently, the temperature is an important variable with regard to the final composition of the gases. Cummer and Brown [9] have reported that addition of steam is effective in increasing cracking rates in the temperature range of 950-1250°C. Oxygen at 600-700°C accelerates the destruction of primary products, and inhibits the formation of aromatics; however, once benzene, the primary component of aromatic compounds in tar, is formed it cannot be destroyed by oxygen. Moreover, according to [12], steam-biomass reaction rates may be enhanced relative to the completing reactions by increasing the gasifier steam concentration. Higher temperature will also result in higher rates of reaction and reduce required residence times [14].

In this paper, results of part of an extensive investigation on thermal conversion concept applying high temperature air and mixture of air and steam carried out at the Division of Energy and Furnace Technology at the Royal Institute of Technology in Sweden are presented. The overall objective of the project is to carry out a parametric study on a fixed bed counter-current gasifier, applying the concept and technology of high temperature air and steam gasification as such, since it can improve the efficiency of wastes and biomass gasification process.

Discussion of the results in this work takes into consideration the following simplified steps of gasification process. During the gasification process the following three phases could take place. Energy release for this process is described by simplified combustion mechanism below. The first phase is a process of pyrolysis during which the biomass is converted by heat into charcoal and volatile compounds. The second phase is an exothermic reaction in which part of the carbon is oxidised to carbon dioxide. In the third phase, part of the carbon dioxide, the volatile compounds and the steam are reduced to carbon monoxide, hydrogen and methane. This mixture of gases diluted with nitrogen from the air and unreduced carbon dioxide is known as fuel gas. If the original feedstock is charcoal, then the gasification process becomes two phased, and the amount of tar produced is cut down [8].

The conversion process from solid biomass to fuel gas resolve around the initial combustion reaction (1), which occurs in the reactor, nearest the air inlet.

$$C + O_2 + 3.76N_2 \rightarrow 3.76N_2 + CO_2 \quad (1)$$

The carbon dioxide produced in this first reaction is then, in the presence of glowing carbon, reduced to carbon monoxide (2).

$$C + CO_2 + 3.76N_2 \rightarrow 3.76N_2 + 2CO \quad (2)$$

When the gasifier is first lit, the level of carbon dioxide is at its highest. The quality of the fuel gas increases with the increase in gasifier reactor temperature. The percentage of the carbon dioxide reduced to carbon monoxide will depend on the temperature in the gasifier.

The ideal reaction that should occur in the gasifier bed is (3);

$$2C + O_2 + 3.76N_2 \rightarrow 3.76N_2 + 2CO \quad (3)$$

Chemical reactions (1), (2) and (3) are regarded as only occurring with carbon, and with wood as the feedstock there are other elements involved. The main constituents of dry wood are cellulose (40-50%), hemicellulose (20-30%), lignin (20-25%), resin, and trace mineral matter [15]. When these residues are heated in an inert atmosphere they decompose into three phases: solid (charcoal), liquid (tars) and gases. The solid phase is generally the most interesting because its porous structure means that it can be exposed to a gasification process with a reactive gas, such as CO_2 or steam, yielding active carbon or a gas with high hydrogen and/or methane content, which might be used as a substitute for natural gas [15].

Steam is one of the most commonly used gasification agent due to the good quality of the gas obtained, with a high percentage of hydrogen [8,10,11,16,17]. In the steam gasification of carbonaceous material, several reactions take place simultaneously; see Equations (4) – (8). The extent to which each reaction is involved depends on the operating conditions. Thus reaction (6) requires high temperature, and reactions (7) and (8) require high pressure. Hence, reactions (4) and (5) are the main reactions of interest in steam gasification.

In biomass gasification, the gas can contain significant amounts of tars which can be eliminated by thermal cracking (high-temperature gasification) or by the use of a catalyst (catalytic gasification). According to Garcia et al [18] high- and low-temperature gasification processes can be distinguished, the demarcation being around 700-750 °C. An increase in the gasification temperature causes an increase in the gaseous product yield and a decrease in the tar yield. Despite this lower tar content, the remaining tars must be eliminated in order to be able to use the gas obtained. These tars are mainly polyaromatic hydrocarbons, which are difficult to convert, and the relative proportion of PAH increases with the gasification temperature [18].

The main secondary reactions involved are:

$$C + H_2O \leftrightarrow H_2 + CO \quad (4)$$
$$CO + H_2O \leftrightarrow H_2 + CO_2 \quad (5)$$
$$C + CO_2 \leftrightarrow 2CO \quad (6)$$
$$CO + 3H_2 \leftrightarrow CH_4 + H_2O \quad (7)$$
$$C + 2H_2 \leftrightarrow CH_4 \quad (8)$$
$$\text{tar} + H_2O \rightarrow H_2 + CO \quad (9)$$
$$\text{tar} + H_2 \rightarrow \text{light hydrocarbons} + \text{gases} \quad (10)$$

Equations (4) and (6) represent the gasification reactions of carbon deposits with H_2O and CO_2. These reactions become significant at 700°C. In addition to being endothermic, reaction (4) is relatively slow compared to carbon-oxygen or carbon-carbon dioxide (Boudouard – reaction 6). Steam-biomass reaction rates may be enhanced relative to the competing reactions by increasing the reactor steam concentration. Higher temperatures will also result in higher rates of reaction and reduce required residence times [12].

EXPERIMENTAL SET UP

High Temperature Air/Steam Gasification (HTAG) test facility (Figure 1) has been built at Royal Institute of Technology (KTH). Figure 2 shows the photograph of the HTAG test facility. This is equipped with the following main equipment:

- Batch type, counter-current and the up draft fixed bed gasifier (Figure 4). Fixed-bed gasifier, is a vertical cylindrical reactor, which consists of two sections: a wind box, the feedstock referred as gasifier bed (fixed bed) on top of which off gas flows out of the fixed bed gasifier.

- Afterburning combustion chamber (shortly called afterburner) to burn completely the produced fuel gas. Afterburner has one inlet for the fuel gas from the gasifier and one outlet for the flue gas. Afterburner is also equipped with a pilot burner and set of additional air nozzles to assure complete combustion of the fuel gas produced.
- Electrical heater to preheat feed gas in the lower range of temperatures (below 600°C).
- Highly preheated air generator (Figure 3) to preheat feed gas up to 1250°C.
- Electrical steam boiler to produce slightly preheated steam. The boiler is equipped with a water preparation unit.
- Air blower supplying air into heaters and next to the gasifier.
- Flue gas fan placed at the outlet of the afterburner to cover pressure losses. In this way also the pressure inside the gasifier is controlled.
- Set of flow meters and pressure regulators to control flow rate of air and steam.
- Micro Gas Chromatograph (GC) configured with four independently controlled modules to measure the fuel gas composition (6),
- Data acquisition system.

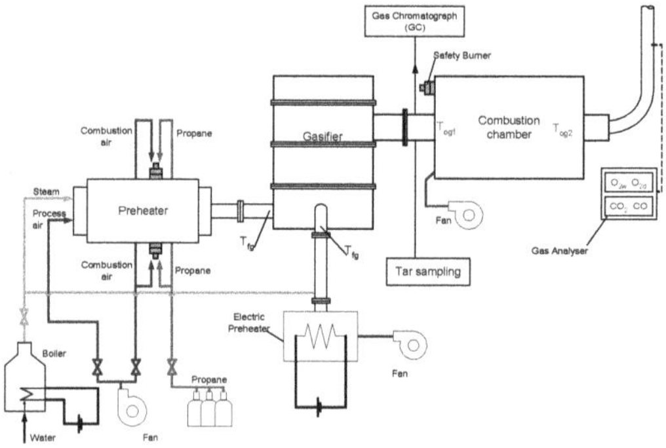

Figure 1: HTAG experimental facility at KTH.

Figure 2: Photograph of HTAG facility stand at KTH with view of the bed (inside gasifier).

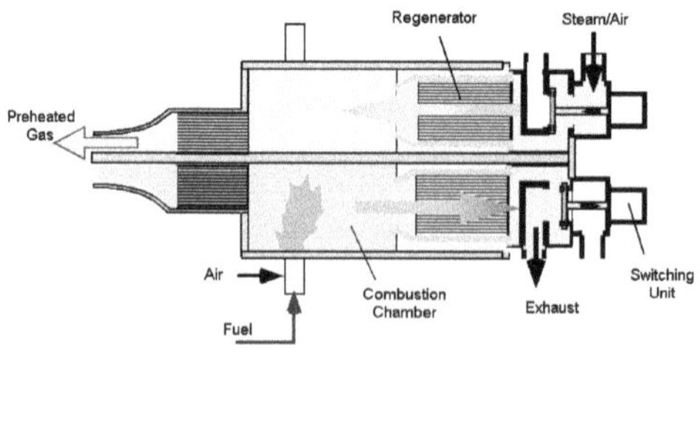

Figure 3: Concept of highly preheated gas generator.

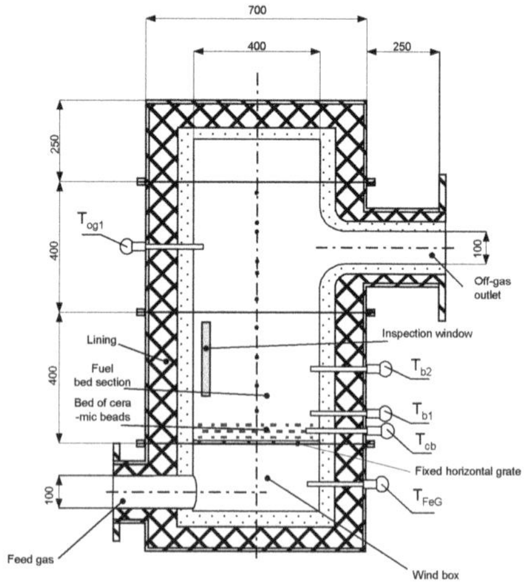

Figure 4: Side view of the KTH gasifier.

A compact high temperature air generator provides the supply of high-temperature air or mixture of air and steam. Figure 3 shows a conceptual diagram of a highly preheated gas generator [4]. While the regenerator located bottom of a combustion chamber heat up by combustion gas, the highly preheated gases go to exit through another chamber. Heat storage and heat release in the regenerators are repeated periodically when combustion gas and low temperature gas are alternately provided by on-off action of switching valve located on the low temperature side. The preheated gas continuously discharge from each exit nozzle at top section, and combustion gas exhaust from bottom section. The features of the facility are that switching valves for regenerator are used only in the low temperature section and the pressure drop at the exit of facility is set larger than that of the combustion gas passing so as to prevent occurring inverse flow of combustion gas in the chamber.

FEEDSTOCK CHARACTERISTICS

Figure 5 shows photographs of the two of feedstock used in studying reforming into fuel gas. High temperature gasification with air and mixture of air and steam was used in gasifying wood pellets, while relatively low temperature air and mixture of air and steam was used in gasifying charcoal.

a) Wood pellets b) Charcoal

Figure 5: Photographs of feedstock used in this study.

As concluded by McKendry [19], key parameters for successful gasification are the feedstock properties (moisture, ash, alkalis and volatiles) and feedstock pre-treatment (drying, particle size, fractionation and leaching). Characteristic values of the wood pellets and of charcoal investigated in this study are presented in Table 1 below.

Table 1: Feedstock analysis.

Proximate/Ultimate Analysis	Charcoal		Wood pellets 12mm	
	As received	Dry weight basis	As received	Dry weight basis
Surface Moisture	3.8			
Inherent Moisture	5.9			
Total moisture [%]	9.7		8.2	
Ash [%]	0.6	0.7	0.4	0.4
Volatile matter [%]	28.6	31.6	77	83.9
Fixed carbon [%]	61.1	67.7	14.4	15.7
Higher Heating Value [MJ/kg] (3)	29.8	33.0	18.8	20.5
Lower Heating Value [MJ/kg] (3)	-	32.5	17.4	19.1
Carbon [%]	82.4	91.3	46.3	50.4
Hydrogen [%]	4.1(1)	3.4	6,6	6.2
Nitrogen [%]	0.2	0.2	<=0.2	<=0.2
Oxygen [%]	12.58(1)	4.39	46.5(2)	42.8
Sulphur [%]	0.03	0.03	0.04	0.04

(1) – Water included, (2) – includes hydrogen in the moisture, (3) – constant pressure

Because wood has a higher hydrogen/carbon ratio (about 0.12) and oxygen/carbon ratio (0.85) than charcoal (respectively 0.037 and 0.048), a higher gasification yield of both gases and hydrocarbons such as tars is expected. Higher chemical efficiencies are thus expected during gasification of charcoal due to its lower H/C compared to wood pellets. This can be explained by the fact that at lower H/C ratios, less product gas has to be recirculated into the gasifier. Instead of this, more carbon of the biomass is transported directly into the combustion zone, which is thermodynamically beneficial [16]. Tars normally represent an important fraction of the heat of combustion of the gas but the amount and composition of the tars is dependent on the fuel, pyrolysis conditions and secondary gas phase reactions.

EXPERIMENTAL PROCEDURE

Start-up of the gasifier includes all operations required until a steady state, whereby the pre-set feed gas temperature is stable. Pre-weighed batches of feedstock are loaded on top on the ceramic beads from the top side of the gasifier. Initial height of the bed was equal to about 0.35 m and diameter was equal to 0.4 m. Fuel is ignited immediately after closing the conversion system. The common feature of the experiments is that ignition occurs on the bottom of the fuel bed.

Shut down procedure includes all the actions to safely seal the gasifier. Computer, steam boiler, heat up process of the feed gas preheater are shut down in an orderly sequence with the 150 kW safety burner on the afterburner remaining on using a propane gas until no combustible gas is produced. Shut down of all fans is the last procedure following cooling of feed gas preheater and of the gasifier. After the reactor cooled down, the top section of the gasifier was opened so that all ash an/or slag were carefully removed and monitored.

The following parameters are measured:

- Temperature of preheated feed gas (T_{FeG}), reacting fixed bed ($T_{b1,}$ T_{b2}), and of the fuel gas (T_{og1}), by means of type N thermocouple, and temperature of flue gas by means of type B thermocouple,
- Fuel gas composition with help of micro Gas Chromatograph (GC) connected to the gas sampling probe,
- Pressure along the air lines,

The following gas components of the fuel gas are measured: hydrogen (H_2), carbon monoxide (CO), carbon dioxide (CO_2), nitrogen (N_2), oxygen (O_2), methane (CH_4) and higher hydrocarbons (CxHy). The composition of fuel gas was determined from gas samples taken at the gasifier outlet. All measurements were performed using a sampling frequency of one sample per 115 seconds. The run time of 115 seconds is built up in a sampling time of 15 seconds and an analysis time of 100 seconds.

In this work, influence of the air to steam ratio in the feed gas was investigated during gasification of wood pellets and charcoal, while the influence of other main parameters, such as feed gas temperature and flow rate of feed gas on the phases of gasification process of wood pellets were investigated and reported elsewhere [5,6,7]. Experimental conditions for the biomass fuels gasification carried out in a fixed bed gasifier whose results are object of study in this paper are presented in Table 2.

Table 2 gives the operation conditions for the case of gasification of charcoal in the experiments with air and mixture of air and steam used as feed gas. Mass of charge was kept constant and equal to 6.11 kg. For the same volume flow rate of the feed (50 Nm^3/h) and same temperature (360°C), the molar fraction of steam was varied from 0 up to 25%.

Table 2 also shows operation conditions of 4 selected gasification runs of wood pellets (sized 12 mm in diameter). The mass of charge was constant and equal to 20 kg. Temperature of the feed gas was slightly varied from 830 up to 920°C whereas its flow rate was slightly varied from 50 to 56 Nm^3/h. Molar fraction of the steam in the feed gas was varied from 0 up to 83%.

The type of biomass largely determines the nature of the tar produced, which is also influenced by the gasification process and the operating conditions. Air gasification produces a low viscosity/low reactivity tar, while steam gasification produces a liquid tar with a low molecular weight. High temperature gasification gives a tar with low oxygen content, consisting mostly of hydrocarbons [19].

Table 2: Operation conditions on the fixed bed gasification of charcoal and wood pellets 12 mm.

Case number	Fuel type	Temperature of the feed gas	Total flow of feed gas	Molar fraction of steam in the feed gas	Mass of fuel
[-]	[-]	[°C]	[Nm³/h]	[%]	[kg]
Ch-1	Charcoal	360	50	0%	6.11
Ch-2	Charcoal	360	50	10%	6.11
Ch-3	Charcoal	360	50	20%	6.11
Ch-4	Charcoal	360	50	25%	6.10
WP-1	Wood pellets	830	50	0%	20
WP-2	Wood pellets	920	55	25%	20
WP-3	Wood pellets	900	53	52%	20
WP-4	Wood pellets	900	56	83%	20

ANALYSIS OF RESULTS AND DISCUSSION

FUEL GAS COMPOSITION

The goal of this study was to study the influence of the amount of steam in feed gas during gasification of two different feedstocks: charcoal and wood pellets. In both feedstocks, temperature of the feed gas was kept constant, being 360°C for charcoal and 900°C for wood pellets. No attempt is made, however, to compare results related to the influence of the feedstock as gasification of charcoal yet corresponds to the initial stage of the ongoing systematic parametric studies. Results of the influence of temperature on the gasification of wood pellets were presented by these authors elsewhere [5,6,7].

Figure 6 and Figure 7 show the evolution of fuel gas composition, lower heating value of the fuel gas (LHV), temperature of the gasifier bed (T_{b1}) and temperature of the fuel gas exiting the gasifier (T_{og1}) during the gasification of charcoal in a fixed bed counter-current gasifier. Similar plots could be seen for the gasification of the wood pellets as well [5,6,7]. Typical composition of the fuel gas comprehend H_2, CO, CO_2, CH_4, N_2, and high hydrocarbons (CxHy).

Comparing gasification time, composition and LHV of the fuel gas, one can clearly see the impact of steam in the feed gas during gasification of charcoal. The gasification time is increased from nearly 45 minutes up to nearly 55 minutes when the molar fraction of steam increases from 0% up to 25%. On the other hand, the results show that the composition of the fuel gas varies and the corresponding maximum LHV is 5.7 MJ/Nm^3, though not showing variation when molar fraction of steam in the feed gas is increased up to 25% (Figure 7).

Main difference in composition of fuel gas between gasification with steam and without steam in feed gas is shown in profiles of hydrogen, carbon monoxide and carbon dioxide. As expected, steam gasification yielded more hydrogen than air alone. Analysing Figure 6 and Figure 7 one can see that molar fraction of H_2 is 15.8% when gasifying charcoal with air as feed gas, and 22.1% when 25% of molar fraction of steam is used in the feed gas. Figure 7 shows that higher molar fraction of hydrogen is observed in the longer range of gasification time. High molar fraction of hydrogen in fuel gas was obtained even when molar fraction of methane has approached zero. This is understandable because using steam on the composition of the feed gas, water gas reaction ($C + H_2O \leftrightarrow CO + H_2$) and water gas shift reaction ($CO + H_2O \leftrightarrow CO_2 + H_2$) take place during gasification process [14]. Also big difference is noticed on profile of carbon dioxide, which is a result of reaction of steam and carbon monoxide. As discussed above the use of air as a feed gas favours first maximum initial pick of molar fraction of CO_2 as shown in Figure 6 and Figure 7. In air gasification, molar fraction of CO_2 subsequently decreases to a value lower than 4% in the expenses of the production of CO due to Bordouard reaction (C + CO2 ↔ CO) which takes place at high temperature (Tb1 > 1000°C). The second pick of CO_2 is observed in Figure 6 occurs after gasification process, indicated by the end of production of CO, and when only combustion takes place in the gasifier.

Impact of steam in the feed gas provides different gasification conditions where molar fraction of CO_2 does not exhibit minimum during gasification process. As shown in Figure 7 temperature of gasifier bed is lower than 1000°C, thus favouring production of CO_2 as well as H_2 from CO and H_2O (water shift reaction). Same figure shows that carbon monoxide has the same shape of profile during gasification with steam and without steam, however molar fraction decreases with increasing amount of steam in fuel gas.

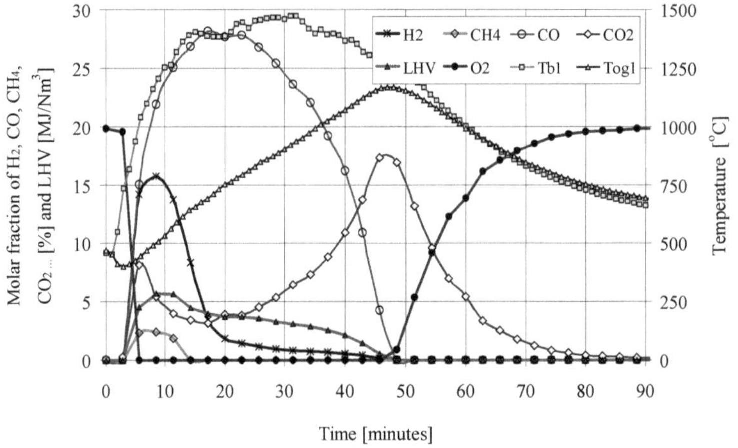

Figure 6: Profiles of composition, lower heating value, and temperature of the fuel bed and of the fuel gas exiting the gasifier during gasification of charcoal with air as feed gas at temperature of 360°C (Case *Ch-1*).

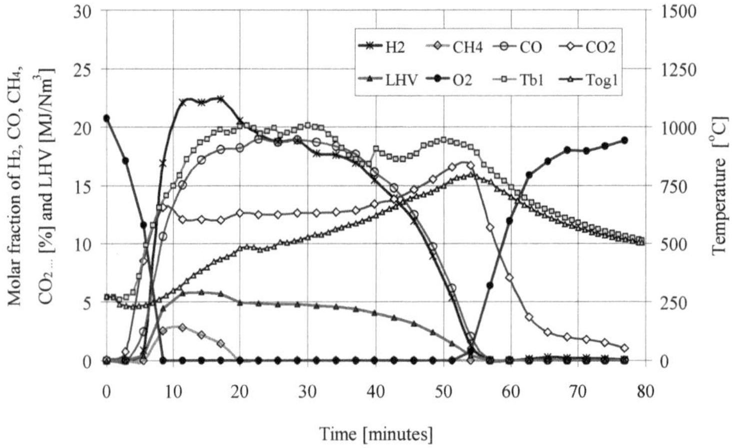

Figure 7: Profiles of composition, lower heating value, and temperature of the fuel bed and of the fuel gas exiting the gasifier during gasification of charcoal with mixture of air and steam as feed gas at of 360°C (Case *Ch-4*).

Figure 8 shows how the composition of fuel gas at the maximum LHV changes with steam during gasification of charcoal. Increase of the molar fraction of steam in the feed gas results in increase in the molar fraction of H_2 from 15.8% up to 22.5%. Slightly linear increase of the molar fraction of CO_2 from 5.4% up to 12.1% is seen in the range 0-25% of molar fraction of steam in the feed gas. Whereas CO decreases from 21.9% down to 17.1%. CH_4 remains fairly constant (~2.3%) whilst higher hydrocarbons (CxHy) remain almost close to zero in the range 0-25% of steam in the feed gas. As to LHV, increases from 5.7 MJ/Nm^3 up to 6.4 MJ/Nm^3 when molar fraction of steam increases from 0% up to 20%. Decrease on the LHV down to 5.8 MJ/Nm^3 is observed with molar fraction of steam in the feed gas at 25%, however.

Composition and maximum lower heating value of the fuel gas during the gasification of wood pellets at 900°C with air and a mixture of air and steam in the feed gas are shown in Figure 9.

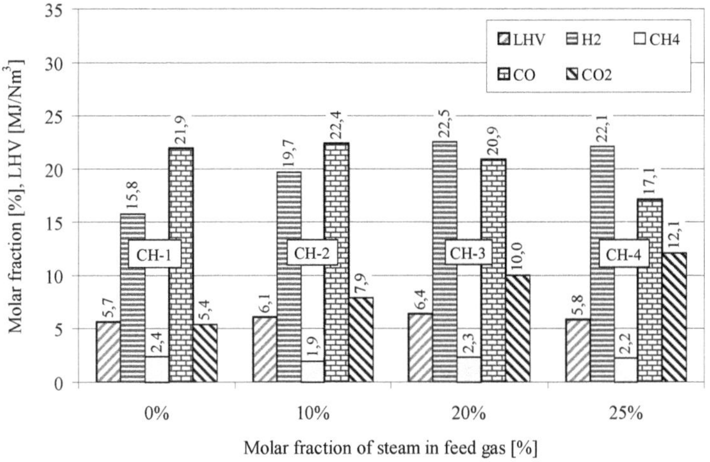

Figure 8: Influence of molar fraction of steam in feed gas on the maximum lower heating value and composition of the fuel gas during gasification of charcoal at maximum LHV.

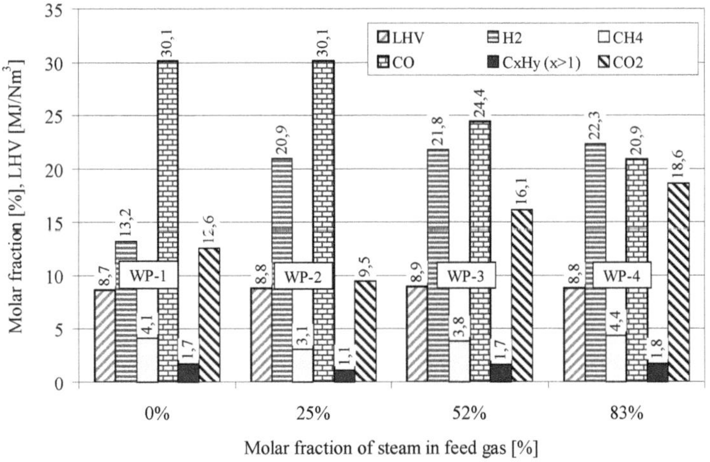

Figure 9: Influence of molar fraction of steam in feed gas on the maximum lower heating value and composition of the fuel gas during gasification of wood pellets at maximum LHV.

As expected, the use of the steam in the composition of the feed gas results in increase of the composition of the hydrogen molar fraction in the fuel gas. Significant increase is observed from 13.2% up to 20.9%, when the molar fraction of steam increases from 0 up to 25%. Beyond 25%, molar fraction of H_2 still increases but slightly from 21.8% up to 22.3% when molar fraction of steam goes up from 52% to 83%. Increase on the composition of carbon dioxide is also shown, ranging from 12.6% up to 18.6%. For the same range of increase of molar fraction of steam in the composition of the feed gas, molar fraction of carbon monoxide exhibits a decrease trend from 30.1% down to 20.9%. Molar fraction of CH_4 decreases from 4.14% down to 3.10% when molar fraction of steam goes up from 0% to 25%. Beyond 25%, CH_4 again increases significantly from 3.10% to 4.38% when molar fraction of steam goes up from 25% to 83%. Similar trend could be seen on the molar fraction of the higher hydrocarbons, going down from 1.72% to 1.13% with increase of the steam from 0% up to 25%, but again increasing up to 1.75% when the molar fraction of steam increases goes up from 25% to 83%.

Mass flow rate and LHV of the fuel gas

A comparison of gasification of charcoal with air as feed gas and with a mixture of air and steam in the feed gas shows that the presence of steam decreases the mass flow rate of fuel gas produced.

As shown in Figure 10, the mass flow rate of dry fuel gas slightly decreases from about 80 kg/h down to about 70 kg/h when the molar fraction of steam increases from 0% up to 25%. The corresponding LHV, however, does show slightly variations towards increase with increase of molar fraction of steam in the same range.

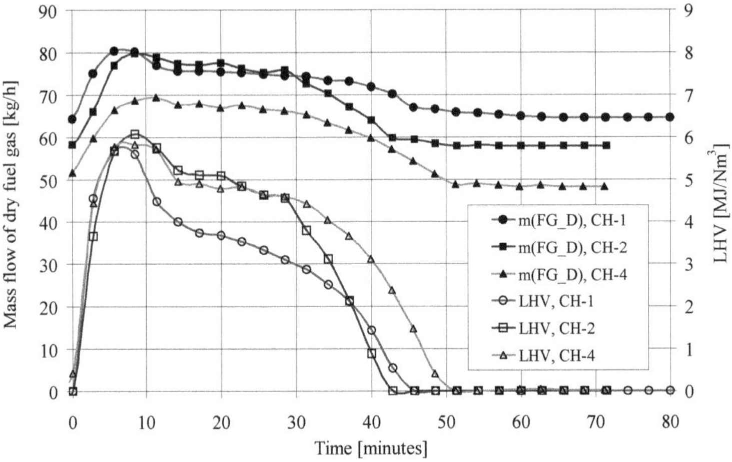

Figure 10: Changes of the mass flow rate and LHV of the fuel gas for various gasification processes of charcoal.

Figure 11 shows mass flow rate as well as the lower heating value of the fuel gas as function of molar fraction of steam obtained during gasification of wood pellets. The figure shows that fuel gas flow rate in the range 125 kg/h is produced with only air as the feed gas compared to flow rate of 35 kg/h when feed gas is having 83% of molar fraction of steam. However, maximum lower heating value of the fuel gas produced by either composition of feed gas shows nearly no variation although the quality of the fuel gas changes due to increase of H_2 and of CO, and due to decrease of CO_2 and tars with increase of steam in the feed gas.

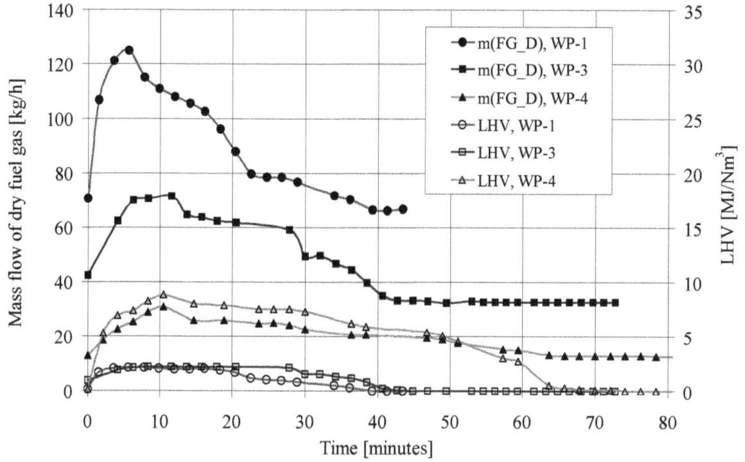

Figure 11: Profiles of mass flow rate and of lower heating value for various wood pellets gasification processes (refer to Table 2).

Process temperature

For the sake of simplification, gasifier temperature (or reaction temperature) and temperature of the fuel gas exiting the gasifier are called process temperature in this study. The importance of both temperatures and the corresponding profiles as a result of variation of molar fraction of steam in the feed gas are discussed in this section.

Figure 12 and Figure 13 show the impact of the preheated feed gas on the reaction temperature referred as temperature of the fuel bed. The sensible heat of the preheated air brought to the reactants induces an increase in the temperature of the fuel bed. During gasification of wood pellets (Figure 13), it goes up to about 1350°C when temperature of the feed gas (only air) is kept constant at 900°C. However, this temperature does reduce to about 830°C and 780°C when molar fraction of steam is increased up to 52% and 83%, respectively.

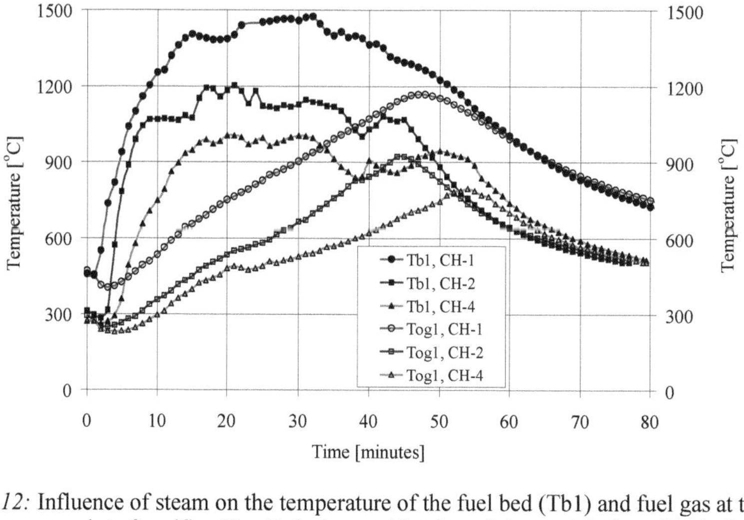

Figure 12: Influence of steam on the temperature of the fuel bed (Tb1) and fuel gas at the exit point of gasifier (Tgo1) during gasification of charcoal (refer to Table 2).

As shown in (Figure 12), temperature of the gasifier bed as higher as 1450°C is seen during gasification of charcoal when the feed gas enters the gasifier with temperature of 360°C. Figure 12 shows that the use of 10% and 25% of molar fraction of steam in the feed gas does significantly decrease the temperature of the fuel bed down to about 1200°C and 1000°C, respectively.

High gasifier bed temperature shown in Figure 12 and 13 favours the endothermic Boudouard reaction:

$$(C + CO_2 \Leftrightarrow 2CO)$$

This reaction requires high temperatures [15], and according to Mathieu et al [20] this reaction produces only CO above 1000°C, which is the reason of high molar fraction of CO using only air and low molar fraction of CO and high molar fraction of CO_2 when mixture of air and steam is used as feed gas.

In some gasifiers, the injection of water in the bed allows to control temperature of the gasifier bed (temperature of reaction) and favours the formation of CO and H_2 according to the following reaction [20]: $C + H_2O \Leftrightarrow CO + H_2$. As discussed above, the increase of the amount of water in the gasifier bed leads to more production of H_2 and CO_2 in expenses of the generation of CO by the following water gas shift reaction $CO + H_2O \Leftrightarrow CO_2 + H_2$ (see Figure 8 and Figure 9). This is caused by the participation of the steam in gas phase reactions and in the char reactions.

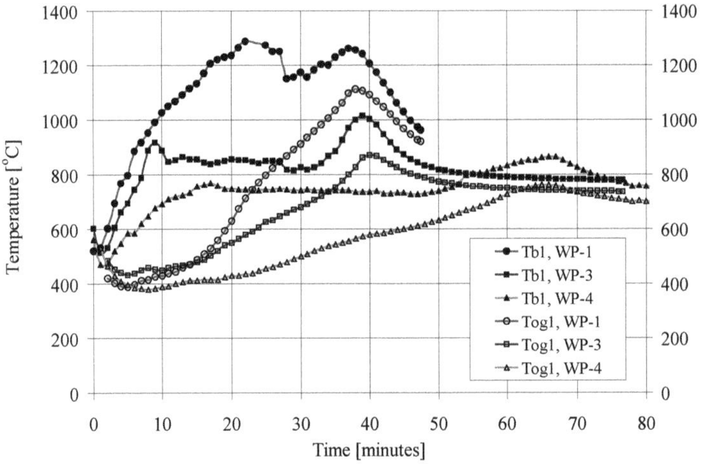

Figure 13: Influence of steam on the temperature of the fuel bed (Tb1) and fuel gas at the exit point of gasifier (Tgo1) during gasification of wood pellets (refer to Table 2).

The results obtained in this study correspond to two levels of temperature of the feed gas. 360°C for charcoal and around 900°C for wood pellets. As for tar removal, it has been stated that a uniform and high gasifier temperature has been found to be the ideal combination to achieve the total cracking of tar [19]. In the previous works [5,6,7], the authors have witnessed that the high temperatures of preheated feed gas increases the temperature of the gasifier bed. It was noticed that little or no tar is produced from this thermal gasification process. From the mass balance performed (not shown in this work), the higher temperature also gave a significant reduction in the yield of carbon into soot, tars and an equally significant reduction in CO_2 production with a corresponding increase in CO yield.

Concerning effect of steam in the gasifier temperature, in both cases (wood pellets and charcoal), the increase of the H_2 content with decrease of T_{bl} is observed because endothermic reaction takes place. The importance of the temperature of the gasifier bed is justified in the study of Cummer et al. [9] who stated that addition of steam and/or oxygen is effective in increasing cracking rates in the temperature range of 950–1250°C.

CONCLUSIONS

Work performed in a fixed bed counter-current gasifier, applying high temperature air and mixture of air and steam gasification concept allows to conclude the fallowing:

- For both charcoal and wood pellets gasification cases it was seen that an increase of steam molar fraction in the feed gas decreases the temperature of the gasification, the gasification rate, the mass yield of the fuel and molar fraction of carbon monoxide but increases molar fraction of hydrogen,
- An increase of the feed gas temperature reduces production of tars, soot and char residue as well as increases calorific value of the fuel gas produced,
- Preheating of the feed gas obtains a higher gasification efficiency and a higher gasification rate,
- High temperature air and mixture of air and steam used as feed gas in a fixed bed gasifier shows the capability of this technology of maximising the gaseous product yield,
- High lower heating value of fuel gas and high molar fraction of hydrogen and hydrocarbons obtained by using high temperature air and steam gasification technology increase the technical possibility of the application of fuel gas.

ACKNOWLEDGMENTS

Authors would like to acknowledge financial support of the grant from the Swedish National Energy Administration, STEM and New Energy and Industrial Technology Development Organization, NEDO as well as industrial partners. The fruitful collaboration particularly with Nippon Furnace Kogyo Kaisha Ltd, Japan is very much acknowledged

REFERENCES

1. Tanaka, R. (1995) *New progress of energy saving technology toward the 21st century*; Frontier of combustion & heat transfer technology, In: 11th IFRF Members Conference, 10-12th May.
2. Blasiak, W., Szewczyk, D., Lucas, C., Tsamba, A. J., Rafidi, N. (2002) *High Temperature Air/Steam Gasification Technical Report no 1: High Exergy Rate Gas Aided Low Grade Fuel Utilization* Technology. Royal Institute of Technology, Stockholm, March. ISRN KTH/MSE - - 02/16 --SE ENERGY/TR.
3. Tsuji, H., Gupta, A. K., Hasegawa, T. Katsuki, M., Kishimoto, K., Morita, M. (2002) *High Temperature Air Combustion: From Energy Conservation to Pollution reduction*, CRC Press.
4. Mochida S., Hasegawa T. (2001) *Development of highly preheated turbulent –jet generator, 3rd International Symposium on Advanced Energy Conversion Systems and Related Technologies*, RAN2001, Nagoya, Japan, December 15-17.
5. Blasiak, W., Szewczyk, D., Lucas, C. (2002) *High Temperature Air/Steam Gasification of biomass wastes,* In: 21st International Conference on Incineration and Thermal Treatment Technologies, IT3 Conference, New Orleans, USA, 13-17 May.
6. Lucas, C., Szewczyk, D. and Blasiak, W. (2002) *High temperature air and steam gasification of densified biofuels*. In: First World Conference on Pellets, Stockholm, 2-7 September.
7. Lucas, C., Blasiak, W. and Szewczyk, D. (2002) *High temperature air and steam gasification of wood pellets*. In: Finnish-Swedish flame days 2002. IFRF. Vaasa, 24-25 September.
8. Coll, R., Salvadó, J., Farriol, X., Montane, D. (2001) *Steam reforming model compounds of biomass gasification tars: conversion at different operating conditions and tendency towards coke formation*. Fuel Processing Technology, **74**, 19–31.

9. Cummer, K. R. and Brown, R. C. (2002) *Ancillary equipment for biomass gasification.* Biomass and Bioenergy, **23**, 113-128.
10. Midilli, A., Dogru, M., Howarth, C. R., and Ayhan, T. (2001) *Hydrogen production from hazelnut shell by applying air-blown downdraft gasification technique*. Int. J. of Hydrogen Energy, **26**, 29-37.
11. Midilli, A., Dogru, M., Akay, G., Howarth, C. R. (2002) *Hydrogen production from hazelnut shell by applying air-blown doundraft gasification technique*. Int. J. of Hydrogen Energy, **27**, 1035-1041.
12. Turn, S., Kinoshita, C., Zhang, Z., Ishimura, D., and Zhou, J. (1998) *An experimental investigation of hydrogen production from biomass gasification.* Int. J. Gydrogen Energy, **23**, 641-648.
13. Encinar, J. M., González, J. F. and González, J. (2002) *Steam gasification of Cynara cardunculus L.: influence of variables.* Fuel, **75**, 27-43.
14. Fagbemi, L., Khezami, L., and Capart, R. (2001) *Pyrolysis products from different biomasses: application to the thermal cracking of tar.* Applied Energy, **69**, 293–306.
15. Encinar, M. J., Gonzaléz, Rodríguez, J. J., and Ramiro, M. J. (2001) *Catalysed and uncatalysed steam gasification of eucalyptus char: influence of variables and kinetic study.* Fuel, **80**, 2025-2036.
16. Schuster, G., Loffler, G., Weigl, K., and Hofbauer, H. (2001) *Biomass steam gasification - an extensive parametric modeling study.* Bioresource Technology, **77**, 71-79
17. Herguido, J., Corella, J., González-Saiz, J. (1992) *Steam gasification of lignocellulosic residues in a fluidised bed at small pilot scale. Effect of the type of feedstock.* Ind. Eng. Chem. Res., **31**, 1274-1282.
18. Garcia, L., Salvador, M.L., Arauzo, J., and Bilbao, R. (1999) *Catalytic gasification of pine sawdust. Effect of catalyst weight/biomass flow rate and steam/biomass ratios on gas production and composition.* Energy & Fuels, **13**, 851-859.
19. McKendry, P. (2002) *Energy production from biomass (part 3): gasification technologies.* Bioresource Technology, **83**, 55-63.
20. Mathieu, P. and Dubuisson, R. (2002) *Performance analysis of a biomass gasifier*. Energy Conversion and Management, **43**, 1291-1299.

Green Diesel from Biomass via Fischer-Tropsch Synthesis: New Insights in Gas Cleaning and Process Design

Harold Boerrigter, Herman den Uil
Energy research Centre of the Netherlands (ECN), Unit ECN Biomass, P.O. Box 1, 1755 ZG Petten, The Netherlands
Hans-Peter Calis
Shell Global Solutions International B.V., The Hague, The Netherlands

ABSTRACT: The interest in biomass is increasing in the light of the growing concern about global warming and the resulting climate change. The emission of the greenhouse gas CO_2 can be reduced when 'green' biomass-derived transportation fuels are used. One of the most promising routes to produce green fuels is the combination of biomass gasification (BG) and Fischer-Tropsch (FT) synthesis, wherein biomass is gasified and after cleaning the biosyngas is used for FT synthesis to produce long-chain hydrocarbons that are converted into 'green diesel'. To demonstrate this route, a small FT unit based on Shell technology was operated for in total 650 hours on biosyngas produced by gasification of willow. In the investigated system, tars were removed in a high-temperature tar cracker and other impurities, like NH_3 and H_2S were removed via wet scrubbing followed by active-carbon and ZnO filters. The experimental work and the supporting system analysis afforded important new insights on the desired gas cleaning and the optimal line-up for biomass gasification processes with a maximised conversion to FT liquids. Two approaches were considered: a front-end approach with reference to the (small) scale of existing CFB gasifiers (1-100 MW_{th}) and a back-end approach with reference to the desired (large) scale for FT synthesis (500-1000 MW_{th}). In general, the sum of H_2 and CO in the raw biosyngas is an important parameter, whereas the H_2/CO ratio is less relevant. BTX (*i.e.* benzene, toluene, and xylenes) are the design guideline for the gas cleaning and with this the tar issue is *de-facto* solved (as tars are easier to remove than BTX). To achieve high yields of FT products the presence of a tar cracker in the system is required. Oxygen gasification allows a further increase in yield of FT products as a N_2-free gas is required for off-gas recycling. The scale of the BG-FT installation determines the line-up of the gas cleaning and the integrated process. It is expected that the future of BG-FT systems will be large plants with pressurised oxygen blown gasifiers and maximised Fischer-Tropsch synthesis.

INTRODUCTION

Although biomass is the oldest energy source known to mankind, its current contribution to the primary energy consumption in the Western world is small. Biomass was gradually substituted after the discovery of huge amounts of cheap fossil fuels. Coal, crude oil, and natural gas were not only cheaper, but also more convenient to use. Furthermore, the 'new' fuels allowed applications that were not possible with biomass: liquid transportation fuels from crude oil and gas for cooking. Biomass appeared to have no future as energy source.

The insight that the fossil fuel reserves are limited, together with concerns over security of supply (*i.e.* the oil crises), initiated the first upraise of interest in biomass (and all other renewable energy forms) in the 1970s. However, continuously low fossil energy prices and the discovery of new fossil fuel reserves impeded the development of biomass technologies. In the 1980s the concern grew that global warming and the resulting climate change were enhanced (if not caused) by CO_2 emissions resulting from fossil fuel consumption. This concern resulted in the Kyoto protocol in which objectives to reduce the anthropogenic CO_2 emissions are documented. To achieve the Kyoto objectives, the share of renewable energy in the primary energy consumption has to increase significantly. The interest in biomass was renewed, as biomass is considered to be one of the most important renewable energy source for this century [1,2].

GREEN DIESEL

Biomass-based energy currently contributes to about 3% of the primary energy consumption in the European Union (EU). This share will have to be tripled within the next decade to meet the EU objective of 8.5% energy from biomass and residues in 2010. An important aspect of biomass is that liquid (bio) fuels can be produced from this renewable source [3]. The emission of the greenhouse gas CO_2 can be reduced when 'green' biomass-derived transportation fuels are used in cars, trucks, and buses. To illustrate this: in the Netherlands 390 PJ (1999) of transportation fuels are consumed annually [4] and replacement of only 5% fossil transportation fuels by renewable fuels would reduce the CO_2 emission by approx. 1.5 million tonnes per year. In addition to the EU objectives for primary energy (*i.e.* electricity and heat), the European Commission (EC) is preparing a directive for renewable transportation fuels. According the directive the target for the share of biofuels is 2% in 2005, increasing to 6% in 2010, and 8% in 2020 [5].

In the Netherlands, the government initiated the so-called GAVE programme to stimulate the demonstration of production chains for alternative renewable gaseous and liquid energy carriers. In the first study phase it was concluded that one of the most promising routes to produce 'green' fuels is the combination of biomass gasification (BG) and Fischer-Tropsch (FT) synthesis [6]. In this route biomass is gasified to yield a producer gas or biosyngas that is rich in H_2 and CO. After cleaning the biosyngas can be used as feed gas for a FT synthesis reactor where the H_2 and CO are converted into long-chain hydrocarbons that are subsequently converted into 'green diesel' in a post-processing step.[1]

FISCHER-TROPSCH SYNTHESIS

In the catalytic Fischer-Tropsch (FT) synthesis one mole of CO reacts with two moles of H_2 to form mainly aliphatic straight-chain hydrocarbons (C_xH_y). Typical FT catalysts are based on iron or cobalt. About 20% of the chemical energy is released as heat in this exothermic reaction:

$$CO + 2H_2 \Rightarrow -(CH_2)- + H_2O \qquad (1)$$

As follows from equation (1), the FT reaction consumes hydrogen and carbon monoxide in a ratio of $H_2/CO = 2$. When the ratio in the feed gas is lower, it can be adjusted with the water-gas shift (WGS) reaction.

$$CO + H_2O \Leftrightarrow CO_2 + H_2 \qquad (2)$$

Iron-based FT catalysts show considerable WGS activity and the H_2/CO ratio is adjusted in the synthesis reactor. In the case of cobalt-based catalysts the ratio needs to be adjusted prior to FT synthesis. Typical operation conditions for FT synthesis, when aiming for long-chain products, are temperatures of 200-250°C and pressures of 25-60 bar [7]. The polymerisation-like chain-growth reaction results in a range of products, comprising light hydrocarbons (C_1 and C_2), LPG (C_3-C_4), naphtha (C_5-C_{11}), diesel (C_9-C_{20}), and wax (>C_{20}) fractions. The distribution of the products depends on the catalyst and the process operation conditions. With respect to the production of green diesel, process conditions can be selected to produce maximum amounts of products in the diesel-range. However, an even higher yield of diesel can be achieved when the FT synthesis is

---

[1] In this paper *"green diesel"* refers to the high-quality ultra-clean diesel-like product produced with Fischer-Tropsch synthesis. This term is used to discriminate from *"bio-diesel"*, which is a liquid product from esterfication of vegetable oils (*e.g.* rapeseed oil and sunflower oil). Furthermore, *"biosyngas"* refers to a gas rich in H_2 and CO obtained by gasification of biomass. Biosyngas discriminates from *"syngas"*, which is a comparable gas but from fossil origin, and from *"biogas"*, which is gas produced by digestion of organic matter and consists mainly of CH_4 and CO_2.

optimised towards production of wax. The wax can be selectively cracked to yield predominantly diesel. For this hydro-cracking additional hydrogen is required, which can be produced from a syngas side-stream that is completely shifted to hydrogen via the WGS reaction (2).

BIOMASS GASIFICATION AND FISCHER-TROPSCH

In a joint research project, ECN and Shell are assessing options for an integrated BG-FT system. Biomass gasification technology has developed to the demonstration stage. Limiting to the most widely used circulating fluidised bed (CFB) gasifier, more than ten installations with a thermal input larger than 10 MW are currently in operation. In nearly all cases the raw product gas is used for firing of cement kilns or co-firing in coal power plants. Furthermore, a variety of smaller CFB installations are in operation as research facility or have the status of pilot-plant. Fischer-Tropsch synthesis is an established technology. Two companies have commercialised their FT technology, *i.e.* Shell (first plant in Malaysia) and Sasol (several plants in South Africa), using natural gas and coal as feedstock to produce the syngas, respectively. The integration of biomass gasification and Fischer-Tropsch synthesis, however, has not yet been demonstrated. The technically most critical and uncertain step in this route is the cleaning of the biosyngas.

The catalysts used in FT synthesis are intrinsically very sensitive to small amounts of poisons. For a number of important impurities in biosyngas, like the sulphur and nitrogen-containing compounds, the effects on the catalyst performance are known. Furthermore, for the removal of these compounds efficient techniques are (commercially) available, although designed for other processes and process conditions. For the removal of the specific biosyngas impurities (*i.e.* the "tars" present in syngas produced by CFB gasification) no technologically mature processes are available in biomass gasification research.

THIS PAPER

One of the targets of the joint ECN/Shell research is to develop the necessary gas cleaning to make biosyngas suitable as a feed gas for FT synthesis. Cleaning of biosyngas for application in gas engines has already been investigated and successfully demonstrated [8]. However, biosyngas cleaning for purpose of catalytic processes, like Fischer-Tropsch synthesis, is much more critical. This paper describes the results of experiments with a BG-FT system and provides new insights regarding the design of integrated BG-FT systems especially with respect to the scale of operation.

INTEGRATED BG-FT SYSTEM

The input of the studied biosyngas cleaning system is a typical biosyngas, while the output is determined by the syngas specifications set by the FT catalyst. The performance of the FT reactor and catalyst are only studied as far as the effect of impurities is concerned. As input for the study a biosyngas produced by a circulating fluidised bed (CFB) gasifier was used. The CFB technology is widely applied in biomass conversion (combustion as well as gasification) as this type of technology is robust, fuel-flexible, and suitable for scales ranging from one to several hundreds of megawatts. The current state-of-the art CFB gasifiers are generally operated at near atmospheric pressures and with air as gasification medium. For FT synthesis a cobalt-based catalyst is used, in view of its superior activity and selectivity relative to iron-based catalysts.

BIOSYNGAS

A typical composition of the raw biosyngas from an air-blown CFB gasifier is shown in Table 1. The syngas components H_2 and CO are present in a ratio of 0.9 and make up only 34 vol%, corresponding to 48.9% of the chemical energy (the LHV of the gas is 8.2 MJ/m_n^3). The remainder of the energy is mainly contained in CH_4, C_2H_4, and benzene (total of 44.4%). The major gas cleaning issue is the presence of 7 g/m_n^3 of tars in the gas. Tars are condensable organic compounds with boiling points between 80-350°C. When the temperature in the system decreases to below ~350°C tars start to condense resulting in fouling and ultimately in failure of the system. Typical inorganic biomass impurities are NH_3, HCl, and H_2S, and in minor quantities COS, CS_2, and HCN are present. Furthermore, 2 g/m_n^3 of solids are present in the raw biosyngas.

Table 1: Typical biosyngas composition for gasification of wood (15% moisture) at 850°C in an atmospheric air-blown CFB gasifier.[2]

Main Constituents	**[vol%, dry]**	**[LHV%]**	**Impurities**	**[mg/m_n^3]**
CO	18	27.8	NH_3	2200
H_2	16	21.1	HCl	130
CO_2	16	-	H_2S	150
H_2O (relative to dry gas)	13	-	all COS, CS_2, HCN, HBr	< 25
N_2	42	-	dust, soot, ash	2000
CH_4	5.5	24.1		
C_2H_2 (acetylene)	0.05	0.4	**Tar classes** [9]	**[mg/m_n^3]**
C_2H_4 (ethene)	1.7	12.4	class 2 (hetero atoms)	350
C_2H_6 (ethane)	0.1	0.8	*- phenol*	*160*
benzene }	0.42	7.9	class 3 (1-ring, excl. BTX)	370
toluene (BTX) }	0.07	1.6	class 4 (2,3-ring)	5300
xylenes }	0.04	1.0	*- naphthalene*	*2250*
sum of tars	0.12	2.8	class 5 (≥ 4-ring)	650
TOTAL	*100*	*100*	class 1 (unidentified)	330

FISCHER-TROPSCH FEED GAS

In commercial operation, catalysts are replaced or regenerated after a certain operational period. The definition of the gas cleaning is therefore based on economic considerations: investment in gas cleaning versus accepting decreasing production due to poisoning of the catalyst. Therefore, there are no 'hard' data on maximum levels for impurities in FT feed gas. For each plant the acceptable levels may be different. As rule-of-thumb a maximum value of 1 ppmV may be used for the sum of the nitrogen-containing and sulphur-containing compounds. For the halides and alkaline metals a lower level of 10 ppbV is assumed(Table 2).

Table 2. Required removal levels for Fischer-Tropsch feed gas.

Impurity	**Removal level**
$H_2S + COS + CS_2$	< 1 ppmV
NH_3 + HCN	< 1 ppmV
HCl + HBr + HF	< 10 ppbV
alkaline metals	< 10 ppbV
solids (soot, dust, ash)	essentially completely
organic compounds (tars)	below dew point
- class 2 (hetero atoms)	*< 1 ppmV*

With respect to the tars in general, there are no limits regarding poisoning of the catalyst. However, as the biosyngas needs to be compressed to 25-60 bar for FT synthesis, the concentration of the organic compounds must be below the dew point at FT pressure to prevent condensation and fouling in the system. Specifically, class 2 tars with S or N hetero atoms (*e.g.* thiophene and pyridine) need to be removed below ppmV level as they are intrinsically poisonous

[2] Calculated for a 5.5 MW_e installation, based on a large collection of experimental experience with the ECN 0.5 MW_{th} CFB gasifier BIVKIN. For information on the BIVKIN see: http://www.ecn.nl/biomass

for the catalyst. Solids must be removed essentially completely, as they foul the system and may obstruct the fixed-bed.

With respect to the other constituents of biosyngas, *i.e.* H_2, CO, CO_2, CH_4, N_2, paraffins (ethane and propane), and olefins (ethene and propene), there are no specifications. The gas is suitable as feed for Fischer-Tropsch synthesis assuming the biosyngas is free of tars and other impurities. This was demonstrated in our previous experimental work, in which FT runs were carried out with mixtures of the mentioned compounds to mimic (clean) biosyngases.

DESIGN APPROACH

In evaluating BG-FT systems, usually the gasifier is taken as starting point and the FT synthesis is considered as a consecutive operation. In this *front-end approach* the scale of the installation is determined by the scale of the gasifier, which is in the range of 10-100 MW_{th} for CFB gasifiers currently in operation. In the view of the ambition to replace a considerable fraction of the fossil-based transportation fuels by renewable transportation fuels, it seems logical to maximise the conversion of (bio)syngas to green diesel. Therefore, in the *back-end approach* the (desired) scale for the Fischer-Tropsch plant is taken as starting point. The relatively small FT plant of Shell in Malaysia (12,500 bbl/day) already corresponds to a thermal input of syngas of more than 1000 MW. For full-scale FT synthesis Shell studies plants with capacities of 75,000 bbl/day [10]. The latter scale would require a BG-FT plant with (several) gasifiers of 500-1000 MW_{th}. Both approaches require different process design and gas cleaning concepts.[3]

FROND-END APPROACH

The principle idea in this approach is that *tri-generation* is the most economic configuration for scales of 1-100 MW_{th} [11]. In a tri-generation system the biosyngas is used for Fischer-Tropsch synthesis, with the off-gas being used to generate electricity and heat in a combined cycle (CC). The fundamental assumptions that rationalise this approach are:

- the gasifier is air-blown as the use of oxygen is not economic for these scales;
- the FT unit is operated in once-through mode to avoid build-up of inerts; this concerns especially N_2 which is present in about 40 vol% in the biosyngas,[4] but also CO_2, CH_4, and gaseous FT products (C_1-C_4);
- no conditioning of the biosyngas (adjustment of H_2/CO ratio via the WGS reaction and CO_2 removal) prior to FT synthesis to reduce units and costs. Especially removal of CO_2 (present in 16 vol%) is an expensive operation that is not essential if the FT off-gas is not recycled.

Importance of H_2/CO ratio

As a consequence of the latter assumption (*i.e.* no syngas conditioning), the H_2 concentration is the limiting factor for the conversion in a once-through Fischer-Tropsch synthesis (*cf.* Table 1). In the case of 90% H_2 conversion, only 40% CO will be converted assuming a consumption ratio of $H_2/CO = 2$. The unconverted H_2 and CO together with the other gas compounds and the gaseous C_1-C_4 hydrocarbons produced in the FT synthesis reaction are used for electricity generation. In Figure 1 an indicative energy balance for the tri-generation system is shown. These data are based on an assumed FT selectivity of 90% to C_5+ products (wax and liquids) versus 10% gaseous C_1-C_4 products, and a CC with 50% electric efficiency. For simplicity it is assumed that the tars and BTX 'pass' the FT reactor as inerts and are converted in the CC.

---

[3] In evaluating processes designs heat integration options and the plant's own electricity consumption are not considered. The net heat and electricity production is dependent of the power balance of the plant.

[4] To produce a nitrogen-free biosyngas an indirect gasifier might be an option. However, as still no recycle can be incorporated due to the presence of other inerts (CO_2 and hydrocarbons) the basic design of the concept will not change.

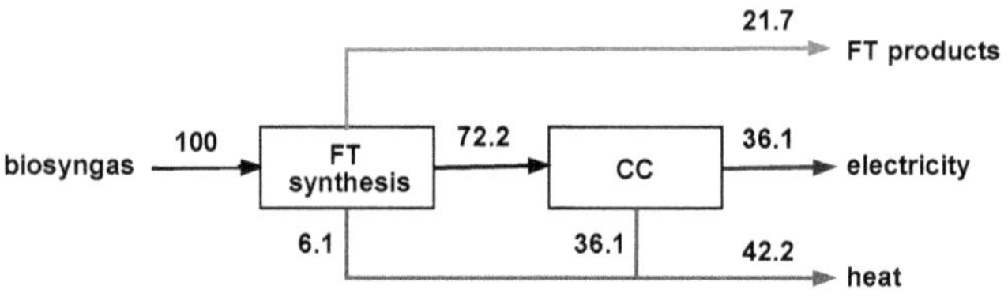

Figure 1: Indicative energy balance for tri-generation without shift. Starting with biosyngas from a CFB (Table 1).

The yield of FT products is low due to the fact that H_2 and CO represent less than one half of the energy content of the syngas, in combination with the fact that the hydrogen deficiency limits the CO conversion to less than 50%. Therefore, in many studies on integrated BG-FT systems, the H_2/CO ratio in the biosyngas is used as an important gasifier selection criterion. The focus is on gasification technologies and operational conditions that afford biosyngases with higher H_2/CO ratios. In general, higher ratios are achieved by adding steam to the gasifier to shift to higher H_2 concentrations already in the gasifier. However, the use of steam is unfavourable for the overall gasification efficiency and results in a net decrease of the total efficiency and should therefore be avoided. Instead of compromising in the gasification efficiency to achieve a higher H_2/CO ratio, which nevertheless remains below the FT consumption ratio of 2, it is more reasonable to use a separate water-gas-shift step in the FT feed gas. When the gas is shifted to the consumption ratio prior to FT synthesis, the energy efficiency to FT wax and liquids is 41% higher (see Figure 2). Also the total energy efficiency to liquids products and power is higher, *viz.* 60.4 versus 57.8%.

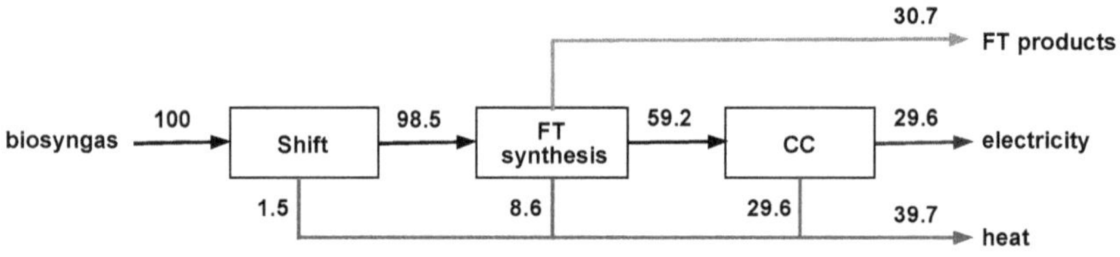

Figure 2: Indicative energy balance for tri-generation with shift to ratio of $H_2/CO = 2$, starting with biosyngas from a CFB (Table 1).

When the purpose of a tri-generation plant is to maximise the production of FT products, a shift step will always be part of the system. Therefore, the H_2/CO ratio in the raw biosyngas is irrelevant as selection criterion. The total yield of H_2+CO in the gas is much more important with respect to the Fischer-Tropsch synthesis.

Tars are the issue in gas cleaning?

In the discussion in the previous section it was implicitely assumed that the gas was sufficiently clean for Fischer-Tropsch synthesis, however, in fact the raw CFB biosyngas does not meet the specifications (*cf.* Table 2). With the current state-of-the-art gas cleaning, biosyngas can be made suitable for application in gas engines [8]. This gas cleaning comprises dust removal by a cyclone, two wet scrubbers to cool the gas and remove tars by condensation and NH_3 and HCl by adsorption, and finally an electrostatic precipitator (ESP) for fine dust and tar aerosols. The 'clean' gas still contains approx. 2 g/m_n^3 tars (mainly class 4), but as the gas has a tar dew point of approx. 30°C, no condensation occurs if the gas is not further cooled. The NH_3 level was above 40 ppmV (removal efficiency is not optimal due to the presence of tars in the water), and the H_2S concentration was 100 ppmV, as this compound was not removed [8].

To meet the FT feed gas specifications the gas needs further cleaning. First of all, the organic compounds must be removed to a level at which no condensation occurs upon compression to FT synthesis pressure (25-60 bar). Taking naphthalene as reference, this would correspond to an allowable tar content of 2 ppmV, while for the BTX compounds this corresponds to 2500 ppmV (for 40 bar at 20°C).

ZnO and active carbon filters are considered suitable for removal of H_2S and remaining traces of other inorganic impurities (*e.g.* NH_3, HCl, HCN, CS_2, and COS), respectively. These two filters can be used as guard beds for the final protection of the catalyst and are are considered a crucial part of the gas cleaning.

Although the active carbon is intended to remove traces of the inorganic impurities, it will also adsorb organic compounds in the gas. Therefore, the active carbon will effectively become a BTX removal step, as the concentration of BTX is still in large excess of the inorganic compounds. This would require frequent regeneration of the guard beds, which is not the preferred approach in the view of process reliability. This implies that it is preferred to remove the tars as well as BTX upstream of the guard beds. Thus, the removal of BTX, and not tars, from the biosyngas is the actual issue in gas cleaning for FT synthesis (as tars are readily removed under conditions for BTX removal).

Gas cleaning for tri-generation

So far no reported studies on BG-FT have considered BTX removal, as these compounds are inert in the FT synthesis. In the tri-generation approach, BTX are even desired for electricity generation in the CC considering the significant contribution to the caloric value of the gas. Conventional gas cleaning techniques used in biomass gasification are not capable of removing tars and BTX to ppmV level. However, recently ECN demonstrated a new powerful technology for tar removal [12]. With this so-called OLGA technology, tars are removed from the gas by an organic washing liquid. As the liquid has an affinity for the tars, the removal is based on absorption as well as liquid collection (the latter is the only mechanism in water scrubbers) [13]. Based on these mechanisms an OLGA unit can also be designed to remove tars as well as BTX to the required levels. Ideally, the tars and BTX are recycled to the gasifier where they are broken down.[5]

The basic process lay-out of the gas cleaning for the tri-generation system is shown in Figure 3. The tars and BTX are removed from the biosyngas with an OLGA unit and the inorganic impurities are removed by (conventional) wet gas cleaning techniques. H_2S and remaining trace impurities are removed with ZnO and active carbon filters, and then the gas is fed to the FT reactor (after compression at some point). The previous discussion on the H_2/CO ratio remains valid even after removal of BTX and the tars, and recycling to the gasifier, however, the numbers will change a little.

Figure 3: Process line-up of gas cleaning for tri-generation.

BACK-END APPROACH

To replace a significant part of the fossil diesel by its green equivalent, a large production capacity is required. Since FT facilities are relatively high capital intensive, the economy of scale is

[5] ECN has demonstrated that tars from the wet cleaning section can be recycled to the gasifier without accumulation of tars (*viz.* the tars are destructed). *Publication in preparation.*

required to produce the green diesel competitively with other alternative renewable transportation fuels, *e.g.* bio-ethanol and bio-diesel (RME) [14]. This implies large-scale FT plants (of more than 1000 MW_{th}) that are dedicated to the production of FT liquids (*viz.* no tri-generation). Fundamental assumptions in this approach are:

- FT off-gas is recycled to the biomass gasifier to achieve maximum syngas conversion;
- electricity is only produced as 'by-product' from the relatively small recycle bleed stream;
- the yield of syngas (*i.e.* the sum of the H_2 and CO content) in the FT feed gas must be as high as possible;

Importance of maximum H_2+CO yield

The importance of a high yield of the syngas compounds H_2 and CO was already mentioned in the discussion of the tri-generation approach. When a maximum yield of FT products is aimed for, all the energy contained in compounds other than H_2 and CO is not available for the FT synthesis. Ideally, all CH_4 (24% of the energy), C_2H_4 (12%), and BTX (11%) should be converted into H_2 and CO. Considering both the demands for high CO+H_2 yield and the necessity to remove all tars and BTX from the FT feed gas, the most optimal process line-up contains a high-temperature tar cracker, in which all hydrocarbons are destructed to yield mainly H_2 and CO.[6] In Figure 4 an indicative energy balance for a system with a thermal tar cracker is shown. For the oxygen-blown tar cracker and the shift (to H_2/CO = 2) together a 15% loss of chemical energy is estimated. As the FT off-gas is recycled (except for a relatively small bleed stream) a 95% overall syngas conversion to wax and liquids is assumed (with 20% heat production and 50% electric efficiency for the CC).

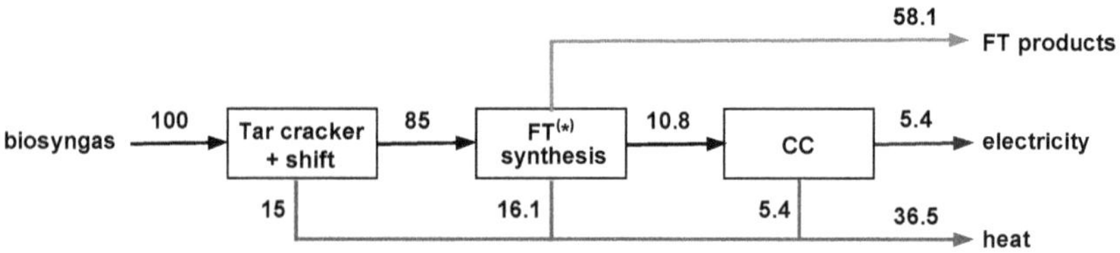

Figure 4: Indicative energy balance for maximum FT production. $FT^{(*)}$ refers to the overall FT synthesis including recycle (for biosyngas from a CFB with a tar cracker).

The increase in the H_2+CO yield results in a much higher production of FT liquids and wax, and also in a higher overall efficiency of 63.5%. The small amount of electricity produced is sufficient to cover the own electricity consumption of the BG-FT plant. In contrast to the other cases described, about 10%-points of the heat is high temperature (>1000°C) heat from the tar cracker and may be used to produce an additional 3 to 4%-points of electricity.

Oxygen gasification?

In the discussion in the previous section it was assumed that recycling of the FT off-gas is possible. To allow economically viable recycling, the concentration of inerts (*i.e.* N_2 and CO_2) in the gas should preferably be below 15 vol% [14]. With already 42 vol% N_2 in the raw biosyngas of an air-blown CFB gasifier this specification can never be met. Therefore, oxygen-blown gasification is required to prevent N_2 dilution of the gas. Although the costs of oxygen are high, they are justified (if optimum yield of FT products is pursued) at large scales by the enormous

[6] Alternatively, a catalytic tar cracker or reformer may be used, which however introduces yet another step that is very sensitive to fouling.

increase in FT yield.[7] As part of the gas conditioning the system will also contain a CO_2 removal step to meet the inert specification.

Gas cleaning for maximised FT production

In Figure 5 the basic process lay-out for the gas cleaning for maximum FT yield is shown. All the tars, BTX, CH_4, and other hydrocarbons are converted into mainly H_2 and CO in the tar cracker. For the tar-free gas the further cleaning is straightforward with wet gas cleaning techniques and ZnO and active carbon guard beds (*cf.* Figure 3).

Figure 5: Process line-up of gas cleaning for maximized H_2+CO yield.

INTEGRATED TEST

To demonstrate the technical feasibility of using biosyngas for Fischer-Tropsch synthesis, in December 2001, a small FT unit based on Shell technology was operated for 150 hours on biosyngas produced by gasification of willow and gas cleaning as shown in Figure 5. This was the first demonstration of the production of Fischer-Tropsch wax from biomass [15].

PROCESS LINE-UP

For the integrated test a process line-up with tar cracker was used, to mimic an oxygen-blow gasifier, a low-temperature externally heated gasifier was used. The process line-up of the system used for the integral test is shown in Figure 6. The biomass (willow) is fed to low-temperature gasification (pyrolysis) reactor with a feed flow of 3 kg/h while argon is used for inertisation. The feed is converted mainly into a gas that is rich in tars and about 25 wt% of the feed (in the case of wood) is converted into char, which is separated. In the high-temperature oxygen-blow tar cracker (operated at >1300°C) all tars in the gas are destructed, while also all BTX and almost all CH_4 are converted into syngas components.

The tar-free syngas is cooled and scrubbed with water in the aqueous scrubber to remove the inorganic impurities (mainly NH_3 and HCl). The clean gas is subsequently compressed to 60 bar in a two-stage compressor with intercooler, after which condensed water is separated. Under these conditions, approximately 8 m_n^3/h of wet raw biosyngas is produced. H_2S is removed from the gas with ZnO filters and final polishing of the gas is achieved with active carbon filters. The compressed gas is stored in gas cylinders to provide a buffer for the Fischer-Tropsch experiments.

For Fischer-Tropsch synthesis a small micro-flow unit with an isothermal fixed-bed reactor is used with a proprietary Shell cobalt-based catalyst. Downstream of the FT reactor the product stream is separated into wax, liquid products, water, and off-gas. The FT off-gas with the lighter gaseous hydrocarbons and the unconverted CO and H_2 is flared in this test. In commercial applications it would be recycled to the gasifier or used for production of power or process heat. Reactor performance and catalyst degradation are monitored via the gas consumption rate, the off-gas composition, and the yield of wax and liquids.

---

[7] A biosyngas with a low N_2 content can also be produced by indirect gasification. This option might be relevant to reduce operational costs, if the use of oxygen can be avoided in this way. However, oxygen is already required to operate the tar cracker.

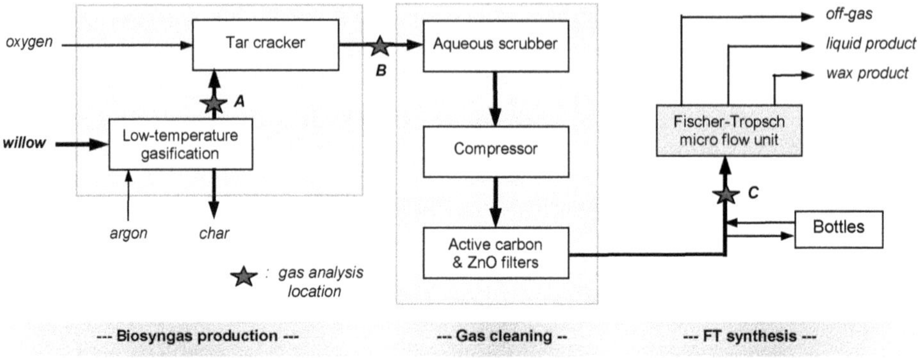

Figure 6: Flow sheet of the integrated system of the biosyngas production, gas cleaning and compression, and Fischer-Tropsch synthesis.

EXPERIMENTAL RESULTS

The integrated test was carried out with willow (dried to 8% moisture) as feed material obtained from a Dutch energy farm [16]. As the micro flow FT unit consumes only ~10 L_n/h of gas, the gasifier was operated periodically to produce additional gas to fill the buffer. The compositions of the biosyngas up- and downstream of the tar cracker, and of the pressurised clean biosyngas are shown in Table 3.

Table 3: Measured gas compositions (dry basis) during 150 hours test with integrated system of gasification, gas cleaning and compression, and FT synthesis.

Gas		**Pyrolysis gas**[a]	**Biosyngas**	**FT feed gas**
Analysis location		**A**	**B**	**C**
Temperature	[°C]	550	485	20
Pressure	[bar]	1	1	40
Moisture	[vol%, wet gas]	-[a]	42.4	0.05
CO	[vol%]	12.5	25.7	27.0
H_2	[vol%]	6.3	28.9	29.4
CO_2	[vol%]	13.1	31.1	28.8
N_2	[vol%]	0.72	0.36	1.08
Ar	[vol%]	56.6	12.0	13.6
CH_4	[vol%]	6.42	0.01	0.01
C_2H_4	[ppmV]	5936	< 5[c]	< 5[c]
C_2H_6	[ppmV]	7359	< 5[c]	< 5[c]
benzene	[ppmV]	736	< 5[c]	< 5[c]
toluene	[ppmV]	530	< 5[c]	< 5[c]
SPA tars[b]	[ppmV]	-[a]	< 10[c]	< 10[c]
NH_3	[ppmV]	~	516	0.02
H_2S	[ppbV]	~	23789	< 10[c]
COS	[ppbV]	~	47578	278
CS_2	[ppbV]	~	207	< 10[c]
TOTAL	*[vol%]*	*100.0*	*100.0*	*100.0*

[a] Composition on dry and tar-free basis; fraction of tars + moisture corresponds to 75 wt% of the pyrolysis gas. [b] Concentration of tars measured by solid phase adsorption (SPA). [c]. Actual concentrations were below detection limit. The reported values are estimated maximum values.

The tar cracker functions very well for the destruction of tars in the pyrolysis gas and conversion of CH_4, C_2H_4, C_2H_6, and BTX into syngas components. In the scrubber all NH_3 is removed, while as expected, the ZnO filters remove H_2S quantitatively. The removal level of COS to ~300 ppbV is sufficient.

During the integrated test, the Fischer-Tropsch catalyst showed no loss of activity. The composition and quality of the wax and light product produced from willow were identical to products from fossil origin as followed from off-line analyses in the Shell laboratories. In January 2002 the test was repeated with an other catalyst sample with a test duration of 500 hours. Again the performance of the reactor was stable over the whole period and the catalyst showed no loss of activity. In Figure 7 samples of the willow and FT wax and light product are shown.

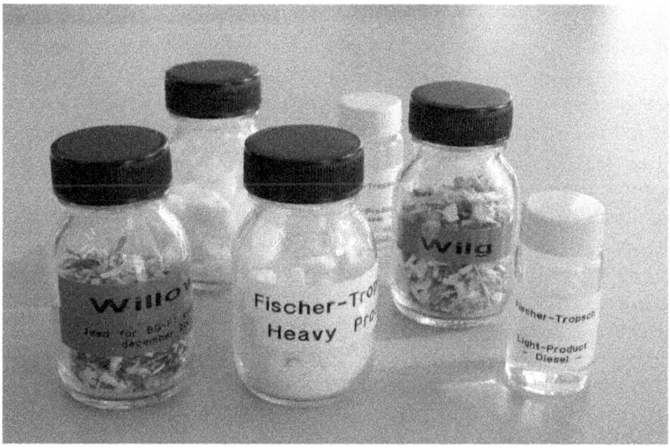

Figure 7: Pictures of the willow used for the integrated test, Fischer-Tropsch wax, and the light product (*viz*. 'green diesel'.

CONCLUSIONS

The emission of the greenhouse gas CO_2 can significantly be reduced by replacing fossil derived transportation fuels by renewable transportation fuels. An interesting route is the production of green diesel via biomass gasification and Fischer-Tropsch synthesis. The critical step is the necessary gas cleaning to make biosyngas suitable as a feed gas for FT synthesis. The experimental work and the supporting system analysis presented in this paper resulted in important new insights on the desired gas cleaning and the optimal line-up for biomass gasification processes with maximised yields of FT liquids. The major insights are:

- BTX are the design guideline for the gas cleaning and not the tars. Organic compounds in the gas will saturate the active carbon guard bed and have to be removed upstream of the guard beds. As BTX are harder to remove than the tars, the tar-issue is *de-facto* solved when the BTX are removed.
- The H_2/CO ratio in a raw biosyngas is irrelevant. To maximise the yield of FT products any process will contain a water-gas-shift step to adjust the H_2/CO ratio to the consumption ratio in the FT feed gas. Thus, it is the sum of H_2 and CO that is the more relevant parameter.
- A high FT yield requires the presence of a tar cracker in the system. The chemical energy contained in hydrocarbons present in biosyngas compounds is lost for FT synthesis unless these compounds are converted into H_2 and CO in a tar cracker.
- A high overall syngas conversion requires gasification with oxygen. Only FT off-gas with a low inert concentration can be recycled to the gasifier. Biosyngas obtained with air-gasification does not meet this specification.

- All inorganic biomass impurities can be removed sufficiently with conventional scrubbing and filtering techniques (when the gas is BTX and tar-free) as was shown in two independent test of 650 h in total.
- The scale of the BG-FT installation determines the line-up of the gas cleaning and the integrated process. A tri-generation system is favourable for smaller scales as no air-separation, CO_2 removal, or FT off-gas recycle are required, however, the yield of FT products is limited. Optimisation of the production of FT liquids (on a large scale) requires oxygen-blow gasification, a tar cracker, CO_2 removal, and an FT off-gas recycle.

OUTLOOK

The technical feasibility of using biosyngas for Fischer-Tropsch synthesis was proven by two integrated test-runs of 650 hours in total. A good understanding was obtained of the essential issues in gas cleaning and process design. The challenge is now to improve the gasification and gas cleaning technologies to achieve higher energy efficiencies and reduce the costs. For example, oxygen-blown biomass gasification, required for high yield FT synthesis, is not yet well developed. Considering the system with an oxygen-blown gasifier and a consecutive tar cracker, one-step gasification at a high temperature in an entrained flow (EF) gasifier might prove to be more energy efficient.

A further issue is the gasification pressure. Pressurised gasification can result in large cost reductions, as higher throughputs are possible with same scale installations, and because compression of syngas from atmospheric pressure to FT synthesis pressure is extremely energy-intensive. Related to gasification at elevated pressure there are also important technical research questions, especially related to the biomass feeding and the use of inert gas, as well as the optimal pressure in view of electricity consumption.

Nevertheless, it is expected that the future of BG-FT systems will be large plants with pressurised oxygen blown gasifiers and maximised Fischer-Tropsch synthesis.

ACKNOWLEDGEMENTS

The work described in this paper was partly sponsored by the *Samenwerkingsverband Duurzame Energie* (SDE) under project number P2000-015. Furthermore, the authors like to acknowledge D.J. Slort, H. Bodenstaff, J. Kuipers, and A.J.M. Kaandorp for carrying out and supporting the experimental work.

REFERENCES

1. Shell International (2001) *Energy Needs, Choices and Possibilities – Scenarios to 2050.*
2. Johansson, T.B.; Kelly, H.; Reddy, A.K.N.; Williams, R.H. (June 1992) *Renewables for fuels and electricity*, UNCED.
3. Larson, E.D.; Jin, H. (1999) *Biomass Conversion to Fischer-Tropsch Liquids: Preliminary Energy Balances.* In: Proceedings of the 4th Biomass Conference of the Americas, Vol. I, pp. 843-853.
4. Energy year report, the Netherlands (1999).
5. Directive of the European Parliament and of the Council, *on alternative fuels for road transportation and on a set of measures to promote the use of biofuels*, Brussels, 7 November 2001, COM(2001) 547, provisional version.
6. GAVE (1999) *Analysis and Evaluation of GAVE chains*, GAVE report 9921. Download from www.gave.novem.org.
7. Dry, M.E. (1981) *The Fischer-Tropsch Synthesis* in Anderson, J.R.; M. Boudart, M. (eds.), Catalysis-Science and Technology, Vol. 1, Springer-Verlag, New York, pp 159-255.

8. (a) Rabou, L.P.L.M.; van der Meijden, C.M.; Brenneisen, L.M.; de Kant, E.; Klein Teeselink, H.; Wubbe, R. (2002) *Gas engine operation on fuel gas from CFB biomass gasifier*, Proceedings of the 12th European Conference and Technology Exhibition on Biomass for Energy, Industry, and Climate Protection, Amsterdam, The Netherlands. (b) Also published as part of ECN report, ECN-RX--02-014, available on www.ecn.nl.
9. Tar comprises a wide spectrum of organic components, which mainly contain a high degree of aromatic rings. For understanding, tar components are organised in classes based on their chemical, condensation and solubility behaviour. See: Van Paasen, S.V.B.; Bergman, P.C.A.; Neeft, J.P.A.; Kiel, J.H.A. (2002) *Primary measures for tar reduction. Reduce the problem at the source*, Proceedings of 12th European Conference and technology Exhibition on Biomass for Energy, Industry and Climate Protection, Amsterdam, The Netherlands, 17-21 June.
10. Gas-to-Liquids News (June 2002) Vol. V, No. 6, p. 11.
11. Daey Ouwens, C.; Den Uil, H.; Boerrigter, H. (2000) *Tri-generation from biomass and residues* in: Proceedings of Progress in Thermodynamical Biomass Conversion (PITBC), Tyrol, Austria, 17-22 September.
12. Netherlands Energy research Centre (ECN), ECN Biomass, *Tar-free producer gas…*, Poster presented on the Nederlandse Duurzame Energie Conferentie (NDEC), Rotterdam, 27-28 February, 2002.
13. Bergman, P.C.A.; Van Paasen, S.V.B.; Boerrigter, H. (2002) *The novel "OLGA" technology for complete tar removal from biomass producer gas*, this conference.
14. Calis, H.P.A.; Haan, H.; Boerrigter, H.; Van den Broek, R.; Van der Drift, A.; Faaij, A.P.C.; Peppink, G.; Venderbosch, R.J. (2002) *Technical and economic feasibility of large-scale synthesis gas production in the Netherlands from imported biomass feedstock*, Preliminary results.
15. Boerrigter, H.; Den Uil, H. (2002) *Productie van Groene Diesel uit Biomassa. Eerste demonstratie van een geïntegreerd systeem van biomassavergassing, gasreiniging en de Fischer-Tropsch synthese* (Dutch). In Proceedings of the Nederlandse Duurzame Energie Conferentie (NDEC), Rotterdam, 27-28 February, 2002.
16. For the composition of willow see the biomass composition database "Phyllis", available on http://www.ecn.nl/phyllis.

CEA Activities on Biomass Gasification to Produce Synfuels

G. Boissonnet, J.M. Seiler, T. Chataing & G. Claudet
CEA Grenoble, DTP/SETEX, 17, rue des Martyrs, 38054 Grenoble Cedex 09, France
M. Brothier & C. Bertrand
CEA Cadarache, DED/SEP, 13108 Saint Paul lez Durance Cedex, FRANCE

ABSTRACT: CEA launched an R&D programme on biomass gasification two years ago, with the objective to develop a process that generates syngas from various biomass resources, to produce fuels or hydrogen. During the first stage of the programme, a process analysis is performed in collaboration with IFP. The key point of this analysis is a detailed comparison between bibliographic data from existing gasification units (heat and material balances) and estimations of these values deduced from thermodynamic calculations. The aim is to get a better understanding of the influence of different parameters (temperature, pressure, reactants, type of reactor, type of primary energy) for the considered application. At the moment, three different processes have been selected for further studies. The second stage of the programme, in progress, consists of modelling tasks (understanding of limitation processes) and preliminary experiments with laboratory pilots (~5 kW), defined according to the previously selected processes and focused on the remaining process difficulties to be solved. This overall approach allows to integrate the existing knowledge, to define which processes and conditions fit the best the requirements and to better understand the physics of the different phenomena involved in the biomass gasification. The third stage of this programme is devoted to the design of a "technical platform" (~500 kW), in collaboration with some European industrial and scientific partners. This facility will be used to test different kinds of reactor technologies and process conditions, in order to prepare a demonstration plant.

INTRODUCTION

France is consuming about 250 Mtoe of primary energy. 100 Mtoe is petroleum, mostly for transport use (about 60%), 50 Mtoe is gas and coal. The share of each primary energy is fossils: 60%, renewables: 10% (hydraulic, bio energies, ...), nuclear: 30%. Fossils primary energy is the most important, even if it remains smaller compared to the world average (Total fossil fuels: 78% (Petroleum: 32,5%, Coal: 27,5%, Gas: 18,0%), Nuclear:4,5%, Hydraulics: 6,5%, Bio energy: 11%) [1]. Three statements can be made:

- The contribution to the reduction of greenhouse effect gases (discussed by politics in Kyoto protocol, EC white book or energy guidelines...), as well as economic dependence on fossil fuels is highly dependent of the reduction of automotive fuels from oil and gas.
- One primary energy cannot be used for all applications. For example, nuclear energy, widely used in France, is dedicated to produce electricity (80%) and cannot be used directly to produce fuels.
- An increased use of Bioenergy in industry usually creates rural development and 2 to 3 net additional employments by 1 000 toe/year substituted [1]. This is true in the transport sector and applicable to the European Union.

Then, strategic and economical as well as environmental and technical motives justify an increased use of renewable fuels in the transport sector.

The most energy efficient way to convert biomass into fuels is by thermal gasification of biomass feedstock to a syngas (H_2 and CO). Such a gas can then be used to synthesise fuels, such

as Fischer Tropsch fuels, methanol, DME and hydrogen and a variety of other chemical compounds. Hence, gasification offers flexibility regarding both feedstock and final product.
While synthesis processes are proven technologies using syngas from natural gas, the biomass gasification requires additional development and proof that produced gas can fit the requirement of synthesis processes.

That is the reason why CEA launched an R&D programme on biomass gasification two years ago, with a team of about 10 persons at the moment. The overall objective for the project is to develop a process that produces a synthesis gas from various biomass resources with the required purity for direct industrial use. More detailed objectives are to:

- Develop syngas production based on steam gasification of biomass, including hot gas filtration, pollutants removal (acid gas, tars, inorganic pollutants such as chlorine, sulphur, nitrogen, alkali), reforming, gas shift and so on.
- Analyse the technical and economical opportunities for the technology and for the fuels.
- Define design criteria for large industrial plants.

The process development is based on the following steps:

- Analytical experimental approach and modelling dedicated to the process objective (fitness for purpose approach and process analysis), which aims are to integrate the existing knowledge and to define which processes and conditions fit the requirements.
- Technical platform, which aim is to test process solutions.
- Pilots and demonstrators, which aims are to validate the process conclusions.

During the first stage of the programme, collaboration with IFP is initiated in order to evaluate more precisely the potential of such a processes and to point out the difficulties and the way to resolve it. Collaborations with CNRS and universities and a contract with ADEME are also in progress. Results presented in the following are based on modelling calculations and process evaluations.

SYNGAS SPECIFIFCATIONS AND TECHNOLOGIES

SYNGAS SPECIFICATIONS

To ensure that the studies and technical choices are adapted to the goal, the main specification for the gas must be kept in mind (see table 1).

Table 1: Syngas specifications for Synfuels.

	Specification
Total Sulphur, Nitrogen, Chlorine or Alkali	< 0.1 mg/Nm3
Dusts	< 0.02 mg/Nm3
Tars	< 0.1 mg/Nm3
CH_4	< 3 %
CO_2	< 5 %
H_2/CO	2

Note that the same order of magnitude is required for hydrogen for fuel cell applications, but the specifications are more drastic. The dust specification requires efficient hot filtration. The issue of tars is one of the most important: catalytic cracking processes seem to have reached their limits with about 0.1 g/Nm3 of tars, which is greater than the specification value with 3 orders of magnitude. Separation processes (quench, gas washing, condensation) or high temperature processes are the possible ways of study.

TECHNOLOGIES

Three technologies have been studied and developed enough to have good chance to become industrial: fixed bed, fluidised bed (bubbling or circulating) and plasma or arc furnace. Table 2 contains relevant information for an overview of these techniques. Note that there are some specific issues left.

Each kind of technology can be adapted to different kind of processes, depending above all on the kind of reactant used to perform the gasification, the pressure and the temperature, as described in table 3.

Table 2: Considered Technologies and Operative Conditions (potential).

	Circulating Fluidised Bed	Fluidised and fixed bed	Plasma or electric arc furnace
Temperature	800 –900 °C	1100°C–1300°C	1500 °C
Tars	Catalysts -cracking	Temperature cracking	
Ashes	Ash separation & Bed replacement		Ash melting
Cleaning	Gas cleaning (volatile species)		
Energy input	From biomass		External energy input
Advantages	Low temperature, Well known technology, Stages		Gas yield, Ashes, Tars
Disadvantages	Gas yield, Amount of steam or air, Small residence time, Ashes agglomeration, Tars	Technical barriers Energy input	Lack of experience, Technical barriers, External energy input (fossil energy excluded)

Table 3: Different kind of process conditions for different uses.

	Steam gasification	Oxygen and steam gasification	Hydrogen Gasification	Air Gasification
Objective	Various use syngas production	Methanol production	Methanol production	Cogeneration (heat, electricity)
Main characteristics	Low pressure[3] High temperature[2]	Low pressure[3] High temperature[2]	High pressure[1] Low temperature[4]	Low pressure[3] High temperature[2]
Products	H_2, CO	H_2, CO, CO_2	CH_4	H_2, CO, CO_2, CH_4
Example of Pilot	FICFB (Austria) [2] SilvaGas (US) [3,4]	Clamecy (France) [5,6]	Hynol (US) [7]	Värnamo (Sweden) [8]
Advantages	Various use of syngas	Autothermal process	Quasi Auto-thermal process	Auto thermal process
Disadvantage	External energy need	Carbon losses	Big amount of fossil fuels	Low energy content gas
Issues	Ext. energy input Tars Ashes Efficient filtration	Tars Ashes Filtration	Ext. energy input Lots of tars Ashes Filtration	Tars Ashes Filtration

(1) Low temperature: ~500°C ; (2) High temperature: ~850°C to 1100°C ; (3) Low pressure: < ~5 bars ; (4) High pressure: > ~30 bars

RESEARCH PROGRAMME AND EXPERIMENTS

EXPERIMENTAL RESEARCHES IN PROGRESS

Experimental studies are performed with three main goals:

- to increase database content by physical and chemical measurement and analysis.
- to better understand the physical and chemical phenomena involved in pyrolysis and gasification.
- to test some technical solutions (e. g. different gasification reactor, hot filtration, tar cracking, gas cleaning).

All experimental results are used as basis of the modelling work described later.

Analysis disposal (Calorimeter)

This facility is used to better understand the process of devolatilization of biomass under different temperature level and different sort of reactant. It is also used to perform energy measurements during devolatilization and evaluate the "enthalpy of formation" of studied biomass (by analogy with the enthalpy of formation of a pure component). Some other analysis disposals are used, coupled with laboratory scale pilots, such as FTIR or GC-MS.

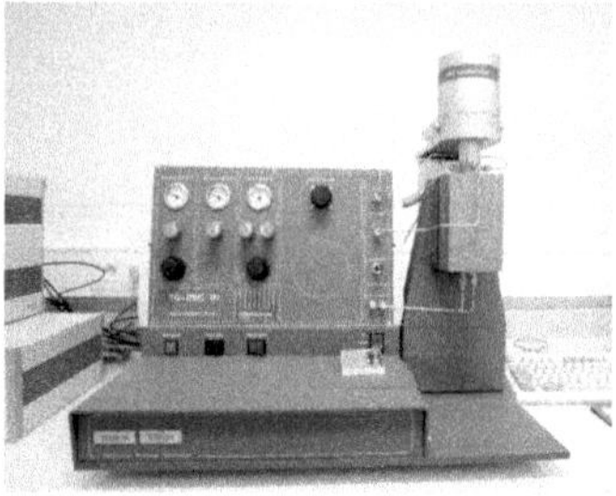

Figure 1: Calorimeter.

Pyrolysis furnace and purification test section

The aim of this facility is to study the pyrolysis of biomass under reducing atmosphere, in the pyrolysis furnace and the gas purification (essentially tar cracking and H_2/CO adjustment) in the purification test facility. The main characteristics of the furnace are:

- Biomass mass flow rate: ~1 $kg.h^{-1}$.
- Temperature: 300 to 600 °C.
- Pressure: 1 bar.
- Max Power: 18 kW.
- Rotating tube: 2 m length, 0.1 m diameter, 1 to 10 rpm rotating speed.
- 3 heating zones with temperature control and power measurements.

Figure 2: Pyrolysis furnace.

The furnace gas outlet is coupled with the purification test section, which main characteristics are:

- Temperature: 800 to 1000 °C.
- Pressure: 1 bar.
- 1.2 m height, 0.15 m inner diameter.
- 3 heating zones with temperature control and power measurements.

- Reactant injection (H_2O, O_2, CO, H_2, CH_4).
- Fill partially or fully by catalyst.

Both facilities are equipped to measure mass and energy balances as precise as possible (thermocouple, flow meters, permanent gas analysis…). Concerning tar studying, a sampling train and an analysis disposal (GC-MS) adapted from the European tar guideline are used, with a modified sampling procedure [9,10]. First results have been obtained in 2002 and the main part of the programme will be in 2003.

Fluidised bed test facility

The aim of the bubbling fluidised bed is to study the steam gasification of biomass or pyrolysis residue in fluidised conditions. Studying hot filtration solutions, ashes problems and bed agglomeration problems are the other goals of this disposal.

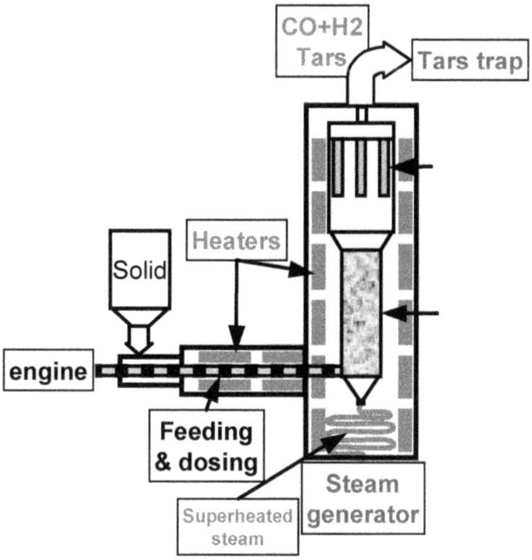

Figure 3: Fluidised bed test facility.

The main characteristics are:
- Biomass mass flow rate: ~1 $kg.h^{-1}$.
- Steam mass flow rate: 0,25 to 2 $kg.h^{-1}$.
- Temperature: 800 to 1000 °C.
- Pressure: 1 to 40 bars.
- Max Power: 15 kW.
- Superficial gas velocity: 0.1 to 0.5 $m.s^{-1}$.
- 1 m height, 0.1 m diameter.
- 6 heating zones with temperature control and power measurements.
- Reactant injection (H_2O, O_2).
- Bed material inert or catalyst.

The fluidised bed is equipped to measure mass and energy balances. A tar sampling train and an analysis disposal are also used. First results will be obtained in 2003 and the main part of the programme will be in 2003 and 2004.

TECHNICAL PLATFORM

In the next two years, a technical platform will be built, which aim is to test and validate, in a representative size, some technical solutions selected from the lab scale laboratory results and conclusions. This platform is not a demonstrator, this new concept consists in building a modular little plant where the main reaction section can be changed without changing the rest of the plant. It will be located on an industrial site (power plant or municipal waste incinerator) in order to be connected to the site facilities. The main characteristics are as follows:

- Type of reaction section: fluidised bed, fixed bed, arc furnace or plasma.
- Pmax: 10 bars.
- Tmax: 1500 °C.
- Biomass flow rate: 100 – 500 $kg.h^{-1}$.

The duration of the program is planned to be 6 to 10 years from 2003.

MODELLING APPROACH

GENERAL MODELLING APPROACH

The modelling work is based on a "fitness for purpose" approach, in order to always keep the final process goal in perspective. This iterative approach is rapidly described hereafter (see fig. 4).

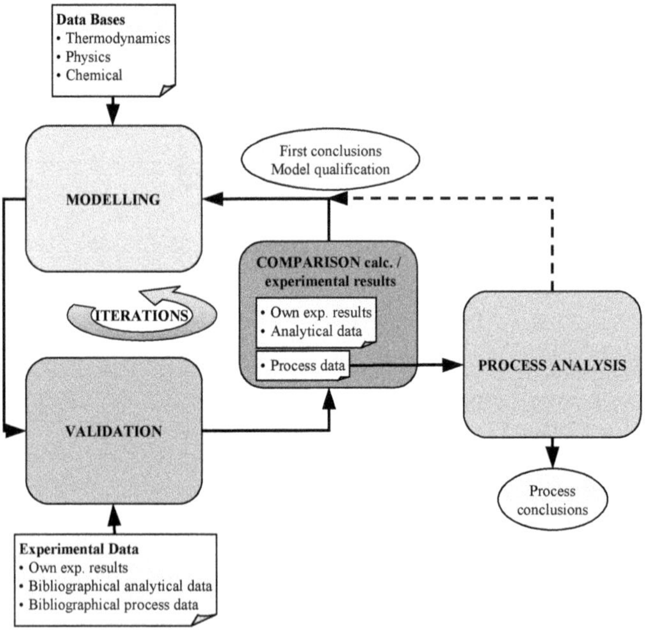

Figure 4: Fitness for purpose modelling approach.

The number of iteration and the level of description depend on the modelling results and their comparison with experimental data from our own disposals or from bibliography. A modelling task is in progress to predict the amount of the main species in the syngas (H_2, CO, CO_2, CH_4, H_2O) under different operative conditions (temperature, pressure, reactants). The first stage of the model is to do thermodynamic calculations with the Coach database and the Gemini solver [11]. Orders of magnitude and comparisons with experimental results has shown that the model is good for the overall predictions, but there is some deviation left, between the equilibrium results and the experimental results especially for the residue and for CH_4. In the following steps, deviation from the equilibrium (physical and chemical limitation processes) and perhaps more precise phenomena description will be considered.

The same kind of iterative approach is performed for tars, inorganic volatile pollutants, ashes composition and for a special task of process simulation. Note that this approach allows physical and chemical interpretations but also a process diagnosis (in terms of operative condition and technologies), using the mass and energy balances.

EXAMPLE

As an example, comparisons between thermodynamic calculations and results of the FICFB process have been made. Experimental results come from a laboratory pilots with the following conditions [12] for 1 "mole" of Biomass described by the formula $C_6H_{8.4}O_{3.6}$:

- Total Steam/Biodry = 7.7 mol/mol (or 1 kg/kg).
- Temperature = 820°C.
- Pressure = 1 bar.
- Biodry mass flow rate = 0.26 kg/h.
- Biomass humidity = 15 %w.
- Bed material: catalyst.

Table 4 contains the main results.

Table 4: Comparison between thermodynamic calculations and FICFB experimental results.

	Units	Experiment	Thermo Results
H_2	Mol	5.54	8.35
CO	Mol	2.66	4.21
CO_2	Mol	2.33	1.78
CH_4	Mol	0.55	0.01
H_2Ogas	Mol	3.95	3.50
$C_{10}H_8$ (tars)	Mol	$7.76\ 10^{-04}$	0
C solid$^{(a)}$	Mol	0,45	0
Orphan H$^{(b)}$	Mol	2.52	0
H_2 $^{(c)}$	Molfrac dry gas	0.50	0.58
CO$^{(c)}$	Molfrac dry gas	0.24	0.29
$CO_2$$^{(c)}$	Molfrac dry gas	0.21	0.12
$CH_4$$^{(c)}$	Molfrac dry gas	0.05	0.01
Tar ($C_{10}H_8$)$^{(c)}$	g/Nm3 dry gas	0.40	0
Dry gas volume$^{(c)}$	Nm3/kg biodry	1.8	2.3
Converted H_2O	%	48.5	54.4
Out. H_2O exp. / Out. H_2O thermo.	-	1.1	1
H_2 Yield	weight % biodry	8	12.1
H_2/CO	-	2.08	1.98
LHV$^{(d)}$	MJ/Nm3 dry gas	10.2	10
LHV$^{(d)}$	MJ/mol biodry	2.53	3.21
LHV$^{(d)}$	MJ/kg biodry	18.4	23.3

(a): hypothesis for the carbonaceous residue is solid carbon
(b): mass balance of H atoms is not totally verified
(c): given data
(d): H2+CO+CH4, calculated from the dry gas composition

Although these results show a very good efficiency of the process, the comparison reveals a gap between experimental results and thermodynamic calculations. The following part proposes several hypotheses to explain it:

- The deviation from the equilibrium is due to transfer limitation and kinetics. The short gas residence time is probably an increasing factor of these deviations. The main limitations seem to come from transfer phenomena in the solid.
- Temperature is too low to have good conditions to produce H_2 and CO under atmospheric pressure. It means that under these conditions, CO_2 and H_2O still remain, even at thermodynamic equilibrium.
- The steam / Biodry ratio is too high. Even at the thermodynamic equilibrium almost 50% of the inlet steam is not converted. Perhaps is it due intrinsically to the fluidised bed technology, which imposes a large amount of gas at the reactor bottom, in order to have good fluidisation conditions.
- The use of catalyst reduces the gap between experimental results and thermodynamic calculations, but not enough to maximize CO and H_2.
- The same observation is valid for the tars, which quantity is too high to fit the syngas specifications.

Note that all results and conclusions are in good agreement with those presented by FICFB [13]. The modelling task will continue to obtain a better representation of the main limitation processes in term of time constant of each phenomenon. The model will allow identification of the parameters of the process or technology that can be varied to improve the syngas quality.

PROCESS EVALUATION

BEST PROCESS CONDITIONS AND PERFORMANCES

As presented before, the modelling approach allows an evaluation of existing processes to be done thanks to the data available in the literature. The main goal of this evaluation is to integrate the existing knowledge, in order to not do studies already performed, and to define which processes and conditions fit the best the requirements. Thanks to this analysis, three different processes have been selected for further studies. The first conclusions about the best process conditions to produce syngas are:

- Steam gasification.
- Steam-Biodry ratio ~ 2 to 5 mol/mol or 0.25 to 0.6 kg/kg.
- Tgasification > 950°C.
- Low pressure < 4 bars.

The first quantitative conclusions about these process performances are gathered in Table 5.

KEY ISSUES FOR THE PROCESS DEVELOPMENT

In addition to these general process conclusions, some technical issues have been found, that are key issues for the success of the process:

- Tars: Catalytic cracking seems to be not adapted for the syngas specifications (max reduction to 0.1 g/Nm3). Solutions: condensation, quench (liquid effluent) or high temperature cracking.
- Ashes and Inorganic Material: As an ultimate residue they must be considered with lots of attention. They are not easy (or even impossible) to separate from the bed material (fluidised bed or fixed bed). Perhaps slagging is a good way of managing them?
- Primary Energy Input: There are still remaining questions about that. What kind of energy and what method of energy injection must be used in order to maximize the mass and energy yields?

Table 5: Best performances under best conditions for three kinds of processes.

	Auto thermal Gasification Processes			**Gasification Processes with External Energy Input**	
	Methanol	**DME**	**FT**	**FT - Plasma**	**FT - Plasma**
Comments	Partial reaction with O_2With external input of O_2			H_2 from gas shift	H_2 external input
Mass Output[a]	~54 %	~41 %	~27%	38 %	58 %
Mass Output[b](Oil Equivalent)	~27 %			~38 %	~58 %
Expected Industrially[b]	15 to 25 %			35 to 55 %	
Max. Energy Yield	~55 %			66 %	-
Expected Industrially	45 to 55 %			-	-
External Energy Input (MJ/kg biodry)	-			~11	~21
Expected Industrially[c]	-			~10 to 17	

(a): Synthesis yield not taken into account and ~50% calorific energy losses
(b): Synthesis yield taken into account (~85%) and ~50% calorific energy losses
(c): With energy recovery in the process

CONCLUSION

The research programme launched by CEA is made of successive steps:

- Process analysis.
- Modelling.
- Lab scale experiments.
- Technical platform experiments and demonstration (later).

The first experimental and modelling results have led to conclusions about:

- Process and operative conditions.
- Technologies.
- Critical points for the success of the process.

Those conclusions allow us to focus on the most appropriate issues for the further experimental work and especially for the technical platform studies. This work is planned to be about ten years from now and will continue with some European industrial and scientific partners.

REFERENCES

1. C. Roy (2002) *Energie Matières Premières et Carbone*. Ademe Report, Ademe.
2. H. Hofbauer, R. Rauch, G. Löffler, S. Kaiser, E. Fercher, H. Tremmel (2002) Six Years Experience with the FICFB-Gasification Process. *12th European Conference and Technology Exhibition on Biomass for Energy, Industry and Climate Protection; Amsterdam*
3. M. A. Paisley, M. C. Farris, J. W. Black, J. M. Irving, R. P. Overend (2000) Preliminary operating results from the Battelle/FERCO gasification demonstration plant in Burlington, Vermont, USA. *1st World Conference and Exhibition on Biomass for Energy and Industry, Sevilla.*
4. M. A. Paisley, J. M. Irving, R. P. Overend (2001) A promising power option – the FERCO SILVAGAS biomass gasification process – operating experience at the Burlington gasifier. *Proceedings of ASME Turbo Expo, New Orleans.*

5. G. Chrysothome, J. M. Lemasle (1986) Production de methanol par gazéification de la biomasse. *VIIth International Symposium on Alcohol Fuels, Paris,* 82-87.
6. J. M. Lesmasle, M. Marcelin (1986) *Production de gaz de synthsèse par gazéification du bois en lit fluidisé sous pression.* ASCAB Internal Report, ASCAB.
7. J. M. Norbeck, K. Johnson (2000) *Evaluation of a process to convert biomass to methanol fuel*, EPA Report NRMRL-RTP-202, US Environmental Protection Agency
8. Sydkraft AB (2001), *VARNAMO demonstration plant.* Sydkraft AB / CEE Report.
9. Neft J. P. A., Knoef H. A. M., Zielke U., Sjöström K., Hasler P., Simell P. A., Dorrington M. A., Abatzoglou N., Deutch S., Greil C., Buffinga G. J., Brage C., Suomalainen M. (2002) *Guideline for sampling and analysis of tar and particles in biomass producer gases (Vesion 3.1).* Energy Project EEN5-1999-00507 (Tar protocol).
10. Brage C., Sjöström K. (2002) *An outline of R&D work supporting the tar Guideline.* www.tarweb.net.
11. B.Cheynet (1989) Complex Chemical Equilibria Calculations with the THERMODATA System. In *Computer Software in Chemical and Extractive Metallurgy*, (Ed. by W.T. Thomson). Pergamon Press, 3144NY.
12. H. Hofbauer, R. Rauch, P. Foscolo, D. Matera (2000) Hydrogen-rich Gas from Biomass Steam Gasification. *1st World Conference and Exhibition on Biomass for Energy and Industry, Sevilla.*
13. G. Schuster, G. Löffler, K. Weigl, H. Hofbauer (2001) Biomass steam gasification - an extensive parametric modeling study. *Bioresource Technology*, **77**, 71-79

Biomass gasification in Supercritical Water, Progress in the Research Centre Karlsruhe

N. Boukis, V. Diem, E. Dinjus, U. Galla, A. Kruse
Institute for Technical Chemistry, Forschungszentrum Karlsruhe
P.O. Box 3640, D-76021 Karlsruhe, Germany

ABSTRACT: Gasification of biomass with a high water content (about 80 wt. %) in supercritical water has several advantages compared with combustion or gasification at ambient pressures. At temperatures and pressures above the critical point of water (T>374°C, p>22.1 MPa), the reaction of the organic substances with water proceeds fast and completely with high space-time yields. Desired product is an hydrogen rich fuel gas. Carbon forms preferentially CO_2, methane is a byproduct. The CO_2 can be easily separated from the product gas, and stored under pressure. Until now, most of the investigations have been performed in laboratory scale equipment with model compounds in continuous flow tubular reactors and biomass in batch experiments. At temperatures of 600°C and about 25 MPa a complete gasification can be achieved. The presence of potassium is important for a high hydrogen yield. Low CO yield in the product gas as well as low tar and char formation are further advantages of the process. Based on the encouraging experimental experiences with laboratory scale equipment and with an 10 kg/h continuous flow plant operated for a similar process, a new continuous flow plant dedicated for the process of biomass gasification in supercritical water has been installed and is now put in operation.

Keywords: hydrogen, agricultural residues, organic waste, gasification, supercritical water

INTRODUCTION

Using biomass for energy generation offers the potential to reduce the rise in atmospheric CO_2 concentration and supports national and European efforts in climate protection.

A lot of the residual biomass appears as wet material containing up to 95 wt. % of water. Wet residual biomass amounts to some 10^7 Mg/a (DOM) corresponding to an energy content (heating value) of several hundreds PJ/a in Germany.

Traditional gasification achieves low efficiencies at these high water contents (in the order of 10% cold gas efficiency at 80% water content). Under high (supercritical) pressure, heat exchange is effective. Estimations on the basis of thermodynamic data give an about six times higher gas efficiency for gasification in supercritical water at a temperature of 600°C and a pressure of 30 MPa. The density of water is approximately one order of magnitude higher than the density of water vapor at 250°C and 1 MPa. This gives rise to an increased space-time yield. Higher thermal conductivity and higher temperature have a positive impact on the endothermal reforming reaction, generally accelerating it. Supercritical water forms a one-phase system with the main product gases H_2, CO_2, CO, CH_4.

Concerning process technology, there is a simple possibility of producing hydrogen at high pressures as required for storage in pressurized tanks. In addition, a cost and energy effective scrubbing step for CO_2 removal under pressure can be integrated. This treatment increases the hydrogen partial pressure. If an applicable long-term storage for the water – CO_2 mixture under pressure is available, and then the process offers a new method to avoid completely CO_2 emissions.

First experiments to convert carbohydrates to liquid products in supercritical water were performed by Modell and Amin in the mid 1970s at MIT [1]. Systematic experimental investigations for the conversion and gasification of biomass, for waste model compounds and for real wastewater under hydrothermal conditions were carried out at the Pacific Northwest Laboratories (PNL), at the university of Hawaii and at the National Institutes for the Resources and Environment [2], Japan (NIRE).

In the research center Karlsruhe numerous experiments with model substances and limited experiments with real biomasses were performed in supercritical water in facilities of various capacities, [3-5]. High pressure experiments in a continuous flow reactor with real biomass slurry represent a special experimental challenge at low flow rates.

At present, a pilot plant with a throughput of 100 kg/h of aqueous biomass (educt) has been installed and is now put in operation. The new facility, VERENA (a German acronym standing for "experimental facility for the energetic exploitation of agricultural matter"), should be expanded into a complete energy generation system on a medium term.

EXPERIMENTAL

PROCESS DESCRIPTION

If necessary, the educt is subjected to a preliminary treatment. Depending on the quality of the educt, this treatment takes place in a separator for the separation of inorganic solid impurities and the decanting of accompanying water or in a mixing tank for educts of variable origin. Furthermore catalytical effective additives are added to the educt, if required. Then, the conditioned educt is passed on to the high-pressure feed system. Depending on the quality of the educt, this system is of variable design. Preliminary crushing is required for solid-containing educts in order to ensure a reliable operation of the high-pressure dosage pump. The operating pressure is typically about 30 MPa.

The educt flow is heated to the required temperature for the gasification reaction (500 – 700 °C) (e.g. $C_6H_{12}O_6 + 6\ H_2O \rightarrow 12\ H_2 + 6CO_2$; $\Delta H = 158.7$ kJ/mol, 882 kJ/kg) and the sensible heat (e.g. H {400 – 600°C} = 1292 kJ/kg). The volume of the reactor is large enough to enable a residence time of at least 1 min.

The effluents are passed on to a separation system, where the product gas is expanded to the pressure required for further treatment. This energy might be recovered by an expansion turbine. Via a level control, the water phase leaves the separator. Depending on the educt quality, the water phase may contain solids that have to be removed in an additional separator. Part of the effluents can be recycled. Qualification of the effluents and solids with regard to management and disposal is an important topic. The objective is to reach the limits specified for direct discharge via sewers.

EDUCTS AND APPARATUS

The model compounds glucose for cellulose, catechol and vanillin for lignin and glycine for proteins were used as aqueous solutions. Real biomass (vegetable foodstuff and vegetable wastes) was treated in batch autoclaves and recently in continuous flow reactors. Experiments with solids have been performed by the addition of a certain amount of the material to the autoclaves. Continuous flow experiments in small apparatuses have been performed by pumping fresh water to the preloaded 20 ml autoclave to entrain product gas continuously and regulate pressure to the desired level.

Experiments with methanol instead of biomass were performed to demonstrate the long time stability of the gasification reaction in supercritical water [6].

For a detailed description of the laboratory equipment used see also [3, 6]. The pilot-plant VERENA is described separately in chapter 4.

RESULTS AND DISCUSSION

METHANOL

The conversion of methanol in or with supercritical water under the experimental conditions (p=25 MPa, Tmax=600 °C, flow rate about 1 ml/min) resulted in a product gas rich in hydrogen. Carbon is converted to a mixture of CO_2, CO and CH_4. In all experiments performed, the conversion rate was higher than 98 %, see Figure 1. Although no catalyst has been used in these experiments, there is evidence that the inner surface of the reactor influences the composition of the product gas. This phenomenon is not clear yet and further work is under progress to clarify this effect.

The apparatus was continuously operated for several weeks. Operation with 5 wt.% methanol lasted over 1000h. The operation, concerning gas production and quality, was stable. After the experiment the reactor has been inspected for residuals. Soot formation was negligible. The overall operating time of the plant in the meantime has reached approximately 3000 h without any major disruptions. All the samples of the product solution were clear, odorless and chemical analysis showed only negligible traces of compounds higher than C1. Remarkable is the high amount of hydrogen in the product gas. The gas composition and especially carbon monoxide and the carbon dioxide concentration is strongly influenced by the methanol concentration of the feed mixture (see Figure 1).

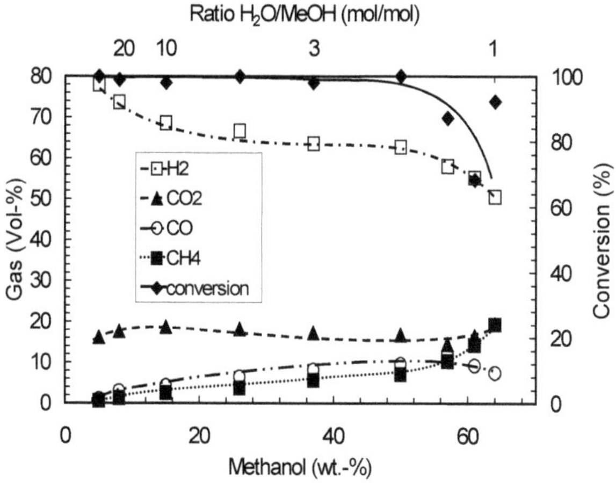

Figure 1: Gas composition and conversion rate as function of the methanol concentration.

MODEL COMPOUNDS

It is known that cellulose hydrolysis proceeds very fast [7], therefore it is better to use glucose instead of cellulose. Glucose can be fed to the reactor as an aqueous solution. Some few mmol of potassium salts have been added to the feed to increase gasification efficiency and to lower CO content; see also [4].

At low temperatures some (in the order of few wt %) soot and tar is formed. At temperatures higher than 550°C complete gasification of glucose occurs. Formation of solid and oily by-products can be suppressed down to trace level. The aqueous effluent (residual TOC) contained mainly phenol and phenol derivatives. Obviously phenol compounds, which are supposed to be less reactive than glucose are formed from glucose.

The main reaction paths for glucose are known [8 - 10] and represented in Figure 2 in a simplified manner. It is expected that a rapid hydrolysis of cellulose to sugar units, e.g. glucose and fructose, first takes place. This hydrolytic depolymerization is followed by decomposition reactions to short-chain aldehydes and acids competes with dehydration and ring closure to furfural derivatives or phenols. The short-chain aldehydes and acids react to the gaseous products desired. The phenols and other aromatic compounds generated are also decomposed to gases under the conditions of biomass gasification in supercritical water. On the other hand, undesired tar, soot or coke may be produced from these aromatic compounds.

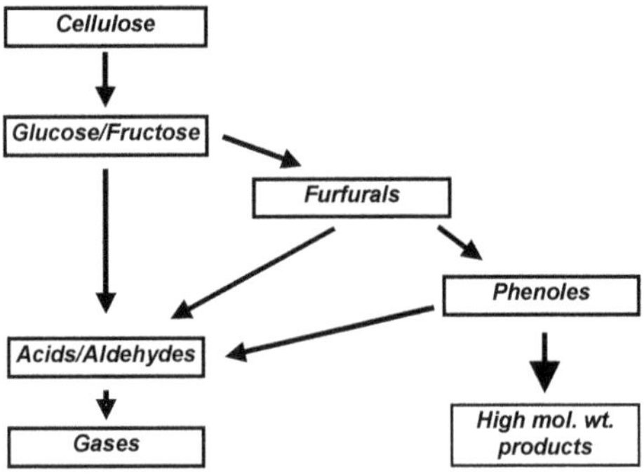

Figure 2: Simplified reaction scheme of cellulose decomposition in near- or supercritical water.

The next step was to study the consequences of the formation of aromatic compounds by the degradation of the aromatic compound, pyrocatechol. Pyrocatechol is also a model substance for lignin. Again KOH was added in the aqueous pyrocatechol feed solution.

Calculations with the computer program equiTherm show an evident temperature dependence of the gas phase composition. The calculated equilibrium value of the relative hydrogen yield (see Figure 3) is nearly reached after two minutes reaction time at 600°C. The hydrogen yield decreases with pressure.

The calculated equilibrium value of methane formation (see Figure 4), which is increasing with pressure is not attained after 4 min. reaction time at 600°C. The reason for this might be that methane formation and methane decomposition are kinetically inhibited reactions at the experimental conditions. Higher hydrocarbons with up to four carbon atoms are formed with a yield of up to 0.6 mole/mole pyrocatechol. Their contents decrease with pressure and reaction time.

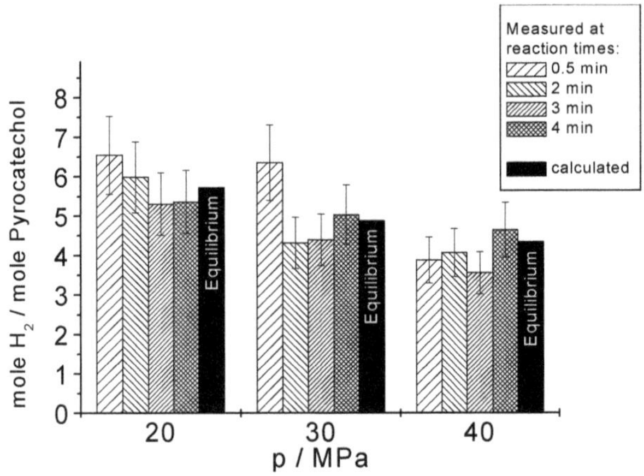

Figure 3: Calculated equilibrium and measured hydrogen yield at 600°C and 20, 30, 40 MPa (concentrations 0.6 mol/l pyrocatechol and 0.005 mol/l KOH).

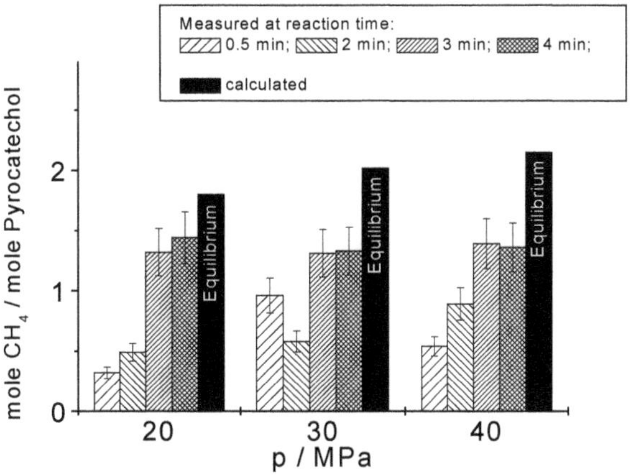

Figure 4: Calculated equilibrium and measured methane yield at 600°C and 20, 30, 40 MPa (concentrations 0.6 mol/l pyrocatechol and 0.005 mol/l KOH).

The residual TOC content decreases evidently with reaction time and slightly with pressure. At 600°C and an initial concentration of 0.6 molar pyrocatechol the TOC of the product solution decreases to less than 0.3% after 2 min reaction time independent of the pressure. The residual TOC content, if there was enough material to analyze, consisted mainly of phenol and small amounts of methyl-, ethyl- and propylphenols, as well as naphthols and indanone. The carbon deposit as solid residue (elementary carbon and organic carbon) after experiment is between 1.7-5.2% of the carbon of the input flow at 20 MPa pressure (subcritical). It decreases at 30 MPa or higher process pressure to less than 0.4%. The solid residue contains also potassium and other metals dissolved from the metal walls of the reactor. This means that even the aromatic compounds react with supercritical water to form gases, see also [4].

The dependence of the gas composition on the dry matter content of foodstuff at a temperature of 500°C has been measured into a continuous stirred tank reactor (lab scale) [5]. Figure 5 shows the gas composition as function of the dry matter content at 500°C. H_2 is the main product followed by CO_2 and CH_4. All other gaseous components, also CO, occur only in very small amounts. This small CO content is in agreement with the thermodynamic calculations. However, it can only be reached in the presence of alkali salts as occurring in the biomass to overcome the kinetic inhibition of the water-gas shift reaction [4]. With increasing the dry matter content, the CH_4 fraction increases too, while the H_2 yield drops [5]. This is also in agreement with the thermodynamic calculation [4, 10].

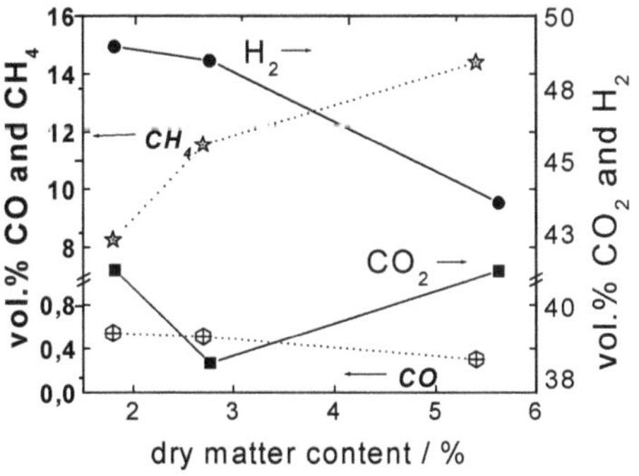

Figure 5: Main gaseous products and CO in the product gas as a function of the dry matter (foodstuff) content (500°C, 5 min. residence time, 30 MPa, continuous stirred tank reactor).

WASTE BIOMASS

Waste vegetable biomass (only 260 mg) was dried, crushed and filled into the autoclave. Water (1 ml/min) was continuously pumped at 25 MPa. Reaction temperature was 500 -700°C and the reaction time was 25 minutes.

Several experiments conducted indicate carbon conversion up to more than 95%; the main constituents of the product gas are hydrogen (> 70 vol.%), carbon dioxide and methane, while higher alkanes and carbon monoxide were found only at lower concentrations. The coke-like residue was always below 5%. Figure 6 shows the gas composition after the gasification of biomass at 600°C.

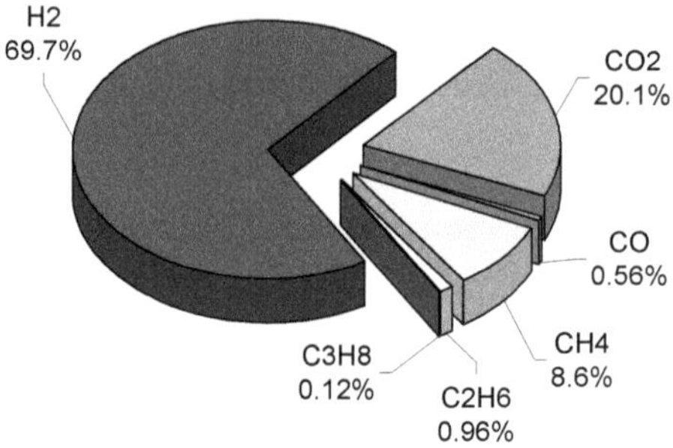

Figure 6: Gas composition for gasification of vegetable waste biomass in supercritical water (feed: 260 mg of dried waste biomass, p = 25 MPa, T = 600°C, reaction time = 25 min, coke -like residue < 5 %).

The dependence of the amount of the gas formed from the reaction temperature show a strong influence of temperature (see Figure 7). The temperature of 700 °C represents a limit for a technical application due to the properties of available reactor materials.

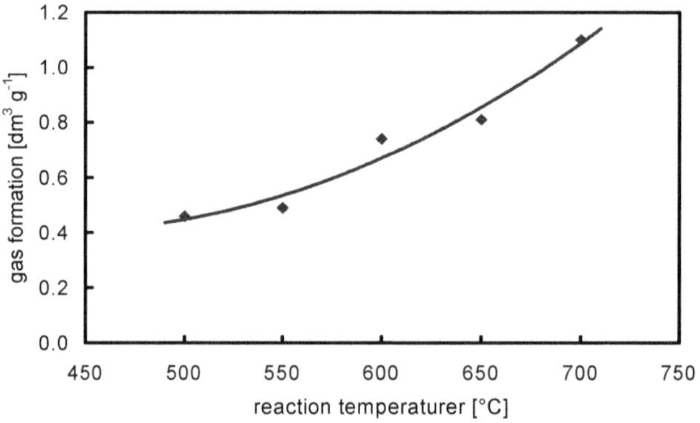

Figure 7: Gas production as a function of the reaction temperature (for the other experimental data see Figure 6).

THE PILOT-PLANT VERENA

Based on the experimental results in laboratory scale a pilot plant (VERENA, see Figure 8), with a total throughput of 100 kg per hour (max. 20 % dry biomass) was planned, constructed and is still under first test operation. The plant is designed for an operating pressure of up to 35 MPa and a maximum temperature of 700 °C [11].

The feed system is qualified for the handling of different educts, such as solid biomass, mud (including sewage sludge) and liquid educts. A measured quantity of the educt is transferred into the stirred preparation tank (volume 1.5 m³) with a connected mill. The water content of the feed and the maximum size of solid particles are adjusted to obtain a homogenous feed suited for pumping. The conditioned educt is transferred to the stirred store tank (volume 3 m³) which has a capacity for 30 hours of continuous operation. The feed is dosed by a mass-flow controlled high pressure metering pump to the reaction system. The feed system was tested successfully with more than one ton of real waste biomass.

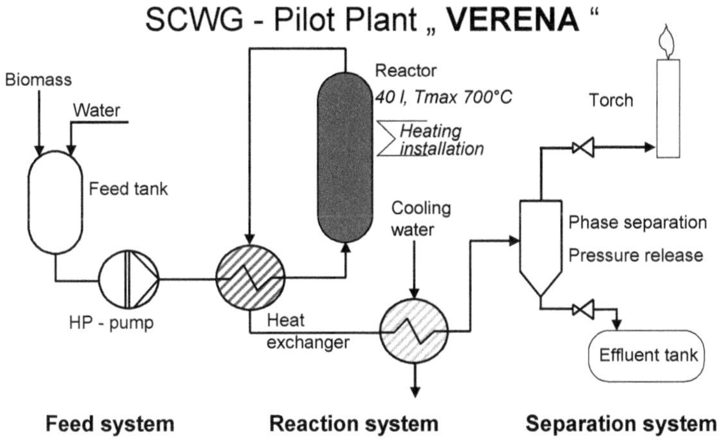

Figure 8: VERENA – simplified process flow sheet.

The educt flow passes a double-pipe heat exchanger and is heated in countercurrent flow by the hot effluent from the reactor (≈600°C). The pre-heated educt is lead into the reactor, which is heated by externally generated flue gas. It has a total volume of 40 dm³, can be operated up to 700 °C and is divided into two parts. The first part is a tubular reactor which acts primarily as heat exchanger, the second part is a slim vessel to guarantee the required residence time of at least 1 min. Under supercritical conditions the gaseous reaction products are soluble in water and leave the reactor as a homogenous phase, pass the heat exchanger and a cooler. By cooling down the product gases separate from the water phase.

The two phases were separated and the pressure can be reduced alternatively in one or two steps. One separator is equipped with an integrated CO_2 scrubber. Here, the CO_2 content in the product gas is reduced by the addition of cold water for increasing the solution capacity for CO_2. The amount of cold water added serves to control the reduction of the CO_2 content and the fraction of burnable gases (H_2, CH_4, CO) of the product gas in the water phase of the scrubber. The product gas is relieved to the pressure required for further treatment. Via a level control, the water phase is passed on to a third separator, operating at ambient pressure, to remove lean gas and eventual solids from the effluent. Then the effluent is lead to a store tank.

The burnable gases can be used principally in different ways for energy generation or chemical reactions. In the VERENA plant the use of product gases is not provided in the first stage. They are passed to a torch and will be burned.

CONCLUSIONS

The process of hydrothermal biomass gasification has been studied in numerous laboratory-scale experiments. The comparison to the traditional gasification process shows some advantages for the hydrothermal gasification especially for wet biomass/organic waste feedstock:

- a much higher thermal efficiency is expected,
- a H_2 rich gas can be produced in one process step,
- soot and tar formation can be suppressed, and
- the heteroatomes (S, N, and halogens) leave the process with the aqueous effluent. An expensive gas cleaning is not necessary.

In the low temperature region between 330-410°C the main gaseous product is CO_2. The formation of gases increases drastically near the critical point of water likely due to a change in the reaction mechanism at decreased density promoting free radical reactions. First experiments at 500°C showed that the main gaseous product is H_2, but with increasing dry matter content, the hydrogen content in the gas phase decreases and the CH_4 content increases. In further experiments, the optimization of the process parameters is focused to the increase of the feed concentration to more than 10 wt% organics (DOM), without losses in the hydrogen yield. In addition, the influence of the type and concentration of the starting material on the composition of the product gas will be investigated further. Technical aspects of the process, such as heat exchanger efficiency, fouling, corrosion, thermal efficiency, solids discharge, and long-term availability, will be studied in the new pilot plant VERENA.

REFERENCES

1. Modell M. (1985) Fundamentals of Thermochemical Biomass Conversion, 95, Eds. R.P.,Milne T.A., Mudge L.K., Elsevier, Amsterdam
2. Xu X., Antal M.J. (1998) Gasification of sewage sludge and other biomass for hydrogen production in supercritical water, Environ. Prog., **17**, 215-220.
3. Schmieder H., Abeln J., Boukis N., Dinjus E., Kruse A., Kluth, M:, Petrich G., Sadri, E., Schacht, M. (2000) J. Supercrit. Fluids **17**, 145-153.
4. Kruse, A., Meier, D., Rimbrecht, P., Schacht, M. (2000) Gasification of Pyrocatechol in Supercritical Water in the Presence of Potassium Hydroxide. Ind.Eng.Chem.Res. **39**, 4842-4848.
5. Henningsen, T. (2002) diploma thesis Fachhochschule Wiesbaden, Germany.
6. Boukis, N., Diem, V., Dinjus, E., Franz, G. and Schmieder, H. (2002) Reforming of Methanol at Supercritical Water Conditions. First Experimental Results. AIDIC Conference Series Editoriale Elsevier, Milano, Italy, Volume 5, 65-69
7. Saka, S.; Ueno, T. (1999) Cellulose Conversion of Various Celluloses to Glucose and its Derivatives in Supercritical Water. Cellulose 6, 177.
8. Kabyemela, B.M.; Adschiri, T.; Malaluan, R.M.; Arai, K. (1999) Glucose and Fructose Decomposition in Subcritical and Supercritical Water: Detailed Reaction Pathway, Mechanisms, and Kinetics. Ind. Eng. Chem. Res. **38**, 2888.
9. Luijkx, G.C.A. (1994) Hydrothermal Conversion of Carbohydrates and Related Compounds. Doctoral thesis, Technical University of Delft.
10. Kruse, A., Gawlik, A., Henningsen, T. (2002) Biomass liquefaction and gasification in near- and supercritical water: Key compounds as a tool to understand chemistry. 4th Int. Symp. on High Pressure Process Techn. and Chem. Eng., 22-25 Sept 2002, Venice, Italy (Proceeding in press)
11. Boukis, N., Diem, V., Dinjus, E., Galla, U., Kruse, A. (2002) Biomass Gasification in Supercritical Water. 12th European Conference and Technology Exhibition on Biomass for Energy, Industry and Climate Protection, 17-21 June Amsterdam, Netherlands (Proceedings in press)

Preliminary Techno-economic Analysis of Large-Scale Synthesis Gas Manufacturing from Imported Biomass

H.P.A. Calis, J.P. Haan
Shell Global Solutions International B.V., Amsterdam, The Netherlands
H. Boerrigter, A. van der Drift, G. Peppink
Energy research Centre of the Netherlands (ECN), Petten, The Netherlands
R. van den Broek
Ecofys B.V., Utrecht, The Netherlands
A.P.C. Faaij
Utrecht University, Utrecht, The Netherlands
R.H. Venderbosch
Biomass Technology Group B.V., Enschede, The Netherlands

ABSTRACT: A techno-economic analysis was made of two concepts for large scale (8 GW_{th}) production of syngas from biomass which, by virtue of the scale of the process, is imported from overseas. The analysis involved projections of technology for the year 2010 and a preliminary estimation of the syngas manufacturing costs, and allows comparison of the two concepts.
The two considered concepts are: 1) production of biomass (energy wood or forestry residues) and overseas transport to a syngas facility, where it is gasified via circulating fluidized bed or entrained flow gasification, and 2) conversion of solid biomass to liquid oil via flash pyrolysis at the biomass production location, followed by overseas transport of the oil to the syngas facility and entrained flow oil gasification. Preliminary deterministic cost models of the complete manufacturing routes, including logistics (transport, storage and transhipment), were developed. For each model input parameter a probability distribution was defined, to account for uncertainties. Via Monte-Carlo simulations and statistical analysis of the cost models, cumulative probability functions were derived for the biosyngas manufacturing costs (€/GJ).

Breakdown of the manufacturing costs indicates that the contribution of the biomass feedstock cost to the syngas manufacturing costs strongly depends on the biomass production scenario. The feedstock element amounts to 15% when forestry residues (@ 0.6 €/GJ) are used, versus 55% when dedicated energy wood (@ 4.0 €/GJ) is used. In the direct route via solid biomass gasification, the remaining part of the syngas manufacturing cost is mainly due to logistics costs and, to a smaller extent, capital charges. For the indirect route, via biomass pyrolysis + oil gasification, the reverse is true. The indirect conversion route has lower logistics costs than the direct route, but this cost advantage is offset by higher feedstock costs (due to a lower overall thermal efficiency) and higher capital costs (two thermal conversion steps, of which pyrolysis is carried out on a much smaller scale than the gasification step). Statistical analysis reveals that the biomass price at the biomass production location, and the scale of operation of the various processes, are the parameters with the biggest impact on the uncertainties in the economic evaluation. The syngas manufacturing costs can be as low as 6 and as high as 13 €/GJ, depending on the biomass price and the selected manufacturing route.

INTRODUCTION

One of the options for transition to sustainable development is to substitute fossil feedstocks (especially coal and fossil oil) by biomass that is produced (as main or by product) in a sustainable way. This increases sustainability by reducing dependence on fossil feedstocks, and also by reducing net CO_2 emissions. Biomass (both dedicated energy crops and forestry/agricultural residues) can be converted into various important commodities, such as power, chemicals and transportation fuels. Learning curves of wind power and photo-voltaics (PV) suggest that in the foreseeable future, wind and PV will probably be less expensive power sources than biomass-fueled power plants. On the other hand, biomass seems a more suitable feedstock than wind or PV for production of chemicals and transportation fuels (except perhaps for hydrogen via photovoltaic water hydrolysis). Furthermore, at present the production of chemicals and especially transportation fuels is responsible for a much larger fraction of fossil feedstock consumption than power generation. These arguments indicate that biomass derived chemicals and transportation fuels could become important options for sustainable development. According to the Shell Energy scenarios [1], biomass derived transportation fuels ("biofuels") may grow at an annual rate of 10% in the period from 2000 to 2050. This is to be compared to projected annual growth rates of only a few percents for fossil coal, oil and gas. In spite of the current modest production capacity of biofuels, such huge growth rates correspond to a global biofuels production capacity in 2050 that may be more than four times the current European gasoline and diesel consumption.

Various technologies are under development for converting biomass into chemicals and transportation fuels. Examples are biochemical conversion (*e.g.*, fermentation of agricultural residues) and chemical conversion (*e.g.*, esterification of vegetable oils) or combinations thereof, at mild temperatures and pressures. Other examples are chemical conversion at elevated temperatures and/or pressures, such as hydrothermal upgrading ("HTU") or flash pyrolysis of agricultural residues or wood. Yet another technology is chemical conversion via synthesis gas as an intermediate product.

The latter route, chemical conversion of biomass into chemicals and synthetic transportation fuels via synthesis gas as an intermediate product, has two attractive features: (*i*) it is an omnivoric route, *i.e.*, a very broad range of biomass feedstocks can be completely converted, and (*ii*) syngas is a universal intermediate to a broad range of commodity chemicals (*e.g.*, ammonia, methanol, fertilizer), transportation fuels (*e.g.*, hydrogen rich synthetic naphtha, kerosene and gasoil) and specialty chemicals (*e.g.*, detergent feedstock, synthetic base oils and specialty paraffinic waxes). For this reason, the present study focuses attention on the conversion of biomass into synthesis gas, in particular syngas of a quality that is suitable for subsequent catalytic conversion, *e.g.*, via Fischer-Tropsch synthesis.

In order for this technology to make a significant contribution towards sustainable development, one needs to consider application of this route on a large scale (in terms of global annual production of syngas from biomass). Two different strategies can be envisaged, which will be referred to as the "local" and "global" strategy. In the "local" strategy, locally available biomass is transported over short distances (less than 100 km) to fairly small local conversion plants (~ hundred kta* of product) and converted into syngas which is further converted into end products. The end product is then transported, typically over longer distances, to the market. Many such facilities would be required to reach a large total production capacity. Alternatively, in the "global" strategy, biomass is collected from various locations and transported over larger distances (overseas) to a large scale conversion facility (~ few thousand kta), that is preferably located close to a market.

The main advantage of the "local" strategy is that biomass transportation costs are lower, especially where appreciable amounts of biomass are available on a single location. Disadvantages are that (*i*) a small-scale conversion plant typically requires a relatively high specific capex, (*ii*) the combination of distance from the market and small production scale suggests manufacturing only a single product, and (*iii*) the plant is heavily dependent on a single or at best a few local suppliers of the biomass. The advantage of the "local" strategy is most apparent in cases where concentrated biomass residue streams are available at almost zero cost and are difficult or expensive to transport. Examples are bagasse from sugar mills or rice husks from rice production.

The alternative option, the "global" strategy, suffers from higher biomass transportation costs. However, in the "global" strategy the biomass feed is purchased on a global biomass market and can thus be purchased where it is most efficiently produced. The use of more efficiently produced biomass may well offset the loss of efficiency due to overseas shipping. Putting it differently, the

---

* kilotons per annum

additional costs of overseas shipping may well be compensated for by lower purchase prices due to normal competition on a global biomass marketplace.

A specific advantage of the "global" strategy is the potentially lower biomass conversion cost due to the economy of scale. In combination with the proximity of the conversion facility to the marketplace, the larger scale also allows production of a wider portfolio of products with a higher market value (*e.g.*, specialty products in the case where the biomass-derived syngas is used for Fischer-Tropsch synthesis of climate neutral transportation fuels). The economic optimization of the product portfolio can be put even further if the biomass conversion plant is located in the proximity of an existing petrochemical refinery or chemicals plant.

Both the "local" and the "global" strategy are interesting strategies to consider in the exploration of sustainable development options involving biomass feedstock. The current study focuses on the "global" strategy, to explore in more detail two technically/logistically different options available under this strategy.

PROBLEM DEFINITION

Key to the development and implementation of biomass-to-syngas routes is the (projected) manufacturing cost of biosyngas. Manufacturing costs of most chemicals and transportation fuels typically depend mainly on feedstock and capital costs. In the case of biomass feedstock, which by its nature needs to be collected from a fairly wide area, also the costs of the feedstock logistics (transport, storage and transhipment) have a significant impact on the manufacturing costs.

Within the aforementioned "global" strategy, two logistically and technically different biosyngas manufacturing routes are envisaged. In the "direct route" biomass is collected (wherever it is most efficiently produced), transported (over land) to an exporting harbor, and shipped (overseas) to the syngas plant, where it is pretreated and gasified to produce syngas. Several biomass gasification technologies are being developed, a.o. circulating fluidized bed (CFB) gasification and (in an early stage of development) entrained flow (EF) gasification. Alternatively, in the "indirect route" the biomass is converted in two steps at the two different locations. First, the biomass is collected and converted into a liquid product ("pyrolysis oil") via a thermal process, at a location close to the biomass production area. Subsequently, the liquid product is transported (over land) to a harbor, shipped (overseas) to the conversion plant and gasified in an oil gasifier.

Each route seems to have distinct advantages and disadvantages. For example, the direct route seems to have the advantage of a higher thermal efficiency, as the thermal conversion takes place in a single step rather than two separate steps. On the other hand, the indirect route seems to have the advantage of concentrating the biomass energy close the source (pyrolysis oil has a four times higher mass and energy (bulk-) density than solid biomass), reducing logistics costs, also by virtue of it being a liquid.

Both routes require research, development and demonstration (RD&D) before they could be implemented on a large scale. A key question of course is, which route to choose to focus further RD&D on? This choice depends on the chances of success of the technical development of either route, but also on the syngas manufacturing costs that are expected under the assumption that the RD&D is indeed successful. In turn, the biosyngas manufacturing costs depend on a large number of parameters, most of which cannot be quantified with certainty but only in terms of probability. An example is the price of biomass, for which a likely bracket can be identified, but which generally varies within this bracket. The differences in technical success chances in combination with the probabilistic nature of most cost variables, hampers a rational choice as to which element of either route RD&D resources could most effectively be allocated to. This defines the problem to be addressed by the current study.

GOAL

The goal of the study is a systematic, quantitative assessment of the technical and economic feasibility of both the direct and the indirect route for large scale production of syngas from biomass, assuming start-up planned for 2010. Questions to be addressed are: what are the chances of technical success of either route; which required RD&D steps are most critical; and if successful, what is the probability distribution of the product (biosyngas) manufacturing cost? Additional questions to be addressed are: which steps in the manufacturing routes have the biggest impact on the all-in manufacturing costs? Which steps add most to the overall *uncertainty* in the projected syngas manufacturing costs?

APPROACH

To reach the stated goal, a strategic decision analysis study (SDA) will be performed for an explicitly defined case. The case concerns the production, at a large production facility, of synthesis gas suitable for application in a catalytic process (manufacturing of chemicals or synthetic transportation fuels), using imported biomass. The production capacity is specified at 8 GW_{th} syngas, which is equivalent to 240 PJ/a, or 40 million boe*/a, or 115,000 boe/day.

Even though the "global" strategy is based on a global biomass market, a deliberate choice was made in the present study to consider a well-defined biomass production location and type of biomass (wood). Although this approach precludes a more general analysis of the issue of availability of biomass, it allows a more pragmatic analysis of the technical and logistic bottlenecks in the production chain.

For this purpose, the three Baltic states were chosen as the biomass production location, a.o. because of the high forest density (about 40%) and the relatively small average distance to the coast is (less than 100 km), which enables efficient transport of the wood to the exporting harbors. Rotterdam was chosen as the location of the syngas manufacturing plant, because of its large port capacity in combination with the presence of a major refinery and proximity to a large market. However, it should be born in mind that these choices for specific locations or a specific type of biomass are not meant to limit the scope of the conclusions drawn from this study to only this specific case.

The study starts with a general description of the two syngas manufacturing routes: the direct route via solid biomass transport and gasification, and the indirect route via biomass pyrolysis, oil transport and gasification. In the direct route, two options for biomass gasification are considered: circulating fluidized bed (CFB) and entrained flow (EF) gasification. Subsequently, critical development paths are defined that visualize the factors that influence the technical feasibility of the different biosyngas manufacturing routes. For each step/factor in the technical development, the chance of technical success is assessed. Multiplication of the success chances of development steps in series results in an estimation of the overall chance that the development will be successful from a technical point of view.

Next, a deterministic model is developed for calculating the syngas manufacturing cost via each of the different routes. A cost-breakdown identifies which cost factors have the biggest impact on the syngas manufacturing costs. Using the model in a Monte-Carlo approach, the probability distribution of the syngas manufacturing costs via each route is determined. Statistical analysis reveals which cost factor, through its probable range of values, has the biggest impact on the probability distribution of the syngas manufacturing costs.

Finally, the aforementioned results are combined to assess (*i*) which of the studied routes has the highest chances of technical success, (*ii*) which route has the lowest expected syngas manufacturing cost, and (*iii*) which are, in either route, the most critical steps, both in terms of technology development success chances and in terms of contribution to the total biosyngas manufacturing costs.

A study like the current one demands judgement and many choices. The presented figures are based on public domain knowledge and in-house knowledge of the various authors. It is beyond the scope of this paper to motivate in detail all choices made or data used.

---

* boe = barrel of oil equivalent = 5.8 GJ

GENERAL DESCRIPTION OF THE BIOSYNGAS MANUFACTURING ROUTES

Figure 1 shows an artist's impression of the syngas manufacturing routes. Biomass (forestry residues or dedicated energy wood) is collected in a number of so-called biomass collection facilities ("b.c.f."), each of which supports an area of about 1000 km^2. In the "direct route", the biomass is transported by truck from the b.c.f. to a nearby exporting harbor ("hub"). In the "indirect route", the biomass is dried, chipped and thermally converted at the b.c.f., to yield a.o. pyrolysis oil. The pyrolysis oil is transported by truck to the hub. In either route, the material is then shipped to the importing harbor. In the "direct route", the solid biomass is dried and chipped/pulverized* and subsequently gasified, via either pressurized (20 bar) oxygen-blown circulating fluidized bed (CFB) gasification plus thermal tar cracker, or pressurized (> 20 bar) oxygen-blown entrained flow (EF) gasification. In the "indirect route", the pyrolysis oil is gasified via pressurized (> 20 bar) oxygen-blown entrained flow (EF) gasification. The raw syngas is then cleaned (removal of solids and inorganic species) and conditioned to meet the specifications for subsequent use [2]. In the studied case, it was assumed that the syngas was to be used for production of predominantly synthetic gasoil (diesel) via Fischer-Tropsch synthesis. For this reason the syngas specifications were set to a certain elevated pressure, a ratio of H_2 to CO of 2, and a maximum inert content of 15% (sum of N_2 and CO_2).

The subsequent steps taken into account in the study are listed in Table 1. The system boundaries were set at the gate of the b.c.f. where the raw biomass is delivered ("front end"), and at the pipe that interfaces the syngas conditioning unit with the syngas conversion unit ("back end").

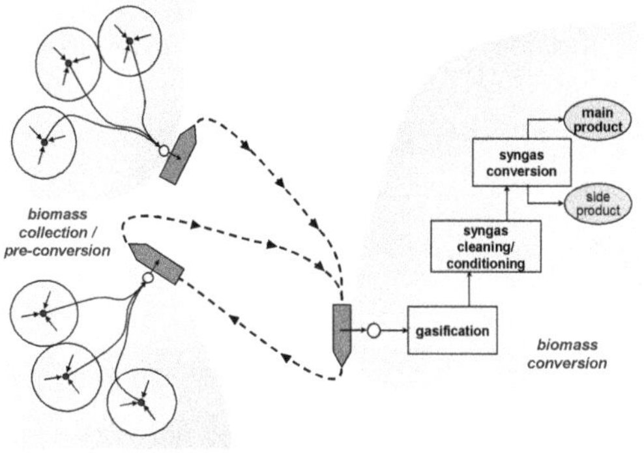

Figure 1: Artist's impression of the biosyngas manufacturing scheme.

CRITICAL PATH DIAGRAMS

To arrive at an assessment of the technical feasibility of a biomass-to-syngas process, one should take into account all major technical hurdles that need to be overcome prior to implementation on a commercial scale. We are dealing with a manufacturing route that contains some elements that have been demonstrated on a modest scale, whereas other elements have already been applied on a commercial scale.

The current study focuses on technology to be available in 2007 (to allow engineering, procurement & construction and ready-for-start-up by 2010), which by that time will have to have been demonstrated on a relevant scale. A preliminary impression of the technical feasibility of the process is obtained by identifying the critical development steps, and associating with them a chance of technical success. Multiplication of the success chances of sequential critical development steps then results in an estimation of the overall chance of technical success of development of the manufacturing route.

* Pulverizing of the biomass is only needed in combination with EF gasification.

Table 1: System steps in biosyngas manufacturing routes.

"direct route"	**"indirect route"**
biomass harvesting	
biomass transport to biomass collection facility ("b.c.f.")	
———*system boundary*———	
biomass storage & transhipment at b.c.f.	
	biomass pretreatment (chipping, drying)
	flash pyrolysis of biomass
	pyrolysis oil storage & transhipment at b.c.f.
biomass transport to exporting harbor ("hub")	pyrolysis oil transport to exporting harbor ("hub")
biomass storage & transhipment at hub	pyrolysis oil storage & transhipment at hub
biomass shipping to importing harbor	pyrolysis oil shipping to importing harbor
biomass storage & transhipment at syngas plant	pyrolysis oil storage & transhipment at syngas plant
biomass preparation (chipping, drying, pulverizing*)	
biomass gasification	pyrolysis oil gasification
syngas cleaning	
syngas compression	
water-gas-shift	
CO_2 removal	
———*system boundary*———	
syngas conversion	

This type of analysis was done for the direct and indirect biosyngas routes. In the direct route, a distinction was made between CFB and EF gasification, as these are in a different stage of development. The results are presented in Figure 2.

Bearing in mind that the estimation of individual success chances is somewhat subjective, the main conclusion drawn from Fig. 2 is that the development and demonstration of pressurized, oxygen-blown CFB gasification of biomass involves significantly greater technical risks than EF gasification of either solid (dried & pulverized) biomass or liquid pyrolysis oil. The implication is that if a cost analysis (under the assumption that technical development & demonstration were successful) were to indicate that all routes give similar biosyngas manufacturing costs, the routes based on EF gasification would be the more attractive routes for further development.

Further observation of Fig. 2 indicates which factors in each of the routes bring along the greatest technical development risk. For the direct route via EF gasification these are slagging in the gasifier and deposition of solids in the syngas cooler. For the direct route via CFB gasification, the greatest risk is fouling of the thermal tar cracker and syngas cooler. For the indirect route, the main critical issue is the reliability of the pyrolysis unit. These observations may be used to direct further RD&D, as resources are most efficiently used by first focusing on the most critical technical hurdles.

---

* Pulverizing of the biomass is only needed in combination with EF gasification.

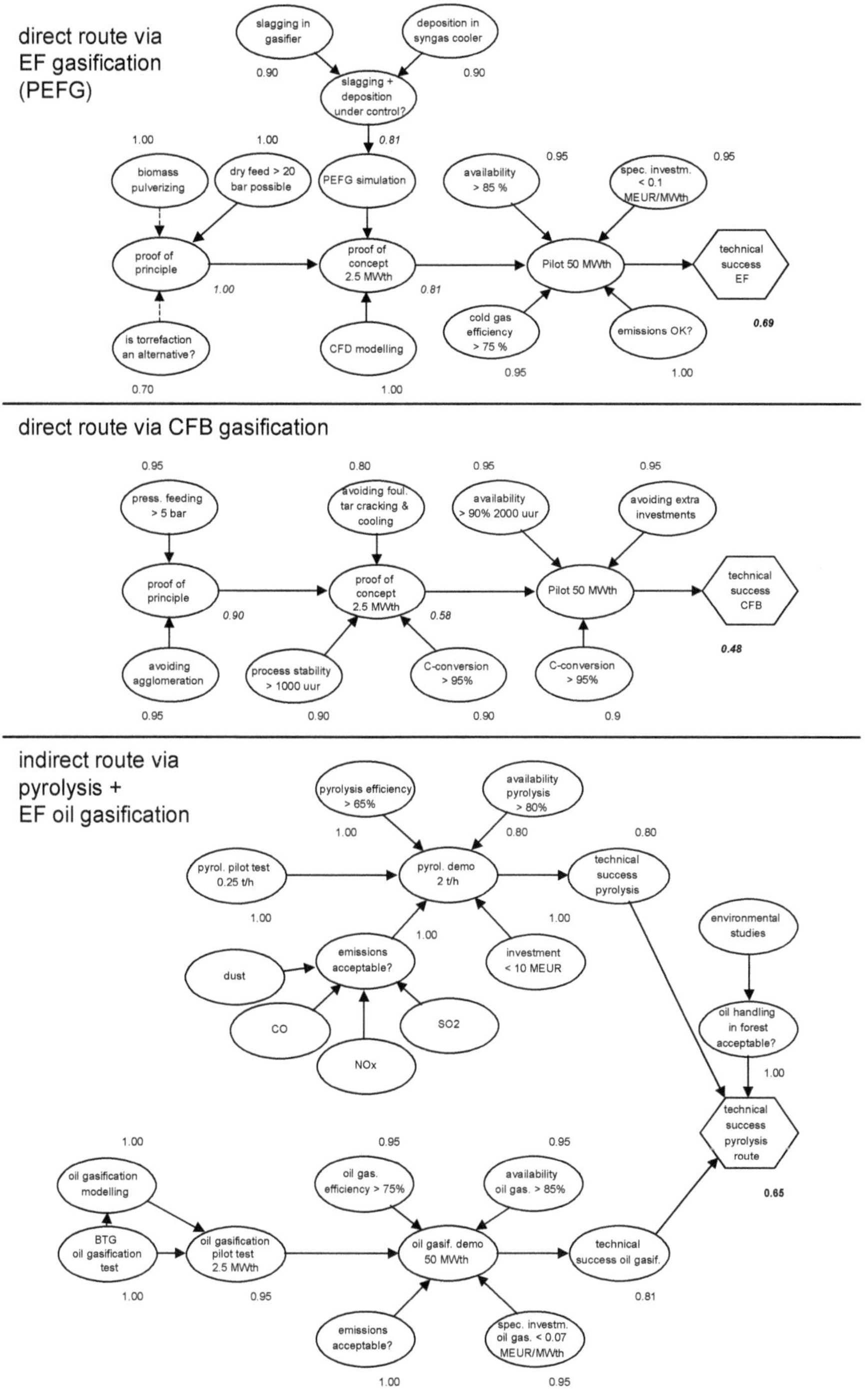

Figure 2: Critical path diagrams visualizing the technical development risks.

COST MODEL

It is beyond the scope of this paper to report the cost model in detail, as the model contains over 500 input parameters and includes fairly elaborate interrelations between the input parameters and model variables. The discussion will therefore be restricted to merely explaining the methodology that was applied, the most relevant input parameters and interrelations, and the main results.

To arrive at a preliminary cost model, the manufacturing chains (from biomass collection up to and including the conditioning of the syngas) were broken down into sequential steps with associated cost drivers, along the lines indicated in Table 1, but with greater detail. Four categories of cost drivers were identified: feedstock, logistics (transport, transhipment and storage), capital, and operation & maintenance. The most relevant elements of the cost model input are briefly discussed in the following sections.

FEEDSTOCK

Feedstock costs are a crucial element in the manufacturing cost of biosyngas. Typical FOB* prices range from a small negative value (*e.g.*, for household waste or demolition wood) up to 6 €/GJ (for virgin wood). These values are to be compared to the market price of gasoil, 4 to 5 €/GJ, as this is one of the possible end-products of the biosyngas.

In the current study, three biomass production scenarios were considered. In the "extensive" scenario, the biomass is assumed to become available as forestry residues at a low price (at the gate of the biomass collection facility) of 0.6 €/GJ. The drawback of this scenario is that a given acreage of forest produces only a limited volume of forestry residues (estimated at 2 otd♣/ha/a). The low productivity implies that the biomass has to be collected from a larger area, on average further away from the coast, increasing transportation costs from the b.c.f. to the exporting harbor.

In the "intensive" scenario, it was assumed that the forests are exploited as production forests, *i.e.*, biomass production is optimized while maintaining a constant average forest density. This yields a higher productivity (estimated at 10 otd/ha/a), but requires additional production resources (*e.g.*, manpower, forestry machinery and fertilizer) and therefore results in a higher biomass price of 4 €/GJ. However, transportation costs from the b.c.f. to the exporting harbor will be lower, as more biomass is produced in a given area close to the coast.

Besides the "extensive" and "intensive" biomass production scenarios, also a "mixed" scenario was considered. In this scenario the overall biomass price depends on the ratio between energy wood and forestry residues that is purchased.

The indicated biomass prices refer to the price at which the material is delivered at a biomass collection facility (b.c.f.), which is a facility in the forest where the biomass is temporarily stored and transhipped. In the indirect biosyngas production route, via pyrolysis oil, the b.c.f. is the location where the conversion of biomass into pyrolysis oil takes place. The b.c.f. supports a biomass production area of about 1000 km^2, which implies an average distance from "tree" to b.c.f. of some 12 km, which is considered a feasible distance for normal forestry machinery.

The biomass delivered at the b.c.f. are wood chunks or branches, with a typical size of 7 cm × 7 cm × 7 cm, and an average moisture content of 30 %wt. The considered biomass pretreatment (at the b.c.f. / pyrolysis plant in the indirect route, and at the syngas plant in the direct route) consists of chipping the biomass to particles smaller than 30 mm, drying of the material from 30 to 7 %wt moisture and (if required) subsequently sizing. The material after drying can be used as such in CFB gasification (ECN). For use in pyrolysis processes, particles sized to 10 mm or less are required (BTG), whereas for EF gasification particles sized to about 1 mm are considered (ECN). The maximum allowable size of biomass particles for EF gasification is still a topic of research; a further reduction of the particle size may prove to be required.

---

* free-on-board, *i.e.*, the price off the supplier's site

♣ oven-dry ton, *i.e.*, dry material

The amount of feedstock required to produce a syngas stream of 8 GW_{th} depends on the overall thermal efficiency of the manufacturing route from biomass to syngas. Given the appreciable fraction of the feedstock cost in the overall manufacturing cost, the thermal efficiencies of the different manufacturing routes are important input parameters in the cost model. The thermal efficiency is expressed as the LHV* of the main syngas components, H_2 and CO, versus the LHV of the biomass feed, weighted by the respective mass flows. Thermal efficiencies of CFB gasification (including a thermal tar cracking step) and EF gasification of solid biomass were estimated at 70% and 77%, respectively (ECN).

The indirect route contains two successive thermal conversion steps; the thermal efficiency of biomass pyrolysis (LHV of the pyrolysis oil relative to the LHV of the biomass feed, weighted by the respective mass flows) was conservatively estimated at 70% (BTG). The thermal efficiency of the EF gasification of the pyrolysis oil was estimated at 81% (Shell). Consequently, the overall thermal efficiency of the indirect route is about 57%, considerably less than that of the direct routes. Assuming that technical development allows an increase of the pyrolysis efficiency from the conservative value of 70% to an optimistic value of 85% (including surplus charcoal in the oil product; BTG), increases the overall efficiency of the indirect route from 57% to 69%, which is still somewhat less than the efficiency of the direct routes. It is noted that such a development affects various pyrolysis oil properties, such as heating value, density, elemental composition and oxygen demand in the oil gasifier. This is part of further research by BTG.

LOGISTICS

Quite unusually for a (petro-)chemical manufacturing process, the feedstock for the considered biosyngas process has to be collected from a very large acreage. Depending on the biomass production scenario (intensive or extensive) and the overall thermal efficiency of the conversion process, the total forest area from which the biomass is to be collected in the considered case is between 18,000 and 115,000 km^2. The actual acreage is even bigger if the forest density is less than 100%, which will usually be the case (in the Baltic countries it is estimated at some 40%). The volumes of biomass transported in the studied project (17-23 mln odt/a) are not exceptionally large when compared to for example coal movements between coal production and consumption locations. However, because of the distributed collection of the biomass, logistics costs form a significant factor in the overall biosyngas manufacturing costs.

In the present study, costs for transport, storage and transhipment were taken into account. Transport consists of transport by truck from the b.c.f. to the exporting harbor, as well as shipping to the importing harbor. The syngas manufacturing facility was assumed to be located close to the Port of Rotterdam, therefore separate over-land transport in Rotterdam was not taken into account. In the estimation of the transportation unit cost rates it was considered that the sizes of trucks and ships is bound to local regulations and infrastructural restrictions such as limited water depths on the route from the Baltics to Rotterdam. For land transport of solid biomass and pyrolysis oil the same unit rates (€/ton/km) were used, since the transport capacity of a single truck is limited by mass rather than volume. For sea transport the same reasoning holds.

Temporary storage of the biomass was accounted for at various locations, a.o. in the forest (along the roadside) and at the b.c.f. In the direct route, additional biomass storage area was defined at the exporting harbors (hubs) and at the importing harbor (Rotterdam). In the indirect route, pyrolysis oil storage tanks were defined at the b.c.f., the hubs and at the importing harbor. The size of the storage capacity was chosen sufficiently large to secure uninterrupted supply (1 week at a b.c.f., 2 ship capacities at a hub, and 3 days at the importing harbor). Biomass storage areas at the b.c.f.s, hubs and the importing harbor were equipped with roofing to allow drying of the biomass during storage, as this relieves the biomass drying duty prior to thermal conversion.
At each storage location, transhipment costs were taken into account, which were based on an analysis of required equipment and manpower. The unit rates for transhipment of pyrolysis oil,

* lower heating value

being a liquid that can be pumped, are much lower than for solid biomass, which has to be hoisted by cranes.

CAPITAL AND OPERATION & MAINTENANCE

To estimate the capital costs, first the installed capital expenditure of the various process steps was estimated on the basis of unit rates (mln € / $MW_{th,\ throughput}$). The unit rates were based on existing capex figures of similar facilities, such as biomass chippers and driers, flash pyrolysis units, CFB gasifiers, EF oil gasifiers, syngas cleaning units, compression stations, water-gas-shift facilities and CO_2-removal units. All equipment was scaled-up to sizes estimated to be a reliable optimum by 2007, using capex scaling exponents of 0.6 to 0.7.

An important assumption (in view of scale advantages) is that both CFB and EF gasification technology can be scaled up to pressurized (20 bar for CFB, even higher for EF), oxygen-blown processes of 1000 MW_{th} by 2007. This requires a significant scale-up effort but is nevertheless considered (ECN/Shell) to have a good chance of technical feasibility, provided that there are sufficient economic incentives for initiating such a development. If, on the other hand, such a scale-up would not be successful, the consequence would be a higher specific capex for the gasification step. Another important assumption is that for EF gasification of solid biomass (which is in an early stage of development, but can derive from commercially proven oil and coal EF gasification technology), substantially larger particles (presumably up to 1 mm) are allowed than for coal EF gasification. This is considered (ECN/Shell) a likely option, because of the much higher reactivity of biomass versus coal; if proven to be feasible, this saves an expensive additional biomass pulverizing step. This type of uncertainty, or differences of opinion between experts concerning the success ratings of technology development, are dealt with in one of the following sections. For example, in the calculation of the specific capex of the CFB gasifier, the possibility that the scale-up might be limited to only 200 MW_{th} is accounted for in the range of input parameter values used.

The projected capacity of the pyrolysis units was based on estimates by BTG and was de-rived from a 2 ton_{wood}/h facility currently being engineered. This unit was scaled up to 10 ton_{wood}/h, or 50 MW_{th} input. That size was considered to be a realistic projection for 2007 (BTG).

Based on the capital expenditure, annual capital charges were calculated using an annuity model, taking into account a ten-year depreciation period for most equipment and assuming a certain required internal rate of return. For the equipment built at the biomass production location a higher internal rate of return (18%) was applied than for equipment built at syngas facility (12%), in view of differences in local risks and inflation.

Operation and maintenance costs were set at about 7% of the capital expenditure, depending on the type of equipment.

RESULTS

The results obtained with the cost model as discussed in the previous sections are summarized in Figure 3, for different biomass production scenarios (extensive, mixed and intensive) and according biomass prices (0.6, 2.3 and 4.0 €/GJ). It shows the cumulative manufacturing costs as a function of the manufacturing stage, providing insight in the contribution of each step to the total costs. The presented cost figures include capital and logistic charges.

The left-hand data points in each of the graphs represent the biomass costs. The lower thermal efficiency of the indirect route (57% overall at 70% pyrolysis efficiency) versus the direct routes (70% for CFB gasification and 77% for EF gasification) imposes a significant feedstock penalty on the indirect route. This penalty becomes more pronounced with increasing biomass price (*i.e.*, going from extensive to intensive biomass production).

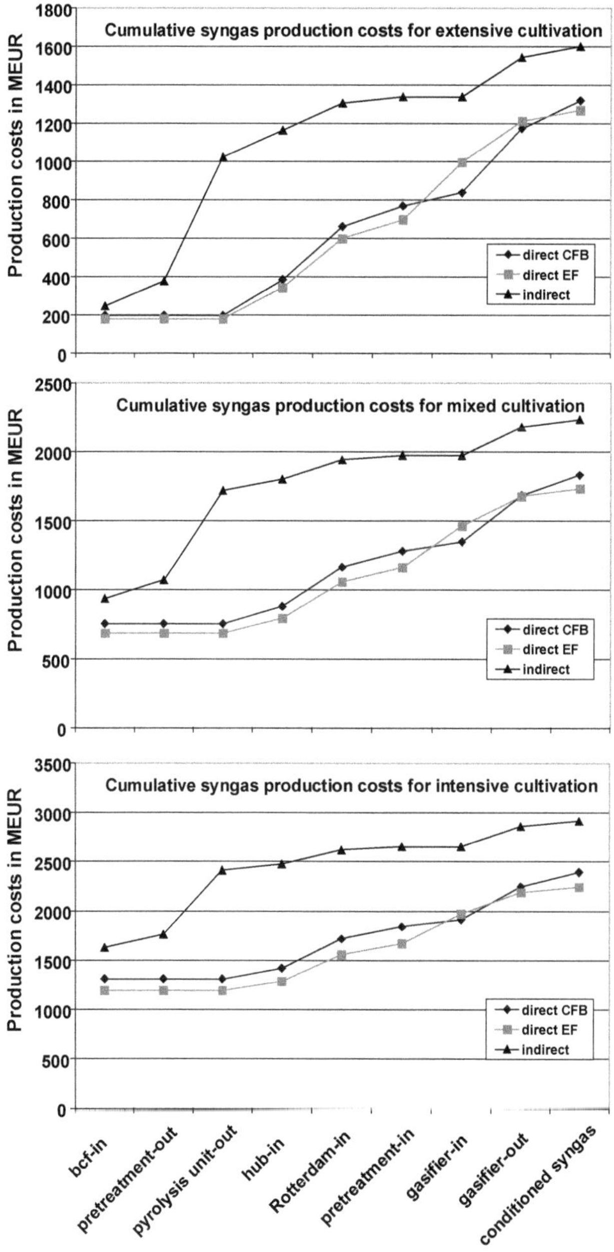

Figure 3: Cumulative annual biosyngas manufacturing costs (mln €/a). Divide manufacturing costs (in mln €/a) by 227 to get the manufacturing cost in €/GJ_{syngas}.
Not all production steps in the graphs are relevant for all manufacturing routes.

Fig. 3 indicates that the conversion of biomass into pyrolysis oil is an important cost element in the indirect route. This is mainly due to the relatively small scale considered for the pyrolysis process (BTG considers 10 ton_{wood}/h to be feasible by 2007), which brings along a high capital expenditure of the pyrolysis units. With a unit capacity of 10 ton_{wood}/h, some 270 units are required to feed the considered syngas facility. Further scale-up in a later stage may well be possible, which would mitigate the impact of the pyrolysis capital costs on the biosyngas manufacturing cost via the indirect route.

The assumption of equivalent land and sea transport costs (€/ton/km) for solid biomass and pyrolysis oil results in a slight cost advantage for the indirect route. This is because the caloric value (GJ/ton) of pyrolysis oil is higher than that of moist wood, and because of the higher gasification thermal efficiency of pyrolysis oil compared to solid biomass: fewer tons of material need to be transported from b.c.f. to the hub and shipped to the importing harbor. In combination with the lower transhipment costs, the effect is a slightly less steep curve of the indirect route at the transport steps.

The impact of the biomass feedstock cost on the syngas manufacturing costs strongly depends on the biomass production scenario. The feedstock element amounts to only 15% when forestry residues (extensive production, 2 odt/ha/a, @ 0.6 €/GJ) are used, versus 55% when dedicated energy wood (intensive production, 10 odt/ha/a, @ 4.0 €/GJ) is used. The remaining part of the syngas manufacturing cost is attributed to mainly capital charges and logistics costs.

Figure 4 presents a manufacturing costs breakdown according to cost category (FOB feedstock cost, logistics cost, capital cost, operation & maintenance and utilities & waste cost). The breakdown is given for three biomass production scenarios (extensive, mixed and intensive). Fig. 4 confirms that in the indirect route (compared to the direct routes) a smaller fraction of the total costs is due to logistics, whereas a higher fraction is due capital. Fig. 4 also shows that in the direct route, with the extensive biomass production scenario (*i.e.*, at a low FOB biomass price of 0.6 \$/GJ), the biomass feedstock cost **at the gate of the syngas plant** (*i.e.*, the FOB biomass costs + the cost of feedstock logistics) is almost four times the FOB biomass price. The contribution of the feedstock gate price to the total manufacturing costs (for the direct routes) ranges from 55% to 75% and increases from extensive to intensive biomass production. In other words, in the studied case the decreased logistics costs associated with intensive biomass production are outbalanced by the higher FOB biomass price.

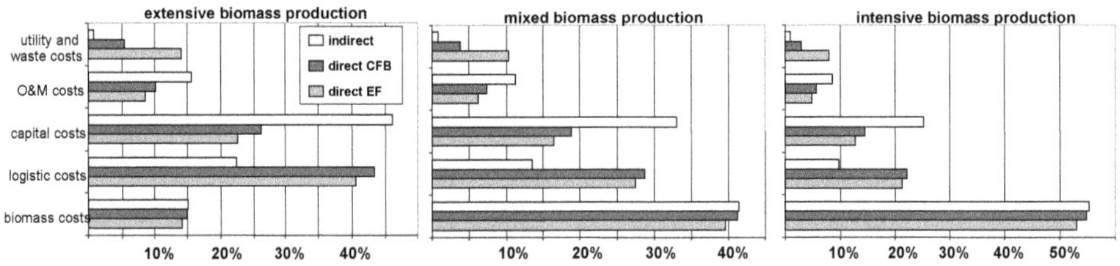

Figure 4: Cost breakdown of the biosyngas manufacturing costs according to cost categories.

EFFECT OF UNCERTAINTIES IN MODEL PARAMETERS ON BIOSYNGAS MANUFACTURING COSTS

All input parameters of the cost model have associated to them a certain degree of uncertainty. Rather than pretending that input parameters are known with an absolute accuracy, one should have the model account for uncertainties in the input parameters. A systematic way of doing so is by defining probability density functions for each input parameter, and running the model repeatedly with random samples taken from the specified range of each parameter. For example, the caloric value of wood, expressed as HHV d.a.f. (higher heating value, dry and ash-free), is known to have a typical value of 20 GJ/ton. However, depending on various conditions, a shipload of wood may have an HHV in the range from 19 to 21 GJ/ton – with 20 GJ/ton being the most likely value. Similarly, the ash content of wood logs may be anywhere in the range from 0.5 to 2.0% – with 1.0% being the most likely value. In that case the parameter is defined to be in the range from 0.5 to 2.0%, and it is specified that the probability of the parameter having a certain value, follows a gaussian distribution with 1.0% as the mean value and which is truncated at 0.5 and 2.0%.

In the present model, such probability density functions (PDFs) were defined for all input parameters for which an appreciable degree of uncertainty exists. For example, the EF gasification efficiency of solid biomass was defined to be in the range from 75 to 79%, biomass storage costs at the b.c.f. were defined in the range from 3.98 to 6.63 €/m^3/year, etcetera.

It was taken into account that some input parameters are linked to each other. For example, the LHV of wood logs, on a dry basis, depends not only on the HHV d.a.f. value, but also on the ash content – each of which follow their own PDF. Another situation is where the probability of one input parameter depends on the probability of another one: similar to the situation that the crude oil price and the electricity spot price both vary in certain ranges, but if the crude oil price happens to be high, the chance that the electricity spot price is high is bigger than if the crude price were low. In this case the values of the parameters or not directly linked to each other like in the previous example, but the PDFs are linked.

After defining the PDFs for the input parameters, the cost model was run typically 5000 times, each time drawing random values of all input parameters according to their respective PDF. For this purpose the software tool @RISK was used. The results are presented in Figure 5.

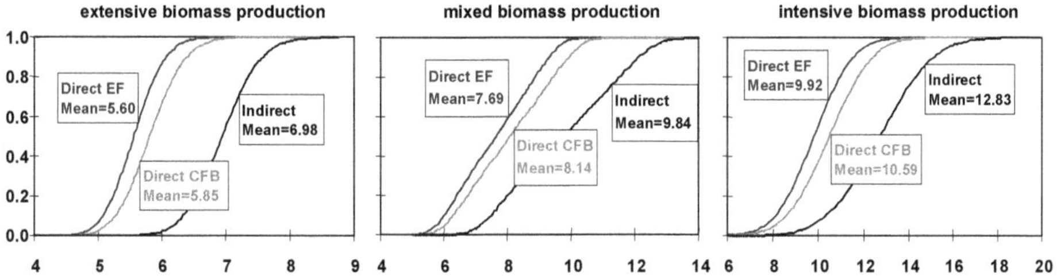

Figure 5: Cumulative probability (vertical axis) of biosyngas all-in manufacturing costs (horizontal axis, €/GJ).

To interpret this graph, we take the case of extensive biomass production (0.6 $/GJ FOB biomass price), in combination with the direct biosyngas manufacturing route via EF gasification. Figure 5 indicates that an all-in biosyngas manufacturing cost of 5.6 $/GJ is estimated, but that the actual value has a 90% chance to be within the range of 5.0 to 6.2 $/GJ. Similarly, it is observed that the chance that the cost exceeds 6.0 $/GJ in this case, is about 14%. This type of information is of course very valuable in the assessment of the economic feasibility of any type of manufacturing process.

Figure 6 presents so-called tornado charts. These illustrate the degree to which the uncertainty of the cost model's output is affected by the uncertainty of the individual parameters put into the model. The longer the bar, the greater the effect that parameter is having on the model's output. The bar length represents the degree of correlation between a certain input parameter and the biosyngas manufacturing costs. The logic is that, the higher the degree of correlation between the input parameter and output variable, the more the input parameter is affecting the output [3].

It is standard practice to plot the parameters from the top down in decreasing size of correlation. If there are positive and negative correlations (a negative correlation means that the output variable decreases upon increasing the input parameter, or vice versa), the result looks a bit like a tornado, hence the name.

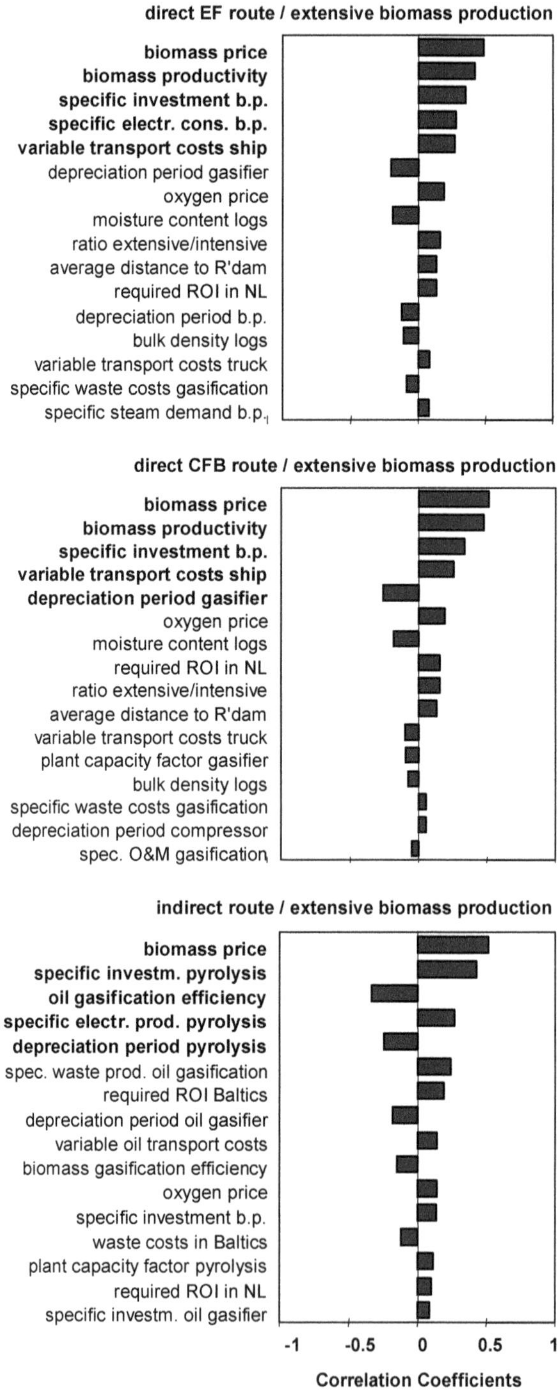

Figure 6: Tornado charts for the different manufacturing routes, pertaining to extensive biomass production. "b.p." means "biomass preparation" (chipping, drying, ...).

The tornado chart is very useful for identifying the key variables and uncertain parameters that are driving the result of the model. It makes sense that, if the uncertainty of these key parameters can be reduced through improved knowledge, or the variability of the problem reduced by changing the system, the total uncertainty of the problem will be reduced too. The tornado chart is therefore very useful for planning any strategy for the reduction of total uncertainty.

Thus, Figure 6 indicates which input parameters have the strongest impact on the uncertainty of the biosyngas manufacturing costs. It is seen that in the case of extensive biomass production (*i.e.*, a low biomass FOB price), variations in the biomass price will have a strong effect on the biosyngas cost. The implication is that if the objective is to minimize the degree of uncertainty concerning the biosyngas manufacturing cost (which is an obvious objective in the assessment of the economic feasibility of the process), one should focus efforts on minimizing the possible variations in the biomass FOB price. In turn, this indicates that it would be important to close long-term biomass supply contracts.

Comparison of the three charts in Figure 6 reveals that in both direct routes, the uncertainty in the variable transportation costs (€/ton/km) per ship have a strong effect on the uncertainty in the overall manufacturing costs. In the indirect route this is less important; uncertainties concerning the capital cost and the thermal efficiency of the pyrolysis unit are more important, suggesting that technology development of biomass pyrolysis is an effective way of reducing uncertainties in the biosyngas manufacturing costs via this route.

CONCLUSIONS

Preliminary critical development path analysis indicates that the development and demonstration of pressurized, oxygen-blown CFB gasification of biomass, on a scale of 1000 MW_{th}, involves significantly greater technical risks than EF gasification of either solid (dried & pulverized) biomass or liquid pyrolysis oil. For the direct biosyngas manufacturing route via CFB gasification, the greatest development risk involves fouling of the thermal tar cracker and syngas cooler. For the direct manufacturing route via EF gasification, it involves slagging in the gasifier and deposition of solids in the syngas cooler. For the indirect route, the main critical issue is the reliability of the pyrolysis unit.

The syngas manufacturing costs can be as low as 6 and as high as 13 €/GJ, depending on the biomass price (ranging from 0.6 to 4.0 €/GJ, at the biomass production location) and the selected manufacturing concept (direct versus indirect). In the studied case the relatively low logistics costs associated with intensive biomass production are outbalanced by the relatively high FOB biomass price for intensive production. On the basis of the parameters used in the cost model, the relatively low logistics costs associated with the indirect route, compared to the direct route, are offset by significantly higher feedstock costs (due to a lower overall thermal efficiency) and higher capital costs (two thermal conversion steps, one of which is carried out on a relatively small scale in the studied case).

In both direct routes, the uncertainty in the variable transportation costs (€/ton/km) per ship have a strong effect on the uncertainty in the overall manufacturing costs. In the indirect route this is less important; uncertainties concerning the capital cost and the thermal efficiency of the pyrolysis unit are more important, suggesting that technology development of biomass pyrolysis is an effective way of reducing uncertainties in the biosyngas manufacturing costs via this route.

REFERENCES

1. Shell International (2001); *Energy Needs, Choices and Possibilities – Scenarios to 2050.*
2. Boerrigter, H.; Den Uil, H.; Calis, H.P. (2002) *Green Diesel from Biomass via Fischer-Tropsch synthesis: new insights in gas cleaning and process design*, Pyrolysis and gasification of Biomass and Waste, Expert Meeting, Strasbourg, France, 30 September-1 October.
3. Vose, D. (2000); *Risk Analysis. A Quantitative Guide*, John Wiley & Sons, Ltd, New York.

Advanced Kinetic Model for the Variation of the Tar Composition in its Catalytic Elimination in Biomass Gasification

J. Corella
Chemical Engineering Department (Faculty of "Quimicas"), University "Complutense" of Madrid, 28040 Madrid, Spain.,
M. A. Caballero, M. P. Aznar
Chemical and Environmental Engineering Department, CPS, 3 María de Luna St., University of Saragossa, 50018 Saragossa, Spain
C. Brage
Kung Tekniska Högskolan (KTH), Chemical Technology, S-100 44 Stockholm, Sweden

ABSTRACT: Biomass gasification in fluidized bed generates tar which can be effectively eliminated with catalysts located downstream from the biomass gasifier. Catalytic tar elimination has been considered till date just as only one or two lumps reaction. In this work tar is considered as composed of six different lumps. Tar is sampled before and after catalytic beds located in a slip-flow downstream from a fluidized bed gasifier, pilot scale. Tar is then completely characterized and its composition is thus determined before and after the catalytic bed. Conversion of each one of the species present in tar, and of lumps thus, is calculated at different temperatures and space-times in the catalytic reactor. A reaction network with six kinetic equations and nine kinetic constants is here presented and used. Once the nine kinetic constants are known, the intrinsic reactivity rates of all lumps are calculated. The species involved in tar are ranked then according to their refractoriness to be destroyed. It is concluded that near all rings behave in a similar way or have the same difficulty to be open (destroyed) by the catalyst.

INTRODUCTION

In biomass gasification in fluidized bed, the produced gas (a mixture of H_2, CO, CO_2, N_2, H_2O, CH_4 and light hydrocarbons) contains some tar (besides of other impurities such as NH_3) which has to be eliminated for most promising and future applications of biomass gasification. Hot gas cleaning methods for tar elimination are usually preferred with respect to wet ones because they really destroy the tar (transferring its energetic content to the flue gas as H_2, CO and CH_4 mainly) instead of transferring it to a liquid waste flow of very difficult disposal. Promising hot gas clean up methods include the use of calcined dolomites, CaO-MgO, (or related materials) or of steam reforming (nickel-based) catalysts, which usually contain CaO or MgO as well. These two kinds of solids catalyze the tar elimination reactions by similar mechanisms (1).

Catalytic tar removal was initially considered as a single reaction in which all tar components were grouped in just only one lump (tar), and assumed to disappear by several simultaneous reactions of (steam-, dry- hydro-, thermal-...) reforming, cracking, etc... The overall rate of tar disappearance was thus given by the sum of the rates of all the elementary individual reactions. The individual reactions were considered as first order with respect to tar and the resulting overall kinetic constant, k_{app}, was first order too (2, 3). This 1st-order single-reaction approach was accepted and used by many (if not all) institutions working world-wide in catalytic hot gas cleaning (in biomass gasification). To compare data on catalyst activities (for tar elimination) from different institutions was then very easy using k_{app} which is directly related to the catalyst activity and to the "reactivity" of the tar to be destroyed.

Since k_{app} is very easy to calculate and handle, a lot of work was made in the last few years using this parameter (k_{app}). Nevertheless, going deeper into catalytic tar removal, problems soon started to appear indicating that the (single, first order reaction) approach was not good enough in many ways. For that reason, Corella and co-workers developed (4) a more advanced kinetic model for the catalytic tar abatement based on two lumps for tar and four kinetic constants. Such advanced approach takes into account the existence of both thermal and catalytic reactions, and of highly reactive or "easy-to-destroy" and of other less reactive or "hard-to-destroy" tar species which only react after relatively long residence times in the catalytic reactor. So, since one lump for tar was not enough in itself, two lumps had to be envisaged and used for a more correct fitting of existing experimental results.

Both one and two lumps based models for tar are easy to understand and handle but they are empirical. Tar conversion depends on tar composition which, in turn, depends on the detailed design and operation of the upstream gasifier (5). Kinetic constants for tar elimination calculated with the above said one or two lumps models are specific for a given plant, then. The tar to be catalytically destroyed downstream from the gasifier may be different from one gasification plant to another one. Extrapolation of the kinetic parameters of the one or two lumps models to different facilities and operating conditions is not safe, thus.

Tar sampling, analysis and definition has been recently agreed world-wide with a new tar protocol (6). Calculation of the amount of tar, and of tar conversion thus, has been standarized. Nevertheless, determining its composition is not standarized yet and probably many years will be needed to reach a consensus in this determination. The different methods used till date for this determination have been compared recently by Padban et al. (7). Among these methods, the one proposed by KTH (Royal Institute of Technology, Stockholm, Sweden) based on SPA (Solid Phase Adsorption), and described in refs. 8-10, is good enough and will be used here to characterize the tar and to determine its composition at the inlet and exit of the catalytic tar elimination reactor.

The effect on tar conversion of catalytic beds located downstream from a biomass gasifier have been published in a lot of papers. Only a few works till date have considered its effect on the variation of the tar composition and none of such previously published works have studied the complex kinetics of the evolution of the tar composition because of the catalytic reactor. This matter will be then the main objective of this paper: the most modern approaches existing for the modelling of the kinetics of the FCC process and of polymers degradation will be here applied to the catalytic tar abatement. The present paper is focussed thus on modelling the kinetics of the evolution of the tar composition, because the existence of a catalytic bed, in the flue gas from a fluidized bed biomass gasifier.

APPROACH TO THE EXISTING REACTING NETWORK.

The basis of the kinetic analysis and modelling presented in this paper is the knowledge of the tar composition, at molecular scale, at the inlet of the catalytic reactor and at its exit for different gas residence times. This information provides the tar composition profiles along the catalytic reactor. Once they are known, the kinetic analysis can be carried out in different ways. Among them, two very different approaches have been selected. The first approach is based on considering the tar as a continuous mixture. The second one is based on lumping together the tar species of a similar chemical structure. Using this 2nd approach, different kinetic models, based on different number of lumps, might be developed. Among the different possibilities in this lumped based modelling, a 6-lump model for tar has been selected (instead of other 4, 5, 7... lumps possible ones) because of the former work of Kinoshita and co-workers (11). Such authors, when studying the tar from a biomass gasifier, defined the following 6 lumps:

1: Benzene
2: 1 ring compounds (excluding benzene): toluene, xylenes, indene, indan.
3: Naphthalene

4: 2 ring compounds (excluding naphthalene): 1- and 2- methylnaphthalenes, biphenyl, acenaphthene, acenaphthylene, fluorene.
5: 3 and 4 ring compounds: phenanthrene, anthracene, fluoranthene, pyrene, benzoanthracene,..
6: Phenolic compounds: phenol, cresols, ethylphenol, xylenols.

These six lumps have been used in a previous work of these authors (20) and will be also used now to study the kinetics of the catalytic tar abatement downstream from the biomass gasifier.

FACILITY USED

All data presented in this paper were obtained in a small pilot plant based on a bubbling fluidized bed of 15 cm i.d., 3.2 m high, continuously fed near the bed bottom with a flow rate of 6 to 10 kg biomass /h. Downstream from the gasifier there are two catalytic reactors connected in-series and located in a slip flow. Results shown in this paper were obtained using the commercial steam reforming catalyst from ICI-Katalystics coded 46-1. All details concerning this pilot plant (gasifier, catalytic reactors, ...) and the biomass used (small chips of pine, Pinus Pinaster) have already been published (12-14). Only a few aspects need now to be described:

1. The kinetic analysis here presented concerns gasification with air, only, with equivalence ratios (ER) from 0.24 to 0.35.
2. These authors, as well as VTT (15, 16) and UMSICHT (17), consider that a very promising overall gasification-gas cleaning process would be based on a gasifier with in-bed dolomite (or a similar tar-elimination material) followed by a nickel-based catalyst (rings or monoliths). For this reason, the data here presented and analyzed correspond to this approach, configuration or process.
3. Intervals of the main experimental conditions in the fixed beds catalytic reactors located downstream from the gasifier were:
   Temperature: between 790 and 860°C.
   Pressure: slightly above atmospheric.
   Space-time (τ): from 0.006 to 0.018 kg catalyst /m^3 (wet, T)/h (low τ values),
   from 0.037 to 0.094 kg catalyst /m^3 (wet, T)/h (relatively high τ values).
4. Tar sampling was made every hour. Several samples of tar were taken thus at the same point during each test run. Results here presented correspond to averaged (regarding length of the test-run or time-on-stream of the catalyst) values.
5. Sampling was made always working in stationary state both in the gasifier and in the catalytic reactors.
6. Samples for tar characterization were taken simultaneously at three different points: a) gasifier exit (inlet of the 1st catalytic reactor), b) exit of the 1st catalytic reactor, and c) exit of the 2nd catalytic reactor. Tar composition was determined before and after each catalytic reactor, thus.
7. Using the same nickel-based catalyst in the two in-series catalytic reactors (in the same test run) is equivalent to having one long catalytic reactor cut in two halves with chemical analysis of the gas at the mid-point. Thus not only the tar composition is known at the inlet and exit of the catalytic reactor but also at its half length. Tar composition measured at three points (inlet, middle and exit), having used besides different space-times, allowed to know the tar composition profiles in the catalytic reactor.

TAR SAMPLING AND ANALYSIS USED

Sampling and characterization of tar have been made following the SPA (solid-phase adsorption) method developed by KTH in Sweden (refs. 8-10). This method allows the determination and quantification of the different tar compounds ranging in molecular weigh from benzene to about coronene.

The SPA method displays several advantages over conventional cold-trapping techniques, such as reduced sampling time (from 60 min. to 1 min.), fast sample preparation (about 3 min.), reproducibility and easy use. On the other hand, some high molecular weight compounds (heavy tar) are not well detected with this method, as discussed by Padban et al. (7). The kinetic analysis presented in this paper will have this limitation. In other words, the accuracy of the following kinetic constants will be the same as that of the SPA method used, thus.

RESULTS. STARTING DATA ON THE TAR COMPOSITION VARIATION BY THE CATALYTIC BED

"Tar conversion" (over nickel-based catalysts) vs. temperature and vs. space-time in the catalytic bed has been already published. The subject now is only the evolution of the "tar composition".

Raw results after the analyses carried out at KTH to the tar, sampled in three different points of the facility and at four different times-on-stream, provide the concentrations (Ci) of each compound or species present in the tar sample. Such results were presented in the past 1st World Conference of Biomass in Sevilla (20).

Once the contents of the species or of the lumps are known at the inlet and exit of a catalytic bed, its conversion is easily calculated. These conversions were calculated for different temperatures of the catalytic bed ($T_{cat,b}$) and for different space-times (τ) or residence times of the gas in the bed.

EFFECT OF SPACE-TIME (τ) OF THE GAS ON CONVERSION OF DIFFERENT SPECIES.

The conversion of the six lumps, at different space-times in the catalytic reactor are shown in Figures 1 and 2. Several important conclusions can be already withdrawn from these Figures:

1) Conversion of 2,3 and 4-rings species (Figure 1) and conversion of 1-ring (excl. benzene) and of phenolic species (Figure 2) are clearly higher than conversions of benzene and of naphthalene (Figure 1) which are relatively low in the interval of τ used.

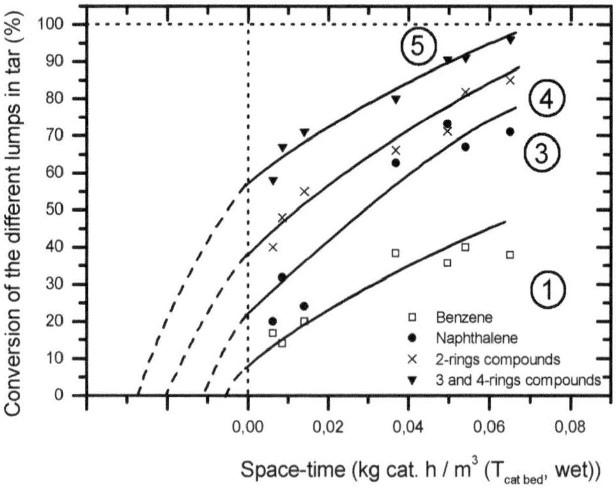

Figure 1: Conversion of benzene, naphthalene, 2, 3 and 4 ring compounds on the ICI 46-1 catalyst for different space times [$T_{cat\ b}$ = 840°C ; ER = 0.24-0.35].

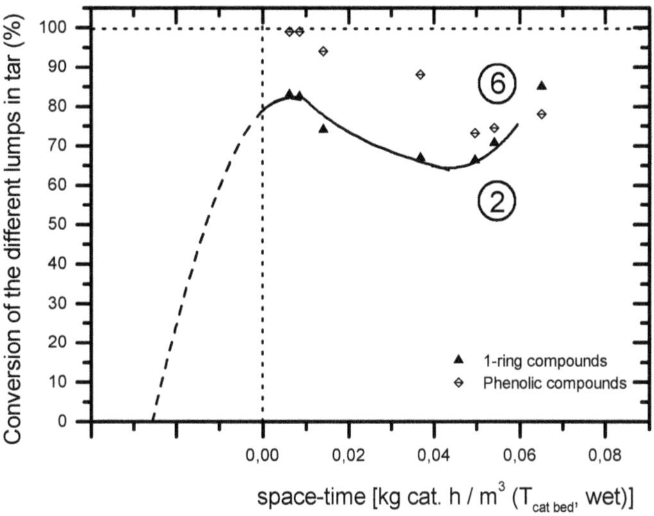

Figure 2: Conversion of 1-ring compounds and phenolic compounds on the ICI 46-1 catalyst for different space-times.

2) Conversions of benzene, naphthalene, two-rings (excl. naphthalene) and of 3 and 4-rings compounds increase with space-time (in the interval used). Yields of phenolic and of one-ring species (excl. benzene) decrease with τ (Figure 2). It means that, on increasing τ, the rate of formation of the species is higher than the one of their disappearance. Their overall yieled decreases with τ (in this interval). It can only be due to their simultaneous formation from another lumps.
3) For space-time =0, when there is no catalyst in the reactor, there exists a quite important conversion for most of the species. It is due both to thermal effects in the gas distributor plate and in the wall and to gas-phase reactions (recalling that these tests were made at 840°C, a quite high temperature).
4) Let's call the conversion at $\tau = 0$, in short and for further studies, "hot wall effect". This "hot wall effect" also appeared and was taken into account in a previous study on the same subject (4). To take into account the hot wall effect, a new space-time (τ_i') may be defined for each i-th lump as

   $\tau'_i = \tau +$ hot wall effect for the i-th lump $= \tau + \tau_{w,i}$, 1.

   in such a way that for each i-th lump, when $\tau_i' = 0$, $X_i = 0$. 2.
5) Each i-th lump has a different refractoriness respect to the hot (840 °C) wall and gas distributor. So, the abscissa for $X_i = 0$ in Figures 1 and 2 (the hot wall effect for the i-th lump) is different for each lump. Extrapolation (below $\tau = 0$) to $X_i = 0$ is doubtful and will limit the accuracy of the kinetic constants further calculated. The chemistry of the process (the nature of the lumps) and the trends of the experimental points help to extrapolate to $X_i = 0$. This extrapolation and the following calculations (mathematical fittings) were carried by selected students of Applied Chemical Kinetics at UCM. To their calculations statistics and errors′ theory (18, 19) were then applied.

If five X_i-τ'_i curves are moved or transported to the left, to get a same origin for the six lumps, only one τ' value can handle then, the same for all lumps (taking into account always eq 1). In such a way, Figure 3 is generated. It has only one (τ') independent variable. When using this Figure, when calculating its abscisa (τ'), it will have to be remembered that the "hot wall effect" ($\tau_{w,i}$) is different for each i-th lump, according to experimental data in Figures 1 and 2.

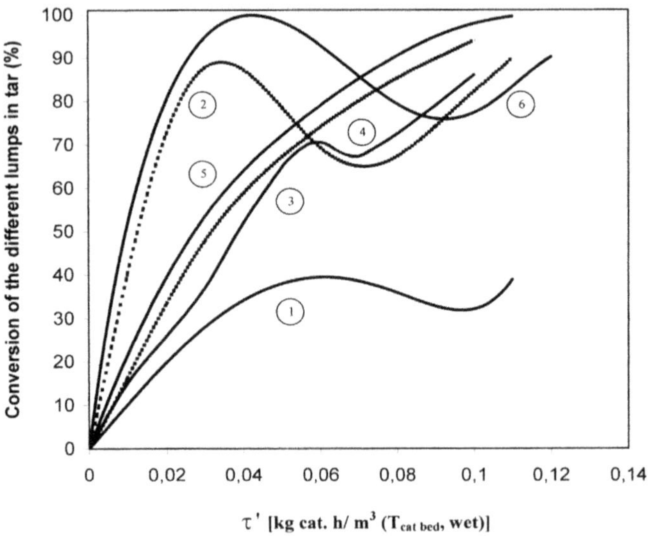

Figure 3: Experimental evolution (in the catalytic bed) of the conversion of the six lumps in the overall tar

KINETIC ANALYSIS AND MODELLING

Proposed reaction network. The above shown results indicate how all tar-species progressively lose (by the combined effect of the catalyst and of H_2O and CO_2) one-by-one all their rings. Once a ring in a tar-species has been open by reaction with an active site in the catalyst or with an adsorbed, and activated thus, H_2O or CO_2 molecule, as shown in Figure 4, the remaining species reacts very soon generating then a tar-species with one ring less. This basic fact, together with results in Figures 1 and 2, motivated to the authors to generate and propose the reaction network shown in Figure 5. Such network explains all the data obtained in this work and is consistent too with all the published data on this subject.

To understand well the reaction network shown in Figure 5, the following facts have to be taken into account:

1) To focus the network on the reacting lumps only and to make Figure 5 easier to understand, the reactants H_2O, CO_2. H_2,.. which convert or destroy all the lumps, as well as the gaseous products (H_2, CO, CH_4,...) and the coke or soot generated in all reactions in the network, have been omitted in such Figure 5. Of course, both gaseous reactants and products (H_2, CO, CH_4,...) exits in the proposed reacting network and can not be forgotten in mass balances, for instance. In other words, the amounts of H_2, CO, CH_4,C_2H_x and coke can not be calculated only by reaction indicated in Figure 5 by k_1, but by all the reactions in such network..
2) The different size of the lumps in Figure 5 represent the relative amount of each lump in the overall tar. Not all lumps are present in the tar in the same concentration or percentage.
3) The network shown in Figure 5 might be enlarged to heavier tars species if they were detected by a tar analysis method different from the SPA one here used. In such case lumps no. 4 and 5 might be considered to be generated too from such heavier tar species, as indicated by dotted lines at the left of the network in Figure 5.
4) Although only one arrow is used in all the reactions in the network, both thermal and catalytic mechanisms are important in every reaction, as demonstrated in ref. 4. Two arrows for each reaction would describe better the real process thus, but would make Figure 5 more complex. One arrow has been omitted then in each reaction, but the two simultaneous mechanisms (thermal and catalytic) should not be forgotten.

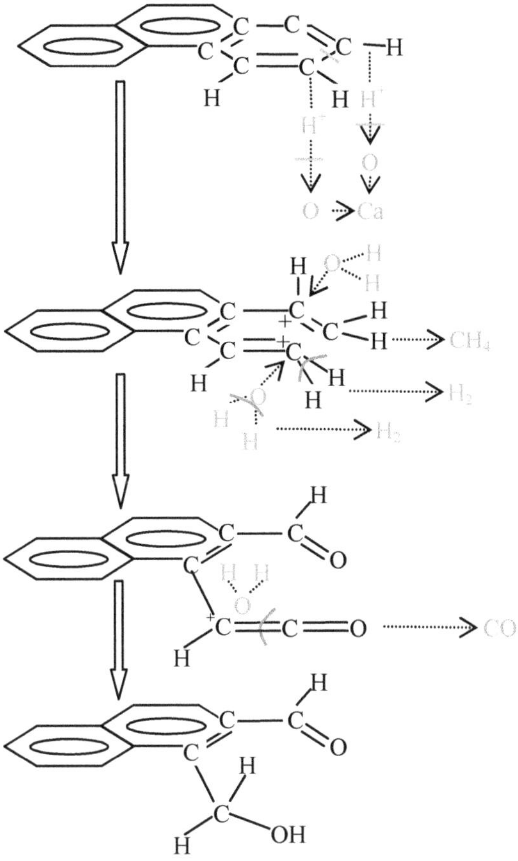

Figure 4: One self consistent mechanism for the catalytic destruction of phenathrene ("typical" tar species): Ring opening by catalytic steam-cracking.

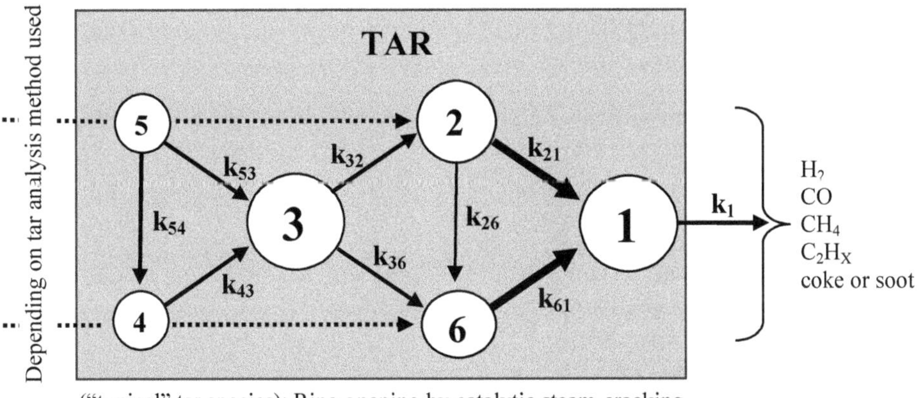

("typical" tar species): Ring opening by catalytic steam-cracking

Figure 5: Proposed 6-lump reacting network. 2nd model here presented for the catalytic tar elimination

Kinetic analysis of the 6-lump reaction network. The reaction network presented in Figure 5 has 9 different kinetic constants. To check the said network and to get its 9 kinetic constants, a set of kinetic equations is proposed after having tried several possibilities. It is based on considering each reaction (from the i-th lump to the j-th one) as first order respect to the concentration of the i-th species considered (Ci). The concentrations of H_2O and CO_2 in the flue gas have also a clear influence in each reaction rate but, since they are much higher than Ci they are considered constant along the reactor and included in the k_{ij} values thus.

If the variation of volume in the gas phase (by the tar elimination reactions) is neglected because it has been determined to be relatively low in the catalytic reactor, the conversion of the i-th lump in the catalytic reactor is given as:
The proposed set of kinetic equations for the network in Figure 5 is, then, the simplest possible one:

$$X_i = \frac{C_{io} - C_i}{C_{io}} = 1 - \frac{C_i}{C_{io}} = 1 - y_i \qquad 3.$$

$$r_1 = \frac{C_{1,0} dX_1}{d\tau'} = k_{61} \cdot C_6 + k_{21} \cdot C_2 - k_1 \cdot C_1 \qquad 4.$$

$$r_2 = \frac{C_{2,0} dX_2}{d\tau'} = k_{32} \cdot C_3 - k_{26} \cdot C_2 - k_{21} \cdot C_2 \qquad 5.$$

$$r_3 = \frac{C_{3,0} dX_3}{d\tau'} = k_{53} \cdot C_5 + k_{43} \cdot C_4 - (k_{32} + k_{36}) \cdot C_3 \qquad 6.$$

$$r_4 = \frac{C_{4,0} dX_4}{d\tau'} = k_{54} \cdot C_5 - k_{43} \cdot C_4 \qquad 7.$$

$$-r_5 = \frac{C_{5,0} dX_5}{d\tau'} = (k_{53} + k_{54}) C_5 \qquad 8.$$

$$r_6 = \frac{C_{6,0} dX_6}{d\tau'} = k_{36} \cdot C_3 + k_{26} \cdot C_2 - k_{61} \cdot C_6 \qquad 9.$$

The units for r_i may seem somewhat strange, but they depend on the units for τ' and C_i which had been previously agreed with several European Institutions such as TPS, KTH, VTT, RUG, ...

Checking of the set of kinetic equations. Calculus of k_{ij}. Estimation of k_{ij} values was made using the whole Xi-τ_i curves. For a first and rough estimate of k_{ij} values, the method of initial rates was applied to eqns. 4 to 9. The initial (for $\tau_i' = 0$, $X_i = 0$) overall rate , $(r_i)_o$, of every lump can be calculated from the slope at $\tau' = 0$ of the X_i-τ' curves shown in Figure 3:

$$(r_i)_0 = C_{i,0} \left(\frac{dX_i}{d\tau'} \right)_{\tau'=0} \qquad 10.$$

In the present experimental conditions, $C_{i,0}$ (at the gasifier exit, inlet of the catalytic bed) had the following averaged values:

$$C_{1,0} = 4400 \pm 400 \text{ mg/m}^3_n \qquad 11.$$
$$C_{2,0} = 2900 \pm 200 \text{ “}$$
$$C_{3,0} = 1320 \pm 120 \text{ “}$$
$$C_{4,0} = 900 \pm 120 \text{ “}$$
$$C_{5,0} = 530 \pm 300 \text{ “}$$
$$C_{6,0} = 200 \pm 100 \text{ “}$$

So, with the method of initial rates, this set is obtained:

$$(r_1)_0 = 44000 \pm 10000 = k_{61} \cdot C_{6,0} + k_{21} \cdot C_{2,0} - k_1 \cdot C_{1,0} \qquad 12.$$

$$(r_2)_0 = 8000 \pm 5000 = k_{32} \cdot C_{3,0} - k_{26} \cdot C_{2,0} - k_{21} \cdot C_{2,0} \qquad 13.$$

$$(r_3)_0 = 10000 \pm 3000 = k_{53} \cdot C_{5,0} + k_{43} \cdot C_{4,0} - (k_{32} + k_{36}) \cdot C_{3,0} \quad 14.$$

$$(r_4)_0 = 8500 \pm 4500 = k_{54} \cdot C_{5,0} - k_{43} \cdot C_{4,0} \quad 15.$$

$$(-r_5)_0 = 15000 \pm 8000 = (k_{53} + k_{54}) C_{5,0} \quad 16.$$

$$(r_6)_0 = 15000 \pm 4000 = k_{36} \cdot C_{3,0} + k_{26} \cdot C_{2,0} - k_{61} \cdot C_{6,0} \quad 17.$$

The relatively big interval of error in the $(r_i)_0$ values come mainly from the uncertainless in extrapolating the X_i-τ' curves to $X_i = 0$. Each selected student of Applied Chemical Kinetics performing these extrapolations calculated a different value for $(r_i)_0$. To the results of the whole team performing the calculations statistics were further applied (18, 19).

X_i-τ_i' curves for some lumps (lumps 3, 2, 6 and 1) have maximums and minimums, as shown in Figure 3, enough well localized (their X_i values known, thus), and in such points $r_i = 0$. So, as a 2nd step another set of algebraic equations was then generated. These two sets were checked then with the whole curve. At the end, the following values of k_{ij} were calculated:

$k_{54} = 80 \pm 20$	[m^3 ($T_{cat.\ bed}$, wet)]/kg cat.·h	18.
$k_{53} = 40 \pm 20$	"	
$k_{43} = 40 \pm 10$	"	
$k_{32} = 30 \pm 8$	"	
$k_{36} = 6 \pm 3$	"	
$k_{26} = 4 \pm 2$	"	
$k_{21} = 8 \pm 2$	"	
$k_{61} = 10 \pm 7$	"	
$k_1 = 3 \pm 1$	"	

Since the overall rate of disappearance of each i-th lump for $C_i = 1$ is the sum of the kinetic constants of the reactions in which such lump disappears, the k_{ij} values provide an index of the reactivity, or of its inverse the refractoriness, of each lump. So, under the present experimental conditions, the rating of the lumps, from more to less reactive for its destruction, or their "intrinsic reactivity rate" is:

lump 5 : (80 + 40 =)	120	mg disappeared/kg cat.·h
lump 4 :	40	"
lump 3 : (30 + 6 =)	36	"
lump 2 : (8 + 4 =)	12	"
lump 6 :	10	"
lump 1 :	3	"

Notice how these "intrinsic reactivity rates" are referred to weight and how the molecular weight and number of rings in the lump also decreases in this ranking. So, if these rates would be referred to "at-gr C" or to the number of rings in the lump, the above shown differences in reactivity would be not so important. It would bring to the easy-to-understand-and-admit conclusion that near all rings, be in a given tar-species or in another one, behave in a similar way or have the same difficulty to be open (destroyed) by the catalyst.

It may be noticed too how the above shown values for k_{ij} are in the same interval that the ones found in previous works (4, 21) with very different models and kinetic analyses. In the authors' opinion, it is a fact in favor of the k_{ij} values above obtained.

With the said k_{ij} values and equations 4 to 9, the y_i-τ' curves shown in Figure 6 have been generated. They are very similar to the experimental curves shown in Figure 3. The calculated y_i-τ' curves (taken always into account is important interval of error) fit enough well the experimental data, then.

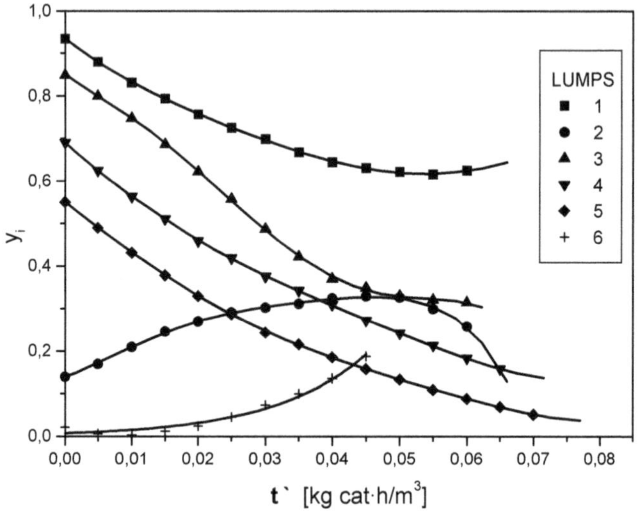

Figure 6: Experimental and predicted curves for the evolution, along catalytic bed height, of the six lumps which compose the tar

Probably, and due to the complexity of the system, other set of kinetic equations and/ or of k_{ij} values might fit well too the experimental data. It does not prevent that it may be concluded that the set of kinetic equations given by equations 4 to 9 together with the above said k_{ij} values are good enough to fit all data. Of course, more experiments would be needed to better localize the maximums of some X_i-τ‘ curves and to obtain more accurate k_{ij} values then, but such more detailed work would require an extensive and costly research program which is out of the scope of this work.

It is concluded thus that the 6-lump based reaction network shown in Figure 5 with the set of kinetic equations given by eqns. 4 to 9 fits well the evolution of the tar composition. This model is consistent with all the data obtained in this work and presented in this paper as well as with the scarce existing data on the matter in the literature.

ACKNOWLEDGEMENT

Javier Gil and Juan-Antonio Martín Berian did an excellent work performing gasification tests and tar sampling. The students of Applied Chemical Kinetics (Chem. Eng. Dept.) of University Complutense of Madrid, Ignacio Bazán, Ana Lepe, Jorge Ruiz-Peinado, David Martín, Alicia Laguna and Daniel de la Torre, helped very much in the kinetic analysis and in statistics calculations. José-Manuel Toledo (UCM) and Christina Hornel and Krister Sjöström (KTH) contributed too in several ways.

The first part of this work (years 1997-2000) was carried out under the project no. JOR3-CT98-0306 and the second part (years 2000-2002), under the NOVACAT Project no. ENK5-CT200-00305 of the EC DG-XII. The authors thank to the European Commission for its financial support.

NOMENCLATURE

Ci	concentration of the i-th species in the produced gas, mg i / m^3_n
$C_{i,0}$	ditto, at the catalytic reactor inlet (equivalent to the gasifier exit), mg i / m^3_n
dp	particle diameter, mm.
ER	equivalence ratio, defined as the air-to-fuel ratio used in the gasifier divided by the air-to-fuel ratio for the stoichiometric combustion, dimensionless.
k_{app}	apparent kinetic constant for a single, 1st order and one lump (for tar) reaction, [(m^3 ($T_{cat\ bed,\ wet}$) /(kg cat.h)].
k_{ij}	kinetic constant for the reaction of the i-th lump to the j-th one, defined by eqns. 4 to 9 [(m^3 ($T_{cat\ bed,\ wet}$) /(kg cat·h)].
Q	gas flow rate [(m^3($T_{cat\ b}$,wet)/h)].
$T_{cat\ b}$ or $T_{cat\ bed}$	temperature, measured in the centre of the catalytic bed (°C).
r_i	overall reaction rate of formation of the i-th lump, [(mg i/ kg cat·h)($m^3_{(Tcat,wet)}$/m^3_n)].
X_i	conversion of the i-th lump, dimensionless.
y_i	fraction remaining of the i-th lump, given as (C_i/C_{io}), dimensionless.
W	weight of catalyst (kg).
Greek symbols	
τ	space-time in the catalytic reactor, defined as W/Q [kg h / m^3 ($T_{cat\ b}$,wet)].
τ_i'	"effective" space-time for the i-th lump, defined by eqn. 1 [kg h / m^3 ($T_{cat\ b}$,wet)].
$\tau_{w,i}$	"hot wall effect" for the i-th lump, defined as: $\tau_{w,i} = \tau_i'$ for $\tau = 0$, and eqn. 2 [kg·h/m^3 ($T_{cat\ b}$,wet)].

REFERENCES

1. Caballero, M.A.; Corella, J.; Aznar, M.P.; Gil, J. (2001). Catalytic hot gas clean up in biomass gasification. Comparison (according to their chemical activities) of dolomites vs. nickel catalysts. 1st World Conference on Biomass for Energy and Industry, Sevilla, 5-9 June, 2000. James & James (Science Publishers) Ltd, 1976-1979.
2. Corella, J.; Narváez, I.; Orío, A. Criteria for selection of dolomites and catalysts for tar elimination from gasification gas; kinetic constants. In "New catalysts for clean environment". VTT Symposium 163. Maijanen, A. and Hase, A. Eds.; Julkaisija and Utgivare Publisher: Espoo, Finland, 1996, 177-184.
3. Corella, J.; Narváez, I.; Orío, A. Fresh tar (from biomass gasification) destruction with downstream catalysts: Comparison of their intrinsic activity with a realistic kinetic model. In "Power Production from Biomass II", VTT Symposium 164. K. Sipila and M. Korhonen, Eds.; VTT Pub: Espoo, Finland, 1996, 269-275.
4. Corella, J.;Toledo, J.M.; Aznar, M.P. (2002). Improving the modelling of the kinetics of the catalytic tar elimination in biomass gasification. Ind. Eng. Chem. Res., 41 (14), 3351-3356.
5. Larsen, E.; Egsgaard, H.; Pedersen, K.; Zielke, U.; Brandt, P. Tar compounds in condensates from different types of gasifiers. Proceed. of the 1st World Conference on Biomass for Energy and Industry, Sevilla, Spain, June 2000, James & James (Science Publishers) Ltd, London, UK, 2001, 1690-1693.
6. Neeft, J. P.A.; Knoef, H.A.M.; Zielke, U.; Sjöstrom, K.; Hasler, P.; Simell, P.A. (2002) Guideline for sampling and analysis of tar and particles in biomass producer gases. Report for Energy Project no. EEN5-1999-00507 (Tar protocol), Brussels, Feb. 2001; presented in the Amsterdam Conference, June 2002.
7. Padban, N.; Wang, W.; Ye, Z.; Bjerle, I.; Odenbrand, I. Tar formation in pressurised fluidized bed in air gasification of woody biomass. Energy & Fuels, 2000, 14, 603-611.
8. Brage C.; Yu, Q.; Chen, G.; Sjöström, K. Use of aminophase adsorbent for biomass tar sampling and separation. Fuel, 1997, 76, ,137-142.

9. Brage, C.; Yu, Q.; Chen, G.; Liliedahl,T.; Rosén, C.; Sjöström, K. Application of solid-phase adsorption (SPA) to monitoring evaluation of biomass tar from different types of gasifiers. In Biomass Gasification and Pyrolysis. M. Kaltschmit and A.V. Bridgwater ,Eds.; CPL Press: Newbury, U.K., 1997, 218-227.
10. Hörnell, C.; Myrén, C.; Brage, C.; Y. Qizhuang,; Björnbom E.; Sjöström, K. Biomass gasification for electricity production: upgrading of tars. In Biomass for Energy and Industry, Proceedings of the International Conference, Würzburg, Germany, 8-11 June 1998, Kopetz, H.; Weber, T.; Palz, W.; Chartier P.; Ferrero, G.L. (eds.). C.A.R.M.E.N., 1662-1664.
11. Kinoshita,C.M.; Wang, Y.; Zhou, J. Tar formation under different biomass gasification conditions. J. Anal. Appl. Pyrolisis, 1994, 29, 169-181.
12. Pérez, P.; Aznar, M.P.; Caballero, M.A.; Gil, J.; Martín, J.A.; Corella, J. Hot gas cleaning and upgrading with a calcined dolomite located downstream a biomass fluidized bed gasifier operating with steam-oxygen mixtures, *Energy & Fuels*, 1997, 11 (6), 1194-1203.
13. Gil,J.; Aznar, M.P.; Caballero, M.A.; Francés, E.; Corella, J. Biomass gasification in fluidized bed at pilot scale with steam-oxygen mixtures. Product distribution for very different operating conditions. Energy & Fuels, 1997, 11(6), 1109-1118.
14. Aznar, M. P.; Corella, J.; Delgado, J.; Lahoz, J. Improved steam gasification of lignocellulosic residues in a fluidized bed with commercial steam reforming catalysts. Ind. Eng. Chem. Res., 1993, 32, 1-10.
15. Simell, P. Catalytic hot gas cleaning of gasification gas. Technical Research Centre of Finland (VTT), Espoo, Finland. VTT publication No. 330; 1997 (p.56).
16. Simell, P.; Kurkela, E.; Ståhlberg, P. Formation and catalytic decomposition of tars from fluidized-bed gasification. In Advances in Therm. Biomass Conversion, Vol. 1, A.V. Brigwater, Ed.; Blackie Academic and Professional: London, 1992, 265-279.
17. Ising, M.; Unger, C.; Gil, J.; Balke, U. Gasification of biomass in a CFB for the generation of power and heat. 1st World Conf. on Biomass for Energy, Sevilla, Spain, June 2000. Ed. by James and James (Sci. Pub.) Ltd., London 2001, 1793-1798.
18. Senent, F. Error Theory and Statistics (Book for students). Ed. by Faculties of Chemistry of Universities of Valladolid (1962-70) and of Valencia (1970-to date), Spain.
19. Neuilly, M. and CETAMA. Modelisation et estimation des erreurs de mesure, (2nd ed), Lavoisier ed., Paris, 1998.
20. Corella, J. ; Caballero, M. A. ; Aznar, M. P. Gil, J. ; Brage, C. (2001). A 6-lump model for the kinetics of the catalytic tar removal in biomass gasification. 1st World Conference on Biomass for Energy and Industry, Sevilla, 5-9 June, 2000. James & James (Science Publishers) Ltd, 1472-1475.
21. Corella, J.; Orío, A.; Toledo, J.M., Biomass gasification with air in a fluidized bed: Exhaustive tar elimination with commercial steam reforming catalysts. *Energy and Fuels*, 1999, 13, 702-709.

Modeling Circulating Fluidized Bed Biomass Gasifiers. 4: A Pseudo–Rigorous 1–Dimensional Model for Stationary State. Some Results

José Corella, Alvaro Sanz and José-Manuel Toledo
Dept. of Chemical Engineering, University "Complutense" of Madrid
28040 – Madrid, Spain

***ABSTRACT*:** Some results from a 1-dimensional and semi rigorous model for atmospheric and circulating fluidized bed biomass gasifiers (CFBBGs) are here shown. The model was presented in a previous paper. Process variables predicted by the model are gas composition (H_2, CO, CO_2, CH_4, C_2H_n, H_2O and O_2 contents), gas yield, tar content in the flue gas and char concentration in the gasifier. Both axial profiles in the riser and values at the gasifier exit are calculated from the model. Variables analyzed are total air flow fed (used as equivalence ratio, ER), percentage of secondary air flow, height (location) of the secondary air flow, biomass moisture, and biomass flow rate (expressed as the biomass weight hourly space velocity in the gasifier). Results here shown in detail correspond to the effect of the equivalence ratio, only. The results from the model agree till date both with known published data from commercial gasifiers and with some tests made to check the model.

INTRODUCTION

Biomass gasification produces a valuable gas (a mixture of H_2, CO, CO_2, CH_4, C_2H_n,...containing some tar and other impurities) by the use of a gasifying agent, usually air, which will be the only one considered in this work. Atmospheric circulating fluidized bed biomass gasifiers (CFBBGs) are very promising reactors for high biomass flow rates or thermal capacities because of their very high throughputs (between 1500 and 4000 kg biomass fed/h·m^2 cross sectional area of the gasifier). Nevertheless, CFBBGs are still facing to some technical and economical problems which make difficult their wide use. A good model for CFBBGs could improve both their design and operation, decreasing then their problems and facilitating the implantation of this technology.

Corella and co-workers at Universities of Saragossa and Madrid (Spain) started to study the modeling of fluidized bed biomass gasifiers in the latest-eighties (i.e., refs. 1 and 2). Later on (3), they studied the reaction network existing in a CFBBG and the problems associated with the accuracy of the corresponding kinetic equations. They also presented (4) a model for bubbling fluidized bed biomass gasifiers which considers only four reactions and kinetic constants, selected among the complex reaction network. Finally, Corella and Sanz (5), part 3 of this modeling, have elaborated and presented a complex model for a whole CFBBG. The main characteristics of the model are:

- 1-dimensional (it takes into account only the longitudinal or axial profiles, and not the radial ones).
- for CFBBGs under stationary state.
- semi-rigorous character because of the assumptions that had to be introduced by lack of accurate knowledge in some parts of the modeling.

A literature review on the field was made in the said previous publications. This paper is now devoted only to show some results from the model presented in part 3. By lack of space, these results will be focused or concentrated on the effect of the equivalence ratio (ER).

PROCESS VARIABLES CALCULATED

With the model detailed in part 3 of this work (5) the *longitudinal or axial profiles in the riser* of a CFBBG can be calculated for the following variables:

(1) Gas composition (H_2, CO, CO_2, CH_4, C_2H_n, O_2, in vol%, dry basis, and H_2O, vol. %).
(2) LHV (MJ/m³n, dry gas).
(3) Tar content (g/m³n, dry gas).
(4) Char concentration [g /kg (sand+dolomite)].

From the longitudinal profiles of these variables, their values at the gasifier exit are further and easily calculated.

The model allows to calculate too the gas yield (at the gasifier exit) and a rough estimation of the carbon (C) content in the fly exit ash. Temperatures at the bottom and in the upper (dilute) zone of the riser (before and after the 2nd air flow) are estimated too. The variables calculated *at the gasifier exit* are, thus:

(1) Gas composition (H_2, CO, CO_2, CH_4, C_2H_n, O_2, in vol%, dry basis and H_2O, vol. %).
(2) LHV (MJ/m³n, dry gas).
(3) "Tar" content (g/m³n, dry gas).
(4) "Char" concentration [g /kg (sand+dolomite)].
(5) C content in fly ash (wt%).
(6) Gas yield (m^3n dry gas/kg biomass daf).
(7) Temperature.

As "tar", all the tar is considered (unreacted "tar 2" + "tar 5"). As "char", the whole char ("char 2" + "char 3") is considered. These lumps ("tar2", "tar5", "char2", "char3") have been defined in part 3 of this work (5). Process variables studied and their intervals, shown in parenthesis, have been:

i) Total ER (0.20 - 0.45)
ii) 2nd air inlet height (6 –10 m)
iii) 2nd air fraction (0 – 40 % of total air)
iv) Biomass moisture (5 – 25 wt%)
v) Biomass flow rate, handled as weight hourly space velocity, WHSV (1.5 – 3 h^{-1}). It is defined as mass (as received) flow rate of biomass fed (Tn/h) to mass of solids or inventory of solids (Tn) in the CFBBG.

Results will be shown here for the following set of process parameters. This set corresponds to a selected and realistic situation in a commercial CFBBG.

It has to be remembered that the process parameters shown in this set are not independent between themselves, but they are related to each other through the heat, mass and hydrodynamic balances and by the stoichiometry of the reactions (shown in ref. 5) existing in the CFBBG. The gasifer topology is shown in Figure 1. Its main dimensions are:

Inner diameter	3.3 m
Total height	14.8 m

RESULTS. EFFECT OF THE TOTAL EQUIVALENCE RATIO (TOTAL ER)

ER is probably the most important and known independent variable in a biomass gasifier. It is a very well known parameter. The effect of ER will be shown for the experimental conditions indicated in Table 1.

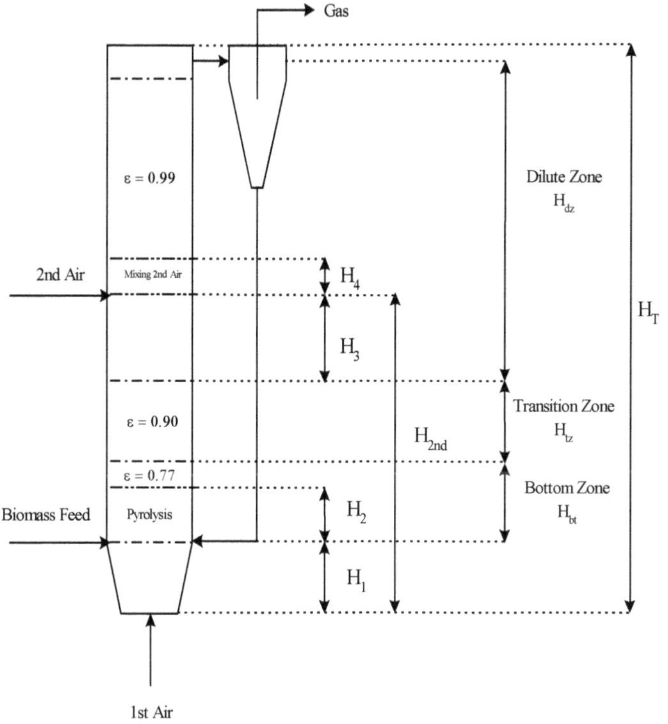

Figure 1: Scheme for the modeled CFBBG.

Table 1: Set of variables used for the results here shown.

Biomass: C (% wt)	50.0
H (% wt)	5.8
O (% wt)........................	44.0
Ash content (wt %)...........	1.0
T (bottom bed) (°C)	825-890
T (dilute zone)(°C)........................	925-990
Air/Biomass (or ER)........................	Variable
Biomass fed (kg a.r./h)........................	15000
WHSV (h^{-1})........................	1.9
Moisture of the biomass (%wt)...........	15
Ash in the biomass (%wt)..................	1
Preheated air temperature(°C)...........	250
% 2nd air (% of total air)..................	20
Inlet 2nd Air (m)........................	6
In-bed dolomite........................	Yes

The *temperature at the bottom of the gasifier*, calculated for different values of ER (from 0.20 to 0.45) and of percentage of secondary air fed, is shown in Figure 2. (In this figure and in following ones the ER value corresponding to 0.30 will be marked with a dotted line to make easier the use of these figures).

Using the temperatures indicated in Figure 2, the calculated longitudinal profiles in the riser of the CFBBG are shown in Figures 3 – 7 for different values of ER.

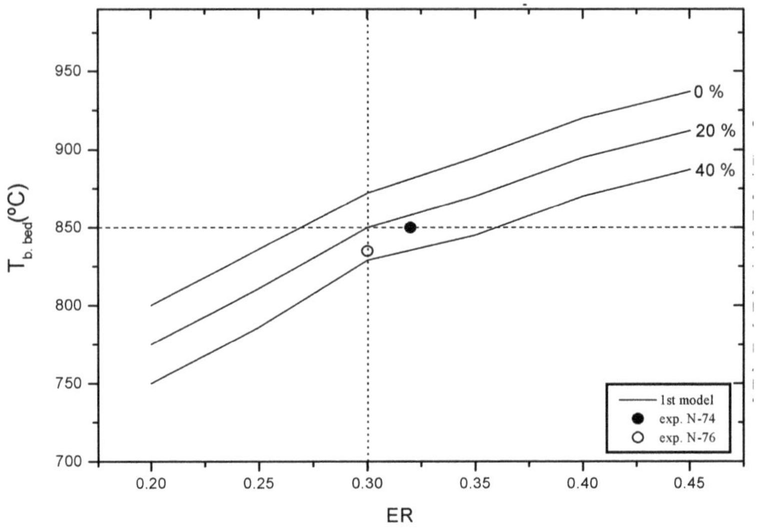

Figure 2: Effect of ER on the temperature at the bottom bed. Parameter: % (defined as % of total ER) 2nd air flow.

All profiles in Figures 3 to 7 have the same shape: The effects of the fast pyrolysis near the biomass feeding point (zone) and of the secondary air are clearly apreciated in such Figures. Although all these profiles are considered useful, the longitudinal profiles of the tar content in the flue gas, shown in Figure 7a, are of particular interest and novelty.

The axial profiles of char concentration shown in Figure 7b correspond to a soft char of very small particle size which is carried upwards by the rising gas. Different axial profiles (for the char) might be obtained for some types of biomass having a high density and particle size. The charred biomass and the char from such biomass could have initially a relatively high particle size and would remain in the bottom bed. The axial profiles would be then somewhat different of those shown in Figure 7b. Nevertheless, since char attrition (in the bottom bed) by the erosive silica sand is very high, the initially big particles of char would be progressively broken to small ones which would be then carried out of the bed by the rising gas. After some time on-stream, under stationary state, the axial profile for the final char would be similar to that shown in Figure 7b. In other words, although some hard chars of initially and relatively big size might have other axial profiles (which remains to be studied in the future), under stationary state such profiles would not be probably very different of those shown in Figure 7b.

The values of the above shown variables have now been calculated at the gasifier exit from results shown in Figures 3–7.

Gas composition (H_2, CO, CO_2, CH_4, C_2H_n, H_2O and O_2 contents) at the gasifier exit is shown in Figures 8, 9 and 10.

The lowest heating value (LHV) of the produced gas is shown in Figure 11. Tar content (in the flue gas) and char concentration (respect to S+D) are in Figure 12. Gas yield and C content in fly ash are finally indicated in Figure 13.

Although all these figures are considered useful and self–explaining, it is pointed out the important quantitative effects of ER on LHV and on tar content shown in Figures 11 and 12b, respectively.

Some test-runs expressly carried out at the small pilot plant, based on a CFBBG, existing at UCM have provided results which agree with predictions of the model. (Experimental points have been omitted in Figures 8-13 shown in this paper to make them easier to understand).

Absolute values and trends from the model, as those shown in Figures 8 to 13, agree quite well too with the published results from pilot and commercial CFBBGs existing worldwide. No important discrepancies have been detected when the same operational parameters existing in the commercial gasifier are introduced and used in the model. The model predicts quite well the effect of ER in a relatively big interval of process parameters.

To make useful this model for gasification community, authors are here inviting to people using a CFBBG to check and compare their results with predictions from the present model. The requested checking or calculations for new conditions would be made quickly and free-of-charge.

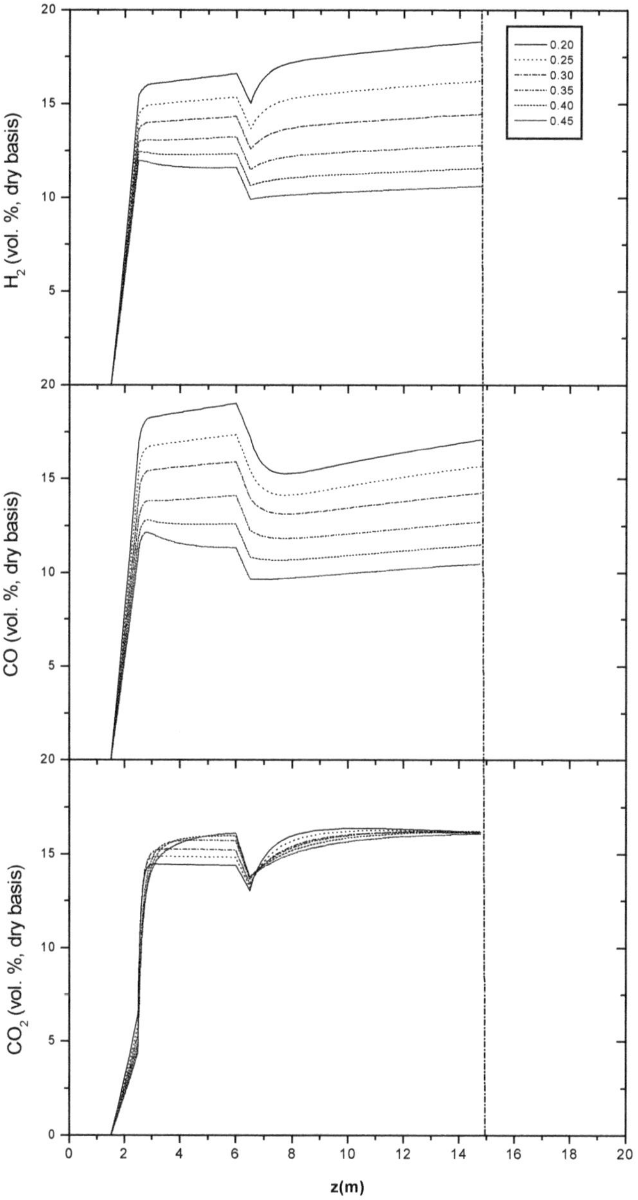

Figure 3: Effect of ER on the longitudinal profiles along the riser of H_2, CO and CO_2 contents (2nd air =20% of total air).

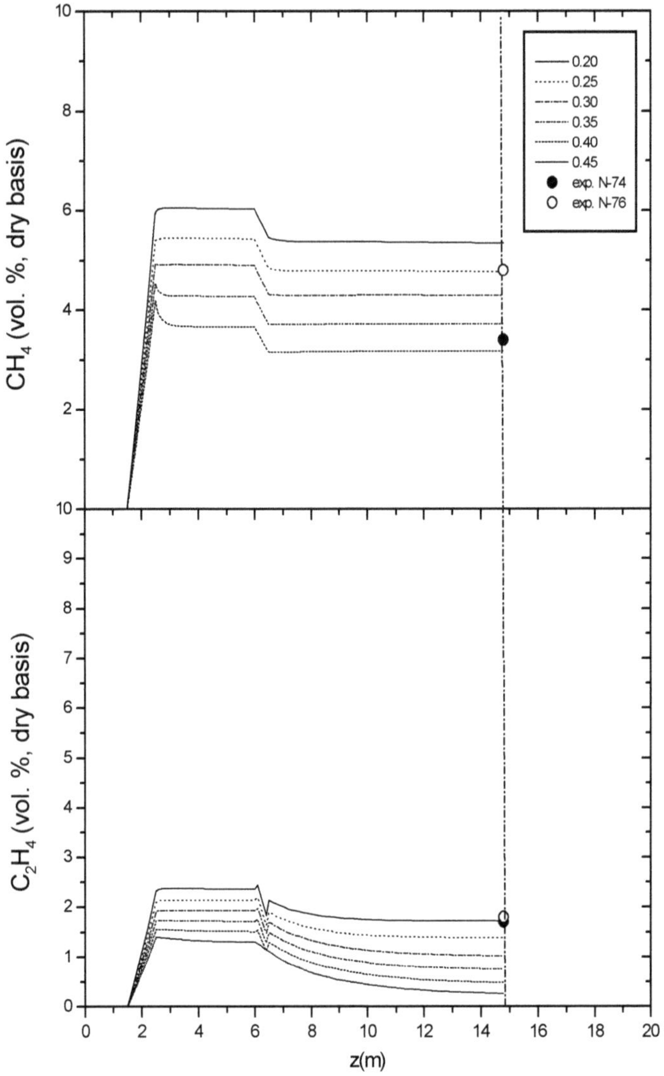

Figure 4: Effect of ER on the longitudinal profiles along the riser of CH_4, C_2H_4 contents (2nd air =20% of total air).

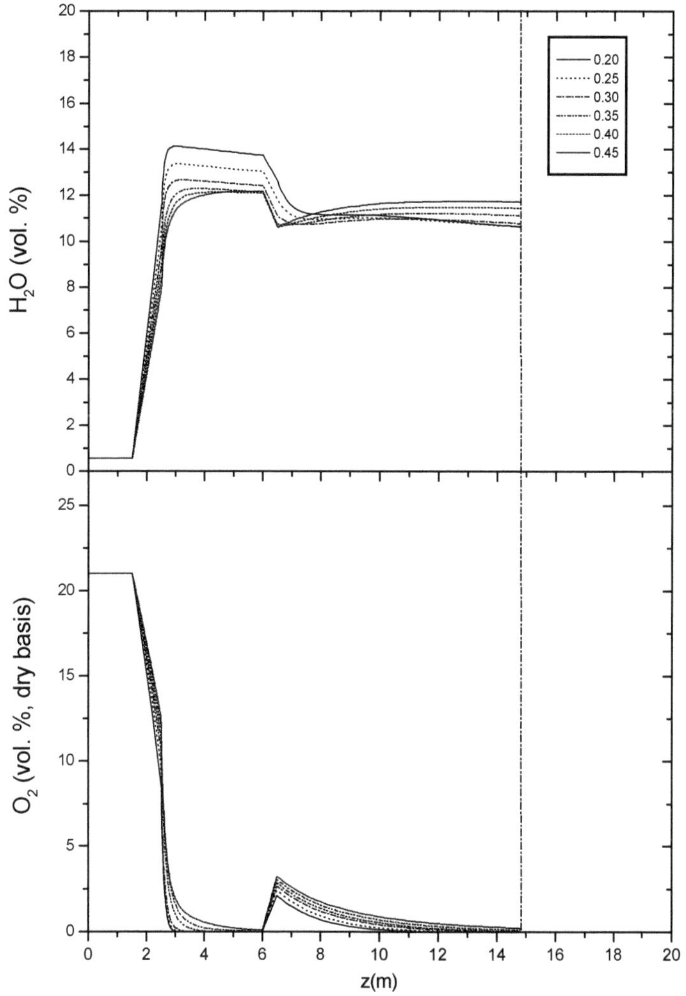

Figure 5: Effect of ER on the longitudinal profiles along the riser of H_2O and O_2 contents (2nd air =20% of total air).

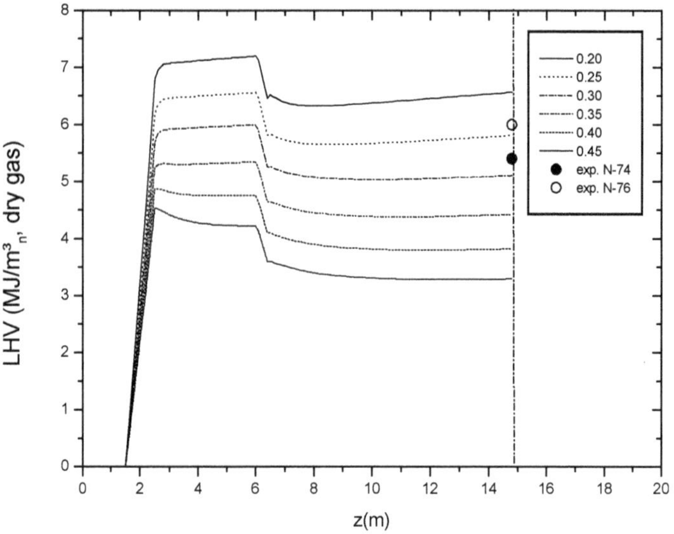

Figure 6: Effect of ER on the longitudinal profiles along the riser of lowest heating value of the flue gas (2nd air =20% of total air).

a)

b)

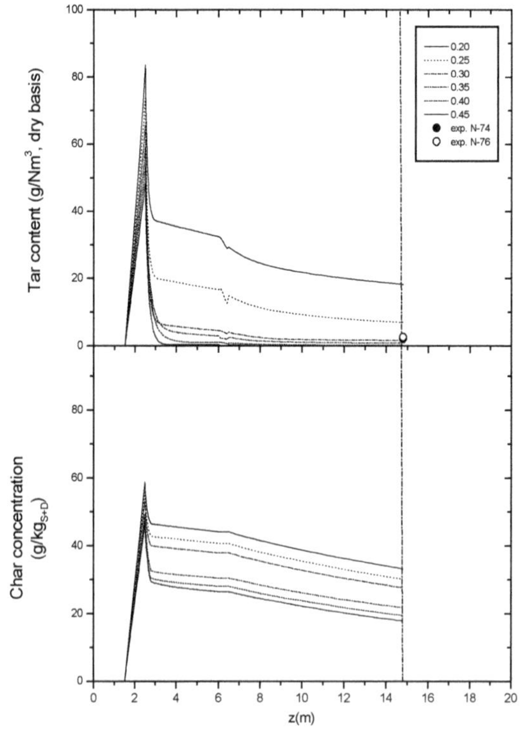

Figure 7: Effect of ER on the longitudinal profiles along the riser of a) tar content and b) char concentration (2nd air =20% of total air).

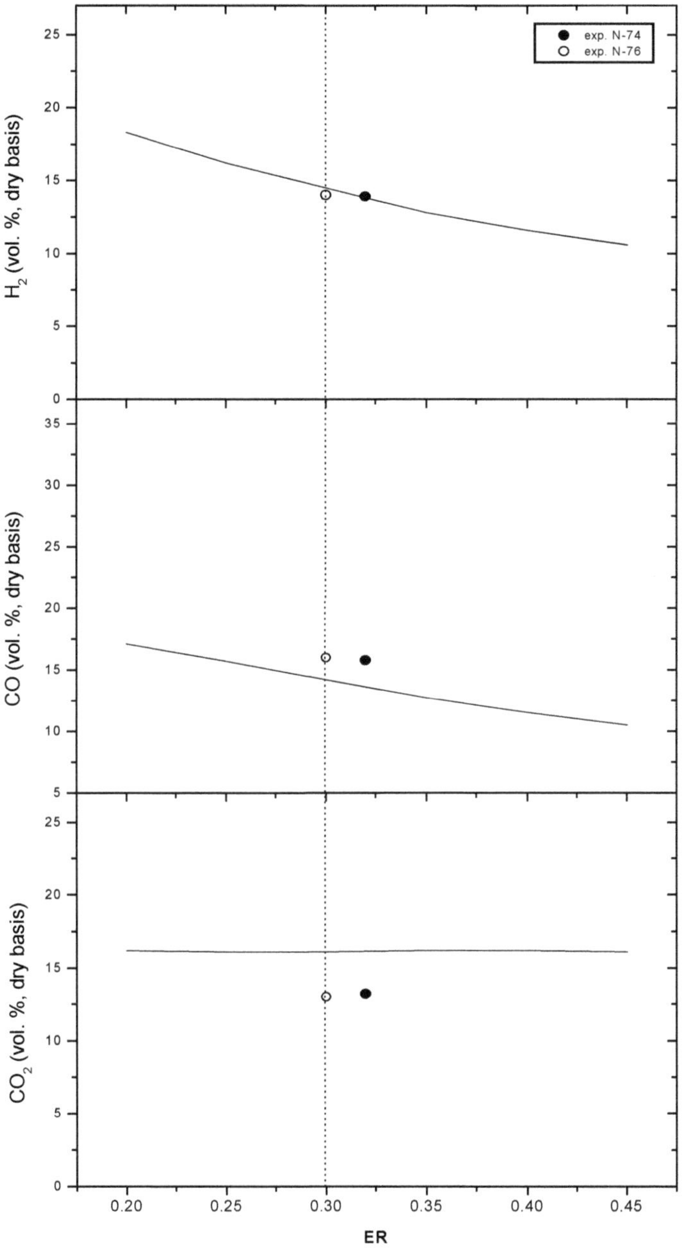

Figure 8: Effect of ER on the H_2, CO and CO_2 contents at the gasifier exit.

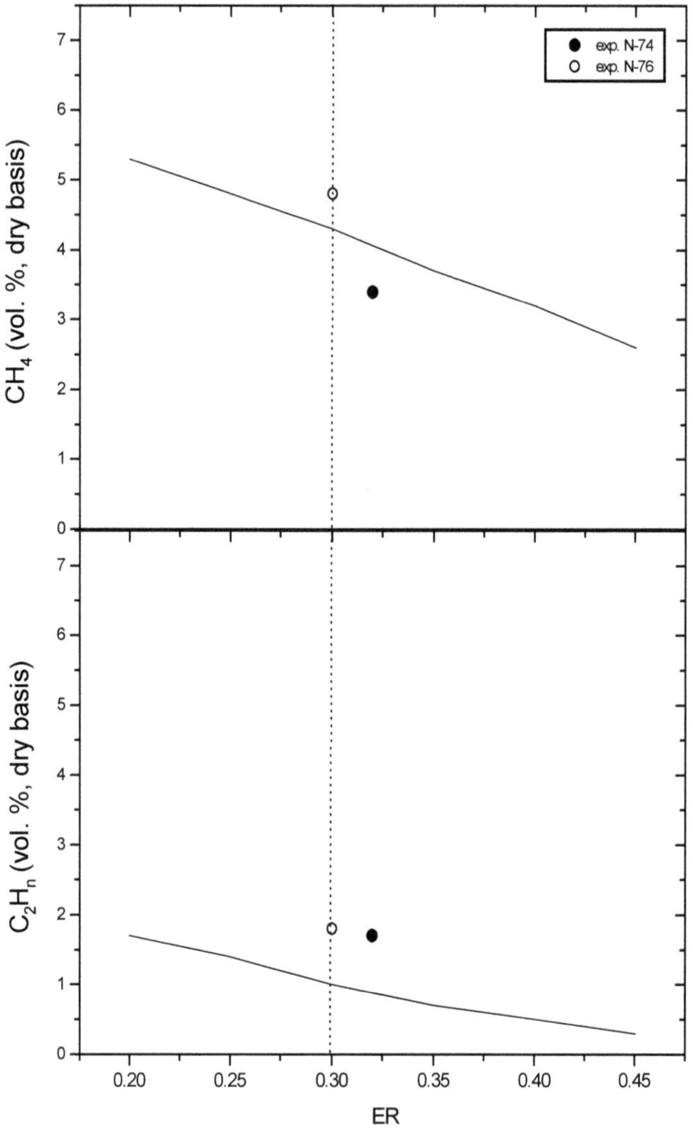

Figure 9: Effect of ER on the CH_4, C_2H_n contents at the gasifier exit.

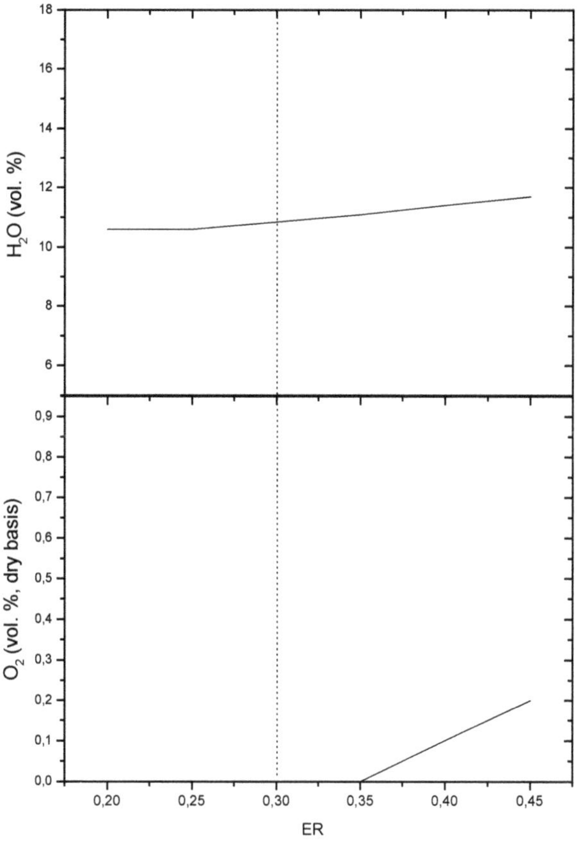

Figure 10: Effect of ER on the H_2O and O_2 contents at the gasifier exit.

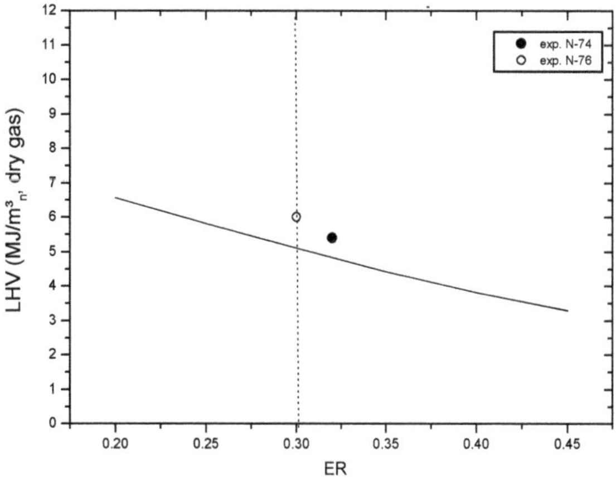

Figure 11: Effect of ER on the lowest heating value in the flue gas at the gasifier exit.

a)

b)

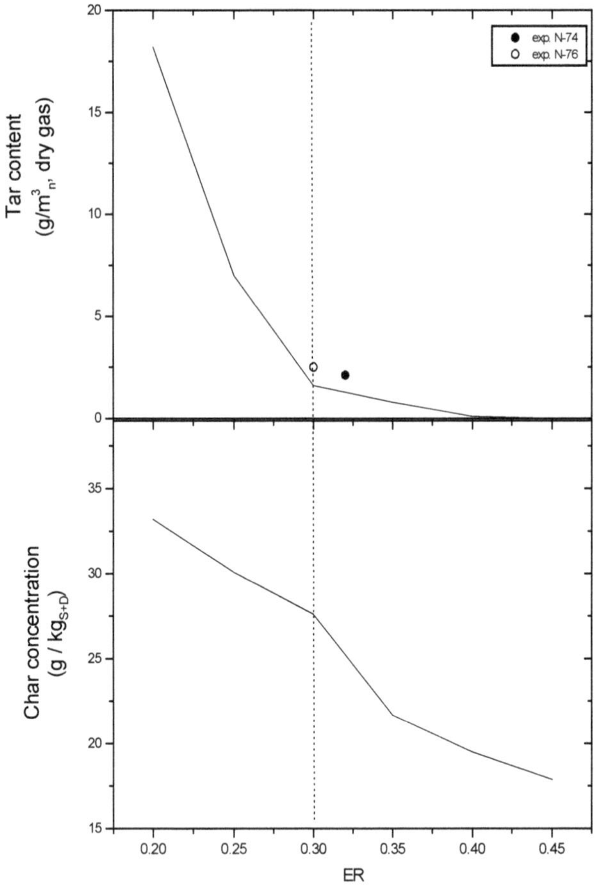

Figure 12: Effect of ER on the tar content and char concentration at the riser exit.

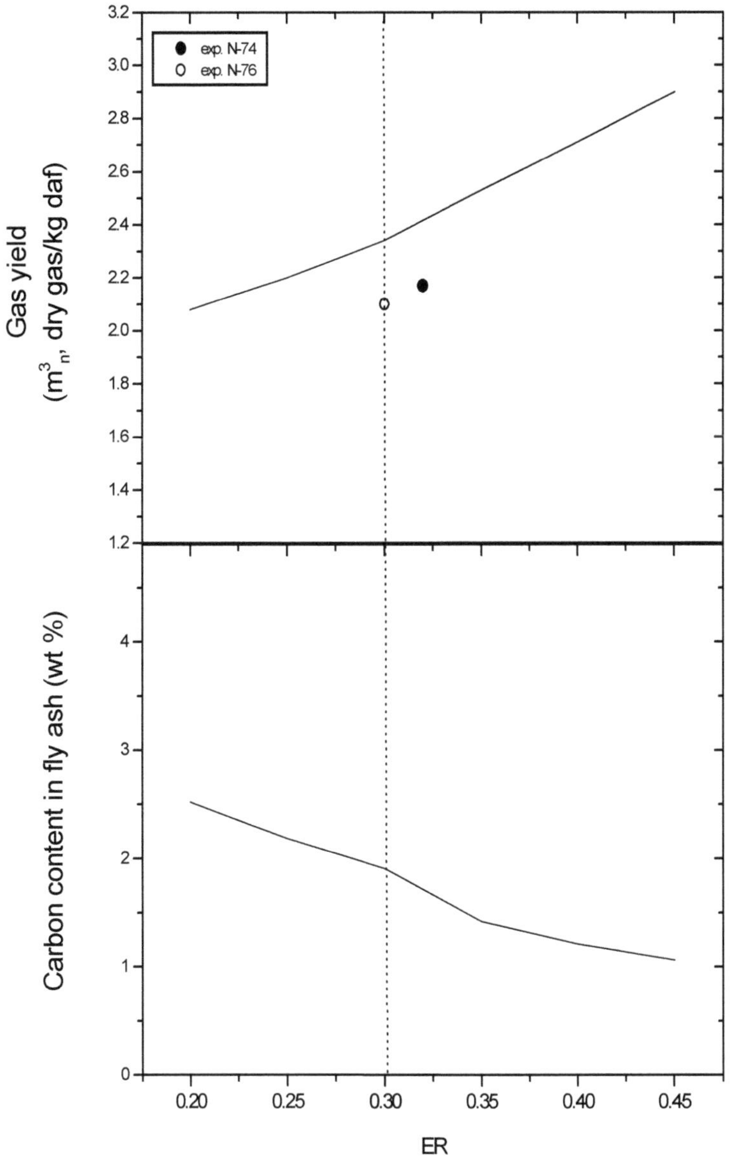

Figure 13: Effect of ER on the gas yield and carbon content in the fly ash at the gasifier exit.

ACKNOWLEDGEMENT

This work was made under the EC-financed project JOR3-CT98-0306. Its Coordinator, Dr. Wennan Zhang, from Mid-Sweden University at Härnosänd (Sweden), did an important effort which is recognized here.

REFERENCES

1. Corella, J.; Alday, F. J. & Herguido, J. (1990). "A model for the non-stationary states of a commercial fluidized bed air gasifier of biomass", En "*Biomass for Energy and Industry*" [Proceed. 5th EC Conference (Lisboa, Portugal, Octobre 1989)], G. Grassi, G. Gosse, G. do Santos (eds.). Elsevier Applied Science, Londres, p. 2804-2809.
2. Corella, J. & Herguido, J. (1994). "Modelling a fluidized bed gasifier of biomass for gasification of biomass in stationary state". En "*Biomass for Energy and Industry*" [Proceed. 7th EC Conference (Florencia, Italia, Octubre 1992)], D. O. Hall, G. Grassi and H. Scheer (eds.). Ponte Press Verlags, Bochum, Alemania, p. 848-856.
3. Corella, J. & Toledo, J. M. (2001) "Modelling a CFB biomass gasifier. Part I: Model formulation". En "*Progress in Thermochemical Biomass Conversión*" [Proceed. Intern. Conference en Tirol, Austria (Sept. 2000)], A. V. Bridgwater (ed.). Blackwell Science Ltd, Oxford, Reino Unido, Vol. 1, p. 333-345.
4. Corella, J.; Herguido, J.; Toledo, J. M.; Gómez-Cívicos, J. I. (2001). "Modeling fluidized bed biomass gasifiers. Part II: gasification with steam in a bubbling fluidized bed". En "*1st World Conference on Biomass for Energy and Industry*" [Proceed. 11th EC and 1st World Conference (Sevilla, España, Junio 2000)], S. Kyritsis, A. A. C. M. Beenackers (eds). James & James (Science Publishers) Ltd., Londres, Reino Unido, vol. 2, p. 1971-1975.
5. Corella, J. & Sanz, A. (2002). "Modeling circulating fluidized bed biomass gasifiers. 3. A pseudo–rigorous model for stationary state". *Chem. Eng. Sci.*, in press.

Large-scale Production of Biomass-derived Fischer-Tropsch Liquids in the Rotterdam Harbor area - A Case Study

C. Daey Ouwens, G.R. Küpers
SDE, PO Box 37107, 1030 AC Amsterdam, The Netherlands
R. van den Broek
Ecofys, PO Box 8408, 3503 RK Utrech, The Netherlands
J.W. Schonewille
Rotterdam Port Authorities, PO Box 6622, 3002 AP Rotterdam, The Netherlands

ABSTRACT: A concept for the large-scale production of "green" liquids from biomass in the Rotterdam Harbour area, have been analyzed in detail. Biomass, imported from Russia is first converted into synthesis gas, which is used in a Fischer-Tropsch synthesis plant to produce the liquids, mainly diesel. The concept is called BIG-FiT (Biomass Integrated Gasification Fischer-Tropsch). For the analysis of the concept a consortium has been set up with ECN, Ecofys, Rabobank, Shell and Volkswagen; SDE coordinated the research project. In cooperation with the consortium, the Rotterdam Port Authorities analyzed the front-end part (acquisition and logistics) of the BIG-FiT chain. Special emphasis has been put on environmental aspects and costs. The environmental aspects considered are: Climate change, abiotic depletion, human toxicity, photochemical oxidation, acidification and eutrophication. The main conclusion of this part is that BIG-FiT liquids (diesel) can be considered as a climate neutral fuel, which significantly reduces most other main emissions; for CO_2 at least 90 %. Concerning the cost, it can be stated that for "expensive" wood of $4/GJ "green" liquids can be produced for about 39 $cent per liter. As cost calculations are based on existing components or subsystems, the numbers obtained are quite reliable. System optimization (higher efficiency) may reduce the cost per liter considerably. For the large scale introduction the development of large gasifiers and cleaning systems are needed. The large scale Fischer-Tropsch process is already "proven technology".

INTRODUCTION

The world's primary energy supply is mainly based on products derived from fossil sources. Roughly 320 EJ[1] or 80% of the total global energy consumption of 400 EJ is based on fossil fuels; 56 EJ is based on renewables and 26 EJ is generated from nuclear. There is no doubt that in the future this has to change to an energy infrastructure that is based on renewable energy sources used in a sustainable way. This process is, according to general accepted understanding, not driven by the scarcity of fossil fuels but due to climate considerations and in a broader sense due to a recognition of the need to meet today's societies energy demands without compromising the needs of the future generations.

From all the primary energy consumed in developed countries, roughly 30% is used for transportion. This share is almost completely supplied by fossil fuels, in particular oil. In view of climate and environmental arguments, it is not surprising that attention for fuels derived from alternative sources is growing. It is the intention of the European Commission to replace a substantial amount of the European transport fuel pool by non fossil based products and to accelerate the production of climate neutral and / or green transport fuels in the near future. Only biomass can be used to replace the fossil derived hydrocarbons.

[1] 1EJ (ExaJoule) equals 10^{18} Joule

To explore the possibilities of the use of biomass for the production of transport fuels a process based on biomass gasification with subsequent conversion of the bio-syngas in a Fischer-Tropsch synthesis unit into hydrocarbons is proposed. The process is called BIG-FiT (Biomass Integrated Gasification – Fischer Tropsch). To explore the feasibility of the process a number of organizations have joined interests and have set-up a consortium. Members of the consortium BIG-FiT are ECN, Ecofys, Rabobank, Shell and Volkswagen. The consortium requested SDE - Samenwerkingsverband Duurzame Energie (Agency for Research in Sustainable Energy) to develop a proposal for a feasibility project and to submit this proposal to Novem with the request for granting a subsidy in the framework of the GAVE program. The proposal has been accepted and the feasibility project [1] was executed under the name: BIG-FiT Prelude. The project has been executed by a project team from representatives of ECN, Ecofys, Shell Global Solutions and SDE. With respect to the front-end part of the BIG-FiT chain (biomass acquisition and logistics), the project team had a creative and fruitful cooperation with the Rotterdam Port Authorities.

BIG-FIT CHAIN

In Figure 1, the complete BIG-FiT chain is represented. The chain has three main parts. The front-end of the chain includes the acquisition of the biomass, the preprocessing and the logistics to transport the biomass to the process plant site (unit number 100). The central part is the processing installations including gasification (unit numbers 200 and 300), gas treatment and gas conditioning (unit numbers 400 and 500), Fischer-Tropsch synthesis (unit number 600) and product upgrading (unit number 700). The back-end site of the plant is mainly the production of electricity and heat (unit number 800) and the end-use of the electricity, heat and fuels (unit number 900). The process scheme in Figure 1 is the basis for all discussions and considerations within the framework of this study. The front-end part of the chain is biomass acquisition and preprocessing of the biomass. Special attention should be paid to biomass crops especially for large-scale applications, i.e. for applications whereby some ten thousands of barrels biomass derived hydrocarbons per day are produced. For such applications, the area needed to cultivate biomass may be many thousands of square kilometers and thus aspects of sustainable forestry need to be taken into account carefully. Preliminary discussions with forestry experts indicated that a robust system of sustainable forestry is feasible [2]. In addition, reference is made to the GRAIN study [3]. In this GRAIN study, biomass availability is investigated for a number of scenario's. In addition, the study discusses criteria for sustainability of biomass acquisition.

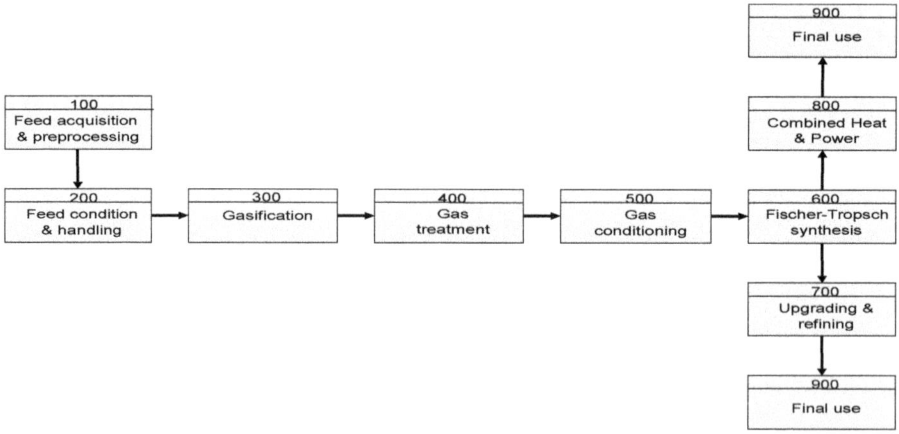

Figure 1: The BIG-FiT chain.

Aspects of biomass logistics (collection, transport, handling, etc) have been studied by the Rotterdam Port Authorities in cooperation with United Nations. The interest in undertaking this work was a result of their ambitions to design sustainable forestry activities in Russia and sustainable harbor activities in Rotterdam. Their results have been made available to the BIG-FiT prelude project and indicate that biomass logistics up to levels of required biomass volumes for a large scale BIG-FiT plant are feasible.

Before the start of the BIG-FiT Prelude project, it was already clear that biomass gasification and gas treatment, to such a specification that the syngas can be converted in the Fischer-Tropsch synthesis unit, are the key aspects in the BIG-FiT process development. Atmospheric air blown biomass gasifiers with thermal inputs in the range of 150 – 200 MW have been operated on a commercial scale [4]. However, for a large-scale bio-refinery, a biomass gasifier with a thermal rating of some Gigawatts (i.e. an order of magnitude larger) will be required. This means that gasifier unit operation must be increased substantially. This is a challenge but comparable technology steps have been made in the past. In the sixties and seventies of the 20th century thermal output of lignite boilers increased an order of magnitude, from 200 MW thermal output to 2500 MW thermal output, only because utilities asked for larger units and technology was ready to be developed. There are other possibilities as well. Circulating Fluid Bed (CFB) gasifiers can be operated under pressure and coal or oil gasification technology may be adopted or modified to use biomass as feedstock. As a reference it can be noted that currently natural gas, oil or coal gasifiers are build in the range from of 500 MW_{th} up to 1250 MW_{th}. Coal combustors are build and operated at unit power levels of 1500 MW_{th}.

Impurities in the raw syngas from the gasifier need to be removed before the syngas is supplied to the Fischer-Tropsch synthesis unit. Within the biomass research program of Dutch Agency for Research in Sustainable Energy (SDE) a large research project is currently underway to define and design gas treatment for the processes like the BIG-FiT chain. Preliminary results of the project indicate that an efficient and cost effective gas treatment process to produce required syngas specifications is feasible. Final project results will be available early 2003.

Fischer-Tropsch (FT) synthesis technology based on syngas produced from natural gas or coal is a mature technology used on a commercial scale for many years. The main drivers for actual applications are the use of stranded or associated gas reserves. Fischer-Tropsch products also exhibit excellent characteristics with respect to combustion in prime movers and low emissions levels to the environment. Additionally there has been a substantial reduction in the capital and operating costs for this kind of projects. The main technology developers nowadays are Shell, Sasol, Exxon/Mobil and BP, followed by a number of small licensors such as Syntroleum and Rentech. Shell and Sasol operate large-scale installations (12.500 – 100.000 bpd).

Gasification and the Fischer-Tropsch synthesis is the basis of BIG-FiT technology. Biomass is gasified and the syngas is used as the feedstock for the Fischer-Tropsch synthesis. As the conversion of natural gas to liquids is called Gas-To-Liquid or GtL, the conversion of biomass to liquid is called Biomass-To-Liquid or BtL.

Related to the syngas feedstock following products are obtained (as percentage of final products in enthalpy content):

- gasoil products (C_{10}-C_{20}) – 60%,
- naphtha product (C_5-C_9) – 12% and
- off gas & light ends (C_1-C_4) – 8%, to be used for CHP production (Unit 800 in Figure 1)
- Around 20% is lost due to thermal inefficiencies, although some of this may be recovered by better heat integration of the process units.

The purpose of BIG-FiT is to yield a maximum premium quality green Fischer-Tropsch diesel. It contains no sulfur or aromatic components and upon combustion, it produces less NOx, particulates and hydrocarbon emissions than fossil diesel. It can be applied as a blending component for fossil diesel to improve the cetane number and emission characteristics of the fossil diesel. The FT diesel can also be applied in higher blending ratios or neat (i.e. pure); in that case the diesel engine needs to be tuned for optimum performance with the synthetic diesel.

Besides FT diesel and naphtha other products with high added value can be produced. The naphtha product stream can be used for blending but is very well suited for use in fuel cells. Fuel cells require a very clean fuel to prevent damage to fuel cell catalyst. FT-naphta can also be used as an intermediate material for the production of "green plastics".

Fischer-Tropsch technology delivers a broad range of high quality products. With biomass as the feedstock the Fischer-Tropsch route represents a way to make a large number of commodity and specialty products sustainable and renewable. In addition to this, a number of other characteristics should be mentioned.

- FT products and wood are excellent storage facilities for chemical products and energy.
- Fischer-Tropsch products are biodegradable.
- Fischer-Tropsch products make use of existing and proven infrastructure.
- Fischer-Tropsch products can be applied for food applications.
- Gasification of biomass in combination with post processing represents a very flexible technology with respect to variety of final products. For instance a smooth transition to hydrogen is possible.

We can only identify the whole BIG-FiT chain sustainable if the environmental aspects and cost show favourable perspectives. For this reason these two issues are elaborated in this paper. For more detailed information we refer to the final report [1].

LIFE CYCLE ASSESSMENT – LCA

The scope and definition of the LCA is defined as follows:

- The goal of the LCA is to understand the drivers that influence the environmental impacts associated with Fischer-Tropsch diesel produced at a large scale so they can be compared to diesel refined from fossil crude oil.
- The functional unit is the amount of diesel needed to fuel 1 km of transport with a heavy-duty (20 ton) truck.
- The final use of the fuel is in the Netherlands in 2020. The biomass is imported from the Baltic States.
- Various types of (woody) biomass are included in the LCA: Dedicated energy crops, primary residues in the form of forest thinnings and secondary residues in the form of wood industry residues.

An important subject to address as part of the inventory phase is allocation. In general, it is difficult to make allocation fully objective, although there is an internationally accepted order of preference in this respect. There are two important allocations addressed in the BIG-FiT LCA:

- Allocation of the environmental impact of the production of wood to forest and industrial residues is on an economic basis since this reflects the societal causing of the environmental impacts in the best way. Since the residues at the production location are assumed to have a zero value, no impact of the production phase is allocated to these residues. This reflects the fact that the main reason the wood is produced is not for production of residues, but the production of main stem wood or saw wood. In the future, if the residues have an economic value at the production location, then part of the impact of the production phase should be allocated to the residues. In this respect, it is important to note that, as a condition, all wood should come from forests that are managed in a sustainable way. This will exclude wood that contributes to deforestation or other forms of non-sustainable forest management.
- Since prices of the end products (e.g. green diesel and green naphtha) are too uncertain to include in the assessment, allocation to the various products that are produced in the BIG-FiT plant is done on an energy basis. In the Fischer-Tropsch synthesis unit of the BIG-FiT plant, losses are not accounted to the syngas flow that will go to the Combined Heat & Power unit, since the Fischer-Tropsch synthesis unit is not necessarily needed to produce heat and power, but only to produce Fischer-Tropsch products like diesel and naphtha.

For all biomass systems, a truck transport distance of 100 km in the exporting country and a ship transport distance of 1000 km are assumed. The BIG-FiT plant is located in Rotterdam and has a thermal input of about 8,5 GW. The allocated efficiency to FT production is based on calculating all losses to the FT products.

As energy crops, cultivation of willow has been taken. Diesel consumption by tractor is assessed in a detailed way accounting for all cultivation steps, in pre-treatment and transport. Relatively modest yields for the year 2020 are assumed (10 tdm/ha/yr) The first commercial plantations with willow have shown that no pesticides are needed to obtain reasonable yields. Fertilisation is applied at a level of 75 kg nitrogen, 60 kg phosphorus and 108 kg potassium per hectare and per year.

For energy crops, detailed data on energy consumption for the different steps in cultivation were used.

The LCA of forest thinnings includes the collection of the wood in the forest, the transport to the road side, pre-treatment and road and sea transport of wood to the BIG-FiT plant.

For the wood industry residues, only (see allocation) pre-treatment, road and truck transport of wood to the BIG-FiT plant was included. Pre-treatment (chipping) is assumed here to be needed in all cases, whereas with some wood industry residues (e.g. sawdust) in practice this is not needed.

The environmental themes and their classification factors used in this study were based on the CML (Centrum voor Milieukunde Leiden) problem-oriented approach which can be found in the recently updated Dutch LCA manual. The focus in this simplified LCA is on 6 environmental themes out of the list of 13 in the CML method. The six themes are: 1 - Climate change, 2 - Abiotic depletion, 3 - Human toxicity, 4 - Photochemical oxidation, 5 - Acidification, and 6 - Eutrophication. The reason for this selection is the fact that from previous LCA's on biomass energy process chains it was found that, after normalisation of the results, these were the most significant environmental themes.

- **Climate change :** All emissions are expressed as CO_2 equivalents. The global warming potential over a 100-year timeframe is used.
- **Abiotic depletion:** Depletion of all types of abiotic raw materials, including all fossil energy carriers.
- **Human toxicity:** Damage to the human being is the end point of the classification here. All emissions are expressed as equivalents of the emission of 1 kg of 1,4-dichlorobenzene
- **Photochemical oxidation:** Photochemical oxidation, better known as the related smog formation during summer, is the result of a reaction between NO_x and VOC (Volatile Organic Compounds). It is expressed here as ethylene equivalents.
- **Acidification:** The score on this environmental theme is based on the potential acid deposition to soil and water. It is expressed as SO_2 equivalents.
- **Eutrophication:** Eutrophication is the increasing nutrient concentration in soil and water. It is expressed as phosphate equivalents.

The findings of the LCA are represented in separate diagrams corresponding to the six environmental themes. Figure 2 provides an explanation of these diagrams. Each diagram is composed of 4 bars. The bars represent Fossil Diesel (FD), FT-diesel from Wood Industry Residues (WIR), FT-diesel from Forest Residues (FR), and FT-diesel from Energy Crops (EC). On top of the bars, the abbreviations FD, WIR, FR and EC are used to indicate which bar belong to which type of biomass.

The y-axis presents the highest scoring diesel system as 100% for each of the environmental themes. The system that scores 100% represents the system with the highest environmental impact. The lower scoring systems are presented as a percentage of this highest impact. This method of presentation allows to show all environmental themes (which are all expressed in different units) on one diagram (Figure 2). Each bar consists out of 5 components which are noted below:

1) *Direct emission FT truck*: The direct emissions from a vehicle that uses the biomass FT diesel as a fuel.
2) *Construction BIG-FIT plant*: The emissions resulting from the production of the main construction material (mainly concrete and steel) for the BIG-FIT plant.
3) *Biomass transport by ship*: The emissions from the ship during transportation of the biomass from the exporting to the importing country. A distance of 1000 km (roughly the distance to the Baltic States) has been used in this respect. The number of 1000 km also facilitates extrapolation to larger distances, since emissions are linear with the transported distance. If this component is left out, the bar that is left would basically represents the impact of FT diesel with Dutch biomass.
4) *Biomass transport by truck*: The local transport in the exporting country in order to get the wood from the site where it is produced (or becomes available) to the harbor where it can be shipped.
5) *Production and pre-treatment biomass*: Production is only relevant for forest residues and energy crops (see allocation). In the case of forest residues, it includes the haulage of the forest residues from the forest to the roadside. In the case of the energy crops, production includes the whole cultivation process of energy crops, including indirect impacts for the production of chemical fertilizer that is used for this cultivation.

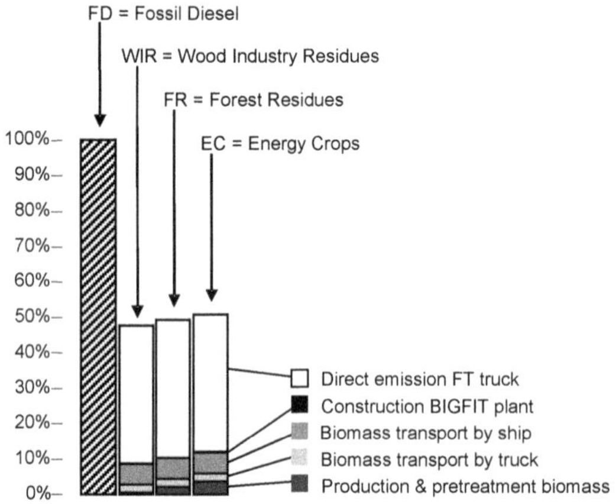

Figure 2: Explanation of result of assessment.

Interpretation is done in three steps: The first step is explaining output from input, the second step is normalisation and the third step is a discussion on uncertainties and subjective elements.

In all four systems, the score on **Abiotic depletion** is almost completely a result of fossil energy carrier depletion. For the biomass systems both national and international transport requires about 2% of the energy output delivered. The energy needed for production is negligible for wood industry residues, and requires about 2% and 7% for forest residues and energy crops respectively. In the case of energy crops about 60% of the energy input required for their production is a result of the energy needed for the production of nitrogen fertiliser and the remaining 40% is for tractor movements during the cultivation process. Overall, biomass FT-diesel requires about 5-11% of the fossil energy input as compared to fossil diesel.

Climate change is to a large extent linked to the score on abiotic depletion by CO_2 caused by fossil energy use. However, in the case of energy crops, the emission of CO_2 equivalents is relatively high (as compared with the score on abiotic depletion) because of N_2O emissions resulting from the use of nitrogen (N) fertiliser in the cultivation process.

Human toxicity impact of fossil diesel is about 90% and is caused by emission of C_xH_y , originating equally from direct and indirect emissions. For the biomass systems, the toxicity impacts are about 80% from direct emissions (72% C_xH_y , 8% NO_x) from the vehicle that uses the FT-diesel. Overall, the human toxicity score of biomass derived diesel is about 50% the score of fossil diesel.

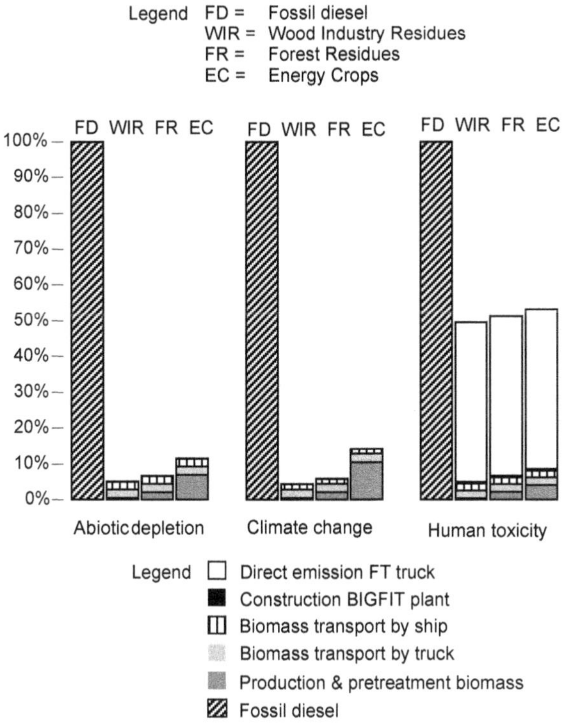

Figure 3: Results of assessment.

In a fossil fuel system, 75% of **Photochemical oxidation** impact is caused by indirect SO_2 emissions and a further 20% by CO emission. For a biomass system, the two main determinants for photochemical oxidation impact are direct CO emissions from the vehicle (35%) and SO_2 emissions from the ship (35%). Overall, the biomass score is about 40% of the fossil energy score.

Some 55% of **acidification** impact of fossil diesel comes from indirect SO_2, a further 35% comes from direct NO_x emissions. For a biomass-based system, the overall score is between 50-70% of that of fossil diesel. The emissions for biomass-based systems are 40-60% from direct NO_x emissions. In case of energy crops, ammonia emissions (because of the use of fertilizer) cause about 25% of the total impact.

Eutrophication in the fossil fuel system, in the wood industry residue and the forest residues systems is for almost 100% caused by direct NO_x emissions from the vehicles. For a biomass system, the overall scores are roughly 10% lower. The highest emission on eutrophication comes from the energy crop system. Beside the direct NO_x emission, the impact here is caused for about 50% from leaching of phosphate and for 30% from nitrate during the energy crop cultivation process. However, these emissions are far within Dutch environmental standards on leaching of nutrients.

Looking back on the process of setting up the LCA and critically reviewing the results some remarks should be made.

- Since the use of FT diesel is not yet widespread in the transport sector, there is only a limited amount of emission measurement data available. Additionally, there are some significant differences between the data in the available open literature. Moreover, emission reduction of FT fuelled vehicles when compared to fossil diesel vehicles may differ with the type of vehicle used. The lower sulphur content of FT diesel will have the additional advantage of providing better possibilities for catalysed flue gas cleaning to reduce other emissions.

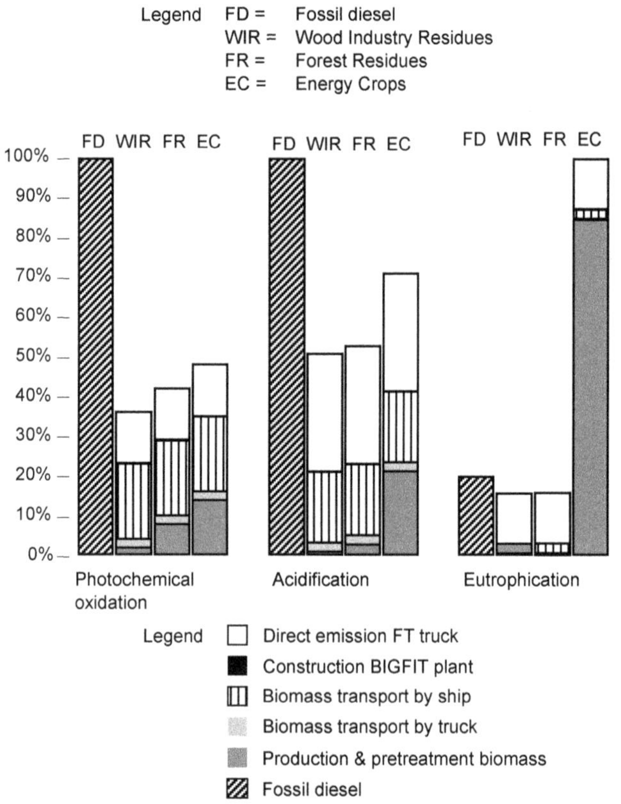

Figure 4: Results of assessment.

- In the definition of the scope of the LCA it was mentioned that the point of time of the LCA is 2020. For the BIG-FiT plant, estimates of the plant size and characteristics starting-up in 2020 are used. To calculate the direct emissions from the vehicles, reference is made to the Euro 5 standard for 2008, being the longest-term estimate available. No data are available on long-term future emissions for the production of fossil diesel, and on direct emissions from ships. In this study these emissions are relative high since present emission data for these applications are used.
- Location is an important factor that influences the inventory phase of the LCA, but location dependency is normally not included in impact assessment methods. This influences the decisiveness of the results obtained. The most obvious example of this is the emission from the sea transport of wood. Emissions that affect local environmental impacts such as human toxicity, photochemical oxidation and acidification, will generally cause less harm when they occur at open sea than when they occur on land. However, this has not been accounted for in the impact assessment. Therefore, the impact caused by sea transport on these environmental themes will be less severe then in reality.

- Allocation between the products produced in the BIG-FiT-plant in this simplified LCA is on an energy basis. It is recognised that a more precise way would be to extend the fossil diesel system to systems that also make products that are produced in a BIG-FiT-plant, but it is not expected that this would lead to significant differences.
- Production of wood industry residues and forest residues (up to the moment of harvest) were left out of the LCA since an allocation incurred on an economic basis is estimated to have a zero value. Although methodologically acceptable, this may not reflect societal acceptance of these types of energy sources in a sufficient way. For instance, if the wood that is used by sawmills comes from unsustainable forestry, it is questionable whether the sawdust that results from cutting this wood would be considered as a sustainable energy resource by the end consumers of the energy.
- In this study, high yield energy crop cultivation with use of chemical fertiliser is selected. However, there are a range of alternatives for this, e.g. no fertilisation and accepting lower yields (like in conventional forestry), using nitrogen-fixing crops, or using animal manure as a fertiliser. These can lower the eutrophication considerably.
- The use of FT diesel for transport will notably reduce the aspect of climate change, but also acidification will diminish.

From the findings of the LCA study the BIG-FiT consortium feels confident to conclude that BIG-FiT diesel scores better than fossil diesel on almost all environmental themes.

- On the environmental themes of climate change and abiotic depletion the score is roughly a factor 10 better. Energy output/input ratios vary between a factor 20 and 9. Even when the wood is imported from much further distances (e.g. 5000 km) than the ones considered in this study (1000 km) energy output/input ratio's would still be between a factor 5 and 10.
- The BIG-FiT diesel scores about a factor 2 better with acidification and photochemical oxidation and human toxicity.
- For eutrophication impact, the score of BIG-FiT-diesel is about 10% better than fossil diesel in the case of biomass from wood industry residues and forest residues, but about a factor 2 worse than fossil diesel in the case of energy crops. This is caused by leaching of nitrate and phosphate, which is a type of emission that does not occur at all in the fossil diesel chain. However, since energy crops do not need very intensive fertilisation, the level of leaching is comfortably within the Dutch emission standards. These emissions can be lowered when the choice can be made for lower fertilisation levels.

The construction of the BIG-FIT plant does not have any significant impact on any of the environmental themes considered.

The main uncertainty in assessing the results is with regard to the future (in the year 2020) values of direct vehicle emissions for FT and fossil diesel. Therefore, the most stringent future emission standards known at this moment, the Euro 5 standard, are used.

Nevertheless BIG-FiT consortium concludes from the LCA study that BIG-FiT diesel can be considered as a climate neutral fuel which significantly reduces most other main emission

COST STUDIES

With respect to a large-scale BIG-FiT refinery it is interesting to know the range of the costs of production for BIG-FiT products. To make an estimate of this range a calculation model has been developed on the basis of information from literature and from information derived from this study.

The large scale demonstration refinery is designed around a Fischer–Tropsch synthesis unit with a thermal rating of 5000 MWth. The hydrocarbons produced are naphtha, diesel and off-gases. The off-gases are feedstock to the Combined Heat and Power installation to generate green power. In case of oxygen blown gasification, approximately 80% of the thermal energy input in the Fischer-Tropsch synthesis unit is converted into hydrocarbons.

With respect to capital expenditure for large-scale installations like a FT- refinery two effects have a significant influence on these costs. One effect is learning from past projects and the other effect relates to economy of scale. In recent years technical developments have enabled the unit capital costs of GTL projects to be significantly reduced. Since the middle of last century, typical costs of 120.000 $ per daily barrel of installed production capacity have been reduced to levels of 20.000 – 40.000 $ per daily barrel.

With respect to the effects of economy of scale reference is made to a recent publication by Corke in Oil & Gas Journal [5]. According to Corke the specific investment costs decrease with the capacity to the power 0.65. For the reference 50.000 bpd FT refinery specific investment costs amounts to $ 21.800 per daily barrel. In general, these investment cost curves apply for installations in the Middle or Far East. For an installation in the Rotterdam Harbor area, a location factor could be applied. In this study a location factor of 1.4 has been taken. Specific investment costs then amount to $ 30.500 per daily barrel.

Next step is to define the distribution of the investment costs over the different process units. This is done on the basis of an engineering study by Foster Wheeler [6] for a 30.000 bpd FT refinery. According to this study, a distribution of investment costs is: syngas production (inclusive oxygen make-up) : 30%; FT-synthesis: 15%; Product upgrading: 10%; Other process units: 10%; Utilities : 15% and off-sites: 20%. As can be observed gasification, inclusive oxygen make-up amounts to a significant portion of the investment costs.

In the cost calculation, the prices for the process units other than the gasification are taken from [6] and multiplied by the specific investment costs. For the gasification including the syngas treating a value of 200$ per kW_{input} has been taken. This value is fixed on the basis of data for conventional thermal conversion installations.

The cost of feedstock contributes substantially to the production cost of biomass derived Fischer-Tropsch products. For this case a value of 4$/GJ has been taken.

With the model described above a price of 39.8 $cent per liter of biomass derived FT-diesel is calculated. This cost price is made up of feedstock costs of 24.7 $cent per liter, CAPEX costs of 10.9 $cent per liter and OPEX costs of 4.2 $cent per liter.

The long-term BIG-FiT case is based on the replacement of 1% of the European transport fuel pool in 2020 by BIG-FiT diesel. As indicated earlier this represents a production capacity of 80.000 barrels diesel per day. Upscaling the 50.000 bpd case of this BIG-FiT study results in following refinery characteristics for the long-term BIG-FiT case:

Table 1: Product range.

Product	production	unit
Diesel	80.000	bpd
Naphtha	15.000	bpd
Total	95.000	bpd

Based on the same cost model, the product price of the BIG-FiT reference refinery can be calculated at 38.4 $cent per liter.

The specific production of CO_2 from diesel and naphtha is assumed to be 73 kg/GJ. The total production of CO_2 by end-users of 95.000 bpd is then 13.9 Mton CO_2. >From the Life Cycle Assessment it can be observed that the biomass derived FT diesel reduces crude based CO_2 production is 90 %. The total amount of avoided CO_2 from the BIG-FiT refinery is 12.5 Mton.

Based on the amount of avoided CO_2 and the cost model, as described for the BIG-FiT refinery, it is possible to make an estimate of the cost per ton of avoided CO_2. However it is difficult to produce one definite value for this number because the reference value, i.e. the cost of conventional crude based diesel is very volatile. In the last years crude prices have varied between 10 and 30 $/bbl. In addition to the fluctuations in the reference costs there is the large influence of the feedstock costs on the BIG-FiT product price. For this reason, the cost of avoided CO_2 has been calculated as a function of the crude price and the feedstock cost is used as parameter. To compare the BIG-FiT diesel price with the crude based diesel price the crude based price must be corrected for the refinery costs. These costs are taken as 6 \$/bbl. The results are given in *Figure 5*. From this figure, it can be observed that in case of a crude price of 25 \$ per barrel and feedstock costs of 4\$/GJ the cost of avoided CO_2 is 75-80 \$ per ton. In case of feedstock cost of 2\$/GJ the costs of avoided CO_2 reduces to 25 – 30 \$ per ton. Further, it can be observed that, at moderate feedstock costs and medium to high fossil derived product prices, a negative value for the cost per ton avoided CO_2 is feasible.

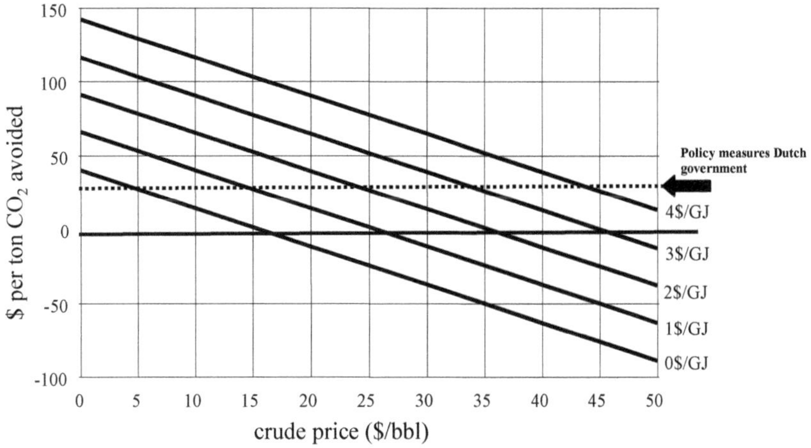

Figure 5: Cost of CO_2 avoided versus crude price.

To position these values in relation to what are considered to be "acceptable" values in the Netherlands, the mean cost of avoided CO_2 for all the policy measures of the Dutch government in 2000 is indicated in Figure 5. This value amounts to 30 € per ton of avoided CO_2.

CONCLUSIONS

Based on these arguments, the BIG-FiT consortium believes that the most rational approach towards progressing of the BIG-FiT concept at present is:

1. To focus attention and resources on fundamental and industrial research aimed at addressing the key technical uncertainties of pressurized oxygen-blown large-scale biomass gasification systems and syngas cleaning.
2. To demonstrate market introduction of FT-diesel for example via a system of CO_2 certificates in combination with FT diesel (out of natural gas) from Bintulu or other future suppliers in addition to other means of promotional activities.

AKNOWLEDGEMENT

This paper is based on a feasibility study executed by the BIG-FiT consortium. Financing of the study has been supplied on a equal base by the consortium and Novem (The Netherlands Agency for Energy and the Environment). The consortium gratefully acknowledge the financial support by Novem.

REFERENCES

1. Projectteam (2002) *Climate neutral transport fuels from biomass – The BIG-FiT concept.* SDE report 2002-10164, Amsterdam.
2. Personnel communication with dr. G.J. Nabuurs from Alterra, Wageningen (2002).
3. Lysen et al (2000),GRAIN: *Global Restrictions for the Availability of Import of biomass to the Netherlands*, Utrecht Centre for Energy Research. GAVE.
4. Personnel communication with Mr. Hiltunen and Mr. Lampenius of Foster Wheeler, Finland (2002).
5. Corke, M.J. (1998), *GTL Technologies focus on lowering costs.* Oil & Gas Journal, September.
6. Ghaemmaghmi, B, Clarke, S.C. (2001), *Design and engineering of a GTL plant,* Foster Wheeler Review, spring.

Long Term Perspectives for Production of Fuels from Biomass: Integrated Assessment and RD&D Priorities

André PC Faaij, Carlo N Hamelinck
Utrecht University, Copernicus Institute, Science Technology Society, Padualaan 14, 3584 CH Utrecht, Netherlands

ABSTRACT: This paper gives some preliminary results of system analyses for a wide range of biofuel production and utilisation options. Analyses are carried out for "well-to-wheel" systems, including biomass production, logistics, conversion, distribution and end-use. Of particular interest is the time-frame considered: about 20-30 years from now. Such a period of time allows for considerable technology development, installing new (distribution) infrastructure and transportation fleets. Performance and cost estimates are therefore made with this long term perspective in mind. Difference is made between the shorter and longer term performance of biofuel chains for all relevant aspects. Hydrogen and ethanol from wood seen to be superior fuels on the long term and could become cheaper than current gasoline driven vehicles when Fuel Cell Vehicles are applied. Fischer-Tropsch liquids and methanol from wood seem very promising for the shorter and medium term.

INTRODUCTION AND RATIONALE

Energy use for transport is one of the main contributors to greenhouse gas emissions and air pollution on a global scale. Seen worldwide, mineral oil derived fuels (gasoline, diesel, kerosine, fuel oil), dominate the worlds transportation fuel supply for almost 100%. Notable exceptions are found in Brazil, where ethanol derived from sugar cane is used for transport fuels on a considerable scale and to a lesser extent ethanol from surpluses of agricultural crops (cereals and maize) in Europe and North America. Fossil energy use for transport accounts for about 25% of the worlds energy consumption and amounts roughly 100 EJ. The global trend is that the share of transport (road and air transport as well as passenger and freight) in the total global energy consumption is increasing, especially in developing countries. This also increases the contribution of the transport sector to climate change.

Besides a strong contribution to greenhouse gas emissions, fossil fuel fired transport systems strongly contribute to urban air pollution. Main contaminants are NO_x, CO, hydrocarbons and dust, being the main cause of formation of photochemical smog. The use of on board catalysts on cars, maintenance programmes, strict legislation and mandated shares of "zero emission vehicles" are reinforced in policies in Western Europe and North America. However, such policies seem still insufficient to meet air quality goals in many situations, especially because the total amount of transport increases at a steady pace in many countries.

Furthermore, mineral oil reserves are concentrated in a limited set of countries (most notably the OPEC member states). With increasing competition on the worlds energy market combined with rapidly decreasing mineral oil supplies of non-OPEC states, the worlds mineral oil production depends on a small number of suppliers, which poses threats for energy security worldwide. Another, related, argument is that many, in particular developing, countries, spend large fractions of their GDP on importing oil, not seldom resulting in large trading deficits.

Many of those countries do however have large biomass production potentials that may be able to off-set oil imports when biofuels could be produced against competitive cost levels.

Transportation fuels derived from biomass could both reducie the dependency on, imported, oil as well as reduce Greenhouse Gas emissions. When biofuels are applied in new vehicle types like fuel cell powered cars, air pollution of transport can also be dramatically reduced. Disadvantages are the current high(er) costs of biofuels compared to gasoline and diesel and the large land areas which are required for substantial amounts of energy. Both should be minimized to be feasible on longer term.

When considering biofuels it should be noted that a large number of options, energy carriers and complete 'energy-chains' are possible for production and their utilisation. A global overview of the most important options is presented in Table 1. Most of those (potential) options are not applied commerically at present. Furthermore, they involve technologies and systems which require substantial development over time. Therefore, the timeframe considered is an extremely relevant factor for estimating and analysing potential performance levels of complete chains.

The main objective of this work is to compare all possible major biofuel chains on an equal basis, determine their potential overall performance from an efficiency and economic point of view on shorter and longer term (i.e. 30 years from now) and identify the main uncertainties and Research & Development needs which are essential to obtain the projected performance levels. Analyses are carried out for "well-to-wheel" systems, including biomass production, logistics, conversion, distribution and end-use.

Table 1: Main biofuel routes and, potential, components.

Carrier	Feedstock	Conversion technology	Distribution & storage	End-use
Hydrogen	Any type of biomass, lignocellulosic preferred	Gasification + shift reaction + H_2 separation	Specially designed pipelines/ liquified or hydrides/ carbon	FCV & ICEV, (Fuel Cell Vehicle & Internal Combustion Engine Vehicle)
Methanol	Any type of biomass, lignocellulosic preferred	Gasification + syngas processing + MeOH synthesis	Conventional, but more bulky than e.g. gasoline	FCV with on boar low termperature reforming & ICEV
Synthetic hydro carbons	Any type of biomass, lignocellulosic preferred	Gasification + syngas processing + Fischer Tropsch synthesis	Conventional, similar to gasoline /diesel	FCV with PROX reactor or ICEV
Ethanol	Lignocellulosic biomass	Hydrolysis + fermentation	Conventional; mixing with gasoline possible	FCV with PROX reactor or ICEV
Ethanol	Sugar / starch crops	Fermentation + distillation	Conventional; mixing with gasoline possible	FCV with PROX reactor or ICEV
Fatty acids (e.g.RME)	Oil seeds (e.g. rapeseed)	Extraction + esterification	Conventional, mixing with diesel possible	FCV with PROX reactor or ICEV
Bio-oil	Any type of biomass, Pyrolysis preferes lignocellulosic	- Pyrolysis + upgrading - HydroThermal Upgrading (gives lower oxygen content than pyrolysis)	Separate; high quality materials and possibly heated piping is required	Raw bio-oil not suited for vehicle use. When fully upgraded (hydrogenated) could be comparable to diesel.

MAIN INPUT DATA

CONVERSION TECHNOLOGIES

Various detailed analysis of conversion routes were used and specifically made for this work. In the case of production of hydrogen, methanol and Fischer-Tropsch liquids via biomass gasification, detailed ASPEN-plus flowsheet modelling and optimisation has been applied (see e.g. [1,2]). For advanced ethanol production systems, similar information is available (see e.g. [3]). For the options concerning annual crops: ethanol from sugar beet and production of bio-diesel from rapeseed, data from operational systems and future projections were used (e.g reported in [4,5]). Production of 'synthetic diesel' via hydrogenation of pyrolysis oil was also included in the analysis, although on forehand it should be recognized that the data quality concerning this option is not as good as for the other routes. Consistenly, difference is made between expected performance on the shorter and longer term.

Long term systems are also assumed to be realized on larger scale. The reference capacity for the short term is 400 MWth input. For the long term this is assumed to be 1000 MWth (taking economies of scale into account for all equipment concerned). Key data concerning conversion are summarized in Table 2.

Table 2: Summary of key data regarding (projected) energetic performance and costs of the conversion systems considered. The capacity of upgrading systems is determined by the output of raw product processed.

Concept	Energy efficiency (HHV) and energy inputs		Investment costs (MU$)		O&M
	Short term (400 MW_{th})	Long term (1000 MW_{th})	Short term (400 MW_{th})	Long term (1000 MW_{th})	
Hydrogen (compressed)	60% (fuel only)	52% (fuel) 6% (power)	191	363	4 %
H_2 liquefaction	0.19 GJ_e/GJ H2		19	32	4 %
Methanol	55% (fuel only)	48% (fuel) 12% (power)	277	526	4 %
Fischer-Tropsch	43% (fuel only)	40% (fuel) 10% (power	287	544	4 %
Ethanol (wood)	46% (fuel) 4% (power)	53% (fuel) 8% (power)	141	180	6 %
Ethanol (sugar)	43% (fuel only) 0.065 GJ_e + 0.24 GJ_{th} /GJ EtOH	43% (fuel only) 0.035 GJ_e + 0.18 GJ_{th} /GJ EtOH	115	173	5 %
Pyrolysis	60%	70%	415	788	4 %
Pyrolysis hydrodeoxygenation	67%; 0.025 GJ_e + 0.48 GJ H_2 /GJ output		140	297	4 %
Biodiesel RME production	88; 0.01 GJ_e + 0.04 GJ MeOH /GJ output		60	106	5 %
Power generation RME	45%	55%	180	250	4 %

BIOMASS PRODUCTION AND SUPPLY

Biomass production under NW European conditions is used as a starting point, although on the long term, in particular wood could be supplied from overseas against lower costs than could be obtained in NW Europe [6]. For main performance data and assumptions see Table 3.

Table 3: Performance characteristics of the energy crops considered on short and long term (HHV per dry tonne is used for calculating energy yields; wood: 19 GJ/tonne, rapeseed: 28 GJ/tonne, straw: 16 GJ/tonne, sugar beet: 19 GJ/tonne).

Crop	Time frame	Yield (odt/ha*yr)	Energy inputs (GJ_{prim}/ha*yr)	Net Yield (GJ/ha*yr)	Net costs (U$/GJ)	Other remarks
SRC-Willow	short term	10	5	180	4-6 (NW Europe)	Perennial crop, low emission levels of agro-chemicals
	long term	15	5	280 (100-300)	1.5 - 3 (incl. tropical regions)	Yield 6-16 depending on land quality; cost strongly depending on location and climate.
Sugar beet	short term	14	13	250	12	Annual crop, high emission levels of agro-chemicals
	long term	20	10	370	8	Shift to fodder beet; assumes good quality land
Rape seed	short term	2.9 seeds 2.6 straw	11	110 (total)	20	Assumes good quality land
	long term	4 seeds 4.5 straw	12	180 (total)	12	Assumes good quality land

* *Biomass logistics:* For woody crops, transport, handling and storage costs add about 10% to the fuel costs in case of road transport in the vicinity of the plant. Energy inputs are about 1-2% of the heating value of the biomass (somewhat higher for sugar beet). When long distance transport (intercontinental) transport is considered, the logistics can add between 0.5-1 U$/GJ. Energy inputs can vary between 6-10% of the heating value of the biomass [6].

FUELS AND DISTRIBUTION

Fuel characteristics have a strong influence on the costs and energy use of the subsequent distribution and marketing systems. Some key data regarding transport and distribution of fuel types are presented in table 4. Apart from hydrogen, all fuels are liquids. The lower energy density of ethanol and methanol compared to gasoline and diesel result in somewhat higher costs. Hydrogen is far more troublesome and expensive; either liquefaction (requiring substantial amounts of electricity) with well isolated storage and transport, or distribution via, specially designed, gas pipelines are required.

Table 4: Key parameters for fuel transport and distribution.

	Energy use (GJ_{prim}/GJ_{fuel})	Transport costs (U$/GJ)	Distribution costs (U$/GJ)
Hydrogen (g)	0.069	0.5	4.4
Hydrogen (l)	0.11	1	2.7
Methanol	0.019	1.9	1.2
Ethanol	0.013	1.5	0.9
Hydrocarbons	0.0084	1	1.6

VEHICLE PERFORMANCE

Last but not least, the vehicle performance is a crucial factor that influences the energy efficiency and costs of the total chain. Two vehicle categories are considered here: Internal Combustion Engine Vehicles (ICEV's) that can be improved on the longer term and Fuel Cell Vehicles, that can be introduced gradually starting about 5-10 years from now. The efficiency of those vehicle categories differ considerably. Furthermore, the efficiency per km driven differs per fuel. FCV's ultimately require hydrogen as a fuel. Therefore on board reforming or partial oxidation is required for non-hydrogen fuels, which in turn results in lower efficiencies and higher costs (and weight) of the vehicle. Most data regarding future vehicle performance and costs are taken from a detailed analysis of Ogden et al. [7], and summarized in table 5.

Table 5: Key data for various vehicle types which represent a middle class car with effective wheel power of 94 kW of 20.000 U$. Performance and maintenance costs valid for an average driving cycle and 17.600 km/year.

	Energy use (MJ/km) (short – long term)	Vehicle costs (kU$) (short – long term)	Maintenance costs (U$/yr)
ICEV-gasoline	3.5 - 2	20	520
ICEV-alcohol & H_2	2.9 - 1.8	20	520
FCV-H_2	0.77	24.3-19.9	400
FCV-MeOH	1.19	23.9-20.4	420
FCV-HxCy+EtOH	1.15	24.5-20.8	420

Regarding ICEV's, alcohols and hydrogen have the benefit that higher compression ratio's are possible, thereby increasing the vehicle efficiency with about 15% compared to hydrocarbon fueling. For all performance and cost data the vehicle category and driving cycles are the same. Therefore, differences in weight, wheel power as well as the price of the vehicle are excluded as disturbing factors in the comparison.

MAIN RESULTS

Some of the main results are depicted in graphs 1 and 2. The first graph shows the calculated costs per GJ of fuel delivered at the car. The second graph shows the number of kilometres that can be driven per hectare. For Figure 1. Number of kilometres that can be driven per hectare (net energy yield) for biofuel chains considered. Calculating this parameter, various energy inputs were taken into account and translated to primary energy inputs in order to calculate net energy yields. From a cost point of view hydrogen and ethanol from wood seen to be superior fuels on the long term and competitive with current gasoline driven vehicles when applied in Fuel Cell Vehicles. However, in order to obtain the projected performance for EtOH production from wood a number of technological breakthroughs is required (most notably in pre-treatment and integrated reactor technology and development of suited enzymes). For hydrogen utilisation, the infrastructure required is the main bottleneck. Fischer-Tropsch liquids and methanol from wood seem very promising for the shorter and medium term.

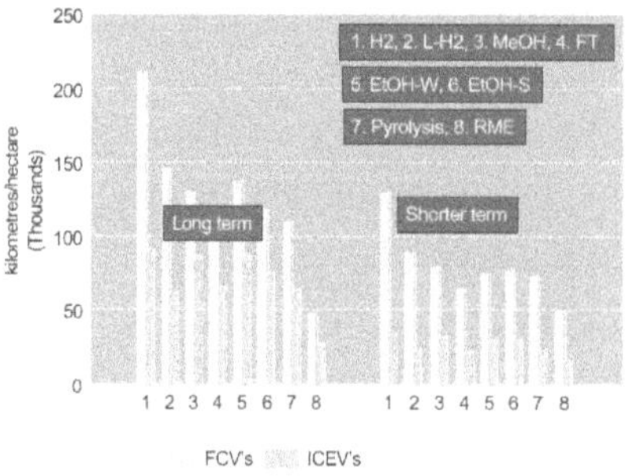

Figure 1: Costs per GJ of fuel delivered at the car.

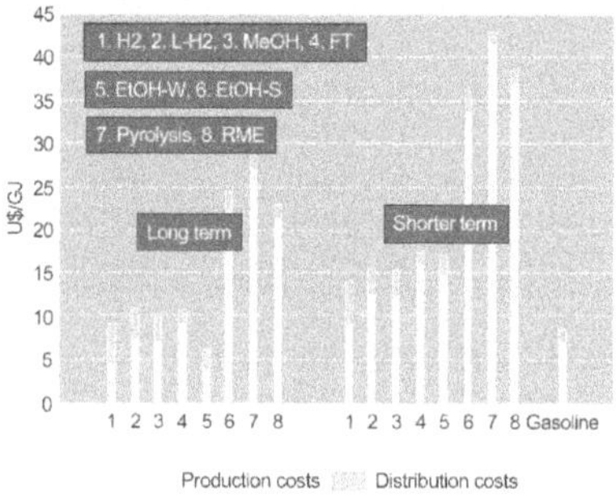

Figure 2: Kilometer yield per hectare.

All these four options require large scale application to obtain the required economomies of scale. Considering the efficiency and minimizing the amount of land needed for transportation fuels, the high efficiency of FCV's and hydrogen fueled FCV's in particular are the decisive factors. The number of kilometers that can be driven *per hectare* is considerably higher for hydrogen when FCV's are used. Methanol and ethanol are second best in this respect.

FINAL REMARKS

Overall, it seems biofuels can offer a feasible alternative for the transportation sector on the longer term, provided sufficient biomass is available to do so. Wood (or perennial crops in general) seem to be the most favourable feedstock. Ethanol from lignocellulosic biomass, hydrogen, methanol and synthetic hydrocarbons produced via Fischer-Tropsch synthesis seem to be competing amongst eachother, especially on shorter term. When breakthroughs in hydrogen storage technology are obtained (possibly avoiding the need for expensive compression and pipeline transport) hydrogen use may become the ultimate energy carrier to be considered.

It is interesting to note that the fuel costs on itself only have a modest influence on the overall driving costs of the vehicle types concerned. The investment costs of the car is the dominating factor. Although the annual fuel costs (for a Figure 2. Costs per GJ fuel delivered at the car for biofuel chains considered.

Fixed number of kilometers) differs widely depending on the (bio)fuel costs and the vehicle's efficiency, the overall driving costs do not differ too much between the chains considered. This conclusion is true when taxes (especially on the fuel itself) are excluded. NW European transport fuel markets are however dominated by taxes which increase the production costs of gasoline by a factor of 4-6, depending on the country in question. This also implies that fuel production costs are not that a decisive factor and that on the shorter term fuel taxation policies could play a crucial role in creating markets for biofuels.

REFERENCES

1. C. Hamelinck, A. Faaij, 2002, *Future prospects for production of methanol and hydrogen from biomass*, Journal of Power Sources 111: 1-22.
2. M. Tijmensen, A. Faaij, C. Hamelinck, M. Hardeveld, 2002, *Exploration of the possibilities for production of Fischer Tropsch liquids and power via biomass gasification*, Biomass & Bioenergy 23: 129-152.
3. L.R. Lynd, 1996, Overview and evaluation of fuel ethanol from cellulosic biomass: technology, economics, the environment and policy, Annual review of the environment 21: 403-65.
4. IEA Automotive Fuels Survey, International Energy Agency, 1996.
5. D. De Jager, A. Faaij, W.P. Troelstra, 1998, Cost effectivity of transport fuels from biomass (in Dutch), EWAB report 9830, Netherlands Agency for Energy and the Environment Novem, Utrecht the Netherlands.
6. A. Agterberg, A. Faaij, 1998, *Bio-energy trade; possibilities and constraints on short and longer term,* EWAB report 9841, EU ALTENER project, co-funded by NOVEM and NUTEK, Utrecht University>Science Technology Society, Utrecht the Netherlands.
7. J.M. Ogden, M.M. Steinbugler, T.G. Kreutz, 1999, A comparison of hydrogen, methanol and gasoline as fuels for fuel cell vehicles: implications for vehicle design and infrastructure development. Journal of Power Sources 79: 143-168.

Steam Gasification of Biomass Using Coprecipitated Nickel-Alumina Catalysts - Effect of Magnesium and Lanthanum as Promoters

L. García, B. Brieva, M.L. Salvador, R. Bilbao, J. Arauzo
Chemical and Environmental Engineering Department, University of Zaragoza
Centro Politécnico Superior, María de Luna, 3, 50015 Zaragoza, Spain

ABSTRACT: The previous work carried out using coprecipitated Ni-Al catalysts have shown a good performance of these catalysts. No attrition problems were observed and a high initial activity of the catalyst was obtained. At longer reaction time or using low values of catalyst weight/biomass flow rate ratio (W/mb $\leq$ 0.4 h), the catalyst loss its activity. The deactivation of the catalyst causes the decrease of H_2, CO, CO_2 and total gas yields. The experimental work carried out incorporating magnesium and lanthanum as promoters in coprecipitated Ni-Al catalysts has shown the improvement of catalyst stability. The $NiMgAl_2O_5$ catalyst prepared at the pH constant technique presented the best results of various magnesium catalysts studied. The Ni-Al coprecipitated catalyst with a 12 % wt of La_2O_3 showed also a good performance.

Steam gasification experiments at 700 °C were carried out in a bench scale installation based on Waterloo Fast Pyrolysis Process technology. The comparative study of the results obtained with $NiAl_2O_4$ (Ni/Al), $NiMgAl_2O_5$ (Ni/Al/Mg) and 12 %wt La_2O_3 coprecipitated Ni-Al catalyst (Ni/Al/La) is presented. Nickel catalyst promoted with lanthanum is the most active catalyst with the highest hydrogen yields.

INTRODUCTION

Hydrogen is widely used in synthesis process like methanol or ammonia, also in refinery processes and metallurgical uses. Nowadays, hydrogen is increasing its interest as a fuel because its pollution effect is almost zero.

Biomass is converting in an alternative energetic resource to conventional fossil fuels with important benefits from an environmental point of view.

Steam gasification of biomass is a suitable process to obtain a rich-hydrogen gas [1-4]. This is an endothermic process and working at low temperatures the input energy for the process decreases but gas yields decreases and liquid product increases. The presence of a catalyst converts liquid products in gases and improves the quality of final gas. The main drawback of the catalyst is its deactivation mainly caused by carbon deposits and sintering.

Different catalytic materials, dolomite, alkali catalysts and nickel catalysts, are being used in biomass gasification [5] with the purpose of destroying tars and removing methane.

The placement of the catalyst in the process is an important fact. Sometimes the catalyst is placed in the same reactor where the thermal decomposition takes places but in other occasions the catalyst is introduced in a secondary reactor downstream from gasifier.

The gasifying temperature and the temperature that the catalyst works are significant variables on tar formation and catalyst stability against deactivation.

In the present work, high severity conditions are employed, the thermal decomposition of biomass occurs at low temperature, 700 °C, and the catalyst is placed in the gasifier. The catalytic materials used are coprecipitated nickel catalysts prepared in the laboratory. The Ni/Al coprecipitated catalyst have shown a good performance with high initial activities [6,4]. However the catalyst deactivates at long reaction times or using low values of catalyst weight/biomass flow rate (W/mb) ratio. With the purpose of increasing the stability of the Ni/Al catalyst magnesium or lanthanum have been incorporated as promoters in its formulation.

Magnesium is presented in several commercial catalysts [7-9] as well as in research formulations [10-13]. The formation of magnesium aluminate spinel increases significantly the strength of the catalyst that is very positive for its use in a fluidized bed reactor. The formation of solid solutions NiO/MgO increases the stability of the catalyst against sintering. Another advantage of the presence of magnesium is its capability of steam adsorption that can be beneficial in coke deactivation.

The addition of lanthanum contributes to catalyst stability and diminishes carbon deposition. Lanthanum is presented in patented formulations on steam reforming of hydrocarbons [14-16] and other catalysts specifically prepared for catalytic conversion of hydrocarbons derived from biomass [12,17,18].

The present work shows a comparative study of $NiMgAl_2O_5$ (Ni/Al/Mg) catalyst, Ni/Al/La catalyst with 12% weight of La_2O_3 and $NiAl_2O_4$ (Ni/Al) catalyst. The experiments have been carried out at low W/mb ratios where the deactivation of the catalyst is higher.

EXPERIMENTAL

EXPERIMENTAL SYSTEM

The experimental system is a bench scale installation that uses WFPP (Waterloo Fast Pyrolysis Process) technology. This technology was developed by Professor D.S. Scott and his co-workers at University of Waterloo in Canada. Flash pyrolysis uses high heating rates and very short residence time of gases in reaction bed [19,20].

A scheme of the installation is presented in Figure 1.

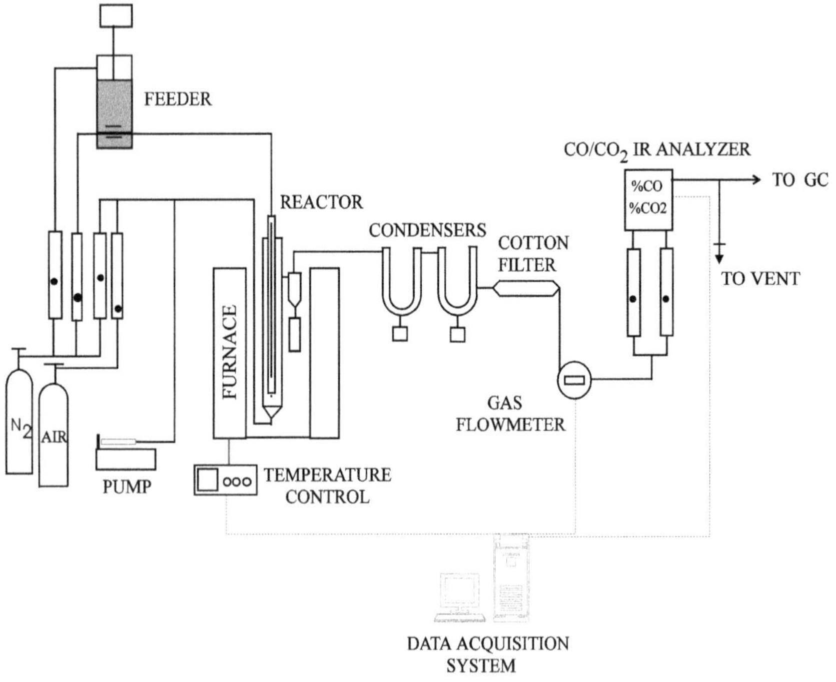

Figure 1: Schematic of the experimental system.

The biomass is fed continuously. A feeder supplies biomass flow rates up to 100 g/h. Two streams are introduced into the fluidized bed reactor. One of nitrogen transports the biomass into the reaction bed, while the other a mixture of nitrogen and water enters at the bottom and reaches the bed through the distributor. Water is measured out in the liquid state with a syringe pump that delivers a constant flow rate (0.1 - 99.9 ml/h). Water becomes steam during its path inside the electrical furnace before entering into the reaction bed.

The product gas is cleaned of char particles using a cyclone. A system of two condensers and a cotton filter allows us to retain liquid products. The gas flow rate is then measured using a dry testmeter, and CO and CO_2 concentrations are continuously determined by an infrared analyzer. In addition, gas samples are taken at regular time intervals and analyzed by chromatography showing the content of H_2, CO, CO_2, CH_4 and C_2 (C_2H_2, C_2H_4 and C_2H_6) in the product gas. Reaction temperature, total gas flow and CO and CO_2 concentrations are registered by a data acquisition system.

Steam gasification experiments have been carried out at 700 °C and atmospheric pressure. Superficial velocity of gas is about 10 times theoretical minimum fluidization velocity. No elutriation and segregation problems were observed with the conditions employed. The reaction bed is composed of 5 or 2 g of catalyst and sand. For all the experiments, the reaction volume was the same, about 34.4 cm^3. The W/mb ratio changes from 0.67 to 0.068 h. The steam/biomass (S/B) ratio ranges from 2.39 to 0.62.

BIOMASS

The biomass used in all the experiments is pine sawdust with a particle size between +150-350 μm. The elemental analysis of this biomass is shown in Table 1. The average moisture content is about 10 %.

Table 1: Elemental analysis of pine sawdust.

	%wt
carbon	48.27
hydrogen	6.45
oxygen*	45.19
nitrogen	0.09

* Calculated by difference.

CATALYSTS

The Ni/Al catalyst employed was prepared in the laboratory by coprecipitation using a similar method to that described by Al-Ubaid and Wolf [21]. Ammonium hydroxide is added to a solution of $Ni(NO_3)_2$ $6H_2O$ and $Al(NO_3)_3$ $9H_2O$ in distilled water until the pH reaches 7.9. The precipitation medium is maintained at 40 °C and moderatly stirred. The salts are mixed in the appropriate proportions to obtain a molar ratio 1:2 (Ni:Al). The precipitate obtained is filtered and washed at 40 °C and dried for about 15 h at 105 °C.

The Ni/Al/La catalyst was prepared using a similar method to that of Ni/Al catalyst. The final pH of precipitation was 8.05.

The Ni/Al/Mg catalyst was prepared at constant pH technique. Two aqueous solutions, one containing the metallic nitrates (1 M) and the other the precipitant, K_2CO_3 and KOH (2.4 M), were simultaneously poured into a glass reaction vessel that initially contained 650 ml of water. The addition was performed dropwise with moderate stirring at 60 °C and a constant pH=10.2±0.2. After precipitation, the final mixture was slowly stirred for 2 h at constant pH while more acid solution was added. The precipitate obtained was filtered, washed at 60 °C, and dried at 70 °C for 12 h.

The precursors were calcined in air atmosphere at low heating rate until the final calcination temperature of 750 °C maintained for 3 h. The catalysts employed in this work were not reduced prior to the biomass reaction. A previous work [22] concluded that the gases generated in the thermal decomposition of biomass, H_2 and CO, have the ability to reduce the Ni/Al catalyst.

The calcined catalysts have been characterized by X-ray diffraction (XRD) and temperature programmed reduction (TPR). The results of XRD indicate that Ni/Al catalyst contained nickel oxide and nickel aluminate with spinel structure phases. These phases are also found in Ni/Al/La and Ni/Al/Mg catalysts. In the latter catalyst $MgAl_2O_4$ phase with spinel structure could be also presented.

The TPR analysis show the maximum reduction peaks for Ni/Al catalyst at 712 °C, for Ni/Al/La catalyst at 730 °C and for Ni/Al/Mg catalyst at 750 °C. The presence of magnesium creates a less reducible catalyst.

RESULTS AND DISCUSSION

GLOBAL EXPERIMENTAL RESULTS

Tables 2 to 4 show global results of steam gasification of biomass using $NiAl_2O_4$ (Ni/Al), nickel catalyst promoted with lanthanum, 12 %wt of La_2O_3 (Ni/Al/La) and $NiMgAl_2O_5$ (Ni/Al/Mg) catalysts respectively. The experiments were performed with 2 or 5 g of catalyst in the reaction bed. The biomass flow rate varied from 7.50 to 29.1 g/h, while the steam flow rate was the same in all the experiments, then different catalyst weight/biomass flow rate (W/mb) and steam/biomass (S/B) ratios resulted.

These tables show experimental variables such as biomass flow rate, catalyst weight, W/mb and S/B ratios, and reaction time. Also indicated are yields of total gas, expressed as mass fraction of the original biomass and steam, the yields of different gases as mass fraction of the original biomass, and the gas composition expressed as molar percentages (N_2 and H_2O free), together with the recovery. The recovery is the ratio between the sum of gases, liquids (tar + H_2O), and the sum of biomass and steam introduced.

Table 2: Results of steam gasification using Ni/Al catalyst.

Run	1	2	3	4
Sawdust feeding rate (g/h)	7.50	15.34	18.11	26.44
W/mb (g catalyst h/g biomass)	0.67	0.33	0.28	0.19
S/B (g steam/g biomass)	2.39	1.17	0.99	0.68
Reaction time (min)	42	58	40.61	61
Catalyst weight (g)	5	5	5	5
Yields (mass fraction)				
gas/biomass	0.97	0.523	0.991	0.96
gas/(biomass+steam)	0.22	0.414	0.461	0.536
recovery	0.962	1.013	0.992	1.008
Gas yields (mass fraction of original biomass)				
H_2	0.045	0.034	0.036	0.036
CO	0.266	0.339	0.415	0.445
CO_2	0.610	0.480	0.455	0.370
CH_4	0.030	0.038	0.040	0.050
Gas composition (% mol, N_2-and H_2O-free)				
H_2	47.31	39.45	37.41	37.64
CO	18.98	28.98	32.21	33.28
CO_2	28.31	23.98	21.71	18.03
CH_4	3.86	5.49	5.44	6.76

Although these tables present experiments with different reaction times and biomass flow rates, some main tendencies can be observed. For each catalyst the simultaneous increase of W/mb and S/B ratios causes the increase of H_2 content while CO and CH_4 content decrease. For the majority of the experiments H_2 and CO_2 yields increase when W/mb and S/B ratio increases while CH_4 yield diminishes.

Table 3: Results of steam gasification using Ni/Al/La catalyst.

Run	5	6	7
Sawdust feeding rate (g/h)	15.41	24.74	29.18
W/mb (g catalyst h/g biomass)	0.32	0.20	0.07
S/B (g steam/g biomass)	1.17	0.73	0.62
Reaction time (min)	48.37	46.40	39.53
Catalyst weight (g)	5	5	2
Yields (mass fraction)			
gas/biomass	1.624	0.950	0.890
gas/(biomass+steam)	0.749	0.555	0.550
recovery	1.077	0.996	0.976
Gas yields (mass fraction of original biomass)			
H_2	0.079	0.024	0.032
CO	0.783	0.278	0.437
CO_2	0.606	0.183	0.318
CH_4	0.074	0.029	0.05
Gas composition (% mol, N_2-and H_2O-free)			
H_2	45.70	37.64	35.94
CO	30.20	34.47	35.61
CO_2	15.80	13.17	16.15
CH_4	4.83	5.99	7.85

Table 4: Results of steam gasification using Ni/Al/Mg catalyst.

Run	8	9
Sawdust feeding rate (g/h)	16.54	29.14
W/mb (g catalyst h/g biomass)	0.12	0.068
S/B (g steam/g biomass)	1.09	0.62
Reaction time (min)	45.2	41.73
Catalyst weight (g)	2	2
Yields (mass fraction)		
gas/biomass	0.92	0.89
gas/(biomass+steam)	0.44	0.53
recovery	0.991	0.985
Gas yields (mass fraction of original biomass)		
H_2	0.027	0.018
CO	0.488	0.470
CO_2	0.261	0.224
CH_4	0.065	0.071
Gas composition (% mol, N_2-and H_2O-free)		
H_2	30.14	23.09
CO	38.22	41.15
CO_2	14.56	12.49
CH_4	9.36	10.58

INFLUENCE OF W/mb AND S/B RATIOS

To analyze the influence of W/mb and S/B ratios on gas yields, for each catalyst the yield of different gases has been represented versus sawdust/catalyst ratio. This ratio allows us to compare experiments carried out with different catalyst weight and biomass flow rate.

For the three catalyst studied, Ni/Al, Ni/Al/La and Ni/Al/Mg, the general tendency shows the increase of total gas, H_2 and CO_2 yields when W/mb and S/B ratios increase. The increase of W/mb ratio favours steam reforming reactions producing higher yields to H_2 and CO, while water-gas shift reaction converts CO to CO_2 and H_2.

The performance of CO yield depends on the catalyst studied. For Ni/Al catalyst it is observed the decrease of CO yield when W/mb and S/B ratios increase. These results can be a consequence of water-gas shift reaction. On the contrary, for Ni/Al/La and Ni/Al/Mg catalysts the increase of W/mb and S/B ratios increases CO yields. This fact can be due to the lower values of W/mb ratios studied. The decrease of W/mb ratio generates lower CO yield due to the minor extension of steam reforming reactions.

Ni/Al catalyst shows the decrease of CH_4 and C_2 yields when W/mb and S/B ratios increases, this can be a consequence of the participation of CH_4 and C_2 steam reforming reactions. For Ni/Al/La and Ni/Al/Mg catalyst a clear tendency is no observed. Another time the minor value of W/mb ratio could justify this result.

Figures 2 to 5 show H_2, CO, CO_2 and CH_4 yields versus sawdust/catalyst ratio using the Ni/Al/La catalyst.

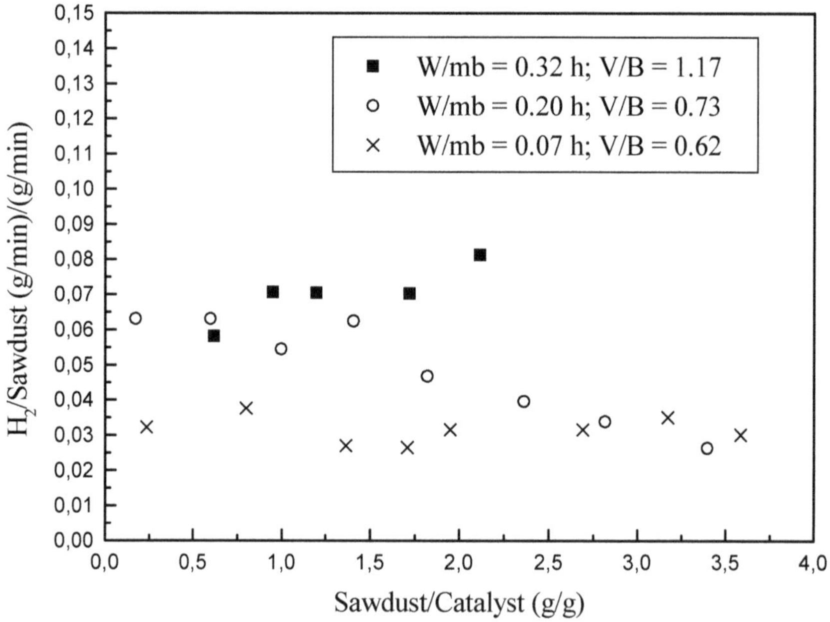

Figure 2: H_2 yield using Ni/Al/La catalyst for different W/mb and S/B ratios.

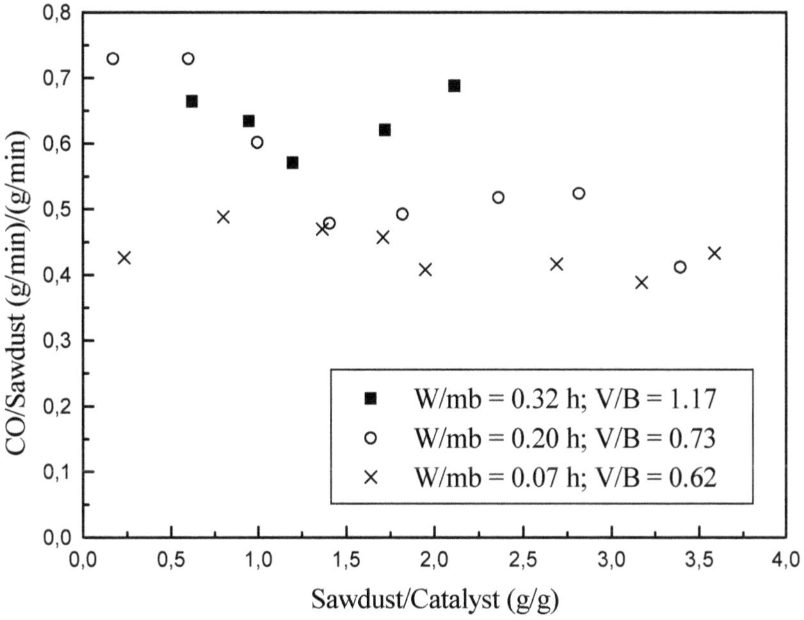

Figure 3: CO yield using Ni/Al/La catalyst for different W/mb and S/B ratios.

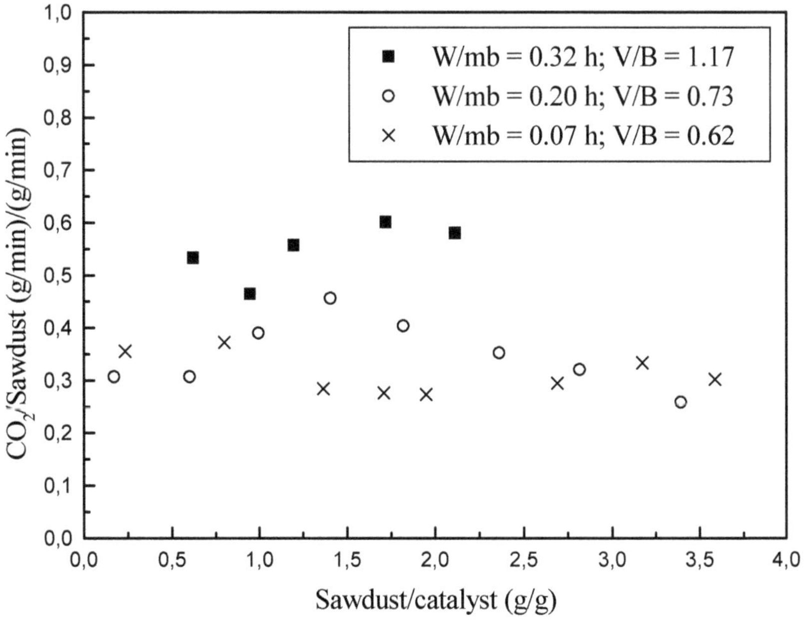

Figure 4: CO_2 yield using Ni/Al/La catalyst for different W/mb and S/B ratios.

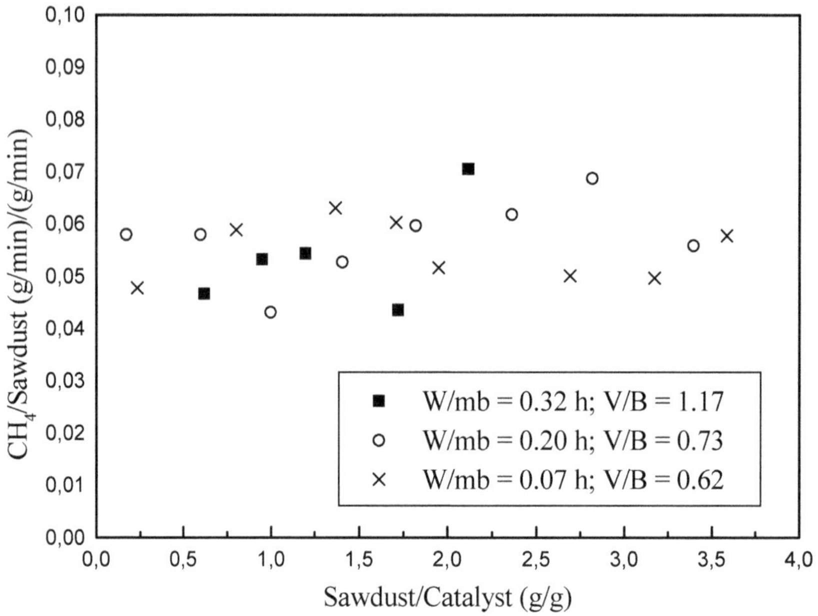

Figure 5: CH_4 yield using Ni/Al/La catalyst for different W/mb and S/B ratios.

COMPARISON OF CATALYSTS

Figures 6 to 9 show H_2, CO, CO_2 and CH_4 yields versus sawdust/catalyst ratio respectively. Figures a) correspond to Ni/Al/La and Ni/Al/Mg catalysts for W/mb ratio of 0.07 h and S/B ratio of 0.62, while figures b) correspond to Ni/Al and Ni/Al/La catalyst for W/mb ratios around 0.30 h and S/B ratios close to 1.1.

For figures a) H_2 and CO_2 yields are higher with the catalyst promoted with lanthanum than for the catalyst promoted with magnesium. For CO and CH_4 yields the results are quite similar, but at high sawdust/catalyst ratios the yields to these gases are higher for Ni/Al/La catalyst. These results indicate that the catalyst promoted with lanthanum is more stable and active for hydrogen production. It must be considered the low values of W/mb ratio studied that correspond to high severity conditions of steam gasification of biomass.

For figures b) H_2 and CO yields are higher for Ni/Al/La catalyst than for Ni/Al catalyst, while CO_2 and CH_4 yields show similar results. These facts indicate the high activity of lanthanum in steam reforming reactions.

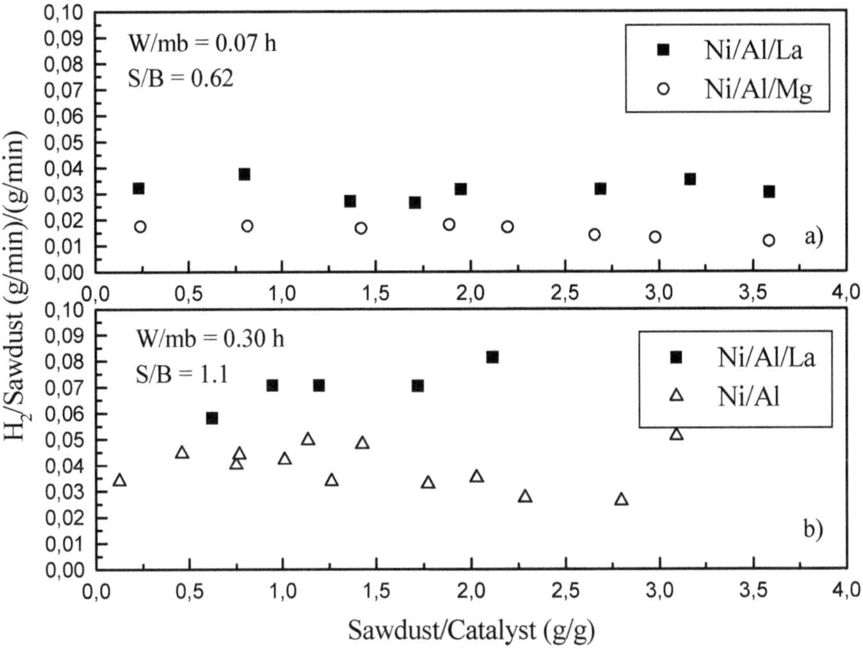

Figure 6: H_2 yield of different compared catalysts.

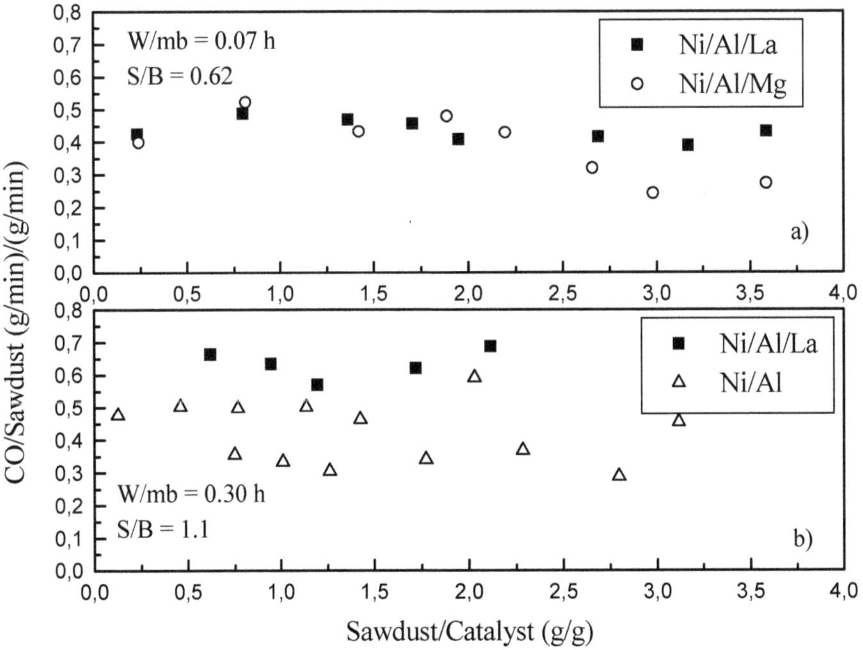

Figure 7: CO yield of different compared catalysts.

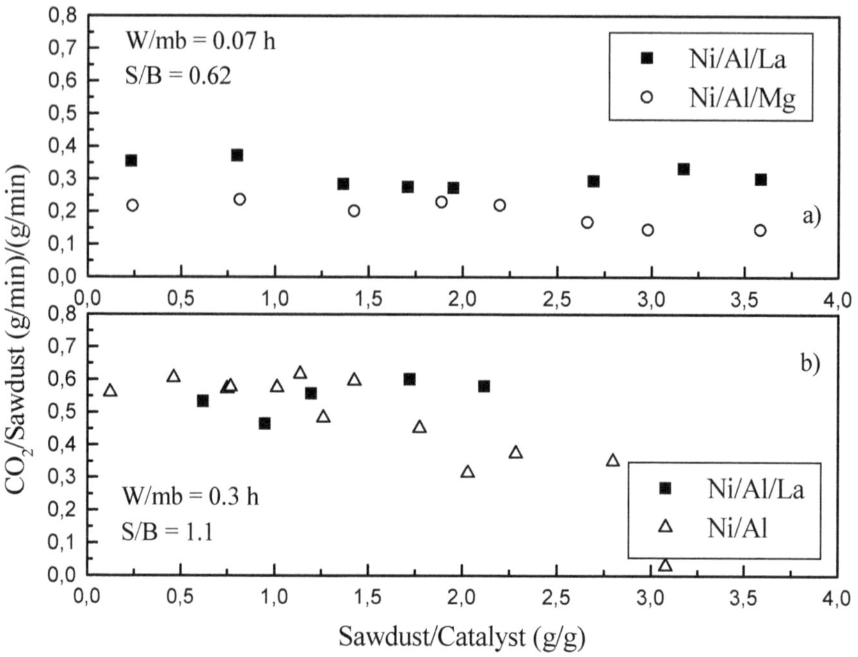

Figure 8: CO_2 yield of different compared catalysts.

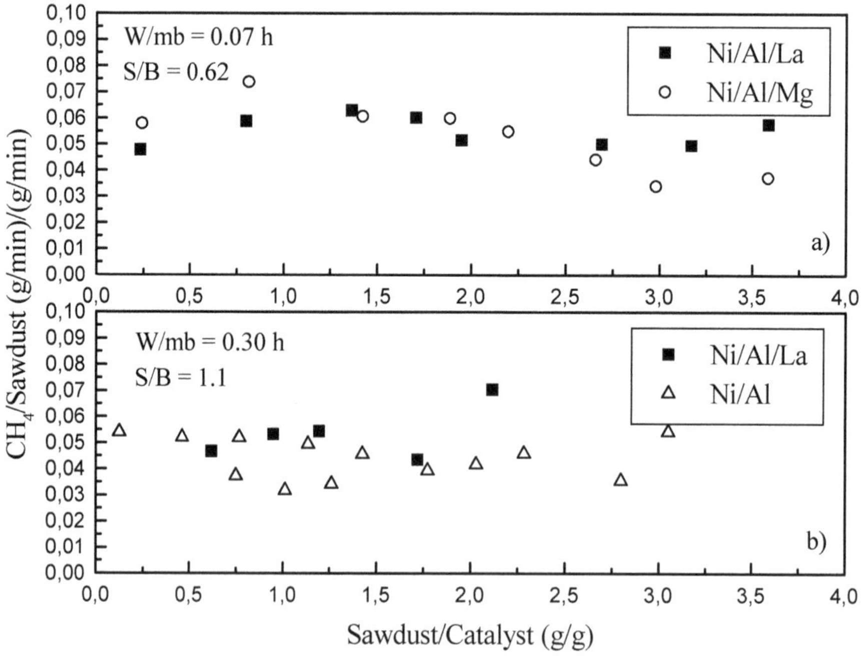

Figure 9: CH_4 yield of different compared catalysts.

CONCLUSIONS

Biomass steam gasification experiments in a fluidized bed reactor have been carried out at 700 °C using $NiAl_2O_4$ (Ni/Al), Ni catalyst promoted with lanthanum (12 %wt La_2O_3) (Ni/Al/La) and $NiMgAl_2O_5$ (Ni/Al/Mg) catalysts showing an adequate performace without attrition problems.

The increase of W/mb and S/B ratios causes the increase of H_2 and CO_2 yields that is explained by the participation of steam reforming and water-gas shift reactions. CO yields depend on the catalyst studied and the W/mb ratio influences significantly.

The comparison of the three catalysts show that nickel catalyst promoted with lanthanum is the most active with the highest hydrogen yields.

ACKNOWLEDGMENT

The authors express their gratitude to the "Comisión Interministerial de Ciencia y Tecnología (C.I.C.Y.T.)" and to the European Commission for providing financial support for the study (Project 2FD97-0890).

REFERENCES

1. Aznar, M.P., Caballero, M.A., Gil, J.; Olivares, A.; Corella, J. (1997) In: *Making a Business from Biomass*, (Ed. By R.P. Overend, E. Chornet), p.859. Pergamon Press, New York.
2. Rapagná, S., Jand, N., Foscolo, P.U. (1998) *Int. J. Hydrogen Energy*, **23**, 551.
3. Turn, S., Kinoshita, C., Zhang, Z., Ishimura,D., Zhou, J. (1998) *Int. J. Hydrogen Energy*, **23**, 641.
4. García, L., Salvador, M.L., Arauzo, J., Bilbao, R. (1999), *Energy and Fuels*, **13**, 851.
5. Sutton, D., Kelleher, B., Ross, J.R.H. (2001), *Fuel Process. Technol.*, **73**, 155.
6. García, L., Salvador, M.L., Arauzo, J., Bilbao, R. (1998), *Ind. Eng. Chem. Res.*, **37**, 3812.
7. Satterfield, C.N. (1991) *Heterogeneous Catalysis in Industrial Practice*, McGraw-Hill, New York.
8. Tottrup, P.B.; Nielsen, B. (1982) *Hydrocarb. Process.* 89.
9. Rostrup-Nielsen, J.R., US Patent RE 028,655 (1975), to Haldor Topsoe A/S.
10. Wang, S., Lu, G.Q. (1998) *Appl. Catal. B*, **16**, 269.
11. Arauzo, J., Radlein, D.; Piskorz, J., Scott, D.S. (1997), *Ind. Eng. Chem. Res.*, **36**, 67.
12. Bangala, D.N.; Abatzoglou, N., Chornet, E. (1998), *AICHE J.*, **44**, 927.
13. Choudhary, V.R., Uphade, B.S., Mamman, A.S. (1997), *J. Catal.,* **172**, 281.
14. Sambrook, R.M., Ross, J.R.H. US Patent 4,469,815 (1984), to Dyson Refractories Limited.
15. Setzer, H.J. US Patent 4,414,140 (1983), to United Technologies Corporation.
16. Bangala, D.N., Chornet, E. US Patent 5,679,614 (1997), to University of Sherbrooke.
17. García, L., French, R., Czernik, S., Chornet, E. (2000), *Appl. Catal. A*, **201**, 255.
18. Rapagná, S., Provendier, H., Petit, C., Kiennemann, A., Foscolo, P.U. (2002), *Biomass and Bioenergy*, **22**, 377.
19. Scott, D.S., Piskorz, J. (1982), *Can. J. Chem. Eng.*, **60**, 666.
20. Scott, D.S., Piskorz, J.; Radlein, D. (1985), *Ind. Eng. Chem. Process Des. Dev.,* **24**, 581.
21. Al-Ubaid,A., Wolf, E.E. (1988), *Appl. Catal.*, **40**, 73.
22. García, L., Salvador, M.L., Bilbao, R., Arauzo, J. (1998), *Energy and Fuels*, **12**, 139.

The First Gasification Tests of Leached Orujillo on a Circulating Fludised-Bed Gasifier

P. García-Ibañez, A. Cabanillas, J.M. Sánchez
CIEMAT, Fossil Fuel Department, Avda. Complutense, 22 28040 Madrid, Spain

ABSTRACT: Nearly a quarter of the world's olive oil production takes place in Spain, where energy companies are starting to exploit the potential of the residues from this industry as biomass fuel for energy production. Approximately 2 million t/yr of orujillo (a residual by-product of the olive oil production industry) are generated in Spain.

Fluidised bed gasification is considered to be the most advanced method for thermochemical conversion of various biomass fuels to energy. Ash-related problems such as sintering, agglomeration, deposition, erosion and corrosion, which are due to the low melting point of ash in the agroresidues, are the main obstacles for economical and viable application of this conversion method for energy explotation of the specific residues. The leaching (washing) of inorganic constituents from biomass leads to changes in inorganic composition and substantial improvements in ash thermal behaviour under gasification conditions.

Leached orujillo has been tested in a 300 kW_{th} atmospheric circulating fluidised-bed (CFB) gasification facility using air as a fluidisation agent.

The preliminary first test series has demonstrated that the CFB test rig operated adequately and makes it possible to carry out gasification experiments with orujillo as a fuel. The lower heating value of the produced gas obtained is 3.8 MJ/Nm^3 at the lowest temperature (780°C), under unfavourable operating conditions. The test runs have also shown a carbon conversion of 86.9% that is reached in the CFB gasification of leached orujillo and a chemical efficiency of up to 91 % has been achieved, even at 800°C.

INTRODUCTION

Renewable energies as a whole accounted for 6.3% of the total primary energy consumption in Spain in 1998. Of this total, just over half corresponds to the use of Biomass Energy. At present, the consumption of energy biomass in Spain is limited to heat applications. Only 4.6% correspond to fuel for generating electricity, in installations linked mostly to large paper-manufacturing plants or areas with a vast residual biomass production [1].

Biomass fuels for heat and power generation are of interest because biomass can be considered an environmentally and climatically sound option due to its almost closed carbon dioxide cycle, low sulphur dioxide emissions, good opportunities of ash utilisation and potential to save fossil energy resources [2,3].

The olive tree has been the source of richness in the Mediterranean area for many years. This area has an olive surface of about 7 million hectares, of which about 2.2 Mha correspond to Spain -over 60% of them in the region of Andalusia-. The production of olive oil in the EU -allowing for the three main producers: Spain, Greece and Italy- is about 80% of the world production. Spain produces more than 30% of the world total production [4].

Only a 21% of the weight of an olive is the oil itself, the reaming 79% consists of water, plus skin, flesh and husk. In the oil mills where the olives are pressed, most of their oil content is extracted. The pulp that results from this extraction process is called "orujo" in Spanish, and it is the raw material used by orujo oil extractors to obtain the residual oil. From the beginning of the 1990s, oil mills have undergone a very important transformation. These changes have always been searching for ways to reduce the environmental impact of the sector. The new method of decanting, using far less water and called the continuous two-phase system, produces one stream of olive oil and a single stream of waste formed of a very wet solid called "alpeorujo" (or two-phase "orujo"). "Alpeorujo" containes an average of 65% water, and approximately 3% fatty. The orujo oil extractors receive the "alperoujo", and dry, extract and refine the orujo olive oil, obtaining a dry solid called "orujillo" -which has a heating value of approximately 4,000 kcal/kg and 8-10% of moisture-.

During the 00/01 season, 700,000 t of olive oil were produced in Spain and over 2 Mt/year of "orujillo" was generated.

Fluidised bed gasification is considered to be the most advanced method and has been used extensively for thermochemical conversion of various biomass fuels to energy offering economical and environmental benefits [5-26]. Ash-related problems including sintering, agglomeration, deposition, erosion and corrosion resulting, due to the low melting point ash of agroresidues, are the main obstacles for economical and viable application of this conversion method for energy exploitation of the specific residues.

Leaching with water of inorganic constituents from biomass offers an approach to avoid extensive ash slagging, deposition and bed agglomeration associated with agroresidues resulting as by-products from various agricultural and/or agro-industrial processes [19-26].

The objective of the present work was to demonstrate the feasibility of the leached orujillo thermal conversion in an air-blown, fluidised bed gasifier and to study the effect of equivalence ratio on the gas quality, carbon conversion and process energy efficiency.

EXPERIMENTAL SECTION

The CFB gasifier test facility is an atmospheric air-blown installation and consists of four main parts: (1) Feeding system; (2) Primary and secondary air supply; (3) Gasifier vessel equipped with a cyclone and a particle recirculation system; (4) Gas analysis equipment.

A schematic diagram of this pilot plant, which has a nominal capacity of 0.3 MW_{th}, is shown in Figure 1 and some of the technical and operational data for the rig are presented in table 1.

Table 1: Technical and operational data of CFB gasifier.

Inside diameter	0.20 m
Reactor height	6.5 m
Gasification temperature	< 900°C
Fluidisation velocity	2-6 m/s
Operation pressure	1.0 bar (abs)
Fuel feed rate (max.)	90 kg/h
Gasification agent	air
MaximumThermal capacity	about 300 kW

The test rig is equipped with a feedstock hopper suitable to low-bulk-density biofuels and the biomass is fed continuously into the sand bed of the reactor via an injector screw which fees fuel at 37 cm above the primary air distributor. The maximum fuel mass flow rate is 90 kg/h. The feeding system consists of two screw feeders in series separated by a rotary valve. The fuel feed rate is controlled on the basis of the calibration curve of the first screw. The second screw has a higher feeding rate thus it will remain almost empty and therefore is not likely to be blocked by pyrolysis products.

Primary gasification air is introduced to the bottom of the reactor via an air distributor plate. Primary air can be preheated to 500°C in the start up. Also secondary air can be fed to the reactor on different levels in the freeboard section. When the gasification air enters into the gasifier below the solid bed, the gas velocity is high enough to fluidise the particles in the bed. At this stage, the bed expands and all the particles are in rapid movement. The gas velocity is so high, that a lot of particles (fly ash and bed material) are conveyed out from the reactor into the cyclone, then returned to the bed.

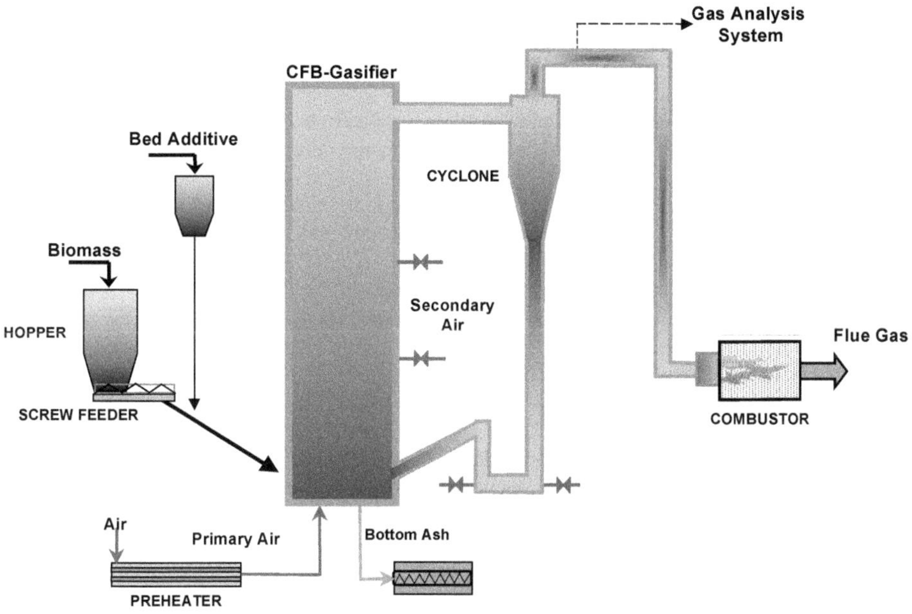

Figure 1: Schematic diagram of the atmospheric CFB gasification test rig.

The heart of this facility is a refractory-lined reactor with 0.2 m in-diameter and a total height of about 6.5 m. The internal refractory forms two concentric cylinders (each one of thickness 6.5 cm) with different thermal conductivity and wearing properties. Additionally, the reactor tube, the cyclone and the recycling line are protected with outside insulation in order to minimise heat losses. The gasifier components (riser, cyclone and standpipe) are equipped with ten thermocouples probes (K-type) and pressure transducers. The thermocouples probes are connected to a computer based data acquisition and control system. In addition, the feed rates of gasification agent are continuously measured and controlled by flowmeters and electronic pressure transducers are used for process evaluation.

After leaving the gasifier, the producer gas enters into a post-combustion chamber where propane gas can be injected to support combustion.

The sampling point for the product gas analyses is located downstream the cyclone. The temperature of the whole sampling system is kept at about 250°C to avoid tar condensation. The main gas constituents, CO, CO_2, H_2, CH_4, and the contents of hydrocarbon components (C_{2+}) are analysed by on-line gas chromatography (GC).

The bed inert material used in all the tests was silica sand (99,5% SiO_2) with an average particle size of 500 μm, a particle density of 2.6 g/cm3) and a bulk density of 1.5 g/cm3. Approximately, 50 kg of silica sand was used as bed inert material during each gasification test.

The leached orujillo samples used in the gasification tests were pre-treated with tap water in order to extract a part of its inorganic constituents, mainly potassium. Table 2 presents the analysis and characterisation of the leached orujillo samples, whereas table 3 gives the ash elemental analysis of the samples.

The start-up of the installation is done with pre-heated hot air and subsequently by combustion of coal the gasifier is heated up to the desired process temperature. The transition from combustion to gasification is done by stopping the feeding of coal and changing the fuel to leached orujillo with increasing the fuel mass flow in one step. Once steady state is reached the air primary pre-heater is cut off.

RESULTS AND DISCUSSION

The experimental conditions and results of these gasification tests are shown in table 4. No secondary air was used in these tests. The gasification temperature was maintained at 800°C in all tests.

The equivalence ratio (ER) is one of the most important operational variables in biomass gasification with air. It is defined as the air-to-fuel weight ratio used divided by the air-to-fuel weight ratio of stoichiometric combustion.

The leached orujillo gasification tests were operated in the ER range of 0.4-0.7 to study its effect on the gas quality, with respect to heating value and raw gas yield, and on the gasification process.

The LHV (dry) of producer gas has been calculated from the fuel gas composition analysed by CG. The hydrogen balance and the nitrogen balance were used to calculate the water vapour content of gas and raw gas flow rate. Carbon conversion (CC) to dry gas was calculated on gas evolution (mass flow rate of carbon in producer gas by the mass flow rate of carbon in fuel). The energy efficiency of the gasification process (PEE) is defined as the rate of energy carried by producer gas divided by the rate of energy supplied to gasifier.

Table 2: Feedstock analyses.

Analysis	Leached orujillo
Moisture content (wt%)	8.9
Bulk density (kg/m^3)	659
Average particle size (mm)	1.89
Proximate (wt%, dry basis)	
Volatile matter	74.4
Fixed Carbon	17.1
Ash	8.5
Ultimate (wt %, daf basis)	
C	52.7
H	7.2
N	1.6
S	0.07
Cl	0.37
O (diff.)	38.1
HHV (MJ/kg)	19.9
LHV (MJ/kg)	18.5

Table 3: Ash composition of the feedstock (wt %).

Ash basis	Na_2O	K_2O	CaO	MgO	SiO_2	P_2O_5	Al_2O_3	Fe_2O_3
Leached orujillo	2.08	24.73	8.01	12.86	16.72	1.45	0.89	2.37

GAS COMPOSITION

The effect of ER on gas composition of the producer gas is presented in figure 2. In the interval studied for ER on this series of experimental tests, on increasing ER, the amount of fuel gases (CO, H_2, CH_4, C_2H_4, C_2H_6) decreases. In Figure 2, at the ER of 0.41 the amount of H_2 (5.4 vol %) is minimum, this result is due to the operating conditions were not steady state by the time the gas composition was analysed by GC. During this set point, operational problems associated with the post-combustion chamber resulted in the end of the gasification test and consequently only one gas sample was analysed. The maximum amount of H_2 (9.3 vol %) is obtained at a ER of 0.59 and an upper zone temperature which is the highest one (893°C). When ER ranges from 0.59 to 0.73, it has little effect on the hydrocarbons.

Table 4: Gasification conditions and results of tests with leached orujillo.

	Units	Tests with leached orujillo			
Test number	-	T2	T5	T7	T8
Test duration (steady conditions)	min	60	120	120	90
Fuel mass flow	kg/h	60	47	64	51
Primary air flow rate	Nm^3/h	109	149	164	149
ER	-	0.41	0.73	0.59	0.67
$T_{gasification}$	°C	780	800	800	800
$T_{upper\ zone}$	°C	753	810	893	824
Gas velocity	m/s	3.7	5.2	5.7	5.2
CO (dry)	vol. %	8.6	6.9	8.4	7.5
CO_2 (dry)	vol. %	21.7	19.9	19.0	19.7
H_2 (dry)	vol. %	5.4	7.3	9.3	7.6
CH_4 (dry)	vol. %	3.0	1.8	1.9	1.8
C_2H_4 (dry)	vol. %	1.6	0.8	0.8	0.9
C_2H_6 (dry)	vol. %	0.3	0.1	0.2	0.2
C_2H_2 (dry)	vol. %	0.0	0.0	0.0	0.0
C_3H_n (dry)	vol. %	0.0	0.0	0.0	0.0
C_6H_6 (dry)	vol. %	0.0	0.0	0.0	0.0
N_2 (dry)	vol. %	59.46	63.16	60.55	62.46
LHV $(dry)_{gas}$	MJ/Nm^3	3.8	2.9	3.3	3.0
Raw gas flow rate	Nm^3/h	173	221	255	226
$Y_{raw\ producer\ gas}$	$Nm^3/kg_{fuel\text{-}daf}$	3.5	5.9	5.0	5.5
Raw gas flow rate (dry)	Nm^3/h	145.2	209.1	237.7	211.5
$Y_{producer\ gas\ (dry\ basis)}$	$Nm^3/kg_{fuel\text{-}daf}$	2.9	5.6	4.7	5.2
CC to dry gas	%	81.0	86.9	83.5	84.7
PEE	$MJ/kg_{fuel\text{-}daf}$	13.4	16.9	16.3	16.7
PEE	%	72.4	91.3	88.1	90.1

LOWER HEATING VALUE OF THE GAS

In Figure 3 the lower heating value of the gas produced at 800°C is shown as function of the equivalence ratio. Due to the dilution of the gas by the nitrogen contained in the air used as gasification agent and due to the increase of carbon dioxide, the heating value decreases with an increasing equivalence ratio. Consequently the highest value is for low equivalence ratio set point (3.8 MJ/Nm^3), in the interval studied for ER.

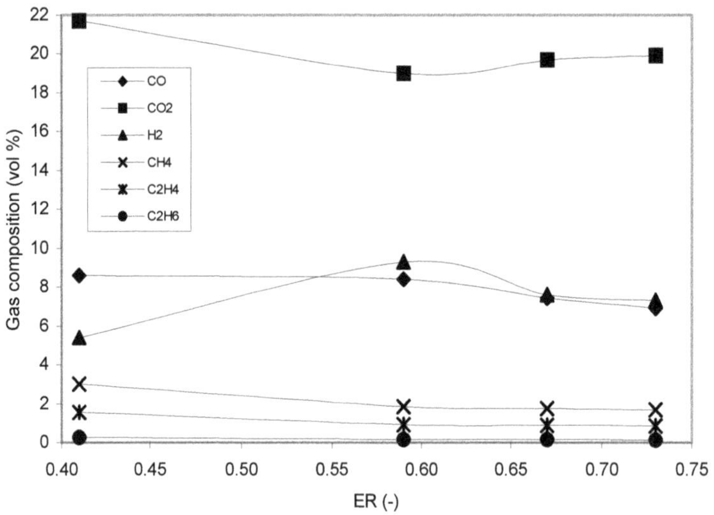

Figure 2: The effect of equivalence ratio on gas composition (dry basis).

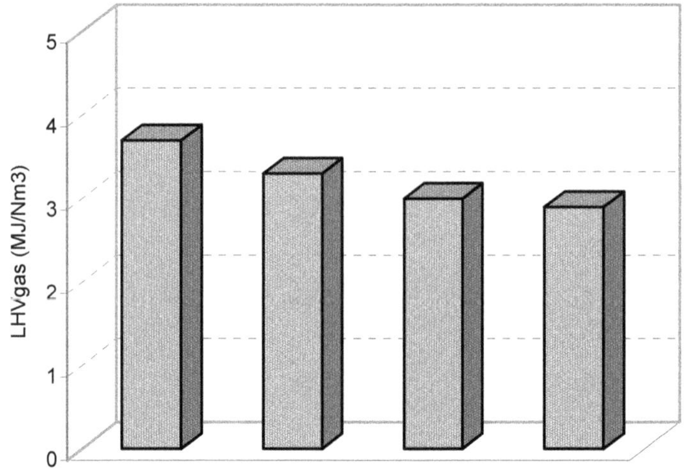

Figure 3: Influence of equivalence ratio on lower heating value of the dry gas.

GAS YIELD

The yield of the product gas and the yields of main gas components are shown in Figure 4. The increase in equivalence ratio generally led also to increased gas yields. This result agrees with the published ones by other authors [5],[7],[12],[13]. From these experimental tests, the yield gas varies from 2.9 $Nm^3/kg_{fuel\text{-}daf}$ at ER of 0.41 to 5.6 $Nm^3/kg_{fuel\text{-}daf}$ at ER of 0.73. Hence, it goes without saying that when ER increase, the gas yield increases, but the heating value of the gas decreases.

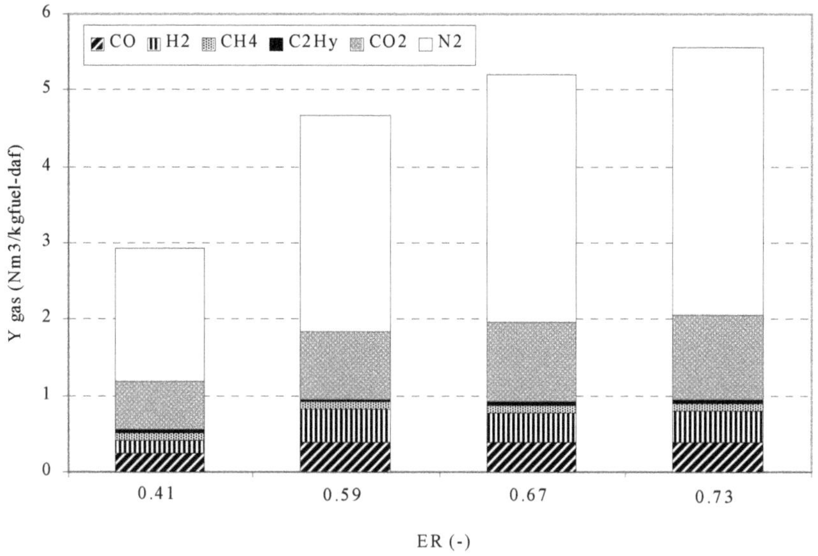

Figure 4: Product gas yield (dry basis) at different equivalence ratios.

CARBON CONVERSION

Figure 5 shows the influence of equivalence ratio on the carbon conversion which, as was said above, that is the amount of carbon converted into a gaseous fuel. At the ER range studied of 0.4-0.7, the carbon conversion varied between 81.0-86.9%. That is to say that an increase in equivalence ratio yields only a slight increase in carbon conversion.

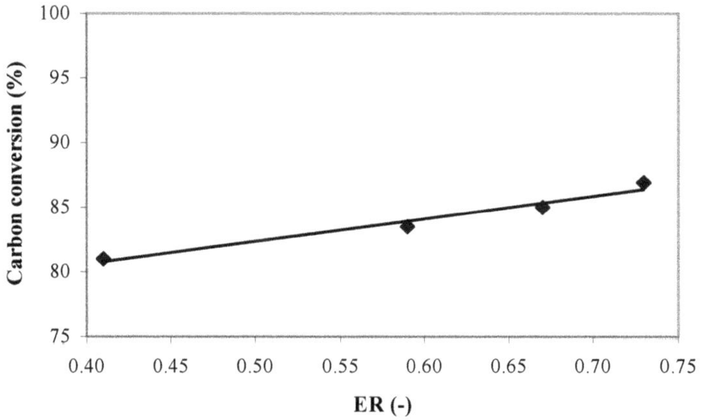

Figure 5: The effect of equivalence ratio on carbon conversion.

CONCLUSIONS

These first test series demonstrated that the CFB test rig operated well and makes it possible to carry out gasification experiments using leached orujillo as a fuel. Consequently, leached orujillo has been found to be well suited for air gasification in a fluidised bed.

The lower heating value of the produced gas obtained, under ill-operating conditions, is 3.8 MJ/Nm^3 at the lowest temperature (780°C). The test runs also showed that a carbon conversion of 86.9% is reached in the CFB gasification of leached orujillo even at 800°C and a process energy efficiency of up to 91 % has been achieved.

Higher equivalence ratios result in higher carbon conversion and higher process energy efficiency, likewise this results in lower gas heating value. Thus, it is obvious that more experimental studies are to be performed in order to achieve a higher calorific value of the product gas.

Gasification tests of 2-3 hours give reliable information on gas yield, LHV and CC, PEE. Nevertheless, phenomena such as ash agglomeration can not be studied adequately in short tests.

REFERENCES

1. Hernández, C. (2000) Biomass in Spain: Activities, Policies and Strategies. The Renewable Energy Promotion Plan. In: *1st World Conference on Biomass for Energy and Industry*, Kyritsis, S., Beenackers, A.A.C.M., Helm, P., Grassi, A., Chiaramonti, D. (Eds.), vol. II, pp. 1213-1216. James & James Ltd.
2. Kaltschmitt M. & Dinkelbach L. (1997) Biomass for Energy in Europe. In *Biomass Gasification and Pyrolysis* (Ed. by M. Kaltschmit & A.V. Bridgwater), pp. 7-23. CPL Press, UK.
3. Bridgwater A.V. (1995) The Technical An Economic Feasibility of Biomass Gasification for Power Generation. *Fuel,* 631-653.
4. Bas, F.J., Colinet, M.J., Lobo, J. & Manso, A. (2000) The olive tree as a Energy Soruce in the Mediterranean Area: Andalusia. In: *1st World Conference on Biomass for Energy and Industry*, Kyritsis, S., Beenackers, A.A.C.M., Helm, P., Grassi, A., Chiaramonti, D. (Eds.), vol. I, pp. 393-395. James & James Ltd.
5. Narváez I., Orío A., Aznar M.P. & Corella J. (1996) Biomass Gasification with Air in an Atmospheric Bubbling Fluidized Bed. Effect of Six Operational Variables on the Quality of the Produced Raw Gas. *Ind. Eng. Chem. Res.*, 35 (7), 2110-2120.
6. Maniatis K., Vassilatos V. & Kyritsis, S. (1994) Design of a Pilot Plant Fluidized bed Gasifier. In *Advances in Thermochemical Biomass Conversion*, (Ed. by A.V. Bridgwater), vol. 1, pp. 403-410. Blackie Academic Proffesional, Glassgow, UK
7. Kurkela E. & Stahlberg P. (1992) Air Gasification of Peat, Wood and Brown Coal in Pressurised Fluidized Bed Reactor. I. Carbon Conversion, Gas Yield, and Tar Formation. *Fuel Process. Technol*, 31, 1-21.
8. Kinoshita C., Wang Y. & Zhou, J. (1994) Tar formation Under Different Biomass Gasification Conditions. *J. Anal. Appl. Pyrolysis,* 29, 160-181.
9. Schenk E.P., van Doorn J. & Kiel J.H.A. (1997) Biomass Gasification Research in Fixed Bed and Fludised Bed Reactors. In *Biomass Gasification and Pyrolysis,* (Ed. by M. Kaltschmit & A.V. Bridgwater), pp. 129-138. CPL Press, UK.
10. Ising M., Hölder D., Backhaus C. & Althaus W. (1998) Fluidized Bed Gasification of Biomass for Cogeneration. In: *Biomass for Energy and Industry,* 10th Conference and Technology Exhibition, Kopetz, H., Weber, T., Palz, W., Chartier, P., Ferrero, G.L. (Eds.) pp. 1549-1551.
11. Gil J., Aznar M.P., Caballero M.A., Francés E. & Corella J. (1997) Biomass Gasification in Fluidized Bed at Pilot Scale with Steam-Oxygen Mixture. Product Distribution for very different operating conditions. *Energy & Fuels*, 11 (6), 1109-1118.
12. Gil J., Corella, J., Aznar M.P. & Caballero M.A. (1999) Biomass Gasification in Atmospheric and Bubbling Fluidized Bed: Effect of the Type of Gasifying Agent on the Product Distribution. *Biomass and Bioenergy*, 17 (5), 389-403.

13. Pinto F., Franco C., André R.N., Gulyurtlu I. & Cabrita, I. (1999) Gasification of waste materials and their co-processing with biomass. *39th IEA Fluidised Bed Conversion: Fluidised Processing of Unconventional Fuels*, (Ed. by A. Cabanillas & M. Miccio), pp. 123-135. Ciemat, Madrid, Spain.
14. Turn S.Q., Kinoshita C.M., Ishimura D.M. & Zhou J. (1998) The fate of inorganic constituents of biomass in fluidized bed gasification. *Fuel,* 77 (3) 135-146.
15. García L., Salvador M.L., Arauzo J., & Bilbao R. (1999) Catalytic Steam Gasification of Pine Sawdust. Effect of Catalyst Weight/Biomass Flow Rate and Steam/Biomass Ratios on Gas Production and Composition. *Energy & Fuels*, 13 (4), 851-859.
16. Pan Y.G., Roca X., Velo E. & Puigjaner L. (1999) Removal of tar by secondary air in fluidised bed gasification of residual biomass and coal. *Fuel,* 78 (14) 1703-1709.
17. Rensfelt E. & Ekstrom C. (1989) Fuel Gas from Municipal Waste in an Integrated Circulating Fluidized bed/Gas Cleaning Processes. *Energy Biomass Wastes*, 12, 811-906.
18. Simell P.A. & Bredenberg J.B.-son (1990) Catalytic Purification of Tarry Fuel Gas. *Fuel*, 69, 1219-1225.
19. Rosén C., Zanzi R. Sjöström K. & Björnbom E. (1998) Pressurized gasification of olive waste in a fluidized bed reactor. In *Biomass for Energy and Industry,* 10th Conference and Technology Exhibition, Kopetz, H., Weber, T., Palz, W., Chartier, P., Ferrero, G.L. (Eds.) pp. 1665-1668.
20. Natarajan E., Öhman M., Gabra M., Nordin A., Liliedahl T. & Rao A.N. (1998) Experimental determination of bed agglomeration tendencies of some common agricultural residues in fluidized bed combustion and gasification. *Biomass and Bioenergy,* 15 (2), 163-169.
21. Ergudenler A. & Ghaly E. (1992) Quality of gas produced from wheat straw in a Dual-Distributor type Fluidised bed Gasifier. *Biomass and Bioenergy*, 3(6), 419-430.
22. Ligthart F.S., van der Drift A. & Olsen A. (1998) Bed-Agglomeration in Fluidized Bed Conversion of Biomass. In: *Biomass for Energy and Industry,* 10th Conference and Technology Exhibition, Kopetz, H., Weber, T., Palz, W., Chartier, P., Ferrero, G.L. (Eds.) pp. 1765-1768.
23. Bakker, R.R., Jenkins, B.M. & Williams, R.B. (2002) Fluidized bed combustion of leached rice straw. *Energy & Fuels* 16, 356-365.
24. Arvelakis, S., Gehrmann, H., Beckman, M. & Koukios, E.G. (2002) Effect of leaching on the ash behaviour of olive residue during fluidized bed gasification. *Biomass and Bioenergy* 22, 55-69.
25. Rosén C., Zanzi R. Sjöström K. & Björnbom E. (1998) Pressurized gasification of olive waste in a fluidized bed reactor. In: *Biomass for Energy and Industry*, 10th Conference and Technology Exhibition, Kopetz, H., Weber, T., Palz, W., Chartier, P., Ferrero, G.L. (Eds.) pp. 1665-1668.
26. Schenk, E.P., van Doorn, J., Bodegom, M., van der Drift, A. (1998) Circulating Fluidized Bed Gasification Experiments at ECN. In: *Biomass for Energy and Industry*, 10th Conference and Technology Exhibition, Kopetz, H., Weber, T., Palz, W., Chartier, P., Ferrero, G.L. (Eds.) pp. 1757-1760.

The Optimizing Design of Downdraft Gasifier to Reducing Tar Content and Preventing Slag

D.Y. Guo, X.L. Yi, S.X. Xin, S.T. Yang, X.S. Lv and C.L. Ni
Energy Research Institute of Shandong Academy of Sciences
19, Keyuan Road, Jinan, China

ABSTRACT: This paper deals with the study of decrease tar content in the gas of down draft gasifier. First, depending on some typical materials' pyrolysis and gasification characteristic, the key to reduce the tar content is to increase the temperature of the combustion zone in the gasifier. The higher the temperature is, the more thorough the tar cracks and the lower the gas contains the tar. But the increase of temperature must be under the ash's (contained in the materials) melting point. Some experiments have been done in the down draft gasifier. Through different shape design of combustion zone to control the temperature of combustion zone so as to crack the tar thoroughly while not to reach the ash's melting point. Thus the gasifier can produce less tar content's gas and operate well. So it gets the optimizing design of down daft gasifier.

BACKGROUND

In China, the biomass is mainly used through direct burning to attain the thermal energy for countryside farmers and forest area residents cooking, which is laborious and unsanitary, takes much time, and has very low conversion efficiency. But biomass energy can be used more efficient, convenient and sanitary while this solid biomass is conversed into gas fuel by pyrolysis gasification, so it is a feasible technique of changing the low quality biomass energy into high quality energy.

Energy research institute of Shandong academy of sciences (SDERI) dedicates themselves to the high efficient usage of biomass. They developed the down draft gasifier unit to supply farmers cooking gas. So far they have set up more than 300 demonstrate units all over the country. With these demonstrates long term's running, it has obtained obviously profits of both economy and surrounding protection.

But there's something trouble with these demonstrates. That is tar. The tar generated in the gasifier mixed with the gas, condense at low temperature and congregate in the pipe and gas cooker, hindering normal flow of the gas. This influences the use of the gas.

Because of some economic and safe reasons, the finely cleaning procedure—static catch tar machine---cannot be used in the tar treatment of the gasifier. The usual way to clean the gas in the down draft gasifier is using mechanical and bathing procedure, etc. But these cleaning procedures cannot eliminate tar from the gas completely.

APPROACH TO REDUCING TAR CONTENT

In the down draft gasifier, the solid biomass first is heated to pyrolysis in the drying and pyrolysis zone. In this process, the tar is generated accompanying with volatile's separating out. The tar consists of complicated aromatic hydrocarbon with big molecules. As it goes down through the combustion zone, it can be cracked due to the high temperature for the second time. The main reaction is that

$$C_2H_6 \xrightarrow{700\sim1000} C_2H_4+H_2$$

$$C_2H_4 \longrightarrow CH_4+C$$

$$CH_4 \longrightarrow C+2H_2$$

As the temperature goes higher, the tar cracks more thoroughly. Thus the tar content in the gas can obviously be reduced.

In the gasifier, it is divided into 3 sections. They are the above section---preheating, drying and pyrolysis section; the middle section---gasification section and the below section---ash section. In the middle gasification section, there's combustion zone with throat and deoxidize zone. See figure 1 below.

If the throat shape of combustion zone is being changed, the air inlet flow velocity must be changed and the combustion temperature will be different and the crack degree of the tar will be different also.

So optimize the throat design to improve the combustion zone's temperature so as to reduce the tar content in the gas is the key mission of this study.

The main test work

A set of biomass gasification is set up in the lab. The device is mainly consisted of a feeder, a gasifier, a gas cleaner, and a blower, etc. In the gasifier, many thermocouples are laid to measure the different zone's temperature. In the gas exit, a tar content, a gas component and a gas heat value instruments are set up (see figure 1).

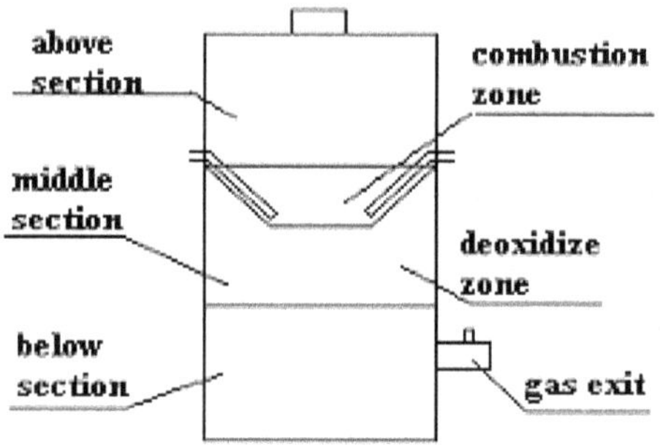

Figure 1: Sketch of the test gasifier

In the test, five different throats model are designed to install in the gasifier respectively. Every model is tested to obtain the influence of the throat dimensions on the temperature and the tar content. During the processes, the valve of return flow keeps the same position. That is in the 5 experiments the gas output of the gasifier is same. The sketches of throat model and air inlet flow shape are showed in figure 2. The different parameters of the throat model are showed in table 1.

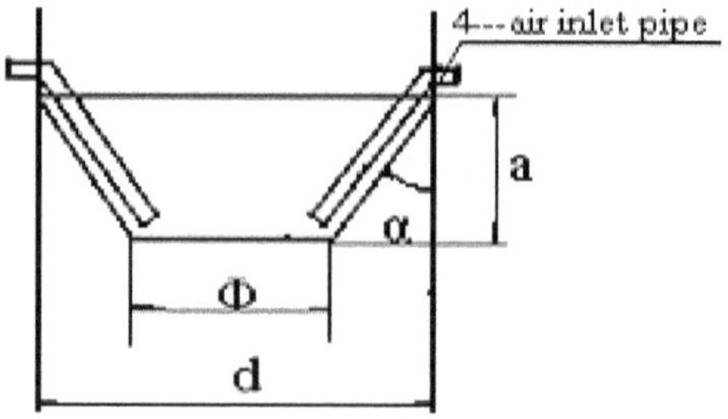

Figure 2: Model sketch of throat and airflow shape.

Table 1: Different parameters of the throat model.

	T1	T2	T3	T4	T5
α (°)	40	33.75	22.5	11.25	0
Φ (mm)	245	379	578	745	900
d (mm)	900	900	900	900	900
a (mm)	390	390	390	390	900

Experiment process. Though it has different combustion zone shape in the gasifier, the test process of the 5 ones is the same. The test stage is as below

(1) Put corn straws cut into pieces to the gasifier up to the throat.
(2) Emblaze corn straws with waste paper.
(3) Turn on the blower and enlarge the output of the blower gradually. Adjust the working condition of the gasifier.
(4) After about 5 minutes try to fire the small flambeau until it's been fired.
(5) Keep the output in 180m3/h and continue operating half an hour.
(6) When the working condition is steady, to measure the operating data every 40 minutes.
(7) Stop testing until there is no material in the gasifier. The whole test last about 3 hours.

Results of the test

The air inlet velocity on the throat section: when we take the velocity in the experiment 5 as the basic one v_0, we can get the following result: the velocity of the experiment 4 v_4 is 1.5 times than the basic one. As the same, the velocity of the experiment 3, 2 and 1 is respectively 2.4, 5.6 and 13.4 times than the basic one. It indicated that: with the reduction of the section of throat, the flow velocities in the combustion zone rise gradually. That is

$V_1=13.4\ v_0$
$V_2=5.6\ v_0$
$V_3=2.4\ v_0$
$V_4=1.5\ v_0$
$V_5=v_0$

Temperature of the combustion zone: the different flow velocity led to the different combustion temperature. With the raise of the flow velocity, the temperature of the combustion zone rises gradually. The changes of the temperature are showed in table 2 below.

Table 2: Combustion zone's temperature of the 5 tests.

	Test 1	Test 2	Test 3	Test 4	Test 5
Temperature (℃)	1225	1160	1060	975	910

The main reaction taken place at the throat is the equation:

$$C+O_2=CO_2+Q_1$$

With the high temperature the reaction rate is high and belongs to the diffusion zone. The coefficient a_{zl} of the mass transfer on the solid surface is the key point if we want to increase the burning rate. It is known from the mass transfer theory that a_{zl} can be enhanced through increasing the relative speed between the air and solid fuel surface. Therefore in the 5 experiments, with the decrease of the throat diameter, the flow speed rises. This means the high relative speed between air and material for gasification reaction. So that the coefficient a_{zl} of mass transfer is high, which enhance the burning rate while the high burning rate indicates the reaction is more intense in a short time, more heat is released, which reversely lead to the rise of the burning temperature.

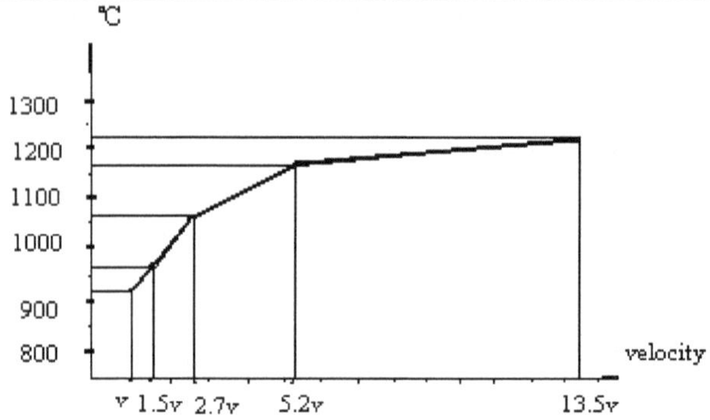

Figure 3: Rise in temperature with rise of flow speed.

The gas component and the gas heat value. The gas component and heat value are also measured and showed in table 3 below. We can find the valid components (H_2 and CO) are increased with the temperature's increasing and the heat value is also increased. This is because the deoxidization zone's temperature increases with the raise of combustion zone's temperature. While the deoxidization reaction belongs to kinetic zone, so if the temperature rises, the reactions will trend to the direction of producing H2 and CO.

Tar content in the gas: from the 5 experiments we get tar content data respective. They are showed in table 4 below. We can find the higher the temperature, the lower the tar content and every 100℃ the reaction temperature goes up, the tar content reduces by 20%. Figure 4 shows the trend of tar content dropping with the rise of combustion temperature.

Table 3: Gas component and heat value.

	Gas outputs	Gas components (%)					Low-heat value
	(Nm^3/h)	CO_2	O_2	CO	H_2	CH_4	(KJ/Nm^3)
Test1	135	11.60	1.10	18.60	17.40	3.23	5383
Test2	135	12.50	1.40	17.20	15.30	2.40	5116
Test3	132	13.00	1.50	16.50	14.10	1.98	4880
Test4	130	13.60	1.60	16.00	13.6	1.80	4750
Test5	132	12.90	1.55	15.20	13.00	1.65	4590

Table 4: Tar content in the gas.

	Combustion temperature (℃)	Tar content (mg/m3)
Test1	1210---1230	1200
Test2	1150---1180	1769
Test3	1050---1070	2332
Tset4	950—980	2769
Test5	900---930	2954

The melting temperature of the ash: in the operation of the gasifier, the resistant characteristic is an important parameter to control the normal operation. The resistance of the gasifier comes from mainly the ash. So if the ash reach its melting temperature point, the resistance will be great enough to influence the operation till to fail in gas output. Because there is not a common standard to measure the corn straw's melting point till now, it can only consult some data of known material like wood and garbage. It is evaluated that the corn straw melting point is about 1000℃. This conclusion is partly be approved by the 5 experiments.

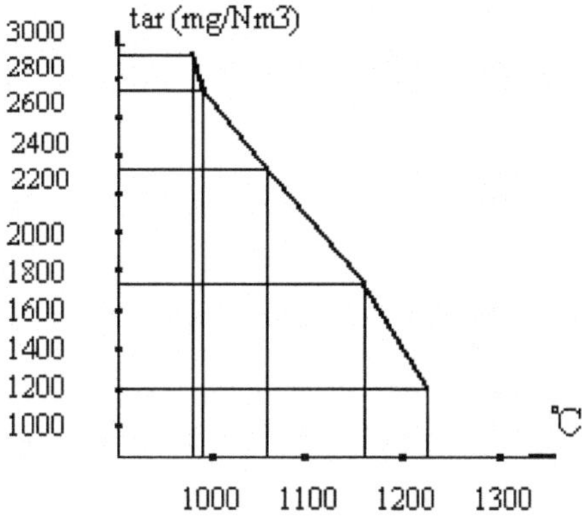

Figure 4: Tar content changing trend.

Table 5 is the minus pressure of each experiment during different stage. We can find in the test 1 and test 2 the minus pressure increase smoothly, while in test 4 and test 5 it does gradient. These changes correspond with the different combustion temperatures.

Table 5: Minus pressure changes of the 5 tests.

	Start	0.5h	1h	1.5h	2h
Test1	15	98	152	370	600
Test2	15	95	145	350	570
Test3	15	92	145	326	520
Test4	15	90	140	300	460
Test5	15	87	133	270	385

In the experimental scope of throat diameter, the size changes affect the minus pressure less. The main factor to influence the minus pressure is ash slag and ash thickness. If the temperature is too much higher than the ash melting point 1000℃ like in test 1. The speed of slag is also too fast and the slag mass will increase rapidly. This will influence the normal operation. While in test 3, the temperature is less high than the melting point and with the shaking of the furnace hearth to drop the small mass of slag, so the operation can last normal. Figure 5 shows minus pressure changing trend of the 5 tests.

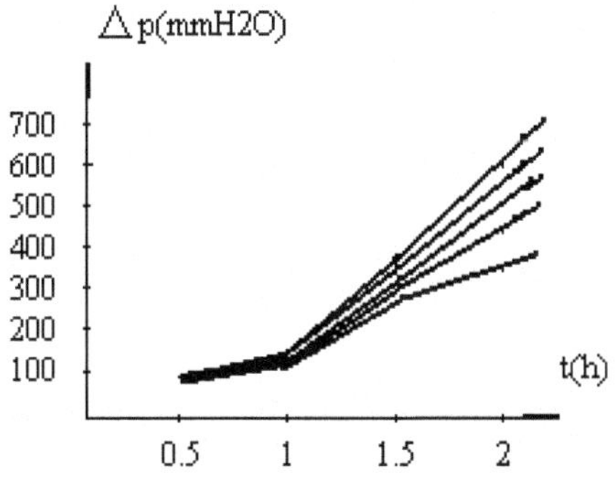

Figure 5: Minus pressure changing trend.

Stability of the gasifier's operation: There is limit for the decrease of the throat diameter. This is the operation stability. Because the straws materials are so less in density, if the throat is too small, the fall speed of straws through the throat will not touch the reaction speed to led unsteady operation of the gasifier. In fact, we have designed a sixth test that the angle α is 45°, corresponding the throat diameter is 120mm, it can not work steadily at all.

Instruction to the design of the gasification stove

Depending on the results of these tests, a new gasification stove with its gas output 200m3/h is designed, built and tested. The angle α is 35°. Its experiment data are that: reaction temperature is 1160℃, the original tar content (tar content in the stove exit) is 1404 mg/Nm3. This type gasification unit has been sold in the market and got a satisfied operating result.

CONCLUSIONS

From the above experiments, we can conclude some items to be useful in the design of gasifier:

(1) The main reaction in the gasification stove is the diffusion reaction, the throat dimensions determine the airflow speed; meanwhile determine the reaction temperature and intensity. As the dimension reduces, the flow speed increases and the reaction temperature rises.

(2) In the down draft gasification stove, the higher the reaction temperature is, the lower the tar content is.

(3) Under the condition that no straw materials jam in the throat happens, no strong slag in the gasifier to influence the normal operation, reducing the throat dimensions and increasing the air flow speed will do good to reduction the tar content in the original gas to reduce the subsequently cleaning work. This should be a principle of the design of the gasification stove.

REFERENCES

1. L.Sun etc. (1993) Crop straw gasifier and village biomass supply network, Proceedings of The 5th International Energy Conference Vol.3
2. Y.Z.Ren and L.L.Zao (2000) Design of lower absorption type of biomass gasification stove, Rural Energy, No.4
3. G.Y.Chen etc. (1999) Experimental Study and Theoretical Analysis On Biomass Pyrolysis In a Fixed-bed, Acta Energiae Solaris Sinica Vol.20, No.2

Future Prospects for Production of Methanol and Hydrogen from Biomass

Carlo N Hamelinck, André PC Faaij
Utrecht University, Copernicus Institute, Science Technology Society, Padualaan 14, 3584 CH Utrecht, Netherlands

ABSTRACT: Technical and economic prospects of the future production of methanol and hydrogen from biomass have been evaluated. A technology review, including possible future components, resulted in a set of promising conversion concepts. Flowsheeting models were made to analyse the technical performance. Results were used for economic evaluations. Overall energy efficiencies are around 55 % HHV for methanol and around 60 % for hydrogen production. Accounting for the lower energy quality of fuel compared to electricity, once-through concepts perform better than fuel only concepts. Hot gas cleaning can contribute to a better performance. 400 MW_{th} input systems produce biofuels at 8 – 12 US$/GJ (current gasoline production price is 4 – 6 US$/GJ). Costs are dictated by capital investments. Outcomes for the various system types are rather comparable, although concepts with little or no electricity co-production perform somewhat better. Hydrogen concepts using ceramic membranes perform well due to their higher overall efficiency combined with modest investment. On the longer term (2020) cost reductions may be obtained by cheaper biomass, technological learning, and application of large scales up to 2000 MW_{th} could bring the production costs of biofuels in the 5 – 7 US$/GJ range, being competitive with current gasoline and diesel costs.

INTRODUCTION

Methanol and hydrogen produced from biomass are promising carbon neutral fuels, both well suited for use in Fuel Cell Vehicles (FCVs). FCVs may have 2-3 times better fuel efficiencies than current gasoline Internal Combustion Engine Vehicles (ICEVs). In addition FCVs are quiet and clean, emitting none of the air pollutants SO_x, NO_x, VOS or dust [1].

Previous analyses [1,2] have shown that methanol can be produced from biomass at 14 – 17 US$/GJ (biomass delivered at 2.3 US$/GJ; all costs are in $US\$_{2001}$), with a net HHV energy efficiency between 54 and 58 %. Hydrogen production costs can be 10 – 14 US$/GJ, with a net HHV energy efficiency of 56 – 64 %. Those levels are not competitive with current gasoline and diesel production costs (4-6 US$/GJ). As these evaluations focused on technologies that are likely to be commercially available on the short term and the scale of production was fixed on about 400 MW_{th}, the key question of the present work [3] is *whether advanced, future technologies, larger scales and alternative concepts may enable competitive production of methanol and hydrogen on longer term.*

BIOFUELS PRODUCTION

In theory a large number of production routes involving conventional and advanced technologies are possible. Methanol or hydrogen production facilities typically consist of the following basic steps (see Figure 1): Pre-treatment, gasification, gas cleaning, reforming of higher hydrocarbons, shift to obtain appropriate H_2:CO ratios, and gas separation for hydrogen production or methanol synthesis and purification. Optional are a gas turbine or boiler, and a steam turbine, for electricity co-production.

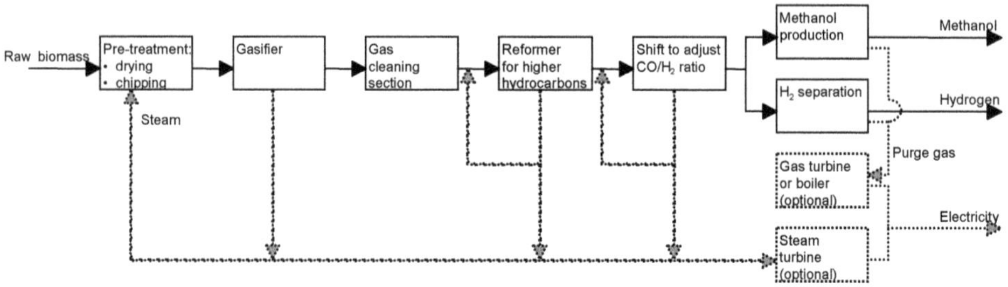

Figure 1: Key components in biomass to methanol/hydrogen production concepts.

Various gasification methods are available for syngas production. Based on throughput, cost, complexity, and efficiency issues, circulated fluidised bed gasifiers are considered most suitable for large-scale application. Two gasifiers are selected: the IGT (Institute of Gas Technology) pressurised direct oxygen fired gasifier, and the BCL (Battelle Columbus) atmospheric indirectly fired gasifier (Table 1). The IGT gasifier can also be operated in a *maximum hydrogen* mode, by increasing the steam input. Both gasifiers produce medium calorific gas, undiluted by atmospheric nitrogen, and represent a very broad range for the H_2:CO ratio of the raw syngas.

Table 1: Key characteristics of selected gasifiers [2].

		IGT	IGT maxH_2	BCL
Bed type		bubbling fluidised bed		Indirect heated CFB
Steam (kg/kg dry feed)		0.3	0.8	0.019
Oxygen (kg/kg dry feed)		0.3	0.38	0
Exit T (°C)		982	920	863
p (bar)		34.5	25	1.2
Composition	H_2O	0.318	0.48	0.199
(mole fraction	H_2	0.208	0.24	0.167
wet basis)	CO	0.15	0.115	0.371
	CO_2	0.239	0.16	0.089
	CH_4	0.0819	0.005	0.126
	C_2H_4	0.0031	0	0.042
	C_2H_6	0	0	0.006
Cold gas	HHV	82.2 %	72.1 %	80.5 %
efficiency	LHV	78.1 %	60.9 %	82.5 %

The produced gas contains tars, dust, alkali compounds and halogens, which can block or poison the catalysts downstream, or corrode the gas turbine. The gas can be cleaned using conventional low temperature technology (e.g. water scrubbing). Alternatively, hot gas cleaning can be considered, using ceramic filters and reagents at 350 – 800 °C. Hot gas cleaning is advantageous for the overall energy balance when followed by a reformer or ceramic membrane (require high inlet temperature). However, not all elements of hot gas cleaning are yet proven technology.

The syngas can contain a considerable amount of methane and other light hydrocarbons, representing a significant part or the heating value of the gas. Reforming converts these compounds to CO and H_2 by steam addition over a nickel catalyst.

Conventional methanol reactors use fixed beds and operate in the gas phase. Conversion is limited by equilibrium and by the high temperature sensitivity of the catalystist. Processes under development at present focus on shifting the equilibrium to the product side to achieve higher conversion per pass, and effective heat removal. In slurry phase processes [4] heat transfer is highly efficient, thereby allowing high conversions per pass without loss of catalyst activity. Moreover the shift reaction occurs at the catalyst surface. Conversion per pass is 15 – 40 % for CO rich gases and 40 – 70 % for balanced and H_2 rich gases, and future CO conversions up to 97 % are predicted. Investment costs for the slurry process are expected to be 5 – 23 % less than for a gas phase process.

After reforming and shifting to a hydrogen rich synthesis gas, hydrogen can be separated and compressed. Pressure Swing Adsorption (PSA) separates components of a gas stream by selective adsorption to a solid at high pressure, and subsequent desorption at low pressure. Membranes are a promising technology for gas separation. They are attractive because of their simple design, may have the ability of combining shift and separation in one reactor, and have potentially better economics than traditional separation technologies [1].

Unconverted fuel gasses that remain after the methanol or hydrogen production section can still contain a significant amount of chemical energy. These gas streams may be combusted in a gas turbine, although they generally have a much lower heating value (4-10 MJ/Nm^3) than most gas turbine combustors have been designed for (35-40 MJ/m^3). Problems for the currently available GTs, can be overcome when designing future GTs. Ongoing developments in gas turbine technology increase efficiency and lower the costs per installed kW over time, application of large scales will give increased turbine efficiency.

Heat is supplied or needed at several points in the biofuel production process (see Figure 1). Supply and demand need to be carefully matched, so that more high quality heat is left to raise and superheat high-pressure steam for electricity production in a steam turbine.

Altogether, in theory, a large number of concepts to produce methanol or hydrogen is possible. Eleven conversion concepts were selected (Table 2) that potentially have low cost and/or high energy efficiency (e.g. unnecessary heat leaps are avoided). The concepts make use of both existing commercially available technologies, as well as (promising) new technologies.

RESULTS

MODELLING

The selected systems are modelled with the Aspen+ process simulation program. For given inputs, product streams can be calculated, or one can evaluate the influence of apparatus adjustments on electrical output. The plant efficiency can be optimised by integrating the heat supply and demand. The pre-treatment and gasification sections are not modelled, their energy use and conversion efficiencies are included in the energy balances, though. The models start with the synthesis gas composition from the gasifiers.

SYSTEM PERFORMANCE

Table 2 summarises the outcomes of the flowsheet models. Based on the overall plant efficiency the methanol concepts lie in a close range: methanol 50 – 57 % and hydrogen 52 – 61 %. Liquid phase methanol production preceded by reforming results in somewhat higher overall efficiencies. After the pressurised IGT gasifier hot gas cleaning leads to higher efficiencies than wet gas cleaning, although not better than concepts with wet gas cleaning after a BCL gasifier. The conventional hydrogen concept 4 has the highest overall plant efficiency, but depends heavily on the import of electricity to the plant. If compared on a fuel only basis, its performance is the worst of the hydrogen concepts. The most advanced concept Hydrogen 3 than is the most efficient.

Several units, such as the gas turbine and reactors may be realised with higher efficiencies than considered here, and the heat integration can still be improved.

ECONOMIC PERFORMANCE

The resulting dimensions of streams and units and the energy balances are used for economic analyses. Plant sizes of 80, 400, 1000 and 2000 MW_{th} HHV are evaluated, 400 MW_{th} being the base scale. The economies of scale of the conversion system is an important factor in the overall economic performance. For BIG/CC systems, scale offsets the increased costs of biomass transport up to capacities of several hundreds of MW_{th} [6]. Large scale biomass conversion is not without precedent: several sugar/ethanol plants in Brazil have a throughput of 1 – 3 Mtonne/year sugarcane.

Table 2: Results of Aspen+ performance calculations, for 430 MW_{th} input HHV systems.

		HHV Output (MW)		HHV Efficiency	
		Fuel	Net electricity[1)] (gross–internal)	Fuel+E	Fuel only[2)]
Methanol					
1	IGTmaxH_2, Wet cleaning, Liquid phase MeOH, Combined Cycle	161	53 (71 – 18)	50%	52%
2	IGT, Hot cleaning, Autothermal reformer, Liquid phase MeOH with steam addition, Combined Cycle	173	62 (82 – 20)	55%	59%
3	IGT, Wet cleaning, Liquid phase MeOH with steam addition, Combined Cycle	113	105 (118 – 14)	51%	58%
4	BCL, Wet cleaning, Steam reformer, Liquid phase MeOH steam addition and recycle	246	0 (25 – 25)	57%	57%
5	IGT, Hot cleaning, Autothermal reformer, Partial shift, Gas phase MeOH with recycle	221	15 (38 – 23)	55%	56%
6	BCL, Wet cleaning, Steam reforming, Partial shift, Gas phase MeOH with recycle	255	-17 (10 – 27)	55%	54%
Hydrogen					
1	IGT, Hot cleaning, dual shift, Pressure Swing Adsorption, Combined Cycle	176	73 (93 – 21)	58%	66%
2	IGTmaxH_2, High temperature dust filter, Ceramic membrane (internal shift), Expansion turbine	259	-1 (25 – 26)	60%	60%
3	IGT, Hot cleaning, Ceramic membrane (internal shift), Combined Cycle	177	84 (103 – 19)	61%	74%
4	BCL, Wet cleaning, Steam reformer, Dual shift, PSA	303	-22 (0 – 22)	65%	63%
5	BCL, Wet cleaning, Dual shift, PSA, Combined Cycle	149	72 (97 – 25)	52%	56%

1) Net electrical output is gross output minus internal use. Gross electricity is produced by gas turbine and/or steam turbine. The internal electricity use stems from pumps, compressors, oxygen separator, etc.
2) The electricity part is assumed to be produced from biomass at η_e = 45 % HHV efficiency [6]. The Fuel only efficiency is calculated by $\eta = \text{Fuel}/(MW_{th,in} - \text{Electricity}/\eta_e)$.

The methanol and hydrogen production costs are calculated by dividing the total annual costs of a system (annual investments, O&M, biomass feedstock, electricity supply/demand) by the produced amount of fuel. The total annual investment is calculated by a factored estimation, based on knowledge of major items of equipment as found in literature or given by experts. Unit investments depend on the size of the components, by scaling from known scales in literature. Various components have a maximum size, above which multiple units will be placed in parallel and the influence of scale on system economy decreases.

Results of the economic analysis are given in Figure 2. The 400 MW_{th} conversion facilities produce methanol at 8.6-12 US$/GJ, and hydrogen at 7.7-11 US$/GJ. Considering the 30 % uncertainty range of this estimation method, one should be careful in ranking the concepts. Some concepts (Methanol 4 and 6 and Hydrogen 2, 3 and 4) perform somewhat better than the other concepts due to an advantageous combination of lower investment costs and higher efficiency.

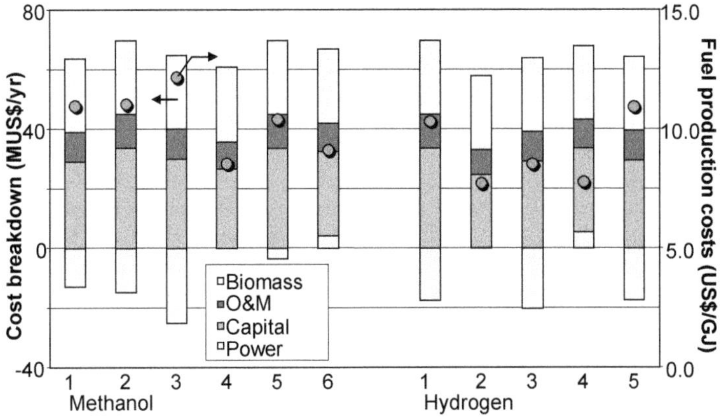

Figure 2: Fuel production costs (spheres refering to right axis) and annual cost breakdown. Interest rate is 10 %, O&M 4 % of total installed investment, biomass costs 2 US$/GJHHV, electricity costs/yields 0.03 US$/kWh.

The lowest methanol production price is found for concepts using the BCL gasifier, having lower investment costs. In hydrogen production the ceramic membrane concepts perform well due to their higher overall efficiency and modest investments. The combination of an expensive oxygen fired IGT gasifier with an combined cycle is generally unfavourable, since the efficiency gain is small compared to the high investment. Only in Hydrogen 3 (using a ceramic membrane) the extra investment seems justified.

Investment redemption accounts for 42 – 76 % of the annual costs and is influenced by the unit investment costs, the interest rate and the plant scale. The build-up of the total investment for all concepts is depicted in Figure 3. Costs for the gasification system (including oxygen production), syngas processing and power generation generally make up the larger part of the investment. Developments in gasification and reforming technology are important to decrease the investments.

Feedstock costs account for 36 – 62 % of the final fuel costs for the mentioned technologies. At a biomass price of 1.7 US$/GJ (realistic for e.g. Brazil), methanol production costs become 8.0-11 US$/GJ, and hydrogen 7.2-10 US$/GJ for 400 MW_{th} concepts. On the other hand, when biomass costs increase to 3 US$/GJ (short term Western Europe), methanol costs will increase to 10-16 US$/GJ, and hydrogen to 9.4-14 US$/GJ.

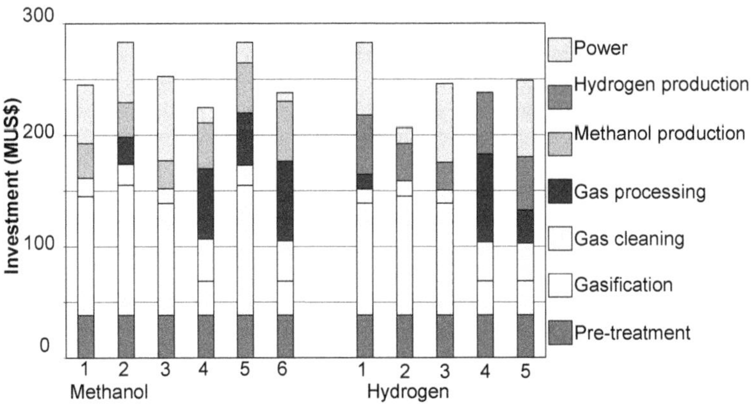

Figure 3: Breakdown of investment costs for 400 MW_{th} concepts.

If the electricity can be sold as green power (*carbon neutral* premium) at 0.08 US$/kWh, the fuel production costs for power co-producing concepts drops: methanol to -0.6-9.5 US$/GJ and hydrogen to 1.9-4.6 US$/GJ. A green electricity scenario, however, may be an atypical scenario, unlikely to be realised concurrent with biomass at 1.7 US$/GJ.

On the long term different cost reductions are possible (Figure 4) because of learning, scale and cheaper biomass. Methanol concepts than produce against 6.1 – 7.4 US$/GJ, and hydrogen concepts against 5.4 – 6.6 US$/GJ. The influence of capital on the annual costs has strongly reduced and the fuel costs of the different concepts lie closer together. Methanol 4 and Hydrogen 3 give the lowest fuel costs. In this long-term scenario, methanol produced from biomass costs considerably less than methanol at the current market, which is priced about 10 US$/GJ. Hydrogen from biomass resides in the cost range of hydrogen at a large natural gas reformer plant 5-9 US$/GJ. Diesel and gasoline production costs vary over time; current gasoline market prices lie in the range 4-6 US$/GJ. Current diesel prices are around 5 US$/GJ, and are projected to rise to 8-11 US$/GJ [7].

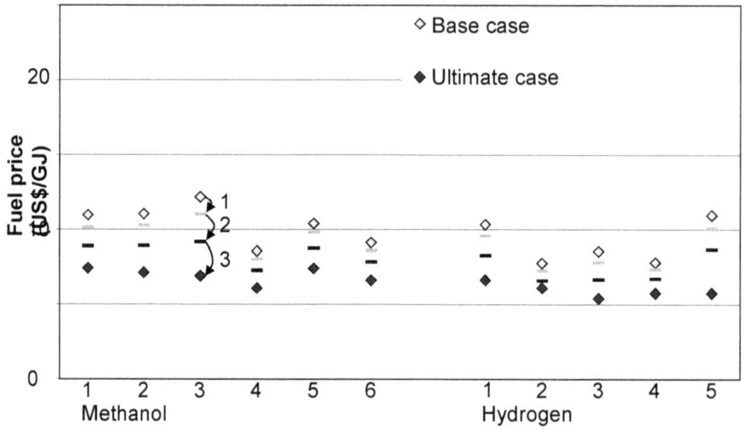

Figure 4: Base case biofuel prices and foreseeable cost reductions: (1) biomass costs 1.7 US$/GJ instead of 2 US$/GJ, (2) technological learning reduces capital investment by 15 % and (3) application of large scale (2000 MW_{th}) reduces unit investment costs.

CONCLUSIONS

Promising conversion concepts for the production of methanol and hydrogen from biomass have been evaluated. The concepts incorporate improved or new technologies for gas processing and synthesis and were selected on potential low cost or high energy efficiency. Overall HHV energy efficiencies remain ~55 % for methanol and ~60 % for hydrogen production. Accounting for the lower energy quality of fuel compared to electricity, once-through concepts perform better than the concepts aiming at fuel only production. Also hot gas cleaning generally shows a better performance. Some of the technologies considered in this study are not yet fully proven/commercially available. Several units may be realised with higher efficiencies than considered here: new catalysts and carrier liquids could improve liquid phase methanol single pass efficiency. At larger scales, conversion and power (especially the combined cycle) may have higher efficiencies.

Methanol can be produced at 9 – 12 US$/GJ and hydrogen at 8 – 11 US$/GJ, slightly above the current production from natural gas prices. The outcomes for the various system types are rather comparable, although concepts focussing on optimised fuel production with little or no electricity co-production perform somewhat better. Hydrogen concepts using ceramic membranes perform well due to their higher overall efficiency combined with modest investment. The present study has shown that biomass-derived methanol and hydrogen can become competitive fuels in the future.

ACKNOWLEDGEMENTS

The underlying study [3] is online available via www.chem.uu.nl/nws.

Integrated research like this cannot be done without information from many different researchers and industries. We would like to thank Eric Larson, Thomas Kreutz, Ryan Katofsky, Daan Jansen, René van Ree, Kees Daey Ouwens, Michiel Tijmensen, Yrjö Solantausta, Mark Paisley, Rick Knight, Peter Tijm, Paul van Ooijen, Hans de Lathouder, Douglas Fain, Robert Williams and Wim Turkenburg.

This work was sponsored by the Netherlands National Research Programme on Global Air Pollution and Climate Change (NOP/MLK II), the Foundation for Technical Sciences (STW), the Netherlands Agency for Energy and the Environment (NOVEM).

REFERENCES

1. Katofsky RE, 1993, The production of fluid fuels from biomass, Center for Energy and Environmental Studies/Princeton University, Princeton.
2. Williams RH, Larson ED, Katofsky RE and Chen J, 1995, Methanol and hydrogen from biomassa for transportation, with comparisons to methanol and hydrogen from natural gas and coal, PU/CEES Report 292, Center for Energy and Environmental Studies/Princeton University, Princeton, 47 pp.
3. Hamelinck CN and Faaij APC, 2001, Future prospects for production of methanol and hydrogen from biomass, Utrecht University>Copernicus Institute>Science Technology and Society, Utrecht the Netherlands, 28 pp + annexes.
4. Cybulski A, 1994, Liquid Phase Methanol Synthesis: Catalysts, Mechanism, Kinetics, Chemical Equilibria, Vapor-Liquid Equilibria, and Modeling - A Review, Catalytic Review - Science Engineering, 36(4): 557-615.
5. Gas Turbine World, 1997, Gas Turbine World 1997 Handbook, Pequot Publishing, Fairfield, Canada, 256 pp.
6. Faaij A, Meuleman B and van Ree R, 1998, Long term perspectives of biomass integrated gasification with combined cycle technology, Netherlands agency for energy and the environment Novem, Utrecht, 93 pp + annexes.
7. Hendriks F, 2000, Natural gas as a feedstock for automotive fuels - An alternative to crude oil (MSc thesis), Department of Science, Technology and Society/Utrecht University, Utrecht, the Netherlands, 85 pp + annexes.

Fischer-Tropsch Products from Biomass: System Study and Routing to Implementation - Preliminary Results

Carlo N Hamelinck, André PC Faaij
Utrecht University, Copernicus Institute, Science Technology Society
Padualaan 14, 3584 CH Utrecht, Netherlands
Herman den Uil, Harold Boerrigter
Energy research Centre of the Netherlands, ECN Biomass, PO Box 1
1755 ZG Petten, Netherlands

ABSTRACT: Fischer-Tropsch liquids produced from biomass via gasification (BIG-FT) may offer a clean and carbon neutral transportation fuel, directly usable in the present transportation sector. This paper reports on system analysis of FT concepts, to investigate techno-economic performance and identify key technical issues for future optimisation and development. System components necessary for FT production from biomass are analysed and combined to a limited set of promising conversion concepts. The main variations are in gasification pressure, the oxygen or air medium, and in optimisation towards liquid fuels only, or towards the product mix of liquid fuels and electricity. Flow sheeting models (Aspen Plus) are made to analyse the energetic performance, these models incorporate the influence of syngas composition, partial pressure and temperature on the FT reactor performance. Results on efficiencies and stream sizes are used for economic evaluations in order to calculated fuel prices by factored estimation. Preliminary results indicate the strong influence of the used gasification medium and pressure on conversion and costs. FT diesel production costs about 16 €/GJ for a 400 MW_{th} input facility, down to 14 €/GJ at very large scale (2000 MW_{th}). Options to further improve this performance are indicated.

INTRODUCTION

Production of Fischer-Tropsch (FT) liquids from biomass via gasification got growing attention in recent years. It offers a clean and carbon neutral transportation fuel, directly usable in the present day transportation sector. Previous research [1] on the combined production of FT liquids and power from biomass, showed that these 'green' FT liquids are not yet competitive with fossil derived FT liquids (16-9 US$/GJ vs. 4 US$/GJ). However, improvement options to bring the production costs of FT liquids from biomass to a competitive level, were indicated. The main system efficiency improvements reside in application of pressurised gasifiers, and in combining high selectivity with high conversion in the FT reactor.

As about 75 % of the investment costs are in the pre-treatment, gasification and gas cleaning section, the influence of the gasification pressure and medium (air / oxygen enriched air / pure oxygen) on economies of both gasifier and downstream equipment is studied. Special attention is given to the technical perspectives for large-scale pressurised biomass gasification. Previous research also indicated that insight is needed in the sensibility of Fischer-Tropsch catalysts to contamination in the biomass-derived synthesis gas, in necessary gas cleaning equipment to avoid catalyst poisoning and ageing, and in reactor design [2].

Several promising FT production configurations are modelled in Aspen Plus, to analyse the energetic performance, and results on efficiencies and stream sizes are used for economic evaluations in order to calculated fuel prices by factored estimation.

BIOMASS TO FISCHER TROPSCH

TECHNOLOGY

The basic steps in producing FT liquids [3; 1] are depicted in Figure 1. After pre-treatment, biomass is gasified to produce synthesis gas (biosyngas). This gas is cleaned and its composition is modified to fit the specifications for Fischer-Tropsch synthesis in the FT reactor. FT offgas is recycled for maximal FT liquids production, or combusted to produce electricity. The liquid FT products are treated to yield gasoline and/or diesel.

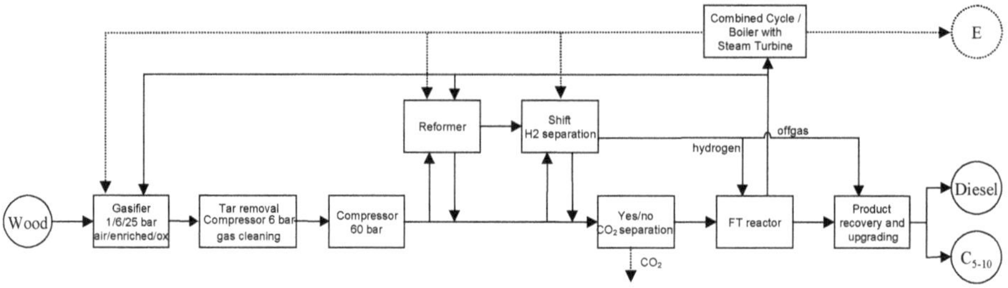

Figure 1: Summary of possible process configurations for FT production.

Several pre-treatment units precede gasification, like storage and size reduction. For drying a significant amount of energy is needed, extracted from the plant offgas, or from the steam cycle. The efficiency of gasification increases with drier biomass [4; 1; 5; 6], however the hydrogen content of the produced syngas decreases, which may be unfavourable for the Fischer Tropsch synthesis downstream. 15 % moisture is assumed.

Many gasification methods are available. Based on throughput, cost, complexity, and efficiency issues, circulated fluidised bed gasifiers are considered for large-scale gas production [4] in this study, entrained flow gasifiers would also be suitable. A direct fired CFB gasifier is selected, to analyse the influence of gasification medium and pressure. Atmospheric CFB gasification up to 300 MW_{th} is technically feasible, but the upper limit of single vessel capacities is not known, as generally fuel supply is considered to be the limiting factor. The economic advantages of pressurised gasification are obvious. Pressurisation after the gasifier is not necessary, or is limited to a single stage, offering the possibility to keep the temperature high between gasifier and FT reactor. The gasifier can be much smaller per unit of throughput, and hence a large single vessel capacity is expected to be possible, thus leading to better economies of scale compared to atmospheric CFB. At pressures above 25-30 bar a heavier construction is needed (or the throughput should be kept small) and feeding becomes complex. The present study limits gasification to 25 bar.

Gasification with oxygen as oxidative medium, offers benefits in downstream equipment size, compression energy, and higher partial pressures of $CO+H_2$, which are of key importance for the FT synthesis. Pure oxygen is expensive, but oxygen enriched air may combine the advances of equipment size reduction and a an inexpensive medium. In the present study, the extremes 'pure' air and pure O_2 are to be studied, as well as one intermediate case assuming O_2 enriched air.
Gas cleaning may be the most critical process section for biomass FT implementation. Gas cleaning can be divided into two parts, tar removal and residual contaminations removal. Tar removal can be done by cracking. At temperatures above 1200 °C, tars are destroyed without a catalyst, usually by addition of steam and oxygen. Drawbacks are the need for expensive materials, soot production [7] and the loss of cold gas efficiency. Catalytic cracking avoids the mentioned problems of thermal cracking. However the technology is not yet fully proven [7] and it is unclear whether the light aromatic compounds (BTX: benzene, toluene, xylene) are converted. To gain more insight in the acceptance of contaminants in FT production, ECN performs tests in a parallel study. Once tars are absent, other impurities, such as NH_3, H_2S, COS, HCl, volatile metals, dust,

and soot, can be removed by standard wet gas cleaning technologies or advanced dry gas cleaning technologies. Maximal acceptable contaminant concentrations for the cobalt FT catalyst are summarised in Table 1, together with the effectiveness of two gas cleaning methods.

Table 1: Contaminant specification for FT synthesis, and cleaning effectiveness of wet and dry gas organization.

	Specs	**Cleaning method & performance**	
		Wet	**Dry**
Particles	0 ppb	Cyclones, Bag filters, Candle filters, Scrubber → Specs met	Cyclones, Moving beds, Candle filters → Specs met
Alkali	< 10 ppb	Scrubber, Active C bed → Specs met	Sorbents → Developing
Halides	< 10 ppb	Dolomite in tar cracker, Scrubber, Active C bed → Specs met	In bed sorbents or in stream + guardbed → < 1 ppm
Nitrogen compounds	< 1 ppmv total N	Scrubber, Hydrolyse HCN to NH_3, Scrubber, Guard bed → Specs met	Catalytic decomposition, Comb. removal NH_3/H_2S, Selective ox. → Developing
Sulphur compounds	< 1 ppmv total S	Scrubber, Hydrolyse COS to H_2S, Scrubber, Guard bed → Specs met	In bed Calcium sorbents, Metal oxide sorbents → <20 ppm

In dry gas cleaning, residual contaminations are removed by chemical absorbents at elevated temperature. In the FT situation, hot gas cleaning has little *energy* advantages, due to the low operation temperature of compressors and the FT reactor. However, dry gas cleaning may have lower operational costs than wet gas cleaning, and decrease the need for expensive water treatment processes [8]. In the present study three gas cleaning trains are applied, as depicted in Figure 2.

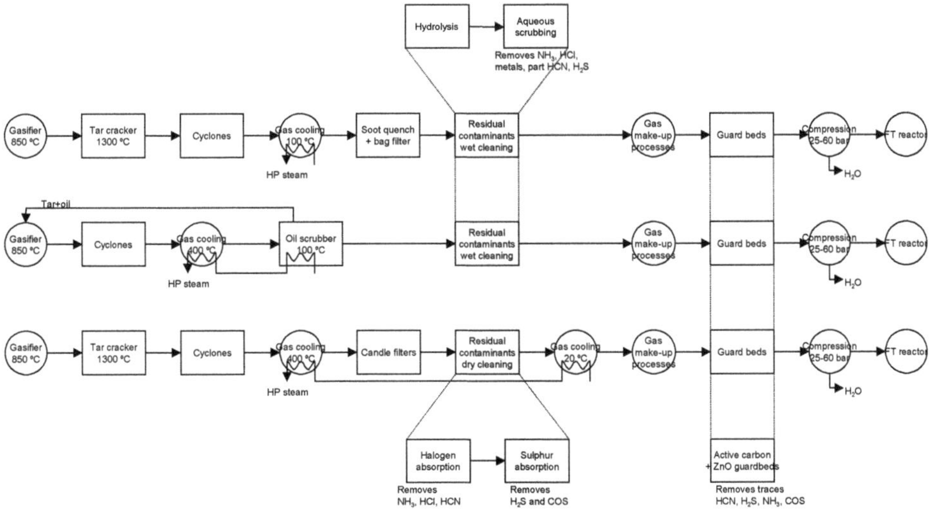

Figure 2: Two gas cleaning trains applied in this study. Top: tar cracking and conventional wet gas cleaning and bottom: tar cracking and dry gas cleaning.

Early compression reduces the size of gas cleaning equipment, however sulphur and chloride compounds condense when compressed and may corrode the compressor. Therefore intermediate compression to 6 bar takes place after bulk removal and 60 bar compression after the guardbed.

The syngas can contain a considerable amount of methane and other light hydrocarbons, representing a significant part of the heating value of the gas. Reforming is optional, to convert these compounds to CO and H_2, driven by steam addition over a nickel catalyst.

The syngas produced by CFB gasification (especially at atmospheric pressures) generally has a low H_2/CO ratio. Although Fischer Tropsch selectivity is favourable at a low ratio [9, p35 and p126; 10, p24 and p103; 11, p183; 12, p306], for high conversion effectively 2 H_2 are needed per CO. This can be realised by water gas shift.

In the FT reactor a high selectivity (α) for liquids should be combined with a high gas to liquid conversion. The growth probability or selectivity, largely depends on the catalyst [13], and further on temperature, local partial pressures of reactants and inertia, and the applied FT technology [14; 15; 1]. For example, α increases with decreasing H_2/CO ratio [16], decreasing reaction temperature and increasing pressure, and α is higher for Ru and Co catalysts relative to Fe catalysts [12]. Conversion depends especially on availability of reactants and the reactor size. Two kinds of FT reactors are considered here: the fixed bed reactor and the slurry phase reactor. Of both reactor types commercial versions exist.

The C_{5+} products are separated by a condensation step and send to the recovery and upgrading section. To maximise the production of FT liquids, the offgas containing unreacted H_2 and CO and produced lower alkanes may be recycled. The recycle can contain a reformer to reconvert $C_{1\text{-}4}$ back into syngas, and a shift reactor. Furthermore the FT off gas may be recycled to the catalytic or thermal (1200-1400 °C) tar cracker, which will work as reformer, but this is only sensible when the gasification pressure is the same as the FT pressure. With higher once through conversions in future, recycling could become obsolete.

Instead of maximised fuel production, the system can also be optimised towards combined fuel and electricity production. In this case the syngas passes 'once through' the FT reactor. The FT offgas is completely purged to a combined cycle for electricity production. Co-production of FT liquids and electricity may combine a high overall efficiency with modes investment costs.

The waxy part of the raw synthesis product is selectively hydrocracked to the desired middle distillate products.

SYSTEM SELECTION

Given all the options for gasification, gas cleaning and conditioning, fuel synthesis and upgrading, and power generation, many concepts to produce FT diesel can be imagined. Figure 1 depicts the many possible combinations and choices.

Not all combinations may be sensible from efficiency or economic point of view. A limited set of configurations is modelled in which maximising FT production is a prime target, especially by optimised once through conversion. In some concepts a recycle is applied. The main variations between these concepts stem from the different gasifier configurations and the gas cleaning train. The performance of different concepts will be evaluated on small, intermediate, and large scale: 80, 400 and 2000 MW_{th} (input). The FT reactor (catalyst) will be modelled with state of the art catalyst performance; a sensitivity analysis is performed towards improved catalysts.

RESULTS

MODELLING

Concepts are modelled using Aspen+ to obtain mass and energy balances. The economic calculations will be based on those balances. To evaluate the chosen configurations, one "master" flowsheet is built in Aspen+, following the scheme of Figure 1. By opening or closing "valves", the different system configurations can be selected. The gasifier, reformer and gas turbine deliver heat, whereas the dryer, gasifier, reformer, and shift require steam. The supply and demand of heat is added to or drawn from the steam turbine, such that any surplus heat is turned into electricity.

Using a gasifier databank, different gasifiers can be 'plugged' into the plant. It is also possible to choose between differently performing FT reactors (catalysts) and different combined cycles, in aid of the sensitivity analysis.

Selection and conversion in FT synthesis are a function of feed stream composition, pressure, temperature, catalyst and reactor type and size. Figure 3 depicts how the influence of various parameters is incorporated in the Aspen Plus model.

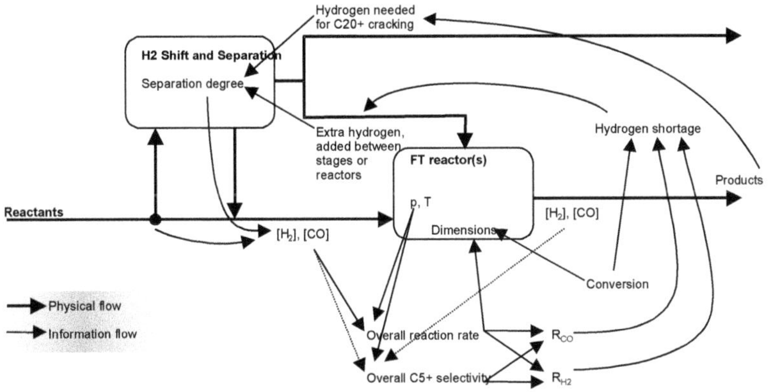

Figure 3: Incorporation of selectivity and conversion in Aspen model. The dashed line illustrates that the C_{5+} selectivity is influenced by either entrance or exit concentrations, depending on the reactor type: solid bed respectively slurry phase.

For high C_{5+} selectivity, the hydrogen/carbon monoxide ratio should be kept below the consumption ratio of these compounds [16]. For the solid bed reactor, which is of plug flow type, this inevitably leads to a decrease of the actual ratio during conversion. As H_2/CO and partial pressures decrease over the reactor length, so change selectivity and conversion rate. The selectivity than remains rather constant, but the reaction rate decreases strongly over conversion: every next reactor needs to be larger. In the slurry reactor (CSTR type) the bulk is assumed totally mixed, so that every local ratio equals reactor exit condition.

PRELIMINARY TECHNO-ECONOMIC RESULTS

Results for 7 once through concepts are given in Figure 4. These concepts apply a tar cracker and conventional wet gas cleaning, but no reforming. Just enough CO is shifted to meet the H_2 demanded for 70 % conversion of $CO+H_2$. Obviously, the production of FT fuel becomes cheaper with increasing gasification pressure and oxygen purity. This coheres with both a decreasing total capital investment (Figure 4) and an increasing plant efficiency. Capital costs decrease for a large part because of decreasing gas volume in the cleaning section, the extra costs for oxygen compression are more than outweighed by reduced compression costs downstream. The plant efficiency increases because at the higher partial reactant pressure, the selectivity for desired product is higher (at same fixed conversion) even though at atmospheric and air gasification significantly more $CO+H_2$ is produced, and also because the internal power needs decrease with higher gasification pressure and less nitrogen dilution.

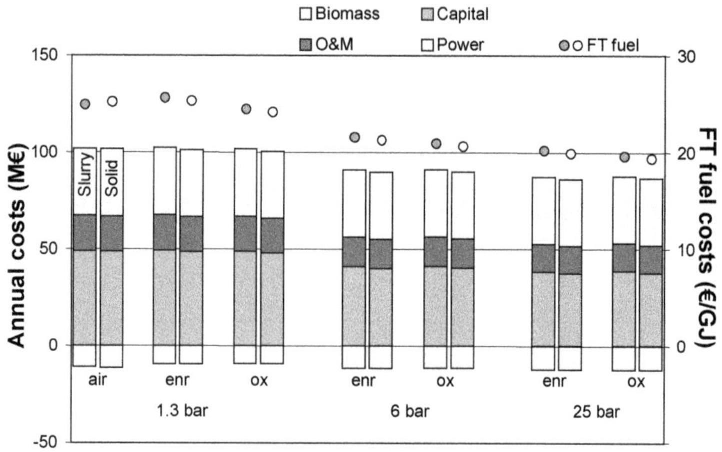

Figure 4: Annual costs (bars) and resulting FT fuel costs (bullets), for different gasification configurations. Fixed settings: 400 MW_{th} input scale, tar cracker, wet gas cleaning, no reforming, once through FT, minimal shift to meet H_2 demand, 70 % conversion $CO+H_2$, biomass price is 3 €/GJ.

Slurry phase and solid bed reactors perform the same, in the rest of this analysis we only discuss solid bed reactors, as similar results can be expected for slurry phase.

The capital build up for the extremities of Figure 4 is given in Figure 5. The gas cleaning costs, which make out a large part in atmospheric air fired concepts, are strongly reduced in pressurised oxygen fired concepts, where – naturally – the costs for oxygen production and compression play an important role. The shares of costs for pre-treatment, gasifier, and power generation slightly grow equally.

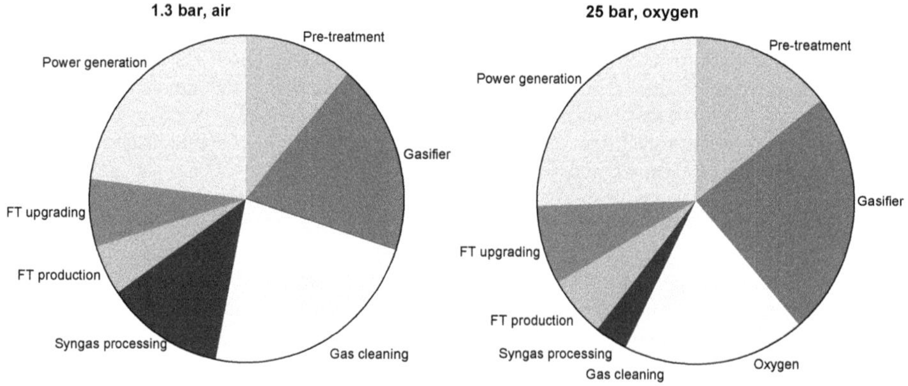

Figure 5: Capital break down for two once through concepts: based on air fired atmospheric gasifier (left) and based on oxygen fired 25 bar gasifier (right). Fixed settings: tar cracker, wet gas cleaning, no reforming, solid bed FT, 70 % conversion.

So far, the $CO+H_2$ conversion in the FT reactor was set to 70 %. A higher conversion can be realised by a larger reactor, leading to higher capital costs. Since the capital costs of the FT reactor are not crucial for the total capital investment, a higher conversion leads to lower FT diesel production costs, see Figure 6.

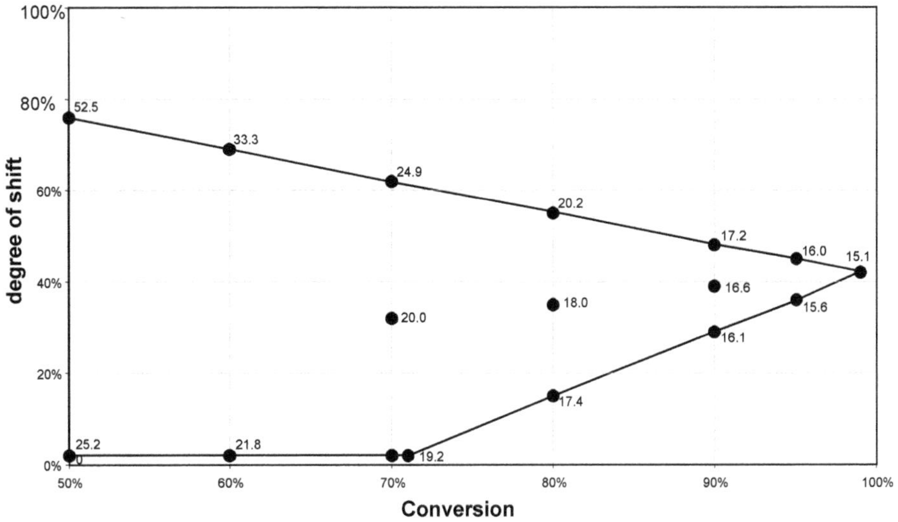

Figure 6: FT Diesel production costs as a function of conversion extent and degree of shift for 25 bar oxygen gasification (purple bullets). Below area is hydrogen shortage, above area is carbon monoxide shortage. Fixed settings: tar cracker, wet gas cleaning, no reforming, once through FT.

According to the graph, the lowest costs are reached at total conversion, this stems from the assumption that reaction rate is constant (the average between entrance and exit reaction speed) over the length of the reactor. Of course, as CO and H_2 disappear over conversion, the reaction rate decreases rapidly when approaching full conversion, leading to exponential reactor size increase for the last converted reactants. We assume that at 90 % conversion, the entrance-exit-averaged reaction rate is just still representative for the volume-averaged reaction speed.

If a reformer is added to the 25 bar oxygen fired wet cleaned concept (see Figure 7), this only increases the costs and reduces the efficiency, as all hydrocarbons were already converted in the tar cracker, and, moreover, the ATR shifts the reaction mixture towards H_2 excess, which slightly decreases the selectivity in the FT reactor. The fraction CO_2 before the FT reactor is about 30 volume %, removing this fraction improves both selectivity and efficiency, but due to the accompanying increase in investment this does not result in lower product costs. The efficiency for the dry gas cleaned concepts is slightly higher than wet cleaned concepts, because the ceramic membrane more effectively shifts (since product is taken away) than the traditional shift reactor, hence less steam is needed. This is outweighed by a slightly higher capital investment, such that the resulting fuel price is the same. The oil scrubber effectively returns BTX and tars to the gasifier where they are cracked, but much energy remains in methane, ethane and ethane, so that the FT yield is low. The performance of concepts with oil scrubber improves when adding a reformer, however, this turns out a more expensive combination than a single tar cracker. The recycle is operated at 70 % conversion of CO+H_2 per pass, 50 % of the off gas is recycled, and the FT pressure is decreased to 25 bar. The continuous temperature leaps and product leaps (biosyngas partly to $C_{1\text{-}4}$ product, and than back to syngas) lead to a both low selectivity and low overall efficiency, and the capital costs are high.

The influence of scale (see Figure 8) is examined for both wet and dry cleaned concepts. It is found that at large scale, FT diesel production costs can drop to about 14 €/GJ. The resulting overall scale factor for both concepts is 0.78. The efficiency advantage of dry cleaned concepts does not lead to smaller costs at larger scales, compared to wet cleaned concepts.

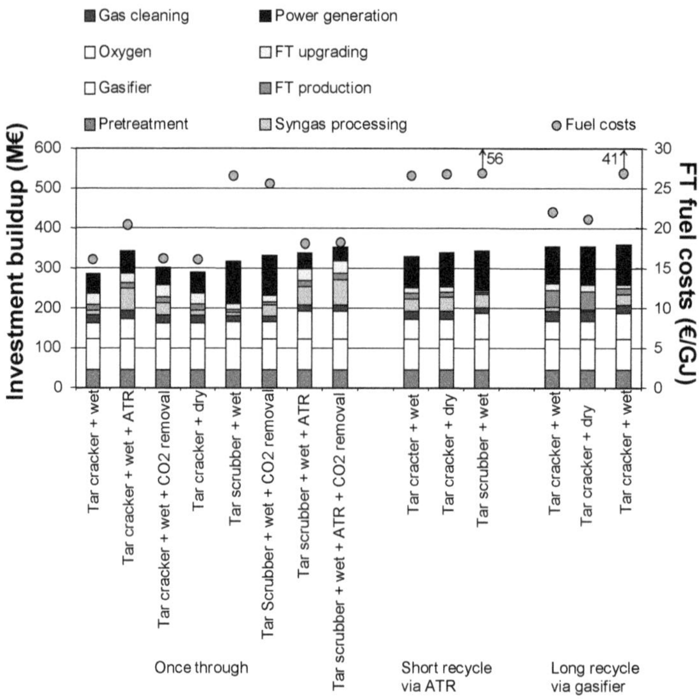

Figure 7: Variations of the key concept (25 bar oxygen gasification) with reforming, or other gas cleaning trains. 90 % conversion.

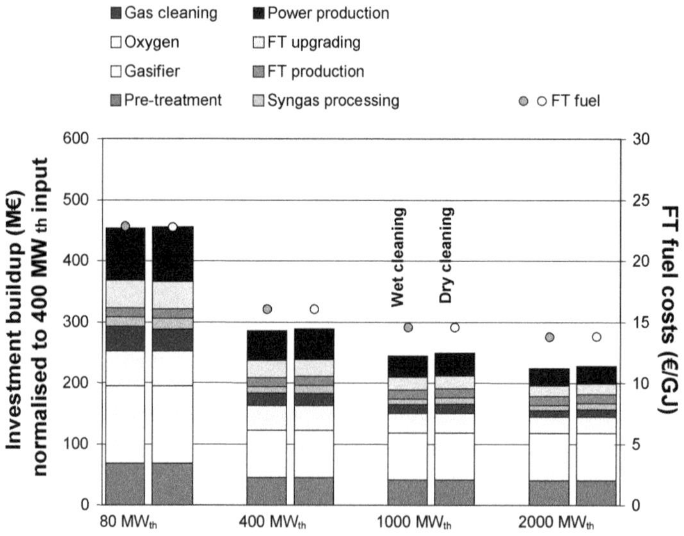

Figure 8: Influence of scale on relative investment costs (bars, left axis), and resulting fuel costs (bullets, right axis) for wet and dry cleaned concepts. Fixed: 25 bar oxygen fired gasification, Tar cracker, no reforming, once through 90 % conversion.

CONCLUSIONS

System components necessary for FT production from biomass are analysed and combined to a limited set of promising conversion concepts. A flowsheeting model with a high level of technical detail has been created. It allows for interaction of parameters as ratio, partial pressures with conversion and selectivity. The results presented here are preliminary, and still need discussion with, and feedback from the project partners. The first results indicate the importance of an undiluted syngas stream, rich in $CO+H_2$. The most promising systems consist of a pressurised oxygen fired gasifier, followed by wet or dry gas cleaning. A reformer is superfluous, and shifting is kept to a minimum, just to satisfy the stoichiometric hydrogen demand. The eventual FT diesel production costs about 16 €/GJ for a 400 MW_{th} input facility, down to 14 €/GJ at very large scale (2000 MW_{th}).

OUTLOOK

Efficiency and cost improvements are still possible, and their influence to the FT production costs will be discussed in the wrap up of this project: cheaper and less energy consuming oxygen production, tar reduction in the gasifier or application of lower temperature tar cracking, developments in catalyst formulation, and efficiency improvements from increasing scale. Finally the study will identify R&D issues essential for the commercialisation of BIG-FT technology and propose development paths towards the ultimate system, via economically attractive and technically feasible systems.

ACKNOWLEDGEMENTS

This work is partly funded by the Foundation Cooperation Sustainable Energy SDE. We would in particular like to thank Cees Daey Ouwens for his infinite enthusiasm and never ending flow of ideas.

REFERENCES

1. Tijmensen MJA, 2000, The production of Fischer Tropsch liquids and power through biomass gasification (MSc thesis), Utrecht University/Science Technology and Society, Utrecht.
2. Wilhelm DJ, Simbeck DR, Karp AD and Dickenson RL, 2001, Syngas production for gas-to-liquids applications: technologies, issues and outlook, Fuel processing technology, 71: 139-148.
3. Geerlings JJC, Wilson JH, Kramer GJ, Kuipers HPCE, Hoek A and Huisman HM, 1999, Fischer-Tropsch technology - from active site to commercial process, Applied Catalysis A: General, 186: pp 27-40.
4. Hamelinck CN, Faaij APC, Larson E and Kreutz T, 2001, Future prospects for the production of methanol and hydrogen from biomass, Science, Technology and Society/Utrecht University, Utrecht, the Netherlands, 28 pp + annexes.
5. Faaij A, Meuleman B and van Ree R, 1998, Long term perspectives of biomass integrated gasification with combined cycle technology, Netherlands agency for energy and the environment Novem, Utrecht, 93 pp + annexes.
6. van Ree R, Oudhuis A, Faaij A and Curvers A, 1995, Modelling of a biomass integrated gasifier/combined cycle (BIG/CC) system with the flowsheet simulation programme ASPEN+, Netherlands Energy Research Foundation ECN and Utrecht University/Science, Technology and Society, Petten.

7. Milne TA, Evans RJ and Abatzoglou N, 1998, Biomass gasifier 'tars': their nature, formation and conversion, report NREL/TP-570-25357, National Renewable Energy Laboratory, Golden, USA, 204.
8. Mitchell SC, 1998, Hot gas cleanup of sulphur, nitrogen, minor and trace elements., IEA Coal Research, London, 84 pp.
9. van der Laan GP, 1999, Kinetics, Selectivity and Scale Up of the Fischer-Tropsch Synthesis (Ph.D. Thesis), University of Groningen, Groningen, 238.
10. Claeys M, 1997, Selektivität, Elementarschritte und kinetische Modellierung bei der Fischer-Tropsch Synthesis (PhD thesis), Technische Hochschule Karlsruhe, Karlsruhe Germany.
11. Anderson RB, 1956, Kinetics and reaction mechanism of the Fischer-Tropsch synthesis, in: Emmett PH (Ed.), Catalysis, volume IV, Hydrocarbon synthesis, hydrogenation and cyclization, Reinhold, New York, USA.
12. Bartholomew CH, 1990, Recent technolgical developments in Fischer-Tropsch catalysis, Catalysis Letters, 7: pp 303-316.
13. Shell, 2001, Shell SMDS Website, http://www.shell.com.my/smds/.
14. Dry ME, 1981, The Fischer-Tropsch process, in: Anderson JR and Boudart M (eds.), Catalysis: Science and Technology, Springer, Berlin, Germany, pp 160-253.
15. Schulz H, 1999, Short history and present trends of Fischer-Tropsch synthesis, Applied Catalysis A: General, 186(1-2): 3-12.
16. Anderson RB, 1984, The Fischer-Tropsch Synthesis, Academic Press Inc, New York.

Tar-free, High Pressure Synthesis Gas from Biomass

E. Henrich, E. Dinjus
For the working group "syngas from biomass" at Forschungszentrum, Karlsruhe (1)
Forschungszentrum Karlsruhe, Institute for Technical Chemistry, CPV,
D-76021 Karlsruhe, Germany

ABSTRACT: A high biomass share in our future energy mix must comprise the agricultural by-products, mainly straw, in addition to wood. Straw and other fast growing herbaceous biomass contains more ash, K and Cl than wood and suitable gasification technologies are not available. A new two-step gasification process is being developed, which considers the special properties of straw and other dry, non-woody biomass. Small or thin dry biomass particles are first liquefied by fast pyrolysis in a number of regional plants. By suspending the smaller amount of pulverised, brittle pyrolysis char in the larger fraction of pyrolysis oil a *pumpable dense slurry* is prepared. The slurries contain up to 90% of the biomass energy, are easily stored in tanks and can be economically transported by rail to a large central plant. A *large* plant with complex technology for the generation and use of synthesis gas is more economic than many small facilities. Biomass is *unique as a renewable carbon resource*. Synfuel or base chemicals production has therefore priority before energy generation. The slurries are pumped into a special, highly pressurised entrained flow gasifier, compatible with much ash and fluctuating ash composition. Efficient slurry conversion is obtained at high temperature and pressure after pneumatic atomisation with oxygen. High p,T-gasification results in a practically tar-free syngas, simplifies gas cleaning and obviates expensive gas compression prior to synthesis. After methanol synthesis or synfuel production in an FT-process, the unconverted syngas and any side- or couple-products are combusted for electricity generation. Only a co-production of synfuel and electric power ensures an efficient and economic use of the chemical energy *plus* the sensible heat in the hot syngas. The selected gasifier type is flexible enough to digest also many other stable slurries from pulverised solid fuels and any organic or aqueous waste solutions, as demonstrated in the SVZ-facilities at "Schwarze Pumpe", Germany.

INTRODUCTION

Substitution of the dwindling fossil fuels by renewable energy and carbon resources is among the key challenges of the century. Development and world-wide exploitation of renewable, environmentally compatible, sufficient and affordable energy sources requires much time and money. Only the rich industrialised nations have the knowledge and the means to manage this huge task in due time.

Global warming and adverse climate changes advocate for a substitution of fossil fuels even before the proven and economically recoverable fossil reserves are exhausted. In the Kyoto protocol, the international community has agreed to limit greenhouse gas emissions, especially CO_2 from the combustion of fossil fuels. Even a temporary shortage of fossil energy supply would cause serious damage to the world economy and distorts social and international relations. Only a partial breakdown of a part of our oil supply would already be a disaster. Guarantee of an uninterrupted supply is not a simple political problem. Substitution of finite fossil fuels, reduction of greenhouse gas emissions and a reliable and sufficient future energy supply are motivating forces for a more efficient and larger use of biomass for energy (2).

SUSTAINABLE ENERGY POTENTIAL OF BIOMASS

About 200 years ago, a world population of a billion relied almost exclusively on renewable biofuels, mainly fuelwood, for the generation low and high temperature heat. After invention of the steam engine and the generator, combustion heat could also be converted into mechanical or electrical power. Biofuel can be transported and stored; the inconvenient fluctuations of wind energy and the fixation to suitable sites for hydropower are thus avoided. The risk of over-exploitation of the forests in the densely populated European countries as a result of the increasing industrialisation, could only be avoided by substituting coal for fuelwood, later on supplemented by oil and gas. Today, the fossil fuel share has increased to 75% in the world energy mix and the biofuel share has dropped to 10%; in Europe to 3+% (3-5). The poorest developing countries have high biomass shares > 50%, e.g. in many African countries south of the Sahara (4).

A total primary energy demand of ~ 20 Gtoe/a – twice the present - is expected in about 50 to 100 years for a stable world population of ~ 10 billion. What could be a sustainable biofuel share in a future energy mix is illustrated in fig. 1. At least half of the biofuel contribution will be fuelwood. The total biomass production in the world forests is about 20+ Gtoe/a, but only 20 to 25% or ~ 5 Gtoe/a are wood (5). The majority of the photosynthetic production is herbaceous biomass, which rots on-site in form of falling leaves etc.. With good forestry practice, more than half of the wood upgrowth in the European forests can be harvested sustainably without over-exploitation. Optimum forestry in the highly productive tropical rain forests is more complex. The present harvest of fuelwood on the one hand and industrial wood for timber and cellulose production on the other hand is about 0.5 and 0.4 Gtoe/a respectively (3, 5). A sustainable increase of the total wood harvest to 2 Gtoe/a of fuelwood and 1 Gtoe/a of timber might be attainable with good forestry methods. This is 3+ times the present consumption and corresponds to ~ 60% of the total wood production. A further extension towards a still higher share could be possible, but will be accompanied with a steep increase of fuelwood cost.

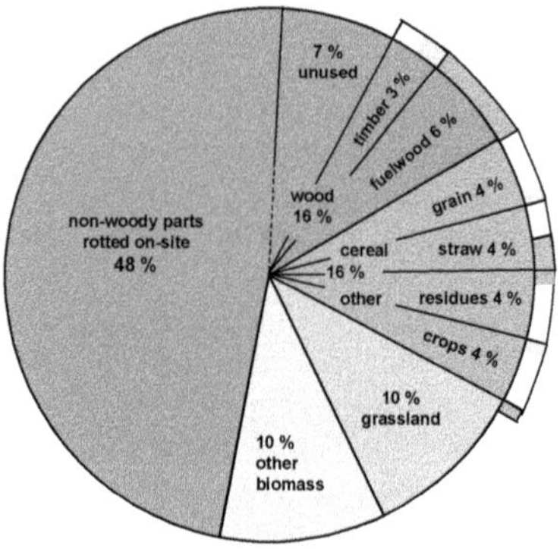

Figure 1: Total land biomass production for a 6 billion population today is ~ 30 Gtoe/a and for 10 billion in ca. 2100 ~ 32 Gtoe/a. The potential percentage share for energy generation from forestry and agriculture is indicated by the grey rim parts. Use of biomass for energy may increase from 1 Gtoe/a today to a maximum of 4 Gtoe/a in the future.

About 1/3 of the earth land area is covered with forests and only about 1/10 is arable land. With a world population growing from 6 to about 10 billion, the total agricultural food and feed biomass production will gradually increase from ~ 3 Gtoe/a today to ~ 5 Gtoe/a in the future. Agricultural food production is assumed to increase in proportion to the population and forest wood production is assumed to be constant. Thus forest wood production and total agricultural biomass production become comparable. (see fig. 1). Cereals are grown on half of the arable land. The world grain harvest amounts to 2.1 Gt/a or 0.7+ Gtoe/a; 90% are wheat, rice, maize and barley (3). The straw harvest is the same, since the average grain-to-straw ratio is about 1. About half of the straw is not used and available for energy generation: ~ 0.4 Gtoe/a today and probably ~ 0.6+ Gtoe/a in the future. Together with dry lignocellulosic by-products from other crops (e.g. sunflower stalks, legumes and rape straw, sugarcane bagasse etc. etc.) as well as some unsuited hay from grassland, this can add to a maximum of about 1 Gtoe/a.

A large diversity of biowastes complements the energetic use of fuelwood and herbaceous by-products from agriculture. Major waste streams are:

wood waste: - manufacturing waste from wooden constructions, furniture etc.
- demolition wood and used wooden articles
- used paper, cardboard (cellulose fibres) and refuse in MSW etc.
- black liquor from cellulose production

crop waste: - waste from human and animal food manufacturing (food industry)

from agriculture: - sewage sludge, animal excrements
- landfill gas (CH_4)

About half of the wood harvest which is used in the industry, is assumed to come back in form of dry wood waste during manufacture and after use of the wooden products. Waste streams from agriculture result from the preparation and use of agricultural crops. Many of these streams are wet and suited for biogas production.

The potential percentage use of ca. 32 Gtoe/a land biomass is out-lined in fig. 1. A land biomass fraction of 12% or ~ 4 Gtoe/a for energy is thought to be a sustainable upper limit. This corresponds to 20% of 20 Gtoe/a primary energy consumption, expected in 50 to 100 years. Fuelwood contributes ~ 50% of the biomass energy, dry agricultural by-products ~ 25% and organic wastes originating from timber and agricultural crops, contribute the rest of 25%. Energy crops like oil seeds for biodiesel or sugar or starch for bioethanol production play a minor role in view of the abundant lignocellulosic biomass. Energy plantations are also more expensive than agricultural by-products, which share the cultivation and harvesting cost with the crops.

PROPERTIES OF AGRICULTURAL BY-PRODUCTS

Herbaceous by-products from agriculture, mainly cereal straw and strawlike residues, are among the cheapest renewables. Contrary to wood, all fast growing plants inevitably contain more ash and heteroatoms, especially much potassium and chloride, and need more expensive fertiliser. The inorganic elements are constituents of biocatalytic systems, which are needed for the faster metabolism. Slowly growing wood contains usually < 1% ash (without bark). Straw or hay contain 5 to 10% ash, and typical K- and Cl-contents are in the 1% range (7). Wood is a relatively clean fuel and traditional technologies for wood combustion and gasification are well developed. Technologies for the use of herbaceous biomass are more complex and not well developed. The large arisings of agricultural by-products for energy correspond to ~ 5% of the world primary energy consumption. This is a significant fraction and justifies the development of a special gasifier suited for agricultural residues.

- A high K-content can lower the ash softening temperature down to < 700°C; this is below a suitable combustion or gasification temperature. The sticky ash can cause reactor slagging and agglomeration of fluidised beds.
- Chlorine is liberated as HCl and causes corrosion, poisoning of catalysts and promotes the formation of toxic polychlorinated "dioxins" and "furans" at unsuitable combustion conditions.
- K-salts like KCl, KOH etc. become volatile at > 650°C and deposition especially of liquid eutectics in downstream gas ducts can cause serious corrosion and plugging.

About 0.2 m of rain in the course of ca. 2 month in central Europe can wash out the K and Cl from the straw swaths left in the field. Delayed baling causes trouble and is undesirable. Another option is a counter-current wash of straw chops with hot water at the plant site. This is an additional technical effort (8). The straw soaks about its own weight of water and must be dried again. Also ~ 2 m^3 of a ~ 1% KCl-waste solution are generated per tonne of straw. After pyrolysis, K and Cl can also be leached from the much smaller amount of pyrolysis char (8). Without all these measures, a large part of K and almost all the Cl will be released into the fluegas or syngas. The latter option is being studied in more detail.

PROCESS SELECTION

Before a rational process can be selected or designed, educts and desired products must be fixed. This is visualised in fig. 2. Concerning the educts, the focus is on herbaceous biomass residues from agriculture, mainly cereal straw, hay or other strawlike products with a high ash, K- and Cl-content. Biomass is the only renewable carbon resource and the use of this unique property has therefore the highest priority. The desired products are organic chemicals and/or high quality synfuel. Since the sustainable amount of 4 Gtoe/a biomass far exceeds the carbon demand of the organic chemical industry, CHO- or CH-synfuels are the products of choice for process design. But an extrapolation of the present world consumption of liquid transport fuel exceeds the available biomass potential.

Figure 2: Aim of the conversion of poor herbaceous biomass via syngas is an economic co-production of valuable products: chemicals, synfuel and electric power.

Synthesis gas, a mixture of CO and H_2, is produced by biomass gasification. Gasification is more flexible, efficient and environmentally compatible than direct combustion. Syngas volumes are much smaller than flue gas volumes from combustion. The easier syngas cleaning procedures are especially helpful for the "dirty" herbaceous biomass.

"Clean" wood in form of logs, chips or pellets is better suited for small residential heating systems, without expensive measures for environmental protection. In medium-sized district heating- and CHP-plants in Denmark a fraction of 1+ Mt/a straw from a total harvest of 5 to 6 Mt/a is combusted (9).

The various synthesis processes have common characteristics: (1.) high operating pressure, (2.) low inert gas content to achieve a high product yield (3.) high syngas purity to prevent catalyst poisoning. This has consequences for the gasifier type and the desired operating conditions. Incomplete conversion and tar production can be avoided at high gasification temperatures > 1000°C. High pressures – above the downstream synthesis pressure – also promote conversion and obviate expensive syngas compression downstream prior to synthesis. The combination of high temperature and high pressure ensures complete gasification in few seconds for pulverised solid fuels, atomised liquids and fuel gases. Various types of pressurised entrained flow gasifiers are suited in principle.

At high gasification temperatures, a fast fuel conversion and a tar-free syngas are obtained at the expense of a large percentage of 20 to 30% of the initial biomass energy present as sensible heat in the hot syngas. Sensible heat is usually less efficiently used than the chemical energy in the combustible CO and H_2 (10). Efficient use of both energy forms is made in IGCC-power plants for electricity generation. If synthesis of CH_3OH, FT-diesel or other organics has priority, an efficient energy use must also comprise the use of sensible heat. The proposed concept is thus the insertion of a synthesis process before electricity generation in a BIG-CC facility: **b**iomass **i**ntegrated **g**asification – **synthesis** – **c**ombined **c**ycle. Similar concepts (GTL = gas to liquid) are proposed for new natural gas power stations.

The consumption of transportation fuel in the rich industrialised countries amounts to about 1 m^3 per capita and year. The large demand is satisfied by large refineries with capacities of several million tons per year. Large plant capacities of the order of a million or more, would also be desirable for the production of biomass derived transport fuels. Many small facilities are more expensive and less competitive.

The energy in 1 Mt diesel fuel for example is equivalent to ~ 3 Mt airdry (15% H_2O) straw. In rural European regions, this plant capacity corresponds to half of the straw harvest from ~ 60000 km^2 or a radius of ca. 140 km. Straw bales have a low bulk density of ~ 100 kg/m^3. Delivery by the local farmers may be acceptable over a maximum distance of 20 to 30 km. A radius of 25 km e.g. corresponds to 2000 km^2 area and a plant capacity of ~ 100 kt/a. In these regional facilities, a simple preparation step should convert the biomass to a compact liquid, which can be stored, transported and handled much more easily. A liquid is also a very suitable feed for an entrained flow gasifier, which must operate at very high pressure.

Fast or flash pyrolysis (11-14) is a relatively simple method, to convert about half of a lignocellulosic biomass or even more into a pyrolysis liquid (raw tar, bio-oil). Char and gas yields are low. The combustible gas can cover the energy requirements of the pyrolysis process. If there is any surplus of pyrolysis gas, it can serve for electricity generation in downstream turbines or internal combustion engines. The brittle pyrolysis char is pulverised and suspended in the oil, to produce a rather dense but still pumpable slurry, which can be transported and stored in tanks. If the pyrolysis oil yield is low and some char will be left, the small surplus can be suspended in water or any other contaminated aqueous or organic waste liquor.

As sketched in fig. 3, the slurry from 20 to 40 regional fast pyrolysis facilities can be transported by rail vessel wagons to a large central plant for syngas production and use. The large facility must always work near full capacity for economic reasons. A large entrained flow gasifier is a guzzler, which is able to digest many different feeds. A gasifier design for a wide spectrum of feedstocks including diverse pumpable waste streams, coal/water slurries and also dry coal dust – which can be conveyed pneumatically with high density from a prefluidised bed – helps to smooth out insufficient biomass-slurry supply, important in the initial stages until routine plant operation is established.

In the slurries from pyrolysis oil plus char, up to 90% of the initial bioenergy can be made available for gasification. Any technical efforts to purify or upgrade the oil are obsolete. On the contrary, the ash still present in the pyrolysis char helps to generate a protecting slag layer in the specially selected gasifier type.

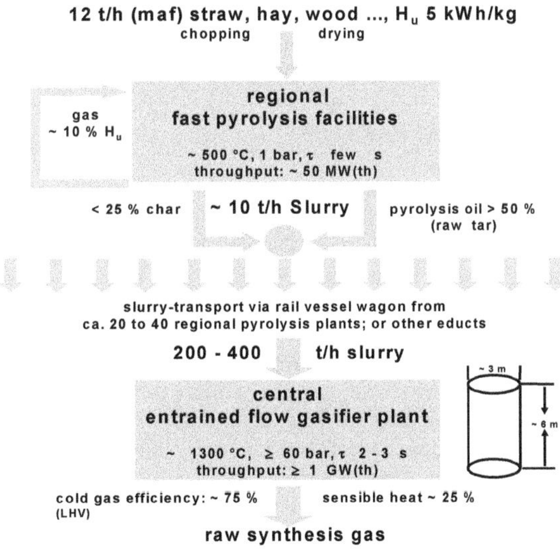

Figure 3: The main characteristic of the proposed concept is fast pyrolysis in many regional plants; slurry transport by rail followed by syngas production and use in few central plants. Extrapolated dimensions of the gasification chamber of a very large 1 GW(th) entrained flow gasifier at 60 bar are ~ 3 m diameter and 6 m height.

PROCESS DESCRIPTION

An application of the fundamental considerations discussed in the previous chapter and some more practical aspects are consistent with the process concept proposed in fig. 4. The multistep process for the conversion of difficult biofuels with high ash K- and Cl-content into synfuel and electricity must be considered as a whole, due to the many mutual interactions between the successive process steps. In the following, "straw" is used as a synonym for all thin-walled and fast drying, non-woody biomass residues from agriculture.

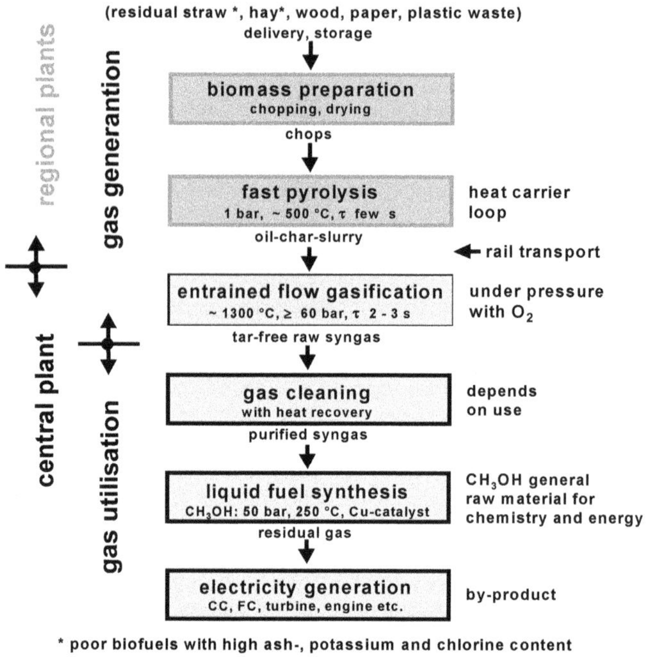

Figure 4: Total process train for the generation and use of syngas from dry lignocellulosic biomass: regional oil/char-slurry production by fast pyrolysis, slurry gasification at high p and T in a large central gasifier, co-production of synfuel and electricity from the hot syngas.

BIOMASS PREPARATION

In-time straw delivery to an intermediate storage facility at the pyrolysis plant must be fixed by contracts with the local farmers, as is known from Danish district heating plants. Large ca. 0.5 tonne square bales with ~ 120 kg/m^3 bulk density are the preferred feedstock for compact transport and storage. A maximum delivery radius of 25 km comprises about 2000 km^2. In rural European regions the unused half of the straw harvest then amounts to ~ 100 000 t/a. This capacity for a regional pyrolysis plant results in a throughput of 12 to 15 t/h of airdry straw (15% H_2O, LHV 4 kWh/kg) or 50 to 60 MW(th) power.

Straw is dried and chopped. Further size reduction is achieved in a hammer mill (28). Especially the stem nodes, which contribute ~ 5% of the weight, must be smashed to ensure a fast heatup during pyrolysis. Moderate size reduction is probably sufficient, since the wall thickness of the straw chops does not change.

FAST PYROLYSIS

The principles of fast pyrolysis are explained in the literature (12-14) and the internet pyne.co.uk. Fast pyrolysis is a special pyrolysis mode aiming at high yields of liquid pyrolysis oil and little char and gas. With an optimum lignocellulosic biomass feedstock and reactor type and operating conditions, liquid yields up to about 75% can be obtained. For the preparation of a suitable slurry for gasification, an extreme optimisation of the fast pyrolysis process is not needed. As explained later in the slurry preparation paragraph, pumpable slurries can be prepared with up to ~ 1/3 char weight, depending on the char properties. A lower liquor yield of 50% and a higher char yield of

25% wt e.g. will be compatible with such requirements. More-over, little surplus char can be suspended in water or – better – aqueous or organic waste solutions from other sources.

The *energy in the pyrolysis gas* is usually in the range of 10% of the initial biomass energy and can be used to supply the energy for the pyrolysis process. If the pyrolysis gas production exceeds the desired low yields, the surplus can be combusted at the pyrolysis plant site in turbines or internal combustion engines for electricity generation. This gases are sufficiently clean, because they have passed a scrubber with cooled pyrolysis oil for quench condensation of the oil vapours. Remaining condensation mist can be removed with an electrofilter. Higher pyrolysis gas yields reduce the energy fraction recovered in the slurry.

Another important aspect is the *purity of the pyrolysis oil.* A relatively high purity and especially an ash free liquor is needed to fuel stationary turbines or engines. The necessary purity is obtained by restrictive production conditions, additional cleaning and sometimes expensive catalysed upgrading procedures. For a gasifier slurry, there are no such purity restrictions. On the contrary, the ash in the char particles is required to produce a molten slag layer, which protects the hot inner gasifier wall from corrosion. Also carry-over of attrited sand from a fluidised pyrolysis bed can be tolerated. The conclusion is, that lower purity and yield requirements permit a simplification of the fast pyrolysis process, which contributes to reliability, availability and economy.

The various reactor types which are being investigated for fast pyrolysis are depicted in fig. 5. Choice of a suitable type based on an analysis of reaction principles (10) is briefly summarised. Usually heat transfer is the rate determining step, not the decomposition reaction. Most reactor types use a solid heat carrier, usually sand, which ensures a fast heat transfer. For dry lignocellulosic biomass, there is an exothermal decomposition section starting somewhat below 300°C and inducing a quick temperature rise to ~ 450°C (29). Since most of the pyrolysis oil is released in this temperature range, also a *slow* pyrolysis under special conditions can exhibit some characteristics of a fast pyrolysis process.

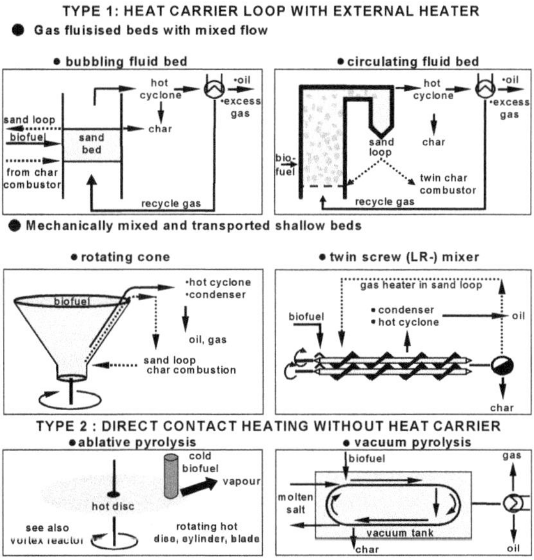

Figure 5: The main reactor types, which are being investigated for fast pyrolysis. The most common types are the gas fluidised beds.

The pyrolysis of a small particle has a definite end. In a bubbling or circulating fluid bed, the residence time spectrum of solids is similar to a CSTR and about exponential. For a desired degree of conversion, this results in a larger residence time and therefore a larger reactor volume compared with a co-current flow and low axial mixing. The larger bed sand volume promotes undesired decomposition reactions of organic pyrolysis oil vapours at the sand surface.

Most fluidised beds are fluidised with a gas, usually with recycled and cooled pyrolysis gas from downstream of the oil condenser. During a quench condensation for pyrolysis oil recovery, only a poor heat recovery can be realised and most of the sensible heat is therefore lost. Since the fluidising gas constitutes a considerable part of the vapour and gas mass flow, the loss of sensible heat will be substantially increased. Much non-condensable gas retards also vapour condensation.

In the rotating cone (15) and the twin screw mixer reactor (16) the fluidisation is obtained by mechanical means. The biofuel particles are mixed with an excess of hot sand at 500+°C and are transported for several seconds in co-current flow with low axial and good radial mixing, which promotes heat transfer. The vapours are quickly removed in cross-current flow from the shallow sand bed. Short sand surface contact helps to reduce vapour decomposition. A small pyrolysis reactor and small sand bed volume is desirable for quick vapour removal. The consequence is, that there is not sufficient space to install a large heat transfer surface to reheat the sand. Therefore the hot sand must be recycled in a closed loop via a larger external sand heater.

In the rotating cone technology the sand is heated by combusting the pyrolysis char particles with air directly in a separate fluid sand bed. The ash is released from the char particles after complete combustion and must be removed from the sand loop. Combustion temperatures must be kept low enough, to prevent agglomeration of the bed sand with sticky ash. For a biomass with much ash like straw, the effort of sand/ash-separation is replaced by a sand/char particle separation. Thus the char can additionally contribute a considerable part to the energy in the slurry The ash is still confined in the char particles, but ash is required anyway for slag formation in the gasifier.

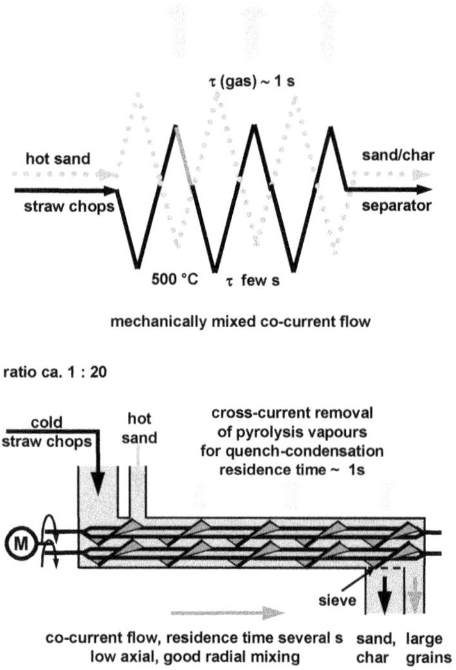

Figure 6: The twin-screw- or LR-mixer reactor has two screws with inter-twining flights to clean each other. There is technical experience for the fast pyrolysis of various refinery products for more than 40 years.

The twin screw or LR-(Lurgi Ruhrgas) mixer reactor in fig. 6 is being applied on the technical scale since more than 40 years for the fast pyrolysis of various refinery products, especially distillation residues. In form of the so-called sand cracker, the reactor has been applied for olefin production by fast pyrolysis of naphtha. The reactor consists of two co-rotating screws with intertwining flights for automatic self-cleaning to ensure reliable operation also with less convenient feedstocks. If the planned experiments with various biomass feedstocks like strawchops or sawdust turn out to be successful, the principal scale-up problems are already solved. A transfer of fast pyrolysis technology into the technical scale for practical demonstration should then be possible in relatively short time.

ENTRAINED FLOW GASIFICATION

The choice of the special gasifier (17, 18) shown in fig. 7 is based on an analysis of available gasification technology (19, 22, 30)). Pressurised entrained flow gasifiers fired with pulverised coal or petcoke are in operation in various IGCC demonstration plants. Gasification temperatures are usually > 1200°C and pressures > 15 bar. The very hot syngas is practically tar-free, but at the expense of 20 to 30% of the slurry energy converted into the less valuable sensible heat of the hot syngas. A partial reconversion into chemical energy by chemical quenching with char powder down to 900 or 1000°C has been proposed (20); it is not clear if the benefit justifies the technical efforts (10).

The high gasification temperature allows almost complete fuel conversion and equilibration of the gaseous products. Crude material and energy balances can therefore be derived from the known feed composition and standard thermodynamic data. The following reaction equation for the gasification of dry lignocellulose is an example. Since the composition of pyrolysis oil/char slurries is not much different from the initial lignocellulose, the reaction equation is also a crude representation of slurry gasification:

dry lignocellulose $\lambda = 0.43$ technical oxygen

$(C_6H_9O_4 + \text{ash, heteroatoms}) + 2.7\,(O_2 + 0.1\,N_2) \rightarrow 1.2\,CO_2 + 4.8\,CO + 2.3\,H_2 + 0.3\,N_2 + 2.2\,H_2O + \text{slag};$

percentage composition of dry syngas:	CO_2	CO	H_2	N_2	
	14	56	27	3	% volume

The equation shows, that only O_2, no water vapour is needed for gasification. The required O_2 amounts to about half of the slurry weight. Since the cost of straw and technical oxygen are similar, use of O_2 instead of air makes a significant contribution to the total manufacturing cost, which is compensated for by other improvements and simplifications.

The gasifier type in fig. 7 has been developed 25 years ago by Deutsches Brennstoff Institut (DBI), Freiberg, Saxony. A 130 MW(th) gasifier of this type is being operated successfully in the SVZ-facilities at "Schwarze Pumpe", Germany for > 15 years with various feeds, including waste slurries and a special brown coal with high NaCl contamination (23). The high KCl content of straw poses similar problems.

The special characteristic of this gasifier type is a thin water-cooled radiation screen inside a cooler steel shell designed for high pressure. A molten slag layer drains down on the inner screen walls and protects the SiC-liner from corrosion. A suitable slag viscosity is adjusted with the O_2-flow (temperature) and – if desirable – suitable inorganic additives in the feed to modify the slag composition. The screen is responsible for the particular flexibility of this gasifier type towards fluctuating ash compositions and larger amounts of ash expected in the large spectrum of herbaceous biomass species.

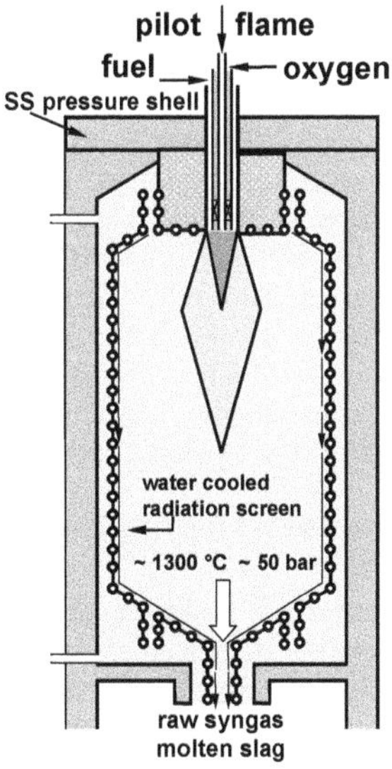

Figure 7: Slagging entrained flow gasifier for operation at high temperature and pressure. Characteristics of this special type is a thin radiation screen with SiC-liner, cooled with pressurised water to keep the pressure resistant steel shell at low temperatures.

GAS CLEANING

Gas cleaning is tuned to the downstream synthesis. The operating pressure corresponds to the gasifier pressure without intermediate compression and the temperature decreases from ~ 1300°C at the gasifier exit to the lower synthesis temperature, e.g. ca. 250°C for CH_3OH. Recovery of the sensible heat from the hot syngas – typically 20+% of the initial biomass energy – must proceed in advance or be integrated into the gas cleaning train. This is essential for a high energy efficiency. Methods applied in IGCC-plants and chemical quenching (20) for the high temperature section etc. are taken into consideration. A nasty problem is the presence of sticky ash particles.

The highly selective synthesis catalysts are rather sensitive towards a number of poisons. The sensitivity towards traces of tar, heavy metals, S- or Cl-compounds etc. extends down to the ppb-level. Raw syngas purification in combination with efficient heat recovery requires a rather complex and expensive sequence of operations. Corresponding technologies and technical experience are available since 60 years e.g. for 0.7 Mt/a synfuel production from coal with the FT-process in Germany during world war 2 and with 10 times larger capacities of > 5 Mt/a at Sasol, South Africa, still in operation today (24, 25). Adjustment of these sophisticated technologies for higher gasifier pressure, biomass feedstocks and other syngas applications is necessary.

In addition to cleaning and heat recovery, also an adjustment of the H_2/CO-ratio is performed at lower temperatures 200 to 400°C via the catalysed shift reaction towards the end of the gas cleaning train: $CO + H_2O \rightarrow CO_2 + H_2$. A complete shift to H_2 for NH_3-synthesis or H_2 use in low temperature fuel cells is possible and practised in some NH_3-production plants with coal-derived syngas.

SYNTHESIS OF LIQUID FUEL

Compared to pyrolysis, only few large synfuel production plants are needed. If they can be integrated into an industrial chemistry plant complex, a more flexible syngas use for various chemicals is possible. Usually, only methanol-production and CH-synfuel production in the Fischer-Tropsch (FT) process are considered, because they confine to syngas components CO and H_2 and do not need additional chemicals. For the two products an optimum H_2/CO-ratio of 2+ must be adjusted with the help of the catalysed shift reaction.

Low pressure methanol synthesis at ~ 50 bar and ~ 250°C proceeds with high selectivity > 99% at sensitive Cu-catalysts (ref. 21, vol. 4, p. 1856). CH_3OH can be used in many different ways: (1.) as direct motor fuel or as an additive, (2.) as compact and convenient storage medium for H_2, which can be generated in the catalysed reforming reaction with water: $CH_3OH + H_2O \rightarrow CO_3 + 3\ H_2$; (3.) Zeolite-catalysed conversion into CH-synfuel in the MTG process (ref. 21, vol. 4, p. 1894); (4.) Zeolite-catalysed conversion into olefins or aromats in the MTO- and MTA-processes. These are base chemicals for the organic chemical industry; methanol synthesis is used here as reference process. High operating pressures in the gasifier above 50 bar are therefore basis of the process design.

The available infrastructure for conventional transport fuels can be used after CH-synfuel production in the Fischer-Tropsch-(FT)-process. The variants of the FT-process are conducted at 200 to 300+°C and 20 to 40 bar with Fe,Co-catalysts (ref. 21, vol. 4, p. 1876). Compared to CH_3OH-synthesis, a higher yield can be obtained in a single pass through the catalyst bed for thermodynamic reasons. In combination with a downstream co-production of electricity from the unconverted syngas, a single pass operation is a desirable option for simplicity and economy reasons.

ELECTRICITY GENERATION

An economic synthesis step can result in ≥ 10% of unconverted syngas for downstream electricity generation. In large central plants this corresponds to ca. 100 MW(th) power or more, a power range suited for combined cycle (CC) systems. The residual gas should be clean enough for a turbine, since the gas purity requirements for the sensitive catalyst upstream are much higher.

CURRENT WORK

Various aspects of the process concept are not reliably covered in the literature and are therefore being studied theoretically and experimentally. The experimental work is accompanied and complemented by modelling of the various processes and a continuous systematic analysis of the technology and the market situation (32). A number of institutions at Forschungszentrum Karlsruhe are involved (26). Up to now, the work has concentrated on syngas generation; the aspects of syngas use will also be covered in the future, together with external collaborators. Similar concepts for biooil and oil/char-slurry gasification in general combined with FT-liquid production are also studied in the Netherlands (30, 31). The following is confined to three major experimental points.

PYROLYSIS

Pyrolysis kinetics are an important basic for reactor design. To guarantee a fast heatup of small thin dry biomass particles to 500 °c in an excess of fluidised hot sand, the characteristic length L = volume/surface-ratio of the particles should be < 0.5 mm better < 0.25 mm. Fibrous cylindrical parts should have a diameter below 1 - 2 mm and flat walls should be only 0.5 - 1 mm thick. Single-walled straw chop sections are equivalent to flat plates and have a characteristic length < 0. 25 mm (thickness 0. 5 mm), the same is true for hay, dry leaves etc.

Dry single-walled straw chops with 10 - 100 mg mass have been plunged from cold lock onto a large moving hot sand bed at the bottom of a vessel with a known volume. In the 0.4 mm size sand with a bulk density of 1500 kg/m^3, the thin chops are rapidly heated until at a center plane temperature of about 300 °C is attained. Then, exothermal pyrolysis reactions induce a fast temperature rise towards the desired 500 °C, accompanied by a fast release of vapours and gases. At 500 °C the volume of pyrolysis vapours and gases amounts to about 0.5 m^3(STP) per kg. Fig. 8 shows, that the vapour release from single-walled chops takes only 1 s; the pyrolysis time for the thick stem nodes is much longer, smashing is therefore an essential preparation step. The measured data are consistent with the theory of unsteady-state heating of solid particles, using the Fourier-number $Fo = \alpha \cdot t/L^2$ with $\alpha = k/(d \cdot C)$ and the Biot number $Bi = h \cdot L/k$ and the following material properties: thermal conductivity k = 0.1 =.3 J/s·m·K, specific heat C = 1800 J/kg·K; φ = 1000 kg/m^3, heat transfer coefficient h ca 100 - 200 J/m^2.

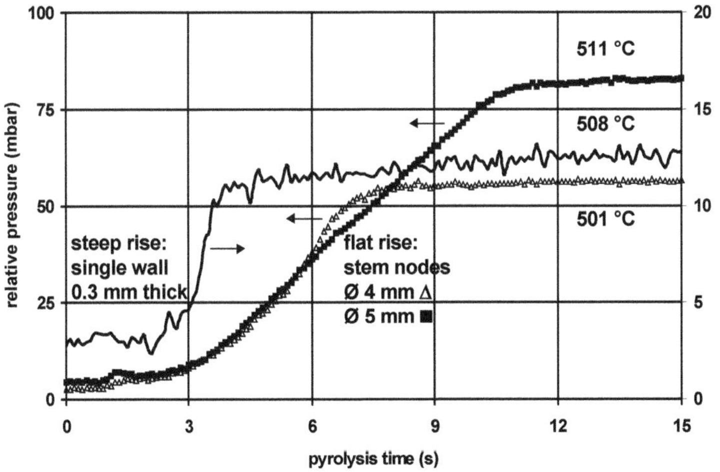

Figure 8: Pyrolysis kinetics of straw chops on a moving sand bed at 500°C. Thin walls are pyrolysed in ~ 1 s; thick stem nodes need order of 10 s and must smashed.

Some design data for a technical 4 kg/s (15 Te/h) twin screw pyrolysis reactor with < 10 s sand residence time can be derived from the data. A large flow of 40 - 80 kg/s of hot sand results in a pyrolysis bed volume of only ~ 0.5 m^3 sand with chops and char. About 6 - 8 m^3/s vapour and gas are released. To prevent carryover of bed sand particles at higher superficial velocities, the bed surface area should be > 4 m^2. this results in a shallow bed height < .2 m and a sufficiently short vapour contact time to reduce decomposition. The volume in the gas phase including the char cyclone upstream from the quench condenser should be kept at < 8 m^3.

SLURRY PREPARATION

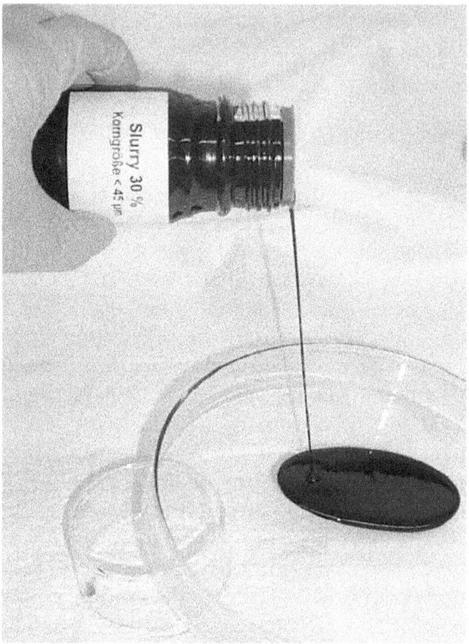

Figure 9: Pumpable slurries can be prepared by mixing pyrolysis oil and up to 33%wt. Powdered pyrolysis char, depending on the char properties. The room temperature viscosity of the highly loaded slurries is several Pas.

The aim is a still free-flowing and pumpable slurry of pyrolysis oil with a high content of pulverised pyrolysis char. Therefore an optimisation of the pyrolysis process with high liquor and low char yields is not necessary. From rheology (36) it is known that freeflowing slurries with spheric particles can be prepared up to a critical volume fraction Φ_{crit} of ~ 60 %; for regular particles with an aspect ratio of ~ 1 Φ_{crit} is ~ 50 %; for needles, platelets etc. Φ_{crit} can be below 0.2. a broad particle size spectrum is more favourable than a narrow one. The brittle pyrolysis chars from biomass are easily pulverised with a broad size spectrum and sauspended in the pyrolysis oil by simple agitation. But the high porosity of the pyrolysis chars, typically between 50 and 75 %, prevents a high loading. The porous particles first soak a lot of the pyrolysis oil, until a sufficient volume is left outside the particles, to serve as a lubricant. The expected true char volume fraction is thus ~ 25 % or even below.

Most experiments have been performed with pyrolysis oil (= raw tar) from commercial charcoal production from beechwood (33): raw tar density of ~ 1200 kg/m^3. the true density of the pulverised char has been 1500 - 1600 kg/m^3. therefore a maximum char particle volume fraction of ~ 50 % corresponds to a weight fraction of ~ 1/3. Still free-flowing oil/char slurries with a maximum char weight of 30 and 33 % have been prepared: Density ~ 1300 kg/m^3. the viscosity at room temperature of several Pas - pyrolysis oil has ~ 0.1 Pas - reduces at 80 °C by more than an order of magnitude; a low viscosity < 0.3 Pas eases pneumatic slurry atomisation in the gasifier. LHV is 6+ kWh/kg, at lower values ≤ 3 kWh/kg a safe self-sustained entrained flow gasification even with O_2 is no more possible. The highly concentrated slurries have been stored over month at room temperature without a stabiliser. Viscosity and other properties did not remarkably change; in the highly concentrated slurries even sedimentation was negligible and they remain pumpable.

SLURRY GASIFICATION

The essential new step in the concept is the atomisation and gasification of highly concentraled slurries with pure pressurised oxygen. The CHO-composition resembles dry lignocellulose with a formal "H_2O"-content of about half. Pressurised O_2 without any water vapour is therefore a suitable gasification agent, which simultaneously serves for pneumatic slurry atomisation. The practical suitability was studied with a 5+ MW(th) pressurised entrained flow gasifier in the Babcock Borsig Power pilot facilities, previously Noell Company, at Freiberg, Germany (see fig. 10) (35).

Slurry preparation: Slurries have been prepared from the easily in larger amounts available by-products of beechwood pyrolysis for commercial charcoal production (33). The pyrolysis oil is termed raw wood tar, has a density of 1184 kg/m^3, a room temperature viscosity of 0.16 Pas and a LHV of 19 MJ/kg. Charcoal dust, LHV 31 MJ/kg, has been further pulverised to <50 μm. Pyrolysis oil in a larger storage tank was recycled via a 1m^3 PE-vessel equipped with a heater and a stirrer. Char powder was slowly added to the stirrer until the desired 20, 23 and 26% wt content was attained; suspension was maintained by continuous circulation. To simulate agricultural residues with high ash content 3% wt of straw ash have been added (district heating plant; Schkölen) and also 0.3% KCI to readjust some Cl-losses. *Slurry properties* are: LHV 21 - 22 MJ/kg, density 1250+ kg/m^3; the viscosity at room temperature 2 - 5 Pas is reduced by an order of magnitude at ~ 50°C temperature rise.

Figure 10: Top of the 5 MW(th) pilot scale pressurised entrained flow gasifier at Babcock Borsig Power (BBP), Freiberg, Saxony.

Gasification conditions: A constant flow of 0.35 te/h (2 MW) warm slurry to the pressurised 26 bar gasifier chamber was maintained with a screw pump (Netzsch company). The slurry was atomised pneumatically with pressurised pure O_2 in a special nozzle. A 0,6 MW(th) pilot flame with natural gas was kept for safety reasons and to compensate for the heat loss in the radiation screen, which becomes negligible in large gasifiers.

Results: With an O_2-stoichiometry of λ=0.4 - 0.5 a gasification temperature of 1300 - 1600°C has been obtained for a smooth slag flow. In large gasifiers with lower surface-to-volume ratio, λ can be reduced. Raw syngas and snolten slag were quenched by water injection directly at the gasification chamber outlet. The *composition of the dry raw syngas* at stationary operation is as expected: practically tarfree with < 0,1 %vol. CH_4; two examples are:

		CO	H_2	CO_2	N_2	CH_4	H_2S	bar
$\lambda(O_2)$ 0.46	%vol.	45	29	18	9	<0,1	29ppmv	-
$\lambda(O_2)$ 0.42	%vol.	47	27	16	9	<0,1	17ppmv	-

In > 100 MW gasifiers, N_2 will be only few percent. Based on the carbon content in the slag and the quench water, which were sampled periodically, the carbon conversion has always been > 99% and ≥ 99.9% at the highest temperatures.

ECONOMIC ASPECTS

A crude rather simplified estimate of the synfuel or electricity manufacturing cost from biomass can be made with the following equation; the basis is explained in (...).

$$\frac{\textbf{cost}}{\textbf{unit}} = \frac{\textbf{1}}{\textbf{unit}}\left[\mathbf{0.2 \cdot A \cdot (S/S_{ref})^{0.7} + P \cdot N \cdot (S/S_{ref})^{0.25} + C \cdot (S/S_{ref})^{1}}\right];$$

An average cost of 60 € per tonne of airdry straw includes transport to a local pyrolysis plant and is equivalent to a crude oil price of 25 €/bbl on a LHV basis. High biomass costs are typical for industrialised countries, in many developing countries biomass is cheaper than oil. The cost of technical grade oxygen ≥ 90 % from a rented on-site production plant is about 8 c€/Nm^3 or 60 €/te and comparable to the straw cost. The amount of O_2 required in the gasifier amounts to about half of the slurry weight or 40 % of the initial straw weight, which corresponds to ~ 24 €/te airdry straw. Slurry transport by rail vessel wagons over a distance of 250 km coa´sts about 16±4 €/te or 12±3 € per tonne airdry straw and assumes, that the pyrolysis plant is connected with the rail net.

The cost for straw or wood, oxygen and transport is almost independent of the plant size and amounts to a total C ~ 96 €/te (see equation) of airdry straw.

The manufacturing costs for the slurries and the synfuel depend on the plant size. Economies of scale are represented by the cost degression exponent of 0.7 for the pilot plant and 0.25 for the personnel respectively. A is the specific plant with a reference size S_{ref} = 50 MW(th) and 500 €/kW(th) for a 1GW(th) gasification and synthesis plant (BIG-FT-CC). Personnel costs of 50 k€ per person and year are assumed and a number N = 36 and 200 of total personnel for the reference plants. Though it is not justified to extrapolate cost estimates over several orders of magnitude plant size since the optimum technology and the equation parameters will change, the approach useful to reveal the dominant cost contributors: personnel for very small plants, investment and operating cost for medium size. Only at a large plant size of ~ 1GW(th) or more, the size independent biofuel (and O_2) costs become the dominant factor and further technology improvement are less important.

OUTLOOK

The various chemical, engineering and economic aspects of the process are being studied by an interdisciplinary working team "syngas from biomass" (1). A process development unit (PDU) for continuous fast pyrolysis with 10 kg/h dry lignocellulose in form of straw chops, hay sawdust, paper/cardboard chips etc. is under construction. The simplified flowsheet is outlined in fig. 11. Central part is a hot sand loop with a twin screw reactor and a sand heater. Various feedstocks, operating conditions but also different component types and design options will be tested and compared. A choice between mechanical and pneumatic sand transport must be based on practical experience. The same is true for the separation of sand and char. The behaviour and separation especially of K and Cl from the process streams will be evaluated. When the twin screw pyrolyser turns out to be suited for fast pyrolysis, a larger > 1 MW(th) pilot facility will be built together with an experienced engineering company, to determine the desired design data for a larger demonstration plant. In view of the available industrial experience, no serious scale-up problems are expected.

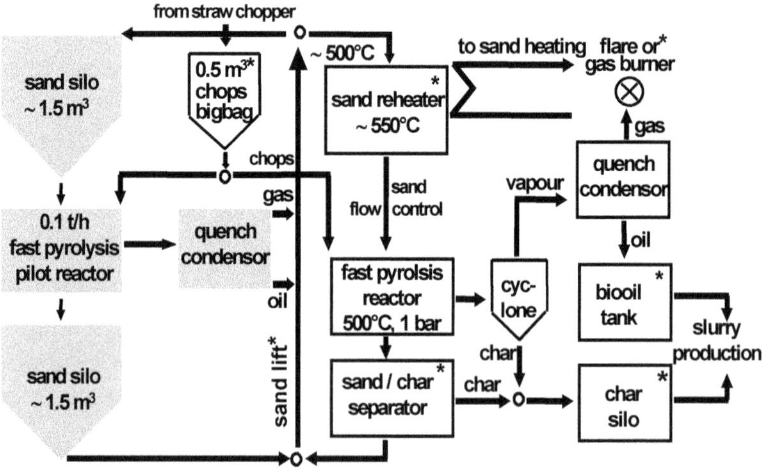

Figure 11: Simplified scheme of the fast pyrolysis process development unit for 10 kg/h straw chop or sawdust under construction at Forschungszentrum Karlsruhe.

Suitable methods will be worked out for the preparation of oil/char slurries. Slurry properties in view to pumping, storage, atomisation and stability will be investigated.

The experimental campaigns at the pressurised (~ 30 bar) entrained flow pilot gasifier in Freiberg will be continued. Various types of biomass slurries but also other fuels including waste slurries and pulverised coal will be tested, to validate a broad feedstock spectrum for a flexible and economic full load operation of a large gasifier. If sufficient experience has been accumulated after few years of operation, an improved pilot gasifier for high pressure of 50 to 100 bar will be designed, built and operated.

This development work will be complemented by a *process development unit for syngas use* with integrated operation of syngas generation, gas cleaning and conditioning, synthesis and eventual electricity generation which comprises all relevant process steps and is able to operate with the real feedstocks. The experimental work will be continuously accompanied by an updated market and technology analysis. A technical scale demonstration of biomass slurry gasification including methanol production and a CC-system for electric power production can in principle be performed in the 130 MW(th) entrained flow gasifier at "Schwarze Pumpe", Saxony, Germany; if sufficient pyrolysis slurries are made available.

ACKNOWLEDGEMENTS

We thank the Ministerium für Ernährung und Ländlichen Raum Baden-Württemberg for financial support.

REFERENCES

1. Working group "syngas from biomass" at Forschungszentrum Karlsruhe; F. Arendt, N. Boukis, W. Breitung, D. G. Cacuci, E. Dinjus, H. Ederer, E. Gasafi, V. Heinzel, E. Henrich, A. Hornung, R. Kirch, L. Krebs, S. Lange, H. Leibold, L. Leible, L. Malcher, L. Meyer, E. Nieke, U. Santo, T. Schulenberg, H. Seifert, J. Starflinger, H. J. Wiemer.
2. EU Greenbook 2000;
3. Der Fischer Weltalmanach 2002", Herausgeber M. von Baratta, Fischer Taschenbuch Verlag Mannheim 2002
4. IEA, „World Energy Outlook, 2001 Insights", OECD/IEA, Paris 2001
5. T. Kojima, „The carbon dioxide problem", Gordon and Breach Science Publishers 1998
6. R. Meyer, J. Börner; Büro für Technikfolgenabschätzung beim Deutschen Bundestag, TAB, Arbeitsbericht Nr. 73, November 2001
7. M. Kaltschmitt, H. Hartmann (Hrsg.); „Energie aus Biomasse", Springer, Berlin 2001, Kap 10, S. 427 ff
8. N. O. Knudsen, P. A. Jensen, B. Sander, K. Dam-Johansen; 10th EU-Conf. Biomass for Energy and Industry, Würzburg, Germany 1998, Proceedings p. 224
9. M. G. Larsen; Proc. 10th EU-Conf. Biomass for Energy and Industry, Würzburg, Germany 1998, Proceedings p. 1196
10. E. Henrich, E. Dinjus, D. Meier; DGMK Tagungsbericht 2002-2, S. 95
11. www.pyne.co.uk;
12. A.Bridgwater et al.; "Fast Pyrolysis of Biomass", a Handbook; cpl-press, UK 1999;
13. M. Kaltschmitt, A. V. Bridgwater (eds.); "Biomass Gasification and Pyrolysis", cpl-press, Newbury 1997
14. A.Bridgwater, C. L. Humphreys, K. Dowden; 10th EU-Conf. Biomass for Energy and Industry, Würzburg, Germany 1998, Proceedings p. 1571
15. M. Wagenaar; „The Rotating Cone Reactor", PhD-thesis, Twente-University (NL). 1994
16. H. Weis, V. F. Pagel; 16th World Petroleum Congress, Calgary (CA), 2000

17. Carl, P. Fritz (eds.); "Noell Konversionverfahren zur Verwertung und Entsorgung von Abfällen", EF-Verlag;Berlin 1997
18. Schingnitz; Chemie-Ingenieur Technik (74) S. 776, 7/2002
19. E. Henrich, F. Weirich; Proc. IT3 ′02 Conf., May 13-17, 2002, New Orleans, LA, USA;
20. B. Wolf, B. Meyer; DGMK Tagungsbericht 2000-1, p. 205
21. G. Ertl, H. Knözinger, J. Weitkamp (eds.): "Handbook of Heterogeneous Catalysis", vol. 1-5
22. M. Kaltschmitt, C. Roesch, L. Dinkelbach (eds.); "Biomass Gasification in Europe", AIR3-CT-94-2284, EUR 18 224 EN, October 1998;
23. W. Seifert, B. Buttker; DGMK Tagungsbericht 2000-1, p. 169-197;
24. W. Teltschik; "Geschichte der deutschen Grosschemie", VCH Weinheim 1992
25. W. Theimer: "Öl und Gas aus Kohle", DTV-Verlag, München 1980
26. Forschungszentrum Karlsruhe: ITC-TAB, ITC-CPV, ITC-ZTS, IKET, ITAS,IRS
27. R. H. Venderbosch, L. van de Beld, W. Prins; 12th EU-Conf. Biomass for Energy and Industry, 17-21 June 2002, Amsterdam, NL in press
28. J. Maier, R. Vetter, V. Siegle, H. Spliethoff; Abschlussbericht zu Forschungsvorhaben Ord. Nr. 22-94.11, Ministerium für Ernährung und Ländlichen Raum Baden Württemberg 1998
29. Winnacker-Küchler; „Chemische Technologie", Bd. 5, p.
30. C. N. Hamelinck, A.P.C. Faaij, H.P. Calis, H. den Uil, H. Boerrigter; 12th EU-Conf. Biomass for Energy and Industry, 17-21 June 2002, Amsterdam, NL in press
31. C. Roesch, W. Wintzer; Vergasung und Pyrolyse von Biomasse, TAB-Arbeitsbericht Nr. 49, Bonn 1997
32. K.D. Ouwens, A. Faaij; 12th EU-Conf. Biomass for Energy and Industry, 17-21 June 2002, Amsterdam, NL in press
33. G. Brocksiepe; „Holzverkohlung", Ullmann Encyklopädie der tech. Chemie, Bd. 12, S. 703
34. L. Leible, A. Arlt, H. Seifert, S. Kälber, E. Nieke, D. Wintzer, B. Fürniss; DGMK Tagungsbericht 2002-2, S. 63
35. E. Henrich, E. Dinjus, D. Meier, F. Weirich; Proc. 12th EU Conf. on Biomass for Energy and Industry, Amsterdam, 17-22 June 2002 (in press)
36. C.W. Macosko; "Rheology", VCH Weinheim 1994;

Biomass CHP Plant Güssing – A Success Story

Hofbauer Hermann, Rauch Reinhard, Bosch Klaus
Renet-Austria, Vienna, AT
Koch Reinhard
Biomasse Kraftwerk Güssing G.m.b.H., Güssing, AT
Aichernig Christian
Repotec, Güssing, AT

ABSTRACT: A steam biomass gasification process has been demonstrated in Güssing, AT. The combined heat and power (CHP) plant has a fuel capacity of 8 MW and an electrical output of about 2 MWel with an electrical efficiency of about 25 %. Wood chips with a water content of 20 - 30 % are used as fuel. The plant consists of a dual fluidized bed steam gasifier, a two stage gas cleaning system , a gas engine with an electricity generator, and a heat utilization system. The start up of the plant was in January 2002 and until September 2002 2500 hours of operation with the gasifier and the gas cleaning system and 750 hours of operation with the gas engine could be reached. The overall performance of the CHP plant is very good, however, some minor problems had to be solved. The most important parts operate quite well only the cooler of the producer gas had to be improved. In Austria and also in other European countries there are currently good conditions for an economical operation of an CHP-plant based on biomass due to high prices for green electricity. Nevertheless, further efforts have to be done to reduce the investment as well as the operating co6sts. The producer gas has a high amount of H_2 and CO and is therefore also well suited for the production of synthesis gas.

BACKGROUND

Güssing is a small town with about 4.000 inhabitants and is located about 200 km south of Vienna near to the Hungarian border. For a long time this border was called the iron curtain and therefore nearly no industry settled there which in consequence led to a lack of jobs for the people living there. Many of them migrated to other regions for ever or at least over week for working. The region was very poor until biomass as source of energy was discovered.

40 % of the region of Güssing is covered with wood. Therefore sufficient raw material is available for the energy supply of the whole city. About ten years ago the major of Güssing and some other visionary people worked out a concept for the supply of Güssing only with energy from renewable raw material from the region. This concept included energy saving as well as energy production by building new and innovative demo-plants. The main fields of energy demand of a city are: fuel for transportation, residential heating, and electricity. Fig. 1 shows these three columns of energy demand and the respective plants in Güssing.

The RME-plant started production already in 1990. This plant is able to produce more biodiesel than necessary for the overall consumption of fuel in the city of Güssing. In 1998 the district heating system was commissioned which was the largest biomass based district heating system in Austria at that time. The grid supplies 95 % of the city of Güssing. The third column was realized in 2002 as a new 2 MWel demonstration plant for CHP production went into operation. This plant is able to produce the whole electricity which is consumed by the people living in Güssing. Nowadays Güssing is supplied by 100 % of renewable energy which is fully based on biomass.

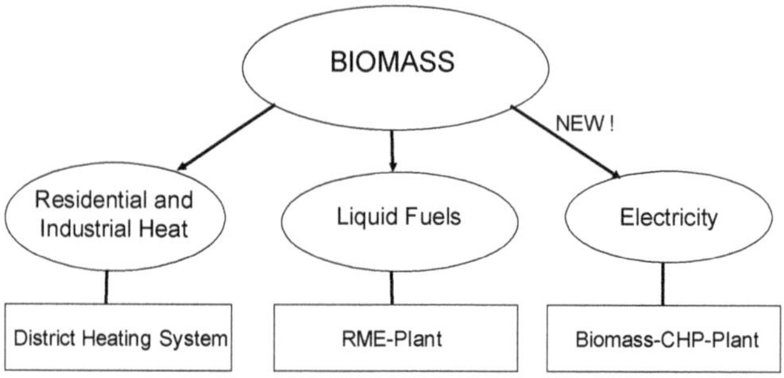

Figure 1: Energy supply of Güssing (100 % renewable).

COMBINED-HEAT AND POWER (CHP) PLANT

SHORT DESCRIPTION OF THE PROCESS

An innovative process for combined heat and power production based on steam gasification has been demonstrated in Güssing. Biomass is gasified in a dual fluidised bed reactor. The producer gas is cooled, cleaned and used in a gas engine. A detailed flow sheet is shown in Fig. 2. The most important characteristic data of the demonstration plant are summarized in Table 1.

Table 1: Characteristic data of the plant.

Type of plant	Demonstration plant	
Fuel Power	8000	kW
Electrical output	2000	kW
Thermal output	4500	kW
Electrical efficiency	25,0	%
Thermal efficiency	56,3	%
Electrical/thermal output	0,44	-
Total efficiency	81,3	%

Biomass chips are transported from a daily hopper to a metering bin and fed into the fluidised bed reactor via a rotary valve system and a screw feeder (Fig. 3). The fluidised bed gasifier consists of two zones, a gasification zone and a combustion zone. The gasification zone is fluidised with steam which is generated by waste heat of the process to produce a nitrogen free producer gas. The combustion zone is fluidised with air and delivers the heat for the gasification process via the circulating bed material.

The producer gas is cooled and cleaned by a two stage cleaning system (Fig. 4). A water cooled heat exchanger reduces the temperature from 850 °C – 900 °C to about 160 °C – 180 °C. The first stage of the cleaning system is a fabric filter to separate the particles and some of the tar from the producer gas. These particles are returned back into combustion zone of the gasifier. In a second stage the gas is liberated from tar by a scrubber.

Spent scrubber liquid saturated with tar and condensate is vaporized and introduced into the combustion zone of the gasifier. The scrubber is used to reduce the temperature of the clean producer gas to about 40 °C which is necessary for the gas engine. The clean gas is finally fed into a gas engine to produce electricity and heat. I the gas engine is not in operation the whole amount of producer gas can be burned in boiler to produce heat. The flue gas of the gas engine is catalytically oxidised to reduce the CO emissions. The heat of the engine´s flue gas is used to product district heat.

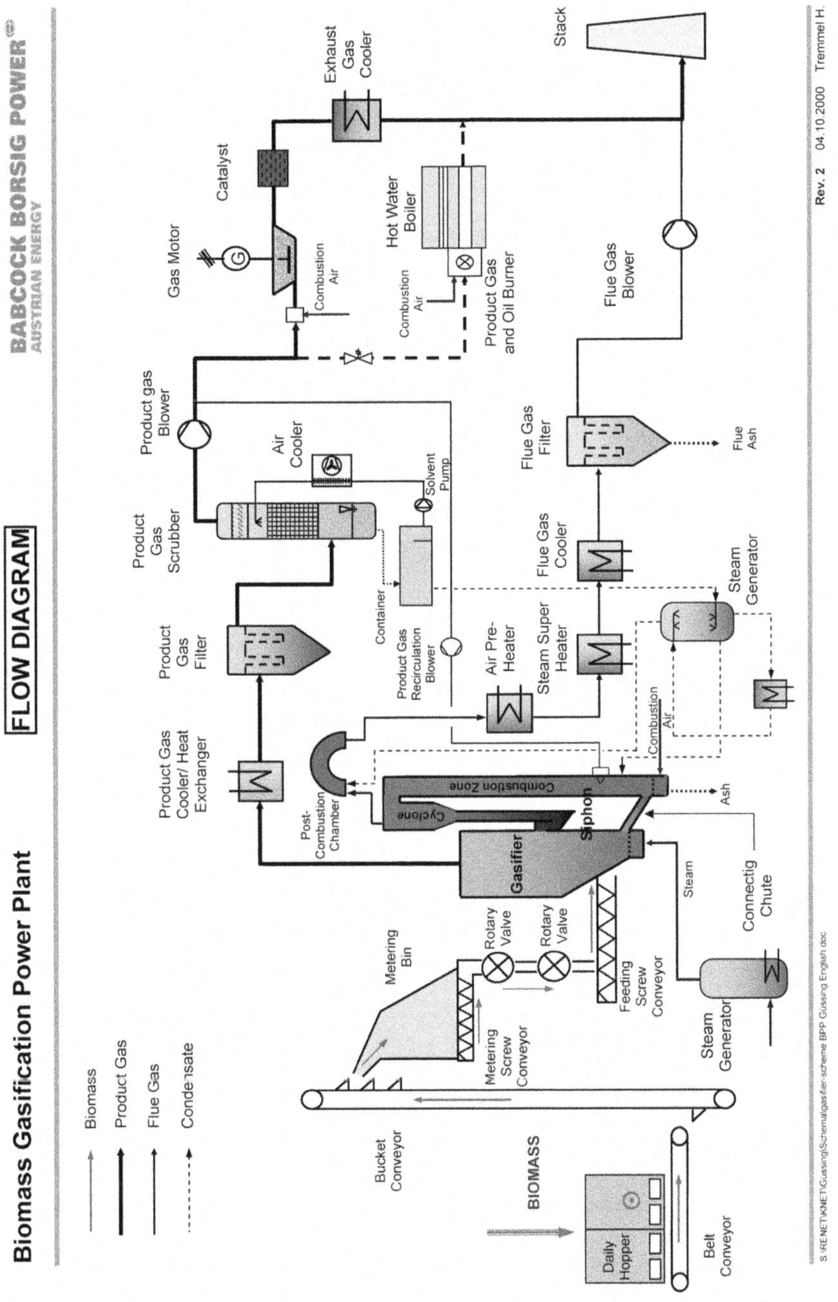

Figure 2: Flow sheet of CHP-plant Güssing.

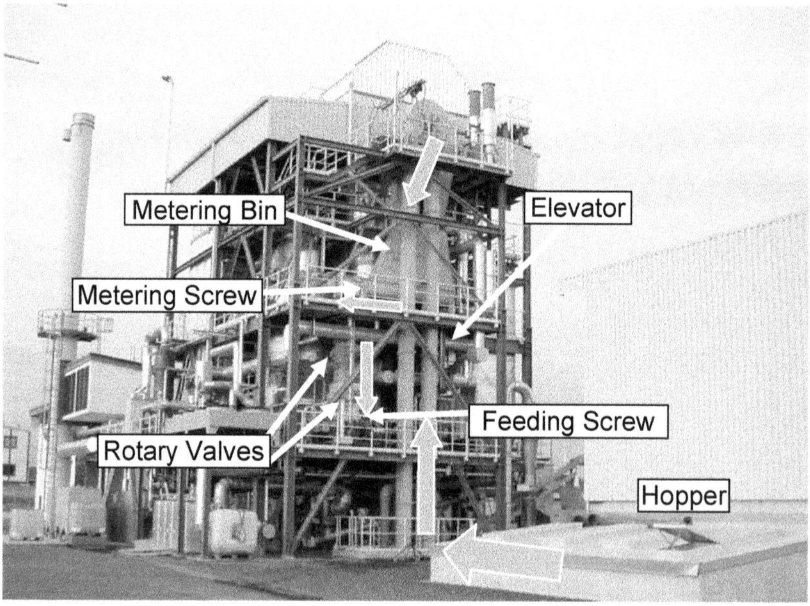

Figure 3: Biomass conveying.

Figure 4: Producer gas line.

The sensible heat of the flue gas for the combustion zone is used for preheating of the air, superheating the steam and also to deliver heat to the district heating system. A gas filter separates the particles before the flue gas is released via a stack to the environment.

DESIGN PHASE

The basic idea of the gasification process originates from the Vienna University of Technology and was published first in 1982 for coal gasification [1]. The gasification process was adapted for biomass during the last ten years. Table 2 shows the main stages of this research and development process.

In 1983 a 10 kW test rig was built to study the fundamental behaviour of a Fast Internal Circulating Fluidised Bed (FICFB) gasifier. The design at that time was a fluidised bed with a daft tube. The basic behaviour of such a system was successfully studied which was the base for the further development. The name FICFB remained the same until now although the design changed to an external circulating fluidised bed system.

A 100 kW pilot plant was used to gasify different fuels (biomass, sewage sludge, coal, etc.) and to carry out a comprehensive parameter study. This pilot plant consists not only of the gasifier but also of the same gas cooling and gas cleaning system which was later realized at the demonstration plant in Güssing. From this pilot plant a lot of design parameters for the demonstration plant could be achieved. A detailed description of the 100 kW pilot plant and the respective results can be found elsewhere [2, 3].

In 2000 a network of competence (Renet-Austria) was established to support the design, construction, commissioning and demonstration phase of the CHP-plant in Güssing. The members of Renet-Austria are the manufacturers of the plant (AE-Energietechnik, Jenbacher AG), the owner of the plant, and a research organisation (TU Vienna). The work of the competence network is funded by the government and two federal states of Austria (Burgenland, Niederösterreich). Due to this support from Renet-Austria many questions during the design phase could be clarified by experiments at the 100 kW pilot plant or by mathematical simulation work [4].

Table 2: Development of the FICFB-Gasification Process.

Stage	Institution	Installation	Period
1	TU Vienna	10 kW test rig	1993 - 1996
2	TU Vienna + AE Energietechnik	100 kW pilot plant + gas cleaning	1997 - 2002
3	Renet Austria	8 MW demonstration plant Güssing	2000 - 2004

CONSTRUCTION PHASE OF THE DEMONSTRATION PLANT

The construction of the demonstration plant started in September 2000 and was finished again in September 2001. This very short time was possible as no serious problems during this phase occurred. The official opening ceremony was hold on 20. September 2001.

COMMISSIONING AND DEMONSTRATION PHASE

The autumn of the year 2001 was very warm in Austria. Therefore the first test runs of the demonstration plant could be carried out November 2001. First the control system of the plant has to be adjusted. This was necessary because no experience was available for such a type of gasification process. Within the time of two month no constant and stable operation was possible. This unsteady operation lead to some deposits in the producer gas cooler due to bad gasification conditions. After this first period a good performance could be reached. From 21. February 2002 to 22. March 2002 the gasification and gas cleaning system was in continuous operation without any problems and also the gas composition was as expected from the results of the pilot plant.

After more than 1500 hours of successful operation with the gasification and the gas cleaning system the gas engine was coupled. Till the end of September 2002 more than 2500 h of operation with the gasifier and gas cleaning and more than 750 h of operation with the gas engine could be reached. Currently a two years demonstration period is carried out to optimise the operation and especially to reduce the operation costs.

OPERATION EXPERIENCE

FUEL SUPPLY

The biomass supply is secured by long term contracts. The fuel for the heat and power production are wood chips delivered by local wood farmers who have established a wood farmers association. The price is fixed for a duration of ten years (within index adaptation) which is currently about 1,6 Cents/kWh. The water content of this wood chips is about 25 %. In future it is intended that 40 % of this fuel should be replaced by a cheaper (0,7 cents/kWh) and dry fuel which are residues of local wood working industries.

HEAT AND POWER UTILIZATION

Heat is delivered to a district heating system which has a length of more than 20 km. The consumers are mainly private houses (300), public offices, schools, and hospital (50), Furthermore, there is a growing demand of industrial heat which is needed over the whole year. The heating station itself operates drying chambers which are additional heat consumers. Currently, first experience could be made with cooling using district heat. Electricity is sold to the electrical grid operator with a feed-in-rate of 12,3 Cents/kWh.

PLANT PERFORMANCE

After optimising the control system a very smooth and stable operation of the gasification and gas cleaning could be obtained. Fig. 5 and Fig. 6 show online measurements of the temperatures, pressures and the gas composition.

The small temperature difference between the combustion and gasification zone indicates that there is sufficient circulation of bed material between the two zones. In this first period of operation olivine is used as bed material. The composition of the producer gas at the demonstration plant was very similar to that measured at the pilot plant. The calorific value of the dry producer gas is constant at about 12 MJ/Nm^3. The nitrogen content originates mainly from the purge gas in the rotary valves and particle filter. Typical ranges of the gas composition can be seen from Table 3.

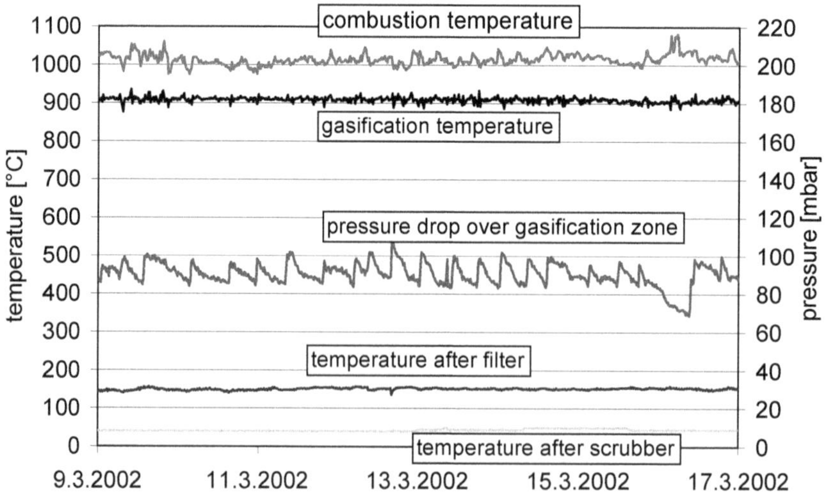

Figure 5: Online measurements of temperatures and the pressure of the gasification bed.

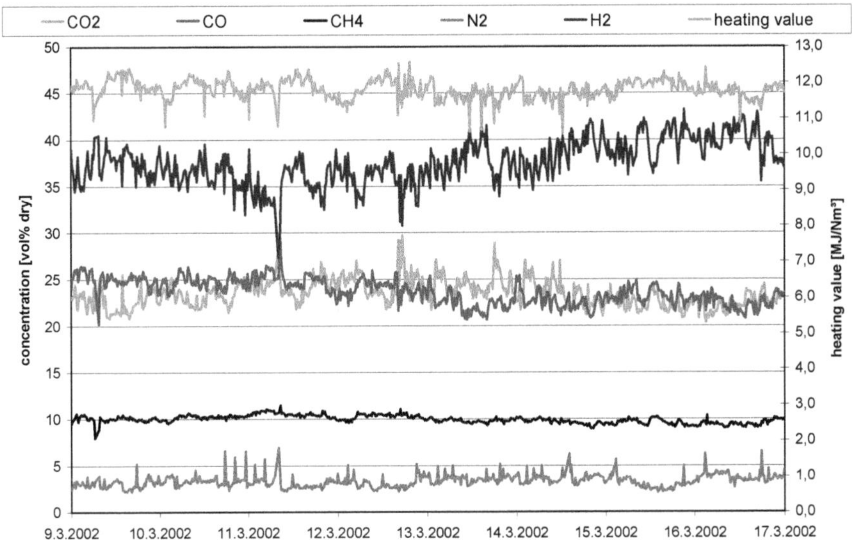

Figure 6: Online measurements of the gas composition and calculation of the heating value.

Table 3: Ranges of the main components in the producer gas (dry gas).

Component	Range	Dimension
hydrogen	35 - 45	Vol-%
carbon monoxide	20 - 30	Vol-%
carbon dioxide	15 - 25	Vol-%
methane	8 - 12	Vol-%
nitrogen	3 - 5	Vol-%

The producer gas leaves the gasifier at the top and enters into the gas cooling and cleaning section. The influence of the tar content on the bed material and gasification agent is shown in Fig. 7. Steam as gasification agent leads to a lower tar content compared with air blown gasifiers. Furtheron, the bed material can act as catalyst for tar cracking. One essential feature of any bed material is a sufficient resistance against attrition. Olivine is a suitable bed material although there are big differences in various olivines. By using a special designed catalyst for tar cracking the primary tar reduction can even improved. In case of Güssing olivine as bed material and steam as gasification agent is used.

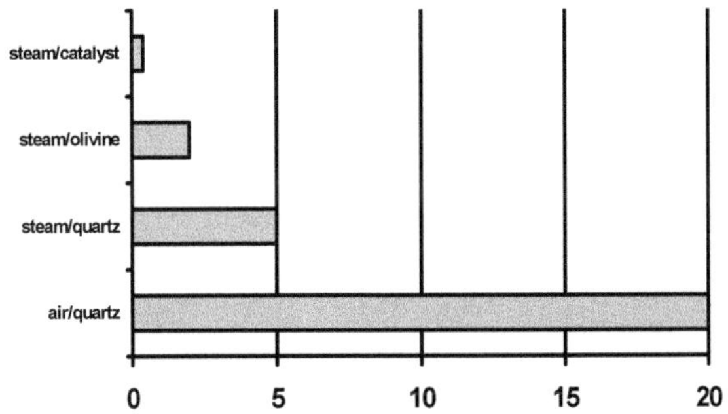

Figure 7: Influence of bed material and gasification agent on tar content (mg/Nm^3 dry gas).

The producer gas is cooled by a heat exchanger from a temperature of 850–900 °C to 150–180 °C. At the beginning of the testing period some fouling problems were observed. This was mainly due to a bad operation of the gasifier. After optimising the operation of the gasifier (temperature >800 °C, steam to biomass ratio >0,6) the fouling disappeared. Corrosion at the entrance of the heat exchanger was observed after 1500 h. It was tried to solve this problem by dividing the heat exchanger into two parts. The first part (hotter part) was manufactured from an higher quality steel better suited to resist the conditions at the entrance.

The particles and some tar is remove from the gas by a precoated fabric filter. This part of the gas cleaning section worked up to now without any problems. After 2500 h of operation no cleaning or exchange of the filter cloths were necessary. The spent precoat material and the separated particles are returned back into the combustion chamber to burn out the carbon and hydrocarbons.

Most of the tar and some ammoina is removed by a scrubber . This scrubber uses a solvent which gives excellent results for tar removal. The spent solvent together with tar components is fed into the combustion zone und combusted. The temperature of the gas is further reduced to about 40 – 50 °C in the scrubber which in necessary for the gas engine. Therefore most of the steam condenses there. This condensate is evaporated and fed into the after burning chamber of the combustion zone. This is a main advantage of the process as no waste water problem arises. Typical values of the tar content along the gas cleaning line can be seen in Fig. 8. The measurements at the demonstration plant were similar to those at the 100 kW pilot plant. The clean producer gas has a very low tar content (10-40 mg/Nm^3 dry gas) which is no problem for the operation of the gas engine.

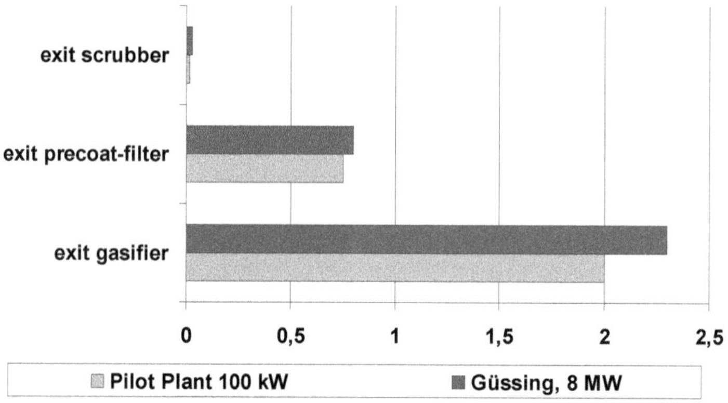

Figure 8: Tar content (mg/Nm^3 dry gas) over the gas cleaning line.

Table 4 contains typical ranges of the minor components in the raw gas as well as the clean producer gas. Hydrogen sulfide was only measured in the clean gas therefore no values are available for the raw gas.

Table 4: Producer gas quality (minor components).

Component	Raw gas	Clean gas	Dimension
tar	1,5 - 4,5	10 - 40	mg/Nm^3
particles	5 - 10	<5	mg/Nm^3
ammonia	1000 - 2000	<400	ppm
hydrogen sulfide	n.m.	20 - 40	ppm

n.m. not measured

ENVIRONMENTAL ASPECTS

The flue gas from the gas engine and the flue gas from the combustion zone are mixed together and released via the stack to the environment. Measurements of the emissions were measured recently and the results are shown in *Table 5*. All measurements are below the emission limits which were set by the local authorities.

As already discussed in the previous chapter there are no liquid emissions from the CHP plant. The condensate from the scrubber is evaporated and fed into the combustion zone where the organic matter is combusted.

The only solid residue is the fly ash from the combustion zone. Therefore the carbon content in this fly ash is very low (<0,5 w-%) and can be handled similar to an ash from biomass combustion. This is an essential advantage compared to the most other gasifiers.

Table 5: Emissions from CHP-plant Güssing (dry gas, ref. 5 % oxygen).

Component	Range	Dimension
CO	900 - 1500[1]	mg/Nm^3
	100 - 150[2]	mg/Nm^3
NOx	300 - 350	mg/Nm^3
dust	< 20	mg/Nm^3

1 without catalyst
2 with catalyst

The CHP plant Güssing is characterised by an excellent environmental performance:
(1) low gaseous emissions
(2) no liquid emissions
(3) ash only from the combustion zone (very low carbon content)

ECONOMY OF THE DEMONSTRATION PLANT

The CHP-plant in Güssing can be operated economically und er the Austrian conditions. The conditions are currently quite well and are characterized on the one hand by high fuel costs and and on the other hand by high feed in tariffs for electricity. Some economic data for the demonstration plant are summarized in Table 6.

Table 6: Economic data of the demonstration plant.

Cost category	Amount	
Investment cost	10	Mio. EURO
Funding (EU, National)	6	Mio. EURO
Operation cost / year	10 - 15	% of investment costs
Price for heat (into grid)	2,0	Cents/kWh
Price for heat (consumer	3,9	Cents/kWh
Price for electricity	12,3	Cents/kWh

For the next plant 25 % reduction of investment cost can be expected due to the experience and learning at the demonstration plant. Furthermore, the operation costs will be reduced essentially. This will be done by aiming at an unmanned operation and an reduction and optimisation of operation means (bed material, precoat material, scrubber liquid).

ACCEPTANCE OF THE CHP-PLANT IN GÜSSING

The acceptance of the CHP-plant by the people of Güssing as well as the local authorities is excellent. The reasons for this acceptance are manifold:

(1) CHP-plant was the missing link for the complete energy supply by biomass for the city of Güssing
(2) The production of heat and electricity only from local raw material end a higher price is paid
(3) Sufficient local biomass is available
(4) Energy supply is now independent from oil price
(5) Local job creation not only by the CHP power plant but also by the settlement of wood working industry

CONCLUSIONS

A CHP-plant based on a steam blown dual fluidised bed biomass gasifier with a capacity of 8 MW has been demonstrated successfully in Güssing. 2500 h of operation of the gasifier and the gas cleaning part of the plant could be reach until September 2002. 750 hours of operation of a Jenbacher gas engine with an electrical output of 2000 kWel showed that the gasification and gas cleaning is working with an excellent performance. All expectations as well as the emission limits could be met. Some minor improvements were necessary especially concerning the producer gas cooler. Compared to other CHP-plants based on gasification of biomass the Güssing plant can be designated as real success story.

The FICFB-gasification process has a large potential for the future as it leads to a high grade producer gas which can be used for various applications:

(1) Combined heat and power production (CHP) with gas engines, gas turbines or fuel cells
(2) Hydrogen production
(3) Synthesis gas for production of substitute natural gas (SNG), methanol, Fischer-Tropsch-diesel

ACKNOWLEDGEMENT

The work described in this paper was funded by the competence centre programme Knet. The financial support of the Ministry of Economy and Labour and of the Federal States of Niederösterreich and Burgenland is gratefully acknowledged.

REFERENCES

1. Hofbauer H. (1982) *Fundamental Investigations on an Internal Circulating Fluidized Bed with a draft tube.* Phd thesis, Vienna University of Technology.
2. Fercher E.; Hofbauer H.; Fleck T.; Rauch R.; Veronik G. (1998) *Two years experience with the FICFB-gasification Process.* Proc. of the 10th European Biomass Conference, Würzburg.
3. Hofbauer H.; Rauch R. (2001) *Stoichiometric water consumption of steam gasification by the FICFB-gasification process.* Proc. Of Progress in Thermochemical Biomass Conversion, Bridgewater A.V. (Ed.), Blackwell Science, Oxford, pp. 199-208.
4. Kaiser S.; Weigl K.; Schuster G.; Tremmel H.; Friedl A.; Hofbauer H. (2000) *Simulation and optimization of a biomass gasification process.* 1st World Conference and Exhibition on Biomass for Energy and Industry, Seville.

Optimisation of Multistaged Biomass Gasification

S. Joseph, K. Wallis, T. Dennis, S. Cowgill, E. De Jonge, P. Klatt, L. Tobiasen
Biomass Energy Services and Technology (BEST),56 Gindurra Rd
SOMERSBY NSW 2250, Australia

ABSTRACT: BEST has been developing a multi-staged biomass gasification process that utilises the waste heat from the electricity generation power plant to improve gasification and overall efficiency. The multistage approach allows for separation of drying/conditioning, torrefaction, pyrolysis and gasification, which in turn allows high control and optimisation of the thermal conversion steps involved.

As part of the development process, BEST has designed and built a simple macro-TGA pyrolysis and char reactor test rig which is used for both design and optimisation of the char reactor part of the process. Initial macro-TGA experiments have been done on Australian rice husks to determine the optimal reaction conditions to produce a low carbon ash. Further studies are to be carried out on green waste and weeds in the near future. In this paper the following will be discussed.

(1) The process that BEST intends to use to ensure high carbon burnout and high gasification efficiencies.
(2) Details on the Macro-TGA test rig.
(3) Further work that will be undertaken.

PROCESS OVERVIEW

There is a considerable body of information that indicates that the carbon conversion and bio-oil oxidation efficiency depends on establishing optimal:

(1) Residence time/ heating rate/ temperature profile in the reactor(s).
(2) Amount, temperature and injection points of air and steam.

The multi-staged approach that BEST is developing allows increased control in the thermal process steps, which in turn allows control of the heating rates and temperature of the biomass. In this way, carbon conversion & tar cracking can be managed and optimised while still focusing on producing high quality ash.

The gasification process can be divided into a series of reaction/process steps. These are indicated in Figure 1.

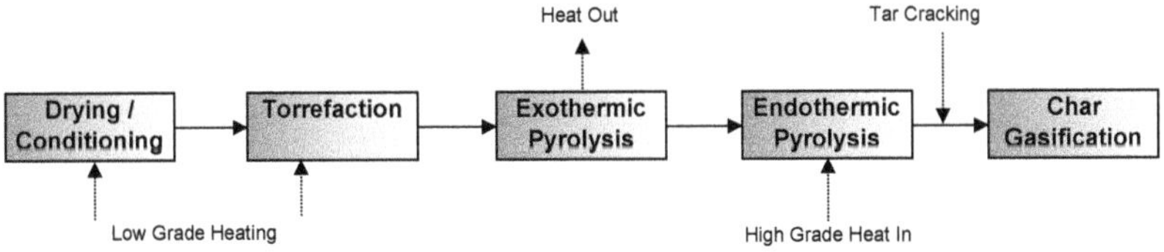

Process:	Heat source:
Drying / conditioning	Low grade waste heat (engine exhaust)
Pyrolysis (including torrefaction, exothermic and endothermic pyrolysis)	Low grade waste heat (engine exhaust) plus heat from recirculated syngas combustion
Char gasification	Heat from char oxidation plus heat from partial bio-oil oxidation
Gas cleaning	N/A
Electricity production	N/A

Figure 1: Process Schematic.

At the time of writing this paper, a pilot scale test rig (~200kW$_{th}$) based on the above process had been constructed, but test results are still pending. The design of this test rig was partly done based on the macro-TGA results described below. Further details of the process steps also follow below.

The energy requirement for each of these process steps is dependent upon the chemical and physical properties of the biomass. The approximate external energy requirements for each of these steps have been estimated for Australian Rice Husks. The requirements per kg of fuel are illustrated in Table 1. The rice husk sample used has 10wt.% moisture and 20wt.% ash on an as received basis.

Table 1: Energy Requirements.

Zone	External Energy Input (Approx.) kJ/kg fuel input (10%m.c., 20% ash)
Drying & Conditioning	470
Torrefaction	130
Exothermic Pyrolysis	0
Endothermic Pyrolysis	350-400
Char Gasification	0 (reducing reactions driven by char oxidation)

DESCRIPTION OF PROCESS STEPS AND EQUIPMENT[1]

It is BEST's contention that slow heating of biomass to approximately 220°C ensures that carbon burnout with steam and/or air is easier to achieve at relatively low temperatures. The literature [5] and the present work on rice husk gasification would indicate that slow heating has the following advantages in respect to reactivity.

(1) There is a greater pore area (both macro and micro) formed in the biomass particles as the steam, volatile organic compounds and CO2 escape and the polymers decompose in the char [5].

(2) The carbon does not react chemically with the ash and/or produce structures that are chemically relatively inert [5].

(3) Due to the long residence times thermal decomposition occurs uniformly throughout the biomass particles [6].

Drying and Conditioning

Drying and conditioning is carried out in a separate reactor which is externally heated using thermal oil[2].The process heat is provided by the waste heat from the energy plant and is carried out over a 5-10minute period to 130-150°C. Removal of almost all of the chemical water has been achieved and the literature would suggest that depolymerisation of the material has commenced [1].

Torrefaction

The material is heated from 130-150°C to 200-240°C in a stirred drum reactor. At the end of the reactor the biomass has darkened and, according to Dielbold [2] this is due to the presence of water, oils and aromatic compounds surrounding the surface of the particles around 200°C in what is termed a melt. The breakdown of lignin produces hydrophilic carboxylic acids, anhydrogluvoses and furanoic compounds [2].The biomass is further depolymerised.

---

1 A plant/process schematic is given in appendix 1

2 For safety reasons the reactor is also internally heated using small amounts of superheated steam

Exothermic Pyrolysis, Endothermic Pyrolysis & partial oxidation of the tars

The next three (3) process steps (exothermic pyrolysis, endothermic pyrolysis and partial oxidation of bio-oils) are also carried out in a swept drum pyrolyser which is externally heated using some of the exhaust from the engine and recycled syngas from the char gasifier. Rapid devolatilisation occurs as soon as the biomass touches the hot walls of the swept drum. Previous experimental work has shown that this devolatilisation, in externally heated stirred drums, takes less than 10 minutes to complete and the final temperature achieved is approximately 400°C [4]. The decomposition is a mainly endothermic reaction, which however has an exothermic region in the temperature range 300-400°C. The reaction produces char, volatile gases and bio-oils (which comprise mainly of alcohols, keytones, aldehydes and phenols) [3].

As the char is slowly moved along the reactor further decomposition of the char occurs from 400°C. A series of reactions that involve the breaking of the O-CH3, OH and C-CO bonds occur. Radicals are formed and these play a major part in the production of more complex hydrocarbons unless reactions capturing the radicals are effectively carried out. These reactions produce biphenols, benzenes and creosoles as well as the gaseous combustible products of H2, CH4 and CO. Unlike the devolatilisation process, this decomposition is solely endothermic.

It is well known that break down of the bio-oils occurs when pyrolysis gas passes through a layer of char into which air/oxygen has been added (in downdraft gasifiers)[3, 5]. As the temperature increases in the char layer, some of the bio-oils can polymerise to produce aromatic and polyaromatic hydrocarbons. Concentrations of so-called secondary and tertiary tars, as low as 50-500 mg/Nm^3, have been measured in downdraft gasifiers [3].

It is the authors' contentions that if moist preheated air (approximately 600°C) is introduced into the reactor in the gas space where endothermic pyrolysis is taking place then the majority of the bio-oils produced will be oxidized or 'cracked' to form lighter combustible gases. Formation of secondary and tertiary tars will be minimised as the introduction of the air and steam will effectively scavenge the carbon, and also because the temperature is intended to be kept relatively low (less than 800°C). Some of the air will react with the char particles that will be in the gas space and flaming pyrolysis will occur. These hot char particles will transfer their heat to the char bed, reducing the amount of external heat required.

Experience, kinetic modeling and experimental results show that a temperature of approximately 400-600°C must be reached before the cracking and oxidation reactions will be initiated. Modeling of the oxidation reactions of representative tar compounds such as phenol, naphthalene and benzene indicates that residence times of approximately 2-5 seconds are required if most of the aromatic compounds are to break down to CO, CH4, CO2 and H2 at 600°C. Reaction times for the lighter oxygenated hydrocarbons are considerably shorter. Initially, modeling would indicate that the excess air ratio should be at least 40-50% (based on the volatiles only). The off gas from this reactor passes through a hot gas cleaning device and is then cooled and residual tars removed.

Char Reactions

Conversion of char occurs in a separate down draft gasification unit into which air and steam are added. The char reactor is distinct from the kiln to facilitate a large residence time without an excessively long kiln. It also enables the units to be controlled separately. The reactor is designed for temperatures up to 800°C. This relatively low temperature is chosen to ensure a high quality residual ash product. The full conversion of char is controlled by the temperatures of the reactor and the components (air, steam, CO_2 concentration and char) as well as the residence time of the air and steam. The principle mechanisms that are responsible for the conversion of char (in the absence of oxygen) are presented in equations (1) and (2).

$$C + CO_2 \xleftrightarrow{r_{11}} 2CO$$
$$C + H_2O \xleftrightarrow{r21} H2 + CO$$

MACRO TGA TEST RIG

BEST has designed and built a macro Thermal Gravimetric Analysis (TGA) reactor to help verify the hypotheses outlined above and thus to assist in the design of a pilot rice husk gasification plant. This rig was designed to determine:

(1) Mass loss as a function of heating rate through the different stages of conditioning, torrefaction/ pyrolysis.

(2) Reaction rates of char as a function of reactor temperature and partial pressures of reactant gases, namely oxygen and steam.

Macro refers to the large sample sizes (10-30g), which differs from conventional TGA reactors, where the main aim is to measure kinetics. The macro TGA inherently measures diffusional affects as well, which allows for the results to be directly comparable to actual reactions in a large scale reactor.

The TGA comprises a 150mm cylindrical reactor which is externally heated by a LPG gas burner. Within the reactor is a mesh tray for the contents of the biomass material. The tray is suspended from the scales via a 10mm rod that passes through the top of the reactor. Water, air, nitrogen and CO2 can be individually metered into the rig. All gases pass through a superheater before they enter the rig. The composition of the gases entering and leaving the rig is measured using a gas chromatograph and I.R. analysers. An illustration of the layout of the rig is given in Figure 2.

It should be noted that the char reacts with steam, air or CO2 directly after pyrolysis. The reaction rates of the char are controlled by both kinetic and diffusion effects since a relative thick bed is being used and because oxidation kinetics are very fast. The rate controlling mechanism depends on the temperature at which the gasification reactions take place.

EXPERIMENTAL PROCEDURE FOR TGA TEST RIG

The tests were carried out using dried rice husks as the feedstock. Ultimate analysis of the husk is provided in Table 2.

Tests to determine weight loss during pyrolysis were carried out. The heating rate for pyrolysis was approximately 10°C/minute. Gasification times of 1 hour were chosen. Gasification was initiated between temperatures of 715-810°C, thereby simulating conditions intended for a larger scale reactor. Figure 3 illustrates the weight loss from rice husks with temperature, for a typical experiment.

Table 2: Ultimate analysis of Rice Husks.

Component	Mass %
Carbon	34.0
Oxygen	31.7
Hydrogen	4.2
Ash	21.5

Literature on downdraft gasification of rice husks was used to approximate the required reactant partial pressures of oxygen and steam to undertake complete gasification of the char [5, 7]. Extracted from this data was a kinetic reactivity for rice husks as well as a reactivity of oxygen with related biomass material (cotton trash). These reactivities formed the basis of a spread sheet model that predicted required partial pressures of reactants for conversion, over the reaction period (30 to 60 minutes). The resulting gasification mixture used is illustrated in Table 3 which equates to a steam to air mass ratio of 2.4:1

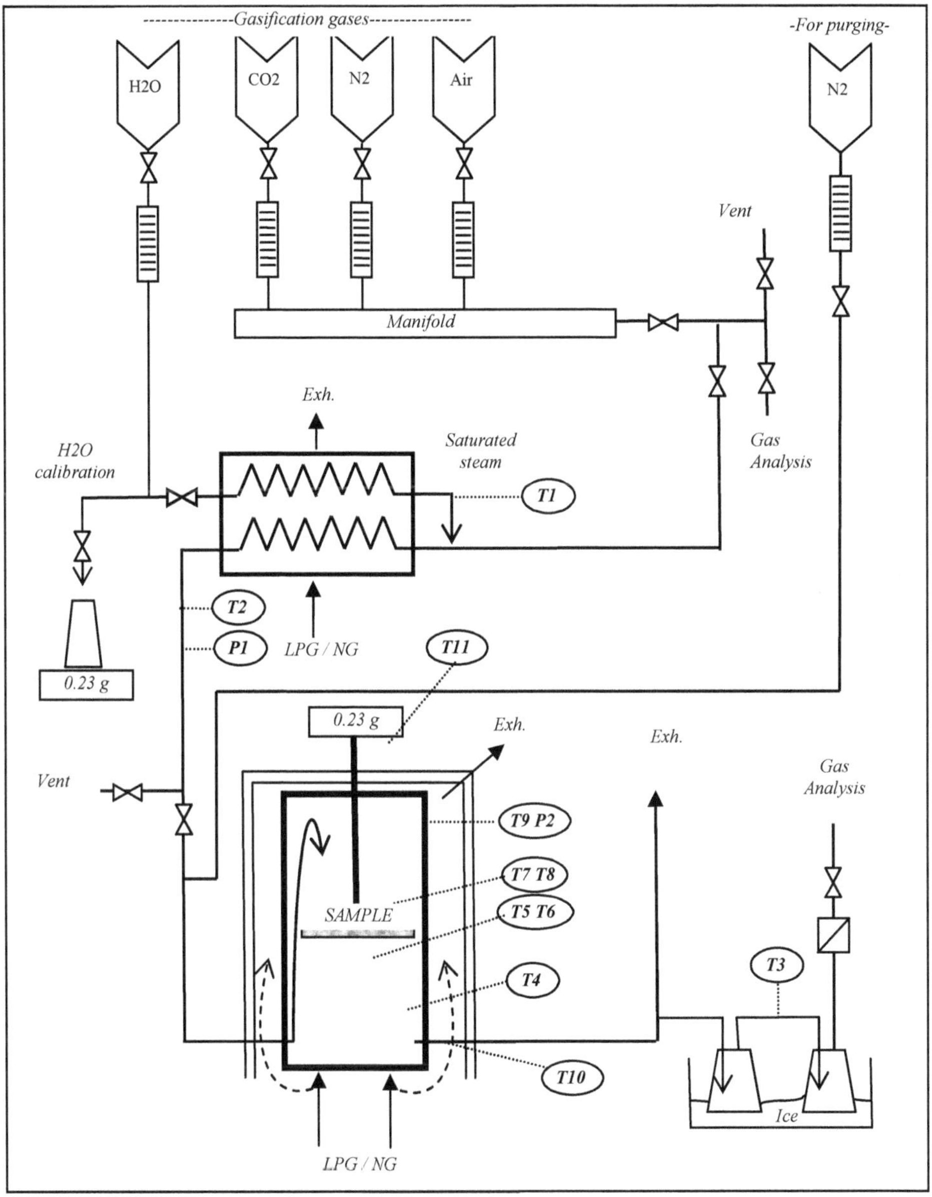

Figure 2: TGA Rig Schematic.

Table 3 Gas Mixture for Gasification

Component	Vol %
Steam	79.3
Oxygen	4.2
Nitrogen	16.5

Table 4: TGA tests on Rice Hulls.

Test Number	Experiment Description	Time (mins)	Temperature (°C)	Gasification Mixture (vol %)
1	Commissioning – Pyrolysis	-	600	-
2	Commissioning – Pyrolysis	-	600	-
3	Pyrolysis	-	645	-
4	Pyrolysis	-	615	-
5	Pyrolysis- Gasification	30	770	100% steam
6	Pyrolysis- Gasification	30	770	100% steam
7	Pyrolysis	-	770	-
8	Pyrolysis- Gasification	30	770	100% steam
9	Pyrolysis- Gasification	30	770	79% H_2O, 17% N_2, 4% O_2
10	Pyrolysis- Gasification	30	770	79% H_2O, 17% N_2, 4% O_2
11	Pyrolysis- Gasification	60	715	79% H_2O, 17% N_2, 4% O_2
12	Pyrolysis- Gasification	60	715	79% H_2O, 17% N_2, 4% O_2
13	Pyrolysis- Gasification	60	715	79% H_2O, 17% N_2, 4% O_2
14	Pyrolysis	60	780	79% H_2O, 17% N_2, 4% O_2
15	Pyrolysis- Gasification	60	810	79% H_2O, 17% N_2, 4% O_2
16	Pyrolysis - Bed temperature test	-	770	-
17	Pyrolysis- Gasification	60	760	79% H_2O, 17% N_2, 4% O_2
18	Pyrolysis- Gasification	60	810	79% H_2O, 17% N_2, 4% O_2
19	Pyrolysis- Gasification	60	760	96% N_2, 4% O_2
20	Pyrolysis- Gasification	60	760	96% N_2, 4% O_2
21	Pyrolysis- Gasification	60	780	79% H_2O, 17% N_2, 4% O_2

TGA TEST RESULTS

Figure 3 illustrates the mass loss with temperature using rice hulls for test number seven (7). Devolatisation occurs mainly between 250°C and 450°C and the residual char at 780°C is 46% of the initial sample. This figure agrees with similar published tests [4].

Examination of the samples indicated that conversion was not uniform across the sample. It was observed that the ash was blacker in the middle and at the edges of the basket, indicating a lack of flow of oxygen and steam in these regions. Since small amounts of rice enter the hull sample, further errors are introduced as the ash composition can change by up to 10%. Measurements of the weight loss under gasification gave a figure of approximately 73 (+/-2)% carbon conversion for temperatures between 715°C and 810°C (Figure 4) and a residence of 60 minutes. However the ash was a very light grey (indicating carbon contents of less than 10%). Samples were directed to a laboratory for loss on ignition (LOI) tests. Loss on ignition is based on the further mass loss when the sample is taken to 1000°C with excess air. The LOI tests gave results of carbon conversion of 84% to 98%. Thus it was decided to assume that the LOI tests are a more accurate measurement of carbon conversion and appropriate corrections for weight change as a function of time were carried out.

Even when taking into account the lack of flow at the edges of the sample and the variable ash composition, the model still over predicts the rate of conversion of the char. The model used to approximate the air/ steam ratio required, was based on kinetics and diffusion of oxygen measured for cotton trash and combined with a kinetics only reactivity for steam. The results indicate that diffusion plays a significant role in determining the reaction rate.

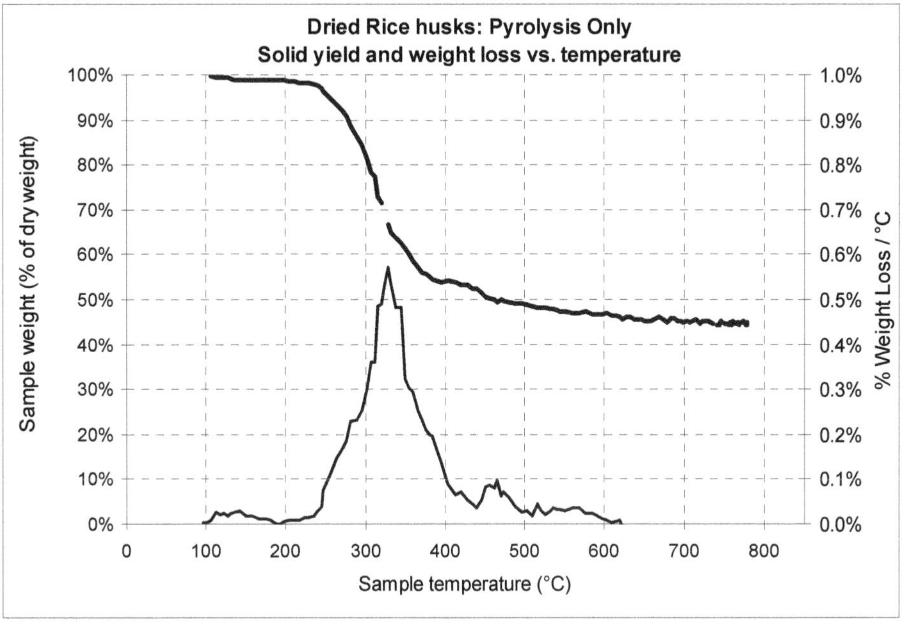

Figure 3: Weight Losses with Temperature of Dried Rice Hulls for Test 7.

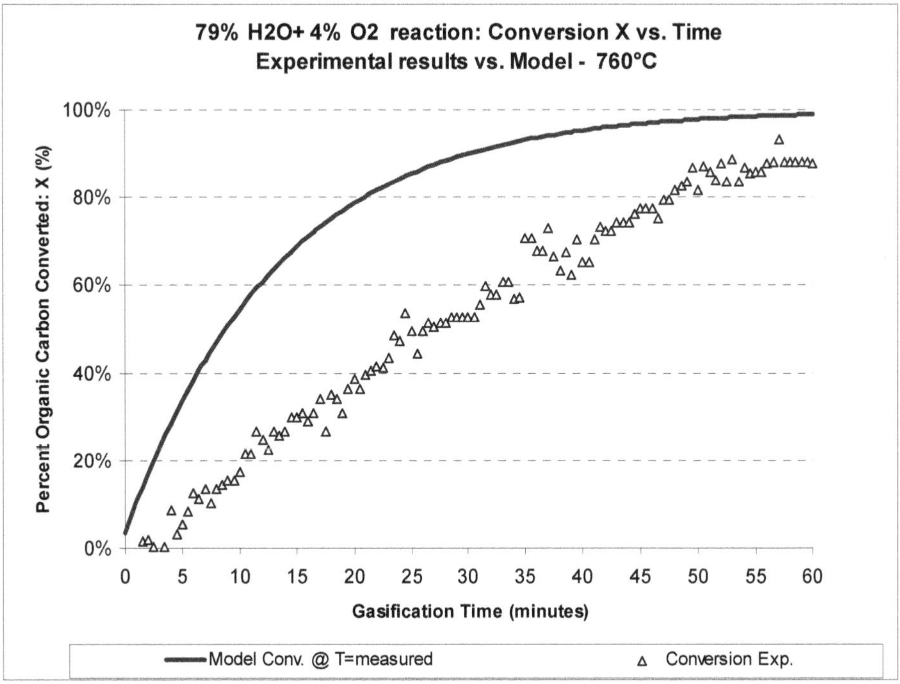

Figure 4: Comparisons between Modelling and Experimental Results for Rice Hulls.

Loss on Ignition (LOI)

Initial LOI results (Table 5) for gasification with steam and air are encouraging, due to the relatively good char conversion in the TGA reactor even at low temperatures. Table 5 illustrates the loss on ignition results received to date.

Table 5: Loss on Ignition.

Test	Sample	LOI
11	Gasification: 79% Steam 4% O2, 715°C	2.3
19	Gasification: 4% O2, 760°C	16.9
18	Gasification: 79% Steam 4% O2, 810°C	1.1

CONCLUSIONS

The test results to date would indicate that the process strategy outlined should produce a low carbon ash if final gasification temperature is maintained below 800°C. Residence times required in the char gasifier are in the order of 1 hour for partial pressures of 4% oxygen and 79% steam.

Further tests are to be undertaken to validate the above results and a second tar cracking reactor may be added to determine the effectiveness of injecting hot air.

FURTHER WORK

BEST has now completed the manufacture of a pilot gasification plant (appendix 1) but experimental results are still pending. This plant consists of a separate reactor for drying and conditioning, pyrolysis and tar oxidation kiln and a down draft char gasification reactor. Hot gas clean up will be achieved using Schumacher ceramic candles and cooling and tar removal using BEST's patented scrubber.

The aim of the test program for the pilot plant is to:

(1) Demonstrate that a clean gas suitable for powering a commercial gas engine can be produced by gasifying a particular biomass waste.

(2) Demonstrate that a hot ceramic candle filter unit can reduce the particulate loading of the gas to less than 50mg/Nm3.

(3) Determine the emissions from the porous burner fired on the syngas produced from the biomass waste.

(4) Provide data for scaling up the design to a larger plant.

ACKNOWLEDGMENT

The authors would like to acknowledge the assistance given by the Sustainable Energy Development Authority (SEDA). SEDA is an agency for the state of New South Wales Government.

REFERENCES

1. Broido A., Javier-Son A.C. & Barrall E.M. (1973) Molecular Weight Decrease in the Early Pyrolysis of Crystalline and Amorphous Cellulose. In: *Journal of Applied Polymer Science,* **17,** 3627-3635.
2. Diebold J.P., Milne T.A., Czernik S., Oasmaa A. & Bridgewater A.V. (1997) Proposed Specifications for Various Grades of Pyrolysis Oils. In: *Developments in Thermochemical Biomass Conversion,* (Ed. By A.V. Bridgwater & D.G.B. Boocock), 433-447. Blackie Academic and Professional,Suffolk.
3. Reed T., Graboski M.S. & Levie B. (1994) Fundamentals, Development & Scaleup of the Air- Oxygen Stratified Downdraft Gasifer. The Biomass Energy Foundation Press, Colaroado
4. Grover P.D. (1989) *Biomass Thermo-Chemical Characterisation for Gasification.* Indian Institute of Technology, Delhi
5. Kaupp A. (1984) *Gasification of Rice Hulls.* Dt. Zentrum fur Techn. Entwicklungstechnologien-GATE in: Dt. Ges.fur Techn. Zusammenarbeit (GTZ) GmbH, Wiesbaden.
6. Di Blasi C., Buonno F. & Branca C. (1998) Reactivities of some biomass chars in air. In: *Carbon 37,* 1227-1238. Pergamon.
7. Manurung R., Continuous Open Core Downdraft Rice Husk Gasification.

Appendix 1

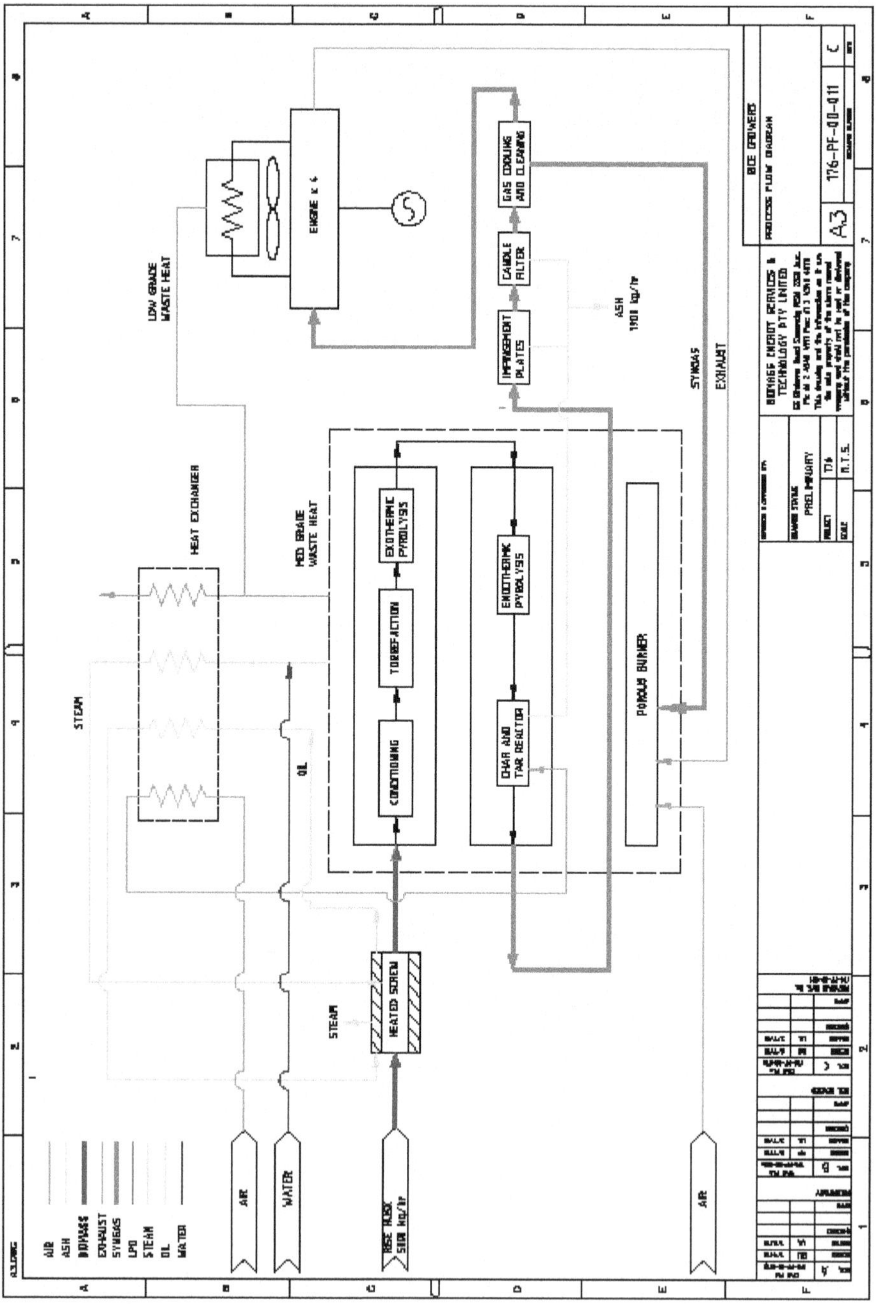

HEAT EXCHANGER
LOW GRADE WASTE HEAT
MED GRADE WASTE HEAT
ENGINE x 4
GAS COOLING AND CLEANING
CANDLE FILTER
IMPINGEMENT PLATES
SYNGAS
EXHAUST
CONDITIONING
TORREFACTION
EXOTHERMIC PYROLYSIS
ENDOTHERMIC PYROLYSIS
CHAR AND TAR REACTOR
POROUS BURNER
HEATED SCREW
STEAM
OIL
AIR
WATER
AIR
ASH
BIOMASS
EXHAUST
SYNGAS
LPG
STEAM
OIL
WATER
PRELIMINARY
N.T.S.
A3
176-PF-00-011
PROCESS FLOW DIAGRAM
BIOMASS ENERGY SERVICES & TECHNOLOGY PTY LIMITED

Gas Cleaning in Biomass Gasification Plants

M. Kleinhappl
Institute of Thermal Engineering, Graz University of Technology, Inffeldgasse 25
A-8010 Graz, Austria

ABSTRACT: The process of biomass gasification is a central aim of nowadays technical and scientific investigations. For medium scale gasifiers for woody biomass usually the technology of air blown atmospheric fixed bed gasifier system is used. The produced gas with low calorific value is burned in gas engines of combined heat and power plants (CHP).

In a general view of the plant the required gas quality must be demonstrated over the way of solving the R&D steps combined with economics efforts. The logical strategy follows technical and availability, environmental and hazard safety and last but not least the valuing measure of economics. The necessary gas quality due to the particulate and tar matter can be reached with just a small number of technologies, but also with a special sorted type of combination of process arrangements. Referencing on the literature and on the own experience just specially grouped processing types can fulfil the increased quality requirements. In evaluation of this some technologies fail as inefficient and not practicable. Good efficiencies have been reached with hot precipitation, controlled cooling, warm or hot gas filtration and wet scrubbing with combined aerosol separation.

The own tests on a 200 kW fixed bed gasification plant have shown a successful operation of a baghouse filter and a rather good operation of a solvent operated single stage scrubber. The basic principle of staged separation of the contaminants is carried on in the waste water treatment device.

INTRODUCTION

The introduction of gasification technologies into the market of energy conversion technologies takes now a several years long period [2,7]. In most cases the general problem of gas cleaning has disabled the widely breakthrough. In general the raised problems can be divided as follows: [3,15]

- Bad or unstable operation of the gasifier.
- Not practicable choice of gas cleaning technology.
- Strong interference between gasification and gas cleaning.
- Consequential damages caused by bad gasifier operation points (initial operation phase and start up).

At the Institute of Thermal Engineering, Graz University of Technology R&D work has been done at a fixed bed gasification plant. During 1998-2001 a special design of gas cleaning with combined waste water treatment for a biomass gasification CHP has been developed [8,9] as shown in Fig. 1. The basic data of this CHP are: [8,10]

- Gasification process: double fire, fixed bed.
- Fuel consumption: 50-55 kg/h (≈ 200 kW).
- Gasification agent: air and steam.
- Producer gas flow: 130-150 m^3/h st.c.[1].
- Gas utilisation with a CHP-plant: gas engine 50 $kW_{el.}$
- Data acquisition equipment for all process parameters.
- Technical devices for quality measurement in gas and process water [9].

[1] St.c.: Standard condition, 1,01325 bar and 0°C

Figure 1: General view of the plant.

GENERAL

GAS UTILISATION TECHNOLOGIES

For combined heat and power production in small and medium scale the following technologies are available: [2,15]

- Gas engines (IC-),
- Gas turbines,
- Fuel cells.

Each of these techniques requires specific qualities of producer gas. Gas engines are today usually available for electric power between 100 and 5000 kW.

GAS QUALITY REQUIREMENTS

For gas utilisation with gas engines some relevant requirements for long term operation are listed in Table 1. These parameters must be reached with an efficient gas cleaning equipment in all cases with stable clean gas quality. For the further strategic operation of gasturbines or fuel cells the limits are much lower [15].

Due to the consumer caused quality requirement in present the fulfilment for engine operation is general aim and for further prospects for realistic chances for operation with fuel cells and other "clean-gas-consumers" must be settled. Our status is shown strategic in Fig. 2.

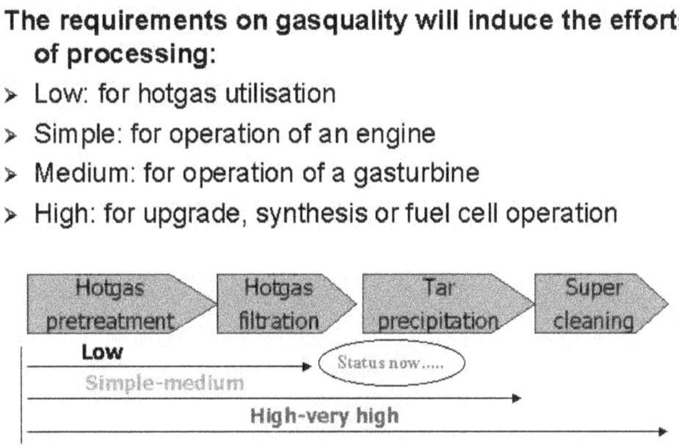

Figure 2: From "clean gas-consumer" induced efforts in gas cleaning [6]

TYPICAL PRODUCER GAS - A CHARACTERISATION - RELEVANCE TO GAS CLEANING

The producer gas can be characterised by useable heating value, the burnability describing numbers, the content of contaminants and the property in gas cleaning and last but not least by the stability of all this parameters.

The ranges of normal contaminants are discussed in Table 1, but each of them must be described in the way of treatability. The summary particulate can be divided into ash, charcoal, soot and condensed salts. And as an example charcoal and ash are very simple in precipitation, but soot as a nucleus carbon particulate is very difficult in precipitation due to the fine sizes [3,4], see Fig 3.

Table 1: Typical values for producer gas and standard limits for operation of IC-engines, in [mg/m³ st.c. dry Producer gas] [2,3,14,16]

Contaminant	Producer gas	Limits, ideal range
Particulate Matter	100-2000	< 50
Tars (heavy)	50-150	< 5
Tars (light)	100-1000	< 50
Phenoles	50-250	? but low
PAH (EPA)	200-600	? but low
Ammonia, as NH_3	150-800	< 55
Chlorides, as HCl	1-30*	<100
Cyanides, as HCN	12-40	? but low
Sulfides, as H_2S	5-25	<1150
Silicates	<1*	< 20
Alkali metals	<20*	? but low (catalyst)
Heavy metals	<1*	< 0.5 (catalyst)

* for uncontaminated woody biomass

Table 2: Typical parameters of producer gas, characterisation for fuelling of a gas engine [2,7,8].

Parameter	Dimension	Producer gas	Limits, ideal range
Temperature	°C	500-800	20-50
Water content	g/m³ st.c. dry	70-150	20-50
Heating value	MJ/m³ st.c. dry	3,5-10	Min. 4,0
Lam. flamespeed	cm/s	39	> 8
Methane number	-	87-92	> 70
Gas pressure	mbar		20-200

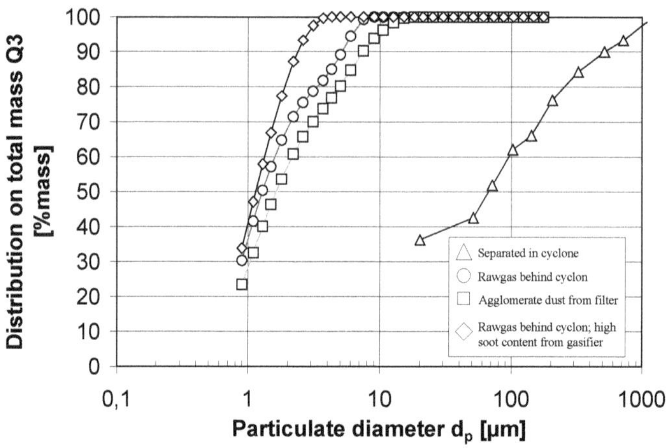

Figure 3: Particulate size distribution on mass for different dust fractions [8,9].

The recognised dust as a mixture out of all particulate components with condensed tar on them has completely differing properties from other kinds of dust. Therefore often the mechanical scrubbers nominated with high separation efficiencies, but not at conditions like gasification dust [3]. Due to the high charcoal and soot content these particulate are always from very low solid densities, and for soot matter with extremely low solid density [4,11,15]. Soot shows in addition the property of extreme electrostatic loading and bad wettability, which is caused by the organic matter and the low moisture content. The tar content as condensable organic matter is present in different component distributions, which are dominated by the "thermal" history where they come from [11,12]. In upstream gasifiers the untreated primary tar components and precursors are present in the producer gas, whilst in downstream gasifiers the ternary components predominate.

In fluidized bed gasifiers a summary of all tar modifications is present, with a serial reaction path through the bed and the freeboard [15]. Due to the age of the tar constituents the behaviour is very different. The relevant partial pressure and saturation pressure gives the reason for condensable and non condensable fractions. Primary tar components condense at much lower temperatures and ternary components at higher, that means earlier, which is important for the precipitation, where also the partial pressure is important. But the thermochemical reactivity is decreasing with status of tar formation, which is very important in catalytic treatment (conversion) [11,12].

The additional contaminants as ammonia, hydrogen sulfide and cyanides have a similar behaviour of different gasification methods. Their reactivity and solubility depends only on the solvent, pH and its temperature, or with chemical sorption to the reactants.

OVERVIEW OF GAS CLEANING TECHNOLOGIES

Gas cleaning methods can be grouped into the following basic principles: [3,8,9,15]

- Particulate precipitation
- Tar precipitation
- Precipitation of others

The different technologies are discussed in the following section and evaluated not as usually by the way of efficiencies, but by the relevant residual contamination after treatment.

For general overview the balance system of such a system should not be neglected. The Contaminants are separated from the producer gas, but disposed in another effluent (shown in Fig 4). We should always pursue the aim of a "zero emission" plant, but not only on the carbon dioxide. By the way of gasification of biomass more harmful effluents produced as in general known. For ecological sustainability always the "on-plant-treatment" of effluents should be included [6].

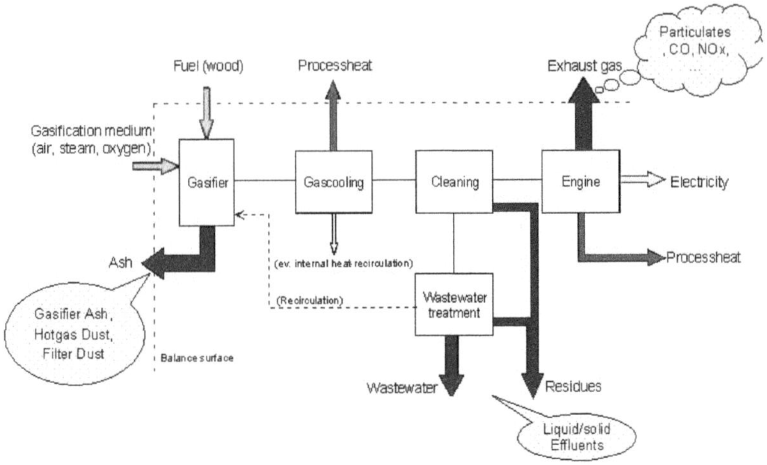

Figure 4: Particulate size distribution on mass for different dust fractions [6].

Particulate precipitation

By primary methods over staging and choice of flow principle the maximum of solid matter particulate contamination can be influenced, as also with the process control and design parameters (velocities as a reason for elutrition of particulates of fuel, charcoal and ash) [10].

With dry process steps like mass force separators (Dust cambers, cyclones) just a pre-treatment can be reached. With electric precipitators a very low residual content of particulate matter can be reached. With different filtering technologies also very low concentrations can be reached.

With the exception of the granular filters, where a bed of solids is used for gathering the dust particulates, all of them a very practicable. Agglomeration of fines on and in the filter cake can be an additional aid, which allows an efficient collection and back flushing.

By the way of wet particulate precipitation that means droplet contacting of particulates, just a medium contamination can be reached, due to the bad scrubbing properties of gasification dust and soot. From mechanical scrubbers like venturi and jet scrubbers just bad efficiencies can be reached [3,8,9,15].

For efficient operation a simple technology for pre-treatment is recommended, where the coarse particulates separated and a second precipitation is necessary to fulfil the end-quality of particulate content. Typical combinations are cyclones and fabric filters or electric precipitators [4,7,15].

Table 3: Residual particulate matter contamination after treatment in [mg/m³ st.c. dry producer] [3,7,8,9,15], with 1000 mg m³/st.c.dry at the inlet.

Technology	Contamination at the outlet
Cyclone	100-500
Wash tower, spray tower	200-300, 400-800
Mechanical scrubber	150-500
Sieve colum	50-200
Packed colum	50-150
Packed colum, solvent	10-100
Venturi scrubbers	20-100
Fabric filter, oneway	10-50
Fabric filter, pulse jetting	20-75
Granular bed filter, (sand)	50-200
Ceramic filter	10-50
Electric precipitator	5-50

Tar precipitation

With primary methods in gasification operation the tar content can be influenced very intensively, but therefore tar generated soot too [10,12]. But also the mentioned tar history and its influence on the composition will have a strong influence to the efficiency of catalytic treatment caused by the reactivity [11,12,15]. Very low and stable residual concentrations have not been reached over long periods. With partial combustion a loss in heating value of about 25 to 30% is often combined, but the tar contents can be reduced over secondary and ternary decomposition reactions [9].

In scrubbing technology just solvent operated columns can reach a low tar content, due to the ideal combination of cooling, condensation, absorption and dissolution of the tarry components. The adsorption on activated carbon is limited in the excess ratio of carbon input, but can't provide an economic operation, due to the high consumption of activated materials [3]. The adsorption in the filtercake of a fabric filter can be an additional effect, but isn't efficient enough to precipitate the light tars [3,15]. The filter temperature must kept over the condensation range of 100-150°C otherwise excessive pre-coating and adding is necessary to prevent cloth condensation [4].

With cooled ESP the tar separation is possible to very low contents, due to the condensation effect and the effective tar aerosol capture [3,4,15].

Table 4: Residual tar matter contamination after treatment in [mg/m³ st.c. dry producer gas] [3,4,7,8,15], with 500 mg m³/st.c.dry at the inlet.

Technology	Contamination at the outlet
In situ catalyis	50-200
Part. combustion	10-100
Post catalyis	50-250
Wash tower	300-400
Spray tower	400-500
Sieve colum	100-200
Packed colum	100-200
Packed colum, solv.	10-100
Venturi scrubber, drop.	40-200
Adsorption	50-200
Co adsorption on filtercake	200-500
Electric precipitator cooled	5-50

For hot gas cleaning a new approach will be the catalytic active ceramic filter candle, which combines dust precipitation and tar conversion [15]. But for cold gas utilisation catalytic processing is just for pre-treatment, the necessary low tar contents can be reached with solvent operated colums or an cooled ESP. The ESP gives the disadvantage that often particulates, tars and condensate will appear as a sticky mixture, which can be disposed only over incineration. Organic solvent agents can be used as additional fuel in the gasification plant [9].

Precipitation of others and cooling

Ammonia, HCl and others can be precipitated with high efficiency only in scrubbers with water contact [9]. Especially for ammonia is due to the saturation a continuous circulation of the scrubbing agent through waste water treatment necessary. For increased volumetric heating value due to dehumidification a cooling with water condensation should be realized, with the advantage that the vapour pressures of contaminants also decrease [8].

OVERVIEW OF WASTE WATER CLEANING TECHNOLOGIES

For an ecological operation of a gasification plant it is necessary to give the pre-purified waste water to the sewer, which can fulfil the emission standard. Due to the strong influence of the used scrubbing agent and it's saturation with ammonia or light tars water treatment is also important for limitation of the contamination of internally circulation scrubbing agents. The following contaminants are present [8,9]:

- Particles, dissolved inorganic and organic matter
- COD, PAH, VOC
- phenols, ammonia, cyanides
- other oxogenates (aliphatic and aromatic)

Several of these contaminants are handled as cancerogenic. For the treatment of the main organic contaminated wastewater several ways of processing are available, but just evaporation, thermal desorption and super cleaning by adsorption have shown the necessary qualities after treatment. For further information see [1,9].

RECENT WORK AND RESULTS

GAS CLEANING SYSTEM

The testplant is configurated for measurements as follows (see in Fig 5 and Fig. 6):

To prevent pollution during the startup phases the first step of gas cleaning occurs in a special burning chamber, in which the startup producer gas can be burned partially. The raw gas, which comes out from the gasifier with a dustload of about 600-1500 mg/Nm³ (dry basis) is precleaned of particles by the way of a cyclone. For a median particle size of 3 µm for ash and 6 µm for charcoal and soot. Then the hot raw gas is cooled down to a temperature of 250-300 °C in a tube steam generator, which produces atmospheric saturated steam used for internal and external heat demands. The operation on temperature profile is shown in Fig. 6 [15].

For precipitation of fines like soot a baghouse filter works at 150-200°C. The backflushing works with nitrogen pulse jets, which are activated over the pressure loss (Fig. 7).

In the scrubbing stage the gas is treated with a venturi scrubber, which exhausts a mixture of liquid and gas into a packed column for droplet separation. The circulated washing agent can be re-cooled. The gas outlet temperature is controlled. But circulating treatment is necessary due to a possible saturation with ammonia and light tars. In this way the scrubbing device is supported in effect.

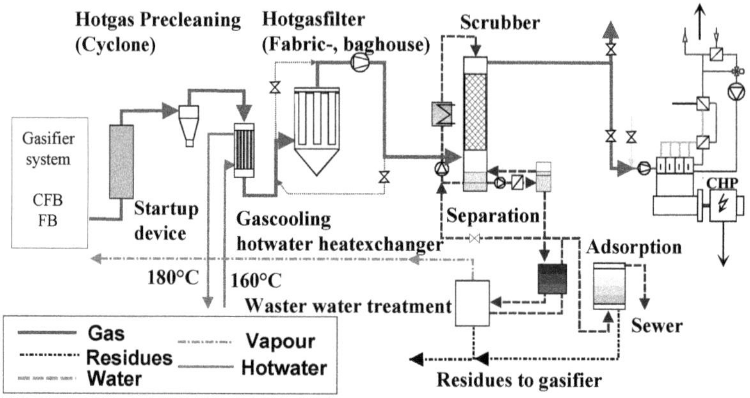

Figure 5: Complete flowsheet of a staged gas cleaning [6,7,8].

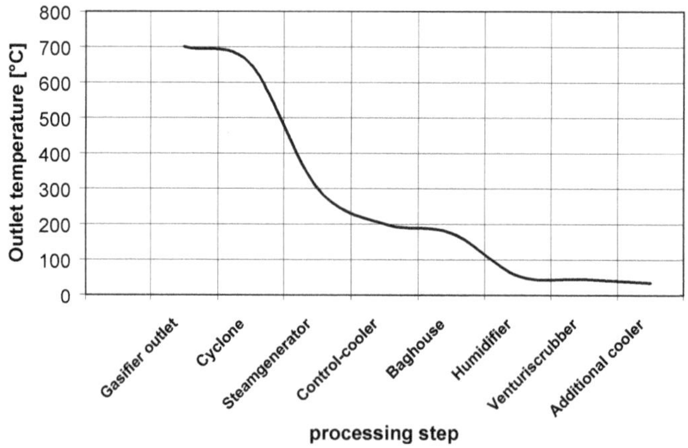

Figure 6: Temperature profile during gas treatment [6].

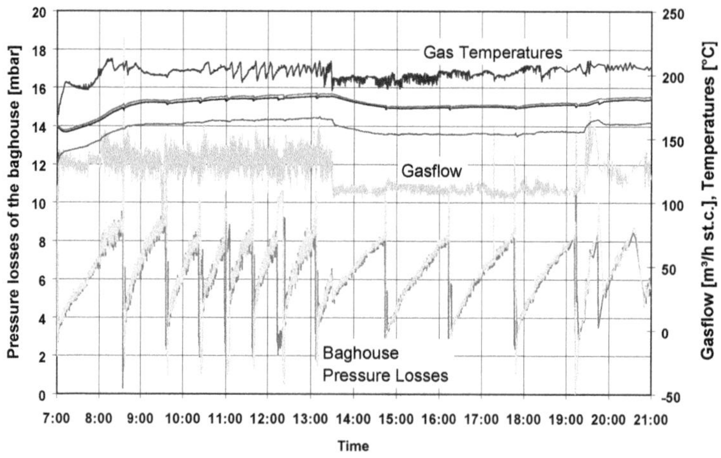

Figure 7: Process look at the working baghouse [6].

For testing the cooling effect on contamination the last stage of the gas treatment consists out of a counter current flow sieve plate column, in which circulates condensate for direct contact cooling and dehumidification.

For compensation of the pressure drops in the gas path of the plant two blowers are integrated. Overall internal power consumption from the gas cleaning 3 kWe (average). The gas flow can be fed to the CHP or alternatively to an externally mounted flare. For water treatment see [9].

GAS QUALITY CONTROL

For measurement and control of the raw gas and the cleaned gas the following parameters are taken into consideration: [6,9,13,14]

- Particle content size >0,42 μm
- extractable matter from the particle load
- condensable and scrubbing tar at 0 °C
- condensable, absorbable and scrubbing tar at –3 °C
- content of ammonia and hydrocyane HCN

The particle and tar measurement is carried out according to the known tar measurement protocols [6,13,14]. As illustrated in Fig. 9, the sampling is carried out with impingers with and without frits (according to the tarexpert meeting 2002 in Amsterdam) and as absorption liquid a mixture of toluene and isopropanol is used. All fractions are sampled and determined on each test run. The margin of detection of this method was determined for components which are as heavy as naphthalene.

The contents of NH_3 and HCN are measured indirectly by absorption in liquid chemicals and wet chemical analysis. For characterising the tar GC-MS technology is used from - 50°C cold trapping samples.

Some representative results are shown in Fig. 8. The ammonia content was found in about 500-600 mg/m³ st.s. on the rawgas and about 20-50 mg/m³ st.c. in the cleaned gas. For ammonia it is very important, that the scrubbing is treated online saturation will be reached [1,6,7].

For basic data of the process water see [9].

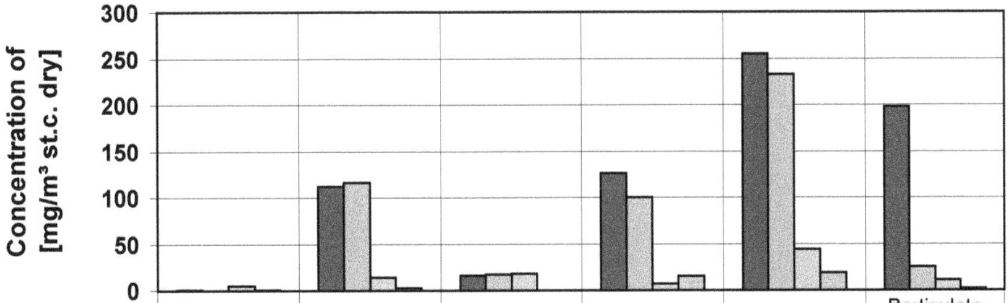

	Filter, tar	Cond., tar	Water, tar	Toluene, tar	Summary, Tar	Particulate Matter
Gasifier, Cyclone	1	112	16	126	255	198
Baghouse	0	116	17	100	233	25
Venturiscrubber	5	14	18	7	44	11
Last Cooling	1	3	0	15	19	2

fractions in tar traping

Figure 8: Representative outlet contaminants of tar and particulates; note gasifier outlet measured after the cyclone.

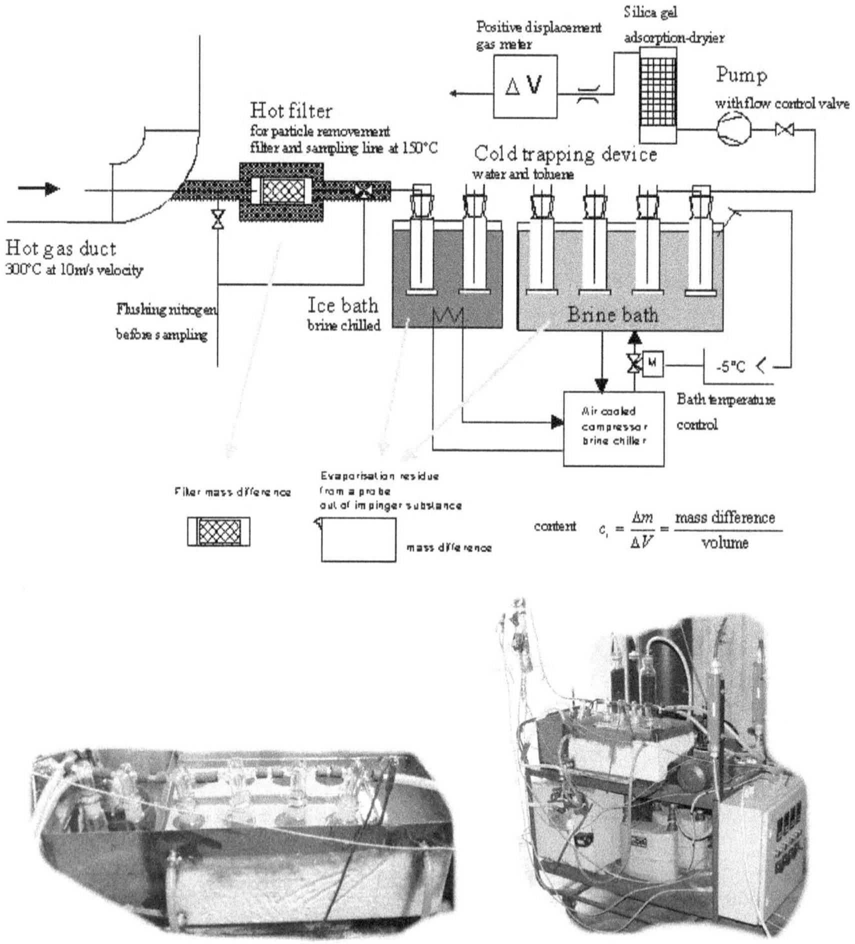

Figure 9: Sampling equipment with impingers and picture.

OTHER RELATED WORK

As above mentioned a lot of other related work is done:

In the related work of my colleagues F. Lettner and Ch. Wallner, serious progress has been made in last time in gasification with staging, improved air supply system and single process optimisation [10]. In the field of safety-needs, automation and process control developments are made for further piloting the process principle. These are necessary preparations for further piloting this principle at a plant site.

CONCLUSIONS

According to the combination of gasification technology and planned aggregate for utilisation the basic design for gas treatment can be defined. Preferring for demonstration of gasification technology the prior available gas engine (suction or turbocharger) is established with defined costs in investment and operation.

The gas quality for disturbance less and efficient engine operation can only be reached with a proper gas cleaning system, which is designed on the characteristics of the producer gas.
Primary and overall measures can decrease the caused efforts in following parts of the chain from biomass to combined heat and power. A logical strategy follows technical durability and availability, environmental aspects and hazards/safety and last but not least the valuing effect of economics.

The necessary gas quality due to the particulate and tar matter can be reached with just a few number of technologies, like dry dust separation and wet scrubbing with solvents. For low environmental impact the treatment of circulation solvent and condensate is fully integrated on plant, in definition "zero emission plant". Our own developments show promising results which gives the base for further development.

For the combined contamination of gasifiers producer gas a coupled dry and wet gas cleaning facility is necessary for the fulfilment of the requirements of gas engine operation. In staged treatment particulates are separated with a baghouse filter at 150-200 °C and the tar is precipitated with a solvent working scrubber. The results of test runs are promising. In the actual work the dimensioning guideline for plant design is developed.

ACKNOWLEDGEMENT

These work has been supported by the Styrian electricity utility STEWEAG/ESTAG/ATP.

ABBREVIATIONS

St.c.: Measurement conditions are defined with dry volume basis at standard conditions like: 1, 01325 bar or 1 atm at 0 °C.

REFERENCES

1. Amann G. (2001) m. thesis *Measurements on a gasification plant for biomass.* Technical University Graz.
2. Bühler R., et al (1997) *Proceedings: IC engines for LCV gas from biomass gasifiers.* IEA Thermal Gasification Seminar, Zürich.
3. Hasler P., et al. (1998) *Gas Cleaning and Waste Water Treatment for Small Scale Biomass Gasifiers.* Report of EU Project JOR3-CT95-0084.
4. Hindsgaul C.(2000) *Low Temperature Particle Filtration of Producer Gas with Low Tar Content.* DTU
5. IEA Working Group Bioenergy (1998) *Provisional Protocol for the sampling and analysis of tar and particulates from Biomass gasifiers* Bioenergy Agreement of Bioenergy Gasification Task.
6. Kleinhappl M. (2002) *Gas cleaning in Biomass Gasification Plants.* Paper and oral Presentation, 12.th European conference and Technology Exhibition on Biomass for Energy, Industry and Climate Protection, Amsterdam.
7. Kleinhappl M., F. Lettner (2001) *Research in Biomass utilisation.* 4.th Internat. Meeting of Producer gas Technology Karlsruhe.
8. Kleinhappl M., M. Pogoreutz (2000) *Gas cleaning in Biomass Gasification.* EFG-report, Technical University Graz
9. Kleinhappl M. (2000) *Gas cleaning in Biomass Gasification Plants and Tarmeasurements at a Doublefire Gasifier.* 1.st World conference on Biomass for Energy and Industry, Sevilla.
10. Lettner F. (2002) *Dimensioning-Guidelines for Biomass-Fixed-Bed –Gasifiers.* 12.th European conference and Technology Exhibition on Biomass for Energy, Industry and Climate Protection, Amsterdam.

11. Milne T.A., et al. (1998) *Biomass Gasifier "Tars": Their Nature, Formation and Conversion,* NREL, Colorado, USA
12. Morf Ph., et al.(2002) "*Mechanisms and kinetics of homogenous secondary reactions of tar from continuous pyrolysis of wood chips*. Fuel, 81, page 843-85
13. Neeft J.P.A., H. Knoef, et al. (1999) *Guideline for Sampling and Analysis of Tar and Particles in Biomass Producer Gases V3.1.* European Energy project EEN5-1999-00507 Tar protocol.
14. Stahlberg P., et al. (1998) *Samling of contaminants from product gases of biomass gasifiers.* VTT Technical Research Centre of Finland.
15. Don Stevens J. (2001) *Hot Gas Conditioning: Recent Progress With Larger-Scale Biomass Gasification Systems.* NREL, Colorado, USA
16. Zielke U. (1997) *Tar and solid Particles in LCV-gas from fixed bed gasifiers.* IEA Seminar: IC Engines for LCV-gas from Biomass gasifiers, Zürich, Switzerland.

Hydrogen Production via Biomass Gasification – A Life Cycle Analysis Approach

C. Koroneos, A. Dompros, G. Roumbas, N. Moussiopoulos
Laboratory of Heat Transfer and Environmental Engineering, Aristotle University of Thessaloniki, P.O. Box 483, 541 24 Thessaloniki, Greece

ABSTRACT: The use of hydrogen as an alternative fuel is gaining more and more acceptance as the environmental impact of hydrocarbons becomes more evident. Hydrogen production is accomplished by steam reforming of natural gas and other fossil primary energy at approximately 97% of total. Thus, less than 3% is based on renewable energy sources. The use of biomass gasification is investigated in the production of hydrogen. The technical and environmental performance of such systems is examined using the method of life cycle analysis (LCA). LCA is a powerful tool in the valuation of the environmental impacts of processes. Its main advantage in environmental decision-making is providing a quantification basis for assessing potential improvements in environmental performance of a system throughout its life cycle.

INTRODUCTION

Hydrogen has the potential to be a clean alternative to the fossil fuels currently used in the transportation sector. This is especially true if the hydrogen is manufactured from renewable resources, primarily sunlight, wind, and biomass.

Use of wood and other forms of biomass as fuels for generating electricity, heat and energy cariers like hydrogen has become a focus of renewed interest in many parts of the world. Biomass is an indigenous, often cheap, and above all renewable fuel. The increasing availability of biomass, combined with the recent development of technologies to use it efficiently and with low levels of emissions, promise to make biomass an increasingly attractive fuel option.

Although biomass is being used in industrial applications in Central and Eastern Europe and in developing countries, the systems in operation are often old, inefficient, and poorly maintained, resulting in waste and pollution. Yet biomass, in contrast to fossil fuels, has a unique potential for making a positive environmental impact.

Biomass can be burned without emitting large amounts of nitrogen oxides (NO_x). Moreover, because the sulfur content of biomass is very low, emissions of sulfur dioxide (SO_2) will also be low, especially compared with the emissions from coal-fired combustion. In addition, using biomass as a fuel would mobilize previously bothersome or environmentally detrimental resources (such as agricultural residues and residues from the wood-working industries)

Using sustainably grown biomass will not result in net CO_2 emissions, as the same amount of carbon released in combustion has been recovered from the atmosphere during biomass growth. The utilization of biomass-derived fuels might, however, result in CO_2 emissions from fossil fuels used for biomass production, transportation, conversion and distribution of final energy carriers. Biomass based energy carriers may however, be used also in these operations and reduce or even eliminate net CO_2 emissions.

Hydrogen can be extracted from biomass sources like wood chips and agricultural waste and utilised as an energy carrier. Biomass has the potential to accelerate the realization of hydrogen as a major fuel of the future. Since biomass is renewable and consumes atmospheric CO_2 during growth, it can have a small net CO_2 impact compared to fossil fuels used for hydrogen production. However, hydrogen from biomass has major challenges. There are no completed technology demonstrations. The yield of hydrogen is low from biomass since the hydrogen content in biomass is low to begin with (approximately 6% versus 25% for methane) and the energy content is low due to the 40% oxygen content of biomass. Since over half of the hydrogen from biomass comes from spitting water in the steam reforming reaction, the energy content of the feedstock is an inherent limitation of the process.

The purpose of this analysis was to assess the environmental feasibility and efficiency of producing hydrogen from biomass via two processes: 1) gasification followed by reforming of the syngas, and 2) gasification followed by electricity generation and electrolysis. In each process, hydrogen is stored in liquefied state. The systems examined are based on the Battelle/FERCO low-pressure indirectly heated biomass gasifier. The main advantage of the first system is the higher efficiency in hydrogen production. The second system has the advantage that because electricity can be transferred easily the biomass to electricity plant can be located near biomass sources and the hydrogen production plant near consumption sources thus minimizing transportation costs.

For the assessment of the environmental efficiency of the two systems the Life Cycle Analysis (LCA) methodology is utilized. LCA is a systematic analytical method to identify, evaluate and help minimize environmental impacts of a specific process or competing processes. Material and energy balances are used to quantify emissions, resource depletion and energy consumption of all processes between transformation of raw materials into useful products and the final disposal of all products and byproducts. The results are then used to evaluate the environmental impacts of the process so that efforts can be focused on mitigating possible effects.

The primary purpose of conducting this LCA study was to answer that are raised about biomass power in regards CO_2 and energy use, and to identify other environmental effects that might become important once such systems are implemented.

The systems under investigation are assumed to be located in an agricultural region of Greece so the biomass potential of Greece is presented first.

BIOMASS POTENTIAL AND CURRENT EXPLOITATION IN GREECE

Biomass, especially agricultural wastes, constitutes an important energy source that could have a great impact in the sustainable development of Greece. Agricultural wastes consist mainly of straw, rice bark, corn shanks, cotton shanks, tobacco shanks, peach cores and olive cores, which are materials with high energy potential. Only a small percentage of this energy potential is exploited through combustion. The available biomass potential in Greece is sufficient to supply of a large percentage of the energy demands in the manufacturing sector. Proper management and exploitation of this potential could lead to economically profitable processes for the agricultural industry. Furthermore, pyrolysis and gasification techniques in the production of liquid and gaseous fuels, could establish a new approach for a more efficient utilization of biomass. The biomass sources in Greece that could be exploited at a regional level are (5):

- *Biomass from cotton cultivation and processing*: The magnitude of cotton cultivation in Greece today has reached the level of 4,000 square kilometers, which produce 1,200,000 tn of teased cotton. 120,000 tn of cotton gin wastes result from the ginning of this quantity, namely 42,000 tn of oil equivalent (TOE). Furthermore, the total amount of cotton shanks that remains after the collection of the seeds is 350 tons per square kilometer with heat content of 14,235 kJ/kg. The production of cotton shanks reaches 1,400,000 tons annually, an amount that corresponds to 408,000 TOE.
- *Olive foot cake waste*: The solids contained in the liquid wastes make up 12 % of the olive crop with heat content reaching 24,409 kJ/kg. It is estimated that 100,000 TOE of olive foot cake waste is being unused in Greece.
- *Biomass from corn*: Corn cultivation in Greece reaches a total area of 2,000 square kilometers and the production of corn biomass is estimated to 1,200-1,500 tons per square kilometer. The above figures reveal that the annual production of dry biomass that is destroyed through combustion in the fields is equivalent to 800,000 TOE.
- *Biomass from rice*: The rice cultivation area in Greece covers 250 square kilometers. Rice production is estimated to 700 tons per square kilometer of cultivated land, namely 175,000 tn. During rice processing, 35,000 tn of rice bark are produced, a quantity that corresponds to 10,000 TOE.

Other biomass sources available in Greece originate from straw, tree branches, peach cores, sawdust etc. Table 1 summarizes the energy levels for each biomass source (5).

Table 1: Biomass energy potential.

Biomass	Tons of oil equivalent (TOE)
Cotton gin	$0.42\ 10^5$
Cotton shanks	$4.08\ 10^5$
Olive foot cake	$1.00\ 10^5$
Corn	$8.00\ 10^5$
Rice	$0.10\ 10^5$
Total	$13.60\ 10^5$
Gross Transport Consumption	$65.00\ 10^5$

Biomass is generally exploited within its area of origin for the partial fulfilment of manufacturing production energy demand using combustion techniques (5).

- *Cotton gin waste*: From the total available quantity of cotton gin waste only a small portion, 35.000 tn, is currently exploited for energy purposes. Consequently, there is an enormous quantity of biomass that remains unexploited since cotton gin waste is considered unsuitable for animal feeding.
- *Cotton shanks*: Cotton shanks remain in the fields and are shattered and dispersed by the majority of the producers. However, the residual of cotton shanks either in their integral or dispersed form, generate pathogenic microorganisms that are harmful to fresh plantations. Consequently, a bigger demand arises for pesticides that have as a result great environmental implications.
- *Olive foot cake waste*: As far as the olive foot cake waste is concerned, there has been no effort to exploit them. Olive foot cake wastes constitute a major environmental problem due to their elevated concentration in organic compounds.
- *Biomass from corn*: No effort is made towards the collection and exploitation of corn biomass and as a result it is burnt in the fields since it is unsuitable for animal feeding.
- *Biomass from rice cultivation and processing*: Rice biomass is partly exploited in the rice mills for energy purposes. Furthermore, the ash that is generated during pyrolysis has a high content in SiO_2 and is profitably exported to European countries.
- *Straw Biomass*: Straw is partly used for animal feed as well as for several other agricultural and industrial uses. However, a substantial quantity remains in the fields and is destroyed by burning.
- *Tree Branches*: Tree branches from lopping are burned on the spot and only a small quantity is used in farmhouses.
- *Peach Kernels*: Peach kernels are exploited in small-scale greenhouses for heat generation.
- *Sawdust*: Sawdust is combusted in large industrial installations for energy production. Nevertheless, it is considered as waste in smaller industries and only in some cases it is utilized for heat generation in the winter.

HYDROGEN PRODUCTION SYSTEMS INVESTIGATED

Block flow diagrams of the two options for producing hydrogen from biomass are shown in Figs. 1, 2. Both systems utilize Battelle/FERCO low-pressure indirectly heated biomass gasifier for the production of synthesis gas (syngas) (1, 3, 6, 7). The first system utilises syngas via catalytic steam reforming and water swift reaction for the production of hydrogen (Fig. 1). The second system utilises syngas for electricity generation via a combine cycle and then by electrolysis hydrogen is produced.

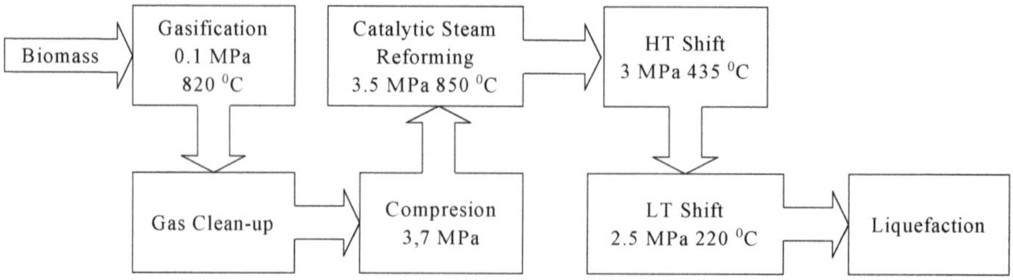

Figure 1: Hydrogen production by reforming synthesis gas (syngas).

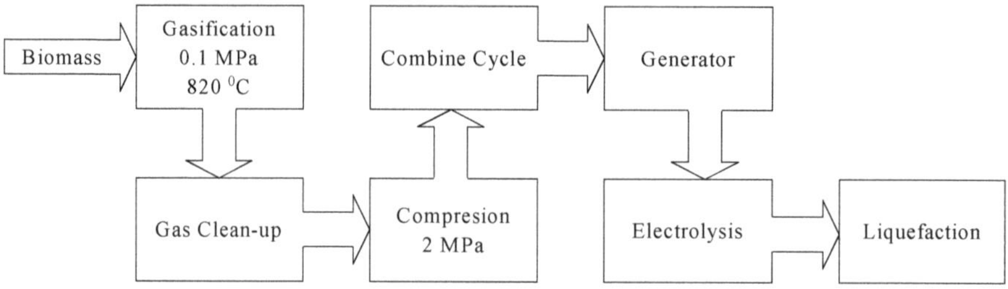

Figure 2: Hydrogen production by electrolysis.

LOW-PRESSURE INDIRECTLY HEATED BIOMASS GASIFIER

The low-pressure indirectly-heated gasifier selected for this study was developed by Bettelle Columbus Laboratory (BCL) specifically for biomass gasification. A schematic of this gasifier is shown in Fig. 3 (1, 6).

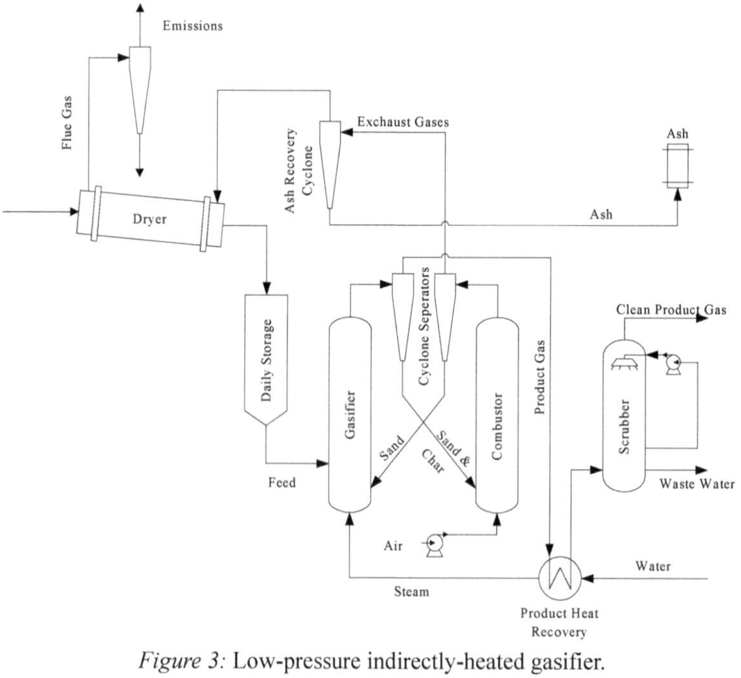

Figure 3: Low-pressure indirectly-heated gasifier.

This system is called indirectly-heated because the heat necessary for the endothermic gasification reactions is supplied by hot sand circulating between the char combustor and the gasification vessel. The distinctive feature of the gasifier is that unlike direct heated gasifiers, which use both steam and air, only steam is injected with the biomass to promote gasification. Therefore the syngas produced has a greater calorific value than that produced by air blown gasifiers. The gasifier operates at nearly atmospheric pressure at 825 °C.

In the present study it is assumed biomass with 50% moisture content, which is reduced in the dryer to 11%. The dried biomass enters the gasifier where it is converted to syngas and char. The biomass composition used for this study is typical of woody biomass (1). The operating conditions of the gasifier are given in table 2 (1). Biomass and composition of the product gas are given tables 3 and 4 (1, 6, 7).

Table 2: Indirectly Heated Gasifier Operating conditions.

Gasifier Temperature (^{0}C)	826	Steam / Biomass Ratio	0.45
Gasifier Pressure (Mpa)	0.17	Sand / Biomass Ratio	0.03

Table 3: Biomass elemental analysis.

Component	% weight dry basis
Carbon	50.88
Oxygen	41.90
Hydrogen	6.04
Nitrogen	0.17
Sulfur	0.09
Ash	0.92
Higher Heating Value (HHV_b) MJ/kg	20.225 MJ/kg

Table 4: Product Gas Composition.

Component	% mol
Hydrogen (H_2)	21.28
Carbon Monoxie (CO)	43.16
Carbon Dioxide (CO_2)	13.45
Methane (CH_4)	15.83
Acetylene (C_2H_2)	0.36
Ethylene (C_2H_4)	4.62
Ethane (C_2H_6)	0.62
Tars	0.40
H_2S	0.08
NH_3	0.37

STEAM REFORMING OF PRODUCT GAS

The reforming process, shown in Fig. 4, is similar to that of used in hydrogen production of natural gas (6). The major unit operations are a primary reformer to convert methane and the higher hydrocarbons present in the syngas to hydrogen, plus shift reaction to convert CO to hydrogen. The reactions governing the reforming process are shown in the following equations:

$$C_nH_m + nH_2O + \text{heat} \rightarrow nCO + (m/2 + n)H_2 \qquad (1)$$

$$CO + H_2O \rightarrow CO_2 + H_2 + \text{heat} \qquad (2)$$

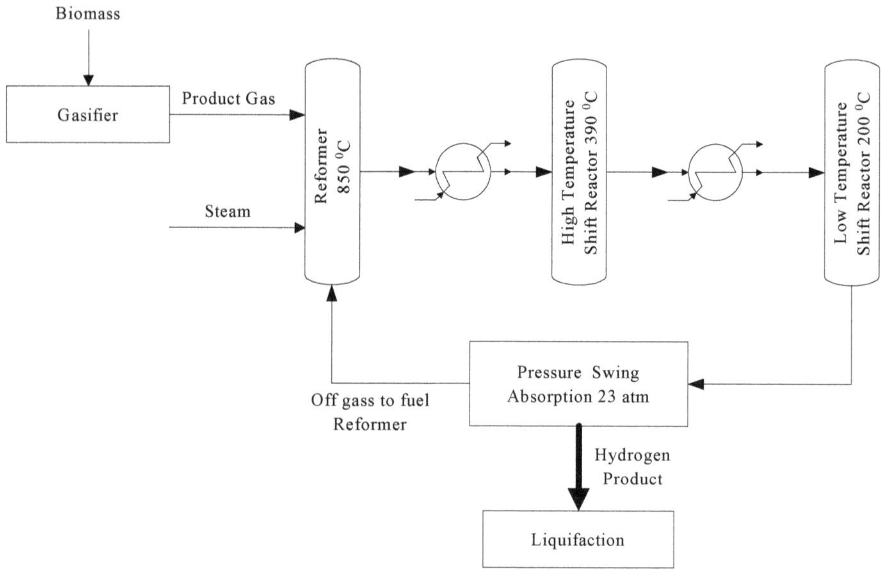

Figure 4: Reforming Process.

The primary reformer converts the methane and the higher hydrocarbons to CO and H_2 (Reaction 1) and performs a significant portion of the water-gas swift reaction (Reaction 2). According to the thermodynamics of the process practically all of the C_2H_x species are consumed, 60% of the CH_4 is converted and there is a 22% net conversion of CO. In the high and low temperature shift reactors nearly complete conversion of CO is accomplished. The produced gas from the shift reactors has a dry-basis composition of 61% H_2, 34.1% CO_2, 2.4% CH_4 and 1.1% CO. The produced gas is purified in the Pressure Swing Absorption (PSA) process, which requires at least 70 mol% hydrogen in the input stream. Thus the incorporation of recycle is necessary. The overall separation recovery is assumed to be 77%. The flue gas from the PSA unit is used for fueling the steam reformer.

BIOMASS TO HYDROGEN VIA ELECTROLYSIS

The biomass to hydrogen via electrolysis is presented schematically in Fig. 5 (1, 7). The product gas is utilized by a combine cycle power generation system to produce the necessary electricity. The clean gas from the gasifier (~97 ^{0}C) is compressed from 172 kPa to 2068 kPa and is heated up to 371 ^{0}C indirectly with process heat from the char combustor and the scrubber. The Hot clean product gas is introduced into the gas turbine combustor along with air from the high-pressure turbine compressor. Gas turbine exhaust is ducted to the heat recovery steam generator (HRSG), which incorporates a superheater, high and low pressure boilers, and economizer. In the steam cycle, superheated steam at 538 ^{0}C and 10 MPa is expanded to the high pressure (HP) turbine. The steam is then combined with steam from the low pressure (LP) boiler, reheated and introduced into the intermediate pressure (IP) turbine. Exhaust from the IP turbine is passed through the LP turbine. Exhaust steam from the LP turbine is used for gasification.

Hydrogen production by electrolysis has been one of the current methods that is applied broadly and has become more mature. The overall energy efficiency of the electrolysis process is assumed to be 77% (4). Construction materials data were taken from GEMIS 2001 (4).

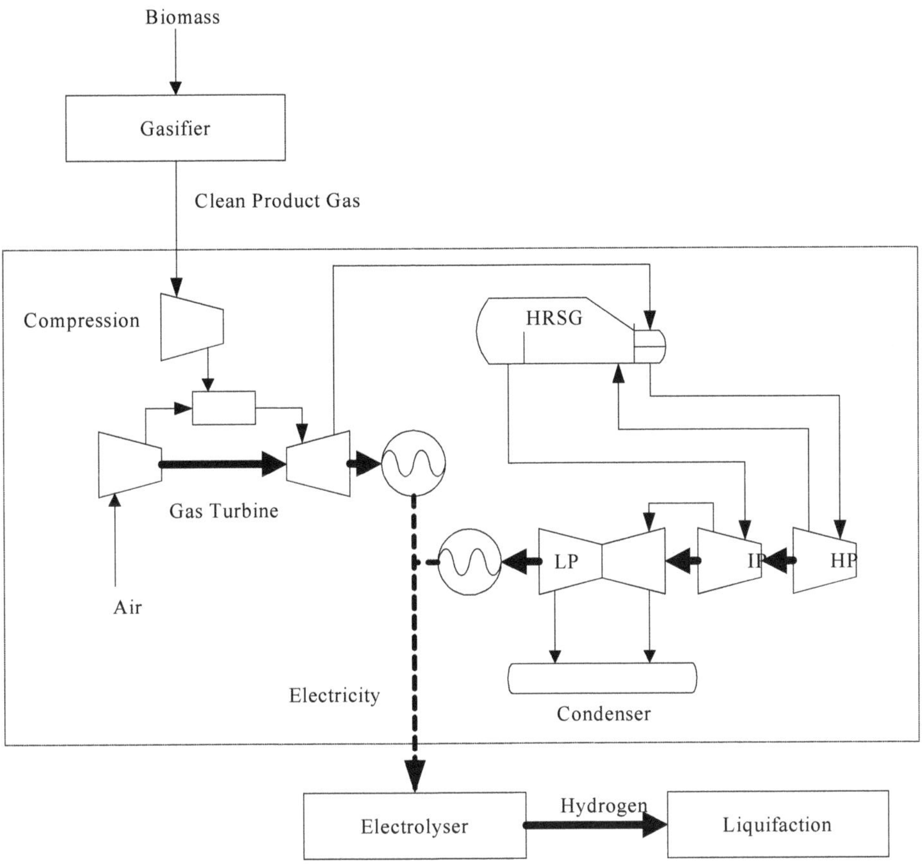

Figure 5: Biomass to Hydrogen via electrolysis.

LIQUEFACTION PROCESS

Hydrogen must be cooled down to –253°Celsius for it to liquefy. From the thermodynamic point of view the best liquefaction process is a combination of isothermic compression followed by adiabatic expansion, whereby the gas cools down due to the Joule Thomson effect. A quantity of 0.97Kwh/kg heat, a condensation enthalpy of 0.13kW/kg and an energy release of 0.2kW/kg due to the Ortho-Para-Conversion has to be withdrawn for liquefaction of hydrogen The theoretical minimum energy requirement is due to the Carnot-efficiency much higher, around 4kWh/kg, depending on process management. In reality, however, none of these ideal processes is reached and therefore plants cool down the gas gradually, usually by pre-cooling it with liquid nitrogen. An electricity requirement of 0.347MJ/MJ is given in the GEMIS-database (4), (refer to 30 bar inflow pressure).

BIOMASS TO HYDROGEN LIFE CYCLE ANALYSIS RESULTS

The complete system investigated is shown in Fig. 6. The functional unit for the evaluation of the cycle is 1 MJ of produced Hydrogen.

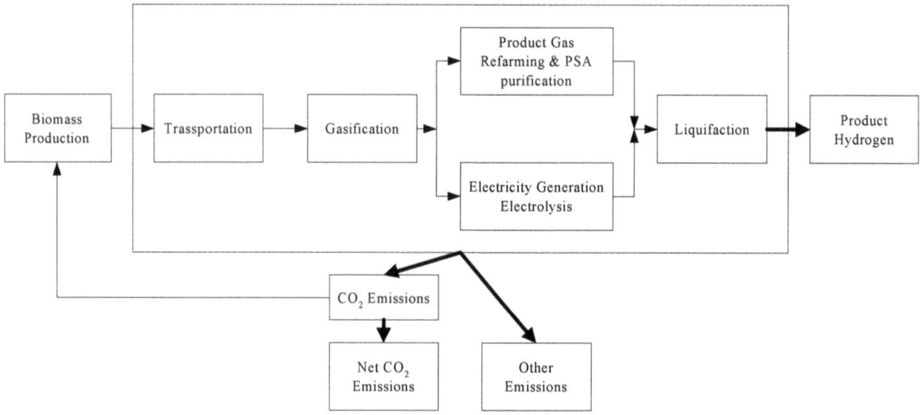

Figure 6: The Life Cycle investigated.

The life cycle can be described briefly as follows:

1. Biomass acquisition and transportation
2. Gasification of biomass
3. Hydrogen production via electrolysis or steam reforming process
4. Liquefaction and storage of hydrogen

The characterisation results of some selected impact categories are shown in Figs. 7, 8 and 9. As it can be seen the gasification to electricity to hydrogen system has better environmental performance in almost all impact categories. This can be seen more clearly in the evaluation results of the impacts.

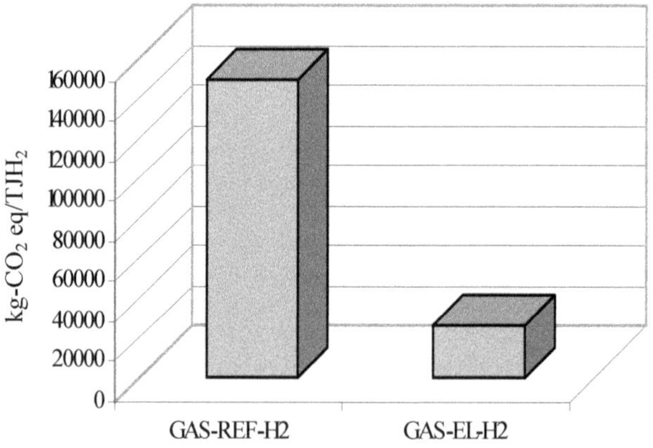

Figure 7: Greenhouse gas emissions, GAS-REF-H2 for the gasification-reformer and GAS-EL-H2 for the gasification-electrical generation system.

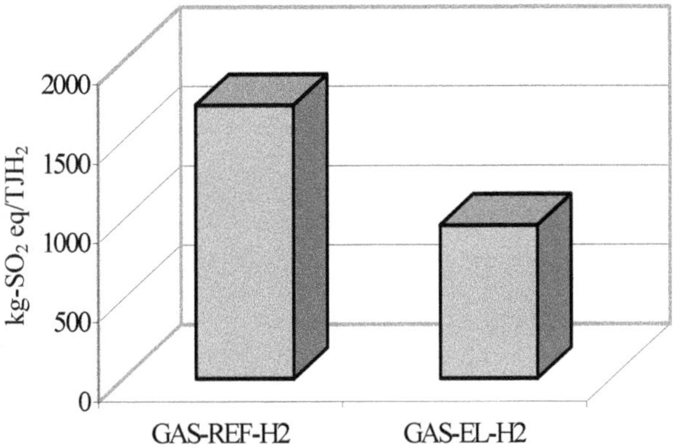

Figure 8: Acidification effect, GAS-REF-H2 for the gasification-reformer and GAS-EL-H2 for the gasification-electrical generation system.

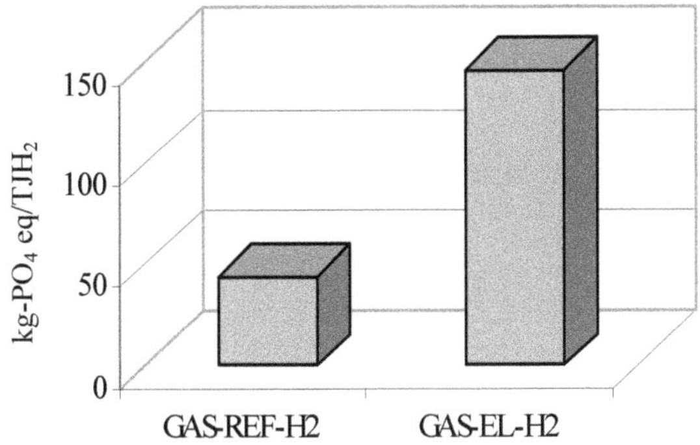

Figure 9: Eutrophication effect, GAS-REF-H2 for the gasification-reformer and GAS-EL-H2 for the gasification-electrical generation system.

The evaluation of the environmental score of each effect is calculated based on the simple formula: Environmental Score = Characterised Value x Normalisation x Weighting factor

Due to lack of such factors for Greece, the classification of the inventory data to impact categories was made using EcoIndicator 95 method. The factors used are presented in Table 5.

Table 5: Normalisation and weighting factors.

Class	**Normalisation**	**Evaluation**
Greenhouse	0,0000742	2,5
Acidification	0,00888	10
Eutrophication	0,0262	5
Heavy metals	17,8	5
Carcinogenesis	106	10
Winter smog	0,0106	5
Summer smog	0,0507	2,5

Valuation factors are used in order to show the relative importance of these effects. The seriousness of the impact of the environmental effects is measured in indicator points. The valuation scores are added up to give a total impact. The gas reformation system has greater impact score (263.4) than the electrolysis system (123.03). The "indicator" graph (Fig. 10) shows that the acidification effect is by far the biggest contributing factor.

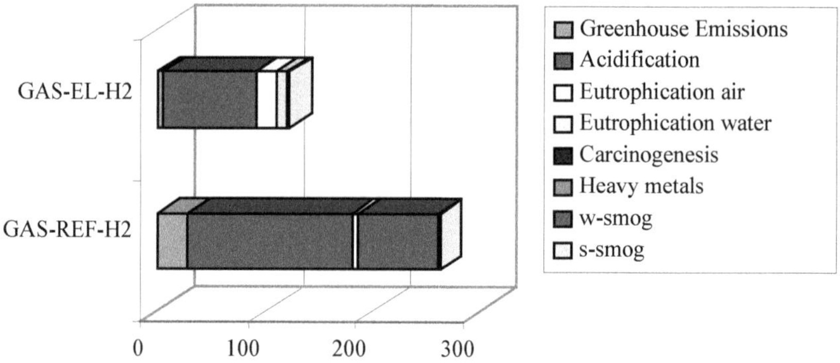

Figure 10: Indicator graph, GAS-REF-H2 for the gasification-reformer and GAS-EL-H2 for the gasification-electrical generation system.

This can also be seen on Fig. 11 where the acidification effect accounts for 69.72% of the total impact for the gasification-electricity generation system while 58.05% for gasification-steam reforming system. The greenhouse effect has larger contribution to the gasification-steam reforming system (28.95%).

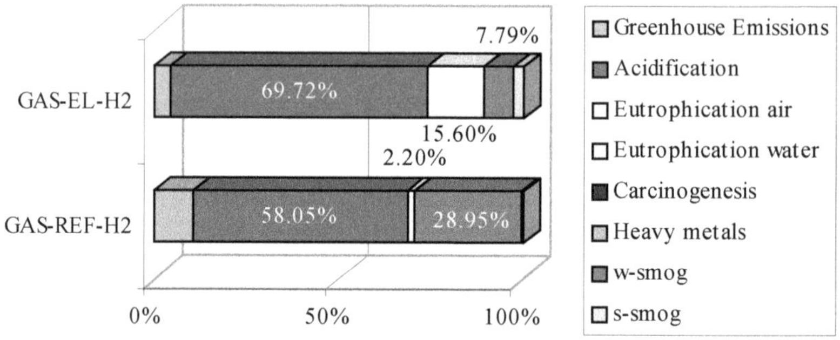

Figure 11: Indicator graph, GAS-REF-H2 for the gasification-reformer and GAS-EL-H2 for the gasification-electrical generation system.

CONCLUSIONS

The Life Cycle Analysis of the two biomass to hydrogen systems indicates that the biomass-gasification-electricity-electrolysis route may have better environmental performance than the biomass-gasification-steam reforming-PSA route. One of the main reasons for this result is that it is assumed that the biomass-gasification-power plant produces all the electricity required for electrolysis and liquefaction steps without need of additional power source. On the other hand the gasification-steam reforming plant requires additional electricity due to compression requirements that involve the steam reforming and PSA processes. The electricity is assumed to come from the local grid, which is mainly fuelled with no renewable energy sources (2).

The biomass-gasification-electricity-electrolysis route requires 4.2 TJ of primary energy input per TJ of hydrogen while for the biomass-gasification-steam reforming-PSA route require only 2.4 TJ per TJ hydrogen produced. But as it can be seen from the Fig 12 gasification-electricity-electrolysis route has 92.9% share of renewable energy in the primary energy input. Thus the gasification-steam reforming-PSA route is the most energy efficient but it requires improvements on its environmental efficiency. This can be achieved by reducing the dependence from non-renewable sources in the steam reforming and PSA process.

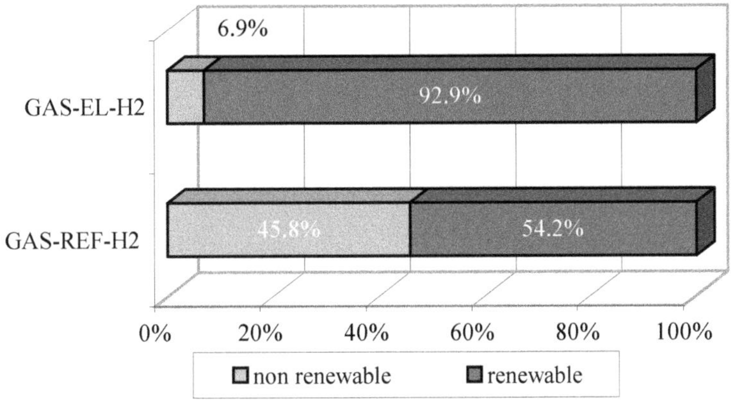

Figure 12: Primary energy input.

REFERENCES

1. Craig K. R., Mann M. K. (October 1996) *Cost and Performance Analysis of Biomass- Based Integrated Gasification Combined-Cycle (BIGCC) Power Systems,* Technical Report NREL/TP-430-21657, US Department of Energy
2. ExternE Project Report (December 1997) *External Costs of Electricity Production in Greece,* Laboratory of Industrial & Energy Economics, National Technical University of Athens, Non Nuclear Energy Programme JOULE III.C.
3. Francis S. L., Zabransky R., Bowen D. A. et al. *Techno-Economic Analysis of Hydrogen Production by Gasification of Biomass,* Proceeding of the 2002 US Hydrogen Program Review, NREL/CP-610-32405.
4. Global Emission Model for Integrated Systems (GEMIS) database (September 2002), Φko-Institut, Gesamthochschule Kasse.
5. Koroneos C.J., Boura A., Moussiopoulos N., Zambaniotou A. and Filippopoulos N. (2000), *Technical, environmental, economical and energy analysis of alternative methods for the exploitation of agricultural wastes in Greece and new perspectives, biomass for energy and industry*, Proceedings of the 1st World Conference and Technology Exhibition, Seville, Spain, 5-9 June.
6. Mann M.K. (August 1995) *Technical and Economical Assement of Producing Hydrogen by Reforming Syngas from th Bettelle Indirectly Heated Biomass Gasifier,* Technical Report NREL/TP-431-8143, US Department of Energy

7. Mann M.K. Spath P.L. (December 1997) *Life Cycle Analysis of a Biomass Gasification Combined-Cycle System*, Technical Report NREL/TP-430-23076, US Department of Energy
8. Spath P. L., Lane J. M., Mann M. K., Amos W. A. (April 2000) *Update of Hydrogen from Biomass-Determination of the Delivered Costs of Hydrogen,* Milestone Report for the US Department of Energy's Hydrogen Program.

Solid Recovered Fuel Gasification for Co-combustion in Pulverised Coal-fired Boilers - Lahti Case Study

E.Kurkela
VTT Processes, Espoo, Finland
J.Palonen
Foster Wheeler Energia Oy, Varkaus, Finland
M.Kivelä & H.Takala
Lahti Energia Oy, Lahti, Finland

ABSTRACT: The replacement of coal in existing large pulverised coal-fired boilers is a cost-effective technology which can significantly lower the CO_2 emissions of power production in many countries already at short term. A power plant concept consisting of a gasifier connected to a large conventional boiler with a high efficiency steam cycle offers an attractive and efficient way to use local biomass and waste sources in energy production. The Kymijarvi power plant in Lahti, Finland, has been a real success story for large-scale biomass/waste gasification. This 60 MW circulating fluidised bed gasifier was commissioned in early 1998 and has since then been in commercial operation replacing hard coal in a 360 MW_{th} boiler producing power and district heat. The annual gasifier availability has been over 95 % in each year and the Lahti plant has clearly demonstrated that the technology is technically proven and is able to reduce the emission of CO_2, SO_2, dust and NO_x compared to coal-alone combustion.

The most simple gasification-based co-combustion concept realised in Lahti without product gas cleaning is suitable to clean biofuels and clean waste-derived fuels, which do not contain high amounts of chlorine or heavy metals. Product gas cleaning makes it possible to utilise also fuels with higher chlorine and metal contents. The dry gas cleaning methods have been tested at pilot scale and will be soon demonstrated in gasification plants designed for solid recovered fuels originating from household and industrial wastes.

INTRODUCTION

Almost all large (> 200 MWe) coal-fired power plants are based on pulverised combustion. These boilers dominate in most European countries. During last two decades many of the large PC-boiler plants have been equipped with efficient environmental control equipment. In addition to electric precipitators, different types of desulphurisation processes are widely utilised. The NOx-emission in old plants is often minimised by low-NOx burners, while catalytic deNOx-technology is utilised in new power plants. The emission limits set by the European LCP-directive to the old power plants can be reached without desulphurisation and deNOx plants by using clean low-sulphur coal and low-NOx burners. The Kymijärvi power plant at Lahti Finland described in this report represents this-type of old pulverised coal-fired steam boiler utilising coal, which has low sulphur and chlorine contents. In this type of boiler, the use gasification gas produced from biomass and waste-derived fuels makes it possible to lower the sulphur and CO_2 emissions compared to 100 % coal-firing.

Presently the international targets for reducing the CO_2 emissions are governing the energy policies in most European countries. The renewable energy sources play a key role in the European CO_2 reduction targets. Biomass and biodegradable part of household and industrial wastes are the main feedstocks, the share of which can be considerably increased already at short and medium term. Biofuels, as well as waste-derived fuels, are local feedstocks. The energy density of these fuels is small and therefore, transporting from long distances is not in an economical sense an attractive solution. This is the main reason why biofuel-based power plants are typically quite small compared to the coal fired power plants. The specific investment and operation costs for small stand-alone biomass power plants are always much higher than in large coal or natural gas fired power plants. In addition, the power production efficiency is typically lower in small plants.

Thus, the replacement of coal in existing large pulverised coal-fired boilers is a cost-effective technology which can significantly lower the CO_2 emissions of power production in many countries already at short term. In Europe, it is typical that about 30 - 100 MW of biofuels and suitable waste-derived materials is available within 50 km from a given power plant, a sufficient amount to gasify and utilise in medium or large size coal-fired boilers. Thus, a power plant concept consisting of a gasifier connected to a large conventional boiler with a high efficiency steam cycle offers an attractive and efficient way to use local biomass and waste sources in energy production.

The fluidised-bed gasification technologies originally developed in 1980's for woody biofuels can be utilised with a wide range of waste-derived fuels. However, source-separation and/or effective pre-treatment of the waste material is required in this technology. The Finnish solution has been the production of Solid Recovered Fuel (SRF), which fulfils quality requirements defined in a fuel standard [1]. The SRF production is based on source-separation at homes, offices and industries followed by crushing, sieving and separation of different impurities at local or regional SRF production plants. The resulting SRF-material is in many respects similar to biofuel and it contains typically ca. 70 wt-% of biodegradable materials (e.g. paper and wood). This feedstock can be utilised together with locally available biomass residues in fluidised-bed gasifiers connected to PC-boilers.

GASIFICATION PROCESS

The leading gasification technology applied in co-firing applications is the Circulating Fluidised Bed (CFB) gasifier. The Finnish Foster Wheeler Energia Oy is the leading technology supplier at present. The first CFB gasifier of Foster Wheeler has since 1983 replaced 35 MW fuel oil in a lime kiln at Wisaforest Oy, Pietarsaari, Finland. Since then similar gasification plants having the same basic technology have been installed at two pulp mills in Sweden and one mill in Portugal. These gasifiers produce lime kiln fuel from bark and waste wood and they also utilise part of the generated gas in drying plants [2,3]. At Lahti this technology is applied to SRF and mixtures of SRF and biomass fuels. CFB gasifiers have also been developed in Europe by other companies: Lurgi GmbH (Germany), Termiska Processer Ab (Sweden) and Austrian Energy/Babkock Borsig (Austria). So far, Foster Wheeler technology has been commercially and technically most successful and all constructed plants have been in successful commercial operation for several years. Presently, Foster Wheeler is constructing a Lahti-type gasifier plant to a coal-fired boiler in Ruien, Belgium.

Atmospheric-pressure bubbling fluidised-bed gasification (BFB) technology has also been developed in recent years in Finland. The BFB technology seems to be economically more suitable to medium size applications (15-40 MW) while the CFB technology is most economic on larger scale (40-100 MW). The first commercial application of the atmospheric-pressure BFB gasification in Finland was realised in Varkaus, central Finland by Corenso United Ltd. This gasifier supplied by Foster Wheeler utilises plastics and aluminium containing reject material coming from the recycling process for used liquid cartoons. In this process, the aluminium is removed from the gas as utilised for recovered aluminium production, while the product gas produced from the plastic material is combusted in a steam boiler. The 40 MW_{th} gasifier has been taken into operation in 2001. Since then it has been in commercial operation with high availability.

BFB gasification technology has also been developed for MSW-derived SRF by Pohjolan Voima Oy and Vapo Oy. The gasification and gas cleaning process has been tested at a 1 MW pilot plant located at the Technical Research Centre of Finland (VTT). So far ca. 500 hours of pilot testing has been carried out and simultaneously the first industrial demonstration plant is being designed.

GAS CLEANING OPTIONS

Figure 1 illustrates the different gasification and gas cleaning options. With woody biofuels and clean waste-derived fuels the most simple connection without gas cleaning is sufficient as will be described in the following for the Lahti gasifier. In this case the efficient gas combustion below the coal flames together with the flue gas cleaning equipment will guarantee excellent environmental performance. This technology has reached fully commercial state.

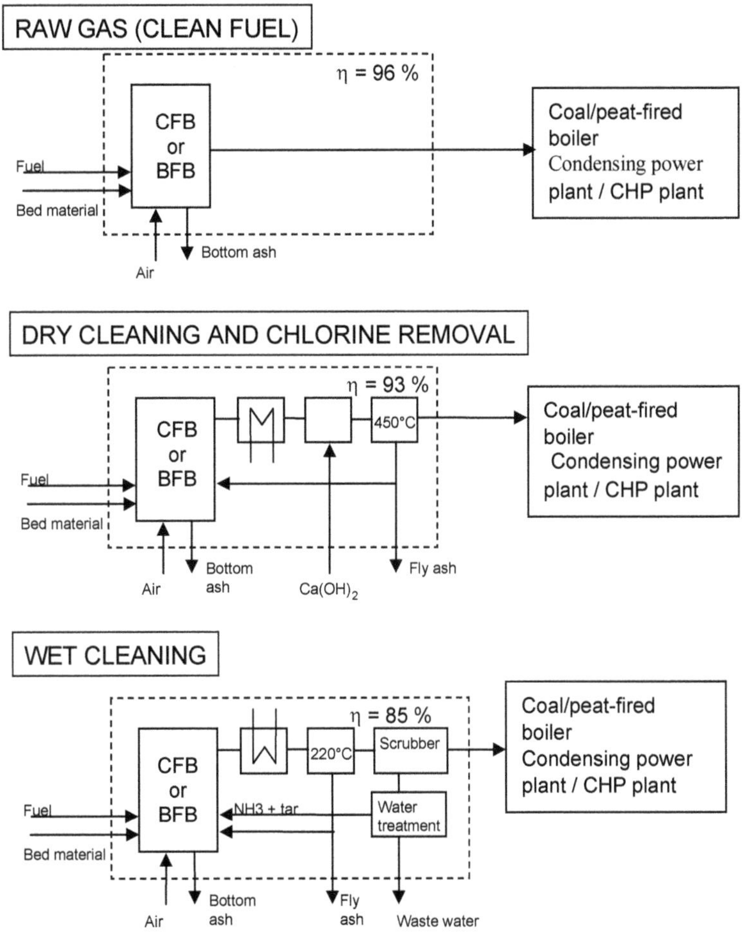

Figure 1: Different principle connections of the gasifier to the coal-fired boiler.

The second process alternative is based on dry gas cleaning prior to the boiler. This alternative makes it possible to utilise high-alkali biofuels (such as straw) as well as SRF with higher chlorine and heavy metal contents. This technology is ready for large-scale demonstration, but has not yet reached commercially proven status. The developed method is based on gas filtration at 200-450 °C temperature in a bag filter unit. With SRF fuels and other fuels containing a lot of plastic material, the filtration temperature has to be higher than 350 °C in order to avoid tar condensation problems. Ceramic filter bags are used. Calcium hydroxide or other alkali sorbents are injected into the gas before the filter unit, if necessary, to improve chlorine capture. However, the inherent fuel alkali-metals and the calcium-based bed materials also react efficiently with HCl during gas cooling and especially in the filter cake. All heavy metals except mercury are also almost completely (> 99 %) removed from the gas during filtration.

The efficiency of the gas cleaning process has been demonstrated at the pilot plants of VTT (both CFB and BFB gasification plants) using different types of SRF and biomass fuels. Foster Wheeler has also operated similar-type of gas cleaning process at their 3 MW CFB gasification pilot plant. Presently Foster Wheeler is constructing a slip stream gas cleaning train to be connected to the Lahti gasifier. Long-term tests are required to demonstrate the durability of the filter media as well as to test the developed operation methods in a real plant site.

The third alternative including wet gas cleaning is a method which can produce a very clean gas, but investment costs are roughly 50 % higher than in the two previous cases. In addition, the waste water treatment may be difficult to design in an environmentally fully acceptable manner.

KYMIJÄRVI POWER STATION AT LAHTI

POWER PLANT DESCRIPTION

The Kymijärvi power plant was started in 1976. Originally, the plant was heavy-oil-fired but in 1982 was modified for coal firing. The boiler is a Benson-type once-through boiler. The steam data is 125 kg/s 540 °C/170 bar/540 °C/40 bar and the plant produces electric power and district heat to the Lahti city. The maximum power capacity is 167 MWe and the maximum district heat production is 240 MW. The annual operating time of the boiler is about 7000 h/a. In the summer, when the heat demand is low, the boiler is shut down. In the spring and autumn, the boiler is operating in low capacity, with natural gas the main fuel. In 1986, the plant was furnished with a gas turbine connected to the heat exchanger preheating the boiler feed water. The maximum electrical output of the gas turbine is 49 MWe, when the outside temperature is -25 °C.

The boiler uses 1850 GWh/a (270,000 ton/a) of coal and about 100 GWh/a of natural gas. The boiler is not equipped with a sulphur removal system. However, the coal utilised contains only 0,3 to 0,4 per cent sulphur. The burners are provided with flue gas circulation and staged combustion to reduce NOx emissions. The biomass/SRF gasifier was connected to the boiler at the end of 1997. The arrangement is illustrated in Figure 2.

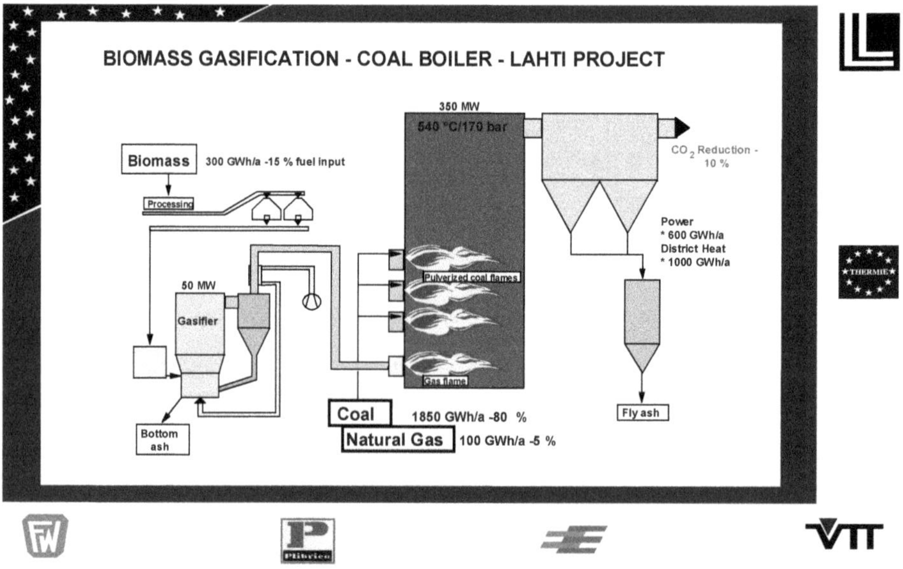

Figure 2: Flow sheet of the Lahden Lämpövoima Oy gasifier.

GASIFIER FUELS AND FUEL HANDLING EQUIPMENT

Approximately 300 GWh/a of different types of biofuels and SRF fuels are available in the Lahti area. On an annual basis this amount is enough to substitute for about 15 % of the fuels burned in the main boiler. Table 1 presents a summary of the available biofuels in the Lahti area. Table 2 gives analysis data for the gasifier fuels and coal used in 2001.

Table 1: The available local fuels on annual basis in the Lahti area (total 300 GWh/a) [2-4].

Fuel	Amount %-weight of total	Moisture %-weight
Saw dust	10	45 – 55
Wood residues (bark, wood chips, wet and fresh wood residues, etc.)	30	45 – 55
Dry wood residues from the wood working industry (plywood, particle board, cuttings, etc.)	30	10 – 20
Solid Recovered Fuel (SRF)	30	10 – 30

Table 2: Analysis data for the fuels used in 2001 [6].

		wood	glue wood	municipal sludge	SRF	Wood+gluelam+ SRF	Coal
LHV, w.b.	MJ/kg	7.61	15.84	7.12	16.04	11.48	25.14
Sulphur, d.b	wt-%	0.03	0.01	0.93	0.08	0.05	0.36
Ash, d.b	wt-%	3.77	1.19	46.92	6.3	3.79	14.66
trace components	d.b						
Cl	ppm-m	87	115	178	2580	1700	93
Na	ppm-m	146	1830	1570	1190	725	1700
K	ppm-m	1710	720	1820	670	910	3110
Br	ppm-m	< 3	< 3	< 3	< 3	< 3	< 3
F	ppm-m	3	< 3	39	43	12	25
As	ppm-m	0.3	0.3	3.2	1.4	0.6	5.5
Cd	ppm-m	0.4	0.1	0.8	0.1	0.2	< 0,02
Cr	ppm-m	15	10	45	21	26	23
Cu	ppm-m	5.2	3.5	160	20	29	7.1
Ni	ppm-m	3.4	2.2	26	8.5	8.2	6.1
Hg	ppm-m	< 0.2	< 0.2	0.6	0.3	0.2	0.2
Tl	ppm-m	0.1	0.1	0.1	< 0,05	0.1	< 0,05
Zn	ppm-m	110	19	390	74	120	15
Pb	ppm-m	1.3	0.9	14	4.3	7.5	3.6

The solid recovered fuel SRF is produced from the in origin classified refuses, which are coming from households, offices, shops and construction sites. The processing of SRF was started by the municipally owned waste management company (Päijät-Hämeen Jätehuolto Oy) in 1997. The SRF consists of 5-15 wt-% plastics, 20-40 wt-% paper, 10-30 wt-% cardboard and 30-60 wt-% wood. In addition to these fuels listed above, demolition wood waste, and shredded tires have also been used as fuels in the gasification plant.

Fuels are transported to the power plant in trucks. There is one receiving hall for SRF and one receiving station for incoming biofuels. The SRF hall is equipped with a receiving pit having a lamella feeder, which controls the fuel flow to a crusher. Coarse biofuel, which originates mainly from the wood working industry is also fed through the SRF system. The trucks tip SRF and coarse biofuels on the floor of the hall or directly into the pit, after which they are crushed in the slowly rotating crusher. The underground conveyor from the first receiving bunker transports SRF and biofuels from the crusher. The other receiving station is used for the finer biofuel and peat. This biofuel is transported to the site in special trucks. The transport platforms of the trucks are furnished with conveyors. These conveyors discharge biofuel and peat from the trucks and the fuel falls through a screen down onto the chain conveyor at the bottom of the bunker. The coarser particles separated by the screen will be moved to the SRF hall for crushing. The underground conveyor lifts the fuels to the belt conveyor, which has a magnetic separator above it. The belt conveyor transports the fuels onto the disk screen. The coarse fuel fractions from the disk screen fall into the final crusher, while the fine fractions from the screen and the crushed biofuel will be transported by a chain conveyor to the fuel storage silos.

The gasification plant is furnished with one storage silo for fuels. Besides serving as a storage silo, this silo is used for homogenisation of the fuel mixture before it is transported into the gasification building. The discharge of the silo has variable speed controls.

ATMOSPHERIC-PRESSURE CFB GASIFICATION PROCESS OF FOSTER WHEELER

The gasifier at Kymijärvi power station is a CFB gasifier (Fig. 3) supplied by Foster Wheeler. The atmospheric CFB gasifier is very simple. The system consists of a refractory-lined reactor where the gasification takes place, of a uniflow cyclone to separate the circulating material from the gas and of a return leg for returning the circulating material to the bottom part of the gasifier. The operating temperature in the reactor is typically 800 to 1000 °C, depending on the fuel and the application.

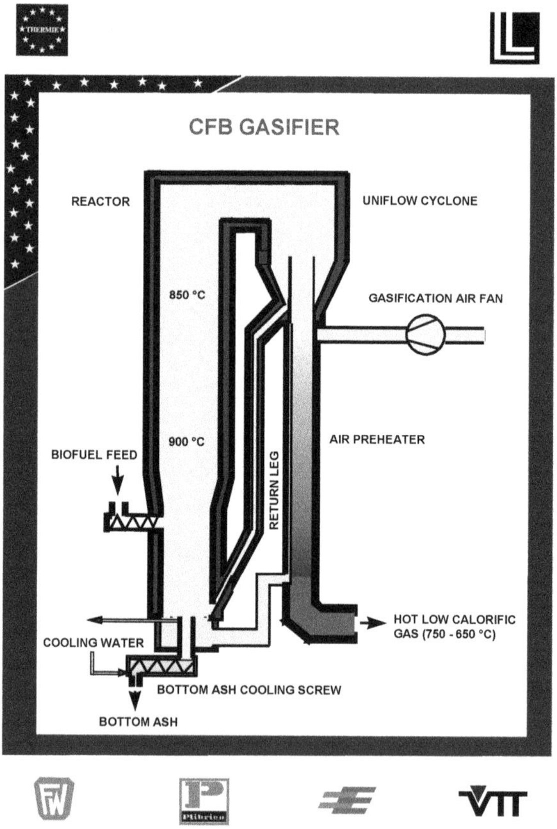

Figure 3: Foster Wheeler CFB gasifier [2].

The fuel is fed into the lower part of the gasifier above a certain distance from the air distribution grid. When entering the reactor, the biofuel particles start to dry rapidly and the pyrolysis also occurs. The gaseous products of drying and pyrolysis flow upwards in the reactor.

Part of the charcoal flows down to the more dense part of the fluidised bed while part of the charcoal flows up together with the circulating media into the uniflow cyclone. Most of the solids are separated from the gas in the cyclone and returned to the bottom of the bed, where the charcoal is combusted with the air that is introduced through the grid nozzles to fluidise the bed.

From the process point of view, the major difference compared to the biomass gasifiers constructed in mid-80's is that fuel will not be dried in this application although the moisture content of fuel can be up to 60 %. Some mechanical changes have been made to accommodate the special nature of the fuel components to be used in the gasifier. For fuels like SRF, some wood wastes and shredded tires which may contain different types of solid impurities (nails, screws, metal wires, concrete), the air distribution grid and the bottom ash extraction system have been specially designed. The fuel feeding was also designed in a different way to achieve stable feeding with an inhomogeneous and low-bulk-density fuels.

The product gas for combustion is led directly from the gasifier through the air preheater to two burners which are located below the coal burners in the boiler. The gas is combusted in the main boiler and it replaces part of the coal. When the fuel is wet, the heating value of the gas is very low. Typically, when the fuel moisture is about 50 % the heat value of the gas is only about 2.2 MJ/m^3n. The design of the product gas burners is rather special and is based on pilot scale combustion tests and computational fluid dynamics (CFD) modelling work.

OPERATION EXPERIENCES

The gasifier was connected to the main boiler on December 7, 1997. The very first gasification tests were carried out on January 14, 1998 and the unit has been in continuous operation since Week 4, 1998. The gasifier was shut down for the summer maintenance on June 2 and because of the extremely low electricity price in Finland in summer autumn 1998, the main boiler was put in operation in the beginning of September and the gasification plant two weeks later, i.e. September 21, 1998. Already during the first operating year approximately 4730 hours of operation in the gasification mode was achieved and the availability of the gasification plant (including fuel reception and handling) was 81.8 % (highest monthly availability up to 93 %). The 1999 operational year was even better, with the plant operating in the gasification mode for approximately 5460 hours with an average availability of 87.6 % (highest monthly availability was 95.3 %). Most of the problems, especially in the beginning, were related to the fuel processing plant. Lack of fuel and operational problems at the fuel processing plant decreased the availability of the whole plant during the first half of 1998.

The operating experience of the gasifier during the years 1998 - 2002 has been excellent. On annual basis, the gasifier availability has been between 96.1 - 99.3 %. With regard to the gasification plant itself, the few problems have been related mostly to the use of shredded tires as a fuel in the gasifier. On several occasions the wire content of tires (there is no additional separation of metal wires with magnet after shredding) was so high that accumulated wires blocked the ash extraction system and the gasifier had to be shut down. Otherwise, with all other fuel fractions, the operation of the gasification process was good.

The annual operation hours and the availability figures are given in Table 3 and the annual statistics for the gasified fuels in Figure 4. The availabilities shown in Table 3 are calculated without taking into account the SRF and biofuel handling section of the plant.

Concerning the gasification process itself, the results have met the expectations. The operating conditions as regards temperatures, pressures and flow rates have been as designed and the process measurements as regards the product gas, bottom ash and fly ash composition have been very close to the calculated values. Due to the high moisture content (up to 58 %) of the gasifier fuels, the heating value of the product gas has been low, typically only 1.6 - 3.2 MJ/m^3n.

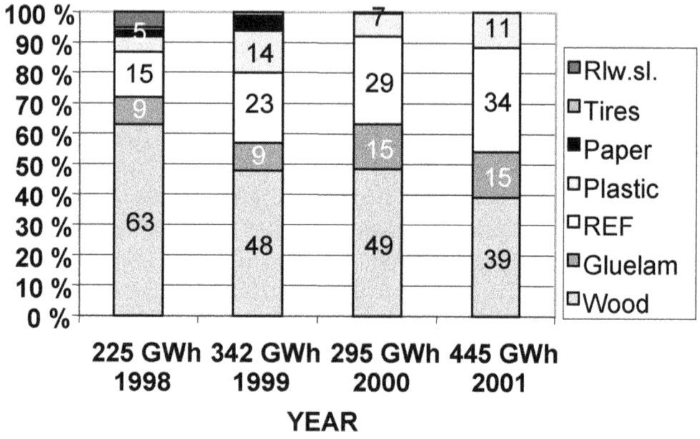

Figure 4: Gasifier fuel distribution and annual consumption in 1998-2001 [5].

Table 3: Operation figures of the Lahti gasifier plant in 1998 - 2001 (design values: operation time 6500 hour/a and energy production 300 GWh/a) [4].

	1998	1999	2000	2001
Operation hours, h	4730	5460	4727	7089
Gasifier availability, %	99.3	98.9	97.1	96.1
Energy produced, GWh	223	343	295	449

The stability of the main boiler steam cycle has also been excellent. The large openings that were made for the low Btu gas burners have not caused any disturbances into the water/steam circulation. Furthermore, as regards the operation of the product gas burners, the product gas combustion has been stable even though the moisture content of the solid fuel has been mostly high and the heating value of the gas very low. The stability of the main boiler coal burners has been normal despite of the fact that the product gas burners were integrated very close to the lowest level coal burners.

PLANT MONITORING AND ENVIRONMENTAL EFFECTS

Simultaneously during the start-up and the commissioning phase of the gasification plant, a thorough one-year monitoring program was started. During this one-year long period the operation of the fuel preparation plant, the gasification plant and the main boiler has been monitored and reported. The monitoring included a) corrosion/deposit formation monitoring in the main boiler with probe testing, b) determination of the gasifier fuel gas composition, c) boiler flue gas emissions, and d) characterisation of the gasifier bottom ash and boiler fly ash. In 1997, a reference test run was carried out, during which the main boiler was fired with 100 % coal. The monitoring results are summarised in the following. More detailed information is given in [2-4] .

Product gas composition

During the monitoring campaigns of 1998, the operating temperature of the gasifier was typically 830 - 860 °C and the gasifier effect varied between 35 MW and 55 MW depending on the gasifier fuel moisture content and on the required gasifier load. During the whole year 1998 the moisture content in the fuel mixture was rather high varying typically between 45 to 58 per cent. Due to that the product gas heating value has been relatively low, typically only 1.6 - 2.4 MJ/m^{3}n. Table 4 summarises typical measured values of the product gas main components. Typical range for the concentration of different trace components are given in Table 5.

Table 4: Average values of the product gas main components.

Gas component	Unit	Average
CO_2	%-vol (wet)	12.9
CO	%-vol (wet)	4.6
H_2	%-vol (wet)	5.9
C_xH_y	%-vol (wet)	3.4
N_2	%-vol (wet)	40.2
H_2O	%-vol (wet)	33.0

Table 5: Typical values of the product gas special measurements.

Component	Unit	Range
NH_3	mg/m^{3}n (dry)	800 – 1000
HCN	mg/m^{3}n (dry)	25 – 45
HCl	ppmv (dry)	30 – 90
H_2S	ppmv (dry)	50 – 80
Benzene	g/m^{3}n (dry)	7 – 12
Tars	g/m^{3}n (dry)	7 – 12
Alkalis	ppmw (dry)	< 0.1
Particulates	g/m^{3}n (wet)	6 – 10

Gasifier bottom ash

The main components of the gasifier bottom ash were bed materials, i.e. sand and limestone. Furthermore, small amounts of solid impurities such as metal pieces, pieces of concrete, glass, etc. were found in bottom ash. Typically, the carbon content in gasifier bottom ash was less than 0.5 per cent. No signs of chlorine were seen in the analyses.

With regard to the trace metals, the following elements were analysed: As, Cd, Cr, Cu, Ni, Pb, Zn and Hg. Elements like chromium (Cr), copper (Cu) and zinc (Zn) were found in the range of hundreds of ppms. When shredded tires were used as a fuel in gasifier, the zinc content in gasifier bottom ash increased to the level of 3000 ppm. All other analysed elements were in the range of a few ppms or tens of ppms. The major part of the elements escaped the gasifier in the gaseous phase or in the fine fly ash particles.

Besides the standard analyses, leachability tests were also made for the bottom ash. According to the tests, the trace metal leachabilities were low. As a result of low trace metal contents and low trace metal leachabilities, the gasifier bottom ash is disposed today as planned at the beginning of the project.

Main boiler flue gas

The main boiler emissions were perhaps under the greatest interest as regards the measurement program of the monitoring phase. In summary, it can be stated that the changes in the emissions were very small. As indicated earlier, the main boiler is not equipped with DeNOx or DeSOx plants and the emission limit values for the emissions were as follows: NOx 240 mg/MJ (as NO_2) and SO_x 240 mg/MJ. Table 6 summarises the effect of the co-combustion of the gasifier product gas on the main boiler emissions.

The dust content in the flue gas after the ESP decreased approximately 10 - 20 mg/m3n. The most probable reason to this has been the increase of the flue gas moisture content, which has enhanced the operation of the ESP. Perhaps the most positive phenomenon has been the decrease in the NOx emission. According to the measurements the NOx content of the main boiler decreased typically approximately 10 mg/MJ, equalling the decrease of 5 to 10 % from the base level. This was evidently due to the cooling effect of the low Btu, high moisture product gas in the bottom part of the boiler. Obviously, due to the cooling effect, the forming of thermal NOx was lower in the coal burners located at the lower part of the boiler. Furthermore, because of the extremely low sulphur content of biofuels, the main boiler SOx emission decreased approximately 20-25 mg/MJ. In contrast, because of the very low chlorine content (0.01 %) of the main boiler coal, the HCl content of flue gas increased approximately by 5 mg/MJ when the gasifier was in operation. The reason for this was the use of SRF fuel and shredded tires in the gasifier. Both of these fuels are known to contain chlorine. As regards the CO emission of the main boiler, no changes could be seen.

Table 6: The effect of gasifier to the main boiler emissions.

Emission	Change caused by gasifier
NOx	Decrease by 10 mg/MJ (= 5 to 10 %)
SOx	Decrease by 20 - 25 mg/MJ
HCl	Increase by 5 mg/MJ *
CO	No change
Particulates	Decrease by 15 mg/m3n
Heavy metals	Slight increase in some elements, base level low
Dioxins	No change
Furans	No change
PAH	No change
Benzenes	No change
Phenols	No change

With regard to the heavy metal stack emissions, increases in some elements could be seen, but because of the very low base levels in coal combustion, the changes that were measured were in practice very small. As regards dioxins, furans, polyaromated hydrocarbons, chlorinated phenols and chlorinated benzenes, no changes could be seen compared to the results from 100 % coal combustion.

Main boiler filter ash

The share of the gasifier fly ash of the main boiler total filter ash is small, only 3 to 5 per cent. Thus, it is obvious that the changes in the main boiler filter ash quality are in practice very small. With regard to the amount of unburnt carbon and alkalis, no change could be seen compared to the 100 % coal combustion. Concerning the heavy metals, the changes in the main boiler filter ash quality were small. Increase in some elements could be seen, but because of the small share of gasifier fly ash, it is obvious that the gasifier fly ash effect on the main boiler filter ash quality is small. As regards the measured content of organic compounds (dioxins, etc.), no changes could be seen when comparing the results to the data of the reference tests.

The leachability tests were made also for the main boiler filter ash. According to the tests, the trace metal leachabilities were low. Because of the small changes in the ash quality and because of the low leachabilities of trace metals, the authorities have given permission to use the main boiler filter ash today as it has been used earlier on construction works (roads, etc.). However, it is noted that permission from the authorities has to be applied for separately for each new application.

Deposit formation monitoring

As regards the results of the corrosion probe monitoring tests no indication of abnormal deposit formation/fouling or corrosion could be seen in the test coupons. Furthermore, in the inspection of the boiler heat transfer surfaces (furnace walls, superheater section, economiser and air pre-heater) during the summer maintenance, no signs of abnormal deposit formation or high temperature corrosion could be found.

ENVIRONMENTAL MEASUREMENTS IN 2001

In 2001, thorough emissions measurements were carried out during a test run, where a small amount of dried sewage sludge was used together with normal gasifier fuels. Results for boiler flue gas emission measurements are shown in Table 7 and data for the ash streams in Table 8. Reference data for 1997 measurements with coal alone are also presented.

Table 7: Comparison of the boiler flue gas emission measurement results from 100 % coal firing (1997) and from co-firing of gasifier gas and coal (2001) [6].

		1997	2001
Fuel consumption:	MW	350	349
coal	MW	350	293
natural gas	MW	0	0
gasifier, total	MW	-	56
- wood	MW	-	29
- glue lam wood	MW	-	8
- SRF	MW	-	16
- sewage sludge	MW	-	3
Flue gas emissions:			
O_2 - content (dry gas)	%	6	6.8
SO_2 , emission	mg/MJ	237	208
SO_2 concentration (dry gas)	mg/m^{3}n	638	636
NOx, emission	mg/MJ	264	187
NOx concentration. (dry gas)	mg/m^{3}n	710	572
Dust emission	mg/MJ	11	7
Dust concentration (wet gas)	mg/m^{3}n	29	19
CO emission	mg/MJ	11	10
CO concentration (dry gas)	mg/m^{3}n	30	30
HCl concentration (dry gas)	mg/m^{3}n	0,4	34
HF concentration (dry gas)	mg/m^{3}n	2,4	7
PAH-compounds total (dry gas)	mg/m^{3}n	0.03	1
Polychlorinated benzenes, -biphenyls and chlorophenols (dry gas)	ng/m^{3}n	56	51
PCDD/PCDF, I-TEQ (dry gas)	ng/m^{3}n	0.005	0.001
Total heavy metal concentrations in dry flue gas:			
Cr	µg/m^{3}n	< 9	4
Ni	µg/m^{3}n	< 9	3
Cd	µg/m^{3}n	< 2	0.2
As	µg/m^{3}n	< 4	2
Cu	µg/m^{3}n	< 9	4
Zn	µg/m^{3}n	< 35	16
Pb	µg/m^{3}n	< 9	< 11
Hg	µg/m^{3}n	< 0,6	1
Sb	µg/m^{3}n	nd	0.4
Co	µg/m^{3}n	nd	0.9
Mn	µg/m^{3}n	nd	22
Tl	µg/m^{3}n	nd	< 8
Sn	µg/m^{3}n	nd	0.8
V	µg/m^{3}n	nd	5
Sum of heavy metals	µg/m^{3}n	-	59

nd = not determined

Table 8: Special environmental analyses for the gasifier bottom ash and for the two boiler ash streams. Samples represent the same operation period described in Table 8 [6].

	unit	boiler fly ash 1997 (only coal)	boiler fly ash 2001	boiler bottom ash 2001	gasifier bottom ash 2001
chlorophenols	ng/g	nd	< 5	< 5	< 5
PAH compounds	ng/g	nd	16	20	10
polychlorinated biphenyls and benzenes	ng/g	nd	< 0,5	< 0,5	< 0,5
dioxins, I-TEQ	pg/g	nd	< 0,5	< 0,5	< 0,5
As	mg/kg	20	1.7	0.9	3.6
Cd	mg/kg	0.4	0.8	0.1	0.1
Cr	mg/kg	119	44	38	86
Cu	mg/kg	50	30	60	640
Ni	mg/kg	56	19	20	35
Hg	mg/kg	n.d	0.7	< 0,2	< 0,2
Zn	mg/kg	150	84	190	1200
Pb	mg/kg	34	10	11	23

Based on the environmental measuring data, the following main conclusions can be draw considering the total environmental effects of the replacement of coal by biomass/SRF-derived product gas at Lahti:

- The emissions of CO_2, SO_2, dust and NO_x are lowered due to the use of gasification gas when compared to coal-alone combustion.
- The effect of the gasifier gas on the heavy metal emissions into the boiler flue gas is negligible compared to the effects caused by changes in coal quality. The total emissions are very low when compared e.g. to the limits set by the Waste Incineration Directive (WID):
  - Hg : 0.001 mg/m^3n, WID limit 0.05 mg/m^3n
  - Cd+Tl: 0.0002 mg/m^3n, WID limit 0.05 mg/m^3n
  - other heavy metals: 0.06 mg/m^3n, WID limit 0.5 mg/m^3n
- The concentration of dioxins was only 0.001 ng/m^3n, while the WID emission limit is 100-times higher (0.1 ng/m^3n)
- The emission of HCl is somewhat increased due to the fact that the chlorine content of the used coal is very low.
- The effect of the gasifier gas co-combustion on the ash composition of the main boiler ash is also very small and cannot be clearly detected below the scatter caused by differences in coal quality.

PLANT ECONOMY

Total costs of the gasification plant at the Kymijärvi power plant were about 12 Milj. €. This included fuel preparation plant, civil works, instrumentation and control as well as the electrification. The project received 3 Milj € support from the THERMIE Program of the European Commission. The estimated payback time of the investment was 5-7 years.

THE EFFICIENCY OF THE GAS CLEANING TECHNOLOGY

The very good environmental performance of the Lahti-type of plant can even be improved by installing the dry gas cleaning system. The use of gas cleaning before leading the gas into the boiler will also make it possible to utilise SRF fuels, which have higher chlorine and heavy metal contents than the rather clean feedstock mixture used at the Lahti gasifier. Table 9 summarises the measured efficiency of the dry gas cleaning methods developed by VTT for chlorine and heavy metals [7]. All heavy metals except mercury are removed almost completely from the gas. The residual HCl content of < 50 ppmv would also result in the Lahti plant to the final flue gas HCl emission of less than 10 ppmv (WID limit for HCl emissions of waste incinerators) when low-chlorine coal is used as the main fuel and the share of gasifier gas is 15-20 %. Presently, VTT is also studying the mercury removal by different sorbents as an further improvement of the developed dry gas cleaning method.

The effects of the co-combustion of cleaned product gas on the emissions of the main boiler have also been estimated by VTT. In these calculations, the effects of coal quality and the use of desulphurisation technology has also been studied. The results are reported in reference [8]. The calculation clearly showed, that only the emissions of mercury have be looked carefully case by case as mercury is not captured by the present gasifier gas cleaning process. However, with most SRF and biomass materials the mercury content is low and similar-type of results can be expected as have been measured at the Lahti gasifier plant, where the mercury emissions have been very low. According to VTT estimation, the mercury emissions are below the WID limit even if the mercury content of the gasifier fuel is 0.75 ppm and up to 30 % of coal is replaced by the gasifier gas. However, this also depends on the mercury emissions of coal combustion and the removal efficiency achieved in flue gas cleaning process.

Table 9: The efficiency of the gasifier gas cleaning for chlorine and heavy metal capture [7].

	removal efficiency %	final concentration in clean gas
HCl (ppm-v)	> 90	< 30-50
Heavy metals (ppm-m):		
V	> 99 %	< 0.01
Cr	> 99 %	< 0.01
Mn	> 99 %	< 0.01
Co	> 99 %	< 0.01
Ni	> 99 %	< 0.01
Cu	> 99 %	< 0.02
Zn	> 99 %	< 0.20
As	> 99 %	< 0.005
Cd	> 99 %	< 0.001
Sn	> 99 %	< 0.01
Sb	> 99%	< 0.01
Pb	> 99 %	< 0.03

CONCLUSIONS

The following conclusions can be drawn on the experiences obtained at Lahti and in the further development work on the gas cleaning:

- The gasification-based co-firing method is a cost-effective and environmentally attractive way of utilising locally available biomass residues and good-quality solid recovered fuels.
- The first commercial plant located at Lahti Finland has demonstrated that the technology is technically proven and is able to reduce the emission of CO_2, SO_2, dust and NO_x compared to coal-alone combustion.
- The emissions measurements clearly show the benefits of this type of waste utilisation technology with respect to dioxin emissions. The reducing gas atmosphere of the gasifier together with the very efficient and high-temperature combustion of the product gas evidently fully eliminate the formation of dioxins. The development of gas cleaning, where over 90 % of chlorine and over 99 % of metals are removed from the gas will even further guarantee the

elimination of dioxins without producing problematic dioxin containing solid wastes as in ordinary incinerators.

- The heavy metal emission in Lahti are very low. The performance will be further improved by using gas cleaning before the boiler.
- The most simple gasification concept realised without gas cleaning is suitable to clean biofuels and clean waste-derived fuels, which do not contain high amounts of chlorine or heavy metals. Product gas cleaning makes it possible to utilise also fuels with higher chlorine and metals contents. However, the gas cleaning still has to be demonstrated before it can be considered to be fully proven technology.

REFERENCES

1. REF standard SFS 5875. 2000. National Finnish Standard. Solid recovered fuel quality control system. Approved in Januray 2000. Helsinki, Finland: Suomen stardardisoimisliitto.
2. Palonen, J., Nieminen, J & Berg, E. Thermie demonstrates biomass CFB gasifier at Lahti. Modern Power Systems, February 1998, pp. 37-41.
3. Nieminen, J., Palonen J. & Kivelä M. Circulating Fluidized Bed Gasifier for Biomass. Thermie demonstration Project in Lahti, Finland. VGB PowerTech, Vol 79, Issue 10/99, pp. 69-74.
4. Palonen, J. & Nieminen, J. Biomass CFB gasifier connected to a 350 MWth steam boiler fired with coal and natural gas - Thermie demonstration project at Kymijärvi CHP power station in Lahti Finland. 2001.
5. Kivelä, M., Nieminen, J. & Palonen, J. Biomass CFB gasifier connected to a 350 MWth steam boiler fired with coal and natural gas - Thermie demonstration project at Kymijärvi CHP power station in Lahti Finland. 2001.
6. Takala, H., Environmental statistics for operation year 2001of the Kymijärvi power plant submitted to the environmental authorities. Lahden Lämpövoima Oy, Lahti, 2002 (unpublished).
7. Kurkela, E. & Nieminen M., Fluidised-bed gasification of waste-derived fuels and subsequent dry gas cleaning from particulates, chlorine and trace metals. VTT project report, October, 2002 (to be published).
8. Hepola, J & Kurkela, E., The effects of gasifier gas co-combustion on the boiler emissions. VTT Prosesses, project report, September 2002 (in Finnish).

Gasification of Paper Reject in a CFB Pilot Plant

H.-T. Lee, K.-T. Wu, C. I. Juch
Energy & Resources Lab., Industrial Technol. Res. Inst., Taiwan, ROC
M. Y. Tsai, C. C. Huang
Hsin Wu Mill, Yuen Foong Yu Paper Mfg. Co., Ltd., Taiwan, ROC
M. Kylä-Sipilä, M. Hiltunen
Karhula R&D Center of Foster Wheeler Energia Oy, Finland

ABSTRACT: In order to evaluate the potential of energy recovery by gasification of refuse derived fuel, all experiments were carried out in a 3 MW atmospheric circulating fluidized bed (CFB) gasifier at the Karhula R&D Center of Foster Wheeler Energia Oy, Finland. The refuse derived fuel used during gasification was the paper reject. The reject from processing of recycled paper for paper production was pelletized before feeding, and it contains approximately 60-70% wood fiber, 10-20% plastics and 10-15% ash. The preliminary results show that the syngas contains around 7.3-10.1 vol% H_2, 5.3-8.0 vol% CO, 4.3-5.8 vol% CH_4, 12.9-16.0 vol% CO_2, etc, at the operating temperature of 850-930oC. It was also observed that the tar contents in the syngas were high, but the tar contents decreased by increasing gasification temperature. In addition, HCl emission can easily be controlled by injecting $Ca(OH)_2$ or $NaHCO_3$ into syngas before the baghouse. Polychlorinated dibentsofurans and dioxins were also measured and analyzed during the experiments.

INTRODUCTION

The R.O.C. government in Taiwan held a national energy conference in 1998 concluding that the uses of new and clean energies should be promoted. A study group was then formed to assess the potential energy supply by new and clean energies. The preliminary assessment showed that the energy supply by various waste-to-energy technologies can reach six million kiloliter oil-equivalent which shares 31% of overall supply of new and clean energies.[1] The gasification, the environmental-beneficial process, then could be considered as a technology to recover energy from biomass and waste in Taiwan.

Gasification can be defined as the conversion of carbonaceous feedstock, e.g. waste, by partial oxidation at elevated temperature.[2] The successful experience of the Lahden Lämpövoima Oy's Kymijärvi power plant gasifcation project has demonstrated that the direct gasifcation of biomass in a commercial atmospheric circulating fluidized bed (CFB) gasifier developed by the Foster Wheeler Energia Oy, Finland.[3] In order to evaluate the potential of energy recovery by gasification of refuse derive fuel (RDF) in Taiwan, a test program is carried out in a 3 MW atmospheric circulating fluidized bed gasifier at the Karhula R&D Center of Foster Wheeler Energia Oy, Finland. The refuse deriver fuel used during gasification was the paper reject. The reject from processing of recycled paper for paper production was pelletized before feeding. The objective of the test is to obtain performance information in the gasifier and efficiency of gas cleaning by injecting additives to bind HCl as in the syngas from RDF gasification. The objective is reached by carrying out the gasification of RDF in determining optimal gasification process conditions for Taiwanese RDF, evaluating the efficiency of additives to capture HCl, and analyzing the composition of the syngas, and the emissions after combustion of the sygas, etc.

EXPERIMENTAL EQUIPMENT

CONFIGURATION OF THE CFB GASIFIER

The main components of the test facility are CFB gasification reactor, syngas cooler and baghouse filter. Part of the syngas producted in gasifier is burned in syngas combustor. The process flow scheme of the test facility is shown in Fig. 1.

The 60 cm ID refractory-lined reactor is inside a 10 m high gas-tight steel shell. The fuel is fed into the fluidized bed, 3 m above the grid. The bed is fluidized by air coming through the air distribution grid at the bottom of the reactor. Air is fed by a Root's blower, the capacity of which is 0.25 m^3/s of 1.4 bar air. The solid matter, entrained from the reactor, is separated in a uniflow cyclone and recycled back to the bottom of the reactor. The returning pipe of the bed material is an insulated steel pipe. A start-up burner is located above the grid. Natural gas is used in the star-up burner.

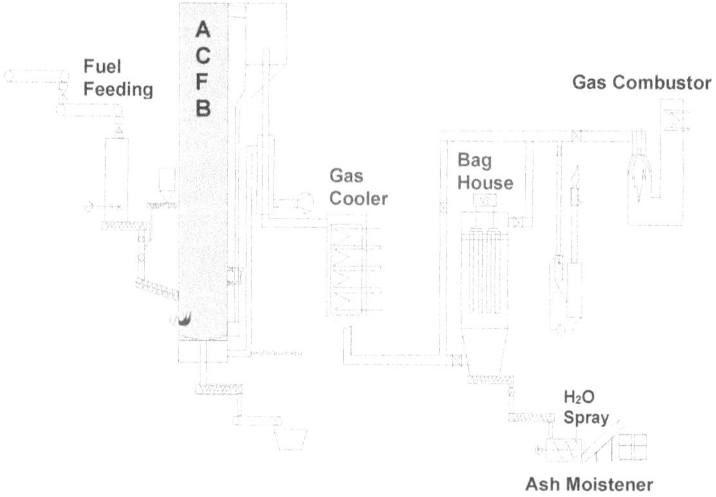

Figure 1: The process flow scheme of the test facility.

The hot syngas, leaving the reactor at 850 - 950°C, is cooled in the gasifier air preheater to 600 - 700 °C. The syngas is cooled further in the water-tube boiler, which is able to operate with maximum values of water 60 bar and temperature 240°C. After the boiler, the gas is being cleaned in a hot filter, in a separately operated baghouse type of a filter unit. After the hot filter, the syngas is led the syngas combustor. The syngas combustor is especially designed for low heat value gas. The objectives of the gas burning are to gain information of the burning properties and emissions of the syngas.

FUELS AND BED MATERIAL

The refuse derived fuel used during gasification test runs was the paper reject from processing of recycled paper for paper production in Hsin Wu Mill, Yuen Foong Yu Paper Mfg. Co., Ltd., Taiwan. The reject was pelletized, and it contains approximately 60-70% wood fiber, 10-20% plastics and 10-15% ash. The analysis of the pre-sample of the paper reject is presented in Table 1. The paper reject contained a high amount of chlorine. The pellets were about 1 to 3 cm of diameter. Silica sand with a mean size of 0.3 mm was employed as the bed material in this test. Bed make-up material consisted of a mixture of limestone and sand.

Table 1: Fuel analyses.

	Proximate analysis	
Moisture	as received	4.24 %
Volatiles	in d.s.	74.7 %
Ash at 575°C	in d.s.	14.1 %
Ash at 815°C	in d.s.	13.9 %
Fixed carbon	in d.s.	11.4 %
	Ultimate analysis	
C	in d.s.	56.3 %
H	in d.s.	7.84 %
N	in d.s.	0.39 %
S	in d.s.	0.17 %
O	in d.s.	21.4 %
	Heat values	
HHV	in d.s.	6,339 kcal/kg
LHV	in d.s.	5,930 kcal/kg
	Halogens	
Cl total	in d.s.	18,700 μg/g

GAS ANALYSIS

The main compounds of syngas, H_2, CO, CO_2, CH_4, C_2H_4 and C_2H_6 were analyzed automatically semicontinuously by a gas chromatograph. The CO, CO_2, O_2 and CH_4 contents in the syngas were also analyzed with continuously operated gas analyzers. The main sampling points located before the gas cooler and after the gas cooler. The flue gas composition after the syngas combustor was also analyzed. The O_2, CO, CO_2, SO_2, NO_x and HCl contents in the flue gas were analyzed by continuously operated gas analyzers. Tar and benzene contents of the syngas were also analyzed. In addition, Polychlorinated dibentsofurans and dioxins were measured and analyzed during the experiments.

COURSE OF THE TEST RUNS

The test matrix is presented in Table 2. The fuel feed rate during the testing varied between 200 and 400 kg/h depending on the plant load and temperature. During Tests 1 - 3, the bed material consisted of sand and limestone. In the rest of the test runs limestone feed rate was increased. Higher amount of limestone feeding was assumed to increase the dust load to the filter and thus decrease the tendency of sticky tar compounds to attach into the filter surface. Limestone is also known to enhance the tar decomposition reactions.

Secondary air feeding into the fuel feeding line was assumed to break tar compounds and some long hydrocarbons in the upper part of the furnace. In addition to this, it was assumed that secondary air into the fuel feeding line prevents the syngas flow from gasifier to the gasifier room via fuel feeding line.

During Tests 4 - 7, the total air feed rate was decided to maintain at a quite a low level (150 - 160 g/s), in order to achieve more residence time for gasification reactions. During Tests 4, 5 and 7, calcium hydroxide ($Ca(OH)_2$) was fed to the syngas line before the baghouse in order to reduce the HCl emissions. During Test 6, sodium bicarbonate ($NaHCO_3$) was fed to the same location.

Table 2: Test matrix.

Test run	Gasification temperature (°C)	Total air flow rate (g/s)	2nd air flow rate (g/s)	Steam flow rate (g/s)	Bed material feed rate (kg/h)	Additive
1	830-860	200	0	20	20-25	
2	880-900	200	0	20	30-35	
3	910-940	200	15	20	18-22	
4	910-930	150	10	0	27-32	$Ca(OH)_2$ feeding
5	900-920	160	10	10	35-45	$Ca(OH)_2$ feeding
6	900-920	160	10	10	40-50	$NaHCO_3$ feeding
7	900-920	160	10	10	45-55	$Ca(OH)_2$ feeding

RESULTS AND DISCUSSION

GASIFICATION TEMPERATURE

Figure 2 shows the typical temperature profile in the gasifier. In the same figure is also shown air feed rate. Typically temperature distribution was very even and temperature changes were smooth. Only if there was disturbance in fuel feeding, temperatures changed quickly.

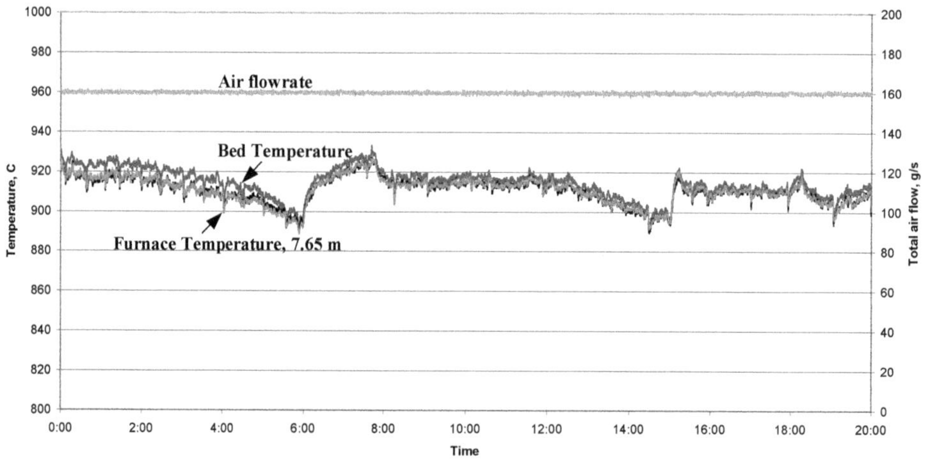

Figure 2: The typical temperature profile in the gasifier (Test 7).

SYNGAS COMPOSITION

The composition of the syngas is shown in Table 3. Carbon conversion was calculated from the material balance using the following equation:

Carbon conversion = [C_fuel – C_ash] ÷ C_fuel

where C_fuel is carbon input (g/s) in the fuel; C_ash is unburned carbon output (g/s) in bottom and fly ashes. The nitrogen (N_2) content of the syngas was calculated as a difference. It can be seen in Table 3 that the syngas contains around 7.3-10.1 vol% H_2, 5.3-8.0 vol% CO, 4.3-5.8 vol% CH_4,

12.9-16.0 vol% CO_2, etc, at the operating temperature of 850-950°C. Compared with other works,[4] the main composition of syngas, such as H_2, CO, CO_2, CH_4, C_2H_4 and C_2H_6, is reasonable. In addition, carbon conversion to gas and tars varied between 94% - 98%, depending on process conditions. Higher carbon conversion was achieved when gasification temperature was increased.

Table 3: The composition of the syngas.

Test run		1	2	3	4	5	6	7
Gasifier average temp (°C)		851	890	915	898	908	915	912
ER (-) [1]		0.269	0.288	0.306	0.331	0.367	0.368	0.340
Syngas HHV, wet (MJ/kg)		6.24	5.80	5.25	4.74	4.30	3.99	4.48
Carbon conversion (%)		94.1	94.1	97.0	95.7	97.3	98.0	98.2
H_2	(vol%, dry)	7.3	10.1	10.1	8.4	8.8	8.8	8.8
N_2	(vol%, dry)	62.3	60.4	60.9	65.7	65.2	64.2	63.4
CO	(vol%, dry)	8.0	7.8	6.7	6.4	5.6	5.3	5.6
CO_2	(vol%, dry)	12.9	13.1	15.1	13.0	14.2	15.9	16.0
CH_4	(vol%, dry)	5.8	5.7	4.9	4.5	4.4	4.3	4.4
C_2H_2	(vol%, dry)	0.41	0.52	0.44	0.31	0.39	0.31	0.42
C_2H_4	(vol%, dry)	3.2	2.2	1.9	1.7	1.4	1.1	1.3
C_2H_6	(vol%, dry)	0.09	0.05	0.04	0.03	0.02	0.02	0.03
C_3–C_5	(vol%, dry) [2]	0.149	0.052	0.034	0.050	0.033	0.016	0.032
HCl	(ppmv, dry)	5,300	3,551	2,755	158	--	68	150
Tars, benzene (g/m^3n, dry)		55	40	41	36	27	24	35

[1] ER: Air equivalence ratio

[2] C_3–C_5 components combined concentrations

In the first 3 test runs, gasification temperatures from 848 to 925°C were tested. It was observed that the tar contents in the syngas were high, but the tar contents were decreased by increased gasification temperature. After the 4th test run, the residence time of syngas in the gasifier was increased to improve the cracking of tars. This was made by decreasing the air flow rate from 200 g/s to 150 – 160 g/s. For the same target, limestone feeding rate to the gasifier (make-up) was also increased significantly. CaO is known to crack tars to lighter components. The lowest tar contents were measured in Test 5 and 6 when the air equivalence ratio (ER) was highest. Although high enough residence time and temperature provided in the gasification process can reduce the tar contents, the concentration of syngas was decrease slightly. Tars and bentzene in the syngas were between 24 and 55 g/m^3n.

It can also be found that the heating value of the syngas decreased with increasing the air equivalence ratio (ER) linearly. The result shown in Fig. 3 is in agreement with the findings of van der Drift et al.[4] Theoretically ER is not an independent parameter, but high ER leads the reaction towards the combustion process. Therefore, the heating value determined by the concentration of syngas decreases with increasing the ER.

To reduce HCl emission, $Ca(OH)_2$ and $NaHCO_3$ was fed to the syngas line before the baghouse. During Tests 4 – 7, it can be seen that HCl emission can easily be controlled significantly by injecting $Ca(OH)_2$ or $NaHCO_3$.

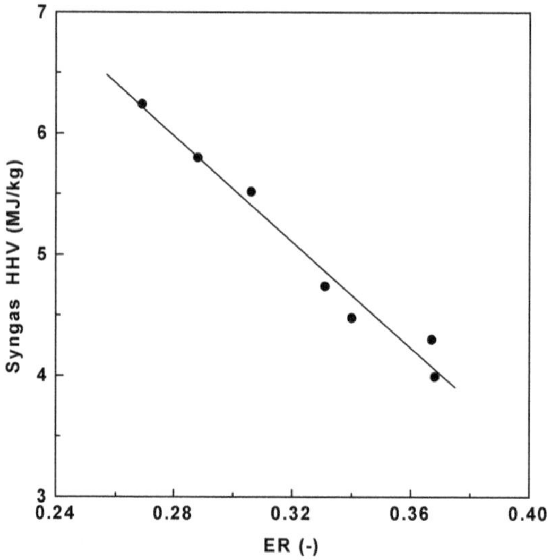

Figure 3: Relationship of heating value of the syngas and the air equivalence ratio.

GASEOUS EMISSIONS

About 1/4 – 1/3 of the total produced syngas was led to the syngas combustor. Table 4 shows the contents of SO_2, NO_x, HCl and O_2 in the flue gas after the syngas combustor.

Table 4 Gaseous emissions after syngas combustor.

Test run	1	2	3	4	5	6	7
Temp in gas burner (°C)	1,107	994	1,111	911	1,022	1,079	1,104
O_2 (vol%, dry) [1]	8.7	11.4	8.6	11.2	9.2	6.9	8.6
CO (ppmv, dry) [2]	16	30	28	21	17	16	30
SO_x (ppmv, dry) [2]	20	7	2	5	6	1	2
NO_x (ppmv, dry) [2]	385	445	233	242	223	208	224
HCl (ppmv, dry) [2]	--	558	694	87	84	39	87

[1] Corrected to 10% O_2 in dry gas

[2] Measured after combustion

Table 4 also shows the burning temperature. All emissios have been corrected to 10% O_2 in dry gas. The measured HCl contents during Tests 1 were on the upper limit of the measurement area, so it is possible that measurements are not representative. As observed in Table 4, HCl content of the flue gas was much higher during Test 3 than that during Tests 4 – 7, when additive material was used. After Test 6, HCl sorbent feeding was decreased; as a consequence the HCl content of flue gas increased more than double. Sulfur content in the syngas can also be reduced by adding limestone into gasifier.

Fig. 4 shows the HCl emission of flue gas after the syngas combustor. After the test was completed, calcium hydroxide feeding was ended and the filter was bypassed. A short test was performed to burn unfiltered syngas in the syngas burner. As seen in the Fig. 4, the HCl content started to increase immediately when the filter was bypassed and calcium hydroxide feeding was stopped.

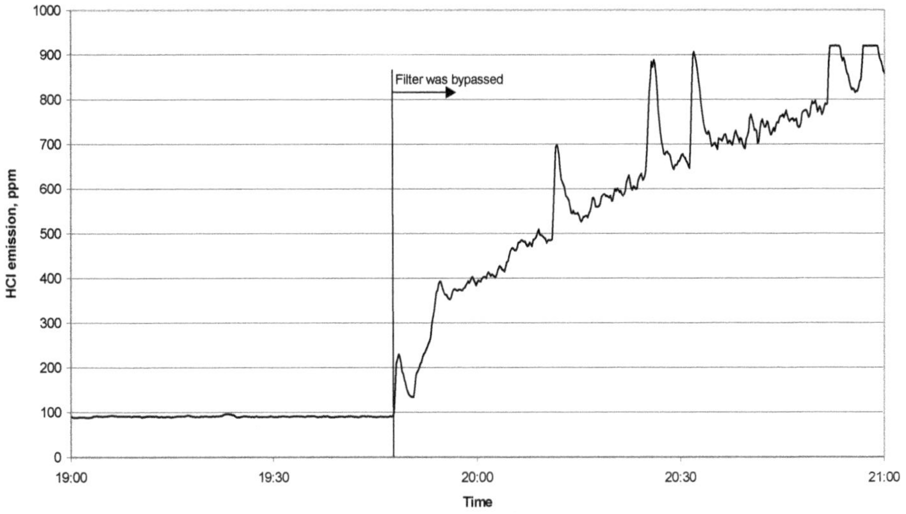

Figure 4: HCl emission after syngas combustion (Test 7).

DIOXINS

During Test 6 and Test 7, the periodical sampling of polychlorinated dibentsofurans and dioxins were taken from syngas and flue gas after the syngas combustor, and also were taken from the bottom and fly ashes. The analyses were carried out by the University of Kuopio, Finland. Table 5 shows the contents of polychlorinated dibentsofurans (PCDF) and dioxins (PCDD) in flue gas, the bottom ash and the fly ash. The values presented as International Toxic Equivalents (I-TEQ) are calculated by using international toxic equivalency factors (I-TEF) for every compound.

Table 5: Dioxins and furans in flue gas and ashes.

Test run	6		7	
	Dioxins	Furans	Dioxins	Furans
	(ng-TQE/m³n)		(ng-TQE/m³n)	
Flue gas [1]				
In gas	1.663	3.488	1.157	2.182
In particles	0.103	0.136	0	0
Ashes				
Fly ash	0.038	0.071	0.011	0.007
Bottom ash	0.003	0.010	0.002	0

[1] At 10% O_2 in dry gas

It can be seen in Table 5 that concentrations of dioxins and furans are low in the bottom and fly ashes of the gasifier, but the emissions of flue gas from the syngas combustor are little higher due to the short residence time in the combustor. Because the syngas combustor load was maintained at the high level, the residence time at the high temperature area in the syngas combustor was quite short (typically less than 1 second). In commercial plant the residence time at the high temperature area is much longer and that causes more complete decomposition of toxic emissions.

CONCLUSIONS

In order to evaluate the potential of energy recovery by gasification of pelletized paper reject, seven tests have been carried out in a 3 MW atmospheric circulating fluidized bed (CFB) gasifier. Some significant results were found during the tests. The preliminary results show that the syngas contains around 7.3-10.1 vol% H_2, 5.3-8.0 vol% CO, 4.3-5.8 vol% CH_4, 12.9-16.0 vol% CO_2, etc, at the operating temperature of 850-930°C. Higher carbon conversion was achieved when gasification temperature was increased. In order to decrease the tar contents, high enough residence time, temperature and make up feed rate must be provided in the gasification. Also heating value of the syngas decreased with increasing the air equivalence ratio linearly. HCl emission can easily be controlled significantly by injecting $Ca(OH)_2$ or $NaHCO_3$ into syngas before the baghouse. Sulfur content in the syngas can also be reduced by adding limestone into gasifier. The concentrations of dioxins and furans are low in the bottom and fly ashes of the gasifier, but the emissions of flue gas from the syngas combustor are little higher due to the short residence time. It is concluded that gasification of the paper reject in Taiwan is feasible.

ACKNOWLEDGEMENTS

The authors H.-T. Lee, K.-T. Wu, and C. I. Juch acknowledge financial support provided by the Energy Commission, MOEA, R.O.C.

REFERENCES

1. Energy Commission (1999) *Report on the RD&D of New and Clean Energy Technologies.* MOEA, R.O.C., Taipei.
2. Reed, T.B. (Ed) (1981) *Biomass Gasification.* Nolyes Data Corp, New Jersey.
3. Raskina, N., Palonenb, J. and Nieminenb, J. (2001) Power boiler fuel augmentation with a biomass fired atmospheric circulating fluuid-bed gasier. *Biomass and Bioenergy*, **20**, 471-481.
4. van der Drifta, A., van Doorna, J., Vermeulenb, J.W. (2001) Ten residual biomass fuels for circulating fluidized-bed gasication. *Biomass and Bioenergy*, **20**, 45-56.

Thermal Treatment of Biomass using Downdraft Gasification

Dr Abdullah Malik
Waste to Energy Ltd, Eyston, Borley Green, Sudbury, Suffolk CO10 7AH, UK

ABSTRACT: Waste to Energy Ltd (WTE) specialises in downdraft gasification for waste destruction and energy capture. In the last few years WTE has made significant progress in successfully treating difficult wastes with a low CV and high ash content. In this paper recent key plant upgrades and performance successes are highlighted, which makes the WTE gasification system to provide a consistent quality gas and unattended operatability, using recently installed plants as case studies.

WTE has developed the downdraft gasification technology to a point where it can handle most combustible solid wastes including all solid renewable resources. Typical feedstock may consist of Willow, Tyres, Waste construction timber, Wastes containing precious metals, Solid biomass, Sorted or processed Municipal Solid Waste, Clinical and other Industrial and hazardous Wastes, Solid wastes from oil industry, Waste packaging etc. WTE has emerged as one of the leading companies offering a complete environmentally sound energy recovery process reducing the environmental impact caused by excessive use of fossil fuels.

WTE plant is designed in a compact form which can be fitted in a standard 40' ISO container offering leaves minimal visual impact, fully portable and can easily be installed at the waste generation source, thereby meeting the proximity principle. This also eliminates the environmental damage caused by transportation of fuel to the power generation plant. Full scale packages come in a range of sizes from 50kg/h to 1000kg/h and for larger capacities multiple modules are employed as appropriate.

WHAT IS DOWNDRAFT GASIFICATION?

A solution to the problem of tar entrainment in the gas stream has been found by designing co-current or downdraft gasifiers. The gasification process can be divided into four distinct zones, each of which is characterised by one important step in the process of converting the solid fuel into a combustible gas. The detail information on the processes which occur in these zones can be found in the literature (1).

The fuel is introduced at the top and the primary gasification air is introduced at the oxidation zone in the gasifier. Hot oxidation zone at 1000-1200°C radiates heat to the fuel sitting above it making it to pyrolyse. The producer gas is withdrawn at the bottom of the gasifier through a reduction zone using an induced draft fan. Gas laden with acidic and tarry distillation products, on their way through the glowing bed of charcoal in the reduction zone gets converted into permanent gases comprising of hydrogen, carbon dioxide, carbon monoxide and methane where more or less complete breakdown of the tars is achieved. The main advantage of downdraft gasifier lays in the production of a tar-free gas suitable for engine application.

The major drawback of this type of gasifier lies in its inability to operate on a number of unprocessed fuels. In particular, fluffy, low density materials give rise to flow problems and excessive pressure drop. Therefore solid fuel must be briquetted. Another important fuel characteristic required to produce consistent quality gas using downdraft gasifier is to maintain moisture below 20%. This is achieved by employing a suitable drying process. High ash content fuels also give immense operational problems in terms of ash fusion thus causing bridge formation.

MAIN WTE PROCESS STAGES

Typically the complete process consists of Feed Handling system, a Dryer (required if the feed moisture is greater than 20%), Gasification and Power Generation as shown in the process flow diagram in Figure 1.

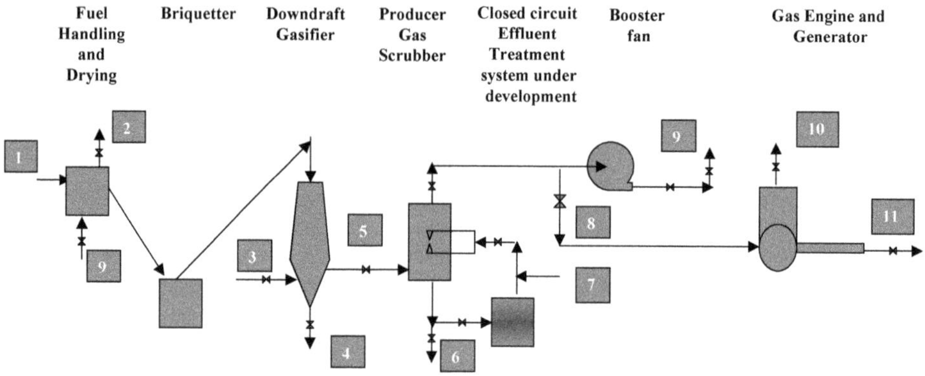

1 After centrifuge sludge enters into the dryer at 20-25% ds
2 Clean dust free exhaust gases after scrubber
3 Air enters the gasifier at about stoichiometric levels
4 White ash (low carbon) with leachable metals but when used as road aggregate, produce zero leachate.
5 Hot producer gas via ~1200°C oxidation zone and 800°C reduction zone leaves the gasifier at ~250°C and enters the scrubber where it is quenched to ~50°C thereby removing dust, acid gases, tars and also eliminate Dioxin reformation.
6 Scrubber residue - Tars, dust
7 Make up water
8 Clean producer gas mainly consisting of H_2, CO, CH_4, CO_2, H_2O, N_2, to duel fuel compression engine
9 Clean gas to dryer or invisible flare

Figure 1: Drying and Gasification - A Completely Emission Free Solution.

FEED HANDLING

The fuel carrying vehicles could discharge into a bulk discharge unit, which would elevate the materials into the hopper. Depending on the type of fuel the bunker may consist of a push-floor discharge located in the bottom with central discharge screw followed by an elevator to handle moist feed and deliver to dryer feed hopper.

DRYING

From the Feed Handling system, moist fuel is conveyed to the dryer system from the storage in a controllable, metered fashion.

Normally the hot air for the drying process is generated mainly by using hot engine exhaust. Depending on the moisture level in the feed any surplus heat requirement is met by a directly fired propane gas air heater. The heat requirement is normally controlled by monitoring of the dryer exhaust gas temperature. As evaporative load increases, the exhaust temperature drops and heat input is proportionally increased. For direct fired dryers, using exhaust gas helps to provide the products of combustion for system inerting. The system would be designed to run at less than 8%

oxygen content by controlled re-cycling of the products of combustion and water vapour, thereby ensuring a safe drying process for the combustible product. The dried product at around 90%ds is then fed to the gasifier.

The hot gases leaving the dryer then pass through a wet scrubber. The scrubber performs two main functions. The first is to condense the water evaporated by the dryer to enable the air to be re-cycled. Its second function is to remove the low level of residual particulate emissions in the case of direct fired dryer.

Up to 20% of the exhaust gases are re-cycled and the remaining gas is discharged to atmosphere by appropriate height stack.

The type of post treatment of exhaust gases depends on the contaminants present which vary with the type of the waste treated and the type of the dryer.

GASIFICATION

The gasification process is designed to meet the UK standard of manufacturing and complies with ISO 9002. The package comprises of a belt conveyor from ground level to the top of the gasifier, a two in one fuel densifier and feeder, a gasification system and a producer gas scrubber.

The 90%ds Fuel is then conveyed to the top of the Densifier Feeding Hopper using an enclosed belt elevator/ conveyor. It densifies the fuel to produce suitable size briquettes and feed the gasifier simultaneously. Densification not only densifies the material but also acts as an air seal for the gasifer.

Fuel Solids at ~90%ds are fed to the gasifier where it is initially ignited to start a self sustainable gasification process running under sub-stoichiometric air supply condition in which part of the heat energy is utilised to keep the gasification process going without any external heat source. It is designed to produce combustible gas of calorific value in the order of $5MJ/m^3$. The produced gas after cooling and cleaning is then supplied to run gas engine. The key feature of the gasifier is that, it has a unique design of throat and air nozzles, which enables a homogeneous final zone temperature distribution. This allows complete combustion of all pollutants (i.e. tar and volatiles) present in the gas prior to leaving the gasifier. The top part of the gasifier is designed to stop any build up of moisture and volatile condensation in the drying and pyrolysis zones.

The gasification system is kept under vacuum by a main process fan to enable the gases to pass through the scrubber. The produced gas at ~250°C generated by the gasifier, after cleaning and cooling using water scrubber will be available at more than 50-60 mbar pressure and the temperature in the range of 40-50 °C with the main combustible constituents of CO 12-18%, $H_2$10-12%, CH_4 2-4% and trace levels of heavy hydrocarbons remainder being CO_2, and N_2. After scrubbing the dust and tar load are expected to be in the order of $30\text{-}70mg/m^3$.

The control plc is programmed to control the gasifier operation. The gasifier operation is flow based. As the pressure drop increases, the booster fan speed increases to reach its maximum set point. This then activates the bottom ash auger relieving the back pressure to a specified set point. The system can be run totally automated in an unattended mode with regular hourly plant checks and is supplied with MMI screen and SCADA link for remote monitoring as shown in Figure 2.

The plant is programmed to start up and shut down automatically with manual option available.

The biogas produced is composed of circa 60-65% inert gases in the form of N_2 and CO_2; this makes handling of the gas inherently very safe. However, a low level enclosed flare system is provided as part of the gasification package to further enhance safe operation, mainly during start-up and shutdown stages. The flare system is fitted with propane gas burners to destroy all pollutants during the flare operation. A picture of flare burning producer gas is shown in Figure 3. A non-soluble dry ash is discharged from the gasifier with Carbon levels in the order of 15-18%.

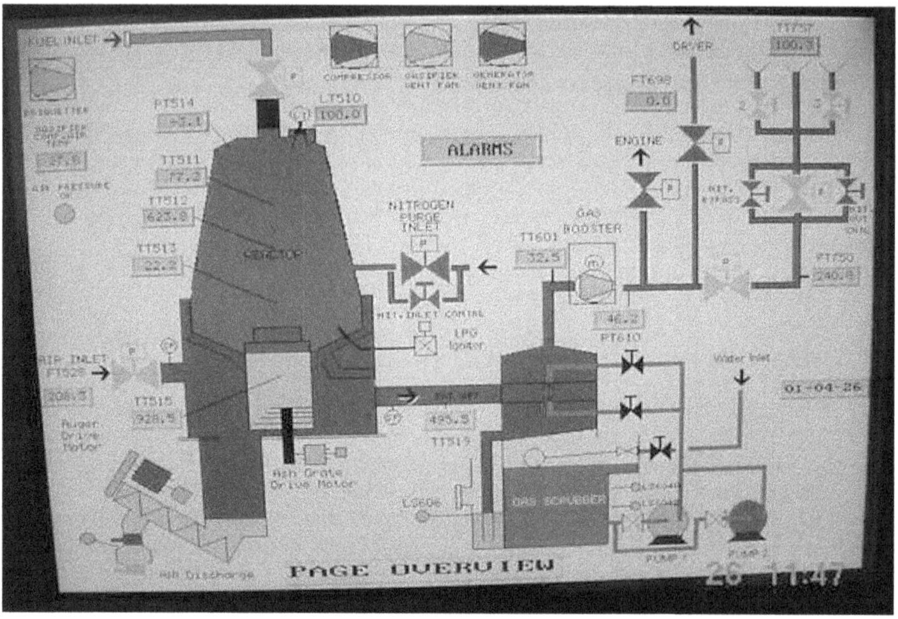

Figure 2: Automatic Control System MMI Screen with SCADA Link for Remote Monitoring.

Typical gas characteristics:

H_2 - 8-12%
CO - 12-18%
CH_4 - 3-6%
CO_2 - 10-12%
N_2 - 45-55%
H_2O - 2-5%
CV - 4.5-5.5MJ/m^3

Figure 3: Gas Burning Inside the Enclosed Flare Clean Emission Free Low CV Gas.

POWER GENERATION

The surplus cooled clean producer gas at ~70°C is then drawn into the gas engine where it will be completely combusted. Hot gases leaving the engine at around 400-450°C will be re-directed to the dryer or heat exchanger for heat recuperation.

EFFLUENT TREATMENT

Although the effluent leaving the system does not exceed the level of contaminants metal content beyond WID limits however, where clean water availability is scarce it is prudent to clean the water and recycle back to the scrubbing system. This is achieved by using state of the art filtration system currently available in the market for tar and dust removal.

KEY PLANT UPGRADES

The WTE gasifier has been upgraded which makes it more or less versatile in handling any type of solid fuel just by minor re-tuning as shown in Figure 4. Some of the upgrades are discussed below:

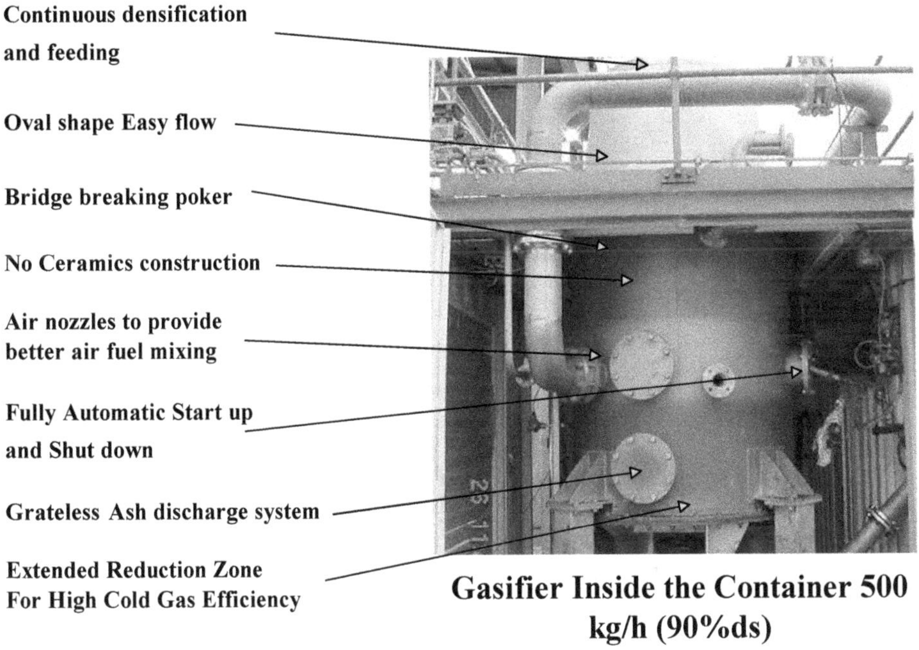

Figure 4: Key Design Upgrades.

CONTINUOUS DENSIFICATION AND FEEDING

The gasifer is fitted with a mechanical level probe which ensures fuel level is maintained during gasifier operation allowing it to run 24 hrs a day 7 days a week continuously without any stoppages.

AIR SUPPLY NOZZLES

Air nozzles are designed to allow air swirl to produce homogeneous temperature distribution in the oxidation zone (A world patent has been awarded for this).

EASY FLOW GASIFIER INTERNAL SHAPE

The gasifier internal wall is fabricated with an oval shape profile to provide free fuel flow without causing any blockage and thereby less likely to form bridging.

REDUCTION ZONE DESIGN

The reduction zone has been specifically designed as a result of recent experience to include a specially designed ash discharge auger to handle difficult fused and clinkered ash; this incorporates increased zone length with a porous zone for complete tar destruction and dust filtration. The gas temperature leaving the gasifier is decreased so as not to exceed 200-250°C. The gas quality is clean enough to directly feed to most types of gas burners.

GAS SCRUBBER

In order to feed the gas engine the gas needs to be cooled to around 75°C. This is achieved by using a water scrubber. This not only cools the gas but also removes any residual tar and dust load, ensuring continuous clean supply. Low thermal and dust load on the scrubber also reduces the water usage producing a low amount of effluent, which is suitable for discharge back to the head of the words without further chemical treatment. However, the water can be cooled and filtered for re-use. A bleed of 0.5-3.0m^3/h effluent with a typical BOD of 200 and COD of 500 depending on the gasifier operating condition and feed type which is treated prior to disposal as appropriate.

BRIDGE BREAKER

The variation in the feed quality in terms of size, ash type and content could lead to briquettes becoming fused together, forming a fuel heap (a bridge) in the pyrolysis zone area. WTE gasifier comes with a bridge breaking system, which gently pokes the pyrolysis zone sufficiently to prevent bridge formation. The bridge breaker linked with plc control to operate in a predefined fashion but adjustable to suit various poking conditions. However in the event a bridge is formed, bridge breakers can be initiated manually at any time during the process. The design and control sequence is to ensure no bridge is formed, however, in the unlikely event of bridge formation or ash clinker fusion the material is moved forward towards the ash discharge via the reduction zone.

ASH CONDITIONING ROTARY KILN

In order to meet the WID requirement of below 5% C in ash levels an additional ash conditioning process is employed where almost the entire residual C is combusted. The hot gases generated are then diverted to become part of the dryer or heat exchanger for heat recuperation. The rotary kiln is fitted with propane burners, which are only used during the start up of the kiln.

PROPANE GAS BURNER

The gasifier is fitted with a propane gas burner. The burner has two functions. Firstly at the start-up condition it is used as an ignition source; secondly during normal operation, in the event of sudden drop in the oxidation zone temperature due to a sudden change in fuel CV, the burner takes over until the gasifier has recovered its normal operating set point.

ENVIRONMENTAL CONSIDERATION

Waste to Energy plant is designed to meet the strictest environmental standards. Gasification process works under reduced condition and oxidation zone operates at >1000°C which not only discourages the formation of Dioxins but also destroys any of the already present in the waste stream. This is followed by a simple zero effluent water scrubbing system (ZEWSS) which cleans the gas from tar, oils, dust, acid gases and other chemical contaminants to be precursors to dioxin/furan formation. The clean gas can then be used in the boiler for steam generation or a standard CHP system for power generation.

The solid residue obtained is inert and reduced to its level of ash present in the waste treated. The gasification process can also be used as Materials Reclamation Facility (MRF) for recycling precious metals.

The physical size of the gasification process plant is small and therefore gives minimal visual impact; is mobile to be installed easily at the waste generation source; and therefore meets the "proximity principle" (See Figure 5).

Figure 5: Skid mounted or mobile unit.

RECENTLY INSTALLED PLANTS

GASIFICATION OF SEWAGE SLUDGE

1) In the first phase of installation, a sewage sludge drying and gasification plant, to treat sludge generated by 60,000pe, was installed during 2000 at one of the leading UK water companies (2, 3). This plant has completed its function, satisfied fundamental operating characteristics. In the phase 2, the same plant will be upgraded to match the site throughput and another full-scale plant will be installed at another site during 2003. The key performance successes are listed as:
   - Sewage sludge 85-90% ds and ash content >30% (iron content of the order of 12% of ash) was successfully gasified. Syngas calorific value was in the range of 3.5-4.9MJ/nm3 and combusted in the dryer burner. Depending on the type and moisture content of the sludge, throughput of 250-450kg/h was achieved.
   - Significant levels of Hydrogen Cyanide in syngas lead to it being scrubbed by the water. The water was treated using Sodium Hypochloride and Sodium Hydroxide by keeping pH >8.5. In order to minimise the cost of chemical treatment scrubber water temperature was raised to keep the cyanide in the gas phase offering increased gas CV.
   - NH_3 was also present in the gas reducing the formation of NO_x to below <100 ppm.
   - Carbon in ash leaving the gasifier was observed of the order of 10-20% which then was left smouldering under controlled thermal condition for 24hrs to bring the levels <3% prior to disposal. The ash analysis showed it to be 99.99% non-leachable.

2) In 2001 a full scale commercial plant delivered to an Israeli organisation and deployed to treat 50-100 kg/h of special waste in Israel. After initial supervised trials, the plant passed the local environmental standards. No further data is in the public domain.
3) Another full scale drying and gasification plant 50-100 kg/h capacity started its operation in 2002 for leather waste destruction, chromium capture and re-cycling and power generation (4). This plant incorporates the latest European Waste Incineration Directive (WID) requirements, has satisfactorily completed its proving trials and is being released for evaluation to key leather industry participants. The key features of this plant are:
   - The plant consists of an air swept dryer, briquetter and gasifier all within a portable ISO container.
   - The plant is fitted with a specially designed grateless reduction zone to handle variable ash contents in the fuel feed when using a number of tannery waste streams i.e. Cr shavings, Veg Shavings, Sludge, Trimming/Tallo, Wet white shavings and EoL with an average CV of 20MJ/kg for 90% ds.
   - The plant is fitted with a three stage poking device to help maintain free fuel flow.

   Some of the key operational data is summarised below:
   - The average moisture level of the wet feed was 50%.
   - Syngas CV was found to be in the range of 4.2-5.5MJ/
   - 35-50% of the energy in the feed was used for drying and remainder was available for power generation.
   - Cold gas efficiency was found to be in the range of 51-65%.
   - The average produced gas composition (%) of H_2 7.8, CO 20, CH_4 1.7 (balance as N_2 and CO_2).
   - While using shavings, Chromium stays in the ash closing the Chromium loop.
   - Ash using other waste streams has shown immense potential for use as constituents for carbon filters.
   - Scrubber water analysis has been completed with ppb levels of aromatics present. These can be easily removed in the conventional treatment systems.
   - The plant has completed its phase one operation and is due to be transferred to a tannery in October.

WTE COMPANY PROFILE

INTRODUCTION

Waste to Energy Ltd (WTE) have a thermal destruction process on a patented advanced gasification principle, combined with most up to date clean up systems available (5). The company's strategy is to provide equipment that will treat all wastes at source eliminating transport costs and the adverse effects transport has on the environment.

WTE have formed a joint venture with Anglian Water and is currently building complete waste-to-energy plant to deal with sewage sludge disposal.

WTE is 25% owned by a financial institution in London, and can offer a variety of financial services including leasing.

R&D

Over the past six years in association with the University of Newcastle, important design data have been obtained from the university's on-site unit provided by WTE on a variety of feedstocks. The information comprises of composition and calorific value of the fuel, produced gas quality, operating temperatures, exhaust emissions, ash analysis, etc.

ENGINEERING CAPABILITIES

WTE philosophy is based on classic process engineering principles. Engineering is based on the preparation of data sheets, calculations, specification, and material requisitions and detail drawings with positive checks and controls to ensure accuracy and consistency.

Our process engineers solve the particular project problem and where necessary produce P&ID's, material selection and other factors to offer the most cost effective solution to the clients needs. Our engineering draws on the extensive design experience gained in advanced ventilation, drying and engine systems. This enables the benefits of past developments to be integrated into every new project.

WTE has developed expertise in Instrumentation and control to offer computer aided SCADA systems for fully automatic & remote control operations.

PROJECT MANAGEMENT CAPABILITIES

WTE places great emphasis on establishing an efficient project organisation in order to ensure smooth contract execution. The project manager is a key individual in the project team. He is the point of contact between the client and WTE. He is totally responsible to WTE management for successful completion of a project. The Project Manager maintains the client interface, manages project team to ensure milestones are met, establishes and oversees a project programme. The Project Manager also oversees the procurement and monitors the project performance to ensure complete client satisfaction.

MANUFACTURING FACILITIES

- Our manufacturing facility at Sudbury on a site of one acre specialises in welded fabrication including storage tanks, pressure vessels and heat exchangers using a variety of materials consisting of carbon steel, stainless steel, incoloy, titanium and copper.
- Our design meets the vessels standard of BS5500, BS2654, BS2594 and BS799.
- We have approved weld procedures and welder qualification to EN287/288. Our welder approval is carried out by Zurich certification Ltd.

QUALITY ASSURANCE PROCEDURES AND SAFETY POLICY

- WTE Ltd aims to operate an integrated quality management system conforming to ISO9002.
- WTE has experience in the control of health and safety hazards associated with design, fabrication and construction of process plants. WTE has always paid high regard to matters affecting health, safety and environment. WTE has an obligation to ensure that the plant and system are designed and constructed to be safe. WTE carries out health and safety reviews of all documents specifying design of our systems.
- Safety reviews such as HAZOP, HAZAN etc. of the design may be performed on all projects in order to identify potential hazards, evaluate consequences and make recommendations to eliminate risks.

CONCLUSIONS

- WTE aims to build on the recent successes achieved by treating difficult wastes.
- It is expected to have installed another five plants by the end of 2003 and the beginning of 2004 using sorted municipal solid wastes, sewage sludge, chicken processing wastes, forestry wastes and leather industry wastes.

REFERENCES

1) Olgun H., Dogru M., Howarth C.R. and Malik A.A., "*Preliminary Studies of Lignocellulosics and Waste Fuels for Fixed Bed Gasification*", International Journal of Global Energy Issues, 15(2001), No 3/4, 264-280.
2) Loram G, *"Local Gasification"*, Energy World April 2001 No.288 pp 16-17
3) Riches S and Evans I, *"Trial Results from the Implementation of Gasification Technology at Broadholme Sewage Treatment*", IQPC, 2002
4) Bowden W, *"Gasifier Operates in Closed Loop on Tannery Wastes*", Journal of BLC Leather Technology Centre September 2002. pp 141-143
5) Waste to Energy Ltd, Web address: www.waste-to-energy.co.uk

Hydrogen from Pressurised Catalytic Conversion of Wet Biomass: Conversion of Organic Matter by a New Catalytic Process

K. Pedersen
Chemidan, Julsoevej 6, DK-8600 Silkeborg, Denmark

ABSTRACT: A new catalytic process is developed, in which all organic matter dissolved in pressurised water is converted into a clean gas mixture of hydrogen and carbon dioxide with the help of a new type of catalyst.

Catalytic experiments with phenol solutions show that it is possible to convert the organic matter into hydrogen-rich gas at 400°C, 10 – 20 MPa and a residence time below 0,5 hour.

The gas comprises 70% hydrogen, 28% carbon dioxide and 2% methane.

Experiments have been conducted with the constituents of biomass:

Glucose represents the fraction of cellulose, xylose from the fraction of hemicellulose and phenol from the fraction of lignin. Phenol turned out to be the least reactive component, for which reason the investigations have been concentrated on this component.

Preliminary calculations on the process economy revealed a fair return on investments.

The production cost is 0.07 €/Nm^3 with a price of biomass of 70 €/dry ton. The European price of hydrogen is 0.20 €/Nm^3.

INTRODUCTION

Most of the hydrogen is currently produced on an industrial scale by the process of steam reforming, or as a by-product of petroleum refining and chemicals production. Steam reforming uses thermal energy to separate hydrogen from the carbon components in e.g. methane, and involves the reaction of these fuels with steam on catalytic surfaces. The first step of the reaction decomposes the fuel into hydrogen and carbon monoxide. Then a "shift reaction" changes the carbon monoxide and water to carbon dioxide and hydrogen. These reactions occur at temperatures of 200°C or higher.

Another way of producing hydrogen is by electrolysis of water. Hydrogen gathers at the cathode and the anode attracts oxygen. A voltage of 1.24 Volts is necessary to separate hydrogen from oxygen in pure water at 25°C. The smallest amount of electricity necessary to electrolyse one mole of water is 65.3 Watt-hours at 25°C. Producing one cubic meter requires 4.8 kWh.

Currently the most cost-effective way to produce hydrogen is steam reforming. According to the U.S. Department of Energy in 1995, the cost was $7.39 per million Btu or $7.00/GJ in large plant production. The assumption is a cost of natural gas of $2.43 per million Btu or $2.30/GJ. This is the equivalent to $0.24 per litre of gasoline. The production of hydrogen by electrolysis using hydroelectricity at off peak rates costs between $10.00 to $20.00/GJ.

The European price (JAN 2001) of hydrogen is 0.20 Euro/Nm^3 or 15.6 €/GJ based on steam reforming of natural gas with a price of 0.238 €/Nm^3 or 6.11€/GJ.

Conversion of wet organic matter into hydrogen normally requires gasification followed by gas cleaning and conditioning before water gas shifting the carbon monoxide containing gas into hydrogen and carbon dioxide. The water gas shift reaction normally requires 2 steps. A high temperature shift reaction followed by a low temperature reaction, since the reaction is exothermal.

In order to gasify the organic material it has to be dried to moisture content of 45 % for downdraft gasifiers and 10 - 15 % for updraft and fluidised bed gasifiers. The drying is most expensive. For this reason it has, until recently, been considered uneconomically to convert wet organic material with a moisture content above 50 % into hydrogen or other valuable products.

High-pressure processes have been investigated earlier after the oil crises in 1973 and 1978 for production of liquid products.

The University of Arizona used the extrusion feed reactor system. The process operated at 21 Mpa and a 20-minute residence time with sodium carbonate as homogeneous catalyst. Wood and product oil was mixed and processed at 375–400°C. The product oil had a heating value around 37 MJ/kg. [1]

Liquefaction of peat and straw was conducted in water at KTH and DTI in autoclaves at 20 Mpa and peat temperatures around 250-400°C and resulted in an oil with an oxygen content around 13% and a heating value of 20 MJ/kg [2].

Other studies were conducted elsewhere with comparison of using water or substitute it with donor liquids like glycols, bio-oils, tetralin, naphthalene a.o

There was also focus on wet gasification into methane rich gases for SNG production. Typical operation conditions were 300-450°C and pressures up to 35 Mpa.

The initial work was related to high temperature gasification by Mudge et al. [3]. The investigation of kinetics and catalysis of producing synthesis gas from biomass was conducted by Sealock et al in 1981-82 [4]. Alkali metals and alkali earth carbonates and supported metals were in the environment of water for production of methane. Other previous work in the field of high temperature pressurised aqueous media includes studies by Model et al (1978)[5], Amin et al (1975)[6] and Model (1985) [7]. Most of this early work consisted of reforming glucose and flour in water at supercritical conditions (>374°C, density >0.31 g/ml). The work presented interesting results, which were attributed to the unique properties of the supercritical water system. The results given by Model and others showed only limited gasification and the catalyst tested had nearly negligible effect.

More recent Antal et al. [8] performed experiments at 385–400°C to study acid and base catalysed systems. Different alcohols were dehydrated to the corresponding hydrogen carbon. Substantial conversion was detected at reaction times below 1 min.

Elliot and Sealock found [9] that nickel was one of the most favourable metals for gasification and that the gasification was increased by addition of alkali metal carbonates (K>Na>Cs) in the temperature region of 350–450°C. Ammonia was also demonstrated as catalyst for the water gas reaction.

Baker [10] conducted additional work by catalytic destruction of p-cresol. Carbon conversion was greater than 99% at 350°C and a residence time below 10 min. In addition methane concentrations were higher at 350°C than at 400°C.

High methane concentrations using reduced nickel have been substantiated in continuous tests with high moisture biomass.

The loss of activity for reduced nickel and others were identified by tests.

The loss of activity was associated with break down of the carrier and depositing of material on the metal surface area [11].

Recent work has been performed by Antal et al, using an active carbon as catalyst at temperatures in the temperature region of 500 - 600°C resulting in a hydrogen and methane rich gas. Salts separates into a powder-like product [8].

When methanation is taking place in the system, it is necessary to operate at a high temperature, in order to obtain a hydrogen rich gas via the equilibrium with steam, methane, hydrogen, and carbon dioxide. Some have found it necessary to ad small amount of oxygen into the system in order to maintain the high temperature.

In order to prevent severe corrosion, Forshungszentrum in Karlsruhe have developed a special reactor [12]. TNO and BTG are currently conducting experiments at 600°C aimed for hydrogen production [13] [14].

The present study enlightens the possibilities for the pressurised process in which the operation is much lower around 400°C. In his region salts are soluble in reaction media and plugging of the reactor system is avoided.

The methanation reaction is blocked on the catalyst surface in the new type of catalyst, which ensures an extreme high yield of hydrogen.

The catalyst has been operating for more than 4000 hours without permanent deactivation in aqueous solutions of organic matter.

EXPERIMENTAL

The experiments were carried out in a plug-flow reactor with a volume of 30 cm^3. Catalysts were crushed down to a size of 0.5–1.0 mm in order to have a uniform catalyst bed in the reactor with an internal diameter of 9 mm.

Thermocouples were placed in the centre of the reactor in order to detect the heat of reaction. Liquids with 10 % organic like methanol, phenol, glucose or xylose were dosed with an HPLC-pump and fed the reactor through a 1/16" pipe with wall thickness of 0.35mm. The small wall thickness was chosen for investigation of any corrosion problems.

The products were cooled down to ambient temperature and left the system through a backpressure valve in a 2-phase flow (liquid and gas). The gas was separated in glass equipment and sent for GC-analysis.

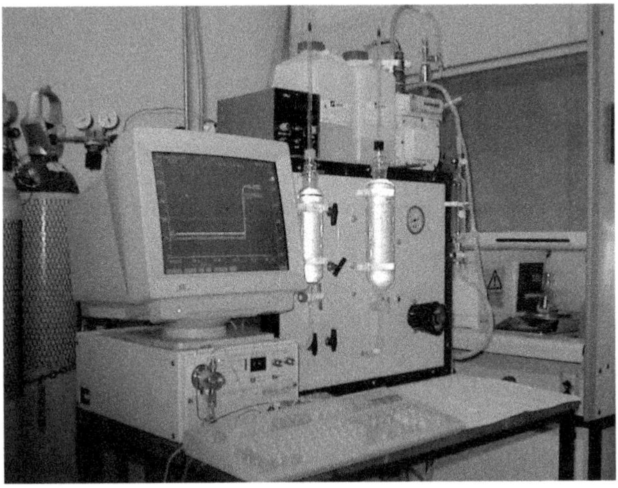

Figure 1: Test unit with control and data logging units.

RESULTS

Methanol (10 w%) decomposes very fast on the catalyst at temperatures above 350°C.

Phenol has the lowest conversion, of the investigated organic matters. However reasonable high conversion around 90% at 10 Mpa, 380 °C and a Liquid Hourly Space Velocity LHSV = 2 (l/l/h).

The rate of reaction is correlated with the rate constant of a first order reaction:

$$\text{Rate} = K^* C$$

The rate constant has the unit h^{-1}

The rate constant is calculated from the conversion (X) of the involved components by using the integrated equation:

$$K = \text{LHSV}^* \text{-LN}(1\text{-X})$$

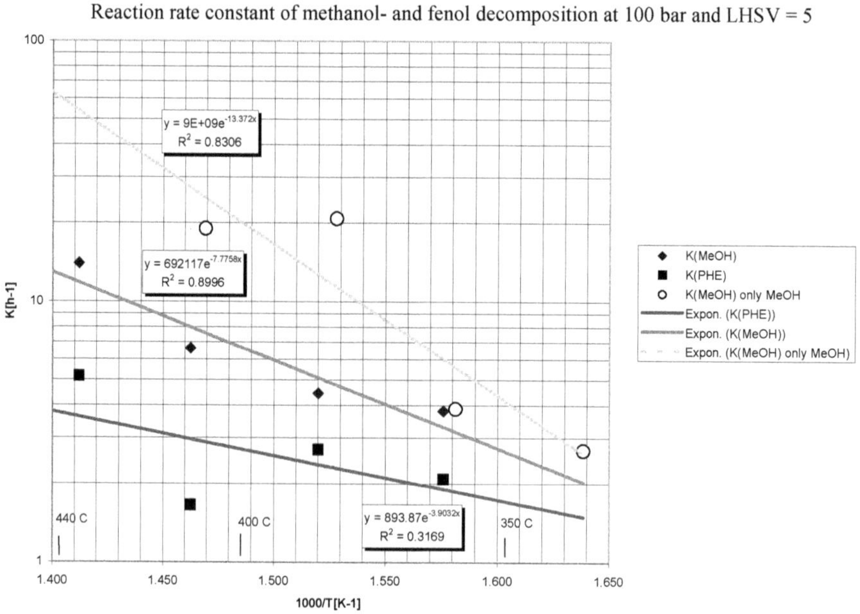

Figure 2: Rate constant versus the reciprocal temperature

At LHSV = 1 we obtain 90% conversion K = 2.3 and 99% conversion with K = 4.6. The activation energy is 111kJ/mole for conversion of methanol and 32 kJ/mole for conversion phenol.

The composition of the gas from conversion of methanol is much like the gas from phenol. The composition of the gas is shown in Fig 3.

The selectivity towards hydrogen is very high at temperatures below 425°C. At higher temperatures the methanation rate increases and lowers the selectivity. However a new modification of the catalyst maintained the high selectivity at higher temperatures.

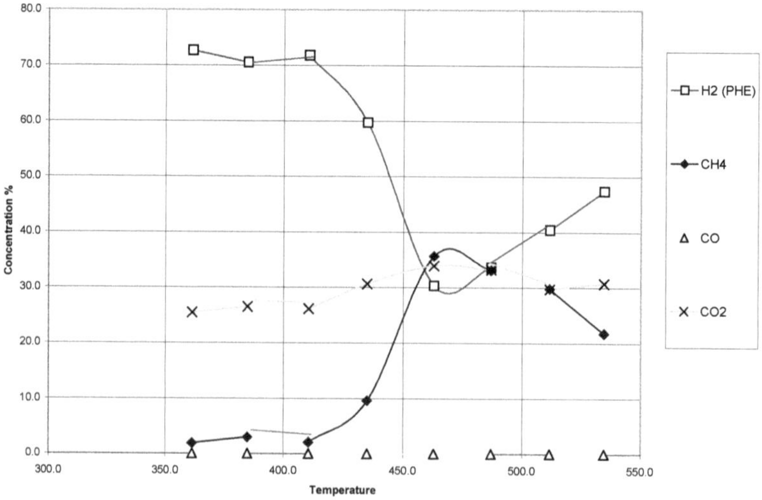

Figure 3: Gas composition by conversion of methanol and phenol at 10 MPz and LHSV = 5.

The selectivity to hydrogen remains constant during long test run at 400°C and 20 Mpa, as is seen in Fig 4.

Prolonged experiments at different pressures were carried out in order to find the pressure dependency and the stability of the catalysts since some deactivation was observed on the fresh catalyst.

The rate constant stabilises at a constant level after approximately 30 hours on stream at a pressure equal or higher than 10 Mpa. It is seen in Fig. 5 that the stabilised rate constant of the phenol conversion is much dependent on the pressure, approximately by the square of the pressure.

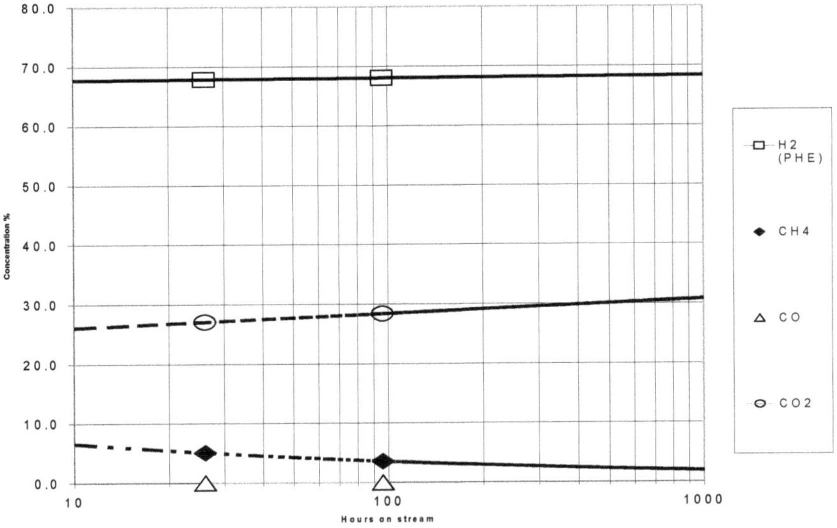

Figure 4: Gas composition by conversion of phenol at 200 bar and LHSV = 2.5.

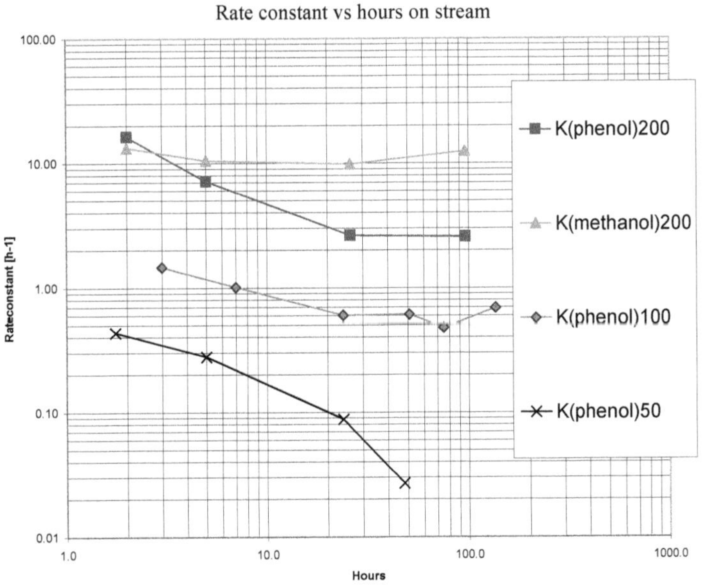

Figure 5: Conversion at different pressure.

The reaction steps are not yet known, but it is supposed to be a sequential steam cracking of C1 molecules from the adsorbed organic molecules. The overall reaction is here illustrated with phenol:

$$C_6H_6O + 11{*}H_2O = 14{*}H_2 + 6{*}CO_2$$

PROCESS STUDY

The study is based on an earlier study of a hydrous pyrolysis process. [15,16]. The process is based on conversion of wet organic material with a high water, typically higher than 50 %.

The organic matter is dissolved in sub- or supercritical water. The dissolving technique is used today in laboratories for extraction of organic matter from samples. At a temperature above 250 – 300°C cellulose dissolves, which is desired in this process. On a larger scale water is used in making pure paper pulp, by dissolving the brown lignin in wood at 200°C and 3 Mpa, by the so-called steam explosion process, which is illustrated below [17].

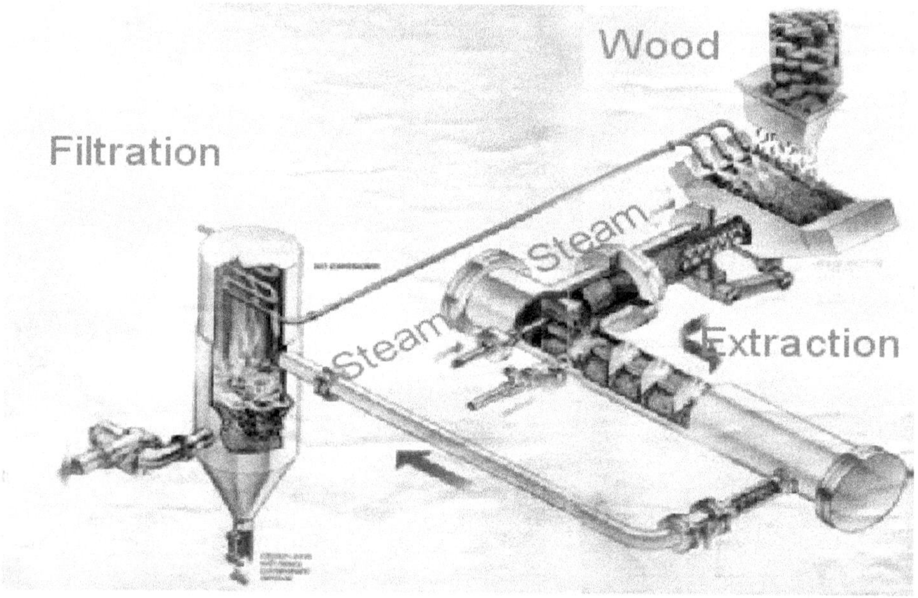

Figure 6: Extraction of organic matter by the steam explosion process.

The diffusion rate in supercritical fluid is comparable to the diffusion rate in gas. With supercritical water it has been possible to extract more than 75% of the dry organic matter, when the conditions involves pyrolysis and thermal breakdown of the polymers of wood [18].

The process scheme of the catalytic conversion of wet organic matter is shown in Fig. 7. The feed composition to the reactor is based on a weight ratio between dry organic matter and steam is 1: 9.

In this example, dry wood with moisture content of 10 % is used. The wood is compressed in an extruder to 15 Mpa, which is close to the normal pressure in an extruder. Superheated steam with carbon dioxide is added the extractor in which 95 % of the carbon is extracted. The rest leaves the extractor with the ashes.

The organic matter is supplied the catalyst, which produces a gas with the same composition as the laboratory experiments.

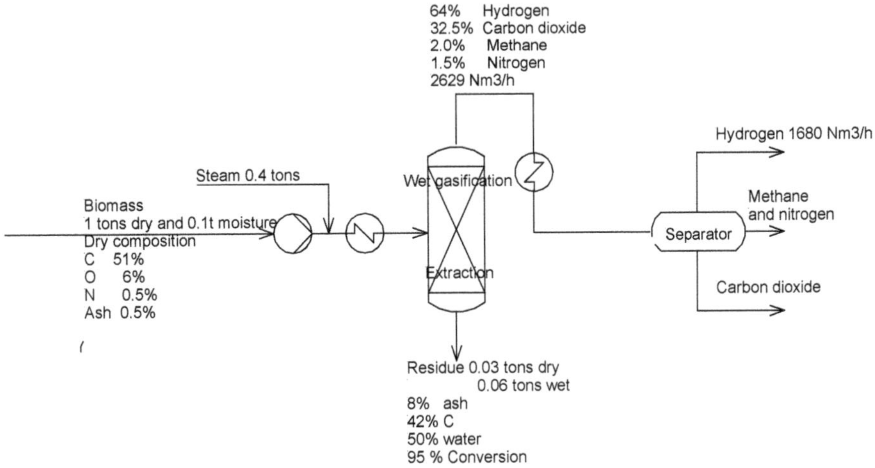

Figure 7: Process scheme of catalytic conversion of wet organic matter.

One ton dry wood is converted into 2629 Nm^3 gas of which 1684 Nm^3 is hydrogen. 8,9 tons steam/water is added the loop, when dry wood is used. Industrial sludge with 90 % water could be used directly without water addition [19].

UTILISATION OF THE PRODUCT GAS

The formed gas could be used ad a hydrogen source as mentioned earlier. The separation is highly favoured by the high process pressure.

The formed gas can with commercial catalysts be converted at the high pressure into other energy products like methanol, dimethylether and hydrocarbons [20]

$$4*H_2 + CO_2 = CH_3OH + 2*H_2O$$

$$2\ CH_3OH = CH_3OCH_3 + H_2O$$

$$n\ CH_3OH = C_nH_n + n*H_2O$$

There are in this way a variety of applications for hydrogen alone and the product gas with carbon dioxide.

PROCESS COSTS

The costs are calculated in accordance to "Plant Design and Economics for Chemical Engineers" [15, 21]. The costs are based on equipment cost. The installed cost is found by using the factor 1–3.5, dependent on the equipment. The interest is 7% and the lifetime and the depreciation period of the plant is 15 years, equivalent to a total interest of 9.4%. Two cases are considered.

- A small district unit processing 1 ton dry or 10 tons wet material with moisture content of 10 %.
- A large plant has a capacity of 70 tons dry corresponding to an annual consumption of 1,000,000 ton biomass with moisture content of 45 % like fresh cut wood.

Capacity t/h dry			1	1	70	70
Produkt mio Nm3 H2/y (1664 Nm3 H2/t dry)			13	13	943	943
Residence time, hours			0.05	0.5	0.05	0.5
Reactor volume m3			0.5	5	35	350
Plant area m2			3500	3500	250000	250000
	Factor	Expon		k€		k€
Buildings and roads	1	1.00	700	700	50000	50000
Inventory	3.5		350	350	350	350
Biomass silos	3.5	0.62	105	105	1463	1463
Compression site, screw type	3.5	0.69	1750	1750	32821	32821
Water separator and steam generator	3.5	0.48	1680	1680	12911	12911
Reaktor	3.5		228	998	3469	15778
Catalyst	1.0	1.00	7.5	75	525	5250
Instrumentation	3.5	0.3	1050	1050	3756	3756
Storage tanks	3.5	0.57	140	140	1577	1577
Total investments k€			6010	6847.5	106871	123906
Capital cost 9.44 % of total investments			567	646	10089	11697
Maintainance 5% of total investments			301	342	5344	6195
Power		1	120	120	8400	8400
Labour	2	0.25	701	701	2027	2027
Administration	2	0.25	200	200	579	579
Total cost			1889	2010	26438	28898
Production cost for different rawmaterials		€/t	€/Nm3	€/Nm3	€/Nm3	€/Nm3
Wood		70	0.182	0.191	0.070	0.072
Straw		35	0.161	0.170	0.049	0.051
Sludge no value		0	0.140	0.149	0.028	0.031
Sludge with deposition tax		-180	0.033	0.042	-0.079	-0.076

The process is based on a feed stream containing 10w % organic matter, for which reason some process water in this case is re-circulated.

The study covers the influence of the residence time on the cost. The residence time in the experiments is around 0.2 hour and the study covers the possible range between 0.05 and 0.5 hour for the industrial plant.

The scaling factor of the production plant is ranging 0.25 –1.0 and listed in the table for the different items.

The feasibility study shows that the hydrogen could be produced at a cost around 0.07 €/Nm^3 in the large industrial plant based on a biomass price of 70 €/ton or 5.6€/GJ.

The influence of the residence time between 0.05-0.5 hour on the production cost is limited.

The study covers different raw materials ranging the most expensive wood to the cheapest sludge, with a deposition tax in Denmark around 180 €/t dry matter. Utilising this sludge in the plant saves the tax.

CONCLUSION AND REMARKS

- It is possible to convert organic matter directly to a clean hydrogen rich gas with a new type of catalyst operating at a temperature around 400°C and a pressure above 10 Mpa.
- The gas contains 70% hydrogen, 2% methane and 28% carbon dioxide.
- The catalyst has been operating for more than 4000 hours without permanent deactivation.
- No corrosion has been observed.
- The process is, when demonstrated, competitive to processes today.
- A pilot unit converting 100 kg biomass per day is being prepared

REFERENCES

1. Sealock, L. John Jr., Elliot, Douglas c., Baker, Eddie g., Buntner, R. Scott (1993), "Chemical Processing in High Aqueous Environments. 1. Historical Perspective and Continuing Developments" Ind. Eng. Chem. Res. 32, 1535-1541
2. Pedersen, K (1989) "Methanol Production based on Straw", The Danish Council of Technology for Renewable Energy. Danish Technological Institute
3. Mudge, L.K., Weber, S.L., Mitchell, D.H., Sealock, L.J., Robertus, RJ (1981), " Investigation on Catalysed Steam gasification of Biomass". PNL-3695 Pacific Northwest Laboratory, Richland
4. Sealock L.J., Elliott, D.C., Hallen, R.T. (1982), " Kinetics and Catalysis of producing synthetic Gas from Biomass. Final report. GRI82/0038, PB83-188644, National technical Information Service Springfield, VA.
5. Modell, M., Reid, R.C., Amin, S (1978), "gasification Process" US Patent 4,113,446
6. Amin, S, Reid, R.C., Modell, M. (1975), Reforming and Decomposition of Glucose in Aqueous Phase) Aerospace Division of ASME Conference on Environmental Systems, San Francisco
7. Modell, M [1985), "Gasification and Liquefaction of Forest Products in Supercritical Water Fundamentals of Thermochemical Biomass p 95-119
8. Michael Antal and Xiaodong Xu (1998), "Total, Catalytic, Supercritical Steam Reforming of Biomass" The Science of Biomass Pyrolysis, PyNe Workshop, Stratford upon Avon 1998, Publ. in Abstracts by A.V. Bridgwater, Aston University
9. Sealock L.J., Elliott, D.C., Hallen, R.T. (1982), " Kinetics and Catalysis of producing synthetic Gas from Biomass. Annual report. GRI80/0116, PB82-214347, National technical Information Service Springfield, VA.
10. Baker, E.G., Sealock, L.J., ((1988), "Catalytic destruction of Hazardous Organic in Aqueous Wastes" Proceedings of the fifth national conference on HWHN 88, Hazardous Materials Research Institute, Silver Springs MD, p 459-462
11. Sealock, L. John Jr., Elliot, Douglas c., Baker, Eddie g., Buntner, R. Scott (1993), "Chemical Processing in High Aqueous Environments. 2. Development of Catalysts for Gasification" Ind. Eng. Chem. Res. 32, 1542-1548
12. Bourkis, Nikolas, Smieder, H., Abeln, J., Dinjus, E., Kruse, A. (2001), "Gasification of High Moisture Waste Biomassin Supercritical Water", Supercritical Fluid, Chromatography, Extraction, and Processing. Symposium Papers, 19-22 August 2001, Myrtle Beach, SC USA.
13. Brem, Gerit (2000), Gasification, pyrolysis and liquifaction at TNO. TNO report TON-MEP P2000/017
14. BTG , Novem, (2001) Superhydrogen. (BTG homepage)
15. Barth, T.; Pedersen, K.; Abbott, A., (2001) "Aqueous Thermal Conversion of Biomass to Hydrocarbons in the Fluid Fuel Range- Hydrocarbons from Biomass". Research funded in part by the European Union in the Framework of the Non Nuclear Energy Programme Joule III under Contract JOR3-CT97-0176. Jan 2001

16. Pedersen, K. (2001), ”Production of Hydrogen by Pressurised Thermal Gasification of Wet Biomass. DTI March 2001. EFP Project No 1383/99-0005. IEA Symposium on Small Stationary Reformers for Hydrogen Production” and ”Biomass to Hydrogen Production”. Novotel, Hammersmith, London 14-15 March 2001
17. http://www.steamexplosion.com/pulp/tech.html
18. Møller, P. (1995), ”Superkritisk Teknologi”, Arbejdsrapport fra miljøstyrelsen nr.7
19. Pedersen, K. (2001) "Production of Clean Combustible gasses by Pressurised Gasification of Wet Biomass" Dansk Kemi 12, Dec. 2001
20. Pedersen, K. (2001), “Production of Clean Combustible gasses by Pressurised Gasification of Wet Biomass” Supercritical Fluid, Chromatography, Extraction, and Processing. Symposium Papers, 19-22 August 2001, Myrtle Beach, SC USA
21. Peters, M.S., Timmerhaus, K.D., (1991), ”Plant Design and Economics for Chemical Engineers, Chemical Engineering Series. McGraw-Hill, Inc

Recycling and Waste Treatment by the Oxygen-Melting-Gasification Method of the imega-s - Process

K. Scheidig
VTI Verfahrenstechnisches Institut für Energie und Umwelt, Box 2005, 07318 Saalfeld, Germany
H. U. Feustel, J. Mallon, M. Schaaf
ingitec Ingenieurbüro für Gießereitechnik GmbH, Box 202, 04167 Leipzig, Germany

ABSTRACT: On the basis of the oxygen cupola furnace technology, ingitec Leipzig, together with its partners, has developed the imega-s – process (ingitec-melting-gasification with special solutions). This process represents a fixed-bed melting-gasification in combined parallel, cross and counter flow, injecting industrial quality pure oxygen. The process is applied to produce synthesis gas for power and heat generation from or biomasses and/or waste materials. The synthesis gas analysis ranges from 40-55% CO and 15-25% H_2, depending on the material gasified and the process control mode.

Depending on the input material the imega-s - reactor produces some slag and hot metal, which must be tapped from time to time or, if the capacity is high enough, even continuously. The technology developed guarantees the complete destruction of any organic pollutants possibly included in the material charged under reducing conditions and at high temperatures up to 2000 °C. The range of charges covers waste wood, natural biomasses, plastic waste, municipal waste, sewage sludge, shredder fractions, old tyres, etc.

Operating in an experimental plant near Leipzig (Germany), the imega-s - process has been developed using a small-scale reactor (hearth diameter: 800 mm; throughput: max. 1.4 t/h). The first commercial plant is under construction, but is very delayed. It should be commissioned in late 2003. The plant is being designed with a capacity of 50,000 t/a of old timber; electrical power: 7.5 MW; thermal power: 9.4 MW.

DESCRIPTION OF THE OXYGEN-MELTING-GASIFICATION METHOD

The imega-s - process is based on the metallurgical process technology of the oxygen cupola furnace [1] - [5]. The reactor developed on this basis is characterised by two tuyere levels, with the gas exhaustion chamber arranged between them. The hearth of the reactor is lined with refractory material up to the height of the bottom tuyere level (level 2). It is provided with a double-wall, water-cooled steel jacket (Fig. 1).

The particular feature of the new process technology is that the reaction gases are passed in parallel flow with the input materials in the shaft of the reactor. Omission of the advantages of material and heat transport in counter flow is intended in order to ensure destruction of organic contaminants under reducing conditions and at high temperature. But within the hearth of the reactor the gaseous reaction products move in counter flow and/or cross flow towards the movement of the packed bed.

The input materials are fed via a hopper and a gate into the gasification shaft. The cold material is dried and pre-heated by hot blast introduced from the top of the reactor. A high-temperature zone is formed at the upper tuyere level (level 1) where industrial quality pure oxygen is injected. Within this zone the organic components contained in the input materials are partially burnt by means of the oxygen. Depending on the material type, extent of pre-treatment, drying and pre-heating, the oxygen offered and other influential factors, different combustion levels are generated with combustion temperatures of up to 2000 °C. The hot gaseous reaction products are exhausted downwards in parallel flow, together with the descending input materials.

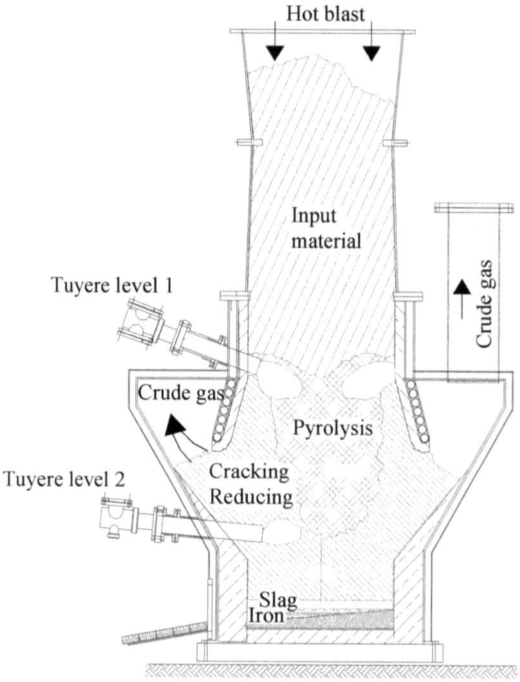

Figure 1: Schematic representation of the imega-s – reactor.

The mineral and metallic components of the input materials melt. They drop or flow downwards and are collected in the hearth. The organic components of the input materials which are not burnt completely in the tuyere level 1, to one extent are gasified, in part being subjected to pyrolytic decomposition resp. cracking.

The pyrolysis coke produced reacts with oxygen in the lower tuyere level (level 2). The heat generated here is used to compensate the heat and temperature losses which are due to the endothermic reactions between the 1st and 2nd tuyere level (gasification, pyrolysis and cracking of organic components as well as reduction of the combustion products from level 1). Thus it is ensured that slag and metal can remain fluid in the hearth and can be tapped or can flow off continuously at larger capacities.

The crude gas exhausted from the zone between the two tuyere levels contains practically no organic contaminated materials because organic compounds as well as the gaseous pyrolysis products produced from the latter are not stable at high temperatures and under reducing conditions. CH_4 contents of < 1 % in the gas produced are a significant indicator for this. The Boudouard equilibrium to high CO contents is provided for the same reason.

The crude gas is exhausted from the reactor within the temperature range of 750 – 850 °C. It is cleaned by conventional processes. The first stage is a dry gas cleaning system consisting of a spray absorber and bag filter system. It is followed by a wet gas cleaning system consisting of quench cooling, aerosol separation and a 2-stage gas scrubber. The necessary underpressure will be provided by an exhauster/compressor arranged downstream, which finally presses the gas then through an activated charcoal filter. Due to the use of oxygen, it is possible to design all these apparatuses with relatively small dimensions.

ADVANTAGES OF THE OXYGEN-MELTING-GASIFICATION METHOD

The new technology is characterised by low investment and operating costs, as well as by operation in an environmentally friendly way. It includes, among others, the following process advantages:

- Preparation of input materials is minimised. Coarse crushing to approx. 300 mm is sufficient. The portion of < 10 mm fines should normally be below 10 %.
- Metallic or mineral components of the input materials must not be removed before charging (no sorting out by hand, no magnetic separation).
- Organic materials as well as contaminated materials are destroyed by destruction integrated in the process itself. No higher molecular hydrocarbons occur in the crude gas of the melter-gasifier. The De-novo synthesis of dioxins and furanes is excluded due to the process described.
- For gas cleaning, well-known processes can be used, meeting highly developed standards. Gas cleaning requires relatively low investment because of the relatively small crude gas volumes per ton input and because of the process-integrated destruction of contaminated materials.
- Process integrated destruction applies also to a number of inorganic and metallic materials. Metal oxides are reduced to a large extent while highly melting heavy metals (e.g. Cu, Cr, Ni) are found completely or to a large extent as alloy elements in iron, which in this way acts as a heavy metal sink.
- The other oxides remaining in the slag are eluate-safe bound in the glassy hardened slag matrix.
- An optimised process and product yield results from the generation of a fuel or synthesis gas which is preferably suited for application in the gas engine or as a substitute for natural gas as the energy carrier; from the production of fluid slag which can be used in form of lumps or granules in the construction of streets and roads; from the production of liquid iron which can then be used as scrap metal.
- A minimum of residues (normally up to 3 - 5 % by weight of the input materials) in the form of dust, sludge and/or salt, which have to be disposed off.
- To a large extent, the single-stage process and the automated operation ensure optimum employment of operators.
- A high specific capacity (approx. 2.5 t/h and m² of hearth area) and an extremely high energy conversion (up to 40 GJ/h and m² of hearth area) allow compact systems with low space requirements.

These advantages offer the possibility to provide low investment costs even for small capacities. For instance, the investment costs of a 50.000 t/a installation can be calculated to be lower than 250 Euro per ton and year. Thus, the imega-s - technology opens the way to profitably working of decentralized small units for the thermal treatment of waste materials.

THE PILOT PLANT

The pilot plant was installed at Leipzig-Holzhausen to test further developments of the process, together with new input materials provided by customers. Hot commissioning was completed in November 2000.

The pilot plant reactor is provided with 3 tuyeres in each of the two tuyere levels. It has an effective hearth diameter of 0.8 m and thus an active hearth area of 0.5 m². At a throughput rate up to 1.4 t/h, its rated capacity is 10.000 t/a. *Fig. 2* shows a partial view of the melter-gasifier of the pilot plant.

From the pilot plant operation in 2001 important information could be obtained for the performance of the process. This includes the verification for the characteristic figures, especially the high throughput rate of 2.8 t/h and m^2 of hearth area, which is more than ten times higher than the corresponding data of grate incineration. In addition to the high throughput rate, there is a high power conversion, too. It ranges up to 40 GJ/h and m^2 of hearth area, depending on the material charged.

Just as with all shaft furnace operation, the use of coke is the prerequisite for normal melting-gasification. Coke provides the permeability of the packed bed, i.e. the gas flow through the material especially at high temperatures, where it is the only solid constituent. The coke consumption runs from 40 to 100 kg per ton of waste dependent of the kind of material processed. Because of the acid constituents of the coke ash and depending on the chemical composition of the waste material, the addition of some limestone is necessary to achieve good flow characteristics of the slag.

The operation of the pilot plant is computer controlled. For this purpose a process model was established and adapted to the operational results achieved with the pilot plant. In this way we gained good experience in the valuation of waste materials using the appropriate chemical analyses.

Figure 2: Partial view of the melter-gasifier of the pilot plant.

Table 1 shows the operating conditions and firing capacity of the generated gas determined for a tested mixture of input materials (melting gasification of old timber, groundwood rejects from the paper industry, bulky waste, and the high calorific fraction of municipal waste). The given electrical and thermal capacities are related to the power generation from synthesis gas in a bio mass block-type heating power station (BHPS) with gas engines.

Table 1: Model calculation of the imega-s – process.

Capacity : 6.58 t/h waste material	7 600 h/a	50 000 t/a
Old timber	304 kg/t	30.4 %
Groundwood rejects	220 kg/t	22.0 %
Bulky waste	290 kg/t	29.0 %
Municipal waste, high calorific fraction	180 kg/t	18.0 %
Metallic parts	6 kg/t	0.6 %
Auxiliary materials:		
Coke	40 kg/t	
Lime stone	40 kg/t	
Oxygen	274 Nm³/t	
Hot blast	300 Nm³/t	
Output materials: Crude gas	8 577 Nm³/h	
Slag	0,7 t/h	
Metal	0,2 t/h	
Synthesis gas parameters		
H_2	21.7 %	
CO	43.4 %	
CH_4	0.9 %	
CO_2	10.8 %	
N_2	18.1 %	
H_2O	5.1 %	
Calorific value	2.39 kWh/Nm3	
BHKW:		
Input	Output	
Gas volume consumed, dry 8 577 Nm³/h	Power, therm. 9.4 MW	
Calorific value 2.39 kWh/ Nm³	Power, electr. 7.5 MW	
Firing power 20.5 MW	Efficiency, therm. 45.9 %	
	Efficiency, electr. 36.6 %	

APPLICATIONS FOR THE imega-s - PROCESS

The imega-s - process is best suited to decentralised systems which are capable of utilising the local available waste potential in an economical manner. The process advantages specified above ensure that the specific investment costs are relatively low. Looking only at the imega-s - reactor with the tapping system for slag and iron, the gas cleaning system and the water treatment plant (without the reactor hall), the investment could be clearly kept below the amount of 250 Euro/t of annual capacity of input material.

The compact design and thus the low space required by imega-s - systems are favourable for tie-in into existing infrastructures. This is particularly applicable in cases where the sale of the heat generated is ensured. Other advantages result, above all, from the installation of relatively small systems at decentralised sites (use of local waste sources, low transport load, easier reached acceptance by people).

An economical alternative should be considered if you can provide that the crude gas generated in the melter-gasifier is cleaned and then burnt as fuel gas in an existing boiler to substitute fossil primary energy carriers. The existing flue gas cleaning unit of the boiler system should then be used for final cleaning in order to ensure the exhaust cleanliness in accordance with the regulations issued by the authorities.

The specific investment costs of a turn-key system can be clearly kept below the amount of 500 Euro/t of annual capacity of input material. These costs should include all major supplies and services, from input material logistics up to the tie-in of a BHPS, including the necessary engineering services, construction and auxiliary construction works, and the installation.

THE ROTHENBURG BIO MASS POWER STATION PROJECT

The Rothenburg project was to be performed by MFU GmbH Leipzig, former partner of ingitec. The order covered the construction of a melting-gasification - system with an annual capacity of 50,000 t. This biomass power station was planned to be taken into operation by the end of 2001, but is very delayed because MFU became insolvent in summer 2001. The investment is partly realized and now should be accomplished in late 2003.

The melter-gasifier in Rothenburg is designed for the thermal utilisation of old wood in accordance with the biomass regulations. A total of 7.5 MW electrical power are to be generated by means of gas engines from the process gas produced. Power is to be fed into the mains according to the Renewable Energy Act (EEG). The waste heat generated when operating the gas engine will be supplied to the Rothenburg long-distance heating network.

MARKETING

FMI Feld Maschinen- und Industriebau GmbH, Oer-Erkenschwick/Germany, is the new Partner of ingitec GmbH Leipzig. FMI offers installations for the melting-gasification of biomass and waste materials using the imega-s – process with capacities of 10 000 t/a to 50 000 t/a.

COOPERATION

For the melting-gasification of dust or sludge it is necessary to agglomerate the material. Special solutions in this field are provided by VTI Verfahrenstechnisches Institut Saalfeld, Germany, which has extensive experience in briquetting and pelletising of dry or wet fine grained material. VTI acts as a partner in cooperation with ingitec and FMI.

REFERENCES

1 Ruschitzka, L., et al. (1994) Gießereiforschung 46 (1994) 2/3, S. 40-50
2 Ruschitzka, L., et al.(1994) Giesserei 81 (1994) Heft 10, S. 297/303
3 Scheidig, K. (1995) Stahl und Eisen Düsseldorf. 115 (1995) Heft 5, S. 59-65 u.177
4 Scheidig, K. (2002) UmweltMagazin Düsseldorf. Nr. 4/5, 2002, S. 80-83
5 Scheidig, K., et al. (2002) Lulea Recycling Conference of MEFOS; TMS & Lulea, University. Lulea/Se, 17-19 June 2002.

Special Gas Utilisation with Gas engines: Requirements and Experiences

Martin Schneider
Product Management, Jenbacher AG, Achenseestr. 1-3; A 6200-Jenbach, Austria

ABSTRACT: Increased ecological consciousness and the knowledge of limited reserves of primary energy in the form of fossil fuels make it necessary to utilise available energy sources economically. Cogeneration plants with gas engines produce electricity and heat at decentralized locations, where they are required. They offer optimal efficiency in the utilisation of energy with minimum environmental burden.

Gas engines are presently powered primarily with natural gas, biogas or propane. The use of "special gases" like producer gases, coke gas, pyrolysis gas or gas from gasification processes gas with low calorific values or changing gas compositions with respect to emission limits of the common air quality requirements is a new challenge for gas engine development. Highly sophisticated gas engines with intelligent engine management systems now allows the utilisation of gases, which could not be burned a few years ago. Jenbacher AG has already installed cogeneration plants running with coke gas (55% H_2 content), pyrolysis gas from domestic waste gasification (35% H_2 content), gas from wood chip gasifier and producer gas from the chemical industry with an extreme low heating value (0.5 kWh/m3).

These experiences has shown, that the gases from gasifier or pyrolysis gas from waste can be used in gas engines, as long the gas fulfils certain requirements.

GAS-CHARACTERISTIC VALUES AND REQUIREMENTS REGARDING THE QUALITY FOR THE USE IN A GAS ENGINE

Gas engines are presently powered mainly with natural gas. But the use of renewable energy sources like landfill gas or sewage gas with low thermal heat values represents a growing market for gas engines all over the world.

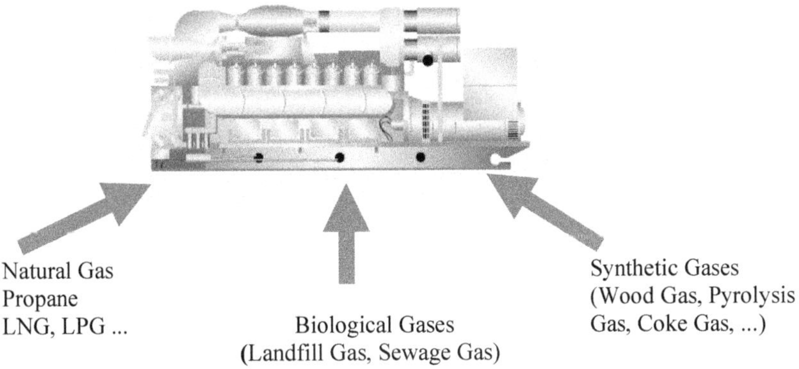

Gases, respectively their constituents, have different properties which can be assessed through their characteristic values, such as methane number, heat value and laminar flame speed, to mention only a few. To be able to achieve an ideal degree of energy conversion, these values must be considered when dealing with the engine.

Table 1: Gas characteristics of different gases.

Gas	Composition	Density [kg/Nm³]	LHV [kWh/Nm³]	Methane number	laminar flame speed [cm/s]
Hydrogen	H_2	0.0899	2.996	0	302
Methane	CH_4	0.717	9.971	100	41
Carbon monoxide	CO	1.25	3.51	75	24
Natural gas (example)	$CH_4 = 88.5\%$ $C_2H_6 = 4.7\%$ $C_3H_8 = 1.6\%$ $C_4H_{10} = 0.2\%$ $N_2 = 5\ \%$	0.798	10.14	80□	41
Biogas (example)	$CH_4 = 65\%$ $CO_2 = 35\%$	1.158	6.5	135	27
Wood gas (example)	$H_2 = 7\%$ $CO = 17\%$ $C_nH = 5\%$ $N_2 = 56\%$ $CO_2 = 15\%$	1.258	1.38		15

CALORIFIC VALUE (LHV) AND THERMAL VALUE (UHV)

Calorific value and thermal value indicate the energy content of a gas. The former can be differentiated from the latter only through the heat of vaporisation of the water resulting from combustion. With regard to the calorific value after combustion, the water is in liquid form after liberating its condensation heat. Figure 1 shows a logarithmic presentation of the thermal value of different gas, which are already used in Jenbacher gas engines.

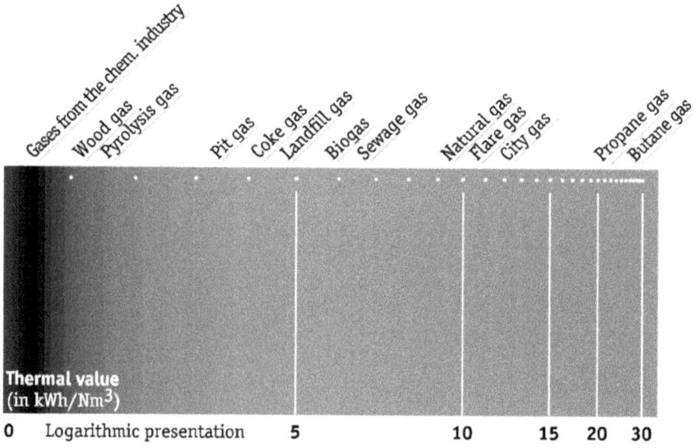

Figure 1: Thermal Value, Heating Value of different gases.

METHANE NUMBER

The determinant parameter for the assessment of the knock resistance of burnable gases is the methane number. It is comparable with the octane number for petrol operation and indicates the percental methane volume ratio of a methane-hydrogen mixture which, in a test engine and under defined conditions, indicates the same knock resistance as the gas to be tested. If, for example, natural gas has a methane number of 85, this signifies that, under specific engine conditions, this natural gas will have the same knock resistance as a mixture of 85 % methane and 15 % hydrogen.

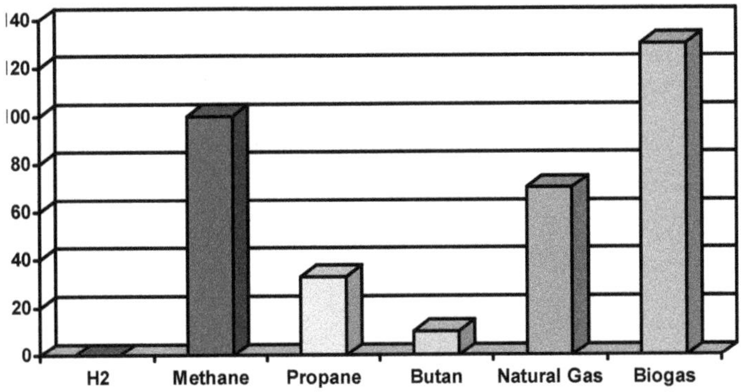

Figure 2: Methane Number of different gases.

LAMINAR FLAME SPEED

Laminar flame speed is the speed at which the oxidation, respectively reaction between the combustible constituents of the gas and oxygen at the surface of the inner zone of the flame takes place. It is dependent on the air ratio Lambda and attains its maximum shortly before Lambda = 1. Figure 3 shows the laminar flame speed for natural gas, landfill gas and wood gas under atmospheric conditions (gas composition according Table 1).

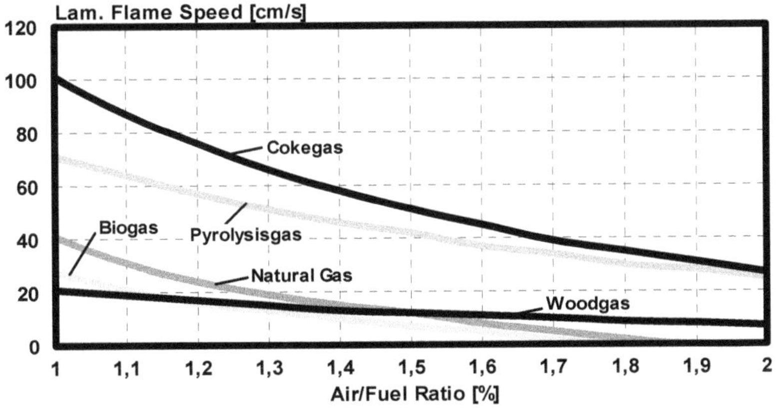

Figure 3: Laminar flame speed. The requirements of the gas engine regarding the composition of the fuel gas can be subdivided into energetic and physical properties, into contaminants, as well as into the variability of properties.

PHYSICAL PROPERTIES

The methane number and the heating value are the determining physical values for standard gases like natural gas, LNG, LPG or biogas. For low concentrated gases, like wood gas, the laminar flame speed is the decisive criterion indicating whether it is possible for the air-gas mixture to be completely burned up in the engine.

The methane number is the crucial physical value for assessing the knock resistance of a burnable gas. The methane number requirement, respectively the knock resistance of the engine, is influenced by constructional and operational factors. In order to be able to guarantee fault-free engine operation, it is necessary to keep the pollutant concentrations in the fuel gas within the prescribed limits. The following table lists the limiting values that Jenbacher AG prescribes for their engines. All quantities refer to the volume of the fuel gas in a standard state (STP = 0 °C; 1.013 bar abs.) and to a thermal value of 10 kWh/m³.

Table 2: Extract "Technical Instruction TI 1000-0300".

	without catalyst	with catalyst
Max. grain size (µm)	3	3
Dust: Max. amount (mg/m³)	50	50
Max. sulphur compounds calculated as H_2S (mg/m³)	2,000	1,150
Max. halogen content (Cl + 2 x sum F mg/m³)	< 100	0
without limitation of guarantee	100 – 400	0
with limitation of guarantee no guarantee	> 400	0
Max. relative silica content [ppm x l/(kW x hrs}		
without limitation of guarantee	< 0.02	0
with limitation of guarantee	> 0.02	0
(for wearing of exhaust valves, bushes and piston rings)		
Max. ammonia content (mg/m³)	55	55
Max. residual oil content in fuel gas (mg/m³)	5	5

MEASURES FOR REDUCING EMISSIONS

Due to increasingly more stringent governmental regulations regarding delimitation of emissions, Jenbacher is working intensively on methods to reduce further pollutants in engine exhaust gas. There are fundamentally two possibilities (Figure 4) to reduce the emission of pollutants from internal combustion engines, namely:

- secondary treatment of exhaust gas and
- measures implemented inside the engine itself.

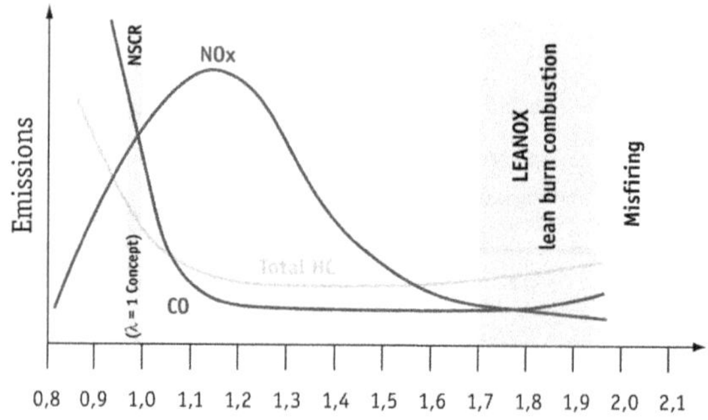

Figure 4: NOx, CO and HC emission dependent on fuel/air ratio.

MEASURES IMPLEMENTED INSIDE THE ENGINE

Lean-burn concept for reduction of NOx

Through operation with a large amount of excess air (lean mixture) the combustion temperature is reduced and hence NOx formation in the combustion chamber is strongly reduced. The "German TA-Luft" limit of NOx (< 500 mg/Nm³) can be complied with reliably and economically by the

Jenbacher lean-burn engines. [NOx emission can be reduced by making the mixture even leaner. This means, however, putting up with a loss of efficiency, a reduction in output (and increased maintenance) to 250 mg/Nm³.] A greater reduction of the emission is not possible at this time without secondary treatment of the exhaust gas. The lean-burn concept was pursued in Jenbach and the LEANOX lean-burn (Figure 5) combustion process developed - a process guaranteeing constant compliance with prescribed NOx emission limits throughout the entire operating time of the engine.

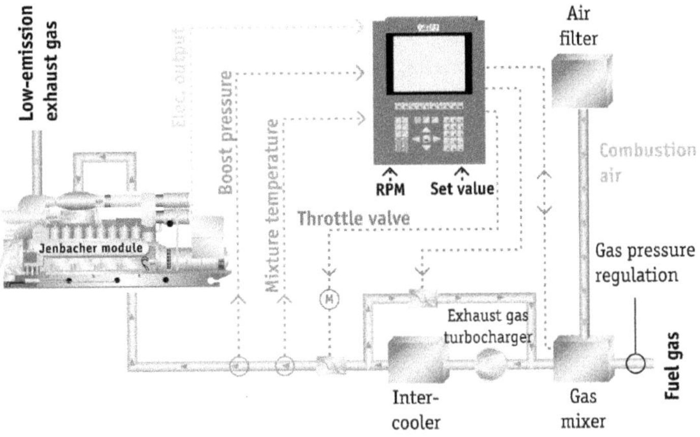

Figure 5: ***LEANOX*** system.

LEANOX-control system

The air ratio Lambda (λ) is linked with thermodynamic engine- and gas-specific parameters. Both constant and variable values are contained in this relationship. Lambda can therefore be represented as a function of:

- engine output,
- mixture pressure (= boost pressure) and
- mixture temperature after the intercooler.

These three values are measured and transmitted to the ***LEANOX*** control system. The signal output by the ***LEANOX*** controller moves the adjusting cone of the air/gas mixer into the desired position, so that the required air ratio is set.

The main advantage of the patented ***LEANOX*** system is the secure and reliable measurement of the input signals for fuel air mixture control. There are no sensors in the hot zones like the combustion chamber or in the exhaust gas. The mixture pressure and temperature sensors are working in low stress, stable environmental conditions and free of deposits and aging. This allows a secure and reliable mixture control. The exact mixture control guarantees not only an accurate emission control, it also controls the whole combustion process and protects the engine against high thermal or mechanical stress.

Reduction of CO by measures inside the engine

The most common way to achieve the "TA-Luft" limit for CO (< 650 mg/Nm³) with biogas is the implementation of internal measures. For example, reducing the compression ratio, adjusting a later ignition point (closer to the TDC) and water cooled exhaust gas manifolds will help to reduce the CO emissions without exhaust gas aftertreatment.

SECONDARY TREATMENT OF EXHAUST GAS FOR THE REDUCTION OF CO

To achieve the "TA-Luft" limit for CO the use of an oxidation catalyst is common for natural gas or propane. Several pollutants in the biogas do not allow the reduction of CO by an oxidation catalyst without a gas pretreatment. These contamination cause deposits or are toxic to the oxidation catalyst and reduces its effectiveness in a short time. But the reduction of CO by an oxidation catalyst allows higher efficiency and higher specific output compared to the measures taken inside the engine.

Jenbacher is following this philosophy (oxidation catalyst and gas pretreatment) for sewage gas and the first pilot plants are running with active carbon gas cleaning systems very successful for several years. One important side effect is the avoiding of silica deposits inside the engine, which can cause problems at the utilisation of biogas in gas engines.

With fuel like landfill gas gases, which are contaminated by changing catalytic poisons, the conversion of CO and HC into CO2 and H2O can be achieved by the Jenbacher thermal aftertreatment system (CL.AIR system).

The CL.AIR system is an installation for the thermal aftertreatment of engine exhaust gases. In heating up the exhaust gas in the CL.AIR system to a temperature of about 800 °C the hydrocarbons (CH4 and NMHC) as well as the CO with the residual oxygen in the exhaust gas are oxidised to form steam and CO2. The nitrogen oxides (NOx) remain unchanged.

With the help of the ***CL.AIR*** system it is also possible for the first time to meet the limiting value prescribed by "TA-Luft" for formaldehyde (HCHO < 20 mg/Nm³). Until now, no manufacturer of engines for sewage- and waste dump gas plants was able to guarantee this value without exhaust gas aftertreatment.

EXAMPLES OF UTILISATION OF SPECIAL GASES

Jenbacher has been working since the middle of the Eighties on the utilisation of special gases like pyrolysis gas or wood gas in gas engines as well. With the target of sufficient performance output and an appropriate cost benefit relation, a turbo-charged gas-otto- lean burn engine has to be used. This requires a certain maximum gas temperature with a low humidity and a low content of heavy hydrocarbon compounds (e.g. tars). Beside the problems of fluctuating gas qualities and the contamination of the gas, it is still one of the challenges, to fulfil the international emission standards (e.g. TA-Luft Standard) at these application. To increase the basic know how and understanding of the utilisation of this kind of gases, Jenbacher participated on several Joule research projects, as well some demonstration projects and also joined some other european and national projects.

Next to these activities, commercial gas engine applications for H2-gases (e.g.: coke gas, H_2 containing weak gas etc.) have been available since 1995 and are in successful operation. All experiences of utilisation of this special gases with H_2 will be directly used for the approach of solutions for problems with the utilisation of woodgas and pyrolysis gas in gas engines. The following examples give a short overview of some commercial special gas applications and some pilot plants operating with pyrolysis gas or wood gas.

GASES FROM THE CHEMICAL INDUSTRY; "KREMS -CHEMIE"/AUSTRIA

Formaldehyde is required for the production of artificial resin which is used throughout the woodworking, paper and rubber industries. As a by-product of this manufacturing process, approx. 12,000 m³/hr of weak, combustible gas (LHV=0.54 kWh/m³ at 18% H_2 and 82% N_2) is in turn used to power Jenbacher cogeneration modules which produce 2 MW electricity and additional process steam.

At the beginning of the 90's the chemical company "Krems Chemie" started to investigate the possibility of utilising the H_2-lean gas, a by-product of the formaldehyde production, in gas engines. Due to the extreme low thermal value none of the engine manufacturer Krems Chemie contacted could give a positive reply. Subsequently one considered to burn the lean gas in a steam boiler and generate electricity with a steam turbine. Shortly before realisation of this concept "Krems Chemie" contacted JES to once again review the possibility of using the gas in a gas engine.

Based on the experience with special gases gathered in the past, especially with wood gas, Jenbacher AG was in the position to use the gas in a gas engine and offered a customised concept. A feasibility study of both concepts put the steam boiler solution out of the running. Approximately 20% less investment costs plus higher electrical efficiency easily compensated the gas engines deficit in steam production. The amortisation time was reduced from the originally planned 3 years to 2 years and due to the short delivery time of the gas engines, the installation was realised in May 1996 instead end of 1996. Since spring 1996 each of the 4 cogeneration modules JMS 320 GS-S.LC generates 588 kW electrical power . From the exhaust gas 1.300 kg/h saturated steam at 10 bar is produced in two steam boilers and the feed water is preheated by the engine jacket water heat. The additional required amount of steam is produced in a steam boiler fired with the same lean gas.

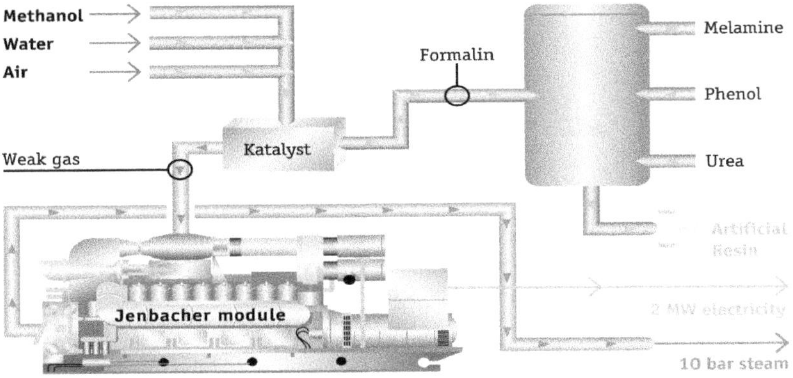

Figure 6: Scheme of cogeneration use at the production of artificial resin at Krems Chemie / Austria.

Assuming that the units operating time is 8.000 h/a, the company's import of electricity can be reduced by 75%. In addition to the economical benefit this installation has as well its benefits for the environment. As a result of the lean burn concept the NOx-emission is below 10 mg/Nm³, whereas 100 mg/Nm³ is the guaranteed value. This " technical" clean gas allows CO-emissions of below 100 mg/Nm³ (guaranteed value CO <150 mg/Nm³) by using an oxidation catalyst. The plant is in commercial operation since commissioning (> 50.000 oh; 7/2002).

COKE GAS: "PROFUSA"/SPAIN

As a waste product of coke production, the company "Profusa" in Bilbao has available a gas with high Hydrogen content. Since August 1995 gas is burned in 12 gas engine generating sets JGS 316 GS-S.L and converted into maximum 6.5 MW electrical power.

The special engine design as well as the appropriate gas mixing equipment make it possible to operate the engines on 100% coke gas, 100% natural gas or on a coke gas/natural gas mixture (60%/40%). The exhaust gas is partly used for waste water treatment and partly for steam production. At night the units operate on 100% coke gas, during the day 40% natural gas is added to the coke gas to increase the output. As an option the engines can operate on 100% natural gas in case of grid failure or emergency conditions. The coke gas from Profusa consists of approx. 55% H_2, approx. 30 % CH_4, and 5% CO, the rest is CO_2 and N_2.

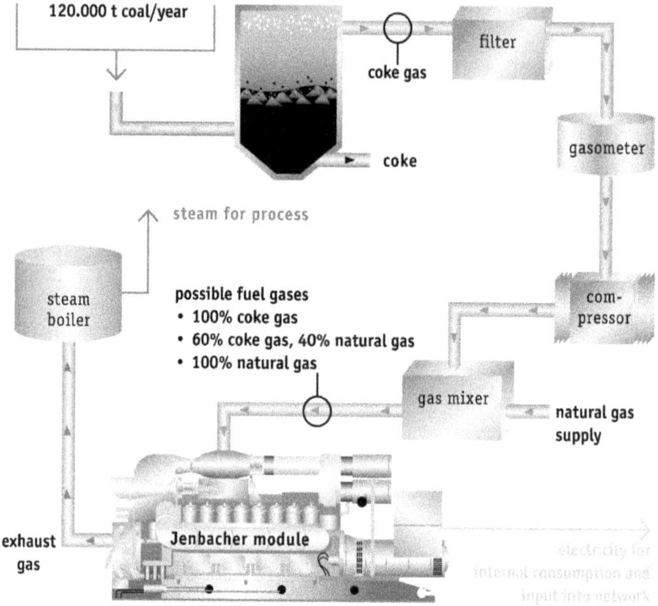

Figure 7: Scheme of coke gas plant at Profusa.

The NOx-emission at coke gas as well as coke gas/natural gas operation is below 300 mg/Nm³(dry exhaust gas, 5%O_2). For operation on pure natural gas "TA-Luft" limits can be achieved. The plant is in commercial operation since commissioning (> 50.000 oh; 7/2002).

Pyrolysis gas: "Thermoselect Plant Fondotoce/Italy"

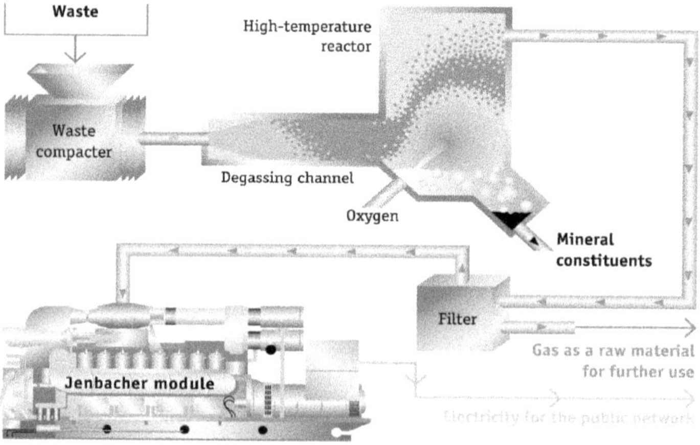

Figure 8: Scheme of the Thermoselect Process.

With the Thermoselect process it is possible to utilise the most varied kinds of waste, for example domestic and industrial waste, shredded material, bulky refuse, sewage sludge, contaminated soil as well as landfill material. By means of compaction of the waste, removal of air, homogenisation and degassing at a temperature of more than 600 °C in a degassing channel, gasification with pure oxygen as the medium, as well as the removal of the mineral and metallic portions in a smelting process, the process represents a closed system for the treatment of waste. The plant functions autonomously on the energy it produces.

The synthetic gas produced in the high-temperature reactor, which is shock-cooled and purified in the course of a number of steps, on the one hand supplies the plant with the necessary energy and on the other hand a Jenbacher gas engine transforms the gas directly into electrical energy. The first pilot plant with a Jenbacher J 612 GS engine was successful in operation from end of 1993 until 1997 in Fondodoce/Italy. Another pilot plant with a J 620 GS engine has been in operation since October 2001 in Chiba/Japan, the first commercial plant should be commissioned end of 2002 in Japan as well.

WOODGAS PLANT: "HARBOORE – VOLUND UPDRAFT GASIFIER"

Already in 1988 the company "Babcock & Wilcox Vølund Aps" decided to take an active part in the development of gasifiers in order to create a gasification principle with the possibility of:

- achieving a stable and continuous gas production
- gasifying fuel with a moisture content of up to 50 %
- producing gas for use in a gas engine
- achieving a high degree of automation for the total plant

To achieve these goals the updraft gasification principle was chosen because of the built-in drying zone. The major advantages of an updraft gasifier are its simplicity, an ability to gasify very wet fuels, high charcoal conversion and internal heat exchange, leading to low gas exit temperature and high gasification efficiency. A disadvantage is the large amount of tar produced. Therefore gas cleaning is required to make the gas usable for gas engines.

After several years of development work the first gasification plant was put into commercial operation in 1993 at the district heating plant Harboøre in Denmark which today supplies heat to approx. 560 heat consumers and to the municipal buildings of the town.

Figures 9 and 10: The Harboøre woodgas CHP plant.

The gasifier has an output of 4 MW(thermal) and the gas has been burnt over the last years in a Low-NOx gas burner built onto a 4 MW hot water boiler. The primary fuel is wood chips but successful tests have been made at the plant with other types of fuel, e.g. chunk-wood, bark and waste wood. The produced gas consists of approx. 15 – 18 % hydrogen, 25 – 28 % carbon monoxide, 7 – 10 % carbon dioxide, 3 – 5 % methane and the rest is nitrogen and water.

During the summer of 1996, a programme to optimise and develop the gasifier was successfully completed. Subsequent research and development activities have concentrated on gas cleaning. Today the gas is cleaned through a complex system of gas scrubbers, heat exchangers and electrostatic filters before it is fed into the gas engines.

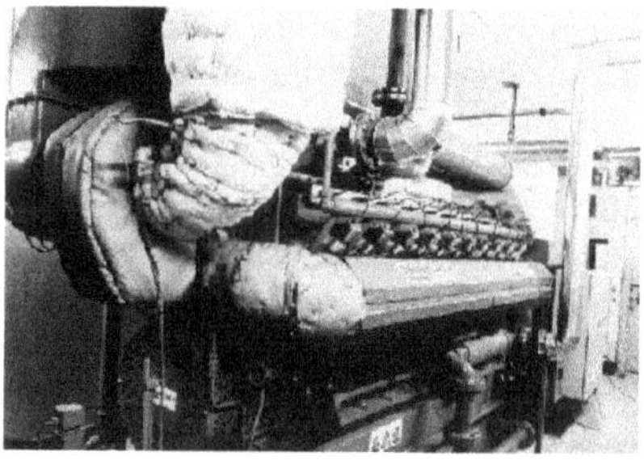

Figure 11: Jenbacher gas engine module JMS 320 GS-S.L.

In the beginning of year 2000 two Jenbacher gas engine modules JMS 320 GS-S.L were installed and the district heating plant was converted into a CHP plant. A significant part of the work has been related to the conditioning of the gasifier product gas for use in gas engines and a reliable solution based on gas cooling, wet electrostatic precipitation and a novel technology for cleaning the resulting tar contaminated water has been demonstrated. The product gas is cooled – using the district heating grid – to about 45°C, during which a considerable amount of water/tar condensate and also aerosols (microscopic water/tar droplets) is released. The aerosols are subsequently removed from the gas stream by means of a wet electrostatic precipitator. After this treatment the gas is clean and applicable for the gas-engines (both tar and dust contents are below 25 mg/Nm3). The gas is boosted to a slightly higher pressure to accomplish engine inlet pressure regulation – by means of a traditional "gas train" – to slightly below atmospheric pressure.

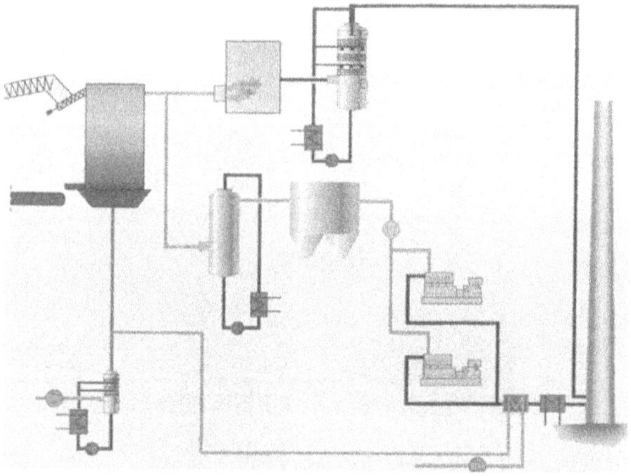

Figure 12 (above): Flow diagram "District Heating Plant Harboøre".

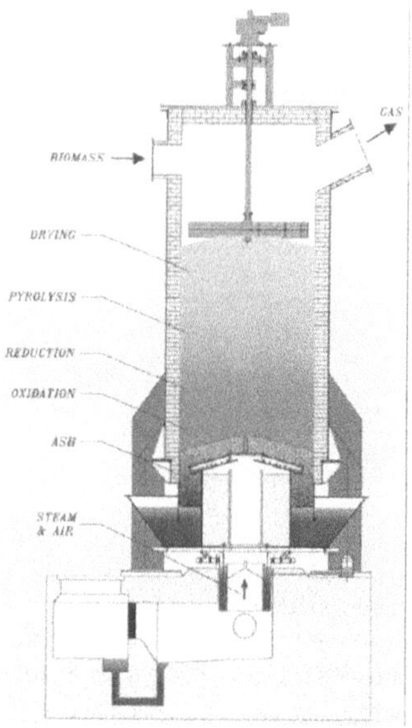

Figure 13 (right): Principle sketch of an updraft gasifier.

At full engine output, the amount of tar contaminated waste water from the Harboøre gasifier amounts to about 1200 litres/hour containing about 18% of various organic acids and tars. Babcock & Wilcox Volund R&D Centre has made several small-scale experiments with the process – first on a laboratory scale system with a capacity of 2 kg/h waste water and during the early Winter 2001/2002 on a pilot scale system with a capacity of about 50 kg/h was operated for a prolonged period. All experiments have proven successful (regarding environmental performance, corrosion and fouling) and BWV is presently in the process of designing and installing a full scale implementation at the Harboore gasifier (Denmark), capable of handling about 1200 litres/h of tar contaminated water from the product gas clean-up system.

Each unit has an electrical output of 648 kW and a thermal output of 883 kW. With these high efficient gas engines the woodgas can be converted with an overall efficiency of up to 90% into electricity and heat, depending on the heating water temperature level. The engines are operated as lean burn LEANOX-controlled gas engines with an air access ratio of approximately 1,6. Therefore low NOx emissions can be obtained. A total of approx. 3000 operating hours of the engines have been successfully achieved by end of summer 2002. Frequently internal inspections of the vital parts of the engines show no sign of tar deposits or other abnormalities up to now.

As a carbon dioxide-neutral energy source, wood is significantly more environmentally friendly than fossil fuels in terms of the greenhouse gas emissions, release in the process of combustion. The Harboøre plant is therefore helping to offset the greenhouse effect by producing heat and electricity with only a limited use of fossil fuels. Based on substitution of coal, the plant's annual heat output saves about 6,700 tonnes/year of CO_2.

WOODGAS PLANT: "GÜSSING – FLUIDIZED BED STEAM GASIFICATION"

In order to make the generation of electricity from biomass possible also in small, decentralised power stations, another new type of power station was realised for the first time in Güssing/Austria.

In the biomass power station of Güssing supplies 4500kW heat for district heating and up to 2000 kW electricity originate from 1760 kg wood per hour. In order to realise this project from the idea to the finished product the partners Austrian Energy, as design engineer, scientists from the Technical University Vienna, the EVN and the District Heating Company Güssing formed the authority network RENET and developed this new, well planned system of cogeneration on the basis of biomass gasification.

Heart of the power station is the fluidized bed steam gasifier. During gasification the biomass is gasified with approximately 850°C under supply of steam. Using steam instead of air as medium of gasification, results in a nitrogen free, tar-poor product gas, with a high heating value. A part of the remaining coke is transported to the combustion chamber via a circulating bed material (sand), which act as heat transfer as well. The heat dissipating to the bed material will used for the process. The flue gas is carried off separately, and the contained heat is recovered for the district heating system.

For the operation of the gas engine, the product gas must be cooled and cleaned. The heat of the gas cooling is used again for district heating. The dust is removed in a woven filter. After this procedure a scrubber reduces the concentrations of tar, ammonia and sour gas components. Due to this special procedure it is possible to feed back all residual substances into the process. As a consequence neither wastes nor waste water result during gas cleaning.

The gas engine converts the chemical energy of the product gas into electrical energy. Beyond that, the waste heat of the engine is used as well for the supply of district heating system. The overall electrical efficiency is 25-28%, the total plant efficiency (electricity and heat) is even more than 85%.

The commissioning of the gasifier started in September 2001, the commissioning of the Jenbacher gas engine J 620 GS E was in February 2002. Jenbacher's aim with this pilot plant is the optimisation of the gas engine for this gas with an relative high H_2 content (30 - 40%) by a new type of gas mixing system and gas supply to the engine. Special test applications for then reduction of the CO-emissions by an oxidation catalyst are under preparation as well. During the first test trial run the engine has achieved the expected output and a the first results of the catalyst test run were positive, but of course long term results are not available yet.

SUMMARY

Applications like "Krems Chemie" or "Profusa" have shown, that a wide range of gases containing hydrogen can be used in gas engines as long the H_2 content achieves a certain limit. The main criteria for the utilisation of these special gases is the laminar flame speed and not the heating value of the gas. The plant "Krems Chemie" has shown that even a gas with a LHV = 0.5 - 0.6 kWh/Nm³ can be used in a gas engine, above this H_2 - limit the influence of the hydrogen content is relative low and can be compensated by different turbo charger tuning. The main criteria for the utilisation of pyrolysis gas or wood gas are the contamination of the gas and the content of condensing hydrocarbons like tars. The common NOx emissions (e.g. TA-Luft: NOx < 500 mg/Nm³) can be achieved by the lean burn combustion concept without any exhaust gas after-treatment. A further challenge for the utilisation of these kind of gases is the often required limits in terms of CO-emission which cannot achieved without exhaust gas after-treatment due to the typical high CO content of the wood gas itself. The utilisation of an oxidation catalyst demands a high standard of gas cleaning and gas pre-treatment to reduce the harmful contamination for the catalyst. Less sensitive but causing higher investment costs would be a thermal after-treatment of the exhaust gas like the Jenbacher Cl.AIR system, which is in successful operation on more than 100 landfill plants to reduce the CO-emissions.

JENBACHER PRODUCT LINE

The Jenbacher product line for gas engines presently comprises four constructional series with a total of nine engines graduated according to cubic capacity and number of cylinders. The program covers a performance spectrum of 330 to 2,734 kW electrical and 363 to 2,790 kW in thermal energy (Table 16). Various versions for fuels such as natural gas, propane and biogases like sewage and waste dump (landfill) gas are available. Jenbacher cogeneration systems achieve a total efficiency of more than 90 % under strict observance of all international emission regulations.

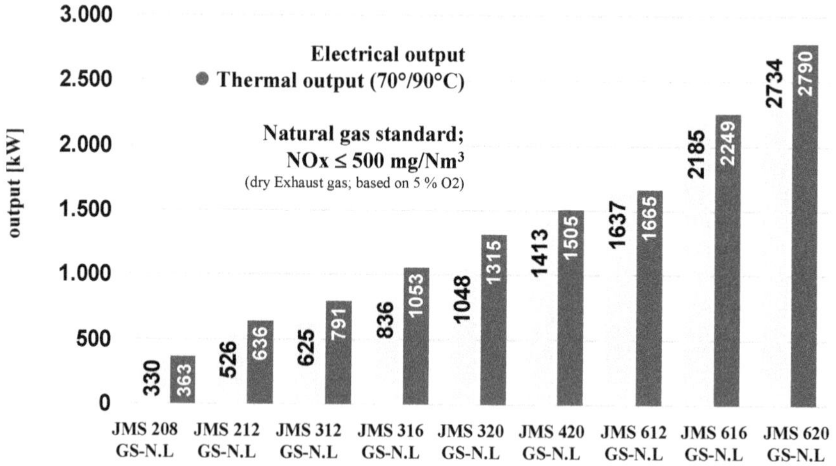

Figure 14: Product Line 2002 - Natural gas; MN = 70.

The integration of planning, design and manufacturing guarantees the most efficient production of standardised components, as well as of tailor-made systems of superior quality. A sophisticated manufacturing process allows a short delivery time of 3 to 5 months depending on the type of engine and the scope of supply. All these engines can be modified to operate with the above mentioned kind of special gases.

Operating Parameters for the Circulating Fluidized Bed (CFB) Processing of Biomass

K. Smolders
Hogeschool voor Wetenschap en Kunst, De Nayer Instituut, Jan De Nayerlaan 5
2860 Sint-Katelijne-Waver, Belgium
J. Baeyens
KULeuven, Chemical Engineering Department, de Croylaan 46
3001 Heverlee, Belgium

ABSTRACT: Growing concern about the global climate change and pollution are driving the development of renewable and sustainable energy sources. Biomass fuels meets these objectives and are said to be "carbon dioxide neutral" because they release a nearly equivalent amount of CO_2 during combustion as plants absorb from the atmosphere. The use of a CFB processing, and associated power generation, could make small scale operations technically and economically viable. If and whenever possible, desk design should define the critical operating parameters and basic design. In the first section of this paper, a design strategy of the CFB for combustion of biomass is developed. Biomass is hereby represented by the idealized formula of $(C_6H_{12}O_5)_n$, with energy content of 16-20 MJ/kg. Kinetic data are taken from thermo-gravimetric measurements and/or literature. The main operating parameters are the solids flux and gas velocity in the riser of the CFB. Isothermal conditions will be guaranteed by circulating inert sand particles of 300 μm at a solids flux of 20 kg/m²s. The gas velocity in the riser is determined at 7.9 m/s. Besides the design of the CFB for biomass combustion, the paper will moreover determine the net combustion capacity (= 4.1 MW_{th}/m^2 riser) of the CFB combustor. Other CFB applications as combustion of coal and sludge will also be illustrated.

INTRODUCTION

Growing concern about the global climate change and pollution is driving the development of renewable and sustainable energy sources. Biomass fuels meet this objective, but being a dispersed and diluted resource, it was for a long time considered as necessary to implement large projects to appreciate the significance and full value of biomass benefits. The circulating fluidized bed (CFB) combustion and associated power generation could make small scale operations technically and economically viable. The CFB has been applied for over 20 years in a variety of processes, mainly involving gas/solid reactions such as coal combustion, coal gasification, minerals' calcination or roasting, etc. The combustion of biomass could yet be another success story for the CFB principle. If and whenever possible, desk design should define the critical operating parameters and preliminary layout.

THE CIRCULATING FLUIDIZED BED

Industrial processes often involve the interaction of a gas and a solid phase. Low gas velocity processes are characterised by limited carry-over and are performed in packed, fluidized or slugging beds. When the gas velocity through the fluidized bed is increased above a critical velocity, entrainment becomes so important that the bed can no longer be maintained in the column unless entrained particles are collected and returned to the bed. Reactors with an external recycling system have a means to control the solids hold-up and are referred to as 'Circulating Fluidized Beds' (CFB). Figure 1 illustrates a CFB coal combustor.

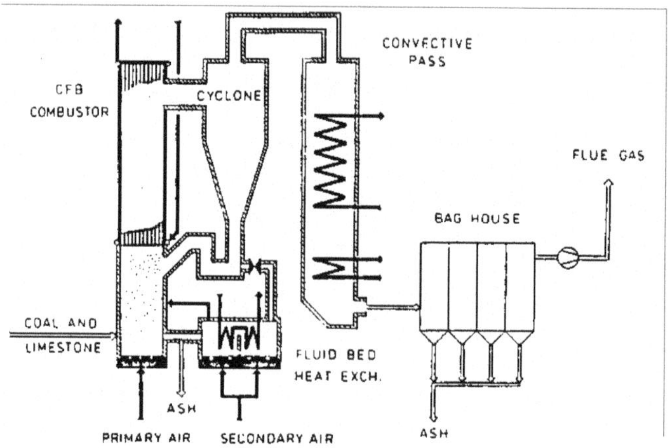

Figure 1: Circulating fluidized bed combustor.

The main goal of the CFB design is the determination of the reactor (riser) dimensions to firstly fulfil requirements of conversion (and thus of residence time) and secondly to achieve the heat exchange rate, if so required. The recycle loop is of additional importance and its design can be based on correlations found in literature (e.g. Smolders et al. [1]) and is not reported in this paper.

THE DESIGN STRATEGY FOR A CFB BIOMASS-COMBUSTOR

Figure 2 illustrates the successive steps in the design strategy of a CFB reactor, which will be applied for the design of a biomass-combustor with a capacity of 2.5 tonnes/hr.

DATA GATHERING

Principles

The design procedure starts with *gathering data* concerning production capacity, particle and gas characteristics, reaction kinetics and heat exchange necessities.

Particle characteristics include the particle size distribution and average size, specific gravities and bulk density, sphericity and heat capacity. Important gas characteristics include gas density, viscosity and heat capacity, which are all temperature dependent. Moreover, the composition of the gas mixture is also an important parameter in reactor design. The reaction kinetics and thermodynamics can be defined by literature data or by experimental work.

Applied to the bio-mass combustor

As is the case for bubbling fluidized bed combustors, CFB-combustors are operated with an inert (sand) bed. At start-up the CFB-combustor circulates sand (ρ_p = 2600 kg/m^3), which is gradually supplemented and replaced by shale and combustion ash (ρ_p = 2500 kg/m^3). The bed characteristics should therefore relate to ash, rather than to sand. The average particle size is approximately 300 μm.

Biomass is represented by the idealised formula of $(C_6H_{12}O_5)_n$. The combustion process is also idealised as:

$$C_6H_{12}O_5 + 6.5\ O_2 \rightarrow 6\ CO_2 + 6\ H_2O$$ with an exothermic reaction heat of 17.5 MJ/kg

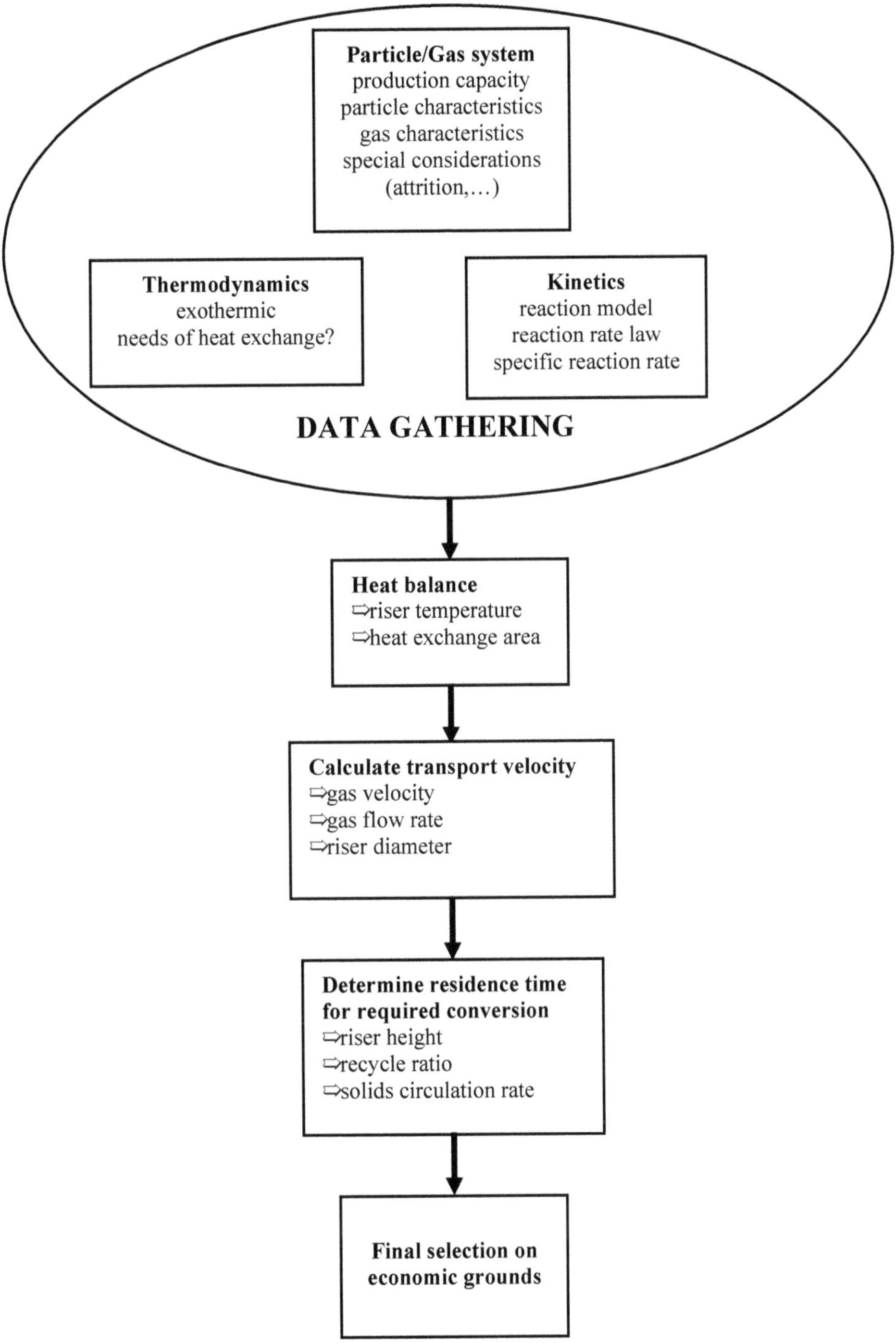

Figure 2: Design sequence of the CFB riser.

Calculations will be made for wood particles between 0.1-2 cm with a particle density of approximately 600 kg/m³. Wood particles contain 40 – 60 wt% moisture, 50 – 30 wt% volatiles, 1 wt% ash and approx. 10wt% fixed carbon (Boroduyla et al. [2]). Kinetics of the devolatilisation and combustion reactions are reported below.

THE HEAT BALANCE DETERMINES THE NET ENERGY YIELD AND FLOW RATES

Principles

A *heat balance* over the riser determines the net energy yield. At sufficiently high solids circulation rates the riser can be assumed to be isothermal (Baeyens [3]). Considerations of the reaction kinetics and heat exchange aspects result in an optimum temperature. Systems for exchange of heat have to be chosen and the exchange surface area has to be fixed for imbedded tubes or for a wall-mounted jacket. Once the operating temperature in the riser is known, the gas characteristics in the riser can be determined.

Applied to the bio-mass combustor

The composition of the biomass $(C_6H_{12}O_5)_n$ determines the flow rate of both the needed air and the flue gases and the composition of the last one. Considering an excess air flow rate of 15%, 4.86 Nm³ air/kg biomass is required, while the flow rate of the flue gases is 5.63 Nm³ flue gas/kg biomass. The total flow rates for a 2.5 tph combustor are then 12,150 Nm³/hr of air and 14,100 Nm³/hr of flue gases.

Biomass combustion occurs mostly at a temperature around 850°C. Heat is produced by the combustion reaction (17.5 MJ/kg) while part of it is consumed by heating the biomass and the air to the reactor temperature. A heat balance around the combustor results in the net energy of 11 MJ/kg biomass. Burning 2500 kg/hr biomass thus corresponds to a 7.6 MW_{th} combustor. Heat can be released by imbedded tubes or by a wall-mounted jacket in the riser or in the recycle loop.

THE TRANSPORT VELOCITY DETERMINES GAS VELOCITY AND RISER DIAMETER

Principles

The *transport velocity* is an important design parameter since it determines the minimum gas velocity required to fully develop a stable fast fluidization mode in the riser. Smolders et al. [4] gathered literature data on the transport velocity of different particles (d_p=24 – 1000 μm, ρ_p = 600 –3000 kg/m³) in CFBs of different size (D = 0.08 – 1 m) They compared these experimental data with the predictions by 8 correlations of several authors. They concluded that the transport velocity can best be calculated by the correlation of Adánez [5] for small particles:

$$Re_{TR} = 2.078 Ar^{0.463} \qquad Ar < 90 \qquad (1)$$

or by the correlation of Perales [6] for larger particles:

$$Re_{TR} = 1.458 Ar^{0.484} \qquad Ar > 90 \qquad (2)$$

To include a safety margin, the operating gas velocity will be chosen to exceed 1.2 U_{TR}. As the minimum gas flow in the riser is determined by the reaction stoichiometry, the *gas velocity* and the *riser diameter* cannot be chosen independently.

Applied to the biomass combustor

CFB-combustors for coal are operated with sand and ash of average particle size 300 μm (Andersson et al. [7]). In view of the small difference of density between both bed materials, transport velocities hardly differ. The biomass combustor will be operated with the same particle size. The transport velocity and operating gas velocity are then:

$$U_{TR} = 6.56 \text{ m/s} \quad (3)$$

$$U_{min} = 1.2\, U_{TR} = 7.9 \text{ m/s} \quad (4)$$

As the the flow rate of incoming air and that of the flue gases are very similar, the bed diameter will be calculated at the average of the two flow rates: 13,000 Nm^3/hr or 53,500 am^3/hr at 850°C. With known gas flow rate and required gas velocity, the riser cross-sectional area A is fixed at 1.88 m^2 which results in a diameter of 1.55 m.

THE RESIDENCE TIME DETERMINES RISER HEIGHT AND RECYCLE RATIO

Principles

The average residence time of the solids, required for the specified conversion can be estimated by applying the reaction rate law. In a riser of reasonable height, this *residence time* cannot be achieved by a single pass of the solids and particles have to be recycled. Figure 3 illustrates the riser system with recycle.

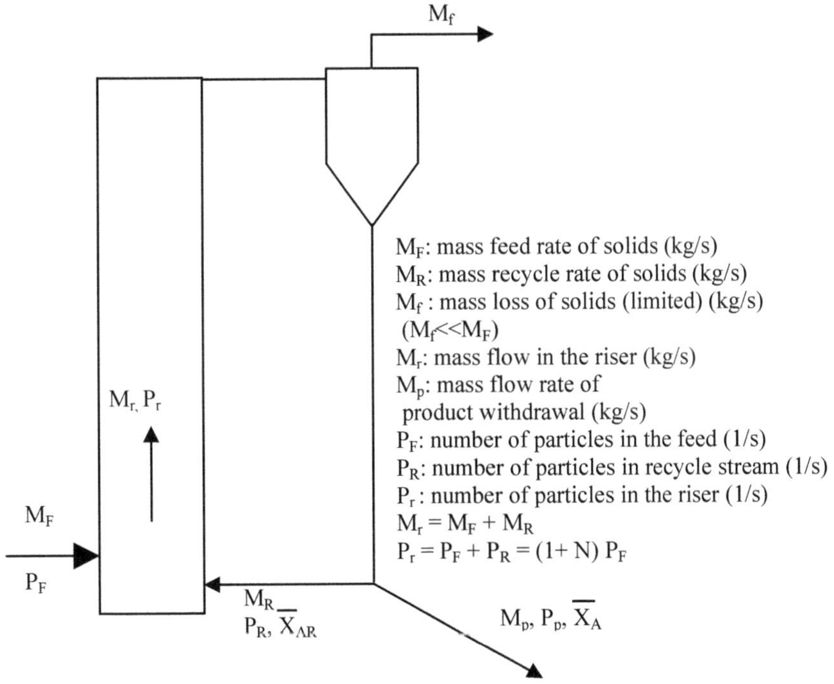

Figure 3: Solid flows in the circulating fluidized bed.

The recycling of solids causes a spread in the residence time, and hence in the conversion of the particles as some rise only once through the riser while others are recycled several times. Partly unreacted feed can hence be discharged, causing insufficient conversion. It can be demonstrated (Smolders et al. [8]) that the conversion in the product stream ($\overline{X}_A$), is equal to:

$$\left(1-\overline{X}_A\right)=\sum_{i=1}^{\infty}\frac{N^{i-1}}{(N+1)^i}\left(1-X_{A,i}\right) \quad (5)$$

with $X_{A,i}$ the conversion of a particle that has a total residence time of i times the average residence time for one pass and with N the recycle ratio and equal to:

$$N = \frac{\text{number of particles in recycle stream}}{\text{number of particles in feed}} = \frac{P_R}{P_F} \qquad (6)$$

If the converted particle is the main reaction product (e.g. for calcination reactions), the recycle ratio N and the height of the reactor determine the average conversion of the discharged particles M_p (and hence the product quality) as illustrated by eqn. (3). Smolders et al. [8] illustrated these principles for a CFB limestone calciner.

In a combustor the mass of products withdrawn from the reactor M_p, represents only the small stream of valueless 'ashes'. The characteristics of this stream are hence less critical in the design of the CFB combustor.

Applied to the bio-mass combustor

Kinetics of the biomass-combustion

In the incinerator, devolatilisation and combustion of volatiles and char take place. Release of vapour from the core cools the particle's surface and keeps its temperature low, so that it can be assumed that char combustion starts only after devolatilisation is completed.

Boroduyla et al. [2] investigated the combustion of wood particles (d_p = 3.5 to 21.5mm) in an fluidized bed of sand particles at temperatures from 1023 to 1123 K The biofuel particle size was found to decrease insignificantly during drying and devolatilisation, with these processes mainly coinciding in time. Furthermore, the volatile extinction time of wood can be estimated from:

$$t_{vo} = 12.1 \times 10^3 \times d_p^{1.3} \qquad (7)$$

To calculate the burnout time of char particles, a reaction model is required. With the assumptions that the reaction is mass transfer limited and that it occurs in two succesive steps with CO as intermediate, the burnout time t of a particle can be theoretically derived as (Smolders [9], Baeyens et al. [10]):

$$t_{ch} = \frac{\rho_M R_0^2}{4 D_{O_2} C_{O_2}} \qquad (8)$$

Biomass combustion in the CFB

When wood particles are introduced in the CFB, they stay in the bottom part of the bed, considered as the dense bed, until their terminal velocity becomes small enough to carry them up the riser with the gas stream. The transport of large particles in the riser occurs not only due to the drag force of the gas but also due to the impact of the fine sand particles on the large wood particles. Geldart et al. [11] demonstrated that particles are elutriated when the gas velocity in the riser is larger than the terminal velocity calculated with an 'effective' gas density. This 'effective' density can be calculated as:

$$\rho_{eff} = \rho_g + \bar{\rho} \qquad (9)$$

This approach was developed using solids ranging in density from 706 to 5000 kg/m³ and in mean size from 62 to 219 μm. To guarantee isothermal conditions in the riser, the circulation rate is set at 20 kg/m² s, corresponding to an 'effective' gas density of 3.8 kg/m³.

The density of the wood particles entering the bed is 600 kg/m³ and will be reduced during devolatilisation. Since the average fixed carbon fraction is approximately 10 wt%, the density of a completely devolatilised particle is 60 kg/m³. It is assumed that char combustion starts only after the end of devolatilisation. The terminal velocity at 850°C for different combinations of particle sizes and densities are given in Table 1.

Table 1: Terminal velocities.

d_p (cm)	0.1	0.5	1	1.1	1.5	2
ρ_p (kg/m³)	600	600	600	600	420	290
U_t (m/s)	1.5	4.9	7.5	7.9	7.9	7.9

Particles up to 1.1 cm can be elutriated from the bed without any devolatilisation. Particles with a size between 1.1 – 2 cm will be partly devolatilised before leaving the bed.

The required time for devolatilisation and the burnout time of the char particles can be calculated by eqn. (7), respectively eqn. (8). The concentration of O_2 varies through the riser from 1.03 mole/m³ after combustion of the volatiles to 0.296 mole/m³ in the flue gases. The concentration of O_2 is assumed to vary linearly during the char combustion. Results are given in Table 2.

Table 2: Theoretical reactions times.

d_p (cm)	0.1	0.5	1	1.1	1.5	2
t_{vo} (s)	1.51	12.3	30	34	51	74
t_{ch} (s)	2.26	56.5	226	273	508	904
t_{total} (s)	3.77	68.7	256	307	559	978

For particles exceeding 1.1 cm, part of the devolatilisation takes place in the dense bed. Taking this into account, the following residence times in the riser are required for complete devolatilisation and char combustion:

Table 3: Required reactions times in the riser.

d_p (cm)	0.1	0.5	1	1.1	1.5	2
t_{vo} (s)	1.51	12.3	30	34	34	32
t_{ch} (s)	2.26	56.5	226	273	508	904
t_{total} (s)	3.77	68.7	256	307	542	936

If the velocity of the particles is estimated as the difference between the gas velocity (7.9 m/s) and their terminal velocity at each moment, the height required for devolatilisation is equal to:

Table 4: Required riser height for devolatilisation.

d_p (cm)	0.1	0.5	1	1.1	1.5	2
H_{vo} (m)	10.4	55	82	82	80	71

The most commonly encountered height for industrial CFB combustors is 20 m. For one pass, the fraction of devolatilisation is given in Table 5.

Table 5: Fraction of devolatilisation in the first pass.

d_p (cm)	0.1	0.5	1	1.1	1.5	2
% devolatised in the first pass	100%	45%	48%	53%	69%	80%

The recycle of wood is thus evidently needed. Since the suspension leaving the riser is quenched in the external steam generating bed, further devolatilisation will be low. Since combustion of volatiles is a fast reaction, volatiles will be completely burned in the outlet duct after the cyclone.

After devolatilisation is completed, char starts to burn. Assuming the char starts to burn at the bottom of the riser, the fraction of char combusted during the first pass in the riser is given in Table 6.

Table 6: Fraction of char combusted in a single pass pass of 20 m.

d_p (cm)	0.1	0.5	1	1.1	1.5	2
% combustion in single pass	100%	5.4%	1.5%	1.3%	0.8%	0.4%

The recycle of the char is thus evidently needed. Since the suspension leaving the riser is quenched in the external steam generating bed where O_2-levels are very low, it is reasonable to accept that no combustion occurs in this recycle loop. In the CFB, char particles of all sizes are found. At equilibrium, the mass flow of char in the recycle loop varies from 375 kg/hr to 29,500 kg/hr when feeding at a constant rate of 2500 kg/hr of the specified particle size. Results are detailed in the Table 7.

Table 7: Char combustion in the riser.

d_p (cm)	0.1	0.5	1	1.1	1.5	2
total height required for char combustion (m)	17.4	384	1390	1645	2865	4695
number of passes before complete burn-out	1	19	70	82	143	235
(+ number of passes for devolatilisation)	(+1)	(+3)	(+4)	(+4)	(+4)	(+4)
total of unburned char in recycle loop (kg/hr)	375	2750	9250	10750	18375	29500

The circulation rate of the inert bed is set at 20 kg/m² s or 135,000 kg/hr. The fraction of unconverted char in the recycle loop is hence max. 22 w% (when feeding 2 mm wood particles).

Only a small stream leaves the combustor as solid i.e. the ashes corresponding to their weight fraction of the biomass (i.e. 1% for wood). This means that 25 kg/hr ash or max. 5.5 kg/hr unburned char (≈ 2 % of the fed char) is lost in the ash stream thus stressing that this loss should not be accounted for in the heat balance and associated calculations.

COMPARISON WITH OTHER COMBUSTION APPLICATIONS

Coal combustion is one of the major applications for CFB reactors. Quoted capacities range from 1 MW_{th} (labscale) to 700 MW_{th} (Soprolif-EDF). It is announced by EDF that capacities will exceed 500-600 MW_e after the year 2000.

The proposed design procedure was used to design a 12 MW_{th} CFB coal-combustor [8]. Comparison of relevant predicted data and typical commercial data demonstrated the validity of the design procedure.

Sludge has significantly higher contents of nitrogen, volatile matter, ash and (for wet sludge) moisture than typical coals and has furthermore a very low fixed carbon content. It has been reported that 11, 25 and 55% of the sludge produced respectively in the European Union, USA and Japan is incinerated. Because of the limitations currently facing the other sludge disposal outlets (landfill and use as fertiliser), it is expected that incineration will gain importance in the next decade. Because of the high moisture content of most sludges, autothermal incineration is only possible if the combustion air is preheated by the flue gases. Sludge combustors are hence mainly built for sludge disposal, rather than for energy production purposes. The proposed procedure was applied for the design of a CFB combustor for a sludge flow of 2 tonnes DS/hr and 4 tonnes water/hr [8]. It was assumed that 50% of the solids were organics with an overall composition of $C_5H_7O_2N$ and a heat content of 22.1 MJ/kg organic DS.

Table 8 illustrates the predicted data for the combustion of biomass, coal and sludge. The particle size of the inert bed, solids circulation rate and gas velocity are the same for the 3 combustors. For the same thermal capacity, the coal and biomass CFB-combustors are comparable in size. The specific combustion load for sludge combustors is lower due to the large amount of water vapour.

Table 8: Comparison of CFB combustors for biomass, coal and sludge.

	biomass	coal	sludge
feeding rate	2500 kg /hr	2200 kg /hr	6000 kg /hr (DS 33%)
surface area of riser	1.88 m^2	2.56 m^2	1.63 m^2
height of riser	20 m	20 m	20m
particle size of bed material	300 μm	300 μm	300 μm
gas velocity	7.9 m/s	7.9 m/s	7.9 m/s
solids circulation rate	20 kg/m^2s	20 kg/m^2s	20 kg/m^2s
combustion capacity	12.2 MW_{th}	17.7 MW_{th}	6.1 MW_{th}
net capacity	7.6 MW_{th}	12 MW_{th}	-
specific combustion load	6.5 MW_{th}/m^2	6.9 MW_{th}/m^2	3.8 MW_{th}/m^2
specific load (net)	4.1 MW_{th}/m^2	4.7 MW_{th}/m^2	-

CONCLUSIONS

A design strategy for a CFB combustor for biomass was presented and illustrated for a 2500 kg/hr combustor. The main operating parameters are the solids flux and gas velocity in the riser of the CFB. Isothermal conditions will be guaranteed by circulating inert sand particles of 300 μm at a solids flux of 20 kg/m^2s. The gas velocity in the riser is determined at 7.9 m/s, exceeding the transport velocity (=6.6 m/s) of the inert sand bed by 20%. About 2% of the char will be lost in the ash stream. The net combustion capacity of the riser is equal to 4.1 MW_{th} per m^2 of the riser, comparable with a coal CFB combustor.

NOMENCLATURE

A	riser surface area	[m^2]
Ar	Archimedes number $\left(=\rho_g(\rho_p-\rho_g)gd_p^3/\mu^2\right)$	[-]
C_{O_2}	concentration of O_2	[mole/m^3]
D_{O_2}	diffusion coefficient for O_2	[m^2/s]
d_p	particle size	[m]
g	gravitational constant	[m/s^2]
H_{vo}	height required for devolatilisation	[m]
M_p	mass flow rate of product withdrawal	[kg/s]
N	recycle ratio	[-]
P_F	number of particles in the feed	[1/s]
P_R	number of particles in the riser	[1/s]
R_0	radius of coal particle	[m]
Re_{TR}	Reynolds number for the transport velocity ($=\rho_g U_{TR} d_p/\mu$)	[-]
t_{ch}	burn-out time for char	[s]
t_{total}	total burn-out time of wood particles	[s]
t_{vo}	volatiles extinction time	[s]
U	superficial gas velocity in the riser	[m/s]

U_{TR}	transport velocity	[m/s]
X_A	conversion	[-]
$\bar{X}_A$	average conversion	[-]
μ	gas viscosity	[Pas]
$\bar{\rho}$	bulk density of fines in the riser	[kg/m³]
ρ_e	'effective' gas density	[kg/m³]
ρ_g	gas density	[kg/m³]
ρ_M	molar density of carbon	[mole C/m³]
ρ_p	particle density	[kg/m³]

REFERENCES

1. Smolders K. & Baeyens J. (1995) The operation of L-valves to control standpipe flow. *Adv. Powder Technology,* **6**, 163-176.
2. Boroduyla V.A., Dikalenko V.I., Palchonok G.I. & Stanchits L.K. (1995) Fluidized bed combustion of solid biological waste and low-grade fuels: experiment and modelling. In: *Fluidization VIII,* Oyland, Florida (U.S.A.), May 1995.
3. Baeyens J. (1998) Heat transfer in fluidized beds, CPA-course, Amsterdam, November 1998.
4. Smolders K. & Baeyens J. (2001) Gas fluidized beds at high gas velocities: a critical review of occurring regimes, *Powder Technology* **119**, 269-291.
5. Adánez J, de Diego L.F. & Gayán P. (1993) Transport velocities of coal and sand particles. *Powder Technology,* **77**, 61-68.
6 Perales J.F., Coll T., Llop M.F., Puigjaner L., Arnaldos J. & Casal J. (1991) On the transition from bubbling to fast fluidization regimes. In: *Circulating Fluidized Bed Technology III,* (Ed. by P. Basu, M. Horio & M. Hasatani), 4-1-1, Pergamon Press.
7 Andersson B.-A. & Leckner B. (1992) Experimental methods of estimating heat transfer in circulating fluidized beds. *Int. J. Heat/Mass Transfer,* **35**, 3353-3362.
8 Smolders K. & Baeyens J. (1999) Design strategy for a gas/solid Circulating Fluidized Bed Reactor. *Powder Handling & Processing,* **11**, 257 –264.
9 Smolders K. (1999) *Hydrodynamics in Circulating Fluidized Beds.* Ph.Dissertation, KUleuven, Belgium.
10 Baeyens J. & Geldart D. (1978) Fluidized bed incineration: a design approach for complete combustion of hydrocarbons. In: *Fluidization,* (Ed. by J.F. Davidson & D.L. Kearns, 264-269, Cambridge University Press.
11 Geldart D., Cullinan J., Georghiades S., Gilvray D. & Pope D.J. (1979) The effect of fines on entrainment from gas fluidized beds. *Trans. Inst. Chem. Eng.* **57**, 269-275

The Tar Analyser - a Suitable Tool for the Development and Control of Gasifiers and Gas Cleaning Systems

B. Staiger, O. Moersch, R. Berger and K.R.G. Hein
Universität Stuttgart, Institut für Verfahrenstechnik und Dampfkesselwesen IVD
Pfaffenwaldring 23, D-70569 Stuttgart, Germany

ABSTRACT: This paper presents a tar analysing method developed at the University of Stuttgart that allows quasi continuous online measurement of the content of condensable hydrocarbons ("tars") in the producer gas from biomass gasification. It considerably simplifies the experimental procedure compared to other methods described in literature. In addition the method also yields the content of gaseous hydrocarbons and can, therefore, replace a conventional analyser with flame ionisation detector.

The method is based on the comparison of the total hydrocarbon content of the hot gas and that of the gas with all tars removed. Hot gas from the gasifier is led directly to the analyser. Hydrocarbons are measured with a flame ionisation detector. Sample loops with a defined volume and high temperature switching valves are used to guarantee high accuracy.

A 19'' rack mounted measurement system is available. Experiments with the gas from a fluidised bed gasifier have been conducted yielding very good results. Tar contents between 200 and 20 000 mg/m³ have been measured reliably. The experimental procedure is simple and repeatability is good. Sampling and analysis time is one minute.

INTRODUCTION

One of the major obstacles for the application of biomass gasification systems is the relatively high tar content of the producer gas that inhibits the use in engines or turbines without further gas cleaning. The determination of the tar content is, therefore, important in research and application. Conventionally tar measurement is performed by cold trapping in wash bottles with a solvent and subsequent off-line analysis by a GC-FID system (gas chromatograph with flame ionization detector) [1][2]. This method is time consuming and does not allow online monitoring of the gas quality. Furthermore, a large amount of solvent is consumed.

At the University of Stuttgart a method for quasi continuous online measurement of the tar content has been developed. With this method, the effect of tar reduction measures can be directly verified. Further, online monitoring of the tar content can help to prevent engine or turbine trouble in commercial plants.

The terminology tar is not well defined in literature. In the context of gasification it is often used to summarize all aromatic and polyaromatic hydrocarbons present in the gas. Since not all of these compounds do cause problems in an application in this paper "tar" is applied to all hydrocarbons that are condensable at ambient temperature. According to this definition BTX components (benzene, toluene, xylene and similar) in concentrations in which they are typically present in the producer gas are not considered to be tars.

TAR SAMPLING SYSTEM

PRINCIPLE

The basic principle of the method is illustrated in Fig. 1. For the determination of the hydrocarbon content flame ionization detectors (FID) are well suited, since they have a high sensitivity and linearity. Further, the signal is with sufficient accuracy directly proportional to the amount of organically bound carbon atoms in the sample independently of the component [3].

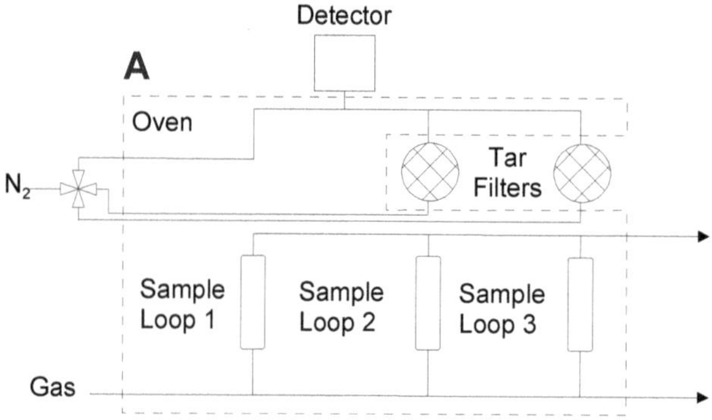

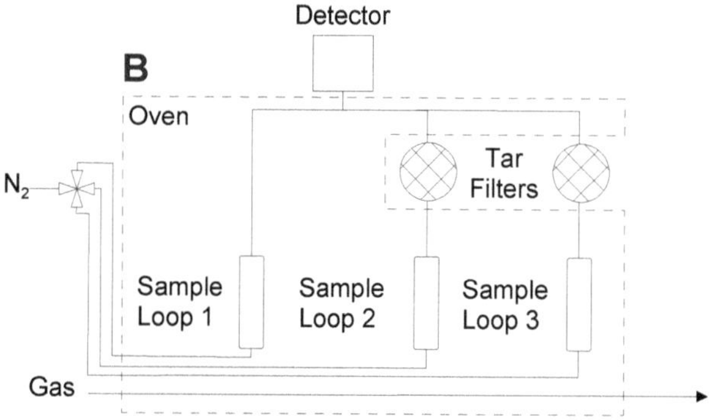

Figure 1: Basic principle of the tar measurement system
A: Sampling Phase, B: Analysing Phase.

At the beginning of the analysis the sample loops are brought in line with the hot gas sampling line from the gasifier. All sample loops are loaded contemporaneously (A-Figure). Fluctuations of the gas concentration do, therefore, not affect the analysis. When the sample loops are filled, high temperature valves switch to analysis and bring the sample loops into the carrier gas flow to the detector (B-Figure). The gases from the sample loops are flushed subsequently to the FID. Gas from loop one and two are led over filters, which can be filled with different filter materials to distinguish the tars. The filter temperature is controlled by peltier elements in the range of 0 - 120 °C. In between runs, the filters are flushed with nitrogen. Commercially available filters of sintered bronce are used for an easy handling. Possible alternative filter materials are quartz wool, activated carbon, SPA material, silica gel and user defined ones. By using the sintered bronce or quartz wool, all condensable substances are removed, by using activated carbon all aromatic hydrocarbons are removed. The gas from sample loop three is led unaltered to the FID to determine the total content of hydrocarbons. The FID output signal is recorded during an analytic cycle and directly evaluated. The area of peak three determines the total hydrocarbon content, the differences in the areas of peak three and one/two determines the tar content.

Figure 2 gives an example for the output signal during one analysis cycle. Analysis time is one minute minimum. The analyzer can run continuously, so that every minutes a present value of tar and gaseous hydrocarbons is available.

Main components of the system are the detector (FID), three sample loops, high temperature switching valves, and two filters for tar condensation. The sample loops guarantee that the reference volume for both flows is identical and contains gas of exactly identical composition. The carrier gas provides the necessary pressure and gas flow rate to the detector.

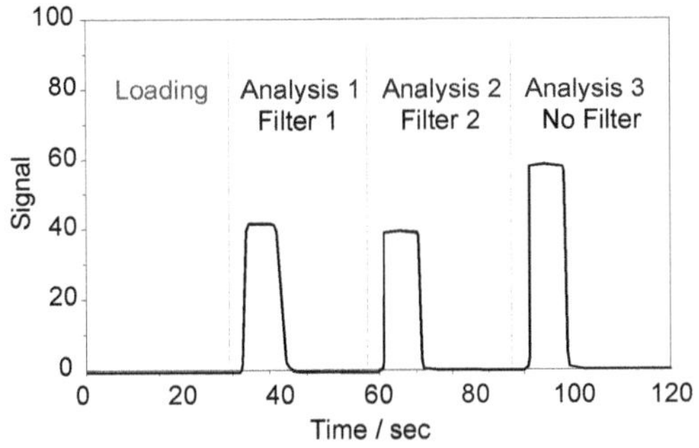

Figure 2: FID-signal during analysis cycle.

If a qualitative analysis of the tar components is desired, the tar filter in the bypass gas flow is removed after a defined volume has been drawn over it. The filter is flushed with dichlormethane. The solution obtained can be analyzed with a GC-MS system (gas chromatograph with mass spectrometer) that allows a detailed identification of single components. An example of such an analysis is shown in Fig. 3. Five liters of gas were drawn over the filter which then was dissolved in 4 ml of dichlormethane.

A 19" rack mounted measurement system is available. The cost will be in the range of a conventional total hydrocarbon analyzer with flame ionization detector. Since the method also yields the content of gaseous hydrocarbons the conventional HC analyzer can be replaced.

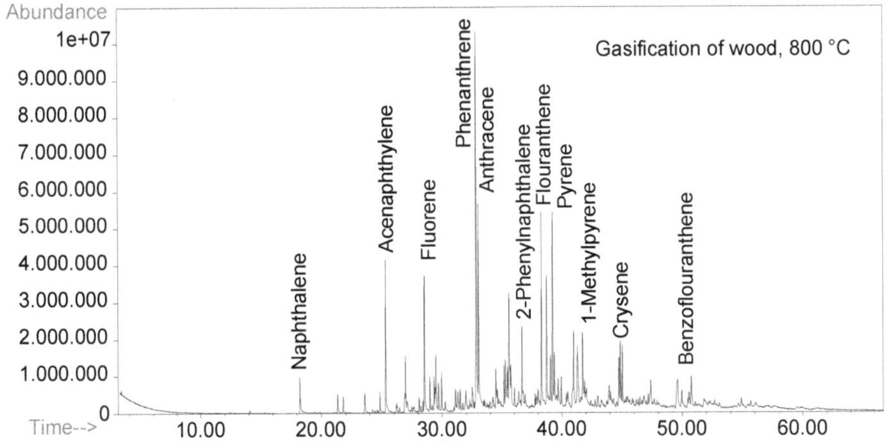

Figure 3. GC-MS analysis of the tar on the bypass filter.

CALIBRATION

As mentioned above, the system takes advantage of the fact that the signal of an FID is almost directly proportional to the amount of organically bound carbon. In Table 1 some response factors are given relative to propane. It can be seen that they are close to one for all compounds, hence the error is small when an average response factor determined with propane is used. The calibration of the system has been performed with a gas containing 37.8 gC/m^3 of propane in nitrogen. For all peaks a separate response factor is determined to compensate small differences in volume or temperature of the sample loops.

Table 1: FID response factors of some aromatic compounds relative to propane.

Substance	Response relative to propane
Propane	1.00
Benzene	0.99
Toluene	1.00
Ethylbenzene	0.92
o-Xylene	0.97
Trimethylbenzene	0.96
Phenol	1.00
Indene	0.92
Naphthalene	1.00

ACCURACY AND DETECTION LIMITS

Since the amount of tars is small (around 0.5 g/m³ for a good gasifier) compared to the amount of total hydrocarbons (in the range of 50 g/m³) high accuracy of the system is crucial to obtain reliable tar values. Flame ionization detectors are quite sensitive to changes in flow rate and temperature. These parameters have to be kept absolutely constant during analysis. Since for both measurements the same detector is used the remaining detector drift has only a minor effect on the result. In Fig. 4 the results of calibration measurements are shown. It can be seen, that for the absolute value of hydrocarbons a high accuracy with less then 0.1 % of scatter can be achieved.

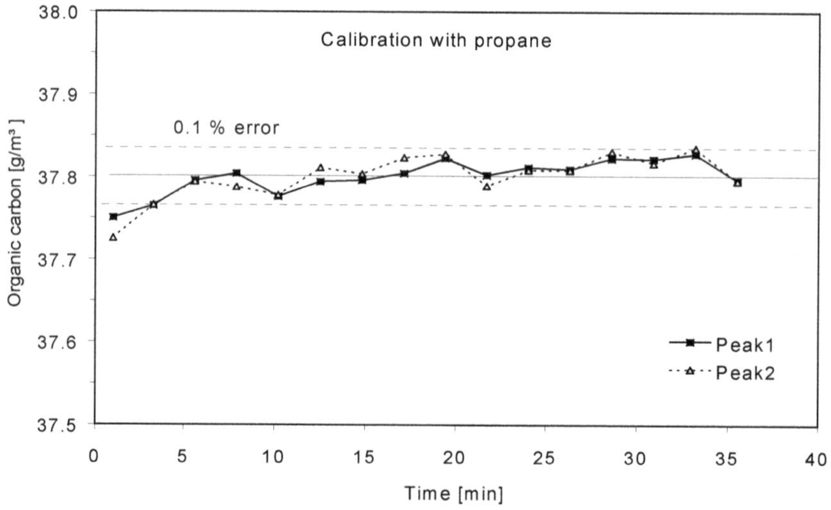

Figure 4: Concentration of organic C calculated from the two peaks during calibration measurements with propane.

Fig. 5 shows the difference of the peaks for the same measurements. The scatter is ±20 mg/m³. The minimum tar concentration that can be detected is, therefore, about 50 mg/m³. The detection limit can be decreased by increasing the volume of the sample loops. This, however, also increases sampling time. To ensure performance periodic calibration runs should be carried out.

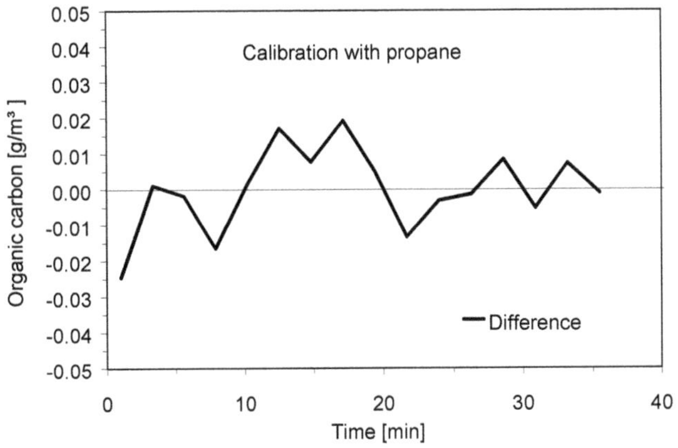

Figure 5: Scattering of zero measurement with propane.

The system determines as "tars" only those components that are condensed and/or removed on the filters. According to the vapor pressure of each component at filter temperature, a certain amount will remain in the vapor phase. However, in a gasification facility this amount should not cause any fouling problems as long as temperature does not decrease further. To ensure an appropriate process monitoring temperature should be chosen below the lowest temperature in the process.

RESULTS OF ONLINE MEASUREMENTS

To test the analysis method a series of gasification experiments have been carried out. The gasifier used is of the bubbling fluidized bed type with a thermal capacity of 25 kW. The facility is equipped with electrical heating that allows to vary air ratio and temperature independently. A detailed description can be found in [4]. In a first test series the results of the online tar measurement have been compared with those obtained with a sampling system based on gravimetric analysis of tar filters [5]. Experiments were conducted with dried beech wood at temperatures of 700, 800, and 950 °C and an equivalence ratio of 0.15-0.25. No additives or catalysts were used. The results are shown in Fig. 6 where tar is plotted over gasification temperature.

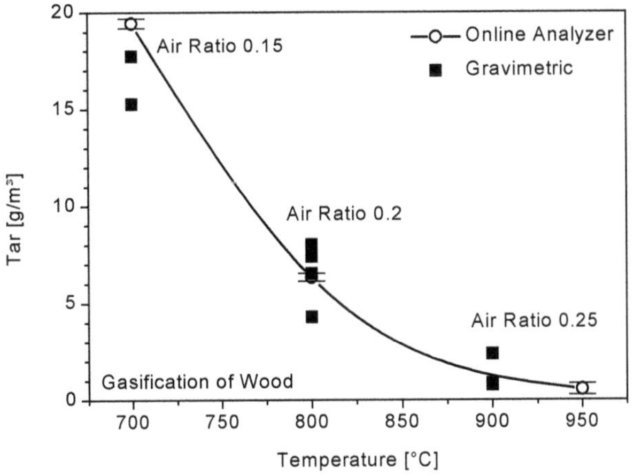

Figure 6: Comparison of results of the online analyzer with results of gravimetric analysis:

The curve shows the typical strong decrease of tar production with increasing temperature and air ratio. Good agreement between the two measuring methods can be observed. The scatter of five to eight measurements performed with the online analyzer at each temperature is much smaller than for the gravimetric method.

Further comparisons were done with a wet chemical method at a downdraft gasifier. The mean value from the tar analyzer during the respective sampling period and the results of the cold trapping are compared in Fig. 7. The sum of heavy tars, PAHs and phenols determined with cold trapping was about 400 mg/m³ compared to 880 mg/m³ of total condensables determined with the analyzer. The filter temperature of the tar analyzer was 10 °C. At this temperature most, probably also BTX-components, are partially retained by the filter. Considering further the insecurities incorporated in any tar measurements, the results confirm the validity of the on-line measurement.

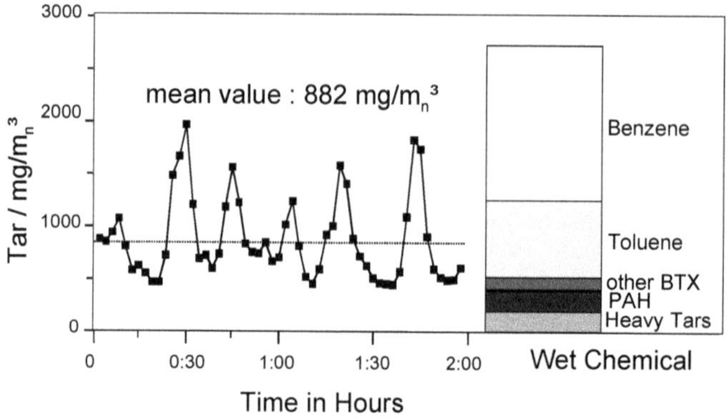

Figure 7: Tar concentration during gasification experiment at a downdraft gasifier.

Measurements where also performed in the clean gas after the sand bed filter of an open top fixed bed gasifier with 200 kW thermal input, coupled to an IC-engine generator set with an output of ca. 40 kW electric. Over a duration of five hours the tar concentration was monitored continuously. As can be seen in Fig. 8, the tar level is decreasing with time after start-up. Further, frequent tar peaks can be observed. These are due to a periodic increase of pressure drop over the gasifier when the burnt out bed collapses and moves down, causing a decrease of air supply and a change of local temperature.

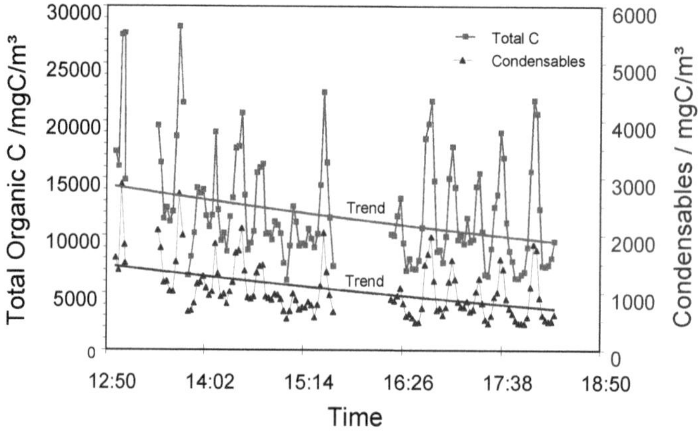

Figure 8: Tar Monitoring at a downdraft gasifier.

In Fig. 9 and 10 parametrical studies to the tar formation are shown exemplary. In the diagrams the tar is plotted over the humidity of the used wood and over the mesh of the sieve, which is corresponding to the particle size.

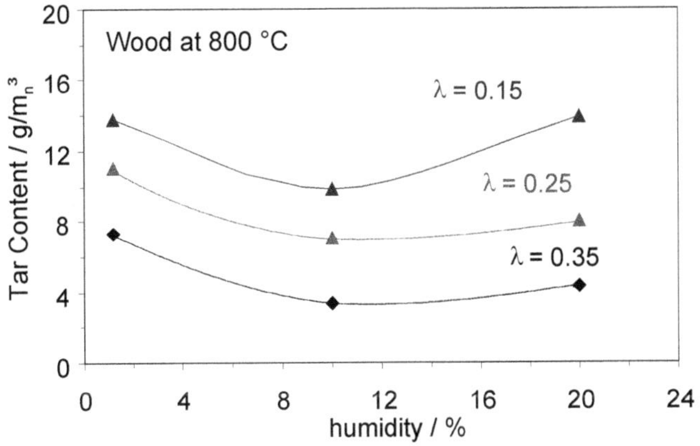

Figure 9: Parametrical Studies (humidity).

As can be seen in Fig. 9 the tar content passes through a minimum with a rising humidity. The tar reducing to up to 10 percent could be explained by the decreasing of the heating value caused by the humidity. Furthermore the water could be a possible reaction partner for the tars.

Figure 10 shows results of a measuring campaign at an industrial downdraft gasifier during its testing phase. Waste wood with high water content was used. The downdraft gasifier was equipped with a scrubber using water as washing liquid and a saw dust filter. The unsuitable fuel caused bad operation conditions and therefore a high value of condensables in the producer gas. The main part of the condesables were washed out in the scrubber, while the further reduction in the saw dust filter was very small.

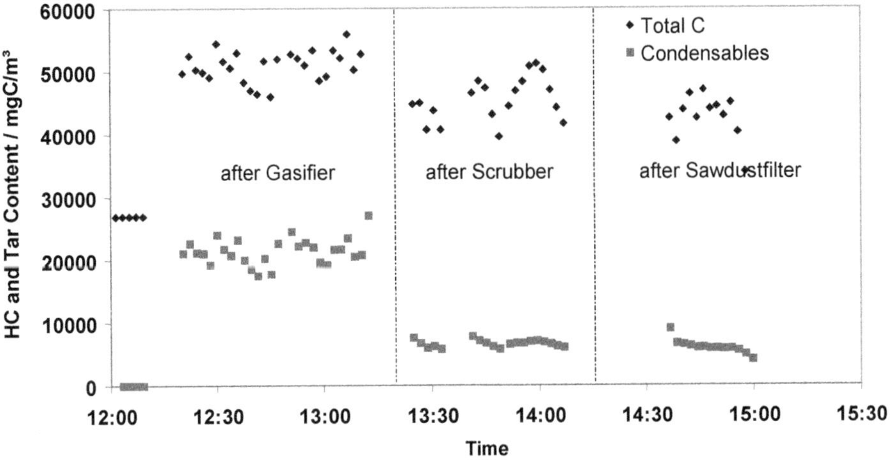

Figure 10: Tar monitoring at a downdraft gasifier fired with waste wood.

REFERENCES

[1] Kurkela E., Stahlberg P. (1992). Air Gasification of Peat,Wood and Brown Coal in a pressurized Fluidized-Bed Reactor. I. Carbon Conversion, Gas Yields and Tar Formation. Fuel Proc.Techn., Vol 31, p. 1-2

[2] Mai M., Struschka M., Baumbach G. (1990). Meßverfahren zur Bestimmung von gas- oder aerosolförmigen organischen Emissionen. Fortschritts-Berichte VDI, Reihe 15, Nr. 75, VDI-Verlag Düsseldorf

[3] Wandinger H. (1997). FID-Meßtechnik, World Wide Web, URL:"http://ourworld.compuserve.com/home pages/Hermann_Wandinger"

[4] Moersch O., Spliethoff H. and Hein K.R.G.(1996). Influence of the Stoichiometry on Gas Quality and Carbon Conversion in Fluidised Bed Gasification of Biomass" in Chartier P., Ferrero G.L., Henius U.M., Hultberg S., Sachau J. and Wiinblad M. (editors): "Biomass for Energy and Environment", Pergamon, p. 1398-1403

[5] Moersch O., Spliethoff H., Hein K.R.G. (1997) A new system for tar sampling and analysis, in Kaltschmitt M. and Bridgewater AV (Editors): Biomass Gasification and Pyrolysis, CPL Press, Newbury UK

Biomass/Coal Gasification for Efficient Embedded Power Generation

J Swithenbank, S Shabangu, B F Chang, A Johari, N V Russell, V N Sharifi
Sheffield University Waste Incineration Centre (SUWIC), Department of Chemical and Process Engineering, University of Sheffield, Sheffield S1 3JD, UK
N Warner
40 High House Drive, Rednal, Birmingham, B45 8ET, UK
F M Lewis
535 East Mariposa Ave, El Segundo, CA 90245- 3013, USA

ABSTRACT: **Gasification** - Our new gasification concept offers the prospect of achieving the goal of high efficiency power generation from various biomass or coal energy sources by utilizing Ultra Superheated Steam (USS). The process uses low-grade steam from sources such as waste incineration, process cooling, and local industry, and then enhances it to a temperature above 1600°C. This is achieved by adding oxygen to the steam to form "artificial air"; gas is then burned in this artificial air to produce an ultra-superheated steam flame. Powdered coal and/or biomass is injected coaxially into the high temperature steam flame where it reacts in about a second to produce a gas that is free of tar and consists largely of CO, H_2 and CH_4. Significantly, the high temperature steam provides the enthalpy for the endothermic gasification reactions. This feature ensures that any ash is cooled below its melting point before it arrives at the reactor walls.

Liquid Metal Based Gas Scrubbing System - Current hot gas clean-up methods focus on the use of metal oxide sorbents for sulphur removal and the utilisation of ceramic barrier filters for particulate cleaning. In contrast, this research project adopts an unconventional approach to fuel gas cleaning technology, incorporating molten tin as the scrubbing medium for the removal of H_2S and solid particulates. The process has been studied using a cold model with fluid circulated by a gas lift to determine the removal efficiency for particulates and gaseous pollutants.

BACKGROUND

Worldwide, there is a progressive change to use a greater proportion of our total energy consumption as electrical energy and citizens in developed countries now use up to about 1kW of electricity per person. This corresponds to about 2.4 tonnes of coal per person per year. At the same time, citizens bring into being up to one tonne of municipal waste each year having an energy equivalent of 300 kg of coal, plus even larger quantities of agricultural and industrial wastes. Bearing in mind that these wastes are largely biomass, the recovery of energy from the waste would not add very much to the net atmospheric CO_2 level. Thus recovering energy from waste helps to mitigate the climate change problem. Furthermore, dedicated agricultural production of biomass for energy generation is increasing in popularity. Sustainable cities therefore require the generation of electrical energy from biomass including suitable fractions of waste that cannot be economically reused or recycled together with clean coal technology.

ULTRA SUPER-HEATED STEAM GASIFICATION

INTRODUCTION

The conversion efficiency of solid fuel via steam raised in a boiler into electricity using steam turbines is only about 20% to 45% depending on the temperature and pressure of the steam. Gasification of solid fuels for IGCC systems produces gas to run gas turbines that are used to produce power together with steam turbines. This enables enhanced efficiency in the conversion of energy from solid fuels and the overall efficiency of a conventional integrated gasification combined cycle is about 51%. In this case, the heat required for the gasification reaction is generated by the reaction of about 15% of the fuel with oxygen. A new process is proposed that uses low-grade steam from an incinerator as the starting point for an ultra superheated steam gasification system for solid fuels. The overall efficiency of the proposed process is about 60%, which is superior to the sum of the separate processes of electricity from waste together with an independent coal/biomass gasification-based electricity generator. When used as an embedded CHP system, the energy complex can produce power and heat from biomass and coal for a sustainable city at an efficiency of about 85% whilst disposing of waste that cannot be effectively reused or recycled.

The aim of this research project is the development of the novel gasification process that utilises ultra-superheated steam.

EXPERIMENTAL INVESTIGATION

The new gasification concept offers the prospect of achieving the goal of high efficiency power generation from various biomass or coal energy sources by utilizing Ultra Superheated Steam (USS). The concept first reduces the coal and/or wastes such as wood waste and biomass to a fine powder that can be reacted in a small residence time and therefore a small volume. The process uses low-grade steam from sources such as waste incineration, process cooling, and local industry, and then enhances it to a temperature above 1600°C. This is achieved by adding oxygen to the steam to form "artificial air"; gas is then burned in this artificial air to produce an ultra-superheated steam flame (Figure 1).

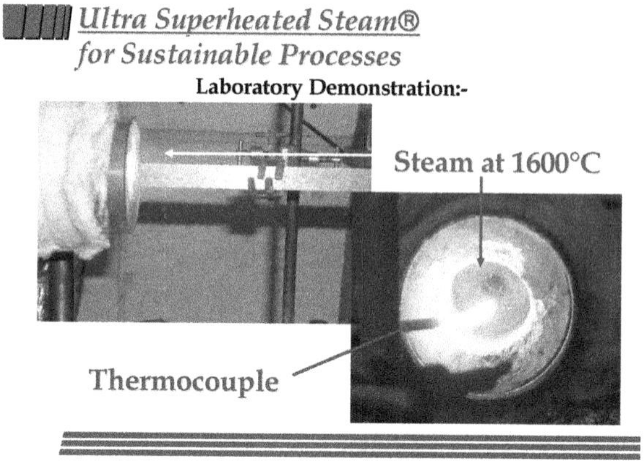

Figure 1: The Ultra Superheated Steam Flame.

The coal and/or biomass is injected coaxially into the high temperature steam flame where it reacts in about a second to produce a gas that is free of tar and consists largely of CO, H_2 and CH_4. Significantly, the high temperature steam provides the enthalpy for the endothermic gasification reactions. This feature ensures that any ash is cooled below its melting point before it arrives at the reactor walls. This is important since the high alkali content of biomass ash results in a low melting point and hence a high tendency to form slag.

The USS gasifier can be envisaged as being composed of two distinct main parts, namely:

1. The USS generator (a conventional burner where the gasification also commences).
2. The gasifier (which is the shell where the reactions proceed).

THE USS BURNER

The North American Manufacturing Company manufactured the burner selected for the generation of USS and subsequent USS gasification (Figure 2). Its design capacities are a maximum air flow of 0.0322 m^3/s and a heat output of 120 kW. Its construction features include air and gas inlets into the burner, a cast iron mounting plate, and a standard quarl that is about 230 mm long.

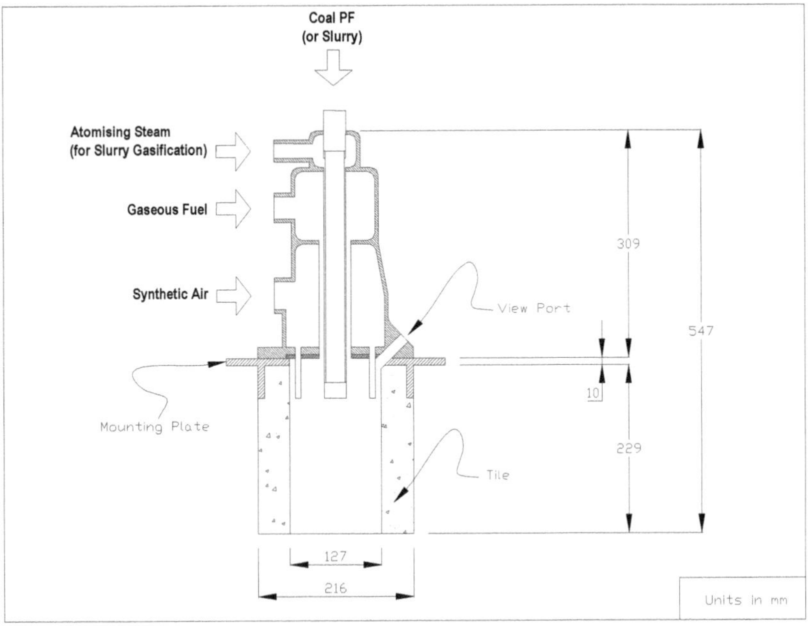

Figure 2: Cross-section of the USS generator.

Figure 3: The Coal Feeding System.

This burner was chosen because, being manufactured for dual fuel operation it could be easily adapted for the USS gasification by replacing the liquid fuel pipe with one suitable for the supply of granular material for the gasification of solids, or by just using the burner as supplied for the gasification of liquid or slurry material.

This burner was adapted to feed in particulate materials by removing the liquid fuel inlet pipe, and replacing it with a pipe with a wider bore, connected to the funnel and vibrating feeder system, (Figure 3).

THE USS GASIFIER

A rational design of coal gasifiers requires the best possible understanding of the fundamental chemical and physical processes that occur:

1. During the transition of the coal to coal-char.
2. During the subsequent gasification of the coal-char itself.

Pressure is one of the operating variables of interest in the investigation of USS gasification. In the gasifier design stage, high pressure USS gasification was considered but the extent of the pressure was limited by the construction and operation feasibilities. Feeding of reactants into a pressurised atmosphere and then maintaining a stable flame at the high pressures presents a difficult operational problem. This then limited the pressure operation of the gasifier to near atmospheric for the preliminary tests, where the material to be gasified is fed into the gasifier by gravity. A photograph of the equipment is shown in Figure 4.

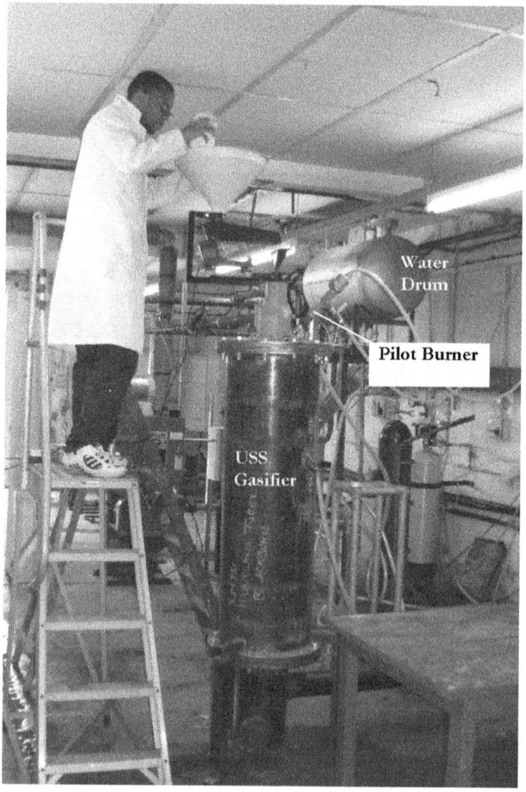

Figure 4: The USS Gasifier.

The Gasification Chamber

The gasification chamber is a cylindrical mild steel shell lined with a 50mm thick fused alumina based castable refractory with high abrasion resistance. The design pressure is 5 bar and the dimensions of this shell are:

- Shell inside diameter = 385 mm
- Length = 1250 mm
- Thickness = 9 mm
- Refractory inside diameter = 285 mm

A catch pot at the bottom collects ash or slag.

The USS gasification was carried out using methane as the gaseous fuel for the generation of USS and propane was used to fire the pilot burner. The yield of gas was measured by collecting samples for GC analysis and also evaluating the ash content.

RESULTS

Two bituminous coals were used in the tests having high volatile (coal A) and semi-anthracite low volatile properties (coal B) respectively.

The theoretical calculations were carried out to estimate the gas yield from USS gasification. The Boudouard, water-gas and hydrogasification reactions occur simultaneously. The predicted yield at equilibrium was used for carbon gasification using steam at 1 atm and 2131K (which is the adiabatic flame temperature for a stoichiometric USS flame). The results are:

Volume fraction of CO = 54.8%
Volume fraction of CO_2 = 0.0%
Volume fraction of H_2 = 45.2%
Volume fraction of H_2O = 0.001%
Volume fraction of CH_4 = 0.007%

Chemical equilibrium **calculations** based on minimising Gibbs free energy are illustrated in Figure 5.

Table 1: The **measured** concentrations of product syngas at the bottom of the reactor.

	Coal A (45-75 μm)		**Coal B (<150 μm)**	
	Sample 1	Sample 2	Sample 1	Sample 2
Time, s	590	1225	843	1110
Syngas Composition, %				
H_2	23.4 (38.9)	22.1 (39.4)	22.1 (37.8)	18.1 (45.3)
O_2	2.8 (0.0)	2.6 (0.0)	2.19 (0.0)	6.2 (0.0)
N_2	28.3 (0.0)	25.2 (0.0)	32.1 (0.0)	38.9 (0.0)
CH_4	6.7 (9.0)	10.1 (12.2)	5.8 (8.3)	2.7 (4.0)
CO_2	14.1 (18.9)	13.9 (16.8)	14.3 (20.4)	13.7 (20.4)
CO	7.1 (9.5)	4.1 (5.0)	11.3 (16.1)	12.5 (18.5)
Balance (H/C)	17.7 (23.4)	22.1 (26.7)	12.3 (17.5)	7.9 (11.7)
TOTAL	100.0	100.0	100.0	100.0

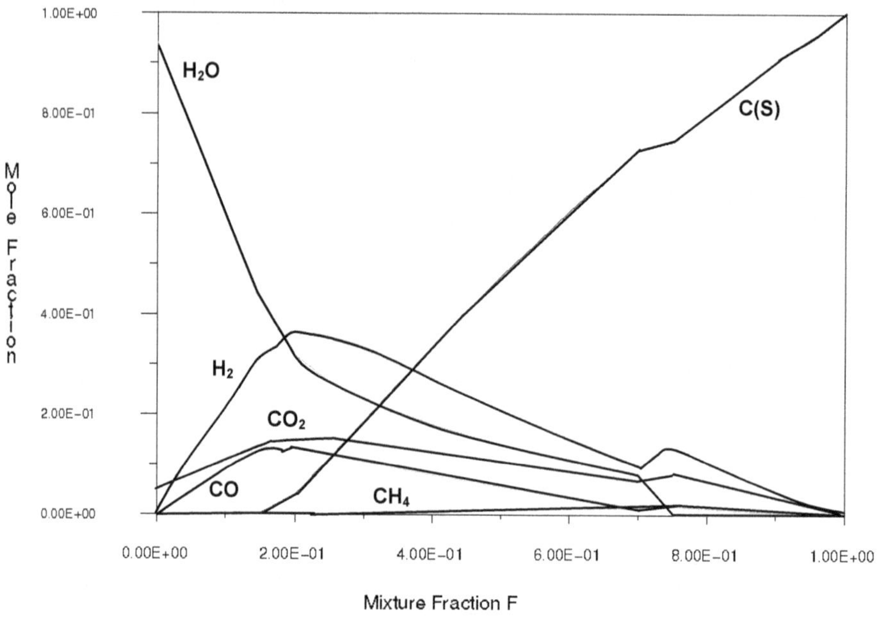

Figure 5: Chemical Equilibrium Instantaneous Species Composition.

It can be seen that there was a significant fraction of nitrogen and oxygen in the raw syngas at the bottom of the reactor. The important source of this is the pulverised coal screw feeder duct that is open to the atmosphere. Due to the poor mixing along the axis of the present chamber some of the oxygen persists to the sampling point. However, the total can be obtained from the nitrogen fraction and the balance would react with the available hydrogen. The initial hydrogen yield is therefore indicated in brackets in the Table 1. Comparing these figures with the theoretical values shows that there are significant two-phase flow phenomena operating in the present reactor.

FUTURE ACTIVITIES FOR USS GASIFICATION

Detailed CFD calculations, combined with modifications to the feeding system are being used to optimise the processes within the USS gasifier chamber. The USS gasifier is envisaged as part of a high efficiency coal gasification power generation system for sustainable cities.

Thermodynamic considerations show that the efficiency of power generation from low-grade steam from a city incinerator is only about 20%. This new gasification concept offers the prospect of achieving the goal of high efficiency power generation from such energy sources by utilizing the synergy possible with a co-incineration system (Figure 6).

The efficiency of gasification can be further improved by preheating the steam before adding the oxygen thus reducing the quantity of oxygen required. Figure 7 shows a regenerative preheating system in which two regenerators operate back-to-back in a cyclic mode. The resulting gas is cool and can be cleaned with conventional technology.

The generation of power from the gas should preferably use a system having a small footprint if the plant is to be installed in a city due to the high cost of land

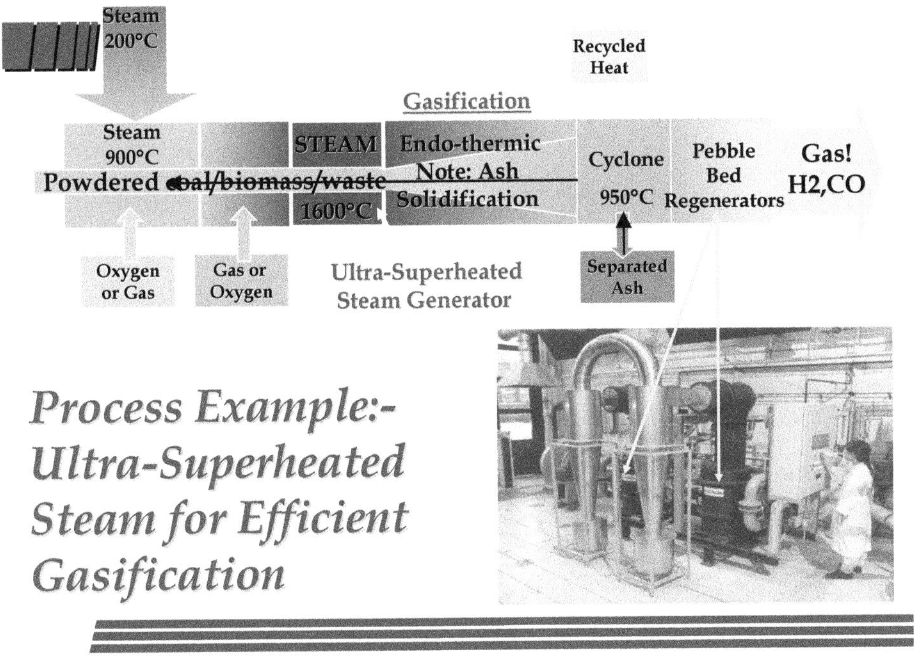

Figure 6: USS gasification process with regenerator.

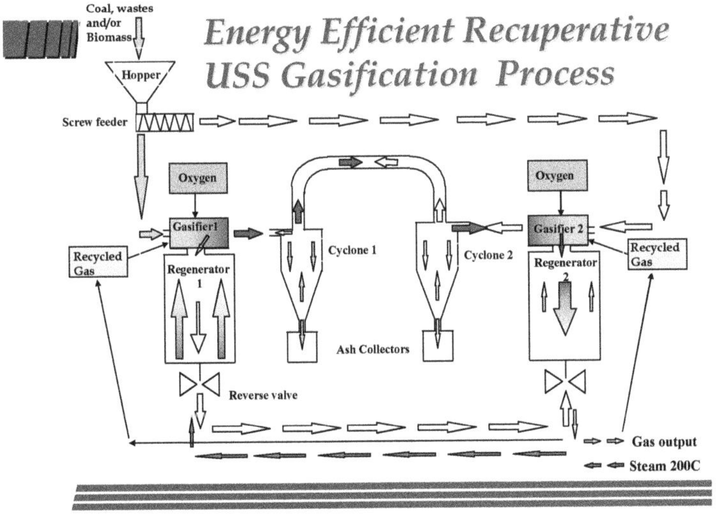

Figure 7: The gasifier regenerator system.

Figure 8 shows the outline of a compact unit that eliminates the Rankine cycle with its attendant cooling towers etc. In this case an advanced recuperative gas turbine is employed. In addition to its high efficiency, the low optimum pressure ratio of this gas turbine system results in lower emissions of oxides of nitrogen than would be obtained from a conventional combined cycle power generation system.

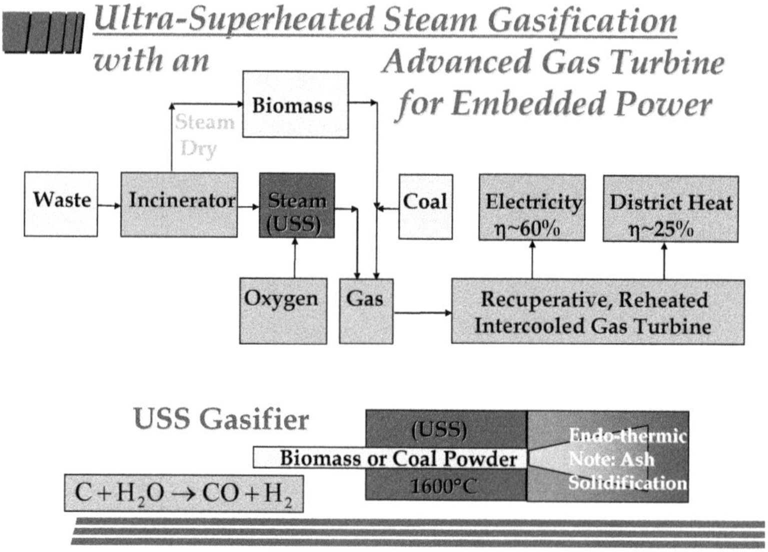

Figure 8: Small Footprint Power Generation System.

More advanced systems can eliminate the emission of carbon dioxide from the power generation by adding a water-gas shift reactor to the process and one such option is shown in Figure 9. In this concept, the hydrogen fuel that is generated can be used directly in a high temperature steam turbine having cooled blades. Because of the high temperature range of the ensuing Rankine cycle, the generation of electrical power is extremely efficient compared to present systems. Alternatively, the hydrogen can be used in a fuel cell to directly produce both electricity and heating.

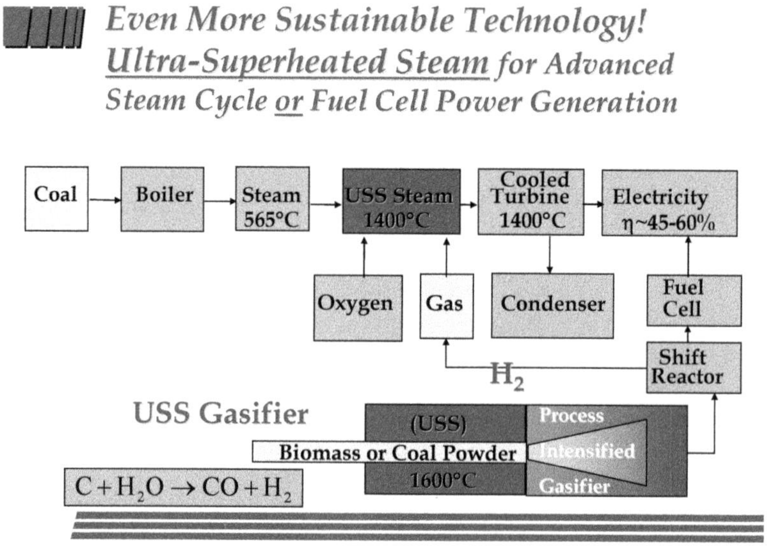

Figure 9: USS Gasification for the Hydrogen Economy.

HOT GAS CLEAN UP BY A LIQUID METAL SCRUBBER SYSTEM

INTRODUCTION

Present integrated gasification combined cycle (IGCC) systems demonstrate high system efficiency and impressive environmental performance, giving these systems an edge over conventional pulverized fuel power stations. However, problems of high capital costs, low reliability and poor operational flexibility (UK DTI, 1998) need to be overcome to make IGCC systems more attractive. This necessitates improvement in various areas of gasification systems, a major one being hot fuel gas cleanup.

Current hot gas clean-up methods focus on the use of metal oxide sorbents for sulphur removal and the utilisation of ceramic barrier filters for particulate cleaning. In contrast, this research project adopts an unconventional approach to fuel gas cleaning technology, incorporating molten tin as scrubbing medium for the removal of H_2S and solid particulates. In the process proposed by Warner, a liquid tin based hot gas scrubbing system addresses this issue. High temperature sulphur removal takes place in a non-wetting, liquid tin irrigated moving packed bed. Sulphur removal occurs via absorption of hydrogen sulphide into molten tin.

$$H_2S + Sn \rightarrow SnS + H_2$$

The reaction should proceed unimpeded as long as the dissolved sulphur in the molten tin is below the saturation level with respect to tin sulphide. To maintain a sufficiently low tin sulphide activity in the molten tin, continuous tin regeneration is effected by external treatment with liquid zinc producing saleable high quality zinc sulphide. In addition to gas desulphurisation, simultaneous solid particulate removal is possible in the packed bed. Molten tin having high surface energy exhibits non-wetting droplet flow on the packing surface, potentially giving good inertial capture of solid particulates by the packed bed from the gas stream.

This process has been studied at SUWIC using cold tests conducted on a non-wetting flow packed bed scrubber and illustrates the application of these data to the design of a small scale molten tin irrigated high temperature scrubber.

EXPERIMENTAL INVESTIGATION

A series of cold tests were conducted on a Perspex model of the gas scrubber operating under non-wetting flow conditions analogous to liquid metal flow. The aims of the cold tests are twofold. Firstly, to study the operational hydraulics of the packed column under non-wetting conditions in order to derive suitable operating parameters and to finalise dimensions for the high temperature system. Secondly, to investigate gas absorption and particulate removal performance under non-wetting flow conditions. In addition, simulation studies on particulate removal in the packed bed under non-wetting flow were carried out using FLUENT. Due to their high surface free energies, liquid metals generally exhibit non-wetting type of flow on solid surfaces. This non-wetting flow behaviour was simulated at room conditions during the cold tests. The room temperature system was therefore irrigated by water for convenience and the selected solid packing material has to be hydrophobic i.e. water repelling to set up an analogous non-wetting room temperature system. This meant that the contact angle of water on the selected solid material has to be greater than 90°. Room-condition tests in the air-water system were performed on an 8 cm internal diameter Perspex column with approximately 19 cm packed bed height. The internal wall of the Perspex column was waxed. Two types of packing were used: 9.5 mm solid high density polyethylene spheres and 10 mm solid soda lime glass spheres treated with paraffin wax.

RESULTS

Flow Visualisation

High-speed digital imaging was performed with a Redlake MotionScope PCI Series monochrome camera to study the non-wetting flow of water on the packing surface. The spherical packing configurations that encourage droplet flow formation were identified. These are sites with a single packing sphere having an immediate space below it and sites at which two or three spheres contact one another. At all these locations, the water droplet grows in size as more water is held up, the additional water flowing either from the top or from within the packed bed. When the droplet diameter reaches approximately 4 mm, it becomes unstable due to its weight and is displaced downwards. This demonstrates the mobility of the static hold-up due to the tendency of water to form droplets rather than a spreading film on the hydrophobic surface of polyethylene and paraffin waxed glass beads. At high water flow, the frequency of droplet formation is sufficiently high such that the individual droplets flow so close to one another that a rivulet flow about 3 to 4 mm width is formed. Generally, rivulet flow became more pronounced at higher fluid flow rates. Since the hot scrubber incorporates spherical ceramic packing dimensionally similar to the polyethylene and waxed glass spheres, the drip tip at the bottom surface of the packing would be dimensionally alike as well. Therefore, the pendent droplet of molten tin at the bottom tip of a ceramic sphere would be expected to be approximately 5 mm. The velocity of the droplets measured in this work assuming vertical downward movement only ranged from 10 mm/s to 37 mm/s.

Flooding Capacity

The flooding capacity of the Perspex column operating under non-wetting flow was investigated. While maintaining a constant water flow, the air flow was increased in a stepwise manner until flooding was achieved, the pressure drops being noted for each air flow after allowing for stabilization of the flow. The flooding points obtained for the water-irrigated packed bed made up of polyethylene spheres as well as paraffin waxed glass spheres were obtained and are plotted in Figure 10. These flooding points are found to be clearly displaced above the now classical generalised flooding correlation of Sherwood, Shipley and Holloway (1938). Sherwood *et al.*'s (1938) correlation was developed using a packed bed of carbon Raschig rings irrigated by water and aqueous organic solutions i.e. essentially wetting systems. On the same plot in Figure 10, the other non-wetting flooding data available in the literature and most importantly, which includes liquid metal irrigated systems has been included for comparison. Good agreement in flooding capacity is demonstrated by all the non-wetting systems. This is despite the fact that the density and surface tension of water (1001 kg/m^3 and 73 dynes/cm respectively at 15°C) are markedly lower than that of the liquid metals (13591 kg/m^3 and 488 dynes/cm for mercury at 15°C). However all the systems have one thing in common i.e. the liquid does not wet the solid surface as the contact angle of the liquid with the solid surface is greater than 90° as follows (Adamson and Gast, 1997): Water on a polyethylene surface: 95°; Water on paraffin surface: 110°; Mercury on glass: 128-148°. This implies that due to the non-wetting condition in the packed bed, the liquid that is held up within the column is lower that that of a wetting system under the same conditions of fluid flow and therefore allowing a greater flooding capacity. This in turn suggests that non-wetting systems can sustain higher fluid velocities than conventional wetting systems before the onset of flooding, hence a smaller column would be viable for a given set of fluid volume flow rates. Also, from Figure 10 it can be seen that the flooding capacity of the non-wetting systems is in general slightly higher for larger contact angles i.e. the more pronounced the non-wetting behaviour.

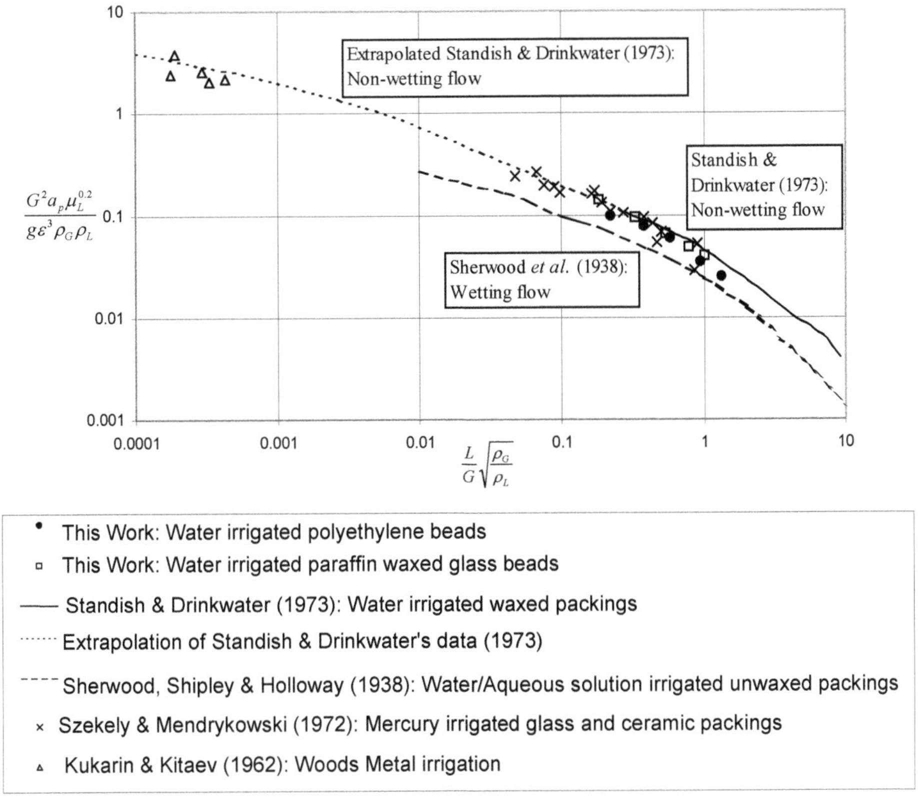

Nomenclature:

G	= gas mass velocity (kg/m^{2}s)
L	= liquid mass velocity (kg/m^{2}s)
g	= gravitational acceleration (m/s^2)
a_p	= specific surface area of packed bed (m^2/m^3)
ε	= bed porosity (-)
μ_L	= liquid viscosity (cP)
ρ_G	= gas density (kg/m^3)
ρ_L	= liquid density (kg/m^3)

Figure 10: Comparison of flooding points under non-wetting and wetting conditions in a packed column.

Liquid Hold-up

In a packed column, the liquid hold-up refers to the liquid retained in the bed of packing i.e. films or droplets on the surface of the packing or the liquid trapped in the interstitial space between the packings. The total liquid hold-up is the total amount of liquid on the packing under dynamic conditions i.e. the sum of the static hold-up and dynamic hold-up. Static hold-up is the volume of liquid per volume of packing that remains in the packed bed that has been fully wetted after the gas and liquid flows have stopped and the bed has drained. The fraction that drains out is the dynamic hold-up.

The static and dynamic liquid hold-ups for the Perspex column were determined. For both hold-up measurements, sufficient time was allowed for the flow through the column to stabilise.

The dynamic hold-up was determined by weighing the amount of water that drained from the bed. The static hold-up was determined by the difference in the weight of the dry packing and that of the wet packing after the bed drained.

The results of the liquid hold-up tests using water-irrigated polyethylene beads are illustrated in Figures 11, 12 and 13. There was a small but definite decrease in static hold-up as water flow increased. The static hold-up varied from an average value of 3.46 to 3.03 vol % of packing as water velocity increased from 0.35 to 0.87 cm/s. The results of this work demonstrate that the static hold-up for a packed column operating under non-wetting flow conditions is actually affected by water flow rates. A likely reason for this difference is that the water droplets held in the interstices of the packing in non-wetting flow are more easily displaced at higher water flow rates than the semi-stagnant pools formed in wetting flow conditions. This hypothesis is further supported by the observed mobility of static hold-up in the high-speed video footage. The results herein suggests that the static hold-up in a non-wetting flow packed bed system has a relatively more effective contribution to mass transfer in a gas absorption process than a conventional wetting system. The dependence of static hold-up on liquid flow rate found in this work is in agreement with the reported static hold-up trend for liquid metal irrigation. The dynamic hold-up for the water irrigated polyethylene packing was found to increase with both the air and water flow rates, with a greater dependence on the liquid rate. The onset of flooding is accompanied by a sudden rapid rise in dynamic hold-up. This observation generally agrees with the dynamic hold-up trend of wetting systems incorporating aqueous solutions and organic liquids.

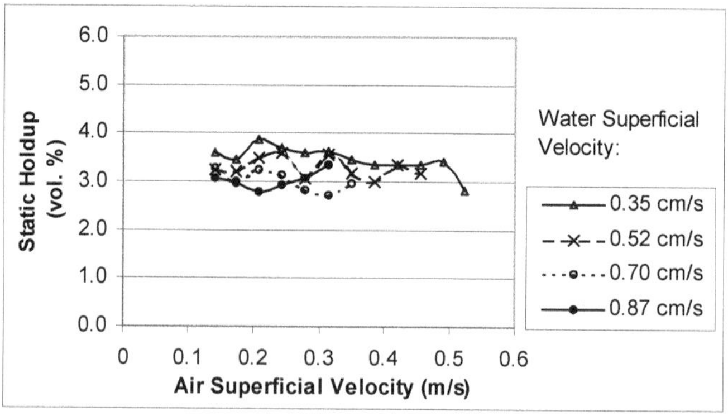

Figure 11: Variation of Static Liquid Hold-up with Air and Water Flow-rate.

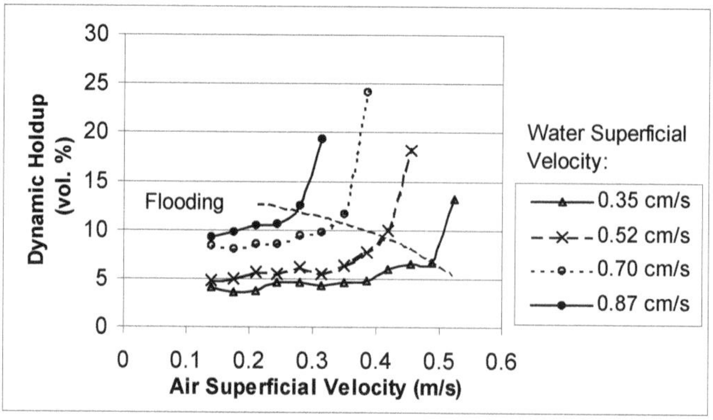

Figure 12: Variation of Dynamic Liquid Hold-up with Air and Water Flow-rate.

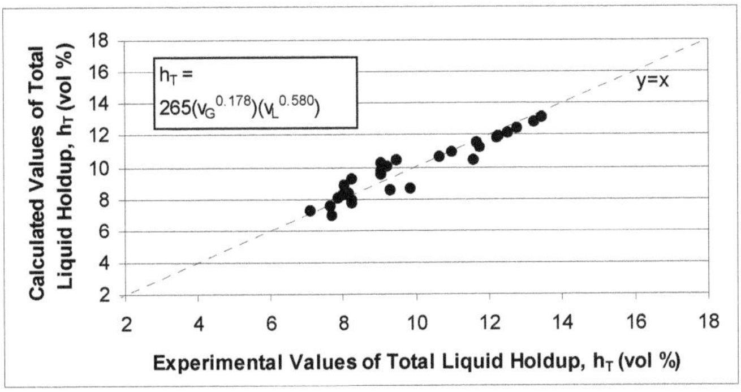

Figure 13: Comparison of Calculated and Experimental Values of Total Liquid Hold-up.

Despite the pronounced differences in liquid density and surface tension, the liquid hold-up trends especially that of the static hold-up found in this work for water irrigated polyethylene packing are in accordance with those reported for liquid metal irrigation. This suggests that for a packed column operating under non-wetting conditions, the liquid-solid packing contact angles exert a dominating influence not only on the flooding capacity but on the liquid hold-up as well.

GAS LIFT

Water circulation through the Perspex packed column via a small air-lift pump was investigated. The dimensions of the air lift are as follows: 1.59 cm internal diameter piping for riser and downcomer, 45 cm lift, 65 cm submergence hence 0.59 submergence ratio (i.e. submergence to total height of air lift). Three different gas injectors were tested: a 5 mm i.d. longitudinal injector, a radial injector with four circumferentially equidistant 2.54 mm air ports and a radial injector with twelve 2.54 mm air ports arranged in 3 rows (4 in each row, circumferentially equidistant). The water discharged at the top of the air lift was collected in a recorded period of time and then weighed. Three readings were taken for each air flow and the average water flow rate determined. It was found that there was negligible difference in the pumping performance of the air lift with these 3 different air injectors as shown in Figure 14. This result is in contrast to the reported improved pump efficiency with multi-port injectors in a larger air lift (Morrison *et al.*, 1987).

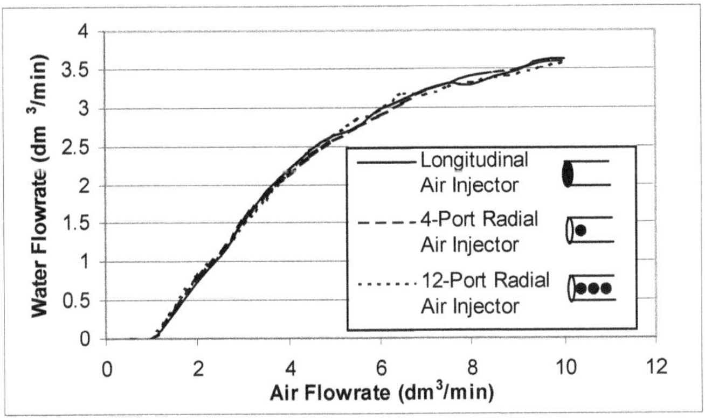

Figure 14: Air Lift Performance Chart with Varying Injectors.

The air lift experimental data obtained in this work is plotted on the axes of Zenz' graphical correlation for the design of air lift pumps (Zenz, 1993). The ordinate includes factors of fluid density, viscosity and surface tension relative to those of water to enable the application of the air lift graphical correlation for other fluids to be pumped apart from water. The experimental data curve obtained from the tested room temperature air lift and plotted on Zenz's axes was used to predict the pumping capacity of the nitrogen gas lift on molten tin in the hot scrubber.

PARTICULATE REMOVAL

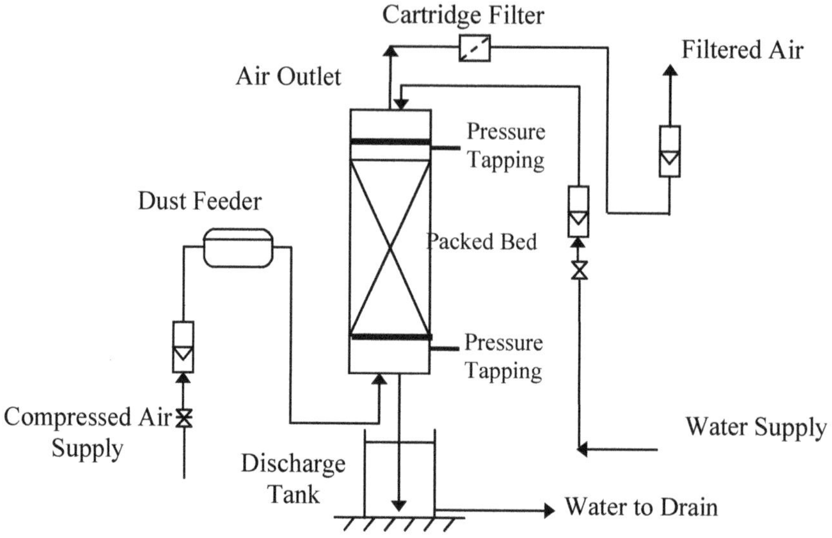

Figure 15: Experimental Setup to Determine Total Efficiency of Dust Removal.

Particulate removal performance of the water-irrigated column packed with polyethylene spheres was investigated using a glass dust laden air stream. The glass dust consists of spherical particles ranging from 1.5 to 35 μm with a uniform density of 1100 kg/m^3. The particle size distribution of the glass dust was determined on the AccuSizer 780A Particle Sizing System. The size distribution of the glass dust resembles very closely a log-normal distribution with mass median diameter of 10.2 μm and geometric standard deviation of 1.8.

Figure 15 illustrates the experimental setup used to determine the total efficiency. A Malvern Instruments' Qspec Dry Powder Feeder was used for dust loading into the inlet air stream. The penetrating glass dust was filtered from the exiting air stream using a Whatman in-line cartridge filter with 1.0 μm particle retention capability. This filter type was chosen for its exceptionally high particle retention and high loading capacity. Isokinetic sampling was not feasible on this small-scale scrubber due to the small flow area of the exit line employed (about 1.5 cm internal diameter). The duration of each test run was approximately 10 to 15 minutes. At the end of the each run, the cartridge filter was removed from its housing, dried in the oven and weighed. The total efficiency of the scrubber, E_T is determined gravimetrically as follows:

$$E_T = 1 - \frac{m_{out}}{m_{in}}$$

where m_{out} = mass of unseparated dust retained in filter (g)
m_{in} = mass of dust fed into the scrubber (g)

The penetrating dust particle size distribution was determined to obtain the grade efficiency. The dust in the exiting air stream was collected into water in an impinger. Samples of the solid in suspension from the impinger are then run through the AccuSizer 780A Particle Sizing System to determine the particle size distribution. The experimental setup used is illustrated in Figure 16.

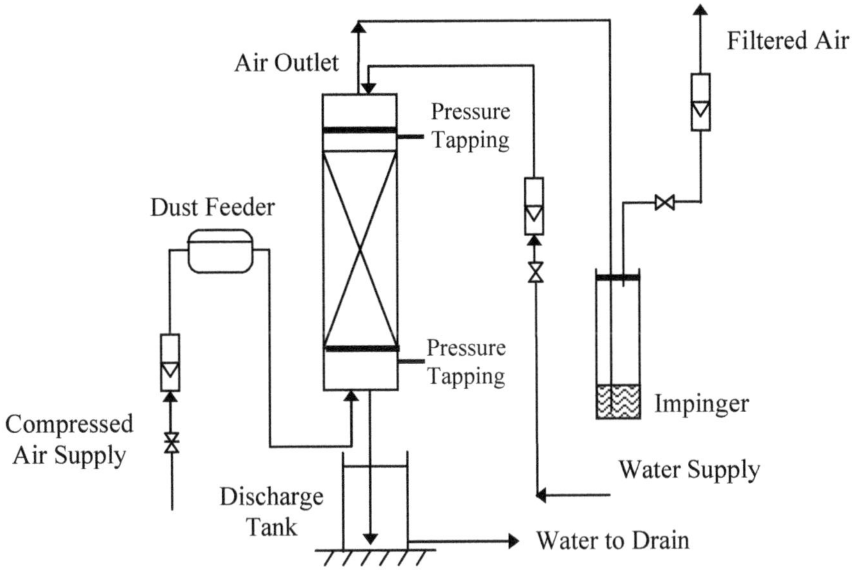

Figure 16: Experimental Setup to Determine Grade Efficiency of Dust Removal.

By controlling the needle valve downstream of the impinger, it was possible to operate at the same column pressure drop with the same operating fluid flow rates as when the filter was in place. This ensured that dust collected in the impinger would be representative of the dust collected on the filter at the same operating fluid flow rate. One impinger was found to be sufficient for full collection of the dust in the exit air stream. The grade efficiency, G(x) is determined gravimetrically as follows:

$$G(x) = 1 - (1 - E_T)\frac{\omega_{out,i}}{\omega_{in,i}}$$

where $\omega_{out,\,i}$ = mass fraction of unseparated dust retained in filter in size range i
$\omega_{in,\,i}$ = mass fraction of dust fed into the scrubber in size range i

The dust feed rate was adjusted at each air flow rate to give an average inlet dust concentration ranging from approximately 500 to 3000 ppmw. The operating air superficial velocity ranged from 18.49 to 29.65 cm/s (measured downstream of the column at atmospheric

Total efficiency E_T increases with air and water flow rate as well as inlet dust concentration. At higher air or water flow rate, the total liquid hold-up in the packed bed increases as shown in the earlier liquid hold-up tests. The droplets and rivulets in the bed increase, therefore improving the deposition of solid particulates through inertial capture and direct interception. High dust loading gives better overall dust separation in the water irrigated packed bed. This is because the moisture encourages agglomeration of dust particles, resulting in larger overall particle sizes and hence improved dust separation from the air stream. The importance of the presence of waterin the packed bed to prevent re-entrainment of dust particles into the air stream is demonstrated by the experimental data curve at zero water flow rate in Figure 18. At a dry bed condition, dust separation decreases with increasing dust concentration. At the start, the dust introduced into the bed will stick to the clean packing surfaces mainly due to electrostatic force generated between the dry dust and the polyethylene spheres. The tendency for electrostatics build up is high due to the relative positions of glass and polyethylene in the triboelectric series. However, the available clean packing surfaces decline rapidly as more dust enters the column. Having found no way to be deposited and retained in the packed bed, the additional dust is swept along with the air stream out of the column. The dry bed has only a certain dust retention capacity. Hence as the inlet dust concentration increases, a higher proportion of the dust penetrates and escapes the packed bed, reducing the overall separation efficiency.

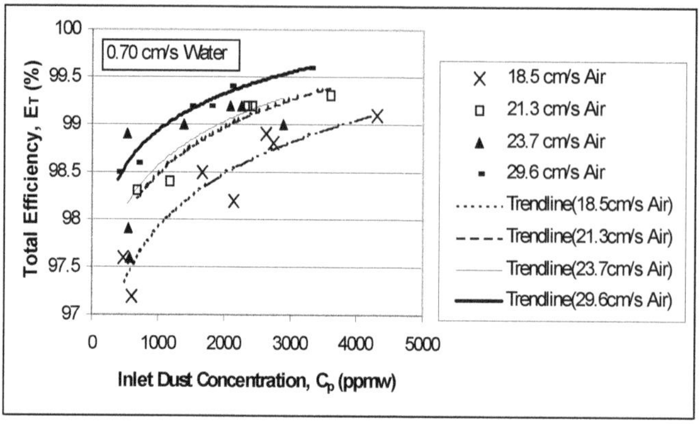

Figure 17: Variation of Total Efficiency with Dust Concentration.

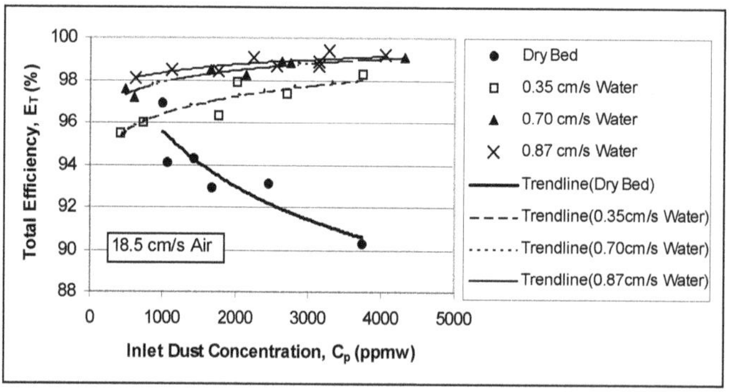

Figure 18: Variation of Total Efficiency with Dust Concentration and Water Flow rate.

Using multiple linear regression analysis, the following correlation is obtained expressing the total efficiency in terms of air and water superficial velocities as well as inlet dust concentration.

$$E_T = 103.20(v_G^{0.025})(v_L^{0.013})(C_p^{0.008})$$

where v_G = air superficial velocity (m/s)
v_L = water superficial velocity (m/s)
C_p = inlet dust concentration (ppmw)

The water-irrigated scrubber demonstrates impressive dust removal performance with total separation efficiency ranging from 95.5 to 99.6%. Cut sizes approaching sub-micron is achievable as shown by the grade efficiency plots in Figure 19.

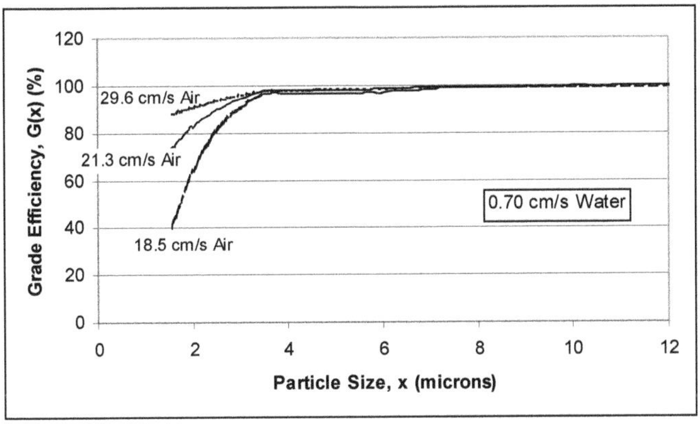

Figure 19: Grade Efficiency Curves.

Compared to the cold scrubber, an even better dust removal performance can be expected of the molten tin irrigated hot scrubber. This is because besides the inertial capture of solid particulates from the gas stream by molten tin droplets and rivulets, the adhesiveness of ash and char particles at elevated temperatures will also encourage deposition and retention on the packing surfaces.

Table 2: Characteristics of packings used.

Packing	Diameter (mm)	Specific Gravity	Specific Surface Area (m^2/m^3)	Porosity	Column to Packing Diameter Ratio
Solid polyethylene beads	9.5	0.96	630	0.45	8.4
Solid glass beads treated with paraffin wax	10.0	2.5	600	0.45	8.0

FUTURE ACTIVITIES

Results from the cold tests have provided insight and understanding for the development of a liquid metal irrigated packed bed scrubber. The design of the hot gas scrubber has been finalised and its construction completed. The next phase of the project will be to investigate the performance of the molten tin irrigated packed bed scrubber in removing H_2S and solid particulates from a simulated hot fuel gas.

REFERENCES

Adamson, A.W. and Gast, A.P. (1997) *Physical Chemistry of Surfaces*. 6th edition. New York, John Wiley & Sons, Inc.

Kukarin, A.S. and Kitaev, B.I. (1962) *Izv. V.U.Z.Chern. Met.*, 12, 20. Cited in: Standish, N. (1973) On the flooding criteria for liquid metals. *Chemical Engineering Science*, 28, 1906-1907.

Morrison, G.L. *et al.* (1987) Experimental analysis of the mechanics of reverse circulation air lift pump. *Ind. Eng. Chem. Res.*, 26, 387-391.

Sherwood, T.K., Shipley, G.H. and Holloway, F.H. (1938) Flooding velocities in packed columns. *Industrial Engineering Chemistry*, 30(7), 765-769.

Szekely, J. and Mendrykowski, J. (1972) On the flooding criteria for liquids with high density and high interfacial tension. *Chemical Engineering Science*, 27, 959-963.

Standish, N. (1968a) Dynamic hold-up in liquid metal irrigated packed bed. *Chemical Engineering Science*, 23, 51-56.

Zenz, F.A. (1993) Explore the potential of air-lift pumps and multiphase. Chemical Engineering Progress, 89(2), 51-56.

Kinetics of Rice Husk, Straw and Bark Gasification

L. Van de steene, P. Girard
Centre de Coopération Internationale de Recherche Agronomique pour le Développement (CIRAD), TA 10/16, 34398 Montpellier, Cedex 5, France
S. Salvador
Laboratoire de Génie des Procédés des Solides Divisés, UMR 2392 CNRS, Ecole des Mines d'Albi-Carmaux (EMAC), Route de Teillet, 81013 Albi Cedex 09, France

ABSTRACT: Rice husk, wheat straw and pine bark behaviour during gasification have been characterised and compared to a bituminous coal. The main reactions involved have been studied:

- devolatilisation (or pyrolysis) of the raw biomass;
- oxidation of the raw biomass;
- charcoal oxidation (attack by O_2);
- charcoal gasification (attack by CO_2).

Two reactors have been used. Dynamic Thermo-Gravimetry reproduces the low heating rates present in reactors such as fixed beds or rotary kilns. An entrained flow reactor was used to simulate the high heating rates encountered in fluidised beds, entrained beds and flash processes. Care was taken to remain in conditions where no heat transfer or mass transfer limitation is effective, so that intrinsic kinetic parameters could be determined. Quantitative values for the reaction rates in temperature domains relevant to industrial applications are derived.

INTRODUCTION

Biomass is one of the most promising renewable energy [1,2], and gasification technology is an attractive route for the production of fuel gases from this natural resource [3]. Up to now, many gasifiers have been designed and tested at different scales, but no significant commercial units have been realised. This situation is mainly due to the fact that design of such reactors is quite complex and relevant information is needed; particularly when biomass is concerned.

As illustrated in figure 1, a gasification process consists of a sequence of chemical reactions, mainly:

- pyrolysis, during which the solid is converted into gases (hydrocarbons, H_2, CO, CO_2, H_2O,...) and a residual charcoal;
- partial heterogeneous oxidation (of the solid fraction) that supplies the energy to heat the reacting system;
- gasifications, that convert the charcoal into CO and H_2 by reactions with CO_2 and H_2O.

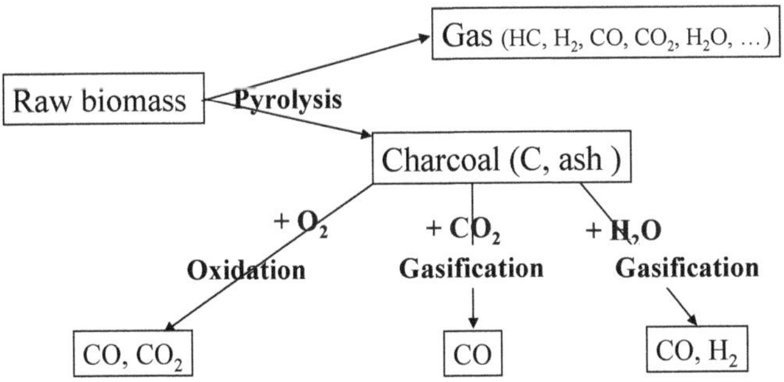

Figure 1: Main chemical reactions and products for a gasification process.

The objectives of this work were to determine the kinetic parameters for three representative biomass fuels, i.e. rice husk, straw and bark, along three different steps:

- pyrolysis, also termed devolatilization;
- heterogeneous (of the charcoal) oxidation by O_2, also termed combustion of the charcoal;
- heterogeneous oxidation by CO_2, here termed gasification.

Experiments were conducted with two apparatus:

- a thermo-balance (Thermo-Gravimetry Analysis, TGA) was used to represent low heating rate processes such as fixed beds or rotary kilns;
- an Entrained Flow Reactor (EFR) in order to reproduce the high heating rates present in processes such as fluidised beds or flash processes.

The experimental results are first compared in terms of the influence of temperature for each biomass during each reaction. Their behaviour is compared to a coal as reference.

The Arrhenius representation is then used in order to put the results together, and analyse the differences in reaction kinetic parameters between the different fuels along the different reactions inside the two reactors. Quantitative practical values for the reaction rates are proposed.

EXPERIMENTAL

BIOMASS

Rice husk and straw are the main agricultural crops in the world; pine bark is an important residue from wood industries. Bark from maritime pine (Pinaster), rice husk from Camargue (south-east of France) and wheat straw (South of France) have been studied. A bituminous coal (Carmaux, France) has also been tested for reference to a "standard" fuel. For all the fuels, the particle sizes were between 50 and 315μm. This fraction was selected by pneumatic sieving.

The upper limit cut-off avoids heat or mass transfer limitations as the reactions progress. This is necessary to achieve "intrinsic" kinetic parameter values;

The lower limit cut-off is to enable easy feeding and pneumatic transport during the experiments.

The proximate and ultimate analysis for the four fuels are given in table 1. Standard NF M03-003, -004 and NF B 55-101 have been used for respectively coal and biomass proximate analysis.

We have checked that the chemical composition of the 50-315μm fraction (that we used) was similar to that of the bulk material.

Table 1: Proximate and ultimate analysis of biomass and coal.

		Rice Husk	Wheat Straw	Pine Bark	Bituminous Coal
Proximate analysis (% dry)	Volatile	61.8	69.0	66.1	23.4
	Ash	15.1	12.0	3.2	19.1
Ultimate analysis (% dry)	C	42.5	40.3	51.7	68.8
	H	5.4	5.3	5.7	4.1
	O	35.7	42.5	38.7	5.7
	N	0.7	1.3	0.6	1.4
	S	0.05	0.0	0.0	0.9

The charcoals used for the characterisation of charcoal combustion and gasification were prepared in a fixed bed of approximately 200g, heated under nitrogen for 2 hours at 800°C. The residual volatile matter content was less than 6%.

ENTRAINED FLOW REACTOR (EFR)

The 80 mm diameter by 2 m long Entrained Flow Reactor is fully described elsewhere [4]. Briefly, the powdered solid fuel was fed through a water cooled probe to a preheated atmosphere gas flow, and crossed the isothermal zone. It was then sampled, together with the gas phase, thanks to a water-cooled nitrogen-quenched probe, that collected most of the solid fraction and stopped the chemical reactions. The controls of the three furnace wall heaters were set to maintain a uniform wall temperature along its length. The solid mass flow-rate was 1g/min. We have shown this value to be the upper limit in order to avoid thermal and chemical interactions between each particle and so to reach an isolated particle situation. This condition was necessary to assess the intrinsic kinetic parameters of chemical reactions [4].

Two types of experiments were carried out in this study for the 3 biomass and the coal listed above.

- A first set of experiments was realised in a nitrogen atmosphere in order to study the devolatilization reaction alone.
- The second set of experiments consisted of combustion tests of the raw biomass in a O_2/N_2 (air) atmosphere.

For each experiment, the mean gas flow velocity was kept constant to 0.55 m/s, which resulted in gas flow rates between 42 and 67 Nl/min depending on the temperature. The particle residence time in the 900mm long reaction zone was estimated to 1.64 s assuming no slip velocity for biomass particles; this is acceptable when considering the very low particle densities. Gas sampled from the reactor was continuously analysed in terms of O_2, CO, CO_2, NO, SO_2, CH_4 and total hydrocarbons.

THERMO-GRAVIMETRIC ANALYSIS REACTOR (TGA)

A Setaram 92-16.18 analyser was used. Experiments were carried out at atmospheric pressure:

- under N_2 atmosphere for pyrolysis experiments;
- under a N_2/O_2 (air) mix for charcoal oxidation experiments;
- under CO_2 for charcoal gasification experiments.

The atmosphere gas flow rate was 1 l/h. The runs were non isothermal (dynamic). Previous experiments have shown that both a low mass of solid and a small heating rate were necessary to avoid heat and mass transfer limitations within the crucible or within the sample bed [5, 6]. The adopted sample mass was between 5 and 7 mg, and the heating rate was 3 K/min. The temperature was increased until no further weight loss was observed. A thermocouple placed directly in contact with the crucible gave a precise registration of the temperature of the sample at any instant.

RESULTS AND DISCUSSION

ENTRAINED FLOW REACTOR (EFR)

Pyrolysis experiments were performed with the 4 solid fuels, in isothermal conditions, for the temperatures of 400, 500, 600, 700 and 800°C. After each test, the weight loss of the solid was measured following the ash tracer method. The results are presented in figure 3 as the particles weight loss (on a dry ash free basis, or "daf"), versus the reaction temperature. The residence time was 1.64 s for all experiments.

Bark, rice husk and straw show very similar weight losses versus temperature. Coal appears to devolatilise more slowly than all biomass. The differences in volatile matter content (table 1) between biomass and coal can partly explain the differences observed in figure 3. However it clearly appears that the pyrolysis reaction starts at lower temperature with biomass, below 400°C, while nothing happens at 500°C with coal.

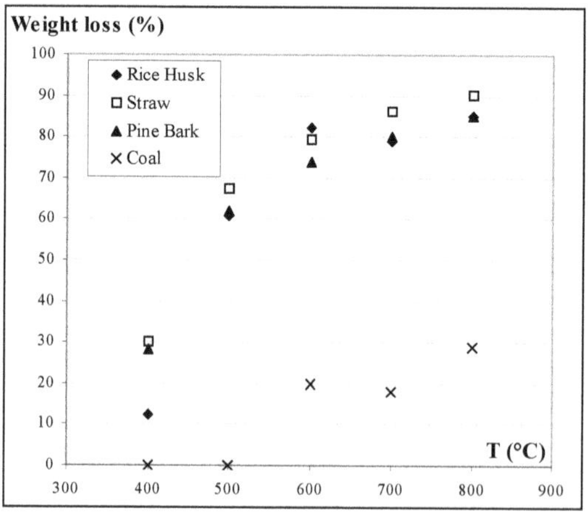

Figure 3: Weight loss (daf) versus temperature during pyrolysis experiments in the EFR (N_2 atm.; 1.64s residence time).

It is also interesting to note that, at 800°C, the total yield of volatile matter is higher in EFR experiments given in table 2 than in the standard tests (note that they have been corrected to an ash free basis). This result has been observed by numerous authors for coal, during flash pyrolysis [7].

Table 2: Volatile yield for EFR experiments and standard tests.

	Standard test* % DAF	EFR** % DAF	EFR / Standard
Rice Husk	73	85	1.16
Straw	78	90	1.15
Pine Bark	68	85	1.25
Coal	22	29	1.32

* 2h at 950°C ** 1,64s at 800°C

Combustion experiments were also performed in isothermal conditions. The solid mass flow rates were so low that the decrease in oxygen content was not significant (3 % max). Results presented in figure 4 show the weight loss of the studied fuels in a 21 % O_2 atmosphere versus temperature. For the three considered biomass (figures 4a, 4b, 4c) the presence of oxygen plays a significant role on the weight loss progress only for temperatures above 500°C, where a difference between pyrolysis and combustion results is observed. The same behaviour exists when coal is considered (figure 4d) but at a higher temperature (600-700°C). Moreover, at 600°C and for 1.64s residence time in the reactor, biomass combustion can be considered as completed (>95%) while only 70 % of conversion was reached at 800°C with the bituminous coal.

THERMO-GRAVIMETRIC ANALYSIS

We have plotted in figure 5 the evolution of the (normalised) mass of the samples versus the sample temperature, along the 3 K/min heating. The oxidation of the raw biomass (fig. 5a) is achieved at temperatures lower than 500°C.

If we consider the charcoal oxidation reactions (fig. 5b), they appear to be completed at temperatures higher by about 50°C than the respective raw biomass oxidation.

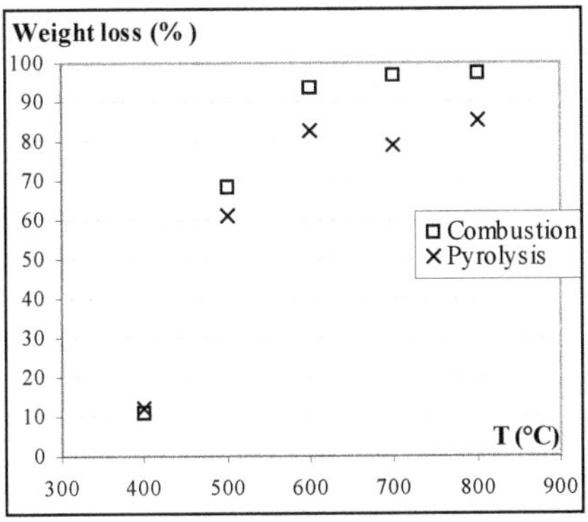

Figure 4a: Rice husk

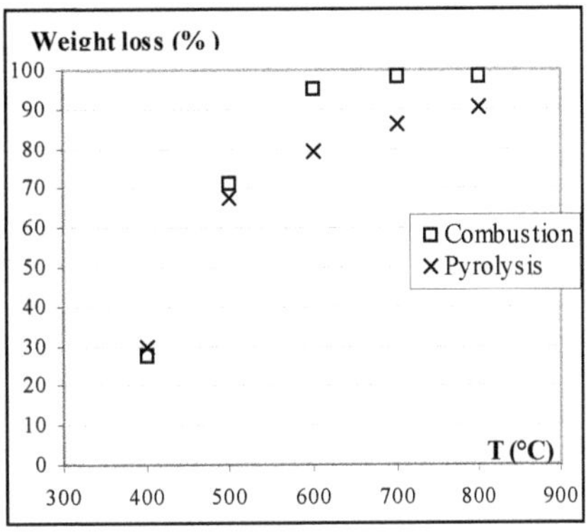

Figure 4b: Straw

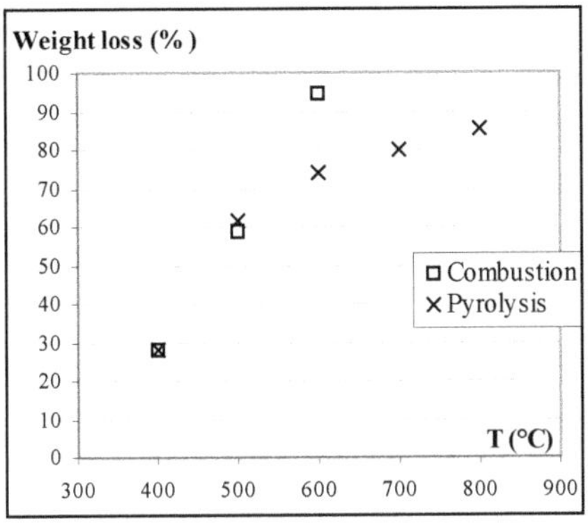

Figure 4c: Pine bark

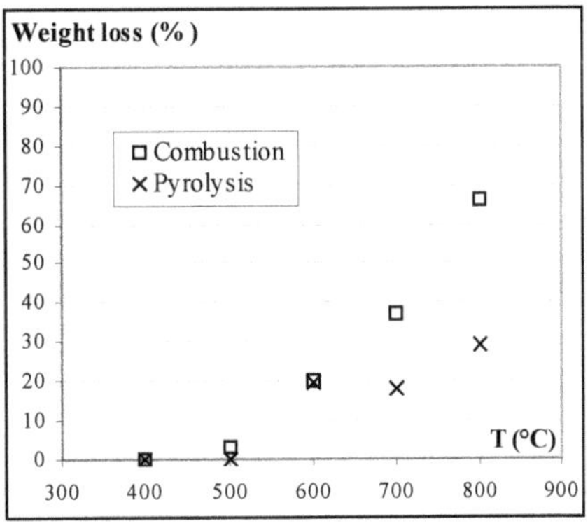

Figure 4d: Coal

Figure 4: Weight loss (daf) versus temperature during coal combustion in EFR ($[O_2]$ = 21 %; Residence time = 1.64s). Comparison with pyrolysis experiments (in N_2, Residence time = 1.64s).

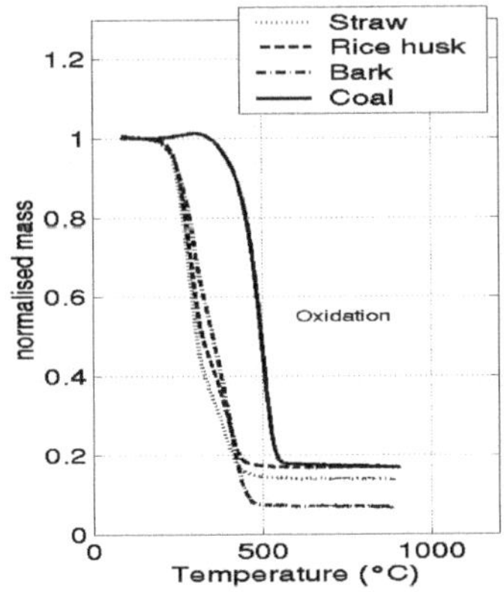

a

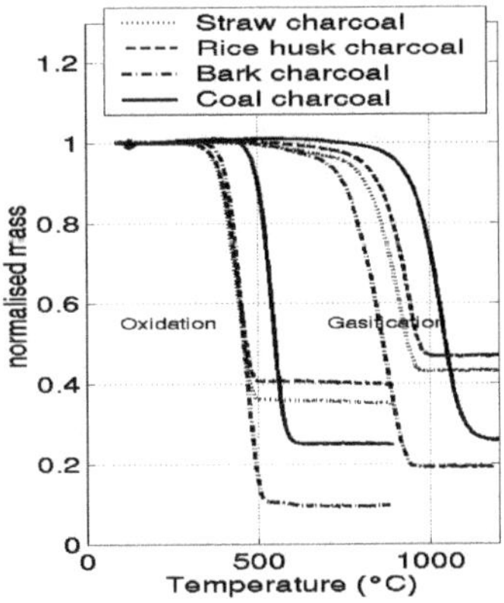

b

Figure 5: Normalised mass versus temperature during TGA experiments (3 K/min).

The attack by CO_2 (gasification) of the charcoals (fig 5b) is completed at higher temperatures than the (respective) attack of the same charcoals by O_2: 900 to 1100°C. If we compare the solid residues after the attack of charcoals by air and CO_2, we can note a slightly higher quantity after CO_2 attack than after air attack.

For the three reactions studied (raw fuel under air, charcoals under air and charcoals under CO_2), biomass again appear to be more reactive than the coal. Bark is less reactive than straw and rice husk during attack by O_2. This order is reversed when CO_2 attack (gasification) is considered.

KINETIC CONSTANT DETERMINATION AND DISCUSSION

It is possible to consolidate the results using the Arrhenius representation. It consists in plotting the reaction rates, on a log scale, versus the reverse of temperature. The reaction rates are to be derived from a model to describe the mass evolution. The model suits the reaction if the plot is a straight line.

Pyrolysis reaction has been described by a single step, following the classical "homogeneous" model. It expresses that the reaction velocity (dm/dt) depends on the remaining mass (m) and a reaction rate (k), that can be expressed from the temperature (T) and the kinetic constants A (frequency factor) and E (activation energy):

$$\frac{dm}{dt} = -k_{pyr} . m \qquad \text{(Eq. 1)}$$

with $k_{pyr} = A_{pyr}\, e^{-\frac{E_{pyr}}{RT}}$

Note that k is in kg lost per kg reacting and per second, or [s^{-1}].

As far as oxidation or gasification are concerned, a classical model has the form:

$$\frac{dm}{dt} = -k_{comb} . S . P_{O_2}^n \qquad \text{(Eq. 2)}$$

with $k_{comb} = A_{comb}\, e^{-\frac{E_{comb}}{RT}}$

where S is the reacting pore surface.

Identifying this surface remains subject to discussions. The BET surface has been used for a long time, but more recent work has shown that the micro-pores did not participate to the reaction [8, 9]. In order to avoid this uncertainty, we will adopt the following "homogeneous" model:

$$\frac{dm}{dt} = -k_{comb} . m . P_{O_2}^n \qquad \text{(Eq. 3)}$$

This model also has the advantage that it will enable the direct comparison between oxidation and gasification reaction rates. Indeed k_{comb} is in [$s^{-1}.atm^{-1}$], which is equivalent to [s^{-1}] at 1 atm.
The reaction order towards oxygen will be taken as 1 for simplification. Values ranging from 0.6 to 1 are found in the literature for various charcoals. Charcoal gasification will be described in the same way following:

$$\frac{dm}{dt} = -k_{gas} . m . P_{CO_2}^n \qquad \text{(Eq. 4)}$$

with $k_{gas} = A_{gas}\, e^{-\frac{E_{gas}}{RT}}$

We have plotted in figure 6 the reaction rates during TGA for the reactions:

- attack of the raw fuels by oxygen (air)
- attack of the charcoals by oxygen in the TGA
- attack of the charcoals by CO_2

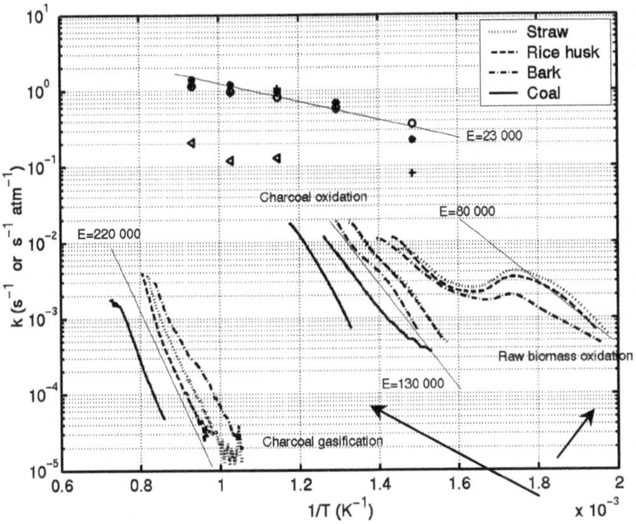

Figure 6: Reaction rates for the reactions of raw biomass devolatilisation, charcoals oxidation and charcoals gasification during DTA. Results from EFR devolatilisation experiments (symbols) have been added: (*) Straw, (+) Rice husk, (o) Bark, (Δ) Coal. E: activation Energy.

The values of k were calculated, for EFR experiments, from the integration of Eq. 1 giving:

$$Ln\left(\frac{m}{m_0}\right) = -k.t$$

with $\frac{m}{m_0} = 1 - \frac{Weight\ loss\ measured\,(\%)}{100}$

For TG experiments, k is calculated along the reaction from dm/dt and m that are continuously recorded during the tests. Only the values of k calculated when the mass loss was between 5% and 95% have been plotted for better readability.

Some additional observations can be made from this representation. The curves for raw biomass attack by oxygen shows 2 distinct straight zones; the low temperature reaction zone (on the right hand side) corresponds to the pyrolysis step, while the higher temperature zone (on the left hand side) results from the residual charcoal oxidation. This last reaction rate is about twice that of the reaction rate of charcoals under the same conditions. An explanation could be that the charcoals from the raw biomass still contain volatiles that are released in parallel. Quantitative comparisons of reaction rates can also be made:

- For all the reactions studied here, the biomass charcoals appear to react - for a given temperature - approximately 10 times faster than the coal charcoal.
- The attack by CO_2 is more than ten thousand times slower than the attack by O_2 (comparing the reaction rates at a given temperature, for a pure oxygen atmosphere and a pure CO_2 atmosphere).

The models we have adopted to describe the attack of the charcoals by O_2 and CO_2 appear to be satisfactory regarding the relatively straight lines we obtain along 5% to 95% mass losses (apart from the experiments of the raw biomass in air during which both devolatilisation and oxidation occurred). The activation energy for charcoal oxidation is circa 130000 $J.mol^{-1}.K^{-1}$ and 220 000 $J.mol^{-1}.K^{-1}$ for charcoal gasification by CO_2.

We have also reported in figure 6 the calculated values of the reaction rates for the flash devolatilisation of the four raw fuels in the EFR. These results can be compared with the soak devolatilisation rates (during DTA, on the right hand side of the curve). A difference in activation energy (slope of the curve) is observed: the activation energy of 80 000 $J.mol^{-1}.K^{-1}$ for soak devolatilisation tends towards 23 000 $J.mol^{-1}.K^{-1}$ in flash conditions. This indicates the probable occurrence of heat or mass transfer limitations during experiments at high temperatures, resulting in a low sensitivity to temperature.

CONCLUSION

The behaviour of bark, rice husk and straw biomass have been quantitatively characterised in terms of intrinsic reaction rates in a temperature range relevant for industrial applications. These rates can be used either directly to predict the reaction velocity of solids in the cases where no heat or mass transfer occur (small isolated particles at low temperatures), or in models solving heat and mass transfer equations in the case of large solids such as pellets, bricks or fixed beds.

During the devolatilisation of the raw biomass, the reaction rates are similar for all of them, despite the fact that their origins are very different. Bark appears to be the less reactive of the three biomass types. The reactivities of biomass are much (10 times) higher than that of our reference coal.

We have observed, despite the small size of the particles and the good dispersion in the atmosphere gas, that fuels devolatilisation is heat and/or mass transfer limited during experiments in the EFR.

Oxidation of the biomass charcoals gives the same trends. A ratio of about 10 is again observed for the reactivities to O_2 of the biomass charcoals and of the coal charcoal. The activation energies are closed to 130.000 $J.mol^{-1}.K^{-1}$.

During the gasification of biomass charcoals, pine bark shows a reactivity to CO_2 two times higher than wheat straw and around five time higher than rice husk, bark charcoal was approximately two times less reactive to O_2 than the two other biomass charcoals. The biomass globally exhibit reactivities 10 times higher than our reference coal. High activation energies (220 000 $J.mol^{-1}.K^{-1}$) were found.

The attack of charcoals (carbon) by pure CO_2 is about 10 000 times less rapid than the attack by pure O_2.

REFERENCES

1. Strehler A. 2000. Potential and global aspects of bioenergy application. 1st world conf. On biomass for energy and industry, Sevilla, Spain: 191-193.
2. Werther J, Saenger M, Hartge EU, Ogada T, Siagi K. 2000. Combustion of agricultural residues. Progress in Energy and Combustion Science 26: 1-27.
3. Reed TB. World state of gasification. Basic gasification studies for development of biomass medium-BTU gasification process. 1997. 3rd biomass conference. of the Americas, Montréal, Canada: 589-595.
4. Van de Steene L, Salvador S, Charnay G. 2000. Controlling powdered fuel combustion at low temperature in a new entrained flow reactor. Combustion Science and Technology 159: 255-279.
5. Morgan PA, Robertson SD, Unsworth JF. 1986. Combustion studies by thermogravimetric analysis 1. Coal oxidation. Fuel 65: 1546-1551.
6. Chen Y., Mori S. Estimating the combustibility of various coals by TG-DTA. Energy and fuels. 1995. 9: 71-74.
7. Therssen E, Gourichon L, Delfosse L. 1995. Devolatilization of coal particles in a flat flame – experimental and modelling study. Comb. and Flame 103: 115-128.
8. Valix MG, Harris DJ, Smith IW, Trimm DL. 1992. The intrinsic combustion reactivity of pulverised coal chars: the use of experimental pore diffusion coefficients. 24th Symposium (International) on Combustion, The combustion Institute. 1217-1223.
9. Valix MG, Trimm DL, Smith IW, Harris DJ. 1992. Mass transfer effects in coal combustion. Chemical Engineering Science 47. 7:1607-1617.

Strategies for Development and Implementation of Fast Pyrolysis of Biomass

A.V. Bridgwater
Aston University Bio-Energy Research Group
Chemical Engineering and Applied Chemistry
Birmingham B4 7ET, UK
M. Grønli
SINTEF Energy Research, Thermal Energy, 7465 Trondheim, NORWAY

The workshop was supported by a panel of the following: T Bridle, ESI; S Czernik, NREL; P Girard, CIRAD; R Guido, Dynamotive; S Gust, Fortum; whose contributions are much appreciated.

FAST PYROLYSIS STATUS

The objective of the workshop was to identify and prioritise what is needed to improve the rate and extent of both penetration and success of fast pyrolysis technology into the market.

A number of fast pyrolysis processes are under industrial development at pilot or demonstration scale or available commercially from BTG in the Netherlands, Dynamotive in Canada, Ensyn in the USA, Fortum in Finland and Wellman in the UK. In addition there is extensive research and development at universities, research institutes and companies around the world. A wide range of applications have been developed and demonstrated for power, heat, fuel and chemicals.

Implementation of fast pyrolysis focuses on three main areas that were discussed and reviewed:

1. Feed materials,
2. Process of fast pyrolysis to liquid,
3. Utilisation of liquid.

FEED MATERIALS

FEED SUPPLY

The importance of the provisions for the supply of feedstocks as forestry materials, energy crops and agricultural by-products and wastes are often underestimated. Long term contracts are important to ensure a secure supply of feedstock. Transport distance is an important consideration to minimise cost and environment impact and feedstock supply and logistics is important when scaling-up. The cost of feed, which is a major contributor to the cost of bio-oil, needs to be minimised. In particular the supply chain needs to be developed and harvesting and storage systems developed.

FEED SELECTION AND AVAILABILITY

Wood residues, forest residues and bagasse are the most interesting short term feed materials. They are widely available, at low cost and with good performance in fast pyrolysis. Straw and agro residues are important in the longer term, but it should be noted that straw has low lignin but high ash content which might cause problems in pyrolysis and use of the bio-oil. Sewage sludge is a significant resource that requires new disposal methods and can be pyrolysed to give liquids. Biomass characteristics should not be a technical barrier in pyrolysis and gasification and most biomass can be successfully pyrolysed.

FEED PREPARATION

Feedstock preparation operations (handling, storage, drying, grinding) are similar to practises in chipboard production (typically 50-200 kt/y) and should not be a problem beyond the tight economic constraints that the energy sector has to meet. However, down-scaling of the commercial equipment used in the wood and pulp industries can be challenging both technically and economically.

FAST PYROLYSIS

PYROLYSIS

Process development and operation is improving with higher quality and more consistent bio-oil being produced. In addition new technologies are reaching demonstration scale with well defined business plans from several companies. However, bio-oil for testing is in short supply. This is preventing long-term performance tests needed for evaluation as well as inhibiting the development of new applications.

Scale-up is an increasingly important issue as pilot and demonstration plants have to be commercially realised. Heat transfer is a critical area in pyrolysis as the pyrolysis process is endothermic and sufficient heat transfer surface has to be provided to meet process heat needs. It was suggested that limits in the short term might be around 100 ton/day, with modular construction of larger plants, although there was also a view that 100 t/d may also be the smallest economic plant size.

Economically attractive operation requires low cost feed, large scale operation and technical development to improve performance and reduce costs.

LIQUIDS

Bio-oil aging should not be a problem since storing longer than 12 months is very unlikely and it should not be forgotten that fossil fuels also age. Phase separation is still seen as the major problem, usually caused by high feed moisture or mixed feed such as forestry residues with a high bark content. However, as companies gain more experience in operating larger-scale units and the fundamentals of pyrolysis process become better understood, this problem should diminish. Good bio-oil quality is a result of process optimisation. Though different feedstocks give oil of somewhat different properties, those liquids can be usually mixed. There was a consensus that the poor reputation of bio-oil was not justified.

Because bio-oil is so much different from petroleum fuels, it requires new standards that have to be developed. For example, bio-oil has low flash-point, but a high ignition temperature. Such differences cause misunderstanding and education and promotion (PR) is an important factor in the development of fast pyrolysis technologies. PR is important and pyrolysis must not be harmed by a negative image, which is where Networks such as PyNe can make a valuable contribution through promotion, training and dissemination of proper information. PyNe is working actively to resolve problems and address priority issues, such as applying for EC projects which has resulted in successful proposals to determine norms and standards for biomass derived pyrolysis liquids, and assess bio-oil toxicity.

APPLICATIONS

The main markets for the liquid products are in heat, power (and CHP) and chemicals and the following applications have been shown to be feasible at various scales of operation:

Application	Scale
Bulk chemicals	Laboratory and demonstration (adhesives)
Energy carrier	Demonstration
Food flavours etc	Commercial
Fuels for transport	Laboratory
Heat	Commercial
Hydrogen	Laboratory
Power and CHP	Demonstration
Speciality chemicals	Laboratory

HEAT

The heat market offers substantial potential but is highly dependent on costs in order to compete with fossil fuels. Bio-oil can substitute for Light Fuel Oil and Heavy Fuel Oil according to specifications. For example, the light fuel oil market in Finland is attractive for intermediate size boilers due to taxation, and burner development is active in Finland to support that market. Co-firing of bio-oil has been demonstrated in 350 MW gas fired power station in Holland, when 1% of the boiler output was successfully replaced. It is in such applications that bio-oil can offer major advantages over solid biomass and gasification due to the ease of handling, storage and combustion in an existing power station when special start-up procedures are not necessary.

POWER

Electricity is an attractive product for its relatively high value, ease of distribution and ease of adapting to local or national market specifications. In addition, the electricity market in Europe offers significant incentives, up to €0.1/kWh, for power gemerated from renewable resources. CHP offers the possibility of enhancing revenues and achieving high overall system efficiencies, although opportunities are site specific and seasonally dependent.

Engines and turbines need further development to provide necessary performance guarantees. Long term demonstration of bio-oil in diesel engines and gas turbine has not been possible due to lack of large quantities bio-oil.

There are similar advantages in storage and transport as for heat applications, although special combustion equipment and start-up and shut down procedures are likely to be needed.

TRANSPORT FUELS

For production of transport fuels there is current interest in gasification of bio-oil/charcoal slurry, which will soon be tested as fuel in a high-pressure gasification unit in Germany, with synthesis of hydrocarbons from the resultant syngas. Use of a bio-oil – char slurry maximises the overall conversion efficiency of the pyrolysis process to a transportable fluid, although solid char contents above around 10% may be difficult to handle. The advantage of this approach is the ease of transport and pressurised feeding of a liquid product. Bio-oil can thus be considered an energy carrier.

CHEMICALS

The only current commercially viable opportunities are in the food flavouring and additives sector (e.g. liquid smoke), however this is a specialised market.

Resins and adhesives for wood panels have been widely investigated with promising results for phenol substitution, but exploitation will require attention to product quality and consistency and defined quality standards.

Another opportunity is in speciality fertilisers in which nitrogen is added to the pyrolysis process or the bio-oil product to give a slow release fertiliser.

Hydrogen production from whole bio-oil is unlikely to be competitive, but the aqueous fraction from phenols recovery or the tar from charcoal production can be reformed to produce hydrogen.

A number of speciality chemicals can be derived such as levoglucosan, anhydrosugars and sugars, none of which yet have a discernible market.

MISCELLANEOUS

INCENTIVES

It is clear that fiscal incentives are essential for successful early commercialisation of any new technology that is competing in a low cost competitive commodity market without the usual economies of scale. While there are European driven incentives and national support programmes for renewable energy, these incentives are often seen as temporary and volatile which inhibits commercial investment.

RESEARCH

It is hard to get funding from EU and national programmes for fundamental research. Fundamental research needs to be linked to applied research, but is essential for successful commercialisation by improving an understanding of the process leading to optimisation and by explaining variations in performance between feed materials, technologies and scales of operation.

INTERACTIONS

Pyrolysis and gasification of biomass has a better reputation than pyrolysis and gasification of waste. There are substantial potential benefits from better interactions between the science and technology of biomass and waste processing both at research levels and commercially. The development of a forum that promotes this interaction, particularly between companies could offer substantial benefits.

CONCLUSIONS

Fast pyrolysis offers a flexible and attractive way of converting solid biomass into an easily stored and transported liquid, which can be successfully used for the production of heat, power and chemicals. Implementation of this technology is, however, hindered by several factors:

- Feedstocks need to be available at the lowest possible cost compatible with suitable physical and chemical characteristics.
- Initially feedstocks are likely to be wood and woody materials and major industrial waste materials with minimum ash and contamination such as bagasse,
- In the future, the feedstock base will include lower-cost biomass materials such as residues, wastes, mixed and contaminated materials,
- Biomass supply chains need to be developed and optimised both technically and economically, and conventional equipment, when possible, should be used to minimize the cost,
- Scale-up of pilot and demonstration plants is an important next step for the sector and support and encouragement is needed,

- There is very limited availability of sufficient quantities of bio-oil at a reasonable cost for long term testing as required to establish performance guarantees for boilers and engines,
- There is very limited availability of bio-oil for exploratory research,
- Support for process optimisation and supply of large quantities of bio-oil for testing should be provided through continued operation of existing pyrolysis plants,
- Continued emphasis on innovation for new processes
- Continued attention needs to be placed on developing bio-oil that is consistent and has improved basic physical properties such as lower viscosity, longer storage life and lower solids content,
- The compatibility of bio-oils from different processes and different feed materials needs to be investigated,
- Encouragement should be offered to explore new ways of using bio-oil for energy, fuels and chemicals,
- The development and publication of norms and standards and all relevant regulations for handling, storage and transport of bio-oil need to be addressed in a centralised and coherent way with widespread dissemination of recommendations and procedures,
- Active promotion of the opportunities offered by bio-oil with sensible and positive publicity is important for the successful implementation of fast pyrolysis in the market place,
- Networks are a very effective of supporting development, identifying and prioritising issues, and directing future directions, particularly if they can be provided with financial, resources to address the most critical problem areas.
- Pyrolysis and gasification should be seen as complementary not competitive.

RECOMMENDATIONS

1. Recognise that plentiful and regular supplies of bio-oil need to be made available for long term testing.
2. Provide resources and support to ensure that adequate supplies of bio-oil are made available at a reasonable cost.
3. Recognise and support fundamental research into production and application of bio-oil to provide support for industrial development.
4. Encourage network activities as a very effective way of supporting development, identifying and prioritising issues and future directions; networks should continue to attract strong support including additional financial resources to address the most critical problem areas and analyze key topics.
5. Fundamental research is vital to the successful optimisation and development of commercial processes and needs to be encouraged in its own right, either through traditional routes or via Networks.
6. The strong emphasis on close collaboration between university and industry should be relaxed. Sometimes such co-operation is essential, but the short term commercial objectives of industry are not always compatible with the creativity and innovation associated with academia, and can lead to imaginative inventions and development being stifled.
7. Pro-active promotion of fast pyrolysis and bio-oil as renewable energy sources will help overcome the sometimes negative image of the technology and the product.
8. Education and training in this and related sectors are essential in order to build up a core of knowledge and expertise on which a viable industrial sector can be built.

Strategies for Development and Implementation of Biomass Gasification

E. Rensfelt
TPS Termiska Processer AB, Studsvik, SE-611 82 Nyköping, Sweden
B. Gobel
Dept of Energy Engineering, Technical University of Denmark, Lyngby, 2800 Denmark

The workshop was supported by a panel of the following: Erik Rensfelt (leader), H Hofbauer, J Swithenbank, E Kurkela, H Knoef, J Corella; whose contributions are much appreciated

INTRODUCTION

The objective of the workshop was to identify and to prioritise what is needed to improve the rate and extent of both penetration and success of gasification technology into the market. A few gasifiers have already been in operation for twenty years and a number of gasification processes are under industrial development at pilot and demonstration scale. Gasifiers are available from from Foster Wheeler in Finland, Lurgi in Germany, Vølund in Denmark, TPS in Sweden and Repotec in Austria. In addition there is extensive research and development at universities, research institutes and companies around the world. Different technologies are more suitable for different scales of operation as shown in Figure 1.

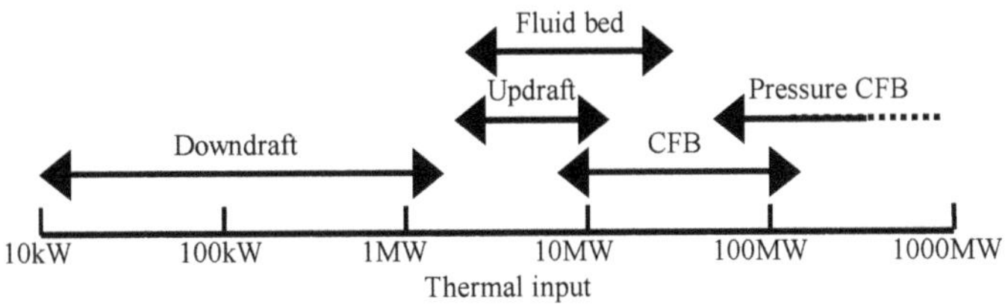

Figure 1: Gasifier Concepts.

A range of feedstock is available for gasification and varies from clean biomass to waste fuels containing different amounts of ash, alkaline and emission precursors like N, Cl and S (Table 1).

Table 1: Biomass and Related Fuels.

Fuel	Price	Ash Properties		Emission precursors		
		Content	Alkaline content	Nitrogen content	Chlorine content	Sulphur content
Clean wood waste	Medium (local)	Low <2%	Low	Low <0.2%	Low <0.1%	Low <0.2%
Green fuel chips					<0.2%	Low
Whole tree	Low	~2%	~0.5%	Medium	Low	<0.2%
SRC	High		~1%			
Agricultural waste	Low			High	High	Low to med
Agro-industrial waste	Low	2-10%	High	0.5-2.5%	1%	<0.5%
Annual energy crops	High					
RDF/MSW (waste fraction)	Low or Negative	High 10-25%		0.5%-1%	0.5%-1%	Medium <0.5%

A wide range of applications has been developed and demonstrated for (co)-firing in boilers/furnaces, gas engines and combined cycles. Comparisons of biomass derived gas with natural gas in terms of basic fuel properties are summarised in Table 2. For some applications, it is important to increase gas-heating value. Low heating value gas has some impact on power generation, which has to be properly considered such as power de-rating and flame stability.

Table 2: Gas inter-changeability.

	Natural gas	Medium heating value		Low heating value
Oxidant		oxygen		air
		Dry	CO_2 free	
LHV (MJ/Nm^3)	35	11-13	16-18	3-7
LHV mix with air at SR=1 (MJ/Nm^3)	3.4	3.2	3.4	2.5
Flame temperature (°C)	1.930	1.700-1.800	1800-1.950	1.400-1.500

A gasification system consists of four main stages and each area was discussed and reviewed with respect to technical barriers in the first part of the workshop:

1. Feeding
2. Gasifier reactor
3. Gas cleaning
4. Utilisation of combustible gas

The second part of the workshop was dedicated on the discussion of non-technical barriers.

TECHNICAL BARRIERS

FEEDING

Pressurized systems

Feeding is important. Especially the feeding of pressurised systems is seen as problematic (mechanical problems, gas tightness problems, etc). Genius engineering solutions are required in this area.

Low-pressure gasifiers can adopt a lot of the technology developed for combustion of biomass.

GASIFIER REACTOR

In addition to the development of rather fuel specific gasifiers, there is currently a strong need for fuel flexible gasifiers that are also able to convert difficult fuels, e.g. high alkali and high ash content fuels. Scale up issues should be considered carefully.

Modelling

The reactor design of gasifiers is not optimised yet and can be improved. In analogy with the development of biomass combustion in the past, a major tool to achieve this is physical and mathematical modelling.

Design and modelling of fluid bed gasifiers have to be done in close co-operation between manufacturers, and universities. However there is a confidentiality problem in providing commercial data and funding is not being spent sensibly in supporting modelling studies when real data cannot be provided. As a result available good models for fluid bed gasification seem to remain unused.

Good models are not available to show what is going on with particles during updraft and downdraft gasification such as shape changes, and flow through the channels between particles. Models and fixed bed tunnelling diagnostic techniques are needed to identify what is happening with particles during their way through the bed.

Heating value
A way to increase the heating value of the gas it was suggested to increase the use of wastes e.g MSW as a feedstock for gasification. The increase of heating value is not necessary for all applications. The most important issue is about the efficiency of the gasifier and not the heating value.

GAS CLEANING

Micropollutants
The consequences of the presence of micropollutants present in biomass and waste, e.g. metals, organo-metals etc., are unidentified and should not be underestimated.

Economical demonstration
Gas cleaning for Cl, Hg, S, N, metals, etc. seems to be sufficient reliable at the moment but its cost should be considered. Niche markets have to be found and further developments should be more towards commercial solutions. This requires economical demonstration plants for long term testing, which should be done in co-operation with research and lab scale experiments, All emission problems have to be solved, before commercial application, even for difficult fuels.

Emphasis on hot gas cleaning
In addition there should be more emphasis on hot gas cleaning such as the removal of sulphur by liquid Ti (99.5 % recovery) or a Zn-absorber, the injection of lime, etc.. However, solutions for hot gas cleaning must be cost effective and not too fancy.

Ashes
Another important point of concern is the quality of the ashes and tendencies for sintering and ash melting.

Tar removal
Tar removal or cracking is a major point of concern in gasification en still needs focus. Nickel catalyst are likely to be used increasingly in the future for tar cracking, although the technology is currently only in the laboratory. Table 3 summarises the main options for removing or cracking tars.

Table 3: Tar conversion and elimination concepts (from 1).

Internal thermal	addition of oxygen and steam
Ultra-high gasification temperature	oxygen and steam
Physical separation	e.g. wet scrubbing, adsorption, centrifugation
Internal catalyst	e.g. nickel or dolomite bed
External catalyst	secondary reactor with e.g. nickel or dolomite
External thermal	secondary reactor with addition or oxygen or air
External reverse flow	secondary reactor with addition of air or oxygen

Filters
Filters like conventional bag house filters have been successfully tested for particle separation at reasonably high tar contents for fluidised bed gasifiers. When a sorbent is used, there is a good reduction in HCl levels in the filter.

Heat exchangers
Corrosion and fouling in hot gas heat exchangers is still a problem. Causes need to be identified and knowledge shared.

APPLICATIONS

The specifications and emissions consequences of prime movers are poorly understood and work is needed to both better define (realistic) specifications and control emissions.
At small scale there are no problems in engines with thermal NOx due to excess air. Fuel NOx is taken out with wet scrubbing of the gas. Instead high emissions of unburnt components such as CO and hydrocarbons are a difficult problem. Catalytic aftertreatment of flue gasa is necessary.

Gas turbine experience at Värnamö has shown that vapour phase tar is not a problem in gas turbines - clean up. At high pressure/high temperature removal of NH_3 is currently not available.

NON-TECHNICAL BARRIERS

COMMERCIAL ASPECTS

Information can be derived from insurance companies, banks and project developers. Thornley and Wright have presented a valuable overview of important items that should be considered during commercialisation of new thermal technologies (2). Particularly important items include reliability, availability, and ease of operation (staffing need and need for automatic operation without staff for small plants).

An EPC contractor is often requested by the customer to avoid gaps in the total contract scope.

CHP is a good opportunity for increased efficiency and lower fuel costs for power.

INCENTIVES

At current biomass and capital cost levels, incentives are essential for a commercially viable biomass projects. Successful examples of incentives include Swedish practices:

- CO_2 tax for heating applications means biomass is commercially very attractive,
- NOx fee. Combustion plants with high emissions pay plants with low emissions of NOx.

These have led to a rapid increase of biomass and steadily lower NOx-emissions based on extensive RD&D.

Financial aspects of gasification projects that require consideration include:

- High initial investment
- Limited private investments
- Remuneration of CHP
- Feedstock availability
- Small subsidies are not stimulating
- Subsidies tend to decrease

Permitting procedures and emission limits are increasingly onerous but necessary:

- Complex, time consuming
- Unknown technology to authorities
- Strict emission limits from incineration
- No one common legislation

Health and safety impacts are summarised in Figure 2 and include:

- Which directive is valid to gasification
- Guidelines for safe operation and construction is needed
- Deforestation
- Fire and explosions
- CO poisoning
- Hot surfaces

- Air pollution
- Noise pollution
- Oil spill
- Air pollution
- Water pollution
- Litter

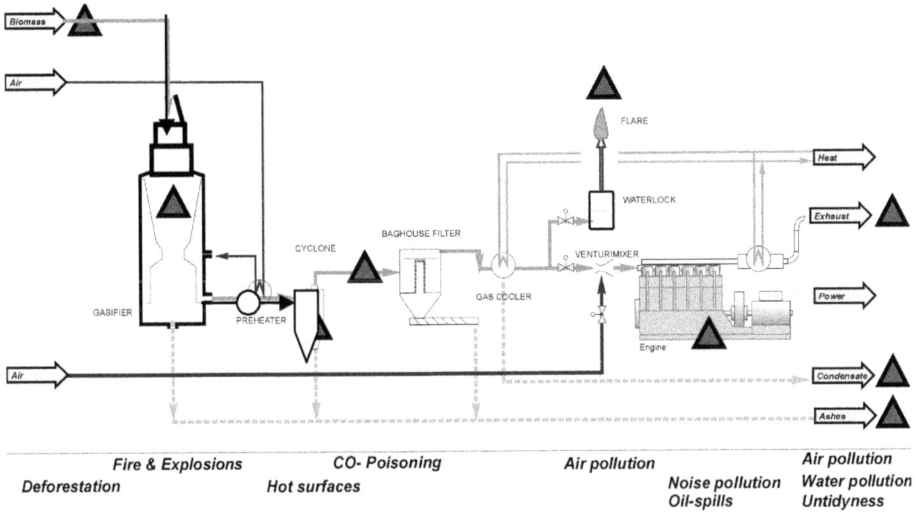

Figure 2: Summary of environment, health and safety aspects of a gasification system.

CONCLUSIONS AND RECOMMENDATIONS

- Demonstration projects are quite risky and should ideally be made by companies with strong finances. In practice, companies that are willing to accept high risk often already have weak finances.
- CHP is a good opportunity for increased efficiency and lower fuel costs for power. But there are high capital costs for piping heat for district heating in combined heat and power systems. CHP is difficult economically and practically in old cities but should carefully be considered in new areas.
- The current strict emissions limits for combustion plants have to be seen as a baseline for new technology e.g. gasification although the difference in oxidizing and reducing atmospheres must be considered. New technology sometimes has an advantage over old technology with bad public reputation.
- At high pressure/high temperature removal of NH_3 is currently not available.
- Economical demonstration plant e.g. in nische markets is required for long term testing. Niche markets need to be identified and exploited in the short term.
- Greater co-operation between industry with such demonstration plants and research organisations through use of laboratory scale experiments would be beneficial for reducing costs and resolving problems. This would also improve the experiences of fundamental researchers.
- For hydrogen, more attention and work is needed on safety aspects including handling. Although many years ago town gas, a mixture of hydrogen and carbon monoxide, was delivered to every household and we all knew how to live with it.

- More emphasis is needed on hot gas cleaning; removing ammonia and sulphur from the hot gas and there are still opportunities to try new concepts.
- It is important and often essential to avoid repeating old mistakes by reading the literature. Wrong approaches have often been adopted because of lack of knowledge of previous work and results from that work, particularly in papers from 1970-2002. Much is already proven and disproven and re-inventing old problems lose time and money. Sometimes old tests should be repeated because they appear to be or might be wrong.
- Supercritical water may offer some useful opportunities.
- Modelling and optimisation can improve reactor vessels and design of gasifiers. Modelling in particular offers much promise for improved design and performance, but there is a problem with validation with data from companies due to confidentiality problems.
- Technology providers need to offer whole gasification plant, otherwise there will be gaps and omission in the performance guarantees. This is the reason why turn-key EPC contracts are important On the other hand the value of guarantees and even turnkey contracts is often overestimated. When problems occur, the customer is often the one that has the major burden.
- It is necessary to demonstrate that a local skilled workforce can operate gasification plants.
- There is a possibility in doing projects in stages if parts of the project have high risk and others low risk. The actual risk is minimised if the initial investment is small, then the potential loss will be small.
- It is important to calculate that the feedstock has to be paid for i.e. it has a cost, even if it can be obtained free initially to ensure there is sufficient profitability over the plant life.
- The question of defining the market and matching the availability of technology to that market is very important:
- Gas cleaning and tar cracking or removal and gas cleaning remain the largest problem in gasification,
- After many years of R&D, workshops and conferences, the bottleneck is still wider exploitation of gasification technology.

A number of important and fundamental questions were posed that could not be fully resolved, particularly:

- What do we want to use the biomass for in the future: energy and/or chemicals? This was considered in several workshops.
- After 15 years of extensive research and development and much funding, there are still only a few commercially operating plants: why?

REFERENCES

1 Knoef, H, "Overview of small-scale biomass gasification", pp 315-324, this publication.
2 Thornley P and Wright E, "Evaluation of Bio-Energy Projects", pp 339-362 in Bridgwater A, (Ed.) "Fast pyrolysis of biomass - a handbook Volume 2", (CPL Press 2002) ISBN 1-872691-47-1

Report of Workshop on Waste Conversion

N. Abatzoglou
Enerkem Technologies Inc. & Université de Sherbrooke, Department of Chemical Engineering, Sherbrooke (Québec), Canada
Panel: K. Pedersen, Chemidan, Denmark; N. Cooper, Compact Power Ltd, UK; P.-D. Oudenne, NESA, Umicore Engineering SA, Belgium.

INTRODUCTION

The purpose of this workshop was to examine the situation worldwide regarding the application of pyrolysis and gasification for the energetic valorization of waste streams. In a world, willing to adhere to the principle of the "Sustainable Development" (see Rio Conference) and to decrease the environmental impacts leading to "Climate Changes" (see Kyoto Conference), it was decided, in Johannesburg Conference, that the implementation mechanisms are not yet as effective as it would be and that new mechanisms and technologies advances are required. It is within this context that Pyrolysis and Gasification represent interesting complementary solutions and could be part of any Integrated Waste Management Strategy, respectful of the "Sustainable Development".

ACTUAL STATUS OF THE HIERARCHY OF THE WASTE MANAGEMENT FILES

The hierarchy of the Waste Management options is well established since many years ago. The symbolism used is 3R—V—D(or E), which means the following:

1. First priority is given to the "Reduction" at source (first R), "Reuse" (second R) and "Recycling" (third R) both physical and chemical.
2. Second priority is given to the "Valorization", either material (i.e. composting of rapidly biodegradable material) or energetic (i.e. thermal conversion for the recovery of the energy content of the waste stream)
3. Last recourse is the "Disposal", usually by landfilling. The letter "E", for "Elimination", is frequently used instead of the letter "D", for "Disposal". In the opinion of the Workshop, "Disposal" is expressing more efficiently the final option because the word "Elimination" means that something is disappearing which is not the case when the Waste is landfilled; it is just a way to set aside and away from view the Waste.

Pyrolysis and Gasification, both thermal conversions are belonging to the "Valorization" file. They produce energy and materials and they can be considered as both energetic and material valorization technologies. Besides, they are more environmentally friendly than incineration. In the opinion of the workshop these two features places Pyrolysis and Gasification at a level of effectiveness higher than "Incineration" which is also part of the "V" file.

WASTE DEFINITION

The workshop thought that it is imperative to have a large definition of what is Waste as feedstock for the Pyrolysis and Gasification technologies. The reason is that the various terms like "Residues", "Waste", "Refuse" are used in literature and there is some confusion to both technical and non-technical people when publications appear.

Despite the fact that Waste" is defined" legally by various countries and organizations, the confusion remains and the communication of the results becomes fuzzy. The definition given to Waste is as follows: A stream of materials is defined as a Waste within a specific context when it is a refuse without objective value; otherwise it constitutes a material at the end of its usefulness. The reason the definition includes the expression "within a specific context" is that spatiotemporarily the same material can have different value, both positive, negative or none depending on many parameters, whose analysis is beyond the scope of this work.

In the opinion of the Workshop the term "Waste" becomes more and more obsolete because of the relative value a waste stream can have within a specific situation. It is proposed to replace the term by the term "Residues" which has no the meaning of "non-usefulness".

WASTE STREAMS FOR PYROLYSIS AND GASIFICATION

Most of Waste streams are appropriate, as they are or after some pre-treatment, conditioning steps, for Pyrolysis or Gasification. In the opinion of this Workshop the following Waste are belonging to this category:

- Urban Waste: this includes what it is called Municipal Solid Waste (MSW) or Household Waste. "RDF" is another term used meaning "Refuse Derived Fuel" and which is used to express a sorted MSW.
- ICI Waste: it includes Waste produced from Institutional, Commercial and Industrial activities.
- Lignocellulosic residues: they include forest, agricultural, demolition and urban activities residues containing mainly lignocellulosic material and which are not anymore useful for the production of added-value products.
- Sludge: this term include sludge produced by wastewater treatment, industrial biological treatment, and agri-food biological activities and whose organic content is sufficiently high to justify thermal treatment.
- All animal waste: this category includes manure, litter and "bone and meat meal". It is material biologically active and it can be characterized normal, special or hazardous depending on biological activity and local legislation.
- Clinical: Clinical waste is divided in three streams:
  a) A stream essentially similar to MSW
  b) A stream of various chemicals, drugs etc. and which is hazardous
  c) A stream containing biologically active and infectious agents.

Although it was tried to have a list of Waste appropriate for thermal treatment, as exhaustive as possible, some other stream could be neglected. Obviously for every Waste reported in this list the pre-treatment steps can vary and this constitutes one of the next topics of this report.

WHAT IS RENEWABLE?

The principle of the Sustainable Development is now well accepted and all our actions must take into account this obligation. Thus, according to the opinion of this workshop, the energetic valorization through Pyrolysis & Gasification adheres to this principle when the Waste treated constitutes a renewable material. To define better which fraction of the Waste feedstock is renewable the Workshop brainstormed on the definition of a renewable material and the deliberation is given below:

Renewable fraction is "all that is biodegradable within a reasonable period of time".

Another point examined throughout the discussions was the relationship between this type of feedstock (renewable) and the production of Greenhouse effect Gases (GHG). It was reported that the petro-chemical industry defines a source as "renewable" when the use of this source is "CO_2-neutral", which means that its use do not produce globally CO_2. This is for example the case of biomass whose use as combustible produces CO_2 equal to the amount of CO_2 captured by the corresponding plant for its growth. The questioning is whether such a relationship is appropriate or if there could be other sources, non CO2-neutral, that fall within the "renewable" sources as well; such an example is the plastic waste streams. The participants gave no definite answer at this point.

THE TECHNOLOGICAL CHALLENGES

There are two levels of technological challenges: the ones at the level of the pre-treatment technologies and those regarding directly Pyrolysis & Gasification.

THE PRE-TREATMENT TECHNOLOGIES

Before accepting a waste as feedstock in a Pyrolysis or Gasification island there are a number of conditioning steps whose nature and extend depends on the nature of the waste stream, including its physico-chemical properties, as well as the needs of the thermal conversion technology (i.e. max. acceptable size and humidity). These pre-treatment technologies are summarized as follows:

- Transport
- Sorting and Separation (Extraction, Heavy media separation, Dewatering, Filtration)
- Baling/Packaging
- Shredding
- Thermal pre-treatment (Drying, Stabilization)
- Granulation/ Pelletizing
- Mixing / Homogenisation

Each of these pre-treatment technologies has some technological challenges to face but they are all mature commercially available technologies. The Workshop has put up a list of these technological challenges without going deeply and examining each technology particularities.

The list is as follows:

- Although each of the pre-treatment technologies is well known and commercially mature their scaling-up for specific feedstock can be difficult and require sometimes "Trial-and-Error" techniques.
- Legislation is changing and EU advances in defining at least three categories of Waste. This can lead to specific pre-treatment needs and might call for normalized pre-treatment steps to be defined technically and financially.
- Work at the Research, Development, Demonstration and Commercialisation (R,D,D&C) levels is required to define pre-treatment modules and trains allowing to produce these types of Waste
- Automation and Health and Safety issues.

THE WASTE VALORIZATION TECHNOLOGIES

The following list of Valorization technologies was set-up by the workshop. It tries to cover all fields including fermentations:

- Incineration/Combustion/Co-combustion
- Thermal or Catalytic Pyrolysis (or Co-pyrolysis)
- Thermal or Catalytic Gasification (or Co-gasification)
- Digestion (known also as Biological Gasification)
  - – Aerobic
  - – Anaerobic
- Wet oxidation
  - – Super-Critical Water Oxidation
  - – Classical WETOX

Among these technologies there are numerous applications at the industrial level. Describing all these applications worldwide goes beyond the scope of this workshop. Nevertheless a short list of Waste Pyrolysis and Gasification units is reported below, showing the type of technology, the waste treated, the products and an approximate number of applications.

For Pyrolysis:

- Rotary kiln: MSW, Industrial Waste, Commercial Waste, Contaminated soils (20 units or more in Europe)
- Positive displacement: All types of Sludge, Tires, Demolition Wood, Clinical, Pharmaceutical, MSW (15 units or more in Europe and worldwide)
- Multistage directly heated system: Waste (several units in Germany, France , Switzerland operating since several years).

The products from these units include: Steam, Electricity, Charcoal, Carbon black, Activated Carbon, Oil for Co-firing.

For Gasification:

- Circulated Fluid-Bed (CFD) + Co-firing in burner for steam to electricity: RDF, Residual lignocellulosics, plastics, other (unit in Finland).
- Two-Stage Melting Gasification: MSW, other varia (unit in Italy).
- Bubbling Fluid-Bed (BFB) with aluminium recovery: MSW and specific RDF (unit in Finland).
- Bubbling Fluid Bed + Internal Combustion engines for electricity production: industrial and commercial plastic residues (unit in Spain).

The products from these units include: Steam, Electricity, Secondary material (i.e. aluminium recovery)

TECHNICAL BARRIERS FOR THE COMMERCIALISATION OF PYROLYSIS & GASIFICATION

Although Pyrolysis & Gasification are already applied technologies in this field (Waste Valorization) there still remain some technical barriers for a large extend commercialization. These barriers are not the same for all particular technologies in this field; for example BFB and CFD Gasification, or Fast Pyrolysis and Indirectly heated Pyrolysis technologies, have not necessarily the same technical barriers. In what follows a list of technical barriers was proposed by this workshop; it includes all technical barriers and the reader must understand that some of them apply and some other not in a specific technology.

- Feeding modules depending on Waste stream
- Bed material handling (for FB gasifiers)
- Syngas clean-up (both cold and hot)
- Products characterization (Standards/Norms)
- Environmental Performances/Health and Safety
- Reliability/Troubleshooting
- Need to mature the technologies under development and develop new ones

NON-TECHNICAL BARRIERS FOR THE COMMERCIALISATION OF PYROLYSIS & GASIFICATION

The commercialization of these technologies depends largely on how the project developers face the non-technical barriers, which are numerous in this field. When trying to apply these technologies to an inherently complex field, like the Environment, all parameters must be taken into account; these include the following factors which are essentially neglected during the technical and technological development steps: social, politic, economic, legal. A list of these barriers is proposed by this workshop:

- Feedstock availability (spatial and temporal): this factor is crucial to the techno-economic feasibility of the project. Projects in this field become economically sound above a marginal treatment capacity, which is usually quite high. If the feedstock is not available locally and must be found at places considerably remote from the application area the cost might become prohibitive.
- Project set-up (all steps): When a promoter set-up a project, there are numerous steps to get through and which are sometimes time consuming (i.e. financing, permitting). The funds available for this step are not inexhaustible and "run-out of money" is a usual situation.
- Bad information / Social acceptability: Environmental pressure groups are very efficient in transmitting and communicating their "buzzwords" to the society, thus creating the perception they defend. Sometimes these perceptions are not expressing the reality and this is the case for Pyrolysis & Gasification. A better vulgarization of the situation is necessary to change some bad perceptions constituting "show-stoppers".
- Permitting: different in each country but generally difficult and time consuming.
- Competition from other technologies (i.e. incineration): At the time being, Incineration has a bad name. This can change in the future and Incineration can become a very important competitor for Pyrolysis & Gasification.
- Environmental Legislation: It is or could become a positive or negative catalyst towards commercializing Pyrolysis & Gasification of Waste streams.
- Energy Policy and other Political issues: Energy is the most important commodity in the World. Energy policies can also become positive or negative catalyst towards commercializing Pyrolysis & Gasification of Waste streams.
- Market fluctuation: Perceptions and realities influence the course of the stock market and consequently the market at large.
- Financing of Demonstration Units: it is usually difficult for new technologies in this field to persuade financiers to invest on demonstration projects.
- Costs at all levels.

THE SOCIAL ACCEPTABILITY ISSUE

The workshop considered that this issue is very important and spent some time in defining it and propose how to proceed during the steps of the project set-up.

The consensus established during the Workshop is described below:

- Social acceptability is created around a set of conditions to be fulfilled in a given project and accepted by a large majority of the 'public opinion'. The meaning of majority meaning that, in case of lack of total consensus, is the large majority which defines the acceptability limits.For environmental projects, in which the issues are complex, the set of conditions is established via an 'informed consensus' among all 'actors' involved in the decision-making process.

"LANDFILLING"

Special thoughts have been expressed during the Workshop on the relation between "Landfilling" and the successful implementation of Pyrolysis & Gasification in the field of waste Management.

It was clearly and consensually established that the eventual penetration of Pyrolysis & Gasification in Waste management is a strong function of the landfill cost. As we know that this cost varies from country to country the possibility of success in implementing Pyrolysis & Gasification is a function of geography and socio-political decisions. It is also thought that governments, through appropriate legislation taking into account the externalities costs, will play a decisive role.

CONCLUSIONS AND STRATEGIES

The Workshop concludes the following:

- Waste is a very promising feedstock for pyrolysis and gasification
- There exist already commercial successes
- Because of the variety of the Waste more R, D, D and C work is needed
- The non-technical barriers are more influential than the technical ones
- Landfill costs is a crucial factor
- Energy and Environmental policies will play a very important role

It proposes the following strategies for a successful implementation of Pyrolysis & Gasification as valuable options within any Integrated Waste Management Project:

- See Pyrolysis & Gasification as the technologies of the future for Waste Valorization.
- Improve the technologies
- Inform people and government efficiently
- Explain that the Valorization is hierarchically at a higher level than landfill and impose restrictions allowing for Waste to have a higher negative tipping fee
- Face systematically and drastically all other non-technical barriers.

PARTICIPANTS IN THIS WORKSHOP (IN RANDOM ORDER)

David Buckle, SITA,
Patricia Thornley, PB Power, UK
Stefan Skrypski-Mäntele, ESI, Germany
Nikolaos Boukis, FZK, Germany
Joan Salvadó, URiV, Spain
André Fontana, Université Libre de Bruxelles, Belgium
Pascal Haxaire, RMD S.A., Spain
Harris Eleftheriadis, TOMI S.A., Greece
Francesco Fantozzi, Université di Perugia, Italy
Lieve Helsen, Katholieke UniversiteitKlaus Eppler, DDH Contractors A/S, Denmark
Xenophon Kakatsios, National technical University of Athens, Greece
Tanja Barth, University of Bergen, Norway
Nic Cooper, Compact Power Ltd., UK
Karsten Pedersen, Chemidan, Denmark
Anastasia Zabaniotou, Aristotle University of Thessaloniki, Greece
Nicolas Abatzoglou, Enerkem Technologies Inc. & Université de Sherbrooke, Canada
Paul-Dominique Oudenne, NESA, Umicore Engineering SA, Belgium.

Syngas for Synfuels – Workshop Report

Carlo N. Hamelinck
Utrecht University, Copernicus Institute, Science Technology Society
Padualaan 14, 3584 CH Utrecht, Netherlands
Richard L. Bain
NREL, 1617 Cole Blvd, MS1613, Golden, CO80401, USA

Contributors to the discussion: Anton Althapp, Harold Boerrigter, Henning Nannen, Edmund Henrich, Hans Haan, Nicolas Abatzoglou, Jan Schelling, Erik Rensfelt, Jorg Schmalfeld, Tanja Barth, Kai Sipila, Karsten Pedersen, Wolter Prins, and Esa Kurkela.

INTRODUCTION

This discussion summarizes an experts workshop on biomass-derived *syngas for synfuels* applications conducted at the "Expert Meeting on Pyrolysis and Gasification of Biomass and Waste," Strasbourg, France, October 1, 2002.

INTEREST IN BIOFUELS

Transportation fuel security and greenhouse gas emissions reduction are the most important drivers for the interest in biofuels. Synthetic fuels–derived from biosyngas–can be produced from a broad range of biomass feedstocks. Examples are methanol, ethanol, hydrogen and Fischer-Tropsch (F-T) liquids. F-T liquids and ethanol can be relatively easily introduced into the existing transportation sector; whereas methanol and hydrogen for fuel cell vehicles are expected to offer a better lifetime economy, but require serious adaptations in fuel distribution infrastructure and automobile technology.

Several automobile manufacturers are interested in high quality synthetic biofuels, partly because synthetic fuels may have improved performance in newly developed engines, compared to traditional oil derived fuels; and also because they see F-T liquids from biomass as an early opportunity to meet society's (the customer's) demand for environmentally sound actions. Manufacturers' representatives are very interested in new developments.

TECHNOLOGY PARALLEL TO FOSSIL

Technology for the production of synthetic fuels from syngas is not new. Methanol and hydrogen are generally produced from natural gas, and there is experience with F-T liquids production from coal. Worldwide, some 85 installations produce chemicals from syngas, for example, methanol, ethanol, mixed alcohols and hydrocarbons. Parallels in fossil fuel technology show that all steps in biomass to synfuels processing, such as pre-treatment, gasification, gas cleaning, make-up, and synthesis are possible at significant scale, and, moreover, that the whole combination is possible as well.

Nevertheless, integrated concepts of synthetic biofuels have been demonstrated only in limited demonstrations, and large-scale developments are on hold. The central question in this workshop is therefore *to identify the most important technological and non-technological barriers to implementation.* This should also lead to recommendations for *direction for governmental programs.*

TECHNOLOGICAL BARRIERS

Required Gasifier

For successful large-scale production of synfuels a fungible–a commercially interchangeable commodity product–syngas is required, *i.e.* a clean gas mix of primarily H_2 and CO with the right composition. It was stated that large gasifiers are required to produce synthetic fuels at the lowest cost (economy of scale). The understanding of downstream requirements of the ultimate process gives direction to the development of large commercial gasifiers, gas cleaning and make-up processes. Some experts hold the view that for the ultimate large-scale processes, oxygen-fired pressurized gasification is a requirement *sine qua non.*

However, current gasification R&D is mainly aimed at decentralized CHP application, which is typically operated at atmospheric pressure and with air as gasification medium. This may mean that gasifiers for CHP are not suitable as demonstration for synthetic biofuels production, and that the development of the required gasification technologies for large-scale fuels production follows a different R&D path. For the eventual large processes, it is therefore suggested that gasification R&D should–from now on–only focus on large-scale syngas production, with the perspective to scale-up up to 500 – 1000 MW_{th} input

For a demonstration plant, the gas composition from the gasifier is not the central aspect, as a shift reactor (with or without CO_2 removal) can make every gas suitable for methanol or F-T production.

Gas Cleaning

Downstream processing requirements can be subdivided into gas cleaning (contaminant specifications) and gas conditioning (process conditions). The latter concerns primarily the H_2/CO ratio–which can be met through the water gas shift reaction–and the concentration of inerts (N_2 and CO_2). The major technological barrier in converting biosyngas into synfuels may thus be in gas cleaning. Some experts suggested that the biomass research community should focus on producing a clean bio syngas and leave the connection to biofuels (*i.e.* gas conditioning and fuel synthesis) to the "synthesis guys".

Requirements for gas cleaning are mainly dictated by the application, and furthermore by the amount and character of the contaminants as a result of feedstock used and gasification technology applied. Catalysts used in the production of methanol and Fischer-Tropsch liquids are easily poisoned by small amounts of alkali metals, halides, and sulphur compounds, which therefore need to be removed to ppm or even ppb level. These levels are much stricter than levels for electricity and heat production, for which most gasifiers are developed at present. There is obviously a break-even point between costs for severe cleaning and costs due to loss of catalyst activity. With ongoing development the acceptance levels for catalyst poisons may go up, or catalysts may become cheaper, and cleaning methods may improve quickly.

Discussion on this topic also revealed an area of conflict between commercial catalyst development and scientific interests. The scientific world complains that much is unclear concerning the requirements of new catalysts, as catalyst manufacturers consider new catalysts proprietary and do not disclose catalyst formulation or poison tolerance details to the public. For example, although the F-T process has been known for 80 years, it is difficult to buy a suitable catalyst.

It is also suggested that one should opt for newly developed gasification processes that produce a cleaner gas to start with. If the economics and scale-up ability of these processes are good, they deserve serious attention. Others share the opinion that the present status of gas cleaning can already fulfill requirements for a demonstration plant. Demonstration will automatically lead to further improvement.

NON-TECHNOLOGICAL BARRIERS

Diversification or Focus?

With the amount of biomass that could be grown in the European Union, a maximum of 20 % of the transportation fuels needed could be produced. Part of the audience prefers that governments make a decision on bioenergy applications today. For instance, some experts feel that decision is needed to use biomass either for CHP or for transportation fuels, and within the transportation fuels area whether the industry should focus on a limited number of fuels, or look at the whole range of fuels. An official focus has the advantage that a clear sign is given to all scientists and developers about the eventual aim of development; and, therefore, that research funding is more effectively spent on the most promising options. Others find that the broad range of applications should remain open, so that the most advantageous option will eventually emerge in the biofuels market.

Although numerous studies have compared the relative economics of biofuels throughout the years, there is no consensus on which biofuel to concentrate developments. Various biofuels from extraction, esterification and hydrolysis-fermentation are named as important alternatives. However, in the long term the preferred process uses the complete energy crop, which may favor a route via the syngas platform. The high temperatures used in gasification processes pose a limit to the maximum efficiency. It may be advantageous to combine a very efficient low temperature process for fuel production from part of the biomass, after which the residue is still suitable for gasification and electricity production.

There are also practical reasons to focus on a particular class of fuels. From an oil company perspective F-T is considered attractive, as it allows the use of existing infrastructure for further processing and distribution. For the long-term, hydrogen from syngas is envisioned, but that route will prove difficult to extend to broad application because of the time and costs associated with infrastructure development. To realize biofuels market penetration within a 10-year timeframe, the industry may need to focus on fuels that are the easiest to produce, even though they may neither be the most economically attractive in the long term, nor be the best options for greenhouse gas emission reductions.

The discussion on whether to focus on one fuel, or to keep all options open, cannot be decided in the experts meeting, but needs to be continued at a higher level.

Company involvement

There is some worry by oil companies' representatives about introducing a green fuel component into their product mix. In some countries, the cooperation between research institutes and major oil companies is not obvious. While some companies seem to be reluctant, others are starting to become interested in alternative fuels, or already have committed themselves to research in this field. This also shows from the representation of several oil companies in this workshop.

The major objections from the oil companies concern the long-term guarantees for attractive economic conditions to sell the fuel produced. For companies to participate they need profitable and stable long-term market conditions; this needs to be created by the European governments. Furthermore, the technology is not yet ready for large-scale application.

The fact, that a lot of natural gas based synfuel projects are under active development, represents an opportunity to get synfuel related research done, which could then also be applied to biomass at some stage in the future.

Feedstock Supply

The supply of inexpensive feedstock is a major barrier for large-scale bioenergy application in Europe. For both the cultivation within Europe, and the import from outside Europe, government support and long term guarantees on tax exemption and regulations are needed. Waste is seen as a very big, almost inexhaustible feedstock. The import of biomass already happens for other applications, and parallels to fossil fuel (coal) transport hint the practical and economic feasibility.

Competition on Two Fronts

In both the feedstock and product arenas, bioenergy operates in competition with other applications and products. Cultivation of bioenergy feedstock competes with the growth of food crops, and, to a lesser extent, with material crops. In Europe, the influence of agricultural policies on the success of bioenergy growth will be huge.

On the product side, there is obvious competition with products from oil and natural gas. Biomass-derived H_2 is more expensive than H_2 from natural gas. Fischer-Tropsch liquids costs are too high, so that production only becomes feasible at very large scales, *i.e.* 1000 MW_{th}. The consensus is that biofuels cannot be produced cheaper than their fossil alternatives within the next 15-20 years. Long-term incentives are needed in the form of tax regulations. The benefits of bioenergy should be taken into account, and be given a monetary value.

Instructive parallels are seen in Brazil and the USA, where large amounts of biomass are already used for energy applications. There, both the large production scale and partial application for other applications decreases the costs.

Bioenergy demonstration projects only cost money. Governments need to pay for not only R&D, but also operation of at least midsize demonstration plants. One of the most difficult challenges facing government and industry is how to manage the transition from government-subsidized R&D to industry-led commercialization. Much technical development fails because of the inability to make this transition. The general opinion is that governments and the public should not expect the demonstration plants to produce synthetic biofuels at a price competitive with fossil fuels.

CONCLUSIONS

The production of biofuels from bio syngas is technically feasible. Many coal-to-chemicals processes show that the integration of gasification, cleaning, make-up and catalysis is possible. What is needed now is the development of similar large-scale biomass based processes.

Some experts feel that gasifier RD&D should focus on large-scale syngas production, using oxygen blown pressurized gasifiers, which have the perspective to scale-up to 500–1000 MW_{th} input. For synthesis of biofuels, a fungible syngas (a commercially interchangeable commodity gas, a clean mix of primarily H_2 and CO with the right composition) is a prerequisite.

There is some disagreement on the gas-cleaning subject. Gas cleaning requirements for synfuel production are very strict; some share the opinion that gas cleaning remains therefore the major barrier in synfuels production, but new voices state that it has been demonstrated that these specifications can be met (for coal, waste, as well as biomass feeds), and that research efforts should be concentrated on gasification research instead.

Non-technical barriers focus on the system economics (plant size, technical risks, dependence on government support), the feedstock availability, costs and infrastructure. It should be realized that small-scale demonstration plants will never produce biofuel at competitive costs. These projects need not only governmental support for RD&D, but for their complete lifetime. Also, large-scale biofuels production facilities will not be economically competitive with fossil based plants if their environmental advantages are not valued, or if the market situation does not change. Society demands that oil companies produce biofuels, but the extent of oil company involvement in research and development depends on the long-term prospects of the biofuels market.

The decision to either focus on a few biofuels or keep all options open is left for later discussion in broader platforms. There was no agreement on which biofuel is the preferred in the short or long term.

For large-scale biomass feedstock supply, parallels are seen in the USA and Latin America where cheap biomass is available. Bioenergy cultivation in Europe can be enhanced through stimulation of multipurpose products, agricultural policies, and by creating a biomass demanding market.

The European government now needs to make a clear decision for stimulating commercial biofuels application–a decision that includes both tax regulations and long term guarantees for green products, and investing in RD&D in order to improve chances for biofuels production.

Power Production Workshop Report

Patricia Thornley
PB Power Ltd., Newcastle upon Tyne, UK
David Chiaramonti
Dept of Energetics "S.Stecco", Faculty of Mechanical Engineering, Florence, Italy

Panel: *Mike Cannon, Alstom, Lincoln, UK; Martin Schneider, Jenbacher, Austria; Timo Nyronen, Vapo Oy, Finland*

BACKGROUND

This report provides a summary of issues discussed at the Power Production workshop of the expert meeting on Pyrolysis and Gasification of Biomass and Waste, held in Strasbourg on 1 October 2002. The objective of the workshop was to identify the technical and non-technical barriers associated with power production in conjunction with pyrolysis and gasification of biomass and waste. Where appropriate, suggestions made by delegates of possible solutions or requirements for further research in order to overcome these barriers have been included.

INTRODUCTION

The workshop focused on methods of producing electricity from syngas, pyrolysis oil or char produced by waste or biomass pyrolysis or gasification technologies. Issues related to receipt and handling of the derived fuels were not discussed, but those relating to fuel delivery to the prime mover, fuel combustion or electrochemical decomposition and supply of electricity to the market were included. Barriers identified were divided as either technical or non-technical. Technical barriers were considered in four categories – those related to the use of:

1. Boilers (for power generation via a conventional steam cycle)
2. Reciprocating engines
3. Gas turbines
4. Fuel Cells

Non-technical barriers are, in many cases, not related to specific technologies and so have been considered generally for all the above power production systems.

TECHNICAL BARRIERS

BOILERS

There are technical challenges associated with the combustion of syngas, char or pyrolysis oil in a boiler to raise steam that can then be used in a conventional steam cycle to produce electricity. Three major areas were identified to which attention must be paid:

1. **The need to achieve good combustion -** which requires, for example, appropriate design of burners and fuel atomisation systems. This is particularly important for pyrolysis oil. However, these barriers are surmountable with appropriate design engineering and examples of operational plants show that appropriate systems can be achieved.

2. **Potential for fouling** – The presence of certain compounds in the biomass or waste derived materials (particularly alkali metal compounds) results in a potential for fouling of boilers. The effects of this can be minimised by appropriate boiler design (tube spacing and ocnfiguration, temperature ranges etc), inclusion of appropriate diagnostic instrumentation for online condition monitoring and, if necessary, implementing an appropriate cleaning regime. These are established techniques, which can be included as design precautions for particular applications and do not represent a fundamental barrier to project implementation.
3. **Potential for corrosion** – The presence of certain chemical components in the original feedstocks (particularly halides in waste) may give rise to potential for corrosion during combustion. This can be minimised by maintaining appropriate temperatures and combustion conditions and is therefore not a substantial barrier.

All of the above barriers have known solutions, which require appropriate design and application. For example there has been development work on burners for boilers using wood-based pyrolysis oils by Oilon (Lahti, Finland) in co-operation with Fortum (Porvoo, Finland) [1, 2].

However, the largest barrier for this method of power production is that it does not achieve a very high efficiency. There are thermodynamic limitations on steam-cycle efficiency, but in addition the problem is exacerbated for these plants by the following:

- It may be necessary to impose low steam conditions to prevent corrosion and other problems
- Temperatures achieved in combustion of these fuels may not be appropriate for high efficiency cycles
- The catchment area of biomass and waste plants is limited by logistics and economics, which limits plant size, necessitaing the use of smaller steam turbines, which tend to be less efficient
- This logistics/size issue does not permit the economies of scale which would make the use of higher steam conditions (giving higher efficiencies) attractive.

The latter two barriers are not strictly technical and, while no-one present had heard of biomass or waste plants operating at ultra super critical steam conditions, there were reports of a 6 MWe plant which operated at 66 bar, 550 °C. Regarding low combustion temperatures there were reports of plants co-firing natural gas or using gas-fired superheaters after the boiler to overcome this problem.

RECIPROCATING ENGINES

There is considerable experience combusting syngas in reciprocating engines and more limited experience of using pyrolysis oil. The engines have to be adapated individually, looking at the particular fuel involved, but the technical feasibility of doing so is proven. Despite this there is not sufficient long term operating experience (30,000 – 40,000 h) to give full confidence in the long term viability and costs of this method of power production. Some long demonstration work (of commisisoning and long term operation) is essential to build commercial confidence.

The biggest technical barrier identified for reciprocating emissions was environmental emissions. NOx emission limits of the TA-Luft standards can be metwith low-heating value biofuels are used, although more stringent standards could be difficult to achieve. However, CO levels tend to be higher than many legislative limits, particularly if low speed diesel engines are used.

The main options available for reducing CO are flue gas treatment technologies: catalytic or post combustion. Catalysts are able to reduce CO emissions, but they are sensitive to flue gas composition and poisons in the gas, so that performance can deteriorate rapidly with time. Also CO-emissions of a gas engine are lower at a high H2-content in the gas than at a low H2-content. One way around this would be to carry out a water-gas shift reaction to change the CO:H2 ratio, but this would be very expensive. The alternative to use of a catalyst to reduce CO emissions is post combustion of CO, of which there is no substantial experience with pyrolysis gas,. but it is already well tested at landfill gas.

There is considerable uncertainty related to the costs of these post combustion units, preventing meaningful financial analysis of overall system economics. For engines adapted to run on landfill gases, the gas treatment accounts for 10-15 % of the complete turn-key engine package cost, with negligible additional operational costs on units which have run for up to 60,000 hours. Verification of the levels of these costs for other fuels is required.

A secondary issue is the presence of tars and condensate in syngas. The associated problems are surmountable by avoiding temperatures near the fuel dew-point, but this results in increased investment costs and it is important to have a reliable fuel analysis to achieve this, which may not be easy with waste and biomass feedstocks.

The use of pyrolysis oil in reciprocating engines presents considerably more challenges than syngas, partly due to the more limited experience. Wartsila have ran trials on a 1.5 MWe engine fed with pyrolysis oil, but the work was then stopped for various reasons [3]. ORMROD (UK) used a pilot injection of diesel oil to ignite pyrolysis oil in a 250 kW engine (one cylinder fed with pyrolysis oil) and work on emulsions of pyrolysis oils and diesel oil has been performed by other groups in Canada and Europe. This experience base and the pyrolysis oil properties identify the following barriers and potential barriers:

- Polymerisation of fuel components in the injection nozzles
- Modifications required to the injection system to cope with the low LHV fuel (one method of addressing this is to use a pre-mix emulsion)
- Corrosion of the injection system (caused by pH of the oil and the presence of char particles)
- Particulate emissions
- Solids in the fuel (which can cause metal erosion)
- Alkalis in the fuel (which can cause fouling and deposition and are mostly contained within the char particles)

Several of the technical barriers are related to the presence of impurities in the fuel. It is possible to engineer a fuel injection system (for example) that is more compatible with pyrolysis oil than the standard systems (e.g fuel delivery and injection systems should be as simple as possible to prevent blockages and materials can be chosen which are appropriate to the pH of the pyrolysis oil), but in the longer term it may make more sense to concentrate on upgrading the pyrolysis oil to make the biofuel more compatible with the currently available technology, as far as possible.

Oil filtration is an available technique, which will reduce char content and thereby also significantly reduce alkali content. Further work is needed on pyrolysis oil filtration to demonstrate its effectiveness in combating some of the above problems. Also, in order to better understand what is going on during combustion of the fuel (and hence appreciate the real reasons for some of the problems reported during trials) improved characterisation of the pyrolysis oil is needed. This work is progressing under the Alterner programme.

Finally it should be noted that in some cases demonstration of combustion of pyrolysis oil in reciprocating engines has been difficult owing to a lack of available pyrolysis oil.

GAS TURBINES

The main attraction of using gas turbines to convert the energy in biofuel to electricity is the potentially high cycle efficiency The average efficiency of biofuel-based gas turbine combined cycles can reach 45 % in case of pressurised gasification and 40 % for atmospheric gasification. This includes the slight increase in efficiency obtained because of the increased mass flow in the turbine when utilising low heating value biofuels. By comparison, the typical efficiency of gas engines are around 38-43 % in case of natural gas (depending on engine size). Using biofuels, if the H_2 content is relatively high and the plant quite large in size, 40 % can be attained, while 35 % is a typical average value for smaller capacity systems. [1]

---

[1] Note these figures do not include the efficiency of the gasification stage.

The choice between gas engines and gas turbines is mainly a matter of size (turbines are more appropriate for larger size plants, gas engines for smaller ones). However, the nature of biomass again constrains larger developments, so that 35-40 MWe would be a reasonable sized gas turbine based plant, which could benefit from the efficiency advantages of increased scale, without suffering the economic disadvantages of increasing the biomass catchment area to supply fuel demands. While microturbines offer promisingly high efficiency cycles at smaller scales there is limited experience with the use of biofuels in microturbines.

Lean-premix combustion technologies have been developed to control CO emissions and low dioxin levels can be simultaneously obtained e.g. in Varnamo, 50 % RDF and 50 % wood fuel gave dioxin emission levels of 0.01 ng/Nm3 (comfortably complying with the Waste Incineration Directive limit of 0.1 ng/Nm3 for dioxins). Thermal NOx is also reduced by the lean-premix technologies. However fuel NOx is a much more significant technical barrier. The efficiency of biomass combined cycle systems could be improved by having a higher inlet temperature to the gas turbine and using hot gas cleaning (although there are limiting factors related to mechanical considerations on materials and component lifetimes) However, if hot gas cleaning is implemented to avail of the related higher cycle efficiency it is difficult to remove ammonia from the syngas and this will form NOx upon combustion in the gas turbine, likely giving rise to NOx levels which exceed current legislative limits. This is a major

There are also technical issues that need to be taken account of in the design of the heat exchanger/gas-cooler for gas turbine exhaust gases. However, these can be addressed with available techniques and appropriate design engineering.

Pyrolysis oil presents more technical barriers than syngas. Many of these are similar to those listed in section 3.2 and the major ones include the following:

- Particles in pyrolysis oil causing erosion
- High alkali content in pyrolysis oil causing deposition
- Possible problems in fuel atomisation

The only significant practical experience of pyrolysis oils and gas turbines of which workshop delegates were aware is the work that ORENDA in Canada has done, which includes gas turbine combustion trials on a 2.5 MWe machine, but not actual gas turbine operation. The published results suggest that deposition on turbine blades does occur and particles in the fuel do cause erosion, but appropriate maintenance schedules may be devisable to keep these under control. This needs further investigation/confirmation, paying particular attention to the actual quality of pyrolysis oil used, specification of maintenance intervention required, feasibility of this and development to actual operation in a gas turbine.

It is likely that further investigation of this area would result in recommendations relating to limits for solid/ash/alkali metal levels in pyrolysis oil for use in gas turbines. The challenge of meeting this would require upgrading of the oil and research and development should be focused in this area.

FUEL CELLS

Fuel cells offer the potential for very high power generation cycle efficiencies with low levels of emissions, but they are not yet readily commercially available and there is little experience of their operation with biofuel systems. Of the four fuel cell categories those of most interest for biofuel power generation are the molten carbonate and Solid Oxide Fuel Cells.

The technical barriers specific to biofuel systems are likely to be purification and conditioning of the syngas. While pyrolysis oil, like many other hydrocarbons, could potentially be steam-reformed to produce a gas for use in a fuel cell, it is not considered likely that it would be attractive to directly use pyrolysis oil for power generation in a fuel cell system.

The major obstacle to deployment of fuel cells in biomass and other systems is the lack of commercially available molten carbonate or solid oxide fuel cells. It is anticipated that, even if these were commercially available, the cost of the system would continue to be a barrier in the near future.

Other technical barriers envisaged include onerous gas purity requirements, variability in gas composition from biomass sources and potentially operation and maintenance requirements/costs. The first technical barrier to be overcome is actual demonstration of a fuel cell system operating with biogas – then we will find out what the other technical barriers are. It was suggested that biogas could be combined with other gases in early demonstration work and possibly in final applications to limit some potential problems.

Some delegates observed that studying the whole cycle from gasification to electric energy generation, the final overall efficiency of the system could be not much better than the more conventional systems discussed above (owing partly to requirements for conditioning of biomass syngas). This would limit the attractiveness of fuel cells for biofuel systems. However, advances in gasification technology (as gasification in supercritical water, producing 75 % H2 rich gas, which is still in its early – but very promising – stage of development in Netherlands and Germany) could modify this.

NON-TECHNICAL BARRIERS

The primary non technical barriers identified by delegates are outlined below.

- **Environmental costs** - The single most important issue facing pyrolysis and gasification of biomass and waste is the economics of power production from these technologies. They are not cost competitive with conventional methods of power production except in certain niche markets. Their advantage is their lower level of environmental emissions, especially CO2 emissions. Until this is recognised by fiscal policies which allow the actual environmental cost of power production by different technologies to be subject to commercial market forces these technologies will remain, by and large, uneconomically attractive.
- **Availability of the feedstock.** Moreover, feedstock market conditions can change, as costs are determined by competition with already well established markets. This is the case of Northern Europe, for instance, where wood by-products have a long history and problems exist in creating new markets
- **Cost of electricity and cost of fossil fuels by comparison** – These are so low in many countries as to make most forms of biomass based generation expensive by comparision. As stated above, unless the environmental benefits of biomass systems are recognised financially there cannot be a level playing field with fossil-fuel based technology.
- **Guarantees:** Many of those involved in technology development at the workshop expressed frustration and concern at the guarantees that are expected from them as a technology provider, with insufficient regard to how other factors beyond their control (e.g. feedstock characteristics) may be influencing performance.
- **Image** - Technology developers also identified the fact that when installing systems, customers tended to want a reduction in capital costs. This would often be achieved by cutting technical corners and the result was a plant that did not perform optimally and therefore did not provide a good base from which to promote or develop the technology.
- **Timescale** necessary for the implementation of bioenergy projects
- **Economy of scale** – The logistics of harvesting and transporting biomass supplies means that plant sizes tend to be small and do not benefit from commercial economies of scale available to competing technologies.
- **Cogeneration** - Biomass systems can be more commercially competitive in a combined heat and power configuration, but requirements to site plants near fuel sources makes the use of heat more difficult than for some competing technologies. District heating is a large potential market for biomass fired plants in many parts of Europe, but it is difficult for new biomass technologies to break into the market when there is already a heat supplier to the system.

- **Operation and maintenance** - For pyrolysis and gasification plants attention needs to be paid to establishing an operational skills base. While it is not unduly difficult to obtain personnel experienced in power generation and control, it is much more difficult to find operators who also have the right balance of chemical engineering skills and specialist knowledge for these plants.
- **Electricity market** - In some countries the electricity market trading mechanisms do not encourage small scale generators, including those using pyrolysis and gasification of biomass and waste. For example in the UK, the risk of penalties under NETA for failure to generate after declaring capacity can be onerous. Similar situations exist in other countries.
- **Public perception/lack of understanding** – This was identified as an obstacle for the advanced technologies, particularly with regard to obtaining necessary permits and licences, even within the relevant authorities.
- **Financial incentivies** - Non-uniformity in EU legislation on bioenergy tariffs and incentives is also a reason for unbalanced development of bioenergy across Europe.

REFERENCES

1. Oasmaa, Anja; Kytö, Matti; Sipilä, Kai. Pyrolysis oil combustion tests in an industrial boiler. - Progress in Thermochemical Biomass Conversion. Bridgwater, A. V. (ed.). Vol. 2. Blackwell Science (2001), s. 1468 - 1481
2. Gust, S. & Kytö, M. Development of combustors for pyrolysis liquids. In: Pyrolysis and Gasification of Biomass and Waste-Expert Meeting 30 September i- 1 October 2002, Strasbourg, France)
3. Jay, D., Rantanen, O., Sipilä, K. & Nylund, N.-O. (1995). Wood pyrolysis oil for diesel engines. Proc. 1995 Fall Technical Conference, Milwaukee, Wisconsin, 24 - 27 Sept. 1995. New York: ASME.

Economics Workshop

Hermann Hofbauer
Renet-Austria, Vienna, AT

Panel: *Yrjo Solantausta, VTT, Finland; Max Lauer, Joanneum Research, Austria*

At the beginning of the workshop the participants defined the topics of common interest. Five topics were selected and discussed during the workshop. The results can be summarized as follows:

PROMOTIONAL DATABASE

The participants felt that there are already success stories but not well documented. The Güssing case was discussed in detail and stated, that it is not only a technical but also an economical success especially due to the benefit of the whole region (wood farmers, job creation, etc.).
The pay back time for bioenergy plants is important for investors. The participants especially fromindustry agreed that the necessary pay back times are:

15 years	for power plants
3-5 years	for pulp and paper industry
even smaller	for risky projects

As a result of this discussion a promotional database was demanded. This database shall include technical as well as economical data and also data from the region where the plant is located. This data base could promote potential investors to invest money into bioenergy projects.

AVAILABILITY OF WASTE FUELS

Due to European and national legislation and improved collecting processes more and more waste fuel are available as fuels at a low prize. Therefore, opportunities exist for operators of bioenergy plants to use such low prize fuels to improve the economy of their plant. But most technologies are not ready yet for such low quality fuels.

As a result of this discussion it was concluded that more attention has to drawn to the to low quality fuels from waste streams during the development of bioenergy plants in order to be able to use these opportunities and to improve the economy of the plant.

FUNDING/SUPPORT

It was common consent that funding and support is necessary for the implementation of bioenergy plants. The question was raised which kind of support should it be.
The majority of the participants shared the opinion that internalising the external costs would give "real costs" and this would be a strong support for bioenergy.

There was a broad discussion about taxation versus "green credits". It was agreed that taxation is the preferable instrument.

The financial funding for bioenergy technologies is limited. Therefore, it was demanded from some participants to select only promising technologies for funding. But no better procedure could be found during discussion for the selection of the promising technologies than that already performed e.g. from European Union. Perhaps some parts could be improved e.g. the selection of the experts.

BIOMASS INDUSTRIES

It was stated that successful biomass industry already exists. The ethanol production in Brazil and US was mentioned and both cases were examined intensively during discussion. For both cases it was seen that the political intention and demand is the most important driving force for these successful implementations.

TRUE NUMBERS FOR ASSESSMENT

Two diagrams showing “specific green house gas emissions” on the one hand and “energy pay back” on the other hand for different renewable energies. The participants did not believe the numbers shown in the diagrams. It was evident that these figures are only used to influence the public in a certain direction (lobby).

It was generally agreed and strongly demanded that the bioenergy society has to evaluate “true numbers” very urgently for the assessments of renewable energy.

Author Index